Vector Algebra and Analysis with Applications

Prasun Kumar Nayak

M.Sc., B.Ed., Ph.D.

Department of Mathematics (UG & PG)

Midnapore College (Autonomous)

Midnapore, West Bengal

Vector Algebra and Analysis with Applications

UNIVERSITIES PRESS (INDIA) PRIVATE LIMITED

Registered Office
3-6-747/1/A & 3-6-754/1, Himayatnagar, Hyderabad 500 029, Telangana, India
info@universitiespress.com; www.universitiespress.com

Distributed by
Orient Blackswan Private Limited

Registered Office
3-6-752 Himayatnagar, Hyderabad 500 029, Telangana, India

Other Offices
Bengaluru, Bhopal, Chennai, Guwahati, Hyderabad, Jaipur, Kolkata,
Lucknow, Mumbai, New Delhi, Noida, Patna, Vijayawada

ISBN 978-93-86235-15-2

Typeset in AGaramondPro 11/12 points *by*
MacroTex Solutions, Chennai 600 091

501322

24022017

Printed at
BB Press, Noida

Published by
Universities Press (India) Private Limited
3-6-747/1/A & 3-6-754/1, Himayatnagar, Hyderabad 500 029, Telangana, India
email: info@universitiespress.com

Preface

Vector is the base of various branches of modern physics, applied mathematics and engineering. So, it is an interdisciplinary subject and a compulsory course for physics, applied mathematics and engineering students. A thorough understanding of vector is thus essential for these students. This book is suitable for both the undergraduate and postgraduate students of this subject.

This book is an outcome of my lectures on vector algebra and analysis over a number of years at the graduate level. The important theories are provided along their proofs, and explained with the help of simple examples. Details of mathematical steps are given to make the subject easy. A variety of worked-out examples are included and a graded set of problems are given to test the knowledge of the readers.

This book contains seven chapters in total. Chapter 1 introduces scalars and vectors. The different types of vectors, addition, subtraction and multiplication by a scalar and linear combination are described. Three important theorems are explained with examples.

Chapter 2 focuses on the product of vectors – scalar product, vector product, scalar triple product and vector triple product, with trigonometric and geometric applications. The concept of reciprocal vector – indispensable for the formulation of geometry and mechanics – is introduced.

The applications of vector algebra (vector equations, geometric problems and mechanics) are presented in Chapter 3. In the next chapter, vector calculus is discussed, with special emphasis on function, limit, continuity and differentiability. A section on curvilinear coordinates, which plays an important role in several branches of physics, is also presented.

Chapter 5 on the vector differential operator provides the framework for a discussion on gradient, divergence and curl, with geometrical significance and applications in several branches of mathematics and physics. Line and multiple integrals are explained in Chapter 6. This is also indispensable for the formulation of Green's problem.

The vector anti-derivative leads us to vector integration, examined in Chapter 7; this is equivalent to integral theorems. Several problems, theorems and diagrams are described herein.

I would now like to express my gratitude to the many people who have inspired and helped me write this book.

Firstly, I would like to thank MR Spiegel, as it was from his book that I began learning this subject. I am indebted to various sources (books, research articles and websites) on this subject and allied topics, which are mentioned in the bibliography. I acknowledge them gratefully.

This book could not have been completed without the moral and loving support and continuous encouragement of my wife Mousumi, my son Soumyadip and my niece Rimjhim.

My sincere respect goes to my teachers, Prof. N Islam, Prof. Bijoy Bera and Prof. N Bhanja of RK Mission Residential College, Narendrapur, who introduced me to the concepts of vector algebra and analysis. I would also like to thank my colleagues, Dr. S Bandyopadhyay, Dr. P Kuila, Dr. M Mandal, Dr. HS Mandal, Dr. S Dey, Dr. Nasreen Khan, Dr. J Majumder, Bandana Bhunia, Avijit Giri, Umasankar Maity, Sudipta Ghorai, Anup Khan, Madhuri Chini and Manjuri Manna, for providing a friendly environment and helping me directly and indirectly, while writing this book.

It gives me pleasure to record my sincere thanks to my students, Ankita Mandal, Riya Dutta, Subhadip Satpati, Riya Basuli and Aditi Chattaraj, who pointed out errors and suggested many improvements.

I express my sincerest gratitude to Prof. (Dr.) Madhumangal Pal, Department of Applied Mathematics with Oceanology and Computer Programming, Vidyasagar University, Midnapore; Prof. (Dr.) S Sen and Dr Swarup Poria, Calcutta University; Dr. Mijanur Rahaman Seikh, Kazi Najrul University, Asansol; and our Principal, Dr. Gopal Chandra Bera for their encouragement.

Special thanks to the honourable reviewers, for their valuable suggestions, careful corrections and kind cooperation.

I feel great reverence for my parents, maternal uncle, brothers, sisters and relatives for their blessings and for being a constant source of inspiration.

I express my gratitude to Thomas Rajesh and Kallol Das of Universities Press, for their cooperation in bringing out this book. Lastly, I am grateful to my editor Madhavi Sethupathi for her assistance.

I welcome constructive criticism that would help me improve the book, from experts in the field as well as from learners.

Prasun Kumar Nayak
prasunkrnayak@gmail.com

Contents

To my beloved wife Mousumi, son Bubai and niece Rimi, for their sacrifice and encouragement

1 Scalars and Vectors

Physics and Applied Mathematics are sciences that deal with observable (quantitatively measurable) physical quantities and the relationships which may exist between the experimentally measured values of the various physical quantities of interest. Physical quantities are divided primarily into two main classes – scalars and vectors, each with characteristic properties and appropriate algebra. Scalar quantities can be fully specified by giving their magnitude alone. The main benefit of using vectors is that they simplify and compress the exposition of physical ideas and make mathematical concepts more perceptible and easy to grasp. Vectors have become an indispensable tool in problems where magnitude and direction need to be dealt with simultaneously, as in elasticity, fluid mechanics, differential geometry, electricity and magnetism.

1.1 SCALAR

A *scalar* is a physical quantity which has only magnitude but no definite direction in space and is completely specified by a single real number (apart from units of measurement which are decided upon in advance). Operations with other scalars are carried out according to the rules of ordinary algebra, and require for its complete specification:

(i) a unit of the same kind, and
(ii) a number specifying how many times the unit is contained in the quantity.

Typical scalar quantities are mass; moment of inertia; length; time, by years, days or seconds; area; volume; density; temperature, by degrees on a thermometric scale; quantity of heat, speed, work, numerical statistics (i.e., birth rates, mortality, population), electric charges, potential or any real number as they are wholly determined when the unit of the scale is fixed. Scalars are represented algebraically by letters in ordinary type like a, b, etc. and as in elementary algebra, denoting the number of units in it. A real number a, which a given quantity bears to the chosen unit, is called the *measure* or *magnitude* of the given scalar quantity. Geometrically, it is represented by a segment of an indirected straight line whose length indicates the number of units.

The positive and negative numbers of ordinary algebra are the typical scalars. For this reason, ordinary algebra is called *scalar algebra* when it is necessary to distinguish it from vector algebra or analysis, which is the subject of this book.

1.2 VECTOR

A *vector* is defined geometrically as a physical quantity having both magnitude and direction in a specific sense in space. Operations with other vectors are carried out according to the rules of vector algebra (discussed later) and require for its complete specification:

(i) a unit of the same kind disregarding direction,
(ii) a number giving the magnitude of the quantity in terms of this unit, and
(iii) a statement of direction.

Typical examples of vectors are displacement, velocity, acceleration, force (gravitational, magnetic, electric), magnetic induction, electric field intensity, position with respect to a fixed origin, etc. The need for a vector concept arose very naturally in geometry and mechanics.

Graphically, a vector is represented by a segment of a directed straight line, say $\vec{OP}$, whose direction is parallel to that of the vector and whose length represents the magnitude of the vector to some convenient scale (Fig. 1.1).

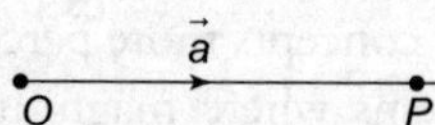

Figure 1.1 Vector as a directed line segment

Analytically, a vector is represented by $\vec{a}$, i.e., $\vec{OP} = \vec{a}$. The point O is called the *initial point* and P is called the *terminal point*, if it is indicated as $\vec{OP}$. The positive number $\overline{OP} = |\vec{OP}| = |\vec{a}| = a$, which is the measure of the length of a vector $\vec{OP}$, is the *magnitude* or *modulus of the vector*. The arrow head indicates the sense of the vector (here, from O towards P). A line of unlimited length of which a directed line segment is a part is called the *support*. Physically, a rectilinear displacement or translation from O to P can be represented by the vector $\vec{OP}$.

Many equations in applied mathematics and physics are of a compact form when written in vector notation. The relationship between the various quantities involved in the equation is revealed immediately when these are written in vector form. Historically, vector notation became widely used with the advent of Maxwell's Electromagnetic Theory, in which the above advantages are clearly seen.

1.2.1 Null and Proper Vectors

A vector whose magnitude is zero can have indeterminate direction associated with it; this is called a *zero vector* (or *null vector*) and is denoted by $\vec{0}$. For a zero vector, the initial point coincides with the terminal point, i.e., $\vec{AA}$ or $\vec{PP}$. The length of the zero vector is zero but it can be thought of as having any line as its line of support. Thus, zero vector has magnitude zero but intermediate direction.

A vector whose magnitude is non-zero is called a *proper vector* or *non-zero vector*. A proper vector has definite length and direction. Thus, a vector $\vec{a}$ is *proper,* if $a = |\vec{a}| > 0$. All vectors will be assumed to be proper unless otherwise stated.

A vector $\vec{a}$ is said to be *spatial* if it lies in a space.

1.2.2 Position Vector

The *position vector* of a point in a space is the vector obtained by joining that point to a pre-assigned point O, taken as the origin in the space. A vector $\overrightarrow{OP}$ is said to be the *position vector* of the point P with respect to the base point or initial point or the origin of reference O (Fig. 1.1). We write it as $\vec{a}$. A point P with position vector $\vec{a}$ is often referred to as the point $P(\vec{a})$.

1.2.3 Unit and Reciprocal Vectors

A vector having unit magnitude or length 1 is called a *unit vector*. If $\vec{a}$ is a given proper vector, then its *unit vector*, denoted by $\hat{a}$, is given by

$$\hat{a} = \frac{1}{|\vec{a}|}\,\vec{a} = \frac{1}{a}\,\vec{a} \tag{1.1}$$

Thus, $\hat{a}$ is a vector having unit magnitude and is drawn in the same direction as that of $\vec{a}$. Also, any vector $\vec{a}$ can be represented as

$$\vec{a} = |\vec{a}|\,\hat{a} = a\,\hat{a} \tag{1.2}$$

The vector which is parallel to the vector $\vec{a}$ and whose magnitude is the reciprocal of the magnitude of the vector $\vec{a}$ is called the *reciprocal vector* of the vector $\vec{a}$. Thus, the reciprocal vector of $\vec{a}$ is $\frac{1}{|\vec{a}|^2}\,\vec{a}$.

1.2.4 Rectangular Unit Vectors

The rectangular unit vectors $\hat{i}, \hat{j}$ and $\hat{k}$ are mutually perpendicular unit vectors having the directions of the positive x, y and z axes, respectively, of the rectangular Cartesian coordinate system (Fig. 1.2).

Such systems derive their name from the fact that a right-handed screw rotated through 90° from ox to oy will advance in the positive z direction.

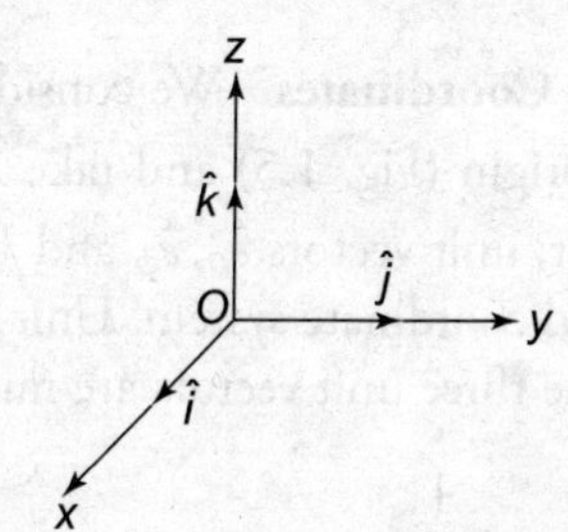

Figure 1.2 Rectangular unit vectors

Polar Coordinates Take O as the base point and OX as the initial line (Fig. 1.3). Here, $\hat{e}_\rho$ and $\hat{e}_\theta$ are the respective unit vectors pointing in the directions of the change in position vector $\vec{r}$, when the coordinates ρ and θ are changed by infinitely small amounts. The two unit vectors are mutually perpendicular and define a right-handed coordinate system. The position vector of the point $P(\rho, \theta)$ can be written as

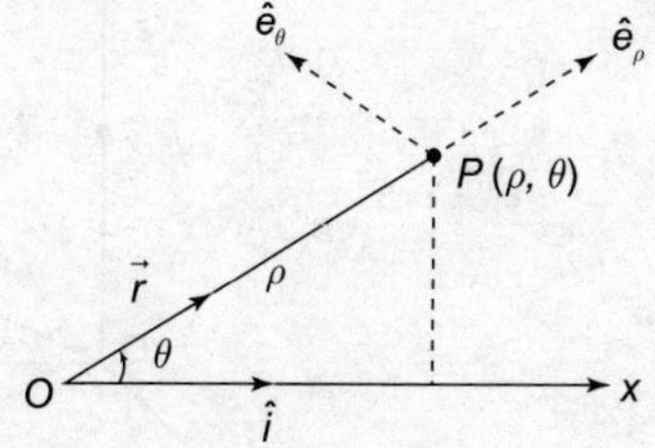

Figure 1.3 Components of a vector (polar coordinates)

$$\vec{r} = \rho\,\hat{e}_\rho + \theta\,\hat{e}_\theta \tag{1.3}$$

Spherical Polar Coordinates We consider a rectangular Cartesian coordinate system *ox*, *oy* and *oz* with *O* as origin (Fig. 1.4) and take $\hat{i}, \hat{j}$ and $\hat{k}$ to be the unit vectors along the respective axes. Here, $\hat{e}_\rho, \hat{e}_\theta$ and $\hat{e}_\phi$ are the respective unit vectors pointing in the directions of the change in position vector $\vec{r}$, when the coordinates ρ, θ and ϕ are changed by infinitely small amounts. The three unit vectors are mutually perpendicular and define a right-handed coordinate system.

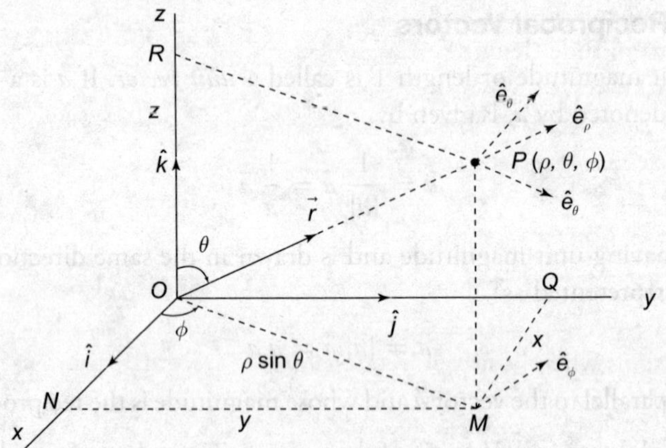

Figure 1.4 Components of a vector (spherical polar coordinates)

Cylindrical Coordinates We consider a rectangular Cartesian coordinate system *ox*, *oy* and *oz* with *O* as origin (Fig. 1.5) and take $\hat{i}, \hat{j}$ and $\hat{k}$ to be the unit vectors along the respective axes. At any point, unit vectors $\hat{e}_\rho, \hat{e}_\phi$ and $\hat{k}$ form a set of mutually perpendicular vectors and define a right-handed coordinate system. Unit vectors $\hat{e}_\rho$ and $\hat{e}_\phi$ are not constant in direction, as shown in Fig. 1.5. The three unit vectors are mutually perpendicular and define a right-handed coordinate system.

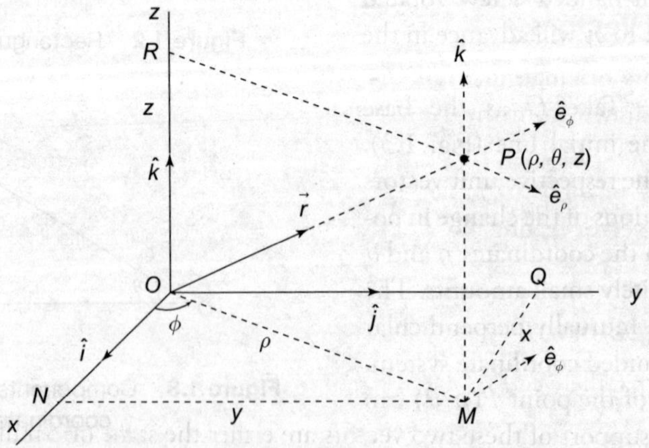

Figure 1.5 Components of a vector (cylindrical coordinates)

1.2.5 Polar and Axial Vectors

To distinguish an ordinary vector from the vector which results from the product of two vectors, we sometimes call the ordinary vector as *polar vector*, and the vector that results from the product of two polar vectors as *axial vector*.

A polar vector can be defined as follows:

(i) In applied mathematics, the term *polar vector* is used to refer to a representation of a vector as a vector magnitude (length) and angle, which is equivalent to specifying its end points in polar coordinates.

(ii) In physics, a *polar vector* is a vector such as the radius vector $\vec{r}$ that reverses sign when the coordinate axes are reversed. Polar vectors are the type of vectors usually simply known as vectors.

Examples of polar vectors include $\vec{r}$, the velocity vector $\vec{v}$, momentum $\vec{p}$ and force $\vec{F}$. The axial vector involves the idea of rotation about an axis perpendicular to the plane area it represents. The essential difference between a polar and an axial vector is that the scalar components of a polar vector change their signs when positive directions of the coordinate axes are reversed, but those of an axial vector do not.

1.2.6 Pseudovector

In physics and mathematics, a pseudovector (or axial vector) is a quantity that transforms like a vector under a proper rotation, but in three dimensions gains an additional sign flip under an improper rotation, such as a reflection. In contrast, pseudovectors do not reverse sign when the coordinate axes are reversed. Geometrically, it is the opposite – of equal magnitude but in the opposite direction – of its mirror image. This is the opposite of a true or polar vector, which on reflection matches its mirror image. Physical examples of pseudovectors include magnetic field, torque, vorticity and angular momentum. The cross-product of two polar vectors is a pseudovector.

1.2.7 Co-initial Vectors

Two or more vectors originating from the same initial point, i.e., vectors drawn from the same point, may be called *co-initial vectors.* In Fig. 1.6, $\vec{a}, \vec{b}$ and $\vec{c}$ are co-initial vectors. It is possible to replace a given vector with another equal vector having any point as the initial point.

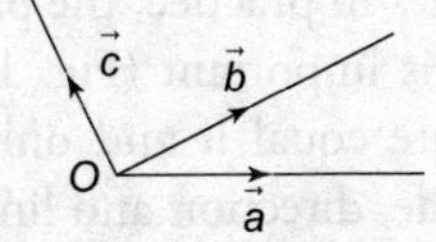

Figure 1.6 Co-initial vectors

1.2.8 Parallel Vectors

Let $\vec{a}$ and $\vec{b}$ be two parallel vectors. Then,

(i) the lines of support of these two vectors are either the same or parallel,
(ii) the senses of them are either the same or opposite, and
(iii) the magnitudes may differ.

Thus, if $\vec{a}$ and $\vec{b}$ are two parallel vectors, then if their senses are the same, there exists a real number $k > 0$ such that $\vec{a} = k\vec{b}$; if their senses are opposite, there exists a real number $k > 0$ such that $\vec{a} = -k\,\vec{b}$. The angle AOB is said to be the angle between the vectors $\overrightarrow{OA}$ and $\overrightarrow{OB}$ (Fig. 1.7). Evidently, this angle does not exceed π. When this angle is $\pi/2$, the vectors are said to be perpendicular and when it is 0 or π, they are said to be parallel or coincident.

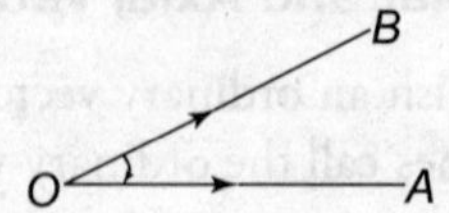

Figure 1.7 Angle between two vectors

1.2.9 Equal Vectors

Two vectors $\vec{a}$ and $\vec{b}$ are said to be *equal* ($\vec{a} = \vec{b}$), if they have the same magnitude ($|\vec{a}| = |\vec{b}|$), same direction and same or parallel supports regardless of their initial points (Fig. 1.8). Thus, two vectors are said to be equal if they have

(i) the same length/magnitude regardless of the position of their initial points,
(ii) the same or parallel supports, and
(iii) the same sense of direction.

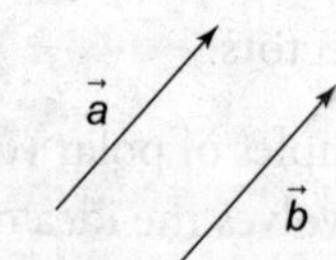

Figure 1.8 Equal vectors

The directions and magnitudes are important in the case of equality of vectors. In Fig. 1.9, $\vec{AB}, \vec{CD}$ and $\vec{EF}$ are equal. Two vectors will not be equal if they have different lengths or inclined supports or even if they have the same lengths and parallel supports but different senses. If $ABCD$ is a parallelogram, we have $\vec{AB} = \vec{DC}$ and $\vec{BC} = \vec{AD}$. Every vector belongs to a class of equal vectors.

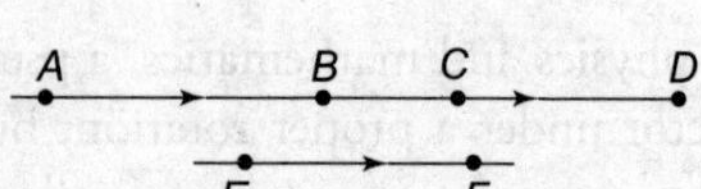

Figure 1.9 Equality of vectors

1.2.10 Sliding Vectors

Sometimes in practice, the particular line of action of a vector is important (Fig. 1.10). In such cases, two vectors are equal if and only if they have the same magnitude, direction and line of action. Such vectors are called *sliding vectors*.

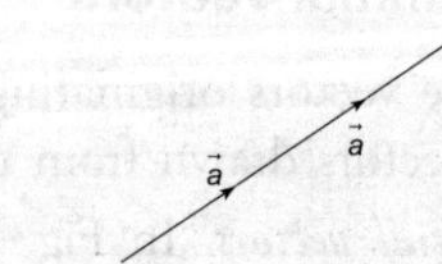

Figure 1.10 Equal sliding vectors

1.2.11 Bound Vectors

Sometimes, it is important to specify the point of action of a vector. Such a vector (Fig. 1.11) is called a *bound vector*. In this case, two vectors will be equal if and only if they are identical.

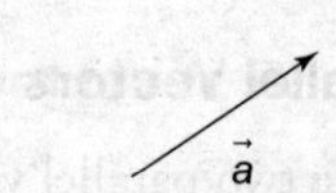

Figure 1.11 Bound vector

1.2.12 Like and Unlike Vectors

Two vectors are said to be *like vectors* if they have the same direction, otherwise they are called *unlike vectors*. The vectors $\vec{a}$ and $\vec{b}$ are said to be *equal and like vectors* if $\vec{a} = \vec{b}$; they will be *equal and unlike vectors* if $\vec{a} = -\vec{b}$.

1.2.13 Free and Localised Vectors

In various physical and geometrical problems, sometimes we come across a vector quantity that is associated with a given point or a given line. In such cases, the vector quantity is said to be *localised* or a *line vector* at that point or line. For example, a force acting at a point on a non-rigid body is a localised vector at that point, whereas a force acting at a point on a rigid body is a localised vector in a line because it can be shifted only along the fixed line (other translations will have other physical effects).

A *free vector* is a directed line segment whose initial point or terminating point can be taken arbitrarily anywhere in space. For example, a vector that arises on the translation of a rigid body is a free force.

A vector is unchanged by pure displacement, and hence, all parallel vectors of the some length and direction ($\vec{a}$, $\vec{b}$ and $\vec{c}$) are equal so that a vector is not localised (Fig. 1.12). Thus, free vectors have no restriction regarding their initial points or terminating points, while a localised vector occupies a definite position in space.

$\vec{a}$ $\vec{b}$ $\vec{c}$

Figure 1.12 Localised vectors

Any two free vectors are equal so long as they have the same magnitude and direction (Fig. 1.13).

A vector which is localised at a point is called a *tied vector.* For example, electric field is a tied vector.

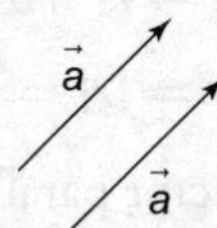

Figure 1.13 Equal free vectors

1.3 VECTOR ALGEBRA

It is possible to develop an algebra of vectors with suitable definition that is capable of extending the algebra of real numbers. This is useful in the study of geometry, mechanics and other branches of physics and applied mathematics. As vectors have their origin in physical problems, their sums and products are defined in such a manner that a number of important physical quantities can be described with the accepted definitions. This section is devoted to the discussion of the algebra of vectors.

1.3.1 Addition of Vectors

The single vector whose effect is equal to the combined effect of two or more vectors is called the *resultant* or *sum* of these vectors. The process of finding the sum is called *vector addition* or *composition of vectors.*

Let the forces $\vec{F} = F_1\hat{i} + F_2\hat{j} + F_3\hat{k}$, $\vec{G} = G_1\hat{i} + G_2\hat{j} + G_3\hat{k}$ and $\vec{H} = H_1\hat{i} + H_2\hat{j} + H_3\hat{k}$ acting on an object be given in terms of the components. Then, the resultant force $\vec{R}$ is given by

$$\vec{R} = \vec{F} + \vec{G} + \vec{H} = (F_1 + G_1 + H_1)\hat{i} + (F_2 + G_2 + H_2)\hat{j} + (F_3 + G_3 + H_3)\hat{k}$$

(a) **Triangle law of addition:** Let two consecutive sides $\vec{AB}$ and $\vec{BC}$ of a triangle ABC represent two vectors. Then, by the triangle law, $\vec{AC}$ is the resultant of the two vectors $\vec{AB}$ and $\vec{BC}$ (Fig. 1.14). $\vec{AB}$ and $\vec{BC}$ are called the *components* of $\vec{AC}$. Borrowing the '+' sign from algebra, it may be written as

$$\vec{AC} = \vec{AB} + \vec{BC}$$

Figure 1.14 Triangle law of addition

This equation is regarded as the definition of *vector addition*. Physically, if a particle is given two successive displacements, one from A to B and a second from B to C, the result is the same as if the particle were given a single displacement from A to C.

Example 1.3.1

Find the unit vector parallel to the resultant of the vectors $\vec{r}_1 = 2\hat{i} + 4\hat{j} - 5\hat{k}$ and $\vec{r}_2 = \hat{i} + 2\hat{j} + 3\hat{k}$.

Solution

Let $\vec{r}$ be the resultant of the two vectors $\vec{r}_1$ and $\vec{r}_2$. Then

$$\vec{r} = \vec{r}_1 + \vec{r}_2 = (2\hat{i} + 4\hat{j} - 5\hat{k}) + (\hat{i} + 2\hat{j} + 3\hat{k}) = 3\hat{i} + 6\hat{j} - 2\hat{k}$$

$$r = |\vec{r}| = |3\hat{i} + 6\hat{j} - 2\hat{k}| = \sqrt{3^2 + 6^2 + (-2)^2} = 7$$

Then, the unit vector parallel to $\vec{r}$ is

$$\frac{1}{r}\vec{r} = \frac{1}{7}(3\hat{i} + 6\hat{j} - 2\hat{k}) = \frac{3}{7}\hat{i} + \frac{6}{7}\hat{j} - \frac{2}{7}\hat{k}$$

(b) **Parallelogram law of addition:** Let two adjacent sides $\vec{AB}$ and $\vec{AD}$ of a parallelogram $ABCD$ represent two co-initial vectors (Fig. 1.15). Then, the parallelogram law of addition states that $\vec{AC}$ gives the resultant of the two vectors represented by $\vec{AB}$ and $\vec{AD}$.

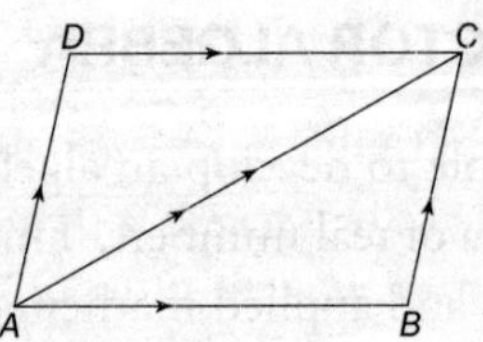

Figure 1.15 Parallelogram law of addition

Thus, the sum of two co-initial vectors $\vec{AB}$ and $\vec{AD}$ is given by $\vec{AC}$, where AC is the diagonal through A of the parallelogram $ABCD$, having AB and AD as adjacent sides.

(c) **Polygon law of addition:** Let $OABCDE$ be a polygon (Fig. 1.16). From triangle OAB, we see that $\vec{OB}$ is the resultant of the vectors $\vec{OA}$ and $\vec{AB}$. If the vectors are drawn one after the other in a chain,

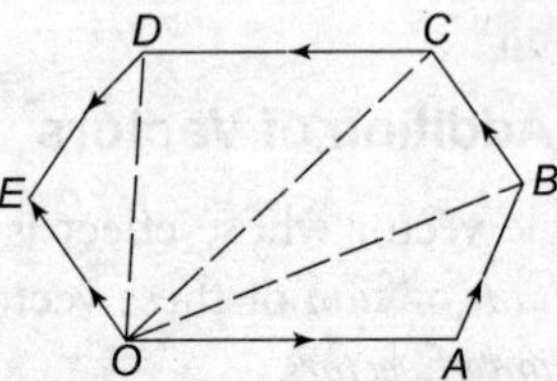

Figure 1.16 Polygon law of addition

each new one from the end of the last one, the straight line joining the origin of the first one to the terminus of the last one gives the vector sum of all.
Therefore, $\vec{OE}$ is the resultant of the vectors $\vec{OA}$, $\vec{AB}$, $\vec{BC}$, $\vec{CD}$ and $\vec{DE}$. This is the *polygon law of addition of vectors*. The vector $\vec{OE}$ is termed the *sum* or *resultant* of the vectors $\vec{OA}$, $\vec{AB}$, $\vec{BC}$, $\vec{CD}$ and $\vec{DE}$, and the polygon *OABCDE* is known as the *vector polygon.*

RESULT 1
From Euclidean geometry, it can be easily shown that:

- $\vec{a}+\vec{b}=\vec{b}+\vec{a}$; commutatively
- $\vec{a}+(\vec{b}+\vec{c})=(\vec{a}+\vec{b})+\vec{c}$; associatively
- $m(\vec{a}+\vec{b})=m\vec{a}+m\vec{b}$; distributively

1.3.2 Subtraction of Vectors

The difference of two given non-collinear vectors $\vec{a}$ and $\vec{b}$ is the addition of the vectors $\vec{a}$ and $-\vec{b}$ (i.e., equal unlike vector of the vector $\vec{b}$) and is denoted by $\vec{a}-\vec{b}$. Evidently, $\vec{a}-\vec{b}=\vec{a}+(-\vec{b})$, i.e., subtraction of $\vec{b}$ from $\vec{a}$ is the same as adding the negative of $\vec{b}$ with $\vec{a}$. From Fig. 1.17, it can be seen that $\vec{OA}=\vec{a}$, $\vec{OB}=\vec{b}$, so $\vec{AB}=\vec{b}-\vec{a}$. In fact

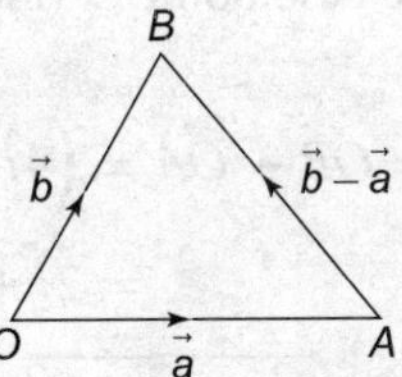

Figure 1.17 Subtraction of vectors

$$\vec{b}-\vec{a}=\vec{b}+(-\vec{a})=\vec{OB}+\vec{AO}=\vec{AO}+\vec{OB}=\vec{AB}$$

Thus, every vector may be expressed as the difference of the position vectors of its end points. $\vec{b}-\vec{a}$ is called the *position vector* of point *B* with respect to point *A*.

RESULT 2
(i) The sum of two vectors is zero when and only when they have the same length and opposite directions. Thus, $\vec{a}-\vec{a}=\vec{0}$.
(ii) If $\vec{a}=\vec{b}$, then $\vec{a}-\vec{b}$ is defined as the null or zero vector and is denoted by $\vec{0}$. This has the magnitude of zero units but its direction is not defined.
(iii) For every vector $\vec{a}$, $\vec{a}+\vec{0}=\vec{a}=\vec{0}+\vec{a}$, where $\vec{0}$ is the zero vector.
(iv) We observe that the negative of the negative of $\vec{a}$ is $\vec{a}$ itself, i.e., $-(-\vec{a})=\vec{a}$.

RESULT 3
We have

$$\left[(-\vec{a})+(-\vec{b})\right]+\vec{b}+\vec{a}=(-\vec{a})+\left[(-\vec{b})+\vec{b}\right]+\vec{a}$$
$$=-\vec{a}+\vec{0}+\vec{a}=(-\vec{a}+\vec{a})+\vec{0}=\vec{0}$$

Therefore, $(-\vec{a})+(-\vec{b})=-(\vec{a}+\vec{b})\Rightarrow-\vec{a}-\vec{b}=-(\vec{a}+\vec{b})$. The proof of the result may also be directly obtained with the help of Fig. 1.18.

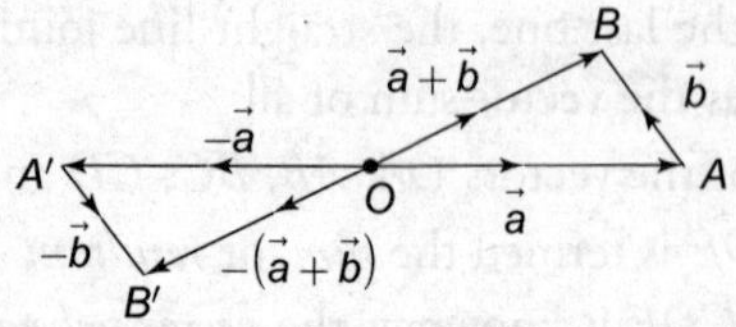

Figure 1.18 Geometrical interpretation

Example 1.3.2

Show that the points A, B and C whose position vectors are, respectively, $2\hat{i}+4\hat{j}-\hat{k}$, $4\hat{i}+5\hat{j}+\hat{k}$ and $3\hat{i}+6\hat{j}-3\hat{k}$ form the vertices of a right-angled triangle.

Solution

The position vector of A, B and C are, respectively, $2\hat{i}+4\hat{j}-\hat{k}$, $4\hat{i}+5\hat{j}+\hat{k}$ and $3\hat{i}+6\hat{j}-3\hat{k}$, with respect to the base point O (Fig. 1.19). We are to show that the Pythagoras theorem is satisfied. Thus

$$\vec{AB}=\vec{OB}-\vec{OA}=\left(4\hat{i}+5\hat{j}+\hat{k}\right)-\left(2\hat{i}+4\hat{j}-\hat{k}\right)$$
$$=2\hat{i}+\hat{j}+2\hat{k}$$

Hence, $\overline{AB}=\sqrt{2^2+1^2+2^2}=3$. Now

$$\vec{BC}=\vec{OC}-\vec{OB}=\left(3\hat{i}+6\hat{j}-3\hat{k}\right)-\left(4\hat{i}+5\hat{j}+\hat{k}\right)$$
$$=-\hat{i}+\hat{j}-4\hat{k}$$

Figure 1.19 Triangle

Therefore, $\overline{BC}=\sqrt{(-1)^2+1^2+(-4)^2}=3\sqrt{2}$. Also

$$\vec{CA}=\vec{OA}-\vec{OC}=\left(2\hat{i}+4\hat{j}-\hat{k}\right)-\left(3\hat{i}+6\hat{j}-3\hat{k}\right)$$
$$=-\hat{i}-2\hat{j}+2\hat{k}$$

Hence, $\overline{CA}=\sqrt{(-1)^2+(-2)^2+2^2}=3$. Now, we see that $BC^2=AB^2+CA^2$ so that A, B and C are the vertices of a right-angled triangle, with $\angle BAC$ as the right angle.

Theorem 1.1

The set of all vectors forms an abelian group with respect to composition as vector addition.

Proof

(i) **Commutative property:** Let the vectors $\vec{AB}$ and $\vec{AD}$ be defined by $\vec{a}$ and $\vec{b}$, respectively (Fig. 1.20).

Using the triangle law of addition, we have

$$\vec{a}+\vec{b}=\vec{AB}+\vec{BC}=\vec{AC}$$
$$=\vec{AD}+\vec{DC}=\vec{b}+\vec{a};\ \text{from } \triangle ADC$$

Thus, the commutative law of vector addition can be proved to hold good.

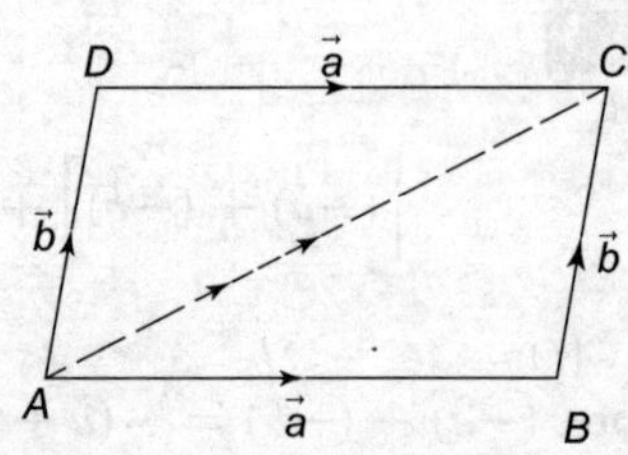

Figure 1.20 Commutative property

(ii) **Associative property:** From Fig. 1.21, let $\vec{AB} = \vec{a}$, $\vec{BC} = \vec{b}$ and $\vec{CD} = \vec{c}$. Then, using the triangle law

$$\vec{a} + (\vec{b} + \vec{c}) = \vec{AB} + (\vec{BC} + \vec{CD})$$
$$= \vec{AB} + \vec{BD} = \vec{AD}$$
$$(\vec{a} + \vec{b}) + \vec{c} = (\vec{AB} + \vec{BC}) + \vec{CD}$$
$$= \vec{AC} + \vec{CD} = \vec{AD}$$

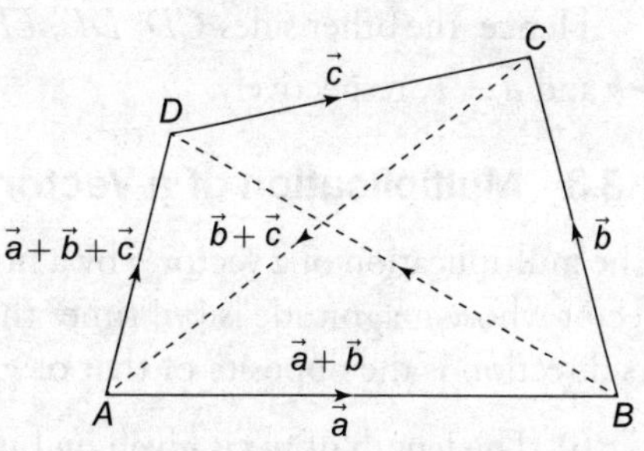

Figure 1.21 Associative property

Since both $\vec{a} + (\vec{b} + \vec{c})$ and $(\vec{a} + \vec{b}) + \vec{c}$ are the same vector $\vec{AD}$, they are equal and are simply written as $\vec{a}+\vec{b}+\vec{c}$. Thus, the vector satisfies the associative law of addition, i.e., $\vec{a} + (\vec{b} + \vec{c}) = (\vec{a} + \vec{b}) + \vec{c}$. The associative property of vector addition shows that the sum of three vectors $\vec{a}$, $\vec{b}$ and $\vec{c}$ is independent of the order in which they are added and of their grouping from partial sums, and is written as $\vec{a} + \vec{b} + \vec{c}$.

(iii) **Existence of identity:** Zero vector has the properties $\vec{AA} = \vec{0}$ and $\vec{BB} = \vec{0}$. Let $\vec{AB} = \vec{a}$. Then, by the triangle law of addition

$$\vec{a} + \vec{0} = \vec{AB} + \vec{BB} = \vec{AB}$$
$$= \vec{AA} + \vec{AB} = \vec{0} + \vec{a}$$

Thus, the vector $\vec{0}$ is such that $\vec{a} + \vec{0} = \vec{0} + \vec{a} = \vec{a}$. Therefore, $\vec{0}$ is the additive identity.

(iv) **Existence of inverse:** A vector $\vec{a}$ is the negative of the vector $\vec{b}$ if $\vec{a} + \vec{b} = \vec{0}$. This is denoted by $(-\vec{a})$. By the triangle law of addition

$$\vec{AB} + \vec{BA} = \vec{AA} = \vec{0}$$

Thus, $\vec{AB}$ is the negative of $\vec{BA}$. The negative of $\vec{AB}$ is a vector $\vec{BA}$ of the same magnitude and support but of opposite direction.

(v) **Closure axiom:** We know that the result of the addition of two vectors is also a vector quantity. Therefore, we conclude that the set of all vectors forms an abelian group with respect to vector addition.

Example 1.3.3

If $\vec{a}$ and $\vec{b}$ be the vectors determined by the two adjacent sides of a regular hexagon, then find the vectors determined by the other sides taken in order.

Solution

The adjacent sides $\vec{AB}$ and $\vec{BC}$ of the regular hexagon $ABCDEF$ (Fig. 1.22) represent the vectors $\vec{a}$ and $\vec{b}$, respectively. Then

$$\vec{AC} = \vec{AB} + \vec{BC} = \vec{a} + \vec{b}, \quad \vec{AD} = 2\,\vec{BC} = 2\,\vec{b}$$
$$\vec{CD} = \vec{AD} - \vec{AC} = 2\vec{b} - (\vec{a} + \vec{b}) = \vec{b} - \vec{a}$$

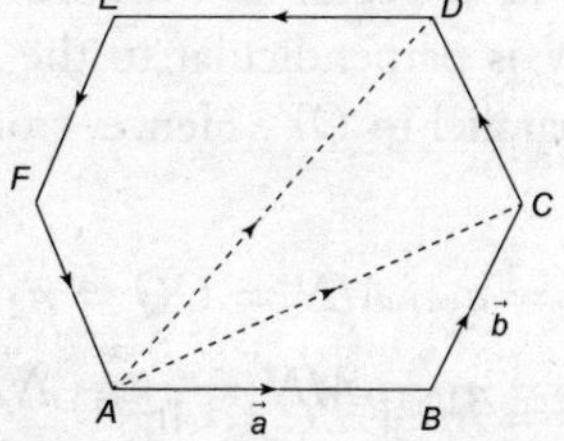

Figure 1.22 Regular hexagon

Since DE is equal to AB but opposite in direction, we have $\vec{DE} = -\vec{a}$. Similarly, $\vec{EF} = -\vec{b}$ and $\vec{FA} = -\vec{CD} = -(\vec{b} - \vec{a}) = \vec{a} - \vec{b}$.

Hence, the other sides $\vec{CD}$, $\vec{DE}$, $\vec{EF}$ and $\vec{FA}$ of the hexagon represent the vectors $(\vec{b}-\vec{a}), -\vec{a}$, $-\vec{b}$ and $\vec{a}-\vec{b}$, respectively.

1.3.3 Multiplication of a Vector by a Scalar

The multiplication of a vector $\vec{a}$ by a non-zero scalar m produces a vector $m\vec{a}$ or $\vec{a}\,m$. This is a new vector whose magnitude is $|m|$ times the magnitude of $\vec{a}$ and parallel to $\vec{a}$. If $m > 0$ and if $m < 0$, its direction is the opposite of that $\vec{a}$.

(i) The length of $m\vec{a}$ is given by $|m\vec{a}| = |m|\,|\vec{a}|$, i.e., the length of the vector $m\vec{a}$ is m times or $-m$ times that of $\vec{a}$, depending of whether m is positive (including zero) or negative. Thus

$$\vec{b} = m\,\vec{a} \Rightarrow m = |\vec{b}|/|\vec{a}|$$

(ii) The support of $m\vec{a}$ is the same or parallel to that of $\vec{a}$.

(iii) The sense of $m\vec{a}$ is the same or opposite of that of $\vec{a}$, depending of whether $m > 0$ or $m < 0$.

(iv) If $m = 0$, $m\,\vec{a} = \vec{0}$, the null vector.

The following results are immediate consequences of the above definition of the multiplication of vectors by scalars:

(i) $-\vec{a} = (-1)\,\vec{a}$ (ii) $\vec{a} = |\vec{a}|\,\hat{a} = a\,\hat{a}$ (iii) $0\,\vec{a} = \vec{0}$; zero vector

(iv) $\vec{a}\,m = \hat{a}(am) = (am)\hat{a} = m(a\hat{a}) = m\,\vec{a}$

(v) $m(n\vec{a}) = mn\vec{a} = n(m\vec{a})$; associative

(vi) $(m+n)\vec{a} = m\vec{a} + n\vec{a}$; (vii) $m(\vec{a}+\vec{b}) = m\vec{a} + m\vec{b}$; distributive

(viii) If $\vec{a} = m\,\vec{b}$, then $\vec{a}$ and $\vec{b}$ are parallel

This shows that vectors obey the associative and distributive laws of multiplication by a scalar. Note that in these laws, only multiplication of a vector by one or more scalars is defined.

1.4 COMPONENTS OF A VECTOR

We consider a rectangular Cartesian coordinate system ox, oy and oz with O as origin (Fig. 1.23). $\hat{i}, \hat{j}$ and $\hat{k}$ are the unit vectors along the respective mutually perpendicular axes. The set $\{\hat{i}, \hat{j}, \hat{k}\}$ is called an *orthogonal triad of vectors*. Let (a_1, a_2, a_3) be the rectangular coordinates of the terminal point P of a vector $\vec{a}$ with initial point at O. Now, PN is perpendicular to the XOY plane and MN is parallel to OY. Hence, from geometry, we have

$$OM = a_1, \quad MN = OQ = a_2 \quad \text{and} \quad NP = a_3$$

$$\Rightarrow \vec{OM} = a_1\,\hat{i}, \ \vec{MN} = a_2\,\hat{j}, \ \vec{NP} = a_3\,\hat{k}$$

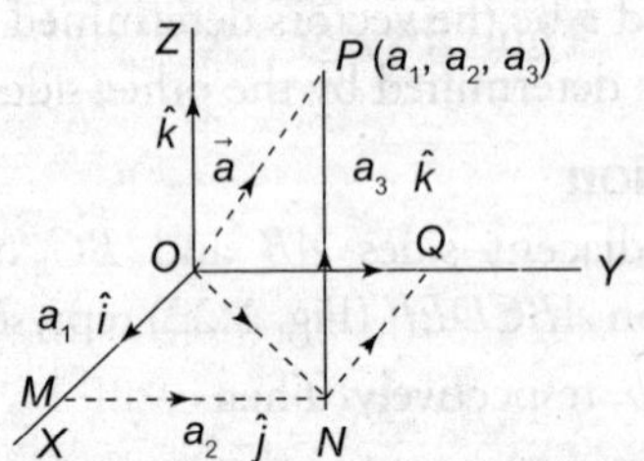

Figure 1.23 Components of a vector

From the triangle OMN, by the triangle law of addition, we get

$$\vec{ON} = \vec{OM} + \vec{MN} = a_1\,\hat{i} + a_2\,\hat{j}$$

From the triangle ONP, we get

$$\vec{OP} = \vec{ON} + \vec{NP} = (a_1\,\hat{\imath} + a_2\,\hat{\jmath}) + a_3\,\hat{k}$$

$$\Rightarrow \quad \vec{a} = a_1\,\hat{\imath} + a_2\,\hat{\jmath} + a_3\,\hat{k} \tag{1.4}$$

Thus, the *position vector* or *radius vector* $\vec{a}$ from O to the point $P(a_1, a_2, a_3)$ can be expressed as a linear combination of $\hat{\imath}, \hat{\jmath}$ and $\hat{k}$ as in Eq. (1.4). In terms of components, we write $\vec{a} = (a_1, a_2, a_3)$. Also, we have

$$OP^2 = ON^2 + NP^2 = OM^2 + OQ^2 + NP^2$$

$$\Rightarrow \quad a^2 = a_1^2 + a_2^2 + a_3^2;\ a = \text{ magnitude of } \vec{a}$$

$$\Rightarrow \quad a = |\vec{a}| = \sqrt{a_1^2 + a_2^2 + a_3^2}$$

Thus, the modulus of a vector is the square root of the sum of the squares of its rectangular components. The unit vector in the direction of $\vec{a}$ is

$$\hat{a} = \frac{1}{|\vec{a}|}\,\vec{a} = \frac{1}{a}(a_1, a_2, a_3) = \frac{1}{\sqrt{a_1^2 + a_2^2 + a_3^2}}\left[a_1\,\hat{\imath} + a_2\,\hat{\jmath} + a_3\,\hat{k}\right] \tag{1.5}$$

whose components along the coordinate axes are

$$\frac{a_1}{a} = \frac{a_1}{\sqrt{a_1^2 + a_2^2 + a_3^2}},\ \frac{a_2}{a} = \frac{a_2}{\sqrt{a_1^2 + a_2^2 + a_3^2}},\ \frac{a_3}{a} = \frac{a_3}{\sqrt{a_1^2 + a_2^2 + a_3^2}}$$

respectively. These are called the *direction cosines* of the line OP and are, respectively, denoted by l, m and n. The vectors $a_1\,\hat{\imath}, a_2\,\hat{\jmath}$ and $a_3\,\hat{k}$ are called the *orthogonal projections* or *rectangular component vectors* or simply *component vectors* of $\vec{a}$ in the x, y and z directions, respectively. The quantities a_1, a_2 and a_3 are called the *rectangular components* or simply *components* of $\vec{a}$ in the x, y and z directions, respectively. Thus, any vector can be resolved into three orthogonal components.

(i) Further, let OP makes angles α, β and γ with rectangular axes at O. Then $\cos\alpha, \cos\beta$ and $\cos\gamma$ are called the *direction cosines* of the line OP, and we write

$$a_1 = |\overrightarrow{OP}|\cos\alpha,\ a_2 = |\overrightarrow{OP}|\cos\beta,\ a_3 = |\overrightarrow{OP}|\cos\gamma$$

Thus, the unit vector is the direction of $\overrightarrow{OP}$ and is given by

$$\hat{a} = \frac{1}{|\overrightarrow{OP}|}\,\overrightarrow{OP} = \cos\alpha\,\hat{\imath} + \cos\beta\,\hat{\jmath} + \cos\gamma\,\hat{k} \tag{1.6}$$

Hence, the coefficients of $\hat{\imath}, \hat{\jmath}$ and $\hat{k}$ form the rectangular resolution of a unit vector and give the direction cosines of that vector.

(ii) Three mutually perpendicular unit vectors are called the *fundamental vectors.*

(iii) Let $\vec{r_p} = (x_p, y_p, z_p)$ be the position vector of several points with respect to some base point O. Then

$$\sum \vec{r_p} = \sum(x_p\,\hat{\imath} + y_p\,\hat{\jmath} + z_p\,\hat{k}) = (\sum x_p)\hat{\imath} + (\sum y_p)\hat{\jmath} + (\sum z_p)\hat{k}$$

as the orientation of the triad $\hat{i}, \hat{j}, \hat{k}$ may be chosen arbitrarily. Thus, we may say that the projection of a sum of vectors on any plane is equal to the sum of the projection of the individual vectors on that plane.

RESULT 4

Let $\vec{r}_1 = a_1\hat{i} + a_2\hat{j} + a_3\hat{k}$ and $\vec{r}_2 = b_1\hat{i} + b_2\hat{j} + b_3\hat{k}$ be, respectively, the position vectors of the two points P and Q with respect to some base point O in component form. Then

$$\vec{PQ} = \vec{r}_2 - \vec{r}_1 = (b_1 - a_1)\hat{i} + (b_2 - a_2)\hat{j} + (b_3 - a_3)\hat{k}$$

Thus, the modulus of $\vec{PQ}$ is

$$\overline{PQ} = |\vec{r}_2 - \vec{r}_1| = \sqrt{(b_1 - a_1)^2 + (b_2 - a_2)^2 + (b_3 - a_3)^2}$$

The unit vector in the direction of $\vec{PQ}$ is

$$\begin{aligned} \hat{PQ} &= \frac{1}{|\vec{PQ}|}\vec{PQ} \\ &= \frac{1}{|\vec{r}_2 - \vec{r}_1|}(\vec{r}_2 - \vec{r}_1) \\ &= \frac{1}{\overline{PQ}}\left[(b_1 - a_1)\hat{i} + (b_2 - a_2)\hat{j} + (b_3 - a_3)\hat{k}\right] \end{aligned}$$

whose components along the coordinate axes are $\frac{b_1 - a_1}{\overline{PQ}}, \frac{b_2 - a_2}{\overline{PQ}}$ and $\frac{b_3 - a_3}{\overline{PQ}}$, respectively, which are the direction cosines of the line PQ.

Example 1.4.1

If the position vectors of A and B with respect to O be $\hat{i} + 3\hat{j} - 7\hat{k}$ and $5\hat{i} - 2\hat{j} + 4\hat{k}$, respectively, then find $\vec{AB}$ and determine its direction cosines and modulus. Find also the unit vector parallel to $\vec{AB}$.

Solution

As in Fig. 1.24, let $\vec{OA} = \hat{i} + 3\hat{j} - 7\hat{k}$, $\vec{OB} = 5\hat{i} - 2\hat{j} + 4\hat{k}$ and let $\vec{AB} = \vec{p}$. Using the triangle law of addition on triangle OAB, we get

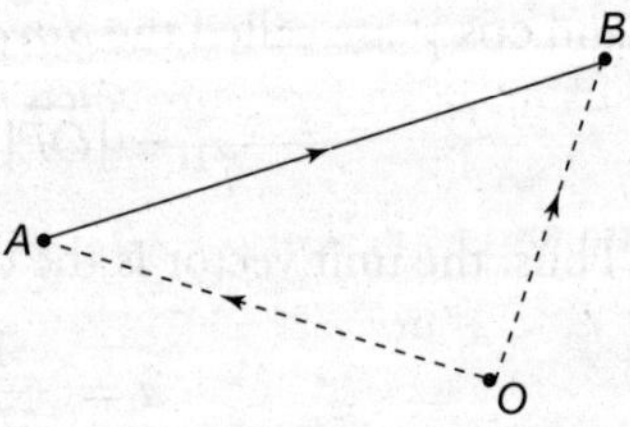

Figure 1.24 Two points

$$\vec{OA} + \vec{AB} = \vec{OB} \Rightarrow \vec{AB} = \vec{OB} - \vec{OA}$$

$$\Rightarrow \vec{p} = (5\hat{i} - 2\hat{j} + 4\hat{k}) - (\hat{i} + 3\hat{j} - 7\hat{k}) = 4\hat{i} - 5\hat{j} + 11\hat{k}$$

$$\Rightarrow |\vec{p}| = \sqrt{4^2 + (-5)^2 + 11^2} = \sqrt{162} = 9\sqrt{2} = |\vec{AB}|$$

Its direction cosines are $\frac{4}{9\sqrt{2}}, \frac{-5}{9\sqrt{2}}$ and $\frac{11}{9\sqrt{2}}$. The unit vector parallel to $\vec{AB}$ is

$$\frac{1}{|\vec{p}|}\vec{p} = \frac{1}{9\sqrt{2}}(4\hat{i} - 5\hat{j} + 11\hat{k})$$

Example 1.4.2

If a vector $\vec{r} = x\hat{i} + y\hat{j} + z\hat{k}$ makes angles α, β and γ with the x, y and z axes, respectively, what are the values of $\cos\alpha, \cos\beta$ and $\cos\gamma$ and show that $\cos^2\alpha + \cos^2\beta + \cos^2\gamma = 1$.

Solution

The vector $\vec{r} = x\hat{i}+y\hat{j}+z\hat{k}$ makes angles α, β and γ with the x, y and z axes, respectively. Referring to Fig. 1.25, $\triangle OAP$ is a right-angled triangle with the right angle at A. Therefore

$$\cos\alpha = \frac{x}{|\vec{r}|} = \frac{x}{r}, \quad \text{where } r = |\vec{r}| = \sqrt{x^2 + y^2 + z^2}$$

Similarly, from right-angled triangles OBP and OCP

$$\cos\beta = \frac{y}{|\vec{r}|} = \frac{y}{r}; \quad \cos\gamma = \frac{z}{|\vec{r}|} = \frac{z}{r}$$

Thus, from the relations $\cos\alpha = \frac{x}{r}$, $\cos\beta = \frac{y}{r}$ and $\cos\gamma = \frac{z}{r}$, the angles α, β and γ can be obtained. From these relations, it follows that

$$\cos^2\alpha + \cos^2\beta + \cos^2\gamma = \frac{x^2 + y^2 + z^2}{r^2} = \frac{r^2}{r^2} = 1$$

The numbers $\cos\alpha, \cos\beta$ and $\cos\gamma$ are called the *direction cosines* of the vector $\vec{OP}$.

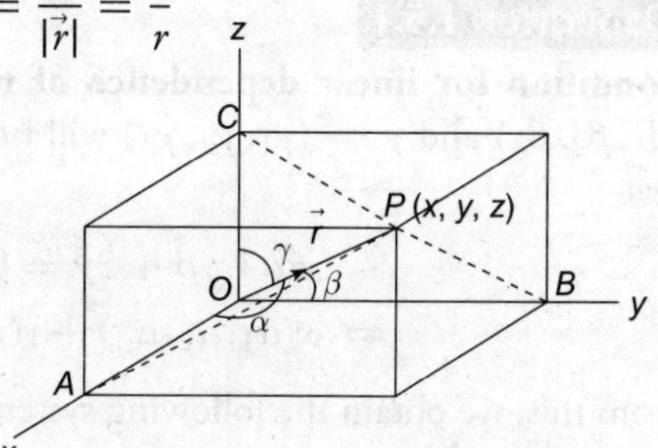

Figure 1.25 Direction cosines

1.5 LINEAR COMBINATION OF VECTORS

A vector $\vec{r}$ is said to be a *linear combination of the vectors* $\vec{\alpha}_1, \vec{\alpha}_2, \cdots, \vec{\alpha}_n$ if there exist scalars $x_1, x_2, \ldots, x_n$ such that

$$\vec{r} = x_1\vec{\alpha}_1 + x_2\vec{\alpha}_2 + \cdots + x_n\vec{\alpha}_n \tag{1.7}$$

The scalars $x_1, x_2, \cdots, x_n$ are called the *coefficients of the linear combination.* For example, $5\vec{\alpha}_1 + 3\vec{\alpha}_2 - \vec{\alpha}_3$ is a linear combination of the vectors $\vec{\alpha}_1, \vec{\alpha}_2$ and $\vec{\alpha}_3$. The linear combinations $x\vec{\alpha}_1$, $x\vec{\alpha}_1 + y\vec{\alpha}_2$ and $x\vec{\alpha}_1 + y\vec{\alpha}_2 + z\vec{\alpha}_3$ will be of special interest to us.

Example 1.5.1

If $\vec{r_1} = 2\hat{i} - \hat{j} + \hat{k}$, $\vec{r_2} = \hat{i} + 3\hat{j} - 2\hat{k}$, $\vec{r_3} = -2\hat{i} + \hat{j} - 3\hat{k}$ and $\vec{r_4} = 3\hat{i} + 2\hat{j} + 5\hat{k}$, find scalars a, b and c such that $\vec{r_4} = a\vec{r_1} + b\vec{r_2} + c\vec{r_3}$.

Solution

Since $\vec{r_4} = a\vec{r_1} + b\vec{r_2} + c\vec{r_3}$, we have

$$3\hat{i} + 2\hat{j} + 5\hat{k} = a(2\hat{i} - \hat{j} + \hat{k}) + b(\hat{i} + 3\hat{j} - 2\hat{k}) + c(-2\hat{i} + \hat{j} - 3\hat{k})$$

$$\text{or,}\quad 3\hat{i} + 2\hat{j} + 5\hat{k} = (2a + b - 2c)\hat{i} + (-a + 3b + c)\hat{j} + (a - 2b - 3c)\hat{k}$$

Since $\hat{i}, \hat{j}$ and $\hat{k}$ are non-coplanar, we have

$$(2a + b - 2c) = 3, \;\; (-a + 3b + c) = 2, \;\; (a - 2b - 3c) = 5$$
$$\Rightarrow a = -2, \;\; b = 1, \;\; c = -3$$

And so, $\vec{r}_4 = -2\vec{r}_1 + \vec{r}_2 - 3\vec{r}_3$.

1.5.1 Linearly Dependent

The set of vectors $\{\vec{\alpha}, \vec{\beta}, \vec{\gamma}\}$ is said to be linearly dependent, if there exist scalars x, y and z, not all zero, such that

$$x\,\vec{\alpha} + y\,\vec{\beta} + z\,\vec{\gamma} = \vec{0} \tag{1.8}$$

DEDUCTION 1.5.1

Condition for linear dependence of three vectors: Three vectors $\vec{\alpha} = (\alpha_1, \alpha_2, \alpha_3)$, $\vec{\beta} = (\beta_1, \beta_2, \beta_3)$ and $\vec{\gamma} = (\gamma_1, \gamma_2, \gamma_3)$ will be linearly dependent if there exist scalars x, y and z such that

$$x\vec{\alpha} + y\vec{\beta} + z\vec{\gamma} = \vec{0}$$
$$\Rightarrow x(\alpha_1, \alpha_2, \alpha_3) + y(\beta_1, \beta_2, \beta_3) + z(\gamma_1, \gamma_2, \gamma_3) = (0, 0, 0)$$

From this, we obtain the following system of equations:

$$x\alpha_i + y\beta_i + z\gamma_i = 0; \;\; i = 1, 2, 3$$

This system will have a non-trivial solution if

$$\begin{vmatrix} \alpha_1 & \alpha_2 & \alpha_3 \\ \beta_1 & \beta_2 & \beta_3 \\ \gamma_1 & \gamma_2 & \gamma_3 \end{vmatrix} = 0 \tag{1.9}$$

which is the condition for dependency of the vectors $\vec{\alpha}, \vec{\beta}$ and $\vec{\gamma}$.

Example 1.5.2

If $\vec{\alpha} = (1, -2, 3)$, $\vec{\beta} = (-2, 3, -4)$ and $\vec{\gamma} = (0, -1, 2)$, show that they form a linearly dependent system.

Solution

Let there be two scalars x and y such that $\vec{\alpha} = x\vec{\beta} + y\vec{\gamma}$, i.e.,

$$(1, -2, 3) = x(-2, 3, -4) + y(0, -1, 2)$$
$$\Rightarrow \;\; -2x = 1, 3x - y = -2, -4x + 2y = 3$$

Solving the first two equations, we get $x = -1/2$, $y = 1/2$, which will satisfy the third equation in x, y. Hence, the given vectors are related as $2\vec{\alpha} + \vec{\beta} - \vec{\gamma} = \vec{0}$, i.e., they are linearly dependent.

1.5.2 Linearly Independent

The set of vectors $\{\vec{\alpha}, \vec{\beta}, \vec{\gamma}\}$ is said to be *linearly independent*, if there exist scalars x, y and z such that

$$x\,\vec{\alpha} + y\,\vec{\beta} + z\,\vec{\gamma} = \vec{0} = \vec{0} \Rightarrow x = y = z = 0 \tag{1.10}$$

DEDUCTION 1.5.2

Condition for linear independence of three vectors: Let three vectors $\vec{\alpha} = (\alpha_1, \alpha_2, \alpha_3)$, $\vec{\beta} = (\beta_1, \beta_2, \beta_3)$ and $\vec{\gamma} = (\gamma_1, \gamma_2, \gamma_3)$ be linearly independent, and the vector $\vec{\delta} = (\delta_1, \delta_2, \delta_3)$ be dependent on $\vec{\alpha}, \vec{\beta}$ and $\vec{\gamma}$. Then, $\vec{\delta}$ can be expressed as a linear combination of $\vec{\alpha}, \vec{\beta}$ and $\vec{\gamma}$ as

$$\vec{\delta} = x\vec{\alpha} + y\vec{\beta} + z\vec{\gamma} \quad \text{i.e., } x\alpha_i + y\beta_i + z\gamma_i = \delta_i; \quad i = 1, 2, 3$$

Now, the above system will have a solution if

$$\begin{vmatrix} \alpha_1 & \alpha_2 & \alpha_3 \\ \beta_1 & \beta_2 & \beta_3 \\ \gamma_1 & \gamma_2 & \gamma_3 \end{vmatrix} \neq 0 \tag{1.11}$$

which is the condition for independence of the vectors $\vec{\alpha}, \vec{\beta}$ and $\vec{\gamma}$.

Example 1.5.3

Show that the vectors $\vec{\alpha} = (1, 2, 3)$, $\vec{\beta} = (-2, 1, -4)$ and $\vec{\gamma} = (1, -1, 2)$ are linearly independent.

Solution

Let there be two scalars x and y such that $\vec{\alpha} = x\vec{\beta} + y\vec{\gamma}$, i.e.,

$$(1, 2, 3) = x(-2, 1, -4) + y(1, -1, 2)$$

$$\Rightarrow \; -2x + y = 1, x - y = 2, -4x + 2y = 3$$

Solving the first two equations, we get $x = -3, y = -5$, which does not satisfy the third equation in x and y. Hence, the given vectors are not related, i.e., they are linearly independent.

1.6 SECTION RATIO (POINT OF DIVISION)

(a) **Internal:** Let the point C divide AB internally in the ratio $m : n$, where m and n are non-zero positive real numbers (Fig. 1.26). Let the position vectors of two points A and B be $\vec{a}$ and $\vec{b}$, respectively, with respect to the same base point O, i.e., $\vec{OA} = \vec{a}$ and $\vec{OB} = \vec{b}$.

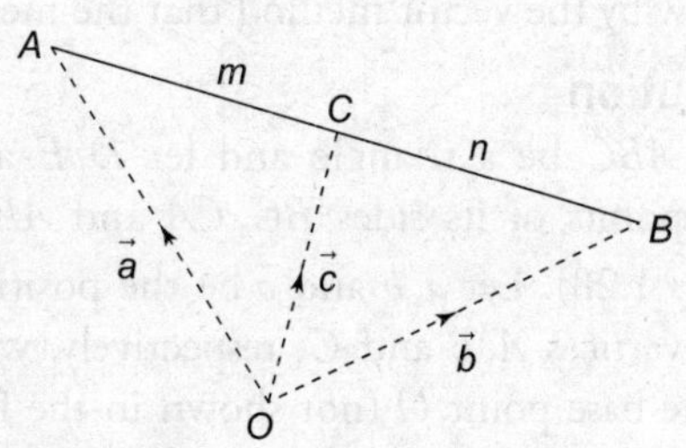

Figure 1.26 Point of division (internal)

In $\triangle OAC$ and $\triangle OCB$, using the triangle law of addition, we have

$$\vec{OA} + \vec{AC} = \vec{OC} \Rightarrow \vec{AC} = \vec{OC} - \vec{OA}$$

$$\vec{OC} + \vec{CB} = \vec{OB} \Rightarrow \vec{CB} = \vec{OB} - \vec{OC}$$

Since C divides AB in the ratio $m : n$, i.e., $AC : CB = m : n$, we have

$$n\,\vec{AC} = m\,\vec{CB} \Rightarrow n(\vec{OC} - \vec{OA}) = m(\vec{OB} - \vec{OC})$$
$$\Rightarrow (m+n)\vec{OC} = m\vec{OB} + n\vec{OA}$$
$$\Rightarrow \vec{OC} = \vec{c} = \frac{1}{m+n} m\vec{OB} + n\vec{OA}$$
$$\Rightarrow \vec{c} = \frac{1}{m+n} m\vec{b} + n\vec{a}, \quad m+n \neq 0 \tag{1.12}$$

This gives the position vector of C with respect to the base point O. In particular, if C is the midpoint or bisecting point of AB $(m = n)$, then from Eq. (1.12)

$$\vec{c} = \frac{1}{m+m} m\,\vec{b} + m\,\vec{a} = \frac{1}{2}(\vec{a} + \vec{b}) \tag{1.13}$$

(b) **External:** Let A and B be two points and $\vec{a}$ and $\vec{b}$ be their position vectors relative to a fixed point O, chosen as origin (Fig. 1.27). Let $\vec{c}$ be the position vector of the point C which divides AB externally such that $AC : BC = m : n$. Then, we have as before

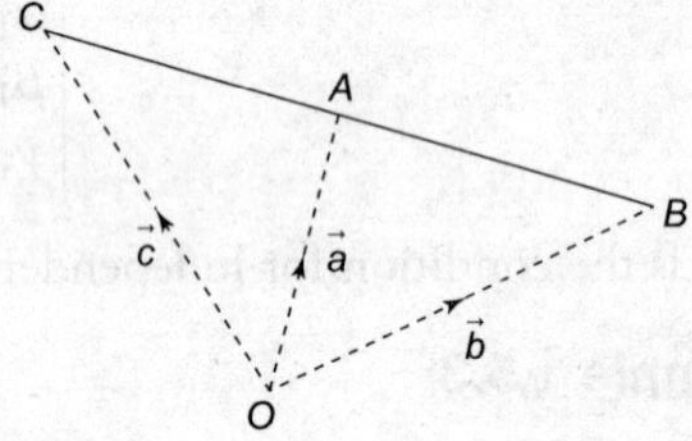

Figure 1.27 Point of division (external)

$$m\,\vec{AC} = n\,\vec{BC}$$
$$\Rightarrow m(\vec{OC} - \vec{OA}) = n(\vec{OC} - \vec{OB})$$
$$\Rightarrow (m-n)\vec{OC} = m\vec{OA} - n\vec{OB}$$
$$\Rightarrow \vec{OC} = \vec{c} = \frac{1}{m-n} m\vec{a} - n\vec{b}, \quad m \neq n \tag{1.14}$$

This gives the position vector of C with respect to the base point O. Thus, the result Eq. (1.14) holds good for $m : n$ regardless of whether it is positive or negative. If the ratio $m : n$ lies between 0 and -1, C is outside AB and closer to A, whereas if $m : n$ lies between -1 and $-\infty$, C is outside AB and closer to B.

Example 1.6.1

Show by the vector method that the medians of a triangle are concurrent.

Solution

Let ABC be a triangle and let D, E and F be the midpoints of its sides BC, CA and AB, respectively (Fig. 1.28). Let $\vec{a}, \vec{b}$ and $\vec{c}$ be the position vectors of the vertices A, B and C, respectively, with respect to some base point O (not shown in the figure). Then, the position vectors of the midpoint D, E and F are, respectively,

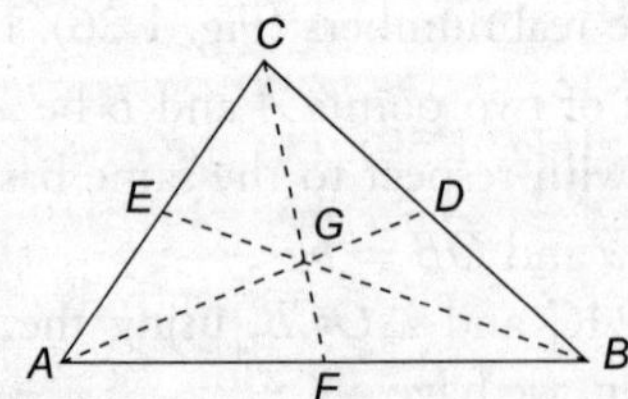

Figure 1.28 Medians of a triangle

$$\vec{OD} = \frac{1}{2}(\vec{b} + \vec{c}), \vec{OE} = \frac{1}{2}(\vec{c} + \vec{a}), \vec{OF} = \frac{1}{2}(\vec{a} + \vec{b})$$

Let us take a point G on the median AD that divides AD in the ratio 2 : 1. Then, the position vector of G is

$$= \frac{1}{2+1}\left[2 \cdot \frac{1}{2}(\vec{a}+\vec{b}) + 1\vec{c}\right] = \frac{1}{3}(\vec{a}+\vec{b}+\vec{c})$$

Similarly, it can be shown that the position vector of the points dividing the median BE and CF internally in the ratio 2 : 1 are each equal to $\frac{1}{3}(\vec{a}+\vec{b}+\vec{c})$. Therefore, the symmetry of the expression shows that all these points coincide with G. Hence, the medians of the triangle are concurrent. The point of concurrence G is the *centroid of the triangle.*

Example 1.6.2

Show that the lines joining the vertices to the centroid of the opposite faces of a tetrahedron are concurrent.

Solution

Let $ABCD$ be a tetrahedron (Fig. 1.29). Let $\vec{a}, \vec{b}, \vec{c}$ and $\vec{d}$ be the position vectors of the points A, B, C and D with respect to some base point O. Let E, F, G and H be the centroids of the faces $\triangle BCD$, $\triangle CDA$, $\triangle DAB$ and $\triangle ABC$, respectively. Then, the position vectors of E, F, G and H are, respectively, $\frac{1}{3}(\vec{b}+\vec{c}+\vec{d})$, $\frac{1}{3}(\vec{c}+\vec{d}+\vec{a})$, $\frac{1}{3}(\vec{d}+\vec{a}+\vec{b})$ and $\frac{1}{3}(\vec{a}+\vec{b}+\vec{c})$. If the point $P(\vec{r})$ divides the median AE internally in the ratio 3 : 1, then the position vector of P is

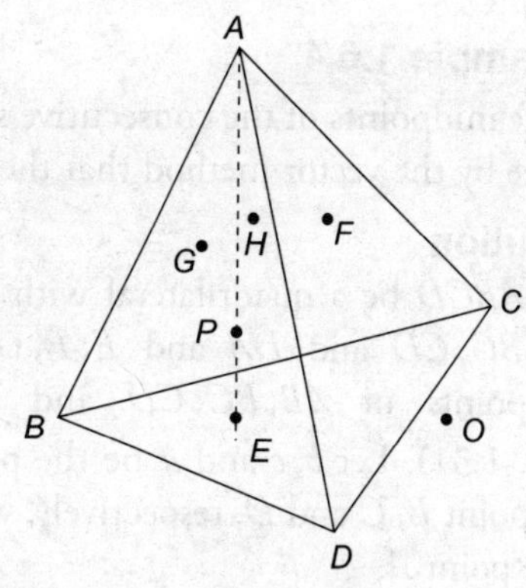

Figure 1.29 Tetrahedron

$$\vec{r} = \frac{1}{3+1}\left[3 \cdot \frac{1}{3}(\vec{b}+\vec{c}+\vec{d}) + 1\vec{a}\right] = \frac{1}{4}(\vec{a}+\vec{b}+\vec{c}+\vec{d})$$

The result is symmetric about $\vec{a}, \vec{b}, \vec{c}$ and $\vec{d}$ and this suggests that the lines BF, CG and DF are also divided internally by the point P in the ratio 3 : 1. Thus, the lines AE, BF, CG and DF are concurrent.

Example 1.6.3

Prove that the straight line joining the midpoints of the two non-parallel sides of a trapezium is parallel to the parallel sides and half of their sum.

Solution

Let $ABCD$ be a trapezium, whose non-parallel sides are AD and BC (Fig. 1.30). Let the middle points of AD and BC be E and F, respectively. Let the position vectors of B, C and D be $\vec{b}, \vec{c}$ and $\vec{d}$, respectively, with respect to the base point A. Thus, the position vectors of E and F are $\frac{1}{2}\vec{d}$ and $\frac{1}{2}(\vec{b}+\vec{c})$, respectively. Since $\vec{AB}$ and $\vec{DC}$ are parallel, we have

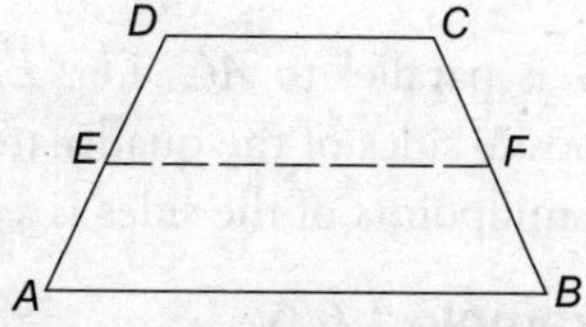

Figure 1.30 Trapezium

$$\vec{AB} = t\vec{DC} \Rightarrow \vec{b} = t(\vec{c}-\vec{d}); \qquad t = \text{ scalar}$$

Now, we are to show that $\vec{EF}$ is parallel to $\vec{AB}$ or $\vec{DC}$.

$$\vec{EF} = \vec{AF} - \vec{AE} = \frac{1}{2}(\vec{b} + \vec{c}) - \frac{1}{2}\vec{d}$$

$$= \frac{1}{2}\left[t(\vec{c} - \vec{d}) + \vec{c} - \vec{d}\right] = \frac{1}{2}(t+1)(\vec{c} - \vec{d})$$

$$\Rightarrow \vec{EF} = \frac{1}{2}(t+1)\vec{DC} = \frac{1}{2}\left(\frac{t+1}{t}\right)\vec{AB}$$

Thus, $\vec{EF}$ and $\vec{DC}$(or $\vec{AB}$) are parallel. Now

$$|\vec{EF}| = \frac{1}{2}(t+1)|\vec{DC}| = \frac{1}{2}\left[t|\vec{DC}| + |\vec{DC}|\right] = \frac{1}{2}\left[|\vec{AB}| + |\vec{DC}|\right]$$

Thus, the straight line EF joining the midpoints of two non-parallel sides AD and BC of a trapezium is parallel to the parallel sides and half of their sum.

Example 1.6.4

If the midpoints of the consecutive sides of any quadrilateral are connected by straight lines, then prove by the vector method that the resulting quadrilateral is a parallelogram.

Solution

Let $ABCD$ be a quadrilateral with consecutive sides AB, BC, CD and DA and E, F, G and H be the midpoints of AB, BC, CD and DA, respectively (Fig. 1.31). Let $\vec{b}, \vec{c}$ and $\vec{d}$ be the position vectors of the point B, C and D, respectively, with respect to the base point A.

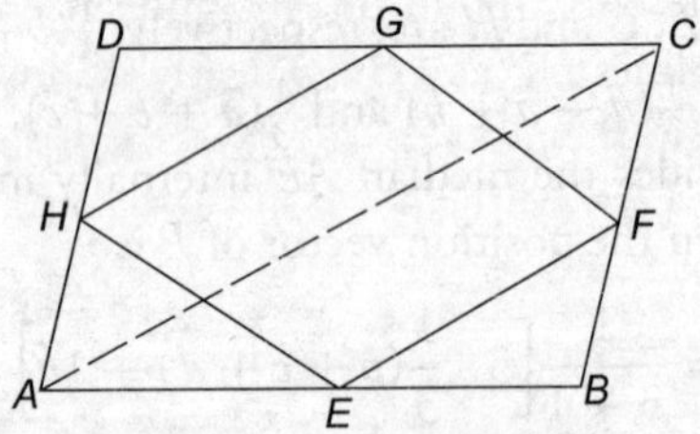

Figure 1.31 Quadrilateral

Then, the position vectors of the midpoints E, F, G and H are $\frac{1}{2}\vec{b}$, $\frac{1}{2}(\vec{b}+\vec{c})$, $\frac{1}{2}(\vec{c}+\vec{d})$ and $\frac{1}{2}\vec{d}$, respectively. Add the diagonal AC. Then, $\vec{AC} = \vec{c}$ and

$$\vec{EF} = \frac{1}{2}(\vec{b} + \vec{c}) - \frac{1}{2}\vec{b} = \frac{1}{2}\vec{c} \Rightarrow \vec{AC} = 2\vec{EF}$$

Therefore, $\vec{EF}$ is parallel to $\vec{AC}$, and similarly

$$\vec{HG} = \frac{1}{2}(\vec{c} + \vec{d}) - \frac{1}{2}\vec{d} = \frac{1}{2}\vec{AC}$$

$\vec{HG}$ is parallel to $\vec{AC}$, i.e., $\vec{EF}$ is parallel to $\vec{HG}$. Also, $|\vec{EF}| = \frac{1}{2}|\vec{AC}| = |\vec{HG}|$. Thus, the opposite sides of the quadrilateral $EFGH$ are equal. Hence, the quadrilateral obtained by joining the midpoints of the sides is a parallelogram.

Example 1.6.5

$ABCD$ is a parallelogram. P and Q are the midpoints of the sides AB and DC, respectively. Show that DP and BQ trisect AC and are trisected by AC.

Solution

Suppose that with reference to a fixed base point O as origin, the position vectors of the vertices A, B, C and D are $\vec{a}, \vec{b}, \vec{c}$ and $\vec{d}$, respectively (Fig. 1.32). Since $ABCD$ is a parallelogram, $\vec{AB} = \vec{DC}$, i.e., $\vec{b} - \vec{a} = \vec{c} - \vec{d}$. P and Q are the midpoints of AB and DC, respectively. The position vectors $\vec{p}$ and $\vec{q}$ of the points P and Q, respectively, are given by

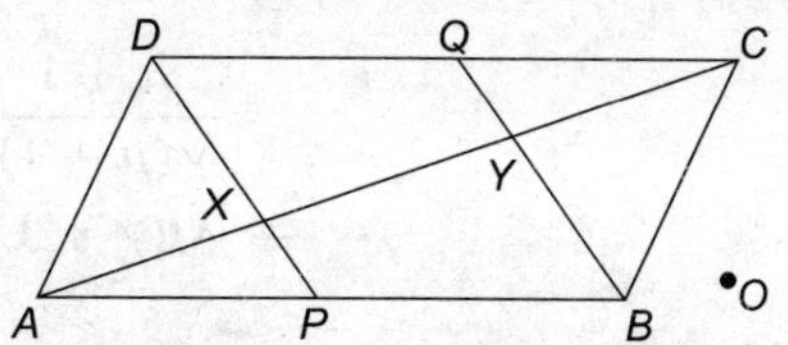

Figure 1.32 Trisection of the diagonal

$$\vec{p} = \frac{1}{2}(\vec{a} + \vec{b}), \quad \vec{q} = \frac{1}{2}(\vec{c} + \vec{d})$$

Let $\vec{x}$ be the position vector of the point X, where DP cuts AC, given by

$$\vec{b} - \vec{a} = \vec{c} - \vec{d}, \quad 2\vec{p} = \vec{a} + \vec{b} \Rightarrow \vec{a} + \vec{c} - \vec{d} + \vec{a} = 2\vec{p}$$

$$\Rightarrow \quad \frac{1}{1+2}(\vec{d} + 2\vec{p}) = \frac{1}{2+1}(2\vec{a} + \vec{c}) = \vec{x}$$

The first number $\frac{1}{1+2}(\vec{d}+2\vec{p})$ represents a point on DP and the second member $\frac{1}{2+1}(2\vec{a}+\vec{c})$ represents a point on AC, and since they are the same, the point is the point of intersection X of the lines DP and AC. Clearly, X divides DP in the ratio 2 : 1 (i.e., X trisects DP) and X divides AC in the ratio 2 : 1 (i.e., AC is trisected by X).

Let $\vec{y}$ be the position vector of the point Y, where BQ cuts AC, given by

$$\vec{b} - \vec{a} = \vec{c} - \vec{d}, \quad 2\vec{q} = \vec{d} + \vec{c}$$

$$\Rightarrow \quad \frac{1}{1+2}(\vec{b} + 2\vec{q}) = \frac{1}{2+1}(2\vec{c} + \vec{a}) = \vec{y}$$

Hence, Y divides BQ in the ratio 2 : 1 and AC in the ratio 2 : 1. This establishes the required results.

Example 1.6.6

If P lies on the line ABC and divides AB, BC and CA in the ratio $\lambda : 1$, $\mu : 1$ and $\gamma : 1$, respectively, then show that $\lambda\mu\gamma + 1 = 0$.

Solution

Let $\vec{a}, \vec{b}$ and $\vec{c}$ be the position vectors of the points A, B and C, respectively, with respect to some origin O. Let $\vec{OP} = \vec{r}$. Hence

$$\vec{r} = \frac{1}{\lambda+1}\left[\lambda\vec{b} + \vec{a}\right] = \frac{1}{\mu+1}\left[\mu\vec{c} + \vec{b}\right] = \frac{1}{\gamma+1}\left[\gamma\vec{a} + \vec{c}\right]$$

$$\text{or,} \quad (\mu+1)\left(\lambda\vec{b} + \vec{a}\right) - (\lambda+1)\left(\mu\vec{c} + \vec{b}\right) = \vec{0}$$

$$\text{or,} \quad (\mu+1)\vec{a} + (\lambda\mu - 1)\vec{b} - \mu(\lambda+1)\vec{c} = \vec{0}$$

Similarly, $\gamma(\mu+1)\vec{a} - (\gamma+1)\vec{b} + (1-\mu\gamma)\vec{c} = \vec{0}$. Comparing these two equations, we have

$$\frac{\mu+1}{\gamma(\mu+1)} = \frac{\lambda\mu-1}{-(\gamma+1)} = \frac{-\mu(\lambda+1)}{1-\mu\gamma}$$

$$\Rightarrow \lambda\mu\gamma + 1 = 0$$

1.7 COLLINEAR AND COPLANAR VECTORS

In this section, we define collinear and coplanar vectors, which are of fundamental importance.

1.7.1 Collinear Vectors

Two or more vectors are said to be *collinear* when they lie in the same line or are parallel to the same line (having the same direction). Two vectors $\vec{a}$ and $\vec{b}$ are collinear when they have the same

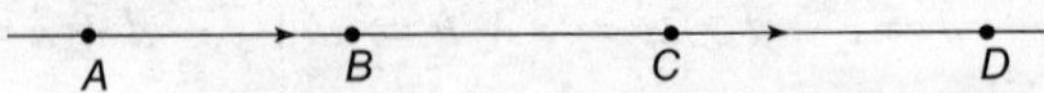

Figure 1.33 Collinear vectors

or parallel supports. In Fig. 1.33, $\vec{AB}$ and $\vec{CD}$ are collinear. If the vectors $\vec{a}$ and $\vec{b}$ are collinear, then there exists an appropriate scalar m such that $\vec{b} = m\,\vec{a}$, i.e., each is a linear combination of the other.

(i) If two collinear free vectors have the same length as well as direction, they are equal.
(ii) It follows that two like vectors are collinear.
(iii) Collinear vectors may or may not be co-initial vectors.
(iv) Collinear vectors are parallel to the same line.

Three vectors ending in the same straight line are called *termino collinear*. The vectors that are not collinear are known as *non-collinear vectors.*

Theorem 1.2 (Collinear vectors)

If $\vec{a}$ and $\vec{b}$ be two proper collinear vectors, then either of them can be expressed as the product of the other by a suitable scalar, the numerical value of the scalar being the ratio of the lengths $\vec{a}$ and $\vec{b}$, and conversely.

Proof: Write $\vec{a} = a\,\hat{a}$ and $\vec{b} = b\,\hat{b}$, where $a = |\vec{a}|$, $b = |\vec{b}|$ and $\hat{a}$ and $\hat{b}$ are unit vectors along $\vec{a}$ and $\vec{b}$, respectively. If $\vec{a}$ and $\vec{b}$ are collinear, then their directions are the same, i.e., $\hat{a} = \hat{b}$. Thus

$$\vec{b} = b\,\hat{b} = \frac{b}{a}\,(a\,\hat{b}) = \frac{b}{a}\,(a\,\hat{a});\ \text{as } \hat{a} = \hat{b}$$

$$\text{i.e.,}\quad \vec{b} = \frac{b}{a}\,\vec{a} = m\,\vec{a},\ \text{where } m = \frac{b}{a} = \frac{\text{length of } \vec{b}}{\text{length of } \vec{a}}$$

Conversely, if $\vec{a} \neq \vec{0}$, then any vector $\vec{b}$, collinear with $\vec{a}$, can be expressed as $m\,\vec{a}$, where m is a suitable scalar. This follows from the definition of multiplication of a vector by a scalar.

Theorem 1.3

If $\vec{a}$ and $\vec{b}$ be two non-zero, non-collinear vectors and m and n be two scalars such that $m\,\vec{a}+n\,\vec{b}=\vec{0}$, then $m=0$ and $n=0$.

Proof: First, suppose that $m \neq 0$. Then

$$m\,\vec{a}+n\,\vec{b}=\vec{0} \Rightarrow \vec{a}=-(n/m)\,\vec{b}$$

i.e., $\vec{a}$ and $\vec{b}$ must be parallel to the same line (collinear), which is a contradiction. Thus, $m=0$ and so $n\,\vec{b}=\vec{0} \Rightarrow n=0$; as $\vec{b} \neq \vec{0}$.

From this theorem, we conclude that if $\vec{a}$ and $\vec{b}$ are non-collinear and

$$m_1\vec{a}+n_1\vec{b}=m_2\vec{a}+n_2\vec{b}$$

$$\Rightarrow (m_1-m_2)\vec{a}+(y_1-y_2)\vec{b}=\vec{0}$$

$$\Rightarrow m_1-m_2=0 \text{ and } n_1-n_2=0 \Rightarrow m_1=m_2 \text{ and } n_1=n_2$$

Theorem 1.4 (Representation theorem)

If $\vec{a}$ and $\vec{b}$ be two non-collinear vectors, then every vector $\vec{r}$, coplanar with $\vec{a}$ and $\vec{b}$, can be uniquely expressed as a linear combination of $\vec{a}$ and $\vec{b}$.

Proof

Existence: Let three vectors $\vec{a}$, $\vec{b}$ and $\vec{p}$ be represented by $\vec{OA}$, $\vec{OB}$ and $\vec{OP}$, repectively, with respect to the base point O, where OA, OB and OP are coplanar (Fig. 1.34). From P, draw PM and PN parallel to BO and AO, respectively, so that they form a parallelogram $OMPN$. Since $\vec{OM}$ and $\vec{ON}$ are parallel to the vectors $\vec{OA}$ and $\vec{OB}$, respectively, there exist suitable scalars m and n, such that

$$\vec{OM}=m\,\vec{OA}=m\,\vec{a} \quad \text{and} \quad \vec{ON}=n\,\vec{OB}=n\,\vec{b}$$

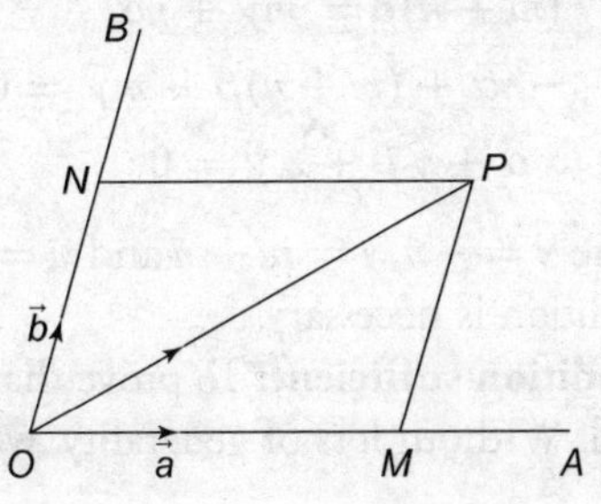

Figure 1.34 Representation theorem

Thus, by the triangle law of addition

$$\vec{OP}=\vec{OM}+\vec{MP}=\vec{OM}+\vec{ON} \Rightarrow \vec{p}=m\,\vec{a}+n\,\vec{b}$$

This shows that $\vec{p}$ is represented as a linear combination of $\vec{a}$ and $\vec{b}$.

Uniqueness: To prove the uniqueness, if possible, let $\vec{p}=m_1\,\vec{a}+n_1\,\vec{b}$. Then, we have

$$m\,\vec{a}+n\,\vec{b}=m_1\,\vec{a}+n_1\,\vec{b} \Rightarrow (m-m_1)\vec{a}+(n-n_1)\vec{b}=\vec{0}$$

Since $\vec{a}$ and $\vec{b}$ are non-collinear, we have $m-m_1=0$, $n-n_1=0$, i.e., $m=m_1$, $n=n_1$. Hence, the uniqueness of the representation.

Result 5

(i) If two equal vectors are expressed in terms of some two collinear vectors, then their corresponding scalar coefficients are equal.

(ii) If $\vec{a}$ and $\vec{b}$ are two non-collinear vectors and $m_1\vec{a} + n_1\vec{b} = m_2\vec{a} + n_2\vec{b}$, then $m_1 = m_2$ and $n_1 = n_2$.

Theorem 1.5 (Condition for the collinearity of three points)

The necessary and sufficient condition for three distinct points A, B and C with positions vectors $\vec{\alpha}, \vec{\beta}$ and $\vec{\gamma}$, respectively, to be collinear is that there exist three scalars x, y and z (not all zero) such that

$$x\vec{\alpha} + y\vec{\beta} + z\vec{\gamma} = \vec{0}; \qquad x + y + z = 0 \tag{1.15}$$

Proof: The position vectors of A, B and C are $\vec{\alpha}, \vec{\beta}$ and $\vec{\gamma}$ with respect to a vector origin O (Fig. 1.35). Then, we have $\vec{AB} = \vec{\beta} - \vec{\alpha}$ and $\vec{AC} = \vec{\gamma} - \vec{\alpha}$. The points A, B and C will be collinear if and only if the vectors $\vec{\beta} - \vec{\alpha}$ and $\vec{\gamma} - \vec{\alpha}$ are collinear.

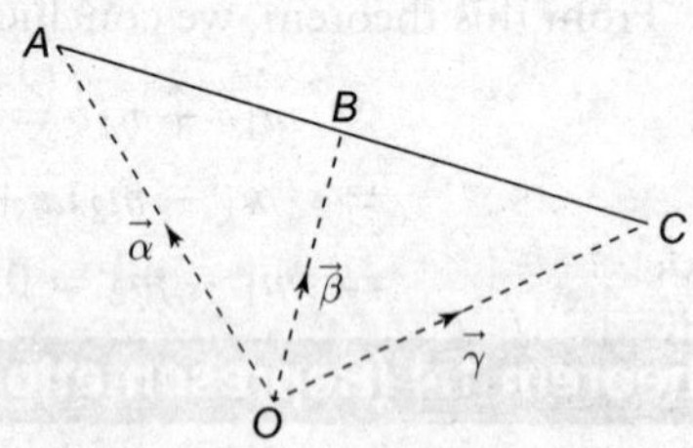

Figure 1.35 Condition for the collinearity of three points

Condition necessary: Let A, B and C be collinear, i.e., they lie on the same line. Let B divide the join of A and C internally in some ratio, say $m : n$. Then

$$\vec{\beta} = \frac{1}{m+n}\left(m\,\vec{\gamma} + n\,\vec{\alpha}\right); \qquad m, n, m+n \neq 0$$

or, $(m+n)\vec{\beta} = m\vec{\gamma} + n\vec{\alpha}$

or, $-n\vec{\alpha} + (m+n)\vec{\beta} - m\vec{\gamma} = \vec{0}$

or, $x\,\vec{\alpha} + y\,\vec{\beta} + z\,\vec{\gamma} = \vec{0}$

where $x = -n, y = m+n$ and $z = -m$, so that $x + y + z = -n + m + n - m = 0$. Hence, the condition is necessary.

Condition sufficient: To prove that the condition is sufficient, let the relation as in Eq. 1.15 hold good. Without loss of generality, we assume that $z \neq 0$ so that $x + y = -z(\neq 0)$. Thus

$$\frac{1}{x+y}\left(x\vec{\alpha} + y\vec{\beta}\right) = \frac{-z}{x+y}\vec{\gamma} = \frac{x+y}{x+y}\vec{\gamma} = \vec{\gamma}$$

This shows that the point C with position vector $\vec{\gamma}$ lies on the line joining the points A and B with position vectors $\vec{\alpha}$ and $\vec{\beta}$, in the ratio $y : x$. Hence, the three points are collinear.

The relations (Eq. 1.15) state that A, B and C divide BC, CA and AB in the ratios $z/y, x/z$ and y/x, whose product is 1.

RESULT 6

If $\vec{a}, \vec{b}$ and $\vec{c}$ be the position vectors of three collinear points A, B and C and x, y and z be three scalars such that $x\vec{a} + y\vec{b} + z\vec{c} = \vec{0}$, then we must have $x + y + z = 0$, provided the origin of reference of the vectors does not lie on the straight line ABC, since $x\vec{a} + y\vec{b} + z\vec{c} = \vec{0}$ may be written as

$$(x+y+z)\vec{a} + y(\vec{b} - \vec{a}) + z(\vec{c} - \vec{a}) = \vec{0}$$

However, as, A, B and C are on the same straight line, the vectors $\vec{a}$ and $y(\vec{b} - \vec{a}) + z(\vec{c} - \vec{a})$ are not collinear. Hence, $x + y + z = 0$.

Example 1.7.1

Prove that the points A, B and C with position vectors $-2\vec{\alpha}+3\vec{\beta}+5\vec{\gamma}$, $\vec{\alpha}+2\vec{\beta}+3\vec{\gamma}$ and $7\vec{\alpha}-\vec{\gamma}$ are collinear. Find the ratio in which C divides AB.

Solution

The position vectors of A, B and C are $-2\vec{\alpha}+3\vec{\beta}+5\vec{\gamma}$, $\vec{\alpha}+2\vec{\beta}+3\vec{\gamma}$ and $7\vec{\alpha}-\vec{\gamma}$, respectively, with respect to the base point O (Fig. 1.26). Therefore

$$\begin{aligned}\vec{AB} &= \vec{OB}-\vec{OA} = (\vec{\alpha}+2\vec{\beta}+3\vec{\gamma})-(-2\vec{\alpha}+3\vec{\beta}+5\vec{\gamma})\\ &= 3\vec{\alpha}-\vec{\beta}-2\vec{\gamma}\\ \vec{AC} &= \vec{OC}-\vec{OA} = (7\vec{\alpha}-\vec{\gamma})-(-2\vec{\alpha}+3\vec{\beta}+5\vec{\gamma})\\ &= 3(3\vec{\alpha}-\vec{\beta}-2\vec{\gamma}) = 3\vec{AB}\end{aligned}$$

Also, the vectors have the same terminal points; hence, the given vectors are collinear. Let the point C divide AB in the ratio $m:n$. The point dividing AB in the ratio $m:n$ being

$$\begin{aligned}&\frac{1}{m+n}\left[m\left(\vec{\alpha}+2\vec{\beta}+3\vec{\gamma}\right)+n\left(-2\vec{\alpha}+3\vec{\beta}+5\vec{\gamma}\right)\right]\\ &=\frac{1}{m+n}\left[(m-2n)\vec{\alpha}+(2m+3n)\vec{\beta}+(3m+5n)\vec{\gamma}\right]\end{aligned}$$

This must be same as $7\vec{\alpha}-\vec{\gamma}$. Therefore

$$\begin{aligned}&\frac{1}{m+n}\left[(m-2n)\vec{\alpha}+(2m+3n)\vec{\beta}+(3m+5n)\vec{\gamma}\right] = 7\vec{\alpha}-\vec{\gamma}\\ &\Rightarrow (m-2n)\vec{\alpha}+(2m+3n)\vec{\beta}+(3m+5n)\vec{\gamma} = 7(m+n)\vec{\alpha}-(m+n)\vec{\gamma}\\ &\Rightarrow (-6m-9n)\vec{\alpha}+(2m+3n)\vec{\beta}+(4m+6n)\vec{\gamma} = \vec{0}\\ &\Rightarrow (-6m-9n) = (2m+3n) = (4m+6n) = 0\end{aligned}$$

These equations are constants in m, n and $\frac{m}{n} = -\frac{3}{2}$. Thus, C divides AB in the ratio $-3:2$.

Also, $x+y+z=0$. Eliminating x, y and z, we get $\Delta = 0$.

Example 1.7.2

Show that for the non-coplanar vectors $\vec{a}, \vec{b}$ and $\vec{c}$, the three points $-2\vec{a}+3\vec{b}+5\vec{c}$, $\vec{a}+2\vec{b}+3\vec{c}$ and $7\vec{a}-\vec{c}$ are collinear.

Solution

If the three given points $\vec{\alpha} = -2\vec{a}+3\vec{b}+5\vec{c}$, $\vec{\beta} = \vec{a}+2\vec{b}+3\vec{c}$ and $\vec{\gamma} = 7\vec{a}-\vec{c}$ are collinear, then using (Eq. 1.15), there exist scalars x, y and z such that

$$x\left(-2\vec{a}+3\vec{b}+5\vec{c}\right)+y\left(\vec{a}+2\vec{b}+3\vec{c}\right)+z(7\vec{a}-\vec{c}) = \vec{0}$$

$$\text{or,}\quad (-2x+y+7z)\vec{a}+(3x+2y)\vec{b}+(5x+3y-z)\vec{c} = \vec{0}$$

Since $\vec{a}, \vec{b}$ and $\vec{c}$ are three non-coplanar vectors, we have

$$-2x+y+7z=0;\quad 3x+2y+0z=0$$

$$5x+3y-z=0 \text{ and } x+y+z=0$$

Let us consider the three equations. The coefficient matrix is given by

$$\begin{pmatrix} -2 & 1 & 7 \\ 3 & 2 & 0 \\ 5 & 3 & -1 \end{pmatrix} \sim \begin{pmatrix} 1 & 0 & 0 \\ 0 & 1 & 0 \\ 0 & 0 & 0 \end{pmatrix}$$

Thus, the rank of the coefficient matrix $= 2 < 3 =$ the number of variables. Thus, the system has non-trivial solutions. Hence, the given three points are collinear.

1.7.2 Coplanar Vectors

A system of vectors is said to be *coplanar* if they lie on the same plane or their supports are parallel to the same plane. A plane parallel to the system of coplanar vectors is called the *plane of the vectors.*

(i) The supports of coplanar co-initial vectors are coplanar.
(ii) Co-initial vectors need not be coplanar.
(iii) Three vectors $\vec{a}, \vec{b}$ and $\vec{c}$ are coplanar, if there exist scalars m and n such that $\vec{c} = m\,\vec{a} + n\,\vec{b}$.

If a system of vectors is not on the same plane, they are *non-coplanar*.

Theorem 1.6

If $\vec{a}, \vec{b}$ and $\vec{c}$ are three proper, non-coplanar vectors and m, n and p are three scalars such that $m\,\vec{a} + n\,\vec{b} + p\,\vec{c} = \vec{0}$, then $m = n = p = 0$.

Proof: Let us suppose, $m \neq 0$. Then, $m\,\vec{a} + n\,\vec{b} + p\,\vec{c} = \vec{0}$ gives

$$m\,\vec{a} = -n\,\vec{b} - n\,\vec{c} \Rightarrow \vec{a} = -\frac{n}{m}\vec{b} - \frac{n}{m}\vec{c}$$

By the representation theorem, the vector $-\frac{n}{m}\vec{b} - \frac{n}{m}\vec{c}$ lies on the plane of $\vec{b}$ and $\vec{c}$, i.e., $\vec{a}$ lies on the plane of $\vec{b}$ and $\vec{c}$, which is clearly a contradiction of the hypothesis that $\vec{a}, \vec{b}$ and $\vec{c}$ are non-coplanar. Therefore, $m = 0$.

By similar reasoning, contradictions are obtained upon supposing $n \neq 0$ and $p \neq 0$. Hence, $m = n = p = 0$.

(i) Three vectors form a linearly dependent system if they are coplanar; they form a linearly independent system if they are non-coplanar.
(ii) If two equal vectors are expressed in terms of the same three non-coplanar vectors, then the corresponding scalar coefficients are equal, and conversely, the two vectors are equal if their corresponding components are equal.

Theorem 1.7 (Representation theorem)

Any vector can be uniquely represented as a linear combination of three non-coplanar vectors.

Proof: Take any point O as origin and let $\overrightarrow{OP} = \vec{r}$ (Fig. 1.36). Let $\hat{a}, \hat{b}$ and $\hat{c}$ be the three unit vectors in the three non-coplanar directions OX, OY and OZ, respectively. On OP as diagonal, a parallelepiped $OAFBC$ is constructed with the edges OA, OB and OC along OX, OY and OZ, respectively. Let the real numbers x, y and z denote the lengths OA, OB and OC such that

$$\overrightarrow{OA} = x\vec{a},\ \overrightarrow{OB} = y\vec{b},\ \overrightarrow{OC} = z\vec{c}$$

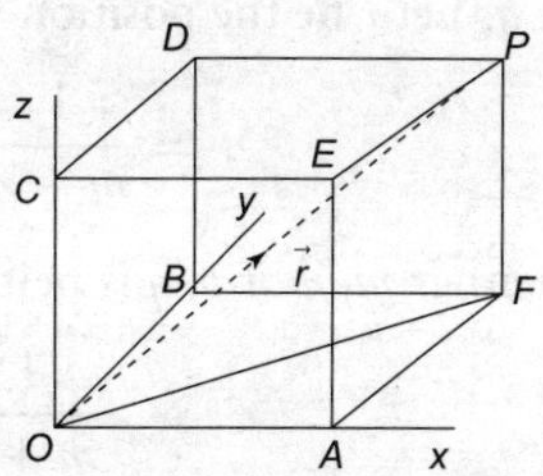

Figure 1.36 Representation theorem

If x, y and z are positive, then the directions of $\overrightarrow{OA}, \overrightarrow{OB}$ and $\overrightarrow{OC}$ are the same as those of $\vec{a}, \vec{b}$ and $\vec{c}$; if any of them are negative, then the corresponding direction will be opposite. Now

$$\begin{aligned}\vec{r} &= \overrightarrow{OP} = \overrightarrow{OF} + \overrightarrow{FP} = (\overrightarrow{OA} + \overrightarrow{AF}) + \overrightarrow{OC} \\ &= \overrightarrow{OA} + \overrightarrow{OB} + \overrightarrow{OC} = x\vec{a} + y\vec{b} + z\vec{c}\end{aligned}$$

Thus, any vector can be expressed as a linear combination of three non-coplanar vectors.

Uniqueness of resolution: If possible, let $\vec{r} = x'\vec{a} + y'\vec{b} + z'\vec{c}$. Then

$$\begin{aligned}x\vec{a} + y\vec{b} + z\vec{c} &= x'\vec{a} + y'\vec{b} + z'\vec{c} \\ \Rightarrow (x - x')\vec{a} + (y - y')\vec{b} + (z - z')\vec{c} &= \vec{0}\end{aligned}$$

But $\hat{a}, \hat{b}$ and $\hat{c}$ are non-coplanar vectors, and hence, they form a linearly independent set. Therefore, by using Theorem 1.3

$$x - x' = y - y' = z - z' = 0 \Rightarrow x = x', y = y', z = z'$$

$\vec{r}$ is called the *resultant* of the three vectors $x\hat{a}, y\hat{b}$ and $z\hat{c}$ which are called the *vector components*. x, y and z are the Cartesian coordinates of P with respect to the axes OX, OY and OZ.

Corollary: The resolution enables us to add two or more vectors by adding their corresponding components. Thus, if

$$\begin{aligned}\vec{r}_1 = x_1\vec{a} + y_1\vec{b} + z_1\vec{c} \text{ and } \vec{r}_2 = x_2\vec{a} + y_2\vec{b} + z_2\vec{c} \\ \text{then } \vec{r} = \vec{r}_1 + \vec{r}_2 = (x_1 + x_2)\vec{a} + (y_1 + y_2)\vec{b} + (z_1 + z_2)\vec{c}\end{aligned}$$

Note: The resolution is unique as one and only one parallelepiped can be constructed with OP as diagonal and OA, OB and OC as the edges.

Theorem 1.8 (Condition for coplanarity of four points)

The necessary and sufficient condition for the four points having position vectors $\vec{\alpha}, \vec{\beta}, \vec{\gamma}$ and $\vec{\delta}$ (no three of them being collinear) to be coplanar is that there exist four scalars x, y, z and t, not all zero, such that

$$x\vec{\alpha} + y\vec{\beta} + z\vec{\gamma} + t\vec{\delta} = \vec{0} \qquad \text{with } x + y + z + t = 0 \qquad (1.16)$$

Proof

Condition necessary: Relative to some base point, $\vec{\alpha}$, $\vec{\beta}$, $\vec{\gamma}$ and $\vec{\delta}$ are the position vectors of the four points A, B, C and D and no three of them are collinear (Fig. 1.37).

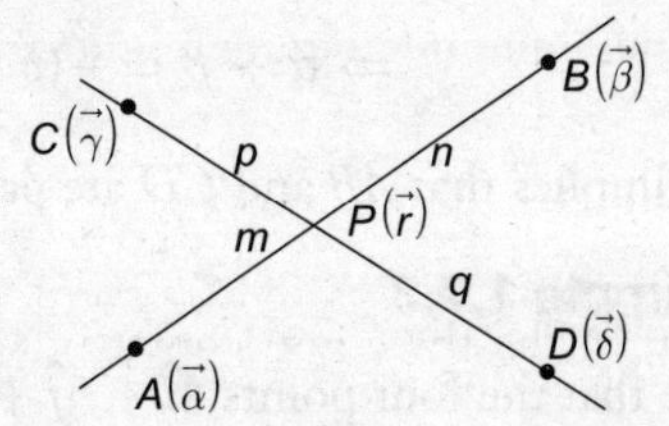

Figure 1.37 Condition for coplanarity of four points

Since no three of the given vectors are collinear, we can select two of them whose join will meet the join of the other two, i.e., AB meets CD at the point P. Let P divide AB in the ratio $m : n$ and CD in the ratio $p : q$. Let $\vec{r}$ be the position vector of P with respect to the base point O. Hence

$$\vec{r} = \frac{1}{m+n}(m\vec{\beta} + n\vec{\alpha}) \text{ and } \vec{r} = \frac{1}{p+q}(p\vec{\delta} + q\vec{\gamma})$$

where neither m/n or p/q is neither zero nor -1. Hence

$$\frac{1}{m+n}(m\vec{\beta} + n\vec{\alpha}) = \frac{1}{p+q}(p\vec{\delta} + q\vec{\gamma})$$

$$\text{or, } \frac{n}{m+n}\vec{\alpha} + \frac{m}{m+n}\vec{\beta} - \frac{q}{p+q}\vec{\gamma} - \frac{p}{p+q}\vec{\delta} = \vec{0}$$

$$\text{or, } x\vec{\alpha} + y\vec{\beta} + z\vec{\gamma} + t\vec{\delta} = \vec{0}$$

where $x = \frac{n}{m+n}$, $y = \frac{m}{m+n}$, $z = -\frac{q}{p+q}$ and $t = -\frac{p}{p+q}$. Therefore

$$x + y + z + t = \frac{n}{m+n} + \frac{m}{m+n} - \frac{q}{p+q} - \frac{p}{p+q} = 1 - 1 = 0$$

Hence, the condition is necessary.

Condition sufficient: To prove that this condition is also sufficient, we assume there exist four scalars x, y, z and t, not all zero, such that the relations (Eq. 1.16) hold. Since x, y, z and t are not all zero, we see that at least one of $x + y$, $y + z$, $z + t$ and $t + x$ is not zero. Without any loss of generality, we assume that $z + t \neq 0$ so that $x + y = -(z + t) \neq 0$. Now

$$x\vec{\alpha} + y\vec{\beta} + z\vec{\gamma} + t\vec{\delta} = \vec{0} \Rightarrow x\vec{\alpha} + y\vec{\beta} = -\left(z\vec{\gamma} + t\vec{\delta}\right)$$

$$\text{or, } \frac{1}{x+y}\left(x\vec{\alpha} + y\vec{\beta}\right) = -\frac{1}{x+y}\left(z\vec{\gamma} + t\vec{\delta}\right) = \frac{1}{z+t}\left(z\vec{\gamma} + t\vec{\delta}\right)$$

This shows that the point P whose position vector is $\frac{1}{x+y}\left(x\vec{\alpha} + y\vec{\beta}\right)$ lies on the join of the points A and B in the ratio $y : x$, and the point whose position vector is $\frac{1}{z+t}\left(z\vec{\gamma} + t\vec{\delta}\right)$ lies on the join of the points C and D, in the ratio $t : z$. Hence, AB and CD intersect, i.e., A, B, C and D are coplanar. Hence, the condition is sufficient.

RESULT 7

When $x + y = 0 = z + t$, we have

$$x\vec{\alpha} + y\vec{\beta} + z\vec{\gamma} + t\vec{\delta} = \vec{0} \Rightarrow x\left(\vec{\alpha} - \vec{\beta}\right) + z\left(\vec{\gamma} - \vec{\delta}\right) = \vec{0}$$

$$\Rightarrow \vec{\alpha} - \vec{\beta} = \frac{z}{x}(\vec{\delta} - \vec{\gamma}) \Rightarrow \vec{BA} = \frac{z}{x}\,\vec{CD}$$

This implies that AB and CD are parallel, and hence, the four points A, B, C and D are coplanar.

Example 1.7.3

Prove that the four points $4\hat{i} + 5\hat{j} + \hat{k}$, $-\hat{j} - \hat{k}$, $3\hat{i} + 9\hat{j} + 4\hat{k}$ and $4(-\hat{i} + \hat{j} + \hat{k})$ are coplanar.

Solution

Let $\vec{\alpha} = 4\hat{i}+5\hat{j}+\hat{k}$, $\vec{\beta} = -\hat{j}-\hat{k}$, $\vec{\gamma} = 3\hat{i}+9\hat{j}+4\hat{k}$ and $\vec{\delta} = 4(-\hat{i}+\hat{j}+\hat{k})$. Now, $x\vec{\alpha}+y\vec{\beta}+z\vec{\gamma}+t\vec{\delta} = \vec{0}$ gives

$$\Rightarrow x(4,5,1)+y(0,-1,-1)+z(3,9,4)+t(-4,4,4)=\vec{0}$$
$$\Rightarrow (4x+3z-4t)\hat{i}+(5x-y+9z+4t)\hat{j}+(x-y+4z+4t)\hat{k}=\vec{0}$$
$$\Rightarrow 4x+3z-4t = 5x-y+9z+4t = x-y+4z+4t = 0$$
$$\Rightarrow x=\frac{5}{2}t,\ \ y=-\frac{3}{2}t,\ \ z=-2t$$
$$\Rightarrow x+y+z+t=\frac{5}{2}t-\frac{3}{2}t-2t+t=0$$

Since the relation (Eq. 1.16) holds, the given four points are coplanar.

Example 1.7.4

If the position vectors of the points A, B, C and D satisfy the equation $x\vec{\alpha}+y\vec{\beta}+z\vec{\gamma}=\vec{\delta}$, where $x+y+z=1$, show that the points are coplanar.

Solution

Let $\vec{OA}=\vec{\alpha}$, $\vec{OB}=\vec{\beta}$, $\vec{OC}=\vec{\gamma}$ and $\vec{OD}=\vec{\delta}$ (Fig. 1.38).

Using the given conditions, $x\vec{\alpha}+y\vec{\beta}+z\vec{\gamma}=\vec{\delta}$ and $x+y+z=1$, we get

$$\vec{\delta}=x\vec{\alpha}+y\vec{\beta}+(1-x-y)\vec{\gamma}$$
$$\vec{\delta}-\vec{\alpha}=(x-1)\vec{\alpha}+y(\vec{\beta}-\vec{\gamma})+(1-x)\vec{\gamma}$$
$$\vec{AD}=y(\vec{\beta}-\vec{\gamma})+(1-x)(\vec{\gamma}-\vec{\alpha})$$
$$=-y\vec{CB}+(1-x)\vec{AC}$$

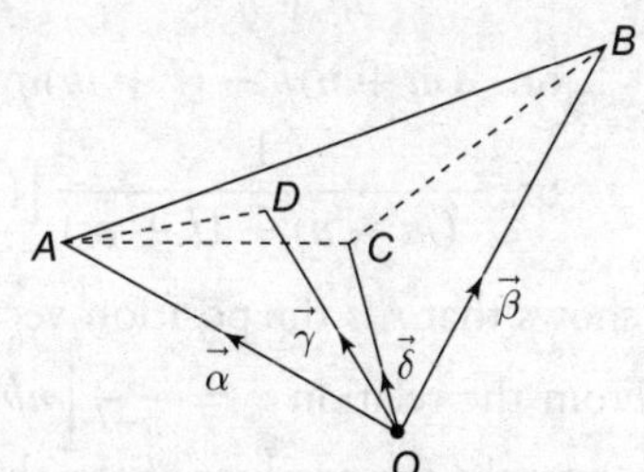

Figure 1.38 Condition for coplanarity of four points

This shows that AD lies on the plane of AC and BC, i.e., the points A, B, C and D are coplanar.

Theorem 1.9

A linear relation $x\vec{a}+y\vec{b}+z\vec{c}=0$ connecting the position vectors $\vec{a}$, $\vec{b}$ and $\vec{c}$ of three points A, B and C will be independent of the origin, if and only if, $x+y+z=0$.

Proof: Let O and O' be the two origins referred to which the position vectors of A, B and C are $\vec{a}, \vec{b}$ and $\vec{c}$ and $\vec{a}', \vec{b}'$ and $\vec{c}'$, respectively. Then, $\vec{a}=\overrightarrow{OO'}+\vec{a}'$, $\vec{b}=\overrightarrow{OO'}+\vec{b}'$, and $\vec{c}=\overrightarrow{OO'}+\vec{c}'$. Thus

$$x\vec{a}+y\vec{b}+z\vec{c}=x(\overrightarrow{OO'}+\vec{a}')+y(\overrightarrow{OO'}+\vec{b}')+z(\overrightarrow{OO'}+\vec{c}')$$
$$=(x+y+z)\overrightarrow{OO'}+x\vec{a}'+y\vec{b}'+z\vec{c}'$$

Now, both $(x\vec{a}+y\vec{b}+z\vec{c})$ and $(x\vec{a}'+y\vec{b}'+z\vec{c}')$ will be null vectors if and only if $x+y+z=0$.

In this section, we shall prove two important theorems: the Theorem of Menelaus and the Theorem of Desargues of elementary geometry. These theorems can be proved by showing the condition for collinearity of three points.

Theorem 1.10 (Theorem of Menelaus)

The product of the ratios in which a transversal cuts the sides of a triangle taken in order is equal to (-1), and conversely.

Proof: Let a transversal cut the sides BC, CA and AB of a triangle ABC at the points P, Q and R, respectively (Fig. 1.39). Let the position vectors of A, B, C, P, Q and R be $\vec{a}, \vec{b}, \vec{c}, \vec{p}, \vec{q}$ and $\vec{r}$, respectively, with respect to some base point O (not shown in the figure). P divides BC in the ratio $m : n$. Now, it is possible to choose a number l such that Q divides CA in the ratio $l : m$. Then

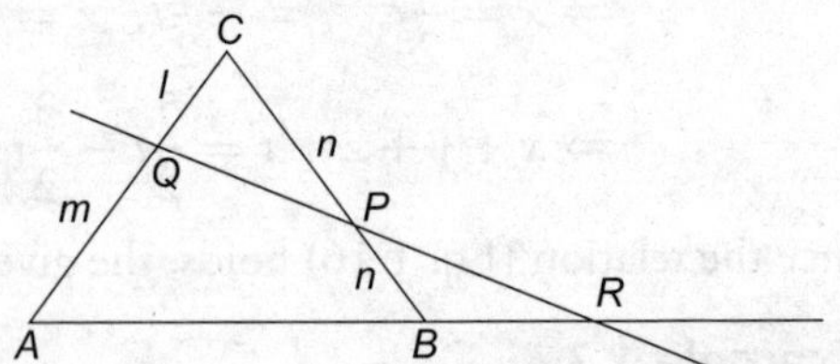

Figure 1.39 Theorem of Menelaus

$$\vec{p} = \frac{1}{m+n}\left(m\vec{c} + n\vec{b}\right) \quad and \quad \vec{q} = \frac{1}{l+m}\left(l\vec{a} + m\vec{c}\right)$$

$$\text{or,} \quad (m+n)\vec{p} - (l+m)\vec{q} = n\vec{b} - l\vec{a}$$

$$\text{or,} \quad \frac{1}{(m+n)-(l+m)}\left[(m+n)\vec{p} - (l+m)\vec{q}\right] = \frac{1}{n-l}\left[n\vec{b} - l\vec{a}\right] = \vec{r} \text{ (say)}$$

This shows that $\vec{r}$ is the position vector of the point R, which is the intersection point of QR and AB. From the relation $\vec{r} = \frac{1}{n-l}\left[n\vec{b} - l\vec{a}\right]$, it is evident that R divides AB in the ratio $n : (-l)$. Consequently, the product of the three ratios is given by

$$\frac{m}{n} \times \frac{l}{m} \times \frac{n}{-l} = -1$$

Conversely, let the product of the three ratios be (-1). We can assume that P, Q and R divide BC, CA and AB in the ratios $n : (-l)$, $m : n$ and $l : m$, respectively. Then

$$\vec{p} = \frac{1}{m+n}\left(m\vec{c} + n\vec{b}\right), \quad \vec{q} = \frac{1}{l+m}\left(l\vec{a} + m\vec{c}\right) \quad and \quad \vec{r} = \frac{1}{n-l}\left(n\vec{b} - l\vec{a}\right)$$

$$\text{or,} \quad (m+n)\vec{p} - (l+m)\vec{q} - (n-l)\vec{r} = \vec{0}$$

$$\text{or,} \quad x\vec{p} + y\vec{q} + z\vec{r} = \vec{0}$$

where the sum of the scalar coefficients is given by

$$x + y + z = (m+n) - (l+m) - (n-l) = 0$$

so that the points P, Q and R lie on a straight line, which is the transversal.

Theorem 1.11 (Theorem of Desargues)

If the straight lines joining the corresponding vertices of two triangles be concurrent, then the points of intersection of the corresponding sides are collinear, and conversely.

Proof: Let the three concurrent straight lines OA, OB and OC be produced to D, E and F, respectively (Fig. 1.40), i.e., the straight lines EB, DA and FC meet at the point O. Hence, our usual symbols for position vectors, the position vector of $O(\vec{h})$, is given by

$$\vec{h} = \frac{1}{m+m_1}\left[m\vec{a} + m_1\vec{d}\right] = \frac{1}{n+n_1}\left[n\vec{b} + n_1\vec{e}\right] = \frac{1}{l+l_1}\left[l\vec{c} + l_1\vec{f}\right]$$

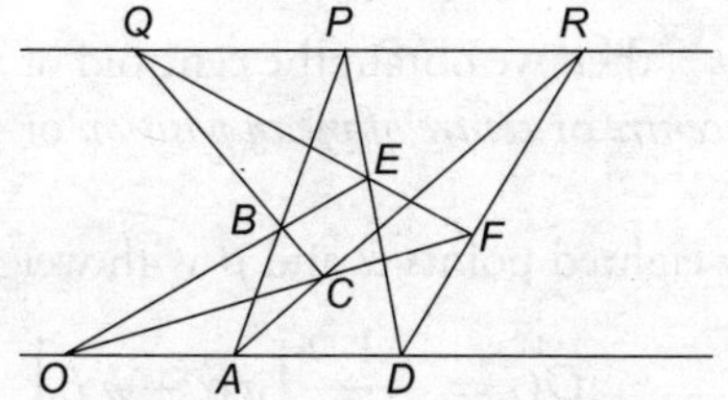

Figure 1.40 Theorem of Desargues

where O divides EB, DA and FC in the ratio $m_1 : m;\ n_1 : n;\ l_1 : l$, respectively. The scalars m, m_1, n, n_1 and l, l_1 have been chosen such that $m + m_1 = n + n_1 = l + l_1 = 1$. Then

$$m\vec{a} + m_1\vec{d} = n\vec{b} + n_1\vec{e} = l\vec{c} + l_1\vec{f}$$

$$\text{or,}\quad n\vec{b} - l\vec{c} = l_1\vec{f} - n_1\vec{e} \quad\text{and}\quad n - l = l_1 - n_1$$

$$\text{or,}\quad \frac{1}{n-l}\left[n\vec{b} - l\vec{c}\right] = \frac{1}{l_1 - n_1}\left[l_1\vec{f} - n_1\vec{e}\right] = \vec{q}$$

Similarly

$$\vec{r} = \frac{1}{m-l}\left[m\vec{a} - l\vec{c}\right] = \frac{1}{l_1 - m_1}\left[l_1\vec{f} - m_1\vec{d}\right]$$

$$\text{and,}\quad \vec{p} = \frac{1}{m-n}\left[m\vec{a} - n\vec{b}\right] = \frac{1}{n_1 - m_1}\left[n_1\vec{e} - m_1\vec{d}\right]$$

where $\vec{p}, \vec{q}$ and $\vec{r}$ are the position vectors of P, Q and R, respectively. Thus

$$(m-n)\vec{p} + (n-l)\vec{q} - (m-l)\vec{r} = \vec{0}$$

$$\Rightarrow x\vec{p} + y\vec{q} + z\vec{r} = \vec{0} \qquad \text{where } x + y + z = 0$$

Hence, the three points P, Q and R are collinear. The converse of this theorem is also true.

1.8 CENTROIDS

Let $\vec{\alpha_1}, \vec{\alpha_2}, \ldots, \vec{\alpha_n}$ be the position vectors of n given points $A_1, A_2, \ldots, A_n$ in space with respect to some base point O. If we associate n real numbers $a_1, a_2, \ldots, a_n$ (called weights) such that $a_1 + a_2 + \cdots + a_n \neq 0$, then the position vector of the point G is given by

$$\vec{OG} = \frac{1}{a_1 + a_2 + \cdots + a_n}\left[a_1\vec{\alpha_1} + a_2\vec{\alpha_2} + \cdots + a_n\vec{\alpha_n}\right]$$

If $m_1, m_2, \ldots, m_n$ be weights of n points $A_1, A_2, \ldots, A_n$ and $\vec{\alpha}_1, \vec{\alpha}_2, \ldots, \vec{\alpha}_n$ be the respective position vectors of these n points relative to the origin O, then the point G, where

$$\vec{OG} = \frac{1}{m_1 + m_2 + \cdots + m_n}\left[m_1\vec{\alpha}_1 + m_2\vec{\alpha}_2 + \cdots + m_n\vec{\alpha}_n\right] \tag{1.17}$$

is called the *centroid* of the given points associated with $m_1, m_2, \ldots, m_n$. The following points are to be noted:

(i) Let $m_1 = m_2 = \cdots = m_n$, then we obtain the centroid of n equally weighted points. This point is called the *mean centre* or *centre of mean position* or simply the *centroid* of the given n points.

(ii) The centroid G of two weighted points $\vec{\alpha}$ and $\vec{\beta}$ with weights a and b is given by

$$\vec{OG} = \frac{1}{a+b}\left[a\,\vec{\alpha} + b\,\vec{\beta}\right]$$

which shows that G divides the line segment AB joining $\vec{\alpha}$ and $\vec{\beta}$ in the ratio $b : a$.

(iii) The centroid G of two given points $\vec{\alpha}$ and $\vec{\beta}$ is given by $\frac{1}{2}(\vec{\alpha} + \vec{\beta})$, i.e., it bisects the join of $\vec{\alpha}$ and $\vec{\beta}$.

1.8.1 Centre of Mass

The centre of mass of a set of particles of masses $m_1, m_2, m_3, \cdots, m_n$ situated at points with position vectors $\vec{\alpha}_1, \vec{\alpha}_2, \vec{\alpha}_3, \cdots, \vec{\alpha}_n$, respectively, is the centroid of these points associated with numbers $m_1, m_2, m_3, \cdots, m_n$. Let $\vec{r}$ be the position vector of the centre of mass. Then

$$\vec{r} = \frac{1}{m_1 + m_2 + m_3 + \ldots + m_n}\left[m_1\vec{\alpha}_1 + m_2\vec{\alpha}_2 + m_3\vec{\alpha}_3 + \ldots + m_n\vec{\alpha}_n\right]$$

$$= \frac{1}{\sum m}\sum m\vec{\alpha} \tag{1.18}$$

The centre of mass of a set of particles is independent of origin and the position is unique.

Theorem 1.12

The centroid of n given points in space is independent of the origin of vectors.

Proof: With reference to the origin O (Fig. 1.41), let the position vectors of n given points be α, β and γ. And with respect to O', the position vector of the points $A, B, C, \ldots$ are $\vec{\alpha}', \vec{\beta}', \vec{\gamma}', \cdots$, respectively, where $\vec{OO'} = \vec{p}$. Then

$$\vec{\alpha}' = \vec{\alpha} - \vec{p},\ \vec{\beta}' = \vec{\beta} - \vec{p},\ \vec{\gamma}' = \vec{\gamma} - \vec{p}, \cdots$$

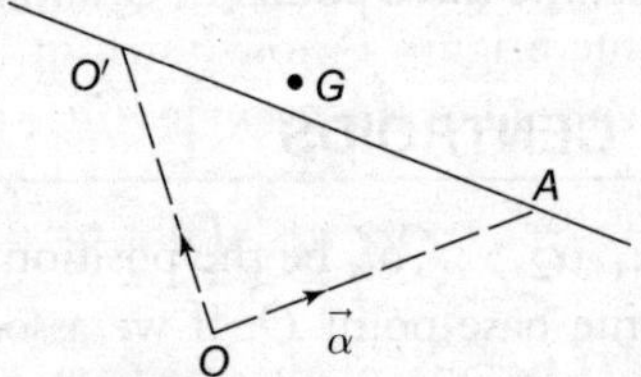

Figure 1.41 Centroid

$\vec{\alpha}, \vec{\beta}, \vec{\gamma}, \cdots$ be the respective position vectors of the points $A, B, C, \cdots$ in space, and $a, b, c, \cdots$ be the associated real numbers such that $a + b + c + \ldots \neq 0$. Let G and G' be, respectively, the centroid with respect to O and O' as origin. Then

$$
\begin{aligned}
\vec{O'G'} &= \frac{1}{a+b+c+\ldots}\left[a\vec{\alpha}' + b\vec{\beta}' + c\vec{\gamma}' + \ldots\right] \\
&= \frac{1}{a+b+c+\ldots}\left[a\left(\vec{\alpha} - \vec{p}\right) + b\left(\vec{\beta} - \vec{p}\right) + c\left(\vec{\gamma} - \vec{p}\right) + \ldots\right] \\
&= \frac{1}{a+b+c+\ldots}\left[\vec{\alpha} + b\vec{\beta} + c\vec{\gamma} + \ldots\right] - \vec{p} \\
&= \vec{OG} - \vec{OO'} = \vec{O'G}
\end{aligned}
$$

This shows that G' coincides with G and the position of G is unique. Thus, relative to any origin, the centroid of a set of weighted points has a position vector that equals the weighted average of the given set of position vectors.

Example 1.8.1

Find the centroid of $3n$ points whose position vectors with reference to an origin O are $\hat{i}, 2\hat{i}, 3\hat{i}, \cdots, n\hat{i}; \hat{j}, 2\hat{j}, 3\hat{j}, \cdots, n\hat{j}; \hat{k}, 2\hat{k}, 3\hat{k}, \cdots, n\hat{k}$.

Solution

Let G be the required centroid. Then

$$
\begin{aligned}
\vec{OG} &= \frac{1}{3n}\left[\left(\hat{i} + 2\hat{i} + 3\hat{i} + \cdots + n\hat{i}\right) + \left(\hat{j} + 2\hat{j} + 3\hat{j} + \cdots + n\hat{j}\right)\right. \\
&\qquad \left. + \left(\hat{k} + 2\hat{k} + 3\hat{k} + \cdots + n\hat{k}\right)\right] \\
&= \frac{1 + 2 + 3 + \cdots + n}{3n}\left(\hat{i} + \hat{j} + \hat{k}\right) \\
&= \frac{n(n+1)}{2}\frac{1}{3n}\left(\hat{i} + \hat{j} + \hat{k}\right) = \frac{n+1}{6}\left(\hat{i} + \hat{j} + \hat{k}\right)
\end{aligned}
$$

Example 1.8.2

Prove that the centroid of the vertices of a triangle bisects the median.

Solution

Let $\vec{\alpha}, \vec{\beta}$ and $\vec{\gamma}$ be the position vectors of the vertices A, B and C of a triangle ABC (Fig. 1.42) with respect to some origin O (not shown in the figure). If G be the centroid of the triangle ABC, then

$$
\vec{OG} = \frac{1}{3}\left(\vec{\alpha} + \vec{\beta} + \vec{\gamma}\right)
$$

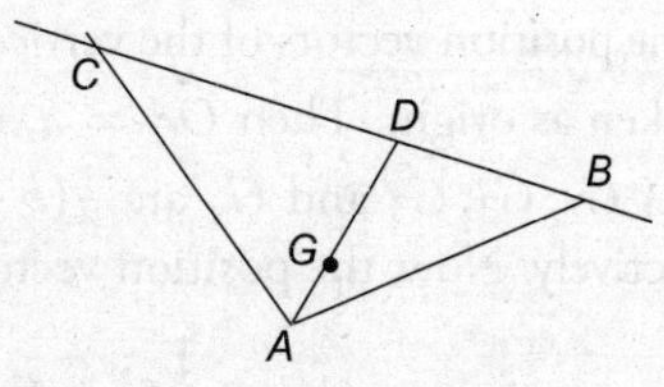

Figure 1.42 Centroid

Let AD be one of the medians. Then, the position vector of D, the midpoint of B and C, is $\vec{OD} = \frac{1}{2}\left(\vec{\beta} + \vec{\gamma}\right)$. Therefore

$$
\vec{OG} = \frac{1}{2+1}\left[2.\frac{1}{2}\left(\vec{\beta} + \vec{\gamma}\right) + \vec{\alpha}\right] = \frac{1}{2+1}\left[2\vec{OD} + \vec{OA}\right]
$$

This shows that G is the point of bisection of the median.

Example 1.8.3

If AD, BE and CF are medians of triangle ABC, show that the triangles ABC and DEF have identical centroids. Also, show that $\vec{AD} + \vec{DE} + \vec{CF} = \vec{AB} + \vec{BC} + \vec{CA}$.

Solution

With respect to the base point O, let $\vec{a}, \vec{b}$ and $\vec{c}$ be the position vectors of the vertices A, B and C, respectively (Fig.1.43). AD, BE and CF are the three medians of the triangle ABC; so the position vectors of D, E and F are $1/2\ (\vec{b} + \vec{c})$, $1/2\ (\vec{c} + \vec{a})$ and $1/2\ (\vec{a} + \vec{b})$, respectively, with respect to the base point O. Let G be the centroid of the triangle ABC. Then

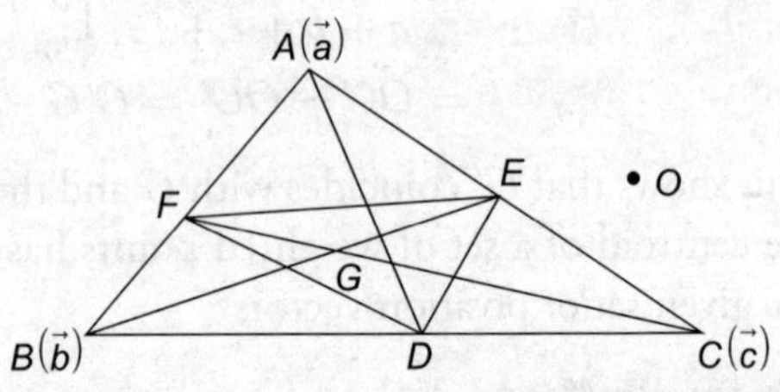

Figure 1.43 Centroid of a triangle

$$\vec{OG} = \frac{1}{3}(\vec{a} + \vec{b} + \vec{c}) = \frac{1}{3}\left[1/2\ (\vec{b} + \vec{c}) + 1/2\ (\vec{c} + \vec{a}) + 1/2\ (\vec{a} + \vec{b})\right]$$
$$= \frac{1}{3}(\vec{OD} + \vec{OE} + \vec{OF})$$

Therefore, G is also the centroid of the triangle DEF. Now

$$\vec{AD} + \vec{BE} + \vec{CF} = (\vec{OD} - \vec{OA}) + (\vec{OE} - \vec{OB}) + (\vec{OF} - \vec{OC})$$
$$= \frac{1}{2}(\vec{b} + \vec{c} - 2\vec{a}) + \frac{1}{2}(\vec{c} + \vec{a} - 2\vec{b}) + \frac{1}{2}(\vec{a} + \vec{b} - 2\vec{c})$$
$$= \vec{0} \text{ and } \vec{AB} + \vec{BC} + \vec{CA} = \vec{AC} + \vec{CA} = \vec{0}$$

Example 1.8.4

Let A, B, C and D be the vertices of a tetrahedron. If G_1, G_2, G_3 and G_4 be the centroids of the triangles BCD, CDA, DAB and ABC, respectively, then show that G divides each of the segments AG_1, BG_2, CG_3 and DG_4 in the ratio 3 : 1.

Solution

Let the position vectors of the vertices A, B, C and D be $\vec{a}, \vec{b}, \vec{c}$ and $\vec{d}$ with respect to the base point O, taken as origin. Then $\vec{OA} = \vec{a}$, $\vec{OB} = \vec{b}$, $\vec{OC} = \vec{c}$ and $\vec{OD} = \vec{d}$. The position vectors of the points G_1, G_2, G_3 and G_4 are $\frac{1}{3}(\vec{b} + \vec{c} + \vec{d})$, $\frac{1}{3}(\vec{c} + \vec{d} + \vec{a})$, $\frac{1}{3}(\vec{d} + \vec{a} + \vec{b})$ and $\frac{1}{3}(\vec{a} + \vec{b} + \vec{c})$, respectively. Now, the position vector of G is given by

$$\vec{OG} = \frac{1}{4}(\vec{a} + \vec{b} + \vec{c} + \vec{d}) = \frac{1}{1+3}\left[1\vec{a} + 3\frac{1}{3}(\vec{b} + \vec{c} + \vec{d})\right]$$
$$= \frac{1}{1+3}\left[1\ \vec{OA} + 3\ \vec{OG}\right]$$

This shows that G divides the line segment AG_1 internally in the ratio 3 : 1. Similarly, for BG_2, CG_3 and DG_4. Hence, G divides each of the other three line segments BG_2, CG_3 and DG_4 internally in the ratio 3 : 1.

Example 1.8.5

ABCD is a quadrilateral and G is the point of intersection of the lines joining the middle points of the opposite sides. Then, for any point O on the plane of the quadrilateral, prove that $\vec{OA}+\vec{OB}+\vec{OC}+\vec{OD}=4\vec{OG}$. Interpret the point G.

Solution

Let P, Q, R and S (Fig. 1.44) be the middle points of pairs of opposite sides AB and CD and AD and BC, respectively. Then

$$\vec{OA}+\vec{OB}=2\vec{OP},\quad \vec{OC}+\vec{OD}=2\vec{OQ}$$
$$\vec{OB}+\vec{OC}=2\vec{OQ},\quad \vec{OA}+\vec{OD}=2\vec{OR}$$

Using commutativity and associativity of vector addition, we get

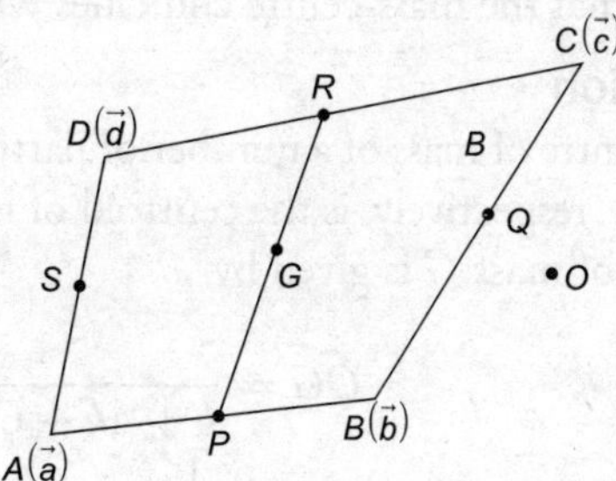

$$\vec{OP}+\vec{OQ}=\vec{OR}+\vec{OS}=\frac{1}{4}(\vec{OA}+\vec{OB}+\vec{OC}+\vec{OD})$$

Figure 1.44 Centroid of a triangle

Since $\vec{OP}+\vec{OQ}$ and $\vec{OR}+\vec{OS}$ are equal and the lines PQ and RS intersect at G, then G must be the middle point of both PQ and RS. Hence

$$\vec{OP}+\vec{OQ}=\vec{OR}+\vec{OS}=2\vec{OG}$$
$$\vec{OA}+\vec{OB}+\vec{OC}+\vec{OD}=\vec{OP}+\vec{OQ}+\vec{OR}+\vec{OS}=4\vec{OG}$$

Interpretation: Let the position vectors of A, B, C and D be $\vec{a}, \vec{b}, \vec{c}$ and $\vec{d}$, respectively, with respect to some base point O, then $\vec{OA}=\vec{a}$, $\vec{OB}=\vec{b}$, $\vec{OC}=\vec{c}$ and $\vec{OD}=\vec{d}$. If $\vec{g}$ be the position vector of the point G with respect to the base point O, then

$$\vec{g}=\vec{OG}=\frac{1}{4}(\vec{a}+\vec{b}+\vec{c}+\vec{d})$$

This shows that G is the centroid of the quadrilateral *ABCD*.

Example 1.8.6

Particles of equal mass are placed at $(n-2)$ of the corners of a regular polygon of n sides. Find their centre of mass.

Solution

If particles of equal mass are placed at all the corners of a regular polygon, the centre of mass would have been the centre O of the polygon. Let two equal masses m and m be placed at the two vacant corners A and B of the polygon; C is the middle point of the line joining A and B. Let $\vec{\alpha}$ and $\vec{\beta}$ be the position vectors of the points O and C with respect to some base point. If G be the required centre of mass whose position vector is $\vec{\gamma}$, then

$$\vec{\alpha}=\frac{1}{(n-2)m+2m}\left[(n-2)m\vec{\gamma}+2m\vec{\beta}\right]$$

$$\text{or,}\quad n\vec{\alpha}-2\vec{\beta}=(n-2)\vec{\gamma}\Rightarrow\vec{\gamma}=\frac{1}{n-2}\left[n\vec{\alpha}-2\vec{\beta}\right]$$

Therefore, the required centre of mass G lies on the join of C and O, such that $CO : OG = (n-2) : 2$.

Example 1.8.7

Masses proportional to $1, 4, 6$ and 8 are placed, respectively, at the extremities of the vectors $\vec{\alpha}_1, \vec{\alpha}_2, \vec{\alpha}_3$ and $\vec{\alpha}_4$ which conform to the relations $5\vec{\alpha}_1+2\vec{\alpha}_2+18\vec{\alpha}_3 = \vec{0}$ and $\vec{\alpha}_1-5\vec{\alpha}_2-12\vec{\alpha}_4 = \vec{0}$. Prove that the mass-centre coincides with the origin of vectors.

Solution

The centre of mass of a number of particles of masses $k, 4k, 6k$ and $8k$ situated at points $\vec{\alpha}_1, \vec{\alpha}_2, \vec{\alpha}_3$ and $\vec{\alpha}_4$, respectively, is the centroid of these points associated with numbers $k, 4k, 6k$ and $8k$. The centre of mass G is given by

$$\begin{aligned}\vec{OG} &= \frac{1}{k+4k+6k+8k}(k\vec{\alpha}_1 + 4k\vec{\alpha}_2 + 6k\vec{\alpha}_3 + 8k\vec{\alpha}_4)\\ &= \frac{1}{19}(\vec{\alpha}_1 + 4\vec{\alpha}_2 + 6\vec{\alpha}_3 + 8\vec{\alpha}_4)\end{aligned}$$

Given $5\vec{\alpha}_1 + 2\vec{\alpha}_2 + 18\vec{\alpha}_3 = \vec{0}$ and $\vec{\alpha}_1 - 5\vec{\alpha}_2 - 12\vec{\alpha}_4 = \vec{0}$, i.e., $\vec{\alpha}_3 = -\frac{1}{18}(5\vec{\alpha}_1 + 2\vec{\alpha}_2)$ and $\vec{\alpha}_4 = \frac{1}{12}(\vec{\alpha} - 5\vec{\alpha}_2)$. So

$$\begin{aligned}\vec{OG} &= \frac{1}{19}\left[\vec{\alpha}_1 + 4\vec{\alpha}_2 - \frac{1}{18}(5\vec{\alpha}_1 + 2\vec{\alpha}_2) + \frac{1}{12}(\vec{\alpha} - 5\vec{\alpha}_2)\right]\\ &= \frac{1}{57}\left[3\vec{\alpha}_1 + 12\vec{\alpha}_2 - 5\vec{\alpha}_1 - 2\vec{\alpha}_2 + 2\vec{\alpha}_1 - 10\vec{\alpha}_2\right] = \vec{0} = \vec{OO}\end{aligned}$$

Therefore, the centre of mass coincides with the origin of vectors.

EXERCISES

Multiple Choice Questions

1. Which one of the following is not a scalar?
   (a) temperature (b) weight (c) density (d) mass
2. Which one of the following is not a vector?
   (a) displacement (b) work (c) centrifugal force (d) gravitational force
3. Electric current is
   (a) scalar (b) vector
   (c) neither scalar nor vector (d) sometimes scalar and sometimes vector
4. Direction of a zero vector
   (a) does not exist (b) is towards origin
   (c) is indeterminate (d) none of these
5. Which of the following is a polar vector?
   (a) torque (b) force (c) angular momentum (d) acceleration

6. Which of the following is an axial vector?
(a) force (b) acceleration (c) linear momentum (d) torque
7. Surface area $\vec{s} = 5\,\hat{k}$ lies on the
(a) *xy* plane (b) *yz* plane (c) *zx* plane (d) arbitrary plane
8. The magnitude of the resultant of two vectors of equal magnitude is equal to the magnitude of one force. Then, the angle between them is
(a) 0° (b) 45° (c) 90° (d) 120°
9. Given that the vectors $\vec{a}$ and $\vec{b}$ are non-collinear, the values of x and y for which the vector quantity $2\vec{u} - \vec{v} = \vec{w}$ holds true if $\vec{u} = x\vec{a} + 2y\vec{b}$, $\vec{v} = -2y\vec{a} + 3x\vec{b}$, $\vec{w} = 4\vec{a} - 2\vec{b}$ are
(a) $x = 4/7, y = 6/7$ (b) $x = 10/7, y = 4/7$
(c) $x = 8/7, y = 2/7$ (d) $x = 2, y = 3$
10. Point A is $\vec{a} + 2\vec{b}$, and $\vec{a}$ divides AB in the ratio 2 : 3. The position vector of B is
(a) $2\vec{a} - \vec{b}$ (b) $\vec{b} - 2\vec{a}$ (c) $\vec{a} - 3\vec{b}$ (d) $\vec{b}$
11. The points O, A, B, X and Y are such that $\vec{OA} = \vec{a}$, $\vec{OB} = \vec{b}$, $\vec{OX} = 3\vec{a}$, $\vec{OY} = 3\vec{b}$, and the point P divides AY in the ratio 1 : 3 internally. Then, $\vec{PB}$ equals
(a) $(\vec{a} - 3\vec{b})/4$ (b) $(3\vec{a} - 2\vec{b})/4$ (c) $(3\vec{a} - \vec{b})/4$ (d) $(2\vec{a} - 3\vec{b})/4$
12. If $\sin\theta\,\hat{i} + \cos\theta\,\hat{j} + \tan\theta\,\hat{k}$ is a unit vector, then the number of values of θ lying in $0, \pi$ is
(a) 1 (b) 2 (c) 3 (d) 4
13. The value of x for which the vector $x(6\hat{i} + 2\hat{j} - 3\hat{k})$ may be of unit length is
(a) 7 (b) −7 (c) 1/7 (d) −1/7
14. The three points $2\hat{i} - \hat{j} + \hat{k}$, $\hat{i} - 3\hat{j} - 5\hat{k}$ and $3\hat{i} - 4\hat{j} - 4\hat{k}$ form a/an
(a) isosceles triangle (b) right-angled triangle
(c) equilateral triangle (d) scalene triangle
15. The distance between two points with position vectors $\hat{i} + \hat{j} + 2\hat{k}$ and $3\hat{i} + 5\hat{j} - 2\hat{k}$ is
(a) 2 units (b) 4 units (c) 6 units (d) 8 units
16. Let $\vec{a}$, $\vec{b}$ and $\vec{c}$ be three non-coplanar vectors, and $3\vec{a} - 7\vec{b} - 4\vec{c}$, $3\vec{a} - 2\vec{b} + \vec{c}$ and $\vec{a} + \vec{b} + x\vec{c}$ be coplanar. The value of x is
(a) 1 (b) 2 (c) 3 (d) 4
17. The position vectors of the four points P, Q, R and S are given by the vectors $\hat{i} + \hat{j} + \hat{k}$, $2\hat{i} + 3\hat{j}$, $3\hat{i} + 5\hat{j} - 2\hat{k}$ and $-\hat{j} + \hat{k}$. Then, $\vec{PQ}$ and $\vec{RS}$ are
(a) parallel (b) non-parallel (c) collinear (d) non-collinear
18. The points whose position vectors are $2\hat{i} + \hat{j} - \hat{k}$, $\hat{i} + 2\hat{j} + \hat{k}$, $2\hat{i} + 2\hat{j} - \hat{k}$ and $3\hat{i} + \hat{j} - 3\hat{k}$ are the vertices of a
(a) rhombus (b) square (c) parallelogram (d) rectangle
19. The end points of the three vectors $\vec{\alpha}$, $3\vec{\alpha} - 2\vec{\beta}$ and $\vec{\beta}$ are
(a) collinear (b) non-collinear (c) coplanar (d) non-coplanar
20. The two vectors $\vec{a} = \hat{i}$ and $\vec{b} = \frac{1}{2}(\hat{i} + \hat{j})$ are
(a) related by rotation (b) related by an inversion
(c) not linearly independent (d) related by reflection through the *xy* plane

21. A rigid body is acted on by two forces: $\vec{F}_1 = a\hat{i} + b\hat{j} - 3\hat{k}$ at the point $(1, 2, -1)$ and $\vec{F}_2 = \hat{i} + a\hat{j} + b\hat{k}$ at the point $(-1, 0, 1)$. If the force system is equipollent to the force $\vec{F}$ and a couple $\vec{G}$, which has no components along $\hat{k}$, then $\vec{F}$ equals
(a) $2\hat{i} + 4\hat{j}$ (b) $2\hat{i} - 4\hat{j}$ (c) $4\hat{i} + 2\hat{j}$ (d) $4\hat{i} - 2\hat{j}$

22. The centroid of the 30 points $\hat{i}, 2\hat{i}, \cdots, 10\hat{i}; \hat{j}, 2\hat{j}, \cdots, 10\hat{j}; \hat{k}, 2\hat{k}, \cdots, 10\hat{k}$ is
(a) $\hat{i} + \hat{j} + \hat{k}$ (b) $11(\hat{i} + \hat{j} + \hat{k})$ (c) $\frac{1}{30}1(\hat{i} + \hat{j} + \hat{k})$ (d) $\frac{11}{6}(\hat{i} + \hat{j} + \hat{k})$

Answers

1. b 2. b 3. a 4. c 5. b 6. d 7. a 8. d 9. b 10. c 11. c 12. b 13. d
14. b 15. c 16. b 17. a 18. c 19. a 20. a 21. c 22. d

Short Answer Questions

1. Give the geometrical representation of a scalar.
2. Define polar and axial vectors. Give one example of each.
3. Define a free vector, line vector and localised vector. Give one example of each.
4. What do you mean by the magnitude and direction cosines of a vector?
5. Prove that the sum of the two vectors is independent of the choice of the origin O.
6. Define position vector. How will you express the displacement of a particle between two points having position vectors $\vec{r}_1$ and $\vec{r}_2$?
7. Prove that the commutative and associative laws hold for the addition of any number of vectors, the sum being independent of the order and grouping of the terms.
8. If $\vec{a}$ and $\vec{b}$ are the vectors representing the diagonals of a parallelogram, find the vectors corresponding to the sides of the parallelogram. *Ans*: $(\vec{a} + \vec{b})/2, (\vec{a} - \vec{b})/2$
9. Given any two vectors $\vec{a}$ and $\vec{b}$, illustrate geometrically the equality $4\vec{a} + 3(\vec{b} - \vec{a}) = \vec{a} + 3\vec{b}$.
10. Show that the point $3\hat{i} + 5\hat{j} + \hat{k}$ bisects the join of the points $4\hat{i} + 3\hat{j} + 5\hat{k}$ and $2\hat{i} + 7\hat{j} - 3\hat{k}$.
11. $OABC$ is a tetrahedron. $\vec{OA} = \vec{a}$, $\vec{OB} = \vec{b}$, $\vec{OC} = \vec{c}$. Express $\vec{BC}$, $\vec{CA}$ and $\vec{AB}$ in terms of $\vec{a}, \vec{b}$ and $\vec{c}$. *Ans*: $\vec{BC} = \vec{c} - \vec{b}$, $\vec{CA} = \vec{a} - \vec{c}$, $\vec{AB} = \vec{b} - \vec{a}$
12. If the position vectors of the points A, B, C and D are $\hat{i} + \hat{j} + \hat{k}$, $2\hat{i} + 3\hat{j}$, $3\hat{i} + 5\hat{j} - 2\hat{k}$ and $-\hat{j} + \hat{k}$, respectively, then prove that $\vec{AB}$ and $\vec{CD}$ are parallel.
13. Deduce the conditions that the vectors $\vec{a} = 5\hat{i} + 7\hat{j} - 3\hat{k}$ and $\vec{b} = 2\hat{i} - p\hat{j} + q\hat{k}$ may be parallel. *Ans*: $p = -14/5, q = -6/5$
14. If $\vec{a} = \hat{i} + \hat{j} - 2\hat{k}$ $\vec{b} = 2\hat{i} + \hat{j} + \hat{k}$ and $\vec{c} = \hat{i} - 2\hat{j} + 2\hat{k}$, find the magnitude and direction cosines of
(a) $\vec{a} + \vec{b} + \vec{c}$ (b) $\vec{a} - \vec{b} + \vec{c}$ *Ans*: (a) $2\sqrt{2}, (0, 2, 2)/2\sqrt{2}$, (b) $\sqrt{17}, (4, 0, 1)/\sqrt{17}$
15. If $\vec{a} = 3\hat{i} - 2\hat{j} + \hat{k}$, $\vec{b} = 2\hat{i} - \hat{j} - 3\hat{k}$ and $\vec{c} = \hat{i} - 2\hat{j} - 2\hat{k}$, then find
(a) $2\vec{a} - 3\vec{b} + 4\vec{c}$ and $|2\vec{a} - 3\vec{b} + 4\vec{c}|$ (b) The unit vector along $\vec{a} - \vec{b}$

(c) The unit vector parallel to $\vec{c}$ *Ans:* (a) $\pm\frac{1}{3}(\hat{\imath}-2\hat{\jmath}-2\hat{k})$, (b) $\sqrt{106}$, (c) $\pm\frac{\sqrt{2}}{6}(\hat{\imath}-\hat{\jmath}+4\hat{k})$

16. Show that the four points $A(6,-7,0)$, $B(16,-19,-4)$, $C(0,3,-6)$ and $D(2,5,10)$ are such that AB and CD intersect at a point $P(1,-1,2)$.

17. The position vectors of the vertices of a plane quadrilateral $ABCD$ satisfy the relation $6\vec{a}-4\vec{b}+\vec{c}-3\vec{d}=\vec{0}$. Find the ratios in which G, the point of intersection of the diagonals AC and BD, divides AC and BD. *Ans:* 6 : 1 and 4 : 3

18. Find the point which divides the join of the points $\vec{a}$ and $\vec{b}$ internally and externally in the ratio 3 : 4. *Ans:* $(4\vec{a}+3\vec{b})/7, (4\vec{a}-3\vec{b})$

19. If two vectors $\vec{a}$ and $\vec{b}$ are added, show that the magnitude of the resultant cannot be greater than $a+b$ or smaller than $|a-b|$.

20. CD is trisected at P; find the point of trisection. *Ans:* $(\vec{c}+2\vec{d})/3$

21. Show that the three points whose position vectors are $\vec{a}$, $3\vec{a}-2\vec{b}$ and $2\vec{b}-\vec{a}$ are collinear, where $\vec{a}$ and $\vec{b}$ are any vectors.

22. Find the unit vector co-directional with the given vector $\vec{a}=4\hat{\imath}+7\hat{\jmath}-3\hat{k}$.
Ans: $\frac{1}{\sqrt{74}}(4\hat{\imath}+7\hat{\jmath}-3\hat{k})$.

23. Prove by the vector method:
(a) The medians of a triangle meet at a point which trisects each of them.
(b) If the diagonals of a quadrilateral bisect each other, then the figure is a parallelogram.

24. Prove that:
(a) Every set of four vectors in three dimensions must be a linearly dependent system.
(b) A triad of non-zero, non-coplanar vectors is a linearly independent system.
(c) A pair of non-zero, non-collinear vectors is a linearly dependent system.

25. A quadrilateral $ABCD$ has masses of 1,2,3 and 4 units located, respectively, at its vertices $A(-1,-2,2)$, $B(3,2,-1)$, $C(1,-2,4)$ and $D(3,1,2)$. Find the coordinates of the centroid.
Ans: $(2,0,2)$

26. The position vectors of points A and B are given by $\vec{r}_1=2\hat{\imath}+3\hat{\jmath}-\hat{k}$ and $\vec{r}_2=4\hat{\imath}-3\hat{\jmath}+2\hat{k}$. Determine $\vec{AB}$ in terms of $\hat{\imath},\hat{\jmath}$ and $\hat{k}$ and find its magnitude. *Ans:* $2\hat{\imath}-6\hat{\jmath}+3\hat{k}, 7$

27. Find the centroid of the 30 points $\hat{\imath}, 2\hat{\imath}, 3\hat{\imath}, \cdots, 10\hat{\imath}; \hat{\jmath}, 2\hat{\jmath}, 3\hat{\jmath}, \cdots, 10\hat{\jmath}; \hat{k}, 2\hat{k}, 3\hat{k}, \cdots, 10\hat{k}$.
Ans: $\frac{11}{6}(\hat{\imath}+\hat{\jmath}+\hat{k})$

Long Answer Questions

1. Find the unit vector parallel to the
(a) xy plane and perpendicular to the vector $4\hat{\imath}-3\hat{\jmath}+\hat{k}$ *Ans:* $\pm(3\hat{\imath}+4\hat{\jmath})/5$
(b) xy plane and perpendicular to the vector $2\hat{\imath}+3\hat{\jmath}+4\hat{k}$ *Ans:* $\pm(2\hat{\imath}-3\hat{\jmath})/\sqrt{13}$

2. If two sides of a triangle are formed by the vectors $\vec{a}=3\hat{\imath}+6\hat{\jmath}-2\hat{k}$ and $\vec{b}=4\hat{\imath}-\hat{\jmath}+2\hat{k}$, determine the third side and all the angles of the triangle. *Ans:* $\pm(7\hat{\imath}+5\hat{\jmath})$

3. Prove that each of the four faces of a tetrahedron subtends the same volume at the centroid.
4. Find the ratio of the volume of a tetrahedron with that of the tetrahedron formed by the centroids of its faces.
5. $ABCD$ is a quadrilateral. The position vectors of its vertices satisfy the relation $2\vec{a}+3\vec{b}-\vec{c}-4\vec{d}=\vec{0}$. Find the ratios in which P divides AB and CD, P being the point of intersection of AB and CD.
6. If $\vec{a}$, $\vec{b}$ and $\vec{c}$ be three non-coplanar vectors, then show that any vector $\vec{r}$ can be uniquely expressed as a linear combination of $\vec{a}$, $\vec{b}$ and $\vec{c}$.
7. Three points whose position vectors are $A(2,4,-1)$, $B(4,5,1)$ and $C(3,6,-3)$ form a triangle.
   (a) Find the lengths of the sides BC, CA and AB. *Ans:* $3, \sqrt{18}, 3$
   (b) Prove that the triangle is an isosceles right-angled triangle.
   (c) Find the direction cosines of $\vec{BC}$, $\vec{CA}$ and $\vec{AB}$.
8. ABC is a triangle. D divides BC in the ratio $l:m$. G divides AD in the ratio $l+m:n$. Find the position vector of D and G. *Ans:* $\vec{d}=\frac{1}{l+m}(l\vec{c}+m\vec{b})$, $\vec{g}=\frac{1}{l+m+n}(l\vec{c}+m\vec{b}+n\vec{a})$
9. ABC is a triangle. D, E and F are the midpoints of the sides BC, CA and AB, respectively. Prove that $\vec{AD}+\vec{BF}+\vec{CF}=\vec{0}$.
10. Prove that the points $(l_i\vec{a}+m_i\vec{b})$; $i=1,2,3$ where $\vec{a}$ and $\vec{b}$ are two non-collinear vectors, are collinear, if and only if

$$\begin{vmatrix} l_1 & l_2 & l_3 \\ m_1 & m_2 & m_3 \\ 1 & 1 & 1 \end{vmatrix} = 0$$

11. Prove that the points $(l_i\vec{a}+m_i\vec{b}+n_i\vec{c})$; $i=1,2,3,4$ where $\vec{a}$, $\vec{b}$ and $\vec{c}$ are three non-coplanar vectors, are coplanar, if and only if

$$\begin{vmatrix} l_1 & l_2 & l_3 & l_4 \\ m_1 & m_2 & m_3 & m_4 \\ n_1 & n_2 & n_3 & n_4 \\ 1 & 1 & 1 & 1 \end{vmatrix} = 0$$

12. The following forces act on a particle $2\hat{i}+3\hat{j}-5\hat{k}$, $-5\hat{i}+\hat{j}+3\hat{k}$, $\hat{i}-2\hat{j}+4\hat{k}$ and $4\hat{i}+3\hat{j}-2\hat{k}$ measured in Newtons. Find
    (a) The resultant of the forces *Ans:* $2\hat{i}-\hat{j}$
    (b) The magnitude of the resultant *Ans:* $\sqrt{5}$
13. (a) Prove that the vectors $\vec{a}=3\hat{i}+\hat{j}-2\hat{k}$, $\vec{b}=-\hat{i}+3\hat{j}+4\hat{k}$ and $\vec{c}=4\hat{i}-2\hat{j}-6\hat{k}$ can form the sides of a triangle.
    (b) Find the lengths of the medians of the triangle. *Ans:* $\sqrt{6}, \sqrt{114}/2, \sqrt{150}/2$

14. $ABCD$ is a plane quadrilateral. The position vectors of its vertices satisfy the relation $2\vec{a}+3\vec{b}-\vec{c}-4\vec{d}=\vec{0}$. Find the ratios in which P divides AB and CD, P being the point of intersection of AB and CD. *Ans*: P divides AB in the ratio $3:2$ and CD in the ratio $4:1$
15. If the position vectors of four points A, B, C and D are $\vec{a}, \vec{b}, \vec{c}$ and $\vec{d}$, respectively, and if $3\vec{a}-2\vec{b}+\vec{c}-2\vec{d}=\vec{0}$, are the points A, B, C and D coplanar? Find P, where AC and BD meet. In what ratios does P divide AC and BD? *Ans*: $1:3$ and $1:1$
16. If the position vectors of n points represent n concurrent forces, then show that the forces will be in equilibrium if the centroid of the points coincides with the origin.
17. Prove that the opposite sides of the equal angles of equiangular triangles are proportional.
18. Find the centroid of the $3n$ points
(a) $\hat{i}, 2\hat{i}, 3\hat{i}, \cdots, n\hat{i}; \hat{j}, 2\hat{j}, 3\hat{j}, \cdots, n\hat{j}; \hat{k}, 2\hat{k}, 3\hat{k}, \cdots, n\hat{k}$. *Ans*: $\frac{n+1}{6}\,(\hat{i}+\hat{j}+\hat{k})$
(b) $\hat{i}, 4\hat{i}, 9\hat{i}, \cdots, n^2\hat{i}; \hat{j}, 4\hat{j}, 9\hat{j}, \cdots, n^2\hat{j}; \hat{k}, 4\hat{k}, 9\hat{k}, \cdots, n^2\hat{k}$. *Ans*: $\frac{(n+1)(2n+1)}{18}\,(\hat{i}+\hat{j}+\hat{k})$
19. Particles of mass proportional to $1, 2, 3, \cdots, n$ are placed, respectively, at the end points of the n vectors $\vec{\alpha}_1+\vec{\alpha}_2; \vec{\alpha}_1+2\vec{\alpha}_2; \vec{\alpha}_1+3\vec{\alpha}_2, \cdots, \vec{\alpha}_1+n\vec{\alpha}_2$; drawn from the origin. Find the vector joining O to the mass-centre. *Ans*: $\vec{\alpha}_1+\frac{1}{3}(2n+1)\vec{\alpha}_2$.
20. Particles of mass $1, 2, 3$ and 4 and $2, 1, 4$ and 3 g, respectively, are placed at the corners A, B, C and D of one face, and the last four at their projections E, F, G and H, respectively, on the opposite face. Find the coordinates of their centre of mass. *Ans*: $\frac{1}{10}(5\hat{i}+5\hat{j}+14\hat{k})$.
21. Show that the centroid of the vertices of a triangle trisects the medians.

2 Product of Vectors

In the previous chapter, we discussed the three main elementary operations, namely, the addition and subtraction of two vectors and the multiplication of a vector by a scalar. Here, we shall discuss the multiplication of two vectors in two different forms:

(i) scalar or dot or inner or direct product
(ii) vector or cross or outer product

Each of these products is proportional to the product of the moduli of the two vectors. We have borrowed from algebra the term multiplication for the process by which a product is formed from two vectors, and these are called the *factors of the product*. The multiplication of two vectors has no interpretation in the ordinary sense of multiplication, as each factor is a concrete quantity, not a mere number.

2.1 SCALAR PRODUCT

The scalar product of two vectors is a scalar which is proportional to the length of each vector. It obeys commutative law, associative law with scalars and distributive law, like the product of real numbers. In mechanics, the application of scalar product is the work done by a force causing displacement of a particle. The length of vectors and the angle between a pair of vectors are expressed in terms of scalar product of vectors. Various formulae of three-dimensional coordinate geometry can be established by using the concept of scalar product of vectors. Many properties of plane geometry and trigonometrical formulae can also be proved using the notion of scalar product.

Definition 2.1.1

The *dot* or *scalar* or *inner product* of two given vectors $\vec{a}$ and $\vec{b}$, denoted by a scalar quantity $\vec{a}\cdot\vec{b}$ (read as vector $\vec{a}$ dot vector $\vec{b}$), is defined by

$$\vec{a}\cdot\vec{b} = |\vec{a}|\ |\vec{b}|\ \cos\theta; \qquad 0 \le \theta \le \pi \qquad (2.1)$$

where θ is the smallest, non-negative angle between the vectors $\vec{a}$ and $\vec{b}$ (Fig. 2.1), and $|\vec{a}|$ and $|\vec{b}|$ represent the magnitude of the vectors $\vec{a}$ and $\vec{b}$, respectively. Sometimes it is denoted by $\langle\vec{a}, \vec{b}\rangle$.

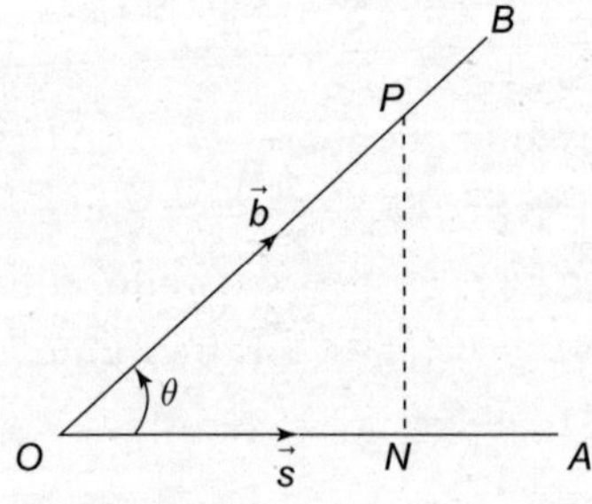

Figure 2.1 Scalar product

The scalar product of any vector with a zero vector is, by definition, a scalar zero. One very important use of the notion of scalar product is that it enables the lengths of vectors and the angles between pairs of vectors to be expressed in terms of the same, and thus provides an analytical tool for the study of metric geometry. Following are the properties of dot products:

Property 2.1.1

The dot product $\vec{a} \cdot \vec{b}$ is positive, zero or negative depending on whether the angle θ is acute, right or obtuse.

Property 2.1.2

Commutativity: The dot product of two vectors obeys commutative law, i.e.,

$$\vec{b} \cdot \vec{a} = |\vec{b}||\vec{a}| \cos(-\theta) = |\vec{a}||\vec{b}| \cos\theta = \vec{a} \cdot \vec{b}$$

for every pair of vectors $\vec{a}$ and $\vec{b}$. Hence, in the case of dot product, the order of the factors may be reversed without altering the value of the product.

Result 8

Since $|\cos\theta| \le 1$ for all vectors of θ, from Eq. (2.1), we have

$$\cos^2\theta = \frac{(\vec{a} \cdot \vec{b})^2}{|\vec{a}|^2 \cdot |\vec{b}|^2} = \frac{(\vec{a} \cdot \vec{b})^2}{(\vec{a} \cdot \vec{a})(\vec{b} \cdot \vec{b})} \le 1$$

$$\Rightarrow (\vec{a} \cdot \vec{b})^2 \le (\vec{a} \cdot \vec{a})(\vec{b} \cdot \vec{b})$$

This is the *Cauchy–Schwarz inequality*.

Property 2.1.3

Length of a vector as a scalar product: If $\vec{a}$ be any vector, then the scalar product $\vec{a} \cdot \vec{a}$ of $\vec{a}$ with itself is given by

$$\vec{a} \cdot \vec{a} = |\vec{a}| \ |\vec{a}| \cos 0^\circ = |\vec{a}|^2 \tag{2.2}$$

Thus, the length $|\vec{a}|$ of any vector $\vec{a}$ is the non-negative square root $\sqrt{\vec{a} \cdot \vec{a}}$.

Property 2.1.4

Condition for orthogonality: Let two proper vectors (i.e., $|\vec{a}| \neq 0$, $|\vec{b}| \neq 0$) be orthogonal. Then, $\theta = \pi/2$, and therefore

$$\vec{a} \cdot \vec{b} = |\vec{a}| \ |\vec{b}| \ \cos\pi/2 = 0$$

Thus, from the above, it is clear that $\vec{a}.\vec{b} = 0$ does not always imply that either $\vec{a}$ or $\vec{b}$ is a zero vector. Conversely, let $\vec{a} \cdot \vec{b} = 0$. When either $\vec{a} = \vec{0}$ or $\vec{b} = \vec{0}$, $\vec{a}$ and $\vec{b}$ are at right angles to each other. Therefore, the necessary and sufficient condition that two proper vectors $\vec{a}$ and $\vec{b}$ are orthogonal is $\vec{a} \cdot \vec{b} = 0$. Thus, for three unit vectors $\hat{i}, \hat{j}$ and $\hat{k}$ along the three mutually perpendicular axes, we have

$$\hat{i} \cdot \hat{j} = \hat{j} \cdot \hat{k} = \hat{k} \cdot \hat{i} = 0 \text{ and } \hat{i} \cdot \hat{i} = \hat{j} \cdot \hat{j} = \hat{k} \cdot \hat{k} = 1 \tag{2.3}$$

A system of three unit vectors, any two of which are mutually orthogonal, is called an *orthogonal triad of vectors*.

Let $\vec{r} = x\hat{i} + y\hat{j} + z\hat{k}$ be any vector. Then, using the above property, we get $\vec{r} \cdot \hat{i} = x\hat{i} \cdot \hat{i} = x$; $\vec{r} \cdot \hat{j} = y\hat{j} \cdot \hat{j} = y$; $\vec{r} \cdot \hat{k} = z\hat{k} \cdot \hat{k} = z$. Therefore

$$\vec{r} = (\vec{r}\cdot\hat{\imath})\hat{\imath} + (\vec{r}\cdot\hat{\jmath})\hat{\jmath} + (\vec{r}\cdot\hat{k})\hat{k}$$

Therefore, $\vec{r}\cdot\hat{\imath}, \vec{r}\cdot\hat{\jmath}$ and $\vec{r}\cdot\hat{k}$ are, respectively, the *rectangular components* of $\vec{r}$.

Following are some examples of dot product using the orthogonality condition:

(i) $\hat{\imath}\cdot\hat{\imath} = |\hat{\imath}|\,|\hat{\imath}|\ \cos 0^\circ = 1\cdot 1\cdot 1 = 1$
(ii) $\hat{\imath}\cdot\hat{\jmath} = |\hat{\imath}|\,|\hat{\jmath}|\ \cos 90^\circ = 1\cdot 1\cdot 0 = 0$
(iii) $\hat{\imath}\cdot(2\hat{\imath} - 5\hat{\jmath} + 7\hat{k}) = 2\hat{\imath}\cdot\hat{\imath} - 5\hat{\imath}\cdot\hat{\jmath} + 7\hat{\imath}\cdot\hat{k} = 2 - 5\cdot 0 + 7\cdot 0 = 2$
(iv) $(3\hat{\imath} - 5\hat{k})\cdot(6\hat{\jmath} + 2\hat{k}) = 3\hat{\imath}\cdot(6\hat{\jmath} + 2\hat{k}) - 5\hat{k}\cdot(6\hat{\jmath} + 2\hat{k})$
$= 18\hat{\imath}\cdot\hat{\jmath} + 6\hat{\imath}\cdot\hat{k} - 30\hat{k}\cdot\hat{\jmath} - 10\hat{k}\cdot\hat{k}$
$= 18\cdot 0 + 6\cdot 0 - 30\cdot 0 - 10\cdot 1 = -10$

Property 2.1.5

For two parallel vectors $\vec{a}$ and $\vec{b}$

(i) $\vec{a}\cdot\vec{b} = |\vec{a}||\vec{b}|\cos 0^0 = |\vec{a}||\vec{b}|$; $\vec{a}$ and $\vec{b}$ have the same sense.
(ii) $\vec{a}\cdot\vec{b} = |\vec{a}||\vec{b}|\cos\pi = -|\vec{a}||\vec{b}|$; $\vec{a}$ and $\vec{b}$ have the opposite sense.

Example 2.1.1

If $\vec{a} = \hat{\imath} + \hat{\jmath} + 2\hat{k}$, $\vec{b} = -\hat{\imath} + 2\hat{\jmath} + 3\hat{k}$ and $\vec{c} = 5\hat{\imath} + 8\hat{k}$, find the scalars x and y such that $\vec{c} - x\vec{a} - y\vec{b}$ is perpendicular to both $\vec{a}$ and $\vec{b}$.

Solution

As $\vec{a} = \hat{\imath} + \hat{\jmath} + 2\hat{k}$, $\vec{b} = -\hat{\imath} + 2\hat{\jmath} + 3\hat{k}$ and $\vec{c} = 5\hat{\imath} + 8\hat{k}$

$$\begin{aligned}\vec{c} - x\vec{a} - y\vec{b} &= 5\hat{\imath} + 8\hat{k} - x(\hat{\imath} + \hat{\jmath} + 2\hat{k}) - y(-\hat{\imath} + 2\hat{\jmath} + 3\hat{k})\\ &= (5 - x + y)\hat{\imath} + (-x - 2y)\hat{\jmath} + (8 - 2x - 3y)\hat{k}\end{aligned}$$

By the given condition, since it is perpendicular to $\vec{a}$, we have

$$\begin{aligned}&\vec{a}\cdot(\vec{c} - x\vec{a} - y\vec{b}) = 0\\ &\Rightarrow (5 - x + y) + (-x - 2y) + 2(8 - 2x - 3y) = 0\\ &\Rightarrow 6x + 7y - 21 = 0\end{aligned}$$

It is perpendicular to $\vec{b}$, so

$$\begin{aligned}&\vec{b}\cdot(\vec{c} - x\vec{a} - y\vec{b}) = 0\\ &\Rightarrow -(5 - x + y) + 2(-x - 2y) + 3(8 - 2x - 3y) = 0\\ &\Rightarrow 7x + 14y - 19 = 0\end{aligned}$$

Solving, we get $x = 23/5$ and $y = -33/35$.

Example 2.1.2

If $\vec{a}$ and $\vec{b}$ are any two vectors such that $|\vec{a} + \vec{b}| = |\vec{a} - \vec{b}|$, prove that $\vec{a}$ is perpendicular to $\vec{b}$, and conversely. Interpret the result geometrically.

Solution

Using the given condition, we have

$$|\vec{a}+\vec{b}|^2 = |\vec{a}-\vec{b}|^2$$

$$\text{or,}\quad (\vec{a}+\vec{b})\cdot(\vec{a}+\vec{b}) = (\vec{a}-\vec{b})\cdot(\vec{a}-\vec{b}); \qquad \text{as } \vec{\alpha}\cdot\vec{\alpha} = |\vec{\alpha}|^2$$

$$\text{or,}\quad \vec{a}\cdot\vec{a}+2\vec{a}\cdot\vec{b}+\vec{b}\cdot\vec{b} = \vec{a}\cdot\vec{a}-2\vec{a}\cdot\vec{b}+\vec{b}\cdot\vec{b}; \quad \text{as } \vec{a}\cdot\vec{b} = \vec{b}\cdot\vec{a}$$

$$\text{or,}\quad 4\vec{a}\cdot\vec{b} = 0 \Rightarrow \vec{a}\cdot\vec{b} = 0$$

Hence, the vectors $\vec{a}$ and $\vec{b}$ are orthogonal (perpendicular). Conversely, we assume that $\vec{a}\cdot\vec{b} = 0$. Now, we have

$$|\vec{a}+\vec{b}| = \sqrt{|\vec{a}+\vec{b}|^2} = \sqrt{(\vec{a}+\vec{b})\cdot(\vec{a}+\vec{b})}; \text{Eq. (2.2)}$$

$$= \sqrt{|\vec{a}|^2+|\vec{b}|^2+2\vec{a}\cdot\vec{b}} = \sqrt{|\vec{a}|^2+|\vec{b}|^2}$$

$$= \sqrt{|\vec{a}|^2+|\vec{b}|^2-2\vec{a}\cdot\vec{b}}, \text{ as } \vec{a}\cdot\vec{b} = 0$$

$$= \sqrt{|\vec{a}-\vec{b}|^2} = \sqrt{(\vec{a}-\vec{b})\cdot(\vec{a}-\vec{b})} = |\vec{a}-\vec{b}|$$

We know that if $\vec{a}$ and $\vec{b}$ are two vectors representing two adjacent sides of a parallelogram, then $\vec{a}+\vec{b}$ and $\vec{a}-\vec{b}$ are the diagonals of the parallelogram. From the above discussion, geometrically it can be concluded that the diagonals of a parallelogram are equal if and only if it is a rectangle.

Example 2.1.3

Find the angles which the vector $\vec{a} = \hat{i}+\hat{j}+\sqrt{2}\,\hat{k}$ makes with the coordinate axes.

Solution

Let θ_1, θ_2 and θ_3 be the angles which $\vec{a}$ makes with the positive x, y and z axes, respectively. Now, the length of $\vec{a}$ is $a = |\vec{a}| = \sqrt{1^2+1^2+(\sqrt{2})^2} = 2$. Thus, we obtain

$$\vec{a}\cdot\hat{i} = a\,|\hat{i}|\,\cos\theta_1 = 2\,\cos\theta_1$$

$$\Rightarrow\ (\hat{i}+\hat{j}+\sqrt{2}\,\hat{k})\cdot\hat{i} = 2\,\cos\theta_1$$

$$\Rightarrow\ \cos\theta_1 = \frac{1}{2} \Rightarrow \theta_1 = \frac{\pi}{3}$$

Proceeding as with the above, we obtain $\theta_2 = \pi/3$ and $\theta_3 = \pi/4$.

Example 2.1.4

If two pairs of opposite sides of a tetrahedron are perpendicular, then show that the third pair is also perpendicular and that the sum of the squares of each pair of opposite sides are equal. Further, show that for such a tetrahedron, the sum of the squares of each pair of opposite edges is the same.

Solution

Let us consider a tetrahedron $OABC$, where the vertex O is chosen as origin (Fig. 2.2). Let $\vec{a}, \vec{b}$ and $\vec{c}$ be the position vectors of the vertices A, B and C, respectively, with respect to O. Also, we consider $\vec{OA}\perp\vec{BC}$ and $\vec{OB}\perp\vec{CA}$; so

$$\vec{OA}\cdot\vec{BC}=0\Rightarrow\vec{a}\cdot(\vec{c}-\vec{b})=0;$$
$$\vec{OB}\cdot\vec{CA}=0\Rightarrow\vec{b}\cdot(\vec{a}-\vec{c})=0$$

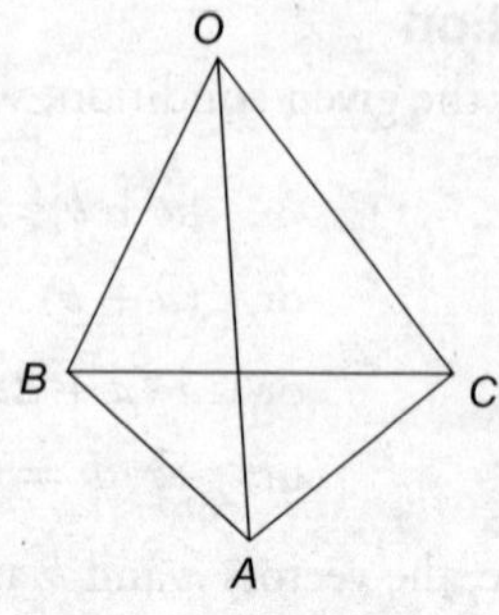

Figure 2.2 Tetrahedron

Adding the results, we obtain

$$\vec{a}\cdot(\vec{c}-\vec{b})+\vec{b}\cdot(\vec{a}-\vec{c})=0$$
$$\Rightarrow\vec{c}\cdot(\vec{b}-\vec{a})=0\Rightarrow\vec{OC}\cdot\vec{AB}=0$$

This shows that $\vec{OC}$ is perpendicular to $\vec{AB}$. Also, the above result gives $\vec{a}\cdot\vec{b}=\vec{b}\cdot\vec{c}=\vec{c}\cdot\vec{a}$. Now, we have

$$\begin{aligned}OA^2+BC^2&=|\vec{a}|^2+|\vec{c}-\vec{b}|^2\\&=|\vec{a}|^2+(\vec{c}-\vec{b})\cdot(\vec{c}-\vec{b})\\&=|\vec{a}|^2+|\vec{b}|^2+|\vec{c}|^2-2\vec{b}\cdot\vec{c}\\OB^2+CA^2&=|\vec{b}|^2+|\vec{a}-\vec{c}|^2\\&=|\vec{b}|^2+(\vec{a}-\vec{c})\cdot(\vec{a}-\vec{c})\\&=|\vec{b}|^2+|\vec{a}|^2+|\vec{c}|^2-2\vec{c}\cdot\vec{a}\\&=|\vec{a}|^2+|\vec{b}|^2+|\vec{c}|^2-2\vec{b}\cdot\vec{c},\text{ as }\vec{c}\cdot\vec{a}\\&=\vec{b}\cdot\vec{c}\\OC^2+AB^2&=|\vec{c}|^2+|\vec{b}-\vec{a}|^2=|\vec{c}|^2+(\vec{b}-\vec{a})\cdot(\vec{b}-\vec{a})\\&=|\vec{c}|^2+|\vec{a}|^2+|\vec{b}|^2-2\vec{a}\cdot\vec{b}\\&=|\vec{a}|^2+|\vec{b}|^2+|\vec{c}|^2-2\vec{b}\cdot\vec{c},\text{ as }\vec{a}\cdot\vec{b}=\vec{b}\cdot\vec{c}\end{aligned}$$

Using the above results, we obtain $OA^2+BC^2=OB^2+CA^2=OC^2+AB^2$. This proves the result.

Example 2.1.5

Show by the vector method that the distance of a corner of a unit cube from the diagonal not passing through it is $\sqrt{6}/3$ units.

Solution

Let $OABCDEFG$ be a unit cube (Fig. 2.3) and also let OA, OC and OE be chosen along the x axis, y axis and z axis, respectively, with respect to O as origin. Further, we assume that $\vec{OA}=\hat{i}$, $\vec{OC}=\hat{j}$ and $\vec{OE}=\hat{k}$. Therefore, the diagonal $\vec{OG}=\hat{i}+\hat{j}+\hat{k}$ and a unit vector along it is $\hat{OG}=\frac{1}{\sqrt{3}}(\hat{i}+\hat{j}+\hat{k})$. Let EP be the perpendicular to OG. Thus, we have

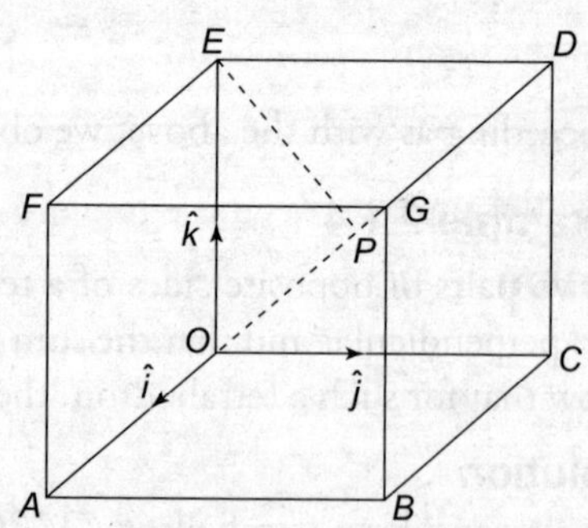

Figure 2.3 Unit cube

$$OP=OE\cos\angle EOP=\cos\angle EOP;\quad\text{as }OE=1\text{ unit}$$
$$=\frac{\vec{OG}\cdot\vec{OE}}{|\vec{OG}|}=\frac{1}{\sqrt{3}}(\hat{i}+\hat{j}+\hat{k})\cdot\hat{k}=\frac{1}{\sqrt{3}}\text{ units}$$

From the triangle OEP, $EP^2 + OP^2 = OE^2$, so

$$EP^2 = OE^2 - OP^2 = 1 - 1/3 = 2/3$$

Thus, $EP = \sqrt{2/3} = \sqrt{6}/3$ units. Similarly, for the other diagonals, the same result can be obtained.

Example 2.1.6

Pythagorian theorem: If $\vec{a}$, $\vec{b}$ and $\vec{c}$ are the position vectors of the vertices A, B and C of a right-angled triangle ABC of which $\angle A$ is the right angle, prove that $|\vec{BC}|^2 = |\vec{AB}|^2 + |\vec{AC}|^2$.

Solution

Since $\vec{a}$, $\vec{b}$ and $\vec{c}$ are the position vectors of the vertices A, B and C with respect to some base point of the right-angled triangle ABC (Fig. 2.4), we have

$$\vec{AB} = \vec{b} - \vec{a} = \vec{p},\ \ \vec{BC} = \vec{c} - \vec{b},\ \ \vec{AC} = \vec{c} - \vec{a} = \vec{q}$$

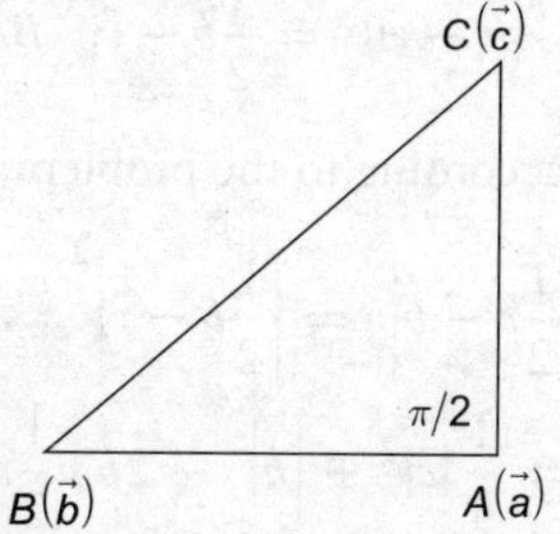

Figure 2.4 Right-angled triangle

Also, as $\angle A$ is right-angled, we have $\vec{AB} \cdot \vec{AC} = 0$, i.e., $(\vec{b} - \vec{a}) \cdot (\vec{a} - \vec{c}) = 0 = \vec{p} \cdot \vec{q}$. Now, we have

$$\begin{aligned}
|\vec{BC}|^2 &= |\vec{c} - \vec{b}|^2 = (\vec{c} - \vec{b}) \cdot (\vec{c} - \vec{b}) \\
&= (\vec{c} - \vec{a} + \vec{a} - \vec{b}) \cdot (\vec{c} - \vec{a} + \vec{a} - \vec{b}) \\
&= (\vec{q} - \vec{p}) \cdot (\vec{q} - \vec{p}) \\
&= \vec{q} \cdot \vec{q} - 2\vec{p} \cdot \vec{q} + \vec{p} \cdot \vec{p} \\
&= \vec{q} \cdot \vec{q} + \vec{p} \cdot \vec{p}; \quad \text{as } \vec{p} \cdot \vec{q} = 0 \\
&= |\vec{q}|^2 + |\vec{p}|^2 = |\vec{AB}|^2 + |\vec{AC}|^2
\end{aligned}$$

Property 2.1.6

Let the vectors $\vec{a}$ and $\vec{b}$ be parallel. Then, $\theta = 0°$, which gives $\vec{a} \cdot \vec{b} = |\vec{a}|\ |\vec{b}|$, i.e., the dot product of two parallel vectors is the product of their magnitudes. For two equal vectors $\vec{a}$ and $\vec{a}$, we have

$$\vec{a} \cdot \vec{a} = |\vec{a}|\ |\vec{a}|\ \cos 0° = |\vec{a}|^2 = a^2$$

Thus, the square of a vector $\vec{a}$, i.e., $|\vec{a}|^2$ is equal to the square of its length or modulus.

Example 2.1.7

If θ be the angle between the two unit vectors $\hat{e}_1$ and $\hat{e}_2$, show that $2\sin\theta/2 = |\hat{e}_1 - \hat{e}_2|$.

Solution

Since $\hat{e}_1$ and $\hat{e}_2$ are two unit vectors, we have $|\hat{e}_1| = 1 = |\hat{e}_2|$. If θ is the angle between them, then

$$\begin{aligned}
\hat{e}_1 \cdot \hat{e}_2 &= |\hat{e}_1||\hat{e}_2| \cos\theta = \cos\theta \\
\therefore\quad |\hat{e}_1 - \hat{e}_2|^2 &= (\hat{e}_1 - \hat{e}_2) \cdot (\hat{e}_1 - \hat{e}_2) \\
&= \hat{e}_1 \cdot \hat{e}_1 - 2\hat{e}_1 \cdot \hat{e}_2 + \hat{e}_2 \cdot \hat{e}_2 \\
&= |\hat{e}_1|^2 - 2\cos\theta + |\hat{e}_2|^2 = 2(1 - \cos\theta) = 4\sin^2\theta/2
\end{aligned}$$

Therefore, $|\hat{e}_1 - \hat{e}_2| = 2\sin\theta/2$.

Example 2.1.8

If the two medians of a triangle are equal, prove that the triangle is isosceles.

Solution

Let ABC be a triangle where BE and CF are two medians of equal length (Fig. 2.5). Let the position vectors of B and C be $\vec{b}$ and $\vec{c}$, respectively, with respect to the base point A. Since BE and CF are medians on the sides AC and AB, respectively, we have

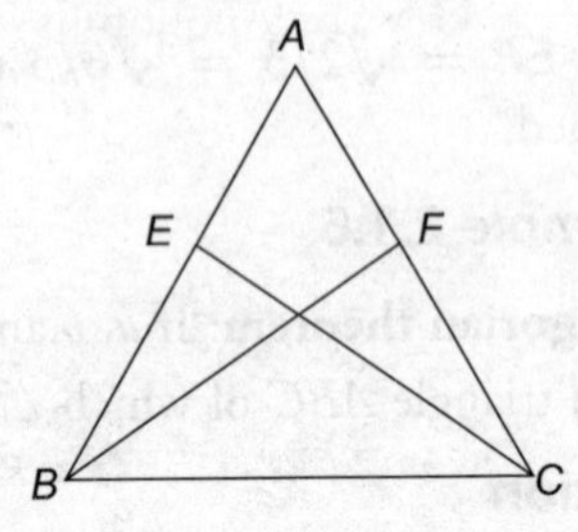

Figure 2.5 Isosceles triangle

$$\vec{CF} = \vec{AF} - \vec{AC} = \frac{1}{2}\vec{b} - \vec{c}, \quad \vec{BE} = \vec{AE} - \vec{AB} = \frac{1}{2}\vec{c} - \vec{b}$$

Now, according to the problem, $|\vec{BE}| = |\vec{CF}|$, so

$$\left|\frac{1}{2}\vec{c} - \vec{b}\right|^2 = \left|\frac{1}{2}\vec{b} - \vec{c}\right|^2 \Rightarrow \left(\frac{1}{2}\vec{c} - \vec{b}\right)\cdot\left(\frac{1}{2}\vec{c} - \vec{b}\right) = \left(\frac{1}{2}\vec{b} - \vec{c}\right)\cdot\left(\frac{1}{2}\vec{b} - \vec{c}\right)$$

$$\Rightarrow \frac{1}{4}|\vec{c}|^2 + \left|\vec{b}\right|^2 - 2\vec{b}\cdot\frac{1}{2}\vec{c} = \frac{1}{4}\left|\vec{b}\right|^2 + |\vec{c}|^2 - 2\frac{1}{2}\vec{b}\cdot\vec{c}$$

$$\text{or,} \quad \frac{3}{4}\left|\vec{b}\right|^2 = \frac{3}{4}|\vec{c}|^2 \Rightarrow AB^2 = AC^2 \Rightarrow AB = AC$$

Hence, the triangle under consideration is isosceles.

Example 2.1.9

Apollonius's theorem: If P be the midpoint of the side XY of the triangle OXY, then show that $OX^2 + OY^2 = 2(OP^2 + XP^2) = 2(OP^2 + PY^2)$.

Solution

We consider the vertex O as the origin and the position vectors of X and Y to be $\vec{a}$ and $\vec{b}$, respectively, with respect to O (Fig. 2.6). So the position vector of P is $\frac{1}{2}(\vec{a} + \vec{b})$. Thus, we have $\vec{XP} = \vec{OP} - \vec{OX} = \frac{1}{2}(\vec{a} + \vec{b}) - \vec{a} = \frac{1}{2}(\vec{b} - \vec{a})$. Now

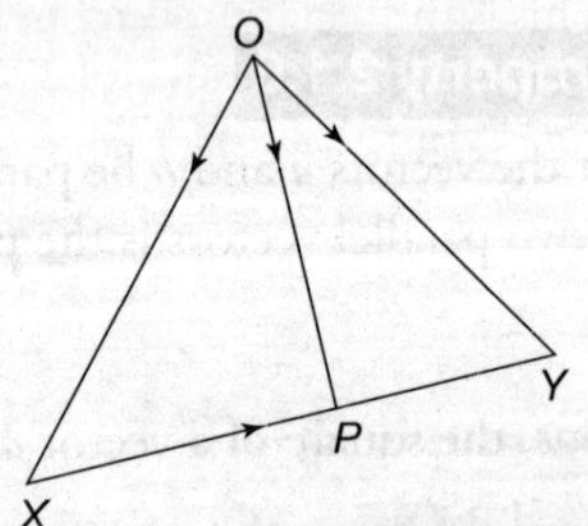

Figure 2.6 Apollonius's theorem

$$\begin{aligned} OP^2 + XP^2 &= OP^2 + PY^2 \\ &= \frac{1}{4}|\vec{a} + \vec{b}|^2 + \frac{1}{4}|\vec{a} - \vec{b}|^2 \\ &= \frac{1}{4}\left\{(\vec{a} + \vec{b})\cdot(\vec{a} + \vec{b}) + (\vec{a} - \vec{b})\cdot(\vec{a} - \vec{b})\right\} \\ &= \frac{1}{4}\left\{|\vec{a}|^2 + |\vec{b}|^2 + 2\vec{a}\cdot\vec{b} + |\vec{a}|^2 + |\vec{b}|^2 - 2\vec{a}\cdot\vec{b}\right\} \\ &= \frac{1}{2}\left\{|\vec{a}|^2 + |\vec{b}|^2\right\} \\ &= \frac{1}{2}(OX^2 + OY^2) \end{aligned}$$

DEDUCTION 2.1.1

If P be the midpoint of the side XY, then $XP = PY = \frac{1}{2}XY = \frac{1}{2}a$, where $XY = a$, $OX = b$ and $OY = c$. Now, by Apollonius's theorem, we find that

$$OX^2 + OY^2 = 2(OP^2 + XP^2)$$
$$\Rightarrow b^2 + c^2 = 2\,OP^2 + \frac{1}{2}a^2$$
$$\Rightarrow OP^2 = \frac{b^2 + c^2}{2} - \frac{1}{4}a^2$$
$$\Rightarrow OP = \sqrt{\frac{b^2 + c^2}{2} - \frac{1}{4}a^2}$$

Property 2.1.7

When the vectors $\vec{a}$ and $\vec{b}$ are parallel or collinear, then $\theta = 0^\circ$ or π so that $\cos\theta = \pm 1$. Hence

$$\vec{a} \cdot \vec{b} = |\vec{a}|\ |\vec{b}|\ \cos\theta = \pm|\vec{a}|\ |\vec{b}|$$

Property 2.1.8

Angle between two vectors in terms of scalar products: Let $\vec{a} = a_1\hat{i} + a_2\hat{j} + a_3\hat{k}$ and $\vec{b} = b_1\hat{i} + b_2\hat{j} + b_3\hat{k}$ be two given vectors in coordinate form. Then, using the result (2.3), we get

$$\vec{a} \cdot \vec{b} = \left(a_1\hat{i} + a_2\hat{j} + a_3\hat{k}\right) \cdot \left(b_1\hat{i} + b_2\hat{j} + b_3\hat{k}\right)$$
$$= a_1\hat{i} \cdot (b_1\hat{i} + b_2\hat{j} + b_3\hat{k}) + a_2\hat{j} \cdot (b_1\hat{i} + b_2\hat{j} + b_3\hat{k})$$
$$+ a_3\hat{k} \cdot (b_1\hat{i} + b_2\hat{j} + b_3\hat{k}) = a_1b_1 + a_2b_2 + a_3b_3$$

This is the scalar product in terms of coordinates. In this case, if θ is the angle between the non-zero vectors, then it is given by

$$\cos\theta = \frac{\vec{a} \cdot \vec{b}}{|\vec{a}||\vec{b}|} = \frac{a_1b_1 + a_2b_2 + a_3b_3}{\sqrt{a_1^2 + a_2^2 + a_3^2}\ \sqrt{b_1^2 + b_2^2 + b_3^2}} \tag{2.4}$$

Example 2.1.10

Find the angle between the vectors $\vec{a} = 2\hat{i} + 2\hat{j} - \hat{k}$ and $\vec{b} = 6\hat{i} - 3\hat{j} + 2\hat{k}$.

Solution

For the given two vectors $\vec{a}$ and $\vec{b}$, we have $|\vec{a}| = 3$ and $|\vec{b}| = 7$. Let θ be the angle between the two vectors $\vec{a}$ and $\vec{b}$, given by (Eq. 2.4). So

$$\cos\theta = \frac{2 \cdot 6 + 2 \cdot (-3) + (-1) \cdot 2}{3 \cdot 7} = \frac{4}{21} \Rightarrow \theta = \cos^{-1}\frac{4}{21}$$

Property 2.1.9

Since $\vec{a} \cdot (\vec{b} \cdot \vec{c})$ is not defined, the associative law between three vectors $\vec{a}$, $\vec{b}$ and $\vec{c}$ need not be considered. We, however, observe that, in general $(\vec{a} \cdot \vec{b})\vec{c} \neq \vec{a}(\vec{b} \cdot \vec{c})$, though both are defined.

Property 2.1.10

Geometrical interpretation of scalar product: Let $\vec{a}$ and $\vec{b}$ be two vectors represented by $\vec{OA}$ and $\vec{OB}$, respectively (Fig. 2.7). Let θ be the angle between $\vec{OA}$ and $\vec{OB}$. Draw $\vec{BL}$ perpendicular to $\vec{OA}$ and $\vec{AM}$ perpendicular to $\vec{OB}$. In Fig. 2.7, let $OP = |\vec{b}|$, then $|\vec{b}|\cos\theta$ is called the *component* or *resolute of* $\vec{b}$ *along* $\vec{a}$ and $|\vec{a}|\cos\theta$ is called the *component* or *resolute of* $\vec{a}$ *along* $\vec{b}$. Now

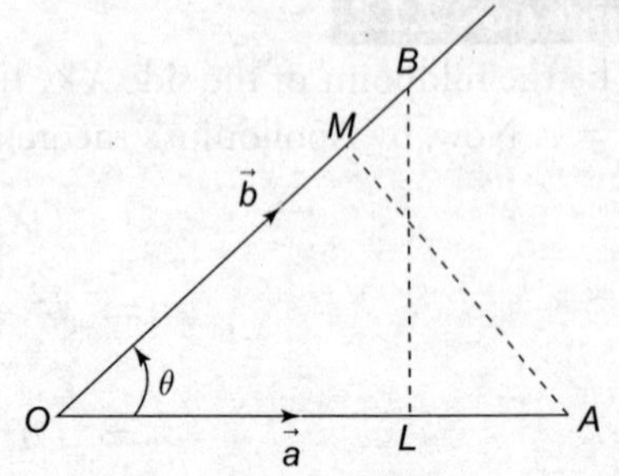

Figure 2.7 Geometrical interpretation of scalar product

$$ON = \text{Projection of } \vec{OP} \text{ on } \vec{OA}$$

$$= |\vec{b}|\cos\theta = \frac{|\vec{a}|\,|\vec{b}|\cos\theta}{|\vec{a}|} = \frac{\vec{a}\cdot\vec{b}}{|\vec{a}|} = \hat{a}\cdot\vec{b} \tag{2.5}$$

Therefore, the projection of $\vec{b}$ on $\vec{a}$ is $\hat{a}\cdot\vec{b}$. Similarly, the projection of $\vec{a}$ on $\vec{b}$ is $\vec{a}\cdot\hat{b}$. Also

$$\vec{a}\cdot\vec{b} = |\vec{a}|\,|\vec{b}|\,\cos\theta = |\vec{a}|\,OB\,\cos\theta = |\vec{a}|\,OL$$

$$= (\text{Magnitude of } \vec{a})\ (\text{Projection of } \vec{b} \text{ on } \vec{a}) \tag{2.6}$$

Similarly, $\vec{a}\cdot\hat{b}$ can also be written in the form

$$\vec{a}\cdot\vec{b} = |\vec{a}|\,|\vec{b}|\,\cos\theta = |\vec{b}|\,OA\,\cos\theta = |\vec{b}|\,OM$$

$$= (\text{Magnitude of } \vec{b})\ (\text{Projection of } \vec{a} \text{ on } \vec{b}) \tag{2.7}$$

Thus, geometrically interpreted, the scalar product of the two vectors is the product of the modulus of either vector and the projection of the other in its direction.

Example 2.1.11

If $\vec{a} = 2\hat{i} + \hat{j} + \hat{k}$, $\vec{b} = \hat{i} - 2\hat{j} + 2\hat{k}$ and $\vec{c} = 3\hat{i} - 4\hat{j} + 2\hat{k}$, then find the projection of $(\vec{a} + \vec{c})$ in the direction of $\vec{b}$.

Solution

For the given vectors $\vec{a}$ and $\vec{c}$, we have

$$\vec{a} + \vec{c} = (2\hat{i} + \hat{j} + \hat{k}) + (3\hat{i} - 4\hat{j} + 2\hat{k}) = 5\hat{i} - 3\hat{j} + 3\hat{k}$$

Hence, the projection of $(\vec{a} + \vec{c})$ on $\vec{b}$ is

$$= \frac{(\vec{a} + \vec{c})\cdot\vec{b}}{|\vec{b}|} = \frac{(5\hat{i} - 3\hat{j} + 3\hat{k})\cdot(\hat{i} - 2\hat{j} + 2\hat{k})}{\sqrt{1^2 + (-2)^2 + 2^2}} = \frac{17}{3}$$

Example 2.1.12

Find the projection of the vector $\vec{a} = \hat{i} - 2\hat{j} + \hat{k}$ on the vector $\vec{b} = 4\hat{i} - 4\hat{j} + 7\hat{k}$. Also, find the component vector of $\vec{a}$ along $\vec{b}$.

Solution

The projection of $\vec{a}$ along $\vec{b}$ is given by

$$\frac{\vec{a}\cdot\vec{b}}{|\vec{b}|}=\frac{1\cdot 4+(-2)\cdot(-4)+1\cdot 7}{\sqrt{4^2+(-4)^2+7^2}}=\frac{19}{9}$$

The component of $\vec{a}$ along $\vec{b}$ is given by

$$\frac{\vec{a}\cdot\vec{b}}{|\vec{b}|}\,\hat{b}=\frac{19}{9}\frac{1}{9}(4\hat{i}-4\hat{j}+7\hat{k})=19(4\hat{i}-4\hat{j}+7\hat{k})/81$$

Property 2.1.11

If either vector of a scalar product of two vectors $\vec{a}\cdot\vec{b}$ is multiplied by a scalar m, then the scalar product is multiplied by that scalar, i.e.,

$$(m\vec{a})\cdot\vec{b}=m(\vec{a}\cdot\vec{b})=\vec{a}\cdot(m\vec{b})$$

This is the *associativity with scalars.* If the angle between $\vec{a}$ and $\vec{b}$ is θ, then the angle between $m\vec{a}$ and $\vec{b}$ is either θ or $\pi-\theta$, depending on whether m is positive or negative.

Property 2.1.12

Let $\vec{a}$ and $\vec{b}$ be any vectors and m and n any scalars. Then

$$(m\vec{a})\cdot(n\vec{b})=mn(\vec{a}\cdot\vec{b})=\vec{a}\cdot(mn\vec{b})=m\left[(n\vec{a})\cdot\vec{b}\right]$$

where m and n are two numbers, positive or negative.

Proof: Let m and n be positive, so that the angle between $m\vec{a}$ and $n\vec{b}$ is the same as that between $\vec{a}$ and $\vec{b}$. Let the angle between $\vec{a}$ and $\vec{b}$ be θ. Then

$$\begin{aligned}m\vec{a}\cdot n\vec{b}&=|m\vec{a}||n\vec{b}|\cos\theta=|m||\vec{a}||n||\vec{b}|\cos\theta\\&=mn|\vec{a}||\vec{b}|\cos\theta=mn(\vec{a}\cdot\vec{b})\end{aligned}$$

Let m now be positive and n be negative so that the angle between $m\vec{a}$ and $n\vec{b}$ is $\pi-\theta$. We have

$$\begin{aligned}m\vec{a}\cdot n\vec{b}&=|m\vec{a}||n\vec{b}|\cos(\pi-\theta)\\&=|m||\vec{a}||n||\vec{b}|(-\cos\theta)\\&=m(-n)|\vec{a}||\vec{b}|(-\cos\theta)=mn(\vec{a}\cdot\vec{b})\end{aligned}$$

Other cases may be similarly disposed of.

Property 2.1.13

Distributivity: The scalar product of vectors is distributive over addition of vectors. For any three vectors $\vec{a}$, $\vec{b}$ and $\vec{c}$, we have

$$\vec{a}\cdot(\vec{b}+\vec{c})=\vec{a}\cdot\vec{b}+\vec{a}\cdot\vec{c};\qquad\text{Left distributive}\qquad(2.8)$$

$$(\vec{b}+\vec{c})\cdot\vec{a}=\vec{b}\cdot\vec{a}+\vec{c}\cdot\vec{a};\qquad\text{Right distributive}\qquad(2.9)$$

Proof: Here, we shall prove the left distributive law. Let $\vec{OA}$, $\vec{OB}$ and $\vec{BC}$ be given as three vectors $\vec{a}$, $\vec{b}$ and $\vec{c}$, respectively (Fig. 2.8). BL and CM are perpendiculars drawn on OA. Then, OL, LM and OM are the respective results or component of $\vec{b}, \vec{c}$ and $(\vec{b}+\vec{c})$ along OA, the line of support of $\vec{a}$. Therefore, $\vec{a}\cdot\vec{b} = |\vec{a}|OL$, $\vec{a}\cdot\vec{c} = |\vec{a}|LM$ and

$$\begin{aligned}\vec{a}\cdot(\vec{b}+\vec{c}) &= |\vec{a}|OM = |\vec{a}|(OL+LM)\\ &= |\vec{a}|OL + |\vec{a}|LM = \vec{a}\cdot\vec{b} + \vec{a}\cdot\vec{c}\end{aligned}$$

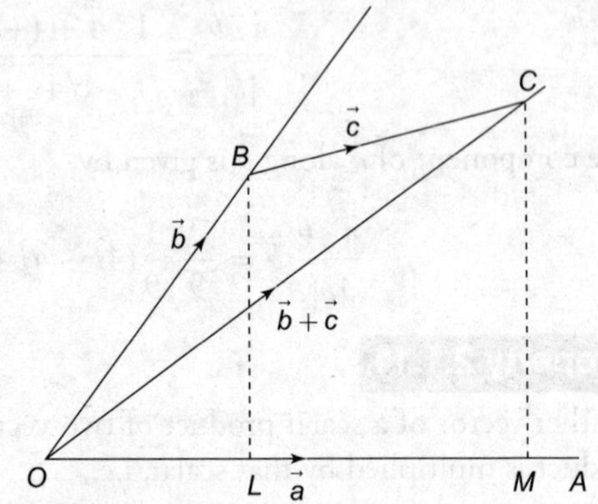

Figure 2.8 Distributive law of scalar product

If one or both of the resolutes are negative, this proof may be ended. We prove the right distributive law in a similar fashion. Therefore, scalar products obey the distributive law over addition. This property is an immediate consequence of the fact that the projection of the sum of two vectors on any line is equal to the sum of their projections on the same line.

RESULT 9

Using the associative and distributive properties of the scalar product of three vectors $\vec{a}$, $\vec{b}$ and $\vec{c}$, we obtain the following identities:

(i) $(\vec{a}+\vec{b})\cdot(\vec{a}-\vec{b}) = |\vec{a}|^2 - |\vec{b}|^2 = a^2 - b^2$

(ii) $|\vec{a}\pm\vec{b}|^2 = (\vec{a}\pm\vec{b})\cdot(\vec{a}\pm\vec{b}) = a^2 \pm 2\vec{a}\cdot\vec{b} + b^2$

(iii) $\vec{a}\cdot\vec{b} = \frac{1}{4}\left[|\vec{a}+\vec{b}|^2 - |\vec{a}-\vec{b}|^2\right]$

Example 2.1.13

Let $\vec{a}$, $\vec{b}$ and $\vec{c}$ be three vectors satisfying the condition $\vec{a}+\vec{b}+\vec{c} = \vec{0}$. If $|\vec{a}| = 3$, $|\vec{b}| = 4$ and $|\vec{c}| = 5$, prove that $\vec{a}\cdot\vec{b} + \vec{b}\cdot\vec{c} + \vec{c}\cdot\vec{a} = -25$.

Solution

Since $\vec{a}+\vec{b}+\vec{c} = \vec{0}$

$$\begin{aligned}&(\vec{a}+\vec{b}+\vec{c})\cdot(\vec{a}+\vec{b}+\vec{c}) = \vec{0}\cdot\vec{0}\\ \text{or,}\quad &|\vec{a}|^2 + |\vec{b}|^2 + |\vec{c}|^2 + 2(\vec{a}\cdot\vec{b} + \vec{b}\cdot\vec{c} + \vec{c}\cdot\vec{a}) = 0\\ \text{or,}\quad &9 + 16 + 25 + 2(\vec{a}\cdot\vec{b} + \vec{b}\cdot\vec{c} + \vec{c}\cdot\vec{a}) = 0\\ \text{or,}\quad &\vec{a}\cdot\vec{b} + \vec{b}\cdot\vec{c} + \vec{c}\cdot\vec{a} = -25\end{aligned}$$

Example 2.1.14

$\vec{a}$, $\vec{b}$ and $\vec{c}$ are three vectors such that $\vec{a}+\vec{b}+\vec{c} = \vec{0}$. If $|\vec{a}| = 3$, $|\vec{b}| = 5$ and $|\vec{c}| = 7$, find the angle between $\vec{a}$ and $\vec{b}$.

Solution

Let θ be the angle between $\vec{a}$ and $\vec{b}$. Here, we have

$$\vec{a}+\vec{b}+\vec{c}=\vec{0};\quad i.e., \vec{a}+\vec{b}=-\vec{c}$$

$$\text{or,}\quad \left(\vec{a}+\vec{b}\right)\cdot\left(\vec{a}+\vec{b}\right)=(-\vec{c})\cdot(-\vec{c})$$

$$\text{or,}\quad \vec{a}\cdot\vec{a}+2\vec{a}\cdot\vec{b}+\vec{b}\cdot\vec{b}=\vec{c}\cdot\vec{c};\ \text{as}\ \vec{a}\cdot\vec{b}=\vec{b}\cdot\vec{a}$$

$$\text{or,}\quad |\vec{a}|^2+2\vec{a}\cdot\vec{b}+|\vec{b}|^2=|\vec{c}|^2$$

$$\text{or,}\quad 2\vec{a}.\vec{b}=|\vec{c}|^2-|\vec{a}|^2-|\vec{b}|^2=15$$

$$\text{or,}\quad 2|\vec{a}||\vec{b}|\cos\theta=15\Rightarrow 2\cdot3\cdot5\ \cos\theta=15$$

$$\text{or,}\quad \cos\theta=\frac{1}{2}\Rightarrow\theta=\frac{\pi^c}{3}\quad\text{or}\quad 60^\circ$$

Hence, the angle between $\vec{a}$ and $\vec{b}$ is 60°.

Example 2.1.15

Using the vector method, prove that for any triangle ABC on a plane,

(i) $\cos A=\dfrac{b^2+c^2-a^2}{2bc}$ (cosine law)

(ii) $c=a\cos B+b\cos A$

in the usual notation of plane trigonometry.

Solution

Let ABC be a triangle on a plane, where $\vec{BC}=\vec{a}$, $\vec{CA}=\vec{b}$ and $\vec{AB}=\vec{c}$ (Fig. 2.9). Let a, b and c be the length of the sides BC, CA and AB, respectively, i.e., $|\vec{BC}|=BC=a$, $|\vec{CA}|=CA=b$ and $|\vec{AB}|=AB=c$. By the triangle law of addition of vectors, we have

$$\vec{BC}+\vec{CA}+\vec{AB}=\vec{0},\quad\text{i.e.,}\ \vec{a}+\vec{b}+\vec{c}=\vec{0}\quad(2.10)$$

(i) Here, the angle between $\vec{b}$ and $\vec{c}$ = angle between CA produced and AB. Using the relation (2.10), we have

$$\vec{a}+\vec{b}+\vec{c}=\vec{0},\quad\text{i.e.,}\ \vec{b}+\vec{c}=-\vec{a}$$

$$\text{or,}\quad (\vec{b}+\vec{c})\cdot(\vec{b}+\vec{c})=\vec{a}\cdot\vec{a}$$

$$\text{or,}\quad \vec{b}\cdot\vec{b}+2\vec{b}\cdot\vec{c}+\vec{c}\cdot\vec{c}=\vec{a}\cdot\vec{a};\ \vec{b}\cdot\vec{c}=\vec{c}\cdot\vec{b}$$

$$\text{or,}\quad |\vec{b}|^2+2|\vec{b}|\ |\vec{c}|\cos(\pi-A)+|\vec{c}|^2=|\vec{a}|^2$$

$$\text{or,}\quad b^2-2bc\cos A+c^2=a^2,\quad\text{i.e.,}\ \cos A=\frac{b^2+c^2-a^2}{2bc}$$

Figure 2.9 Plane triangle

In a similar manner, we have $\cos B=\dfrac{c^2+a^2-b^2}{2ac}$ and $\cos C=\dfrac{a^2+b^2-c^2}{2ab}$. These are called *cosine laws of a plane triangle ABC*.

(ii) Here, the angle between $\vec{b}$ and $\vec{a}$ is $\pi - C$ and the angle between $\vec{c}$ and $\vec{a}$ is $\pi - B$. Using the relation (2.10), we have $\vec{c} = -\vec{a} - \vec{b}$, and so

$$\begin{aligned}
&\vec{c}\cdot\vec{c} = -\vec{c}\cdot\vec{a} - \vec{c}\cdot\vec{b}; \text{ distributive property}\\
\text{or,}\quad &|\vec{c}|^2 = -|\vec{c}|\ |\vec{a}|\cos(\pi - B) - |\vec{c}|\ |\vec{b}|\cos(\pi - A)\\
\text{or,}\quad &c^2 = ca\cos B + cb\cos A\\
\text{or,}\quad &c = a\cos B + b\cos A
\end{aligned}$$

This is known as the *projection formula for triangles.*

Example 2.1.16

Prove by using the concept of scalar product of two vectors that in a plane,

$$\cos(\alpha+\beta) = \cos\alpha\cos\beta - \sin\alpha\sin\beta$$

Solution

Let $\hat{\imath}$ and $\hat{\jmath}$ denote two unit vectors along two rectangular axes OX and OY, respectively (Fig. 2.10), so that $\hat{\imath}\cdot\hat{\imath} = 1 = \hat{\jmath}\cdot\hat{\jmath}$ and $\hat{\imath}\cdot\hat{\jmath} = 0 = \hat{\jmath}\cdot\hat{\imath}$. Let two vectors $\vec{OL}$ and $\vec{OM}$ make angles α and β with OX, as shown in Fig. 2.10, where $|\vec{OL}| = r_1$ and $|\vec{OM}| = r_2$.
Therefore, $\angle XOL = \alpha$, $\angle XOM = \beta$ and $\angle LOM = \alpha + \beta$. Now

$$\begin{aligned}
\vec{OL} &= r_1\cos(-\alpha)\ \hat{\imath} + r_1\sin(-\alpha)\ \hat{\jmath}\\
&= r_1\hat{\imath}\ \cos\alpha - \hat{\jmath}\ \sin\alpha\\
\vec{OM} &= r_2\cos(\beta)\ \hat{\imath} + r_2\sin(\beta)\ \hat{\jmath}\\
&= r_2\hat{\imath}\ \cos\beta + \hat{\jmath}\ \sin\beta
\end{aligned}$$

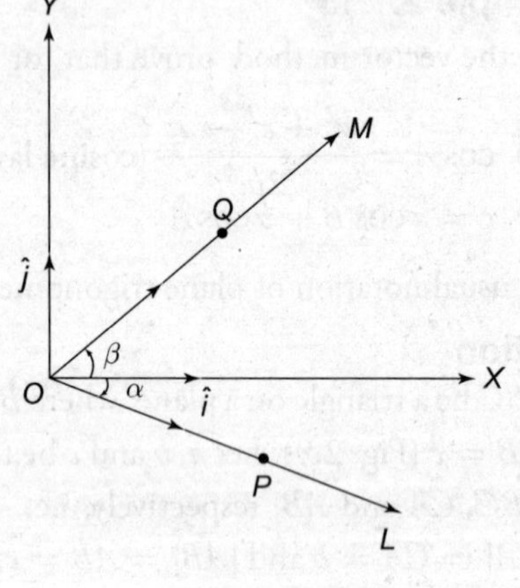

Figure 2.10
$\cos(\alpha+\beta) = \cos\alpha\cos\beta - \sin\alpha\sin\beta$

Taking the dot product of the two vectors $\vec{OL}$ and $\vec{OM}$, we get

$$\begin{aligned}
&\vec{OL}\cdot\vec{OM} = r_1r_2(\hat{\imath}\ \cos\alpha - \hat{\jmath}\ \sin\alpha)\cdot(\hat{\imath}\cos\beta + \hat{\jmath}\sin\beta)\\
\text{or,}\quad &\left|\vec{OL}\right|\left|\vec{OM}\right|\cos(\alpha+\beta) = r_1r_2\,(\cos\alpha\cos\beta - \sin\alpha\sin\beta)\\
\text{or,}\quad &r_1r_2\cos(\alpha+\beta) = r_1r_2\,(\cos\alpha\cos\beta - \sin\alpha\sin\beta)\\
\text{or,}\quad &\cos(\alpha+\beta) = \cos\alpha\cos\beta - \sin\alpha\sin\beta
\end{aligned}$$

Hence, the result follows.

Example 2.1.17

Prove that in a triangle, the perpendiculars drawn from the vertices to the opposite sides are concurrent.

Solution

Let ABC be a triangle (Fig. 2.11). Let AD and BE be the perpendiculars to BC and CA from the vertices A and B, respectively, for the triangle ABC, and they meet at O. Join CO and produce it to meet AB at F.

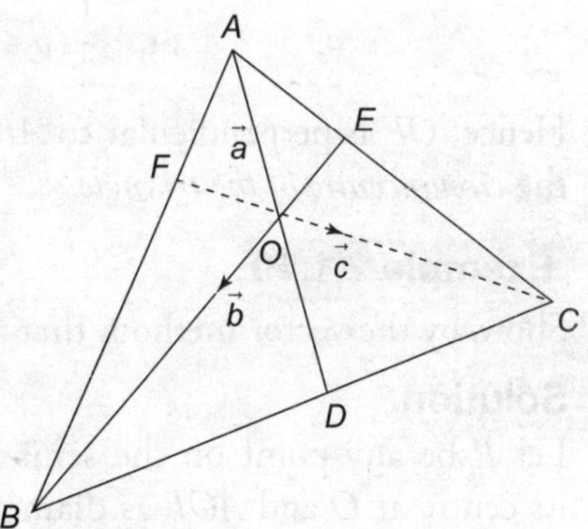

Figure 2.11 Orthocentre

With respect to the base point O, let the position vectors of A, B and C be $\vec{a}, \vec{b}$ and $\vec{c}$, respectively; then $\vec{OA} = \vec{a}$, $\vec{OB} = \vec{b}$ and $\vec{OC} = \vec{c}$. Now, $\vec{AD} = x\vec{a}$ and $\vec{BE} = y\vec{b}$, where x and y are some scalars. Since $\vec{AD}$ is perpendicular to $\vec{BC}$, we have

$$\vec{AD} \cdot \vec{BC} = 0, \quad \text{i.e., } x\vec{a} \cdot (\vec{c} - \vec{b}) = 0$$

$$\text{or,} \quad \vec{a} \cdot \vec{c} = \vec{a} \cdot \vec{b}; \text{ using distributive law} \qquad (2.11)$$

Again, since $\vec{BE}$ is perpendicular to $\vec{AC}$

$$\vec{BE} \cdot \vec{AC} = 0, \quad \text{i.e., } y\vec{b} \cdot (\vec{c} - \vec{a}) = 0$$

$$\text{or,} \quad \vec{b} \cdot \vec{c} = \vec{b} \cdot \vec{a} = \vec{a} \cdot \vec{b}; \text{ using distributive law} \quad (2.12)$$

From Eq. (2.11) and Eq. (2.12), we get

$$\vec{a} \cdot \vec{c} = \vec{a} \cdot \vec{b} = \vec{b} \cdot \vec{c}$$

$$\text{or,} \quad (\vec{a} - \vec{b}) \cdot \vec{c} = 0; \text{ using distributive law}$$

$$\text{or,} \quad z\vec{c} \cdot (\vec{a} - \vec{b}) = 0; \ z \text{ is scalar}$$

$$\text{or,} \quad \vec{CF} \cdot \vec{AB} = 0; \text{ i.e., } \vec{CF} \text{ is perpendicular to } \vec{AB}$$

Therefore, the perpendicular CF from C on the opposite side AB to the triangle ABC also passes through O. Hence, the three perpendiculars from the three vertices A, B and C to the opposite sides of the triangle ABC are concurrent at O; this is known as the *orthocentre of the triangle.*

Example 2.1.18

Prove that the perpendicular bisectors of the sides of a triangle are concurrent.

Solution

Let ABC be a triangle in which the perpendicular bisects DO and EO of BC and CA, respectively, meet at O (Fig. 2.12).Join O to the middle point F of AB. Let $\vec{a}, \vec{b}$ and $\vec{c}$ be the position vectors of A, B and C, respectively, with reference to O as origin. Since D, E and F are the midpoints, we have

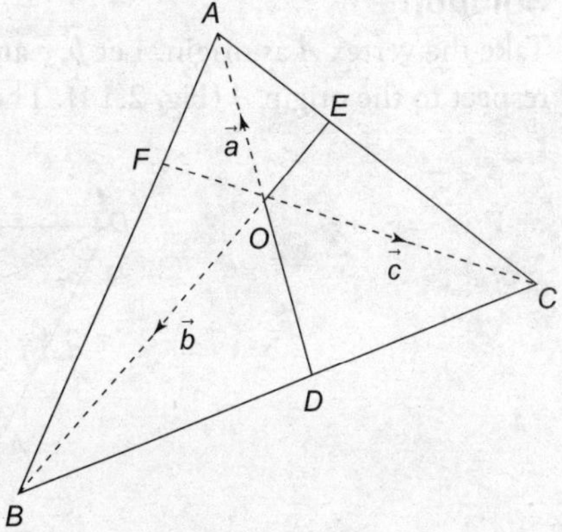

Figure 2.12 Circumcentre

$$\vec{OD} = \frac{1}{2}(\vec{b} + \vec{c}), \quad \vec{OE} = \frac{1}{2}(\vec{c} + \vec{a}), \quad \vec{OF} = \frac{1}{2}(\vec{a} + \vec{b})$$

Since OD is perpendicular to BC, we have $\vec{OD} \cdot \vec{BC} = 0$, i.e.,

$$\frac{1}{2}(\vec{b} + \vec{c}) \cdot (\vec{c} - \vec{b}) = 0 \Rightarrow |\vec{c}|^2 - |\vec{b}|^2 = 0$$

Similarly, since OE is perpendicular to AC, we have $\vec{OE}\cdot\vec{AC}=0$, i.e.,

$$\frac{1}{2}(\vec{c}+\vec{a})\cdot(\vec{c}-\vec{a})=0\Rightarrow|\vec{c}|^2=|\vec{a}|^2=|\vec{b}|^2$$

$$\text{or,}\quad\frac{1}{2}(\vec{b}+\vec{a})\cdot(\vec{b}-\vec{a})=0;\quad\text{i.e., }\vec{OF}\cdot\vec{AB}=0$$

Hence, OF is perpendicular to AB. Hence, the theorem. This point of concurrence is known as the *circumcentre of the triangle.*

Example 2.1.19

Show, by the vector method, that an angle in a semi-circle is a right angle.

Solution

Let P be any point on the semi-circle APB, having its centre at O and AOB as diameter (Fig. 2.13). We are to show that $\angle APB=90^\circ$. Let $\vec{a}$, $\vec{b}$ and $\vec{r}$ be the position vectors of the points A, B and P, respectively, with respect to the base point O. Hence, $\vec{b}=-\vec{a}$ and $|\vec{b}|=|\vec{a}|=|\vec{r}|=$ radius of the semi-circle. Also, $\vec{AP}=\vec{r}-\vec{a}$ and $\vec{BP}=\vec{r}-\vec{b}$. Hence

Figure 2.13 Angle in a semi-circle

$$\begin{aligned}\vec{AP}\cdot\vec{BP}&=(\vec{r}-\vec{a})\cdot(\vec{r}-\vec{b})\\&=(\vec{r}-\vec{a})\cdot(\vec{r}+\vec{a});\qquad\text{as }\vec{b}=-\vec{a}\\&=|\vec{r}|^2-|\vec{a}|^2=0,\qquad\text{as }|\vec{r}|=|\vec{a}|\end{aligned}$$

Hence, $\vec{AP}$ is perpendicular to $\vec{BP}$, i.e., $\angle APB$ is a right angle.

Example 2.1.20

Show that the sum of the squares of the sides of any quadrilateral (plane or skew) equals the sum of the squares of its diagonals together with four times the square of the line joining their middle points.

Solution

Take the vertex A as origin. Let $\vec{b}$, $\vec{c}$ and $\vec{d}$ be the position vectors of the vertices B, C and D with respect to the origin A (Fig. 2.14). The position vector of the midpoints P and Q of the diagonals

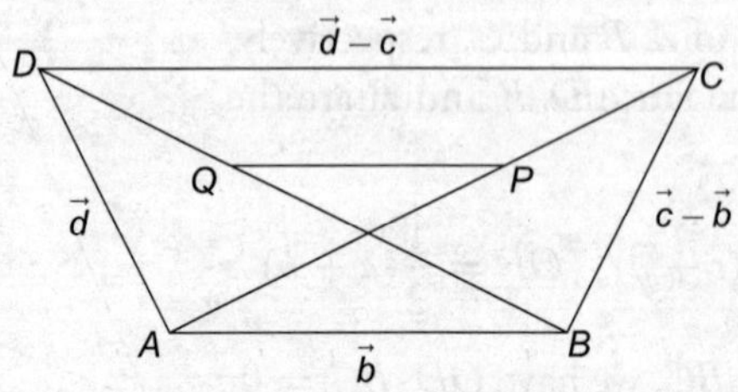

Figure 2.14 Quadrilateral

AC and BD are $\frac{1}{2}\vec{c}$ and $\frac{1}{2}(\vec{b}+\vec{d})$, respectively. We have

$$\begin{aligned} AB^2 + BC^2 + CD^2 + DA^2 &= |\vec{b}|^2 + |\vec{c}-\vec{b}|^2 + |\vec{d}-\vec{c}|^2 + |\vec{d}|^2 \\ &= b^2 + (\vec{c}-\vec{b})\cdot(\vec{c}-\vec{b}) + (\vec{d}-\vec{c})\cdot(\vec{d}-\vec{c}) + d^2 \\ &= b^2 + c^2 - (\vec{b}\cdot\vec{c} + \vec{c}\cdot\vec{d}) + d^2 + c^2 - (\vec{b}\cdot\vec{c} + \vec{c}\cdot\vec{d}) + d^2 \\ &= 2(b^2 + c^2 + d^2) - 2(\vec{b}\cdot\vec{c} + \vec{c}\cdot\vec{d}) \end{aligned}$$

Also

$$\begin{aligned} AC^2 + BD^2 + 4PQ^2 &= |\vec{c}|^2 + |\vec{d}-\vec{b}|^2 + 4\left|\frac{1}{2}(\vec{b}+\vec{d}) - \frac{1}{2}\vec{c}\right|^2 \\ &= c^2 + (\vec{d}-\vec{b})\cdot(\vec{d}-\vec{b}) + (\vec{b}+\vec{d}-\vec{c})\cdot(\vec{b}+\vec{d}-\vec{c}) \\ &= 2(b^2 + c^2 + d^2) - 2(\vec{b}\cdot\vec{c} + \vec{c}\cdot\vec{d}) \end{aligned}$$

Hence, the result.

Example 2.1.21

If the perpendiculars from two vertices B and C to the opposite faces of a tetrahedron $ABCD$ intersect, then BC is perpendicular to AD and the perpendiculars from A and D to the opposite faces also intersect.

Solution

Take A as the origin of reference. Let $\vec{AB} = \vec{b}, \vec{AC} = \vec{c}$ and $\vec{AD} = \vec{d}$. Let the perpendicular from B and C to the opposite faces ACD and ABD meet at a point H, whose position vector is $\vec{h}$. Thus, we have

$$\begin{aligned} &\vec{BH} \perp \vec{AC} \Rightarrow (\vec{h}-\vec{b})\cdot\vec{c} = 0; \quad \vec{BH} \perp \vec{AD} \Rightarrow (\vec{h}-\vec{b})\cdot\vec{d} = 0, \\ &\vec{CH} \perp \vec{AB} \Rightarrow (\vec{h}-\vec{c})\cdot\vec{b} = 0; \quad \vec{CH} \perp \vec{AD} \Rightarrow (\vec{h}-\vec{c})\cdot\vec{d} = 0 \\ \therefore\ &\vec{h}\cdot\vec{c} = \vec{b}\cdot\vec{c}, \vec{h}\cdot\vec{d} = \vec{b}\cdot\vec{d}; \quad \vec{h}\cdot\vec{b} = \vec{b}\cdot\vec{c}, \vec{h}\cdot\vec{d} = \vec{c}\cdot\vec{d} \\ \Rightarrow\ &\vec{b}\cdot\vec{d} = \vec{c}\cdot\vec{d} \Rightarrow (\vec{b}-\vec{c})\cdot\vec{d} = 0 \Rightarrow \vec{BC} \perp \vec{AD} \end{aligned}$$

Let K be the foot of the perpendicular from A to the opposite face BCD, and let k be its position vector. As AK is perpendicular to the plane BCD, we have

$$\begin{aligned} &\vec{k}\cdot(\vec{c}-\vec{b}) = 0; \quad \vec{k}\cdot(\vec{b}-\vec{d}) = 0 \\ \Rightarrow\ &\vec{k}\cdot\vec{c} = \vec{k}\cdot\vec{b} = \vec{k}\cdot\vec{d} \end{aligned}$$

We shall show that there is a point L on AK such that DL is perpendicular to the face ABC. If $\lambda\vec{k}$ be the position vector of this point, the two equations

$$(\lambda\vec{k}-\vec{d})\cdot\vec{b} = 0, \quad (\lambda\vec{k}-\vec{d})\cdot\vec{c} = 0$$

must be consistent in relation to λ. These equations are really the same for

$$\mathrm{k}\cdot\vec{b} = \vec{k}\cdot\vec{c} \text{ and } \vec{b}\cdot\vec{d} = \vec{c}\cdot\vec{d},$$

so that λ is determined. Hence, the result.

Example 2.1.22

The base BC of $\triangle ABC$ is divided at D so that $mBD = nCD$ (Fig. 2.15). Show that $mAB^2 + nAC^2 = mBD^2 + nCD^2 + (m+n)AD^2$.

Solution

With A as origin, let the position vectors of the points B and C be $\vec{b}$ and $\vec{c}$.

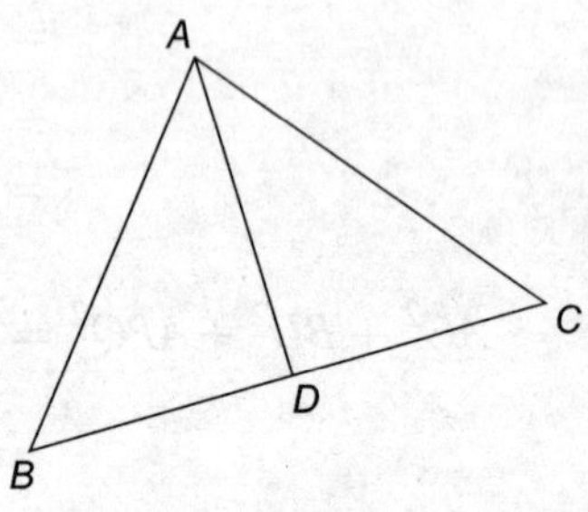

Figure 2.15 Triangle ABC

$$m\vec{BD} = n\vec{DC} \Rightarrow m\vec{DB} + n\vec{DC} = 0$$

Now

$$\begin{aligned} AB^2 &= |\vec{AB}|^2 = |\vec{AD} + \vec{DB}|^2 \\ &= (\vec{AD} + \vec{DB}) \cdot (\vec{AD} + \vec{DB}) \\ &= AD^2 + DB^2 + 2\vec{AD} \cdot \vec{DB} \\ AC^2 &= |\vec{AC}|^2 = |\vec{AD} + \vec{DC}|^2 \\ &= (\vec{AD} + \vec{DC}) \cdot (\vec{AD} + \vec{DC}) \\ &= AD^2 + DC^2 + 2\vec{AD} \cdot \vec{DC} \end{aligned}$$

Multiplying by m and n, respectively, and adding, we get

$$\begin{aligned} mAB^2 + nAC^2 &= (m+n)AD^2 + mBD^2 + nDC^2 + 2\vec{AD}.(m\vec{DB} + n\vec{DC}) \\ &= (m+n)AD^2 + mBD^2 + nDC^2 \end{aligned}$$

Example 2.1.23

Prove that if a point is equidistant from the vertices of a right-angled triangle, its join to the midpoint of the hypotenuse is perpendicular to the plane of the triangle.

Solution

Let ABC be a right-angled triangle (Fig. 2.16) and let D be the midpoint of its hypotenuse BC. Let P be any point and $\vec{a}, \vec{b}, \vec{c}$ and $\vec{p}$ be the position vectors of A, B, C and P, respectively. Given that $PA = PB = PC$, i.e., $PA^2 = PB^2 = PC^2$. Therefore, $|\vec{a} - \vec{p}|^2 = |\vec{b} - \vec{p}|^2 = |\vec{c} - \vec{p}|^2$. Now

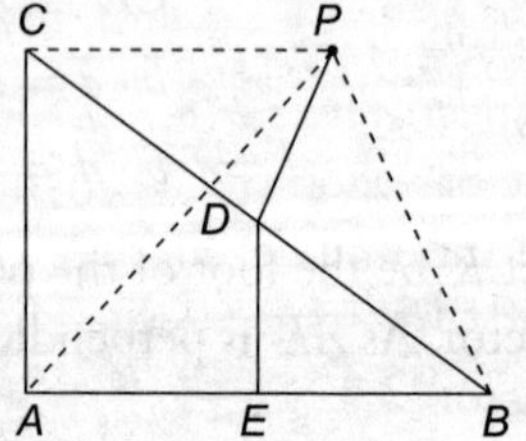

Figure 2.16 Right-angled triangle

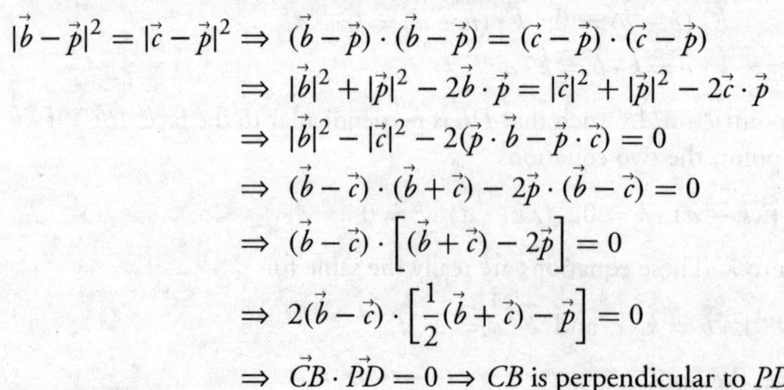

$$\begin{aligned} |\vec{b} - \vec{p}|^2 = |\vec{c} - \vec{p}|^2 &\Rightarrow (\vec{b} - \vec{p}) \cdot (\vec{b} - \vec{p}) = (\vec{c} - \vec{p}) \cdot (\vec{c} - \vec{p}) \\ &\Rightarrow |\vec{b}|^2 + |\vec{p}|^2 - 2\vec{b} \cdot \vec{p} = |\vec{c}|^2 + |\vec{p}|^2 - 2\vec{c} \cdot \vec{p} \\ &\Rightarrow |\vec{b}|^2 - |\vec{c}|^2 - 2(\vec{p} \cdot \vec{b} - \vec{p} \cdot \vec{c}) = 0 \\ &\Rightarrow (\vec{b} - \vec{c}) \cdot (\vec{b} + \vec{c}) - 2\vec{p} \cdot (\vec{b} - \vec{c}) = 0 \\ &\Rightarrow (\vec{b} - \vec{c}) \cdot \left[(\vec{b} + \vec{c}) - 2\vec{p} \right] = 0 \\ &\Rightarrow 2(\vec{b} - \vec{c}) \cdot \left[\frac{1}{2}(\vec{b} + \vec{c}) - \vec{p} \right] = 0 \\ &\Rightarrow \vec{CB} \cdot \vec{PD} = 0 \Rightarrow CB \text{ is perpendicular to } PD \end{aligned}$$

Let E be the midpoint of AB. Then

$$|\vec{a}-\vec{p}|^2 = |\vec{b}-\vec{p}|^2 \Rightarrow 2(\vec{a}-\vec{b})\cdot\left[\frac{1}{2}(\vec{a}+\vec{b})-\vec{p}\right]=0$$

$$\Rightarrow 2\vec{BA}\cdot\vec{PE}=0 \Rightarrow BA \text{ is perpendicular to } PE$$

Since D and E are the midpoints of BC and AB, respectively, DE must be parallel to AC. Again, AC is perpendicular to BA. Hence, DE is perpendicular to BA, i.e., $\vec{AB}\cdot\vec{DE}=0$. Also, $\vec{AD}=\vec{AB}+\vec{BD}$. Therefore

$$\begin{aligned}
\vec{AD}\cdot\vec{DP} &= (\vec{AB}+\vec{BD})\cdot\vec{DP} = \vec{AB}\cdot\vec{DP}+\vec{BD}\cdot\vec{DP}\\
&= \vec{AB}\cdot\vec{DP}+\frac{1}{2}\vec{BC}\cdot\vec{DP}, \text{ as } D \text{ is the midpoint of } BC\\
&= \vec{AB}\cdot\vec{DP}, \text{ as } \vec{CB}\cdot\vec{PD}=0\\
&= \vec{AB}\cdot(\vec{DE}+\vec{EP}) = \vec{AB}\cdot\vec{DE}+\vec{AB}\cdot\vec{EP}=0\\
\Rightarrow &\quad DP \text{ is perpendicular to } AD
\end{aligned}$$

It follows that PD is perpendicular to both CB and AD, and hence, PD must be perpendicular to the plane of triangle ABC.

Example 2.1.24

Euler's theorem: In any triangle, prove that the circumcentre, orthocentre and centroid are collinear and that the centroid divides the join of the other two in the ratio 2 : 1.

Solution

Let H, G and P be, respectively, the circumcentre, centroid and orthocentre of the triangle ABC (Fig. 2.17). Let the position vectors of the points A, B, C, H, G and P be, respectively, $\vec{a}, \vec{b}, \vec{c}, \vec{h}, \vec{g}$ and $\vec{p}$ with respect to the base point O as origin. Let D, E and F be the middle points of the sides BC, CA and AB, respectively. Then, their position vectors are $\frac{1}{2}(\vec{b}+\vec{c})$, $\frac{1}{2}(\vec{c}+\vec{a})$ and $\frac{1}{2}(\vec{a}+\vec{b})$, respectively. Since H is the circumcentre, $\vec{HF}$ is perpendicular to $\vec{AB}$. So

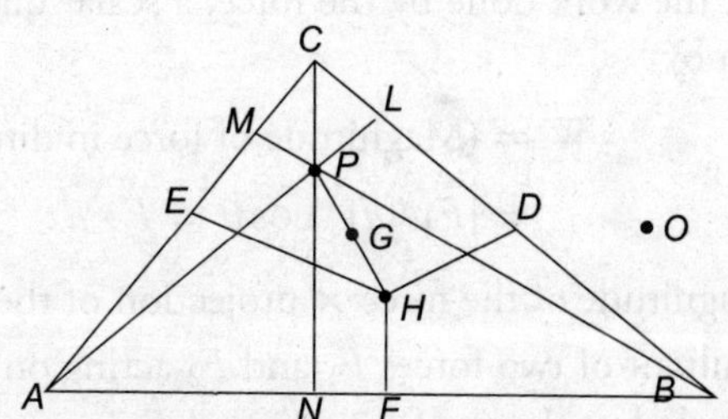

Figure 2.17 Circumcentre, orthocentre and centroid of a triangle

$$\left[\frac{1}{2}(\vec{a}+\vec{b})-\vec{h}\right]\cdot(\vec{b}-\vec{a})=0$$

Also, like the orthocentre P, $\vec{CP}$ is perpendicular to $\vec{AB}$, and thus $(\vec{p}-\vec{c})\cdot(\vec{b}-\vec{a})=0$. Subtracting, we get

$$\left[\frac{1}{2}(\vec{a}+\vec{b})-\vec{h}\right]\cdot(\vec{b}-\vec{a})-\frac{1}{2}(\vec{p}-\vec{c})\cdot(\vec{b}-\vec{a})=0$$

$$\Rightarrow \left[\frac{1}{2}(\vec{a}+\vec{b})-\vec{h}-\frac{1}{2}\vec{p}+\frac{1}{2}\vec{c}\right]\cdot(\vec{b}-\vec{a})=0$$

$$\Rightarrow \left[\vec{a}+\vec{b}+\vec{c}-2\vec{h}-\vec{p}\right]\cdot(\vec{b}-\vec{a})=0$$

This relation holds good when $(\vec{b}-\vec{a})$ is replaced by $(\vec{c}-\vec{b})$ or by $(\vec{a}-\vec{c})$. So we conclude that $\vec{a}+\vec{b}+\vec{c}-2\vec{h}-\vec{p}=\vec{0}$. For otherwise, $\vec{a}+\vec{b}+\vec{c}-2\vec{h}-\vec{p}$ would be perpendicular to three sides of $\triangle ABC$, which is not possible. Therefore

$$\vec{a}+\vec{b}+\vec{c}-2\vec{h}-\vec{p}=\vec{0}\Rightarrow \frac{1}{3}(\vec{a}+\vec{b}+\vec{c})=\frac{1}{3}(2\vec{h}+\vec{p})$$

Since G is the centroid

$$\vec{g}=\frac{1}{3}(\vec{a}+\vec{b}+\vec{c})=\frac{1}{3}(2\vec{h}+\vec{p})\Rightarrow 2\vec{h}+\vec{p}-3\vec{g}=\vec{0},$$

where the sum of the coefficients $2+1-3=0$. Therefore, P, G and H are collinear and G divides HP in the ratio $2:1$.

Property 2.1.14

Physical illustration of the scalar product: Let $\vec{OA}=\vec{F}$ represent a force acting on a particle at O, and $\vec{OB}=\vec{d}$ represent the displacement of the particle from O to B, as shown in Fig. 2.18.

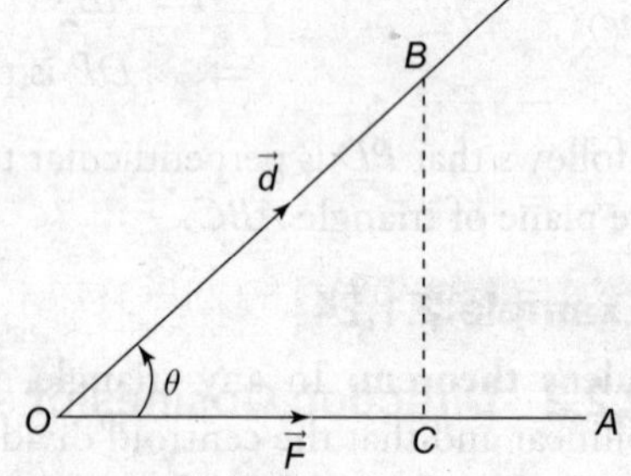

Figure 2.18 Work done

Draw BC perpendicular to OA. Then, the displacement in the direction of the force $= OC = d\cos\theta$. Hence, the work done by the force, a scalar quantity, is given by

$$W = \text{(Magnitude of force in direction of motion)(Distance moved)}$$
$$= |\vec{F}|\ |\vec{d}|\ \cos\theta = \vec{F}\cdot\vec{d} \tag{2.13}$$

i.e., magnitude of the force × projection of the displacement in the direction of the force. If $\vec{F}$ is the resultant of two forces $\vec{F_1}$ and $\vec{F_2}$ acting on a particle subjected to a displacement $\vec{d}$, then the total work done by the forces $\vec{F_1}$ and $\vec{F_2}$ is

$$=\left(\vec{F_1}+\vec{F_2}\right)\cdot\vec{d}=\vec{F}\cdot\vec{d}=\vec{F_1}\cdot\vec{d}+\vec{F_2}\cdot\vec{d}$$

Example 2.1.25

Find the work done in moving an object along a vector $\vec{d}=3\hat{i}+2\hat{j}-5\hat{k}$, if the applied force is $\vec{F}=2\hat{i}-\hat{j}-\hat{k}$.

Solution
Using Eq.(2.13), we get

$$W=(2\hat{i}-\hat{j}-\hat{k})\cdot(3\hat{i}+2\hat{j}-5\hat{k})=6-2+5=9 \text{ units}$$

2.2 APPLICATIONS TO CARTESIAN GEOMETRY

We shall consider some preliminary applications to Cartesian geometry. The main results obtained in this section will be of help for the scalar product of vectors.

2.2.1 Distance Between Two Points

Let the position vectors of the two points P and Q (Fig. 2.19) be $\vec{r}_1 = x_1\hat{i} + y_1\hat{j} + z_1\hat{k}$ and $\vec{r}_2 = x_2\hat{i} + y_2\hat{j} + z_2\hat{k}$, respectively, so that $\vec{OP} = \vec{r}_1$ and $\vec{OQ} = \vec{r}_2$. Therefore

$$\begin{aligned}\vec{PQ} &= \vec{OQ} - \vec{OP} = \vec{r}_2 - \vec{r}_1\\ &= (x_2 - x_1)\hat{i} + (y_2 - y_1)\hat{j} + (z_2 - z_1)\hat{k}\end{aligned}$$

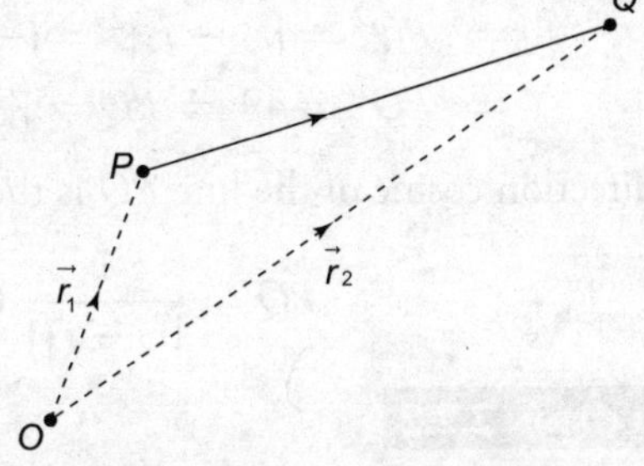

Figure 2.19 Distance between two points

Using the definition of dot product, we get

$$\begin{aligned}PQ^2 &= \vec{PQ} \cdot \vec{PQ} = (\vec{r}_2 - \vec{r}_1) \cdot (\vec{r}_2 - \vec{r}_1)\\ &= |\vec{r}_2 - \vec{r}_1|^2\\ &= \left|(x_2 - x_1)\hat{i} + (y_2 - y_1)\hat{j} + (z_2 - z_1)\hat{k}\right|^2\\ &= (x_2 - x_1)^2 + (y_2 - y_1)^2 + (z_2 - z_1)^2\end{aligned}$$

2.2.2 Direction Cosines of a Line

Consider any directed line and let $\vec{OA}$ denote the vector of unit length parallel to the line PQ (Fig. 2.20). Let $\vec{OA} = l\hat{i} + m\hat{j} + n\hat{k}$. Therefore

$$\begin{aligned}\vec{OA} \cdot \hat{i} &= (l\hat{i} + m\hat{j} + n\hat{k}) \cdot \hat{i}\\ &= l\hat{i} \cdot \hat{i} + m\hat{j} \cdot \hat{i} + n\hat{k} \cdot \hat{i}\\ &= l \cdot 1 + m \cdot 0 + n \cdot 0 = l\end{aligned}$$

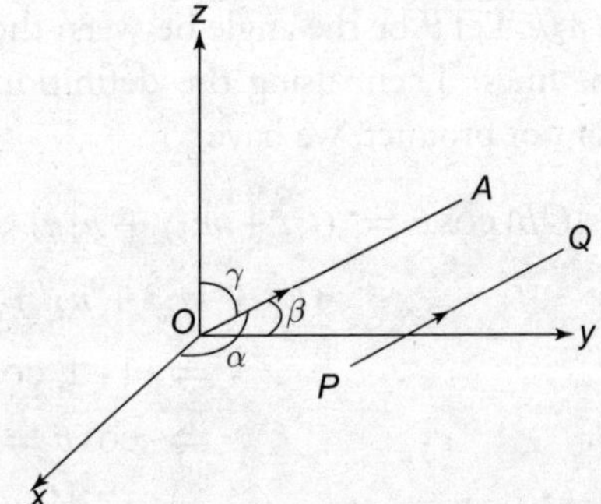

Figure 2.20 Direction cosines of a line

where α denotes the angle that the given line OA makes with the x axis. Similarly, $m = \cos\beta$ and $n = \cos\gamma$, where β and γ are the angles between the given line OA and the y axis and z axis, respectively. Thus, if $l\hat{i} + m\hat{j} + n\hat{k}$ denotes a vector of unit length parallel to a given line, then l, m and n are the *direction cosines of the line.*

DEDUCTION 2.2.1

Let us consider the dot product $\vec{OA} \cdot \vec{OA}$. It is given by

$$\begin{aligned}\vec{OA} \cdot \vec{OA} &= (l\hat{i} + m\hat{j} + n\hat{k}) \cdot (l\hat{i} + m\hat{j} + n\hat{k})\\ \Rightarrow\ 1 &= l^2 + m^2 + n^2\end{aligned}$$

i.e., the sum of the squares of the *direction cosines* of any line is unity.

Example 2.2.1

Find the direction cosines and the distance of the line joining the points, whose position vectors are $3\hat{\imath} + 2\hat{\jmath} - 4\hat{k}$ and $\hat{\imath} - \hat{\jmath} + 2\hat{k}$.

Solution

Let $\vec{r}_1 = 3\hat{\imath} + 2\hat{\jmath} - 4\hat{k}$ and $\vec{r}_2 = \hat{\imath} - \hat{\jmath} + 2\hat{k}$ be the position vectors of the two given points P and Q. Thus

$$\vec{PQ} = \vec{r}_2 - \vec{r}_1 = (\hat{\imath} - \hat{\jmath} + 2\hat{k}) - (3\hat{\imath} + 2\hat{\jmath} - 4\hat{k}) = -2\hat{\imath} - 3\hat{\jmath} + 6\hat{k}$$
$$\Rightarrow PQ^2 = |\vec{r}_2 - \vec{r}_1|^2 = (-2)^2 + (-3)^2 + 6^2$$
$$\Rightarrow PQ^2 = 49 \Rightarrow PQ = 7 \text{ units}$$

The direction cosine of the line PQ is the unit vector in that direction and is given by

$$\hat{PQ} = \frac{1}{|\vec{r}_2 - \vec{r}_1|}(\vec{r}_2 - \vec{r}_1) = \frac{1}{7}(-2\hat{\imath} - 3\hat{\jmath} + 6\hat{k})$$

DEDUCTION 2.2.2

Angle between two straight lines: Let l_1, m_1, n_1 and l_2, m_2, n_2 be the direction cosines of the two given lines OA and OB (Fig. 2.21). The vector of unit length along the given lines is $l_1\hat{\imath} + m_1\hat{\jmath} + n_1\hat{k}$ and $l_2\hat{\imath} + m_2\hat{\jmath} + n_2\hat{k}$. Let θ be the angle between the straight lines. Then, using the definition (2.1) of dot product, we have

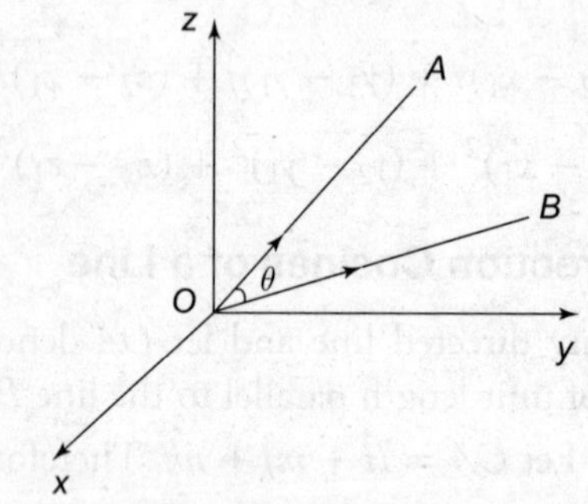

Figure 2.21 Angle between two straight lines

$$|\vec{OA}|\,|\vec{OB}|\cos\theta = (l_1\hat{\imath} + m_1\hat{\jmath} + n_1\hat{k}) \cdot (l_2\hat{\imath} + m_2\hat{\jmath} + n_2\hat{k})$$
$$\Rightarrow 1 \cdot 1 \cos\theta = l_1 l_2 + m_1 m_2 + n_1 n_2$$
$$\Rightarrow \cos\theta = l_1 l_2 + m_1 m_2 + n_1 n_2$$

(i) If the lines are perpendicular, $\theta = \pi/2$, and hence, $l_1 l_2 + m_1 m_2 + n_1 n_2 = 0$.

(ii) If the lines are parallel, then $\frac{l_1}{l_2} = \frac{m_1}{m_2} = \frac{n_1}{n_2}$.

Example 2.2.2

Find the angle between the lines AB and AC where A, B and C are the three points with rectangular coordinates $(1, 2, -1)$, $(2, 0, 3)$ and $(3, -1, 2)$, respectively.

Solution

In terms of usual notation, the position vectors of the given points are $\vec{OA} = \hat{\imath} + 2\hat{\jmath} - \hat{k}$, $\vec{OB} = 2\hat{\imath} + 3\hat{k}$ and $\vec{OC} = 3\hat{\imath} - \hat{\jmath} + 2\hat{k}$, where O is the origin. Thus, we have

$$\vec{AB} = \vec{OB} - \vec{OA} = \hat{\imath} - 2\hat{\jmath} + 4\hat{k},\quad \vec{AC} = \vec{OC} - \vec{OA} = 2\hat{\imath} - 3\hat{\jmath} + 3\hat{k}$$
$$\vec{AB} \cdot \vec{AC} = (\hat{\imath} - 2\hat{\jmath} + 4\hat{k}) \cdot (2\hat{\imath} - 3\hat{\jmath} + 3\hat{k}) = 20$$

$$AB^2 = |\vec{AB}|^2 = \vec{AB} \cdot \vec{AB} = (\hat{i} - 2\hat{j} + 4\hat{k}) \cdot (\hat{i} - 2\hat{j} + 4\hat{k}) = 21$$
$$AC^2 = |\vec{AC}|^2 = \vec{AC} \cdot \vec{AC} = (2\hat{i} - 3\hat{j} + 3\hat{k}) \cdot (2\hat{i} - 3\hat{j} + 3\hat{k}) = 22$$

Let θ denote the angle between AB and AC. Therefore

$$\cos\theta = \frac{\vec{AB} \cdot \vec{AC}}{|\vec{AB}|\,|\vec{AC}|} = \frac{20}{\sqrt{21}\sqrt{22}}$$

Example 2.2.3

A line makes angles α, β, γ and δ with the diagonals of a cube; show that $\cos^2\alpha + \cos^2\beta + \cos^2\gamma + \cos^2\delta = 4/3$.

Solution

Since the angle will remain unchanged for any size of cube, consider a unit cube (Fig. 2.22). Represent the coterminous edges $\vec{OA}, \vec{OB}$ and $\vec{OC}$ by unit vectors $\hat{i}, \hat{j}$ and $\hat{k}$. Then, $\vec{OA} = \hat{i}$, $\vec{OB} = \hat{j}$ and $\vec{OC} = \hat{k}$. So

$$\vec{OP} = \vec{OA} + \vec{AQ} + \vec{QP} = \hat{i} + \hat{j} + \hat{k}$$
$$\vec{CQ} = \hat{i} + \hat{j} - \hat{k}, \quad \vec{AR} = -\hat{i} + \hat{j} + \hat{k}, \quad \vec{BS} = \hat{i} - \hat{j} + \hat{k}$$

Therefore, $OP = CQ = AR = BS = \sqrt{3}$. Let any line OL be given by $\vec{OL} = a_1\hat{i} + a_2\hat{j} + a_3\hat{k}$. Let α be the angle between the lines OL and OP. Then

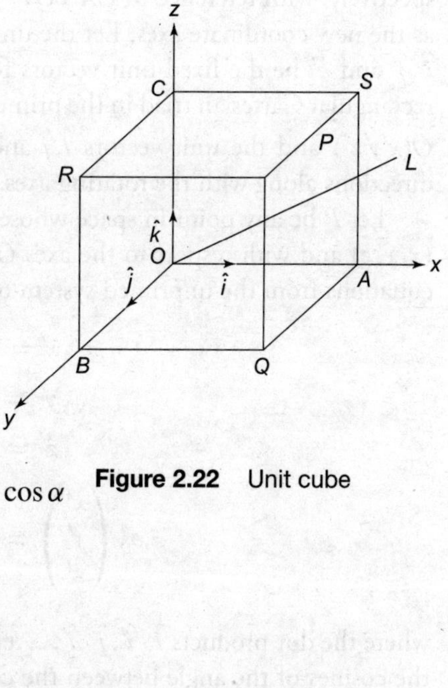

Figure 2.22 Unit cube

$$\vec{OL} \cdot \vec{OP} = |\vec{OL}|\,|\vec{OP}|\,\cos\alpha$$
$$\Rightarrow\ (a_1\hat{i} + a_2\hat{j} + a_3\hat{k}) \cdot (\hat{i} + \hat{j} + \hat{k}) = \sqrt{a_1^2 + a_2^2 + a_3^2}\,\sqrt{3}\,\cos\alpha$$
$$\Rightarrow\ \cos\alpha = \frac{a_1 + a_2 + a_3}{\sqrt{3}\sqrt{a_1^2 + a_2^2 + a_3^2}}$$

In a similar manner, we get

$$\cos\beta = \frac{a_1 + a_2 - a_3}{\sqrt{3}\sqrt{a_1^2 + a_2^2 + a_3^2}}, \cos\gamma = \frac{-a_1 + a_2 + a_3}{\sqrt{3}\sqrt{a_1^2 + a_2^2 + a_3^2}}, \cos\delta = \frac{a_1 - a_2 + a_3}{\sqrt{3}\sqrt{a_1^2 + a_2^2 + a_3^2}}$$

Squaring and adding, we get

$$\cos^2\alpha + \cos^2\beta + \cos^2\gamma + \cos^2\delta$$
$$= \frac{(a_1 + a_2 + a_3)^2 + (a_1 + a_2 - a_3)^2 + (-a_1 + a_2 + a_3)^2 + (a_1 - a_2 + a_3)^2}{3(a_1^2 + a_2^2 + a_3^2)}$$
$$= \frac{4(a_1^2 + a_2^2 + a_3^2)}{3(a_1^2 + a_2^2 + a_3^2)} = \frac{4}{3}$$

2.2.3 Orthogonal Transformation

We consider two rectangular coordinate systems of axes OX, OY and OZ fixed in body, and OX', OY' and OZ' fixed in space, having the same origin O but with the axes rotated with respect to each other (Fig. 2.23). Let l_1, m_1, n_1; l_2, m_2, n_2 and l_3, m_3, n_3 be the direction cosines of OX', OY' and OZ', respectively, with reference to OX, OY and OZ as the coordinate axes. Then, l_1, l_2, l_3; m_1, m_2, m_3 and n_1, n_2, n_3 will be the direction cosines of OX, OY and OZ, respectively, with reference to OX', OY' and OZ' as the new coordinate axes. Let the unit vectors $\hat{i}', \hat{j}'$ and $\hat{k}'$ be the fixed unit vectors forming a rectangular Cartesian triad in the primed system $O(x'y'z')$ and the unit vectors $\hat{i}, \hat{j}$ and $\hat{k}$ in the unprimed system $O(xyz)$ which change their directions along with the rotating axes.

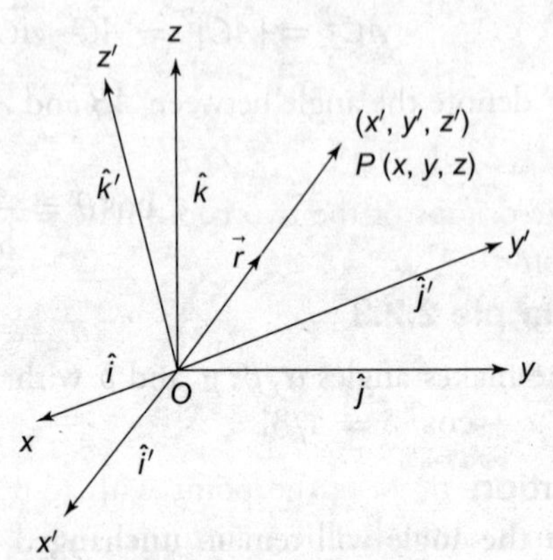

Figure 2.23 Orthogonal transformation

Let P be any point in space whose coordinates with respect to the axes OX, OY and OZ are (x, y, z) and with respect to the axes OX', OY' and OZ' are (x', y', z'). Now, the transformation equations from the unprimed system to the primed system are

$$x' = \vec{r} \cdot \hat{i}' = x\hat{i} \cdot \hat{i}' + y\hat{j} \cdot \hat{i}' + z\hat{k} \cdot \hat{i}'$$

$$y' = \vec{r} \cdot \hat{j}' = x\hat{i} \cdot \hat{j}' + y\hat{j} \cdot \hat{j}' + z\hat{k} \cdot \hat{j}'$$

$$z' = \vec{r} \cdot \hat{k}' = x\hat{i} \cdot \hat{k}' + y\hat{j} \cdot \hat{k}' + z\hat{k} \cdot \hat{k}'$$

$$\text{i.e.,} \quad \begin{pmatrix} x' \\ y' \\ z' \end{pmatrix} = \begin{pmatrix} \hat{i}\cdot\hat{i}' & \hat{j}\cdot\hat{i}' & \hat{k}\cdot\hat{i}' \\ \hat{i}\cdot\hat{j}' & \hat{j}\cdot\hat{j}' & \hat{k}\cdot\hat{j}' \\ \hat{i}\cdot\hat{k}' & \hat{j}\cdot\hat{k}' & \hat{k}\cdot\hat{k}' \end{pmatrix} \begin{pmatrix} x \\ y \\ z \end{pmatrix} \tag{2.14}$$

where the dot products $\hat{i} \cdot \hat{i}', \hat{j} \cdot \hat{i}', \ldots$etc. on the right-hand side of the above equation are simply the cosines of the angle between the corresponding axes. Thus, the transformations between the coordinates are given by

$$\left.\begin{aligned} x' &= l_1\, x + m_1\, y + n_1\, z \\ y' &= l_2\, x + m_2\, y + n_2\, z \\ z' &= l_3\, x + m_2\, y + n_3\, z \end{aligned}\right\} \tag{2.15}$$

The transformation equations (2.15) define a *pure rotation*. The transformation (2.15) is also called an *orthogonal transformation*. Also

$$\left.\begin{aligned} \hat{i}' &= l_1\, \hat{i} + m_1\, \hat{j} + n_1\, \hat{k} \\ \hat{j}' &= l_2\, \hat{i} + m_2\, \hat{j} + n_2\, \hat{k} \\ \hat{k}' &= l_3\, \hat{i} + m_2\, \hat{j} + n_3\, \hat{k} \end{aligned}\right\} \tag{2.16}$$

Using Eq. (2.16), can obtain the equations for inverse transformations as

$$\left.\begin{aligned}\hat{i} &= l_1\,\hat{i}' + l_2\,\hat{j}' + l_3\,\hat{k}'\\ \hat{j} &= m_1\,\hat{i}' + m_2\,\hat{j}' + m_3\,\hat{k}'\\ \hat{k} &= n_1\,\hat{i}' + n_2\,\hat{j}' + n_3\,\hat{k}'\end{aligned}\right\} \quad (2.17)$$

If the origins of the two coordinate systems are not coincident, the equations of transformation become

$$\left.\begin{aligned}x' &= l_1\,x + m_1\,y + n_1\,z + \alpha\\ y' &= l_2\,x + m_2\,y + n_2\,z + \beta\\ z' &= l_3\,x + m_2\,y + n_3\,z + \gamma\end{aligned}\right\} \quad (2.18)$$

where (α, β, γ) is the point with respect to the coordinate axes OX, OY and OZ, when the origin is shifted. The transformation (2.18) defines a rotation plus translation. A general linear transformation is called an *affine transformation.*

2.3 CROSS PRODUCT

The *cross* or *vector* or *outer product* of two given vectors $\vec{a}$ and $\vec{b}$ denoted by $\vec{a} \times \vec{b}$ (read as vector $\vec{a}$ cross vector $\vec{b}$) is defined by the vector quantity

$$\vec{a} \times \vec{b} = |\vec{a}|\ |\vec{b}| \sin\theta\ \hat{n};\quad 0 \le \theta \le \pi \quad (2.19)$$

where $\hat{n}$ is the unit vector perpendicular to the plane of $\vec{a}$ and $\vec{b}$, whose direction coincides with the direction in which a right-handed screw moves when it is rotated from the direction of $\vec{a}$ to the direction of $\vec{b}$ (i.e., $\vec{a}, \vec{b}$ and $\vec{c}$ form a right-handed or dextral system) and θ is the angle between $\vec{a}$ and $\vec{b}$ (Fig. 2.24).

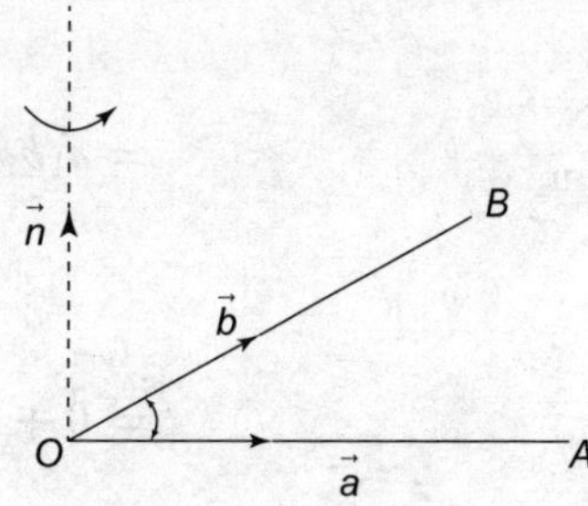

Figure 2.24 Cross product

Based on the definition of vector product, we can easily deduce the following properties.

Property 2.3.1

Commutative property: The cross product of two vectors $\vec{a}$ and $\vec{b}$ is anti-commutative, for

$$\vec{b} \times \vec{a} = |\vec{b}||\vec{a}| \sin(-\theta)\ \hat{n} = -\vec{a} \times \vec{b}$$

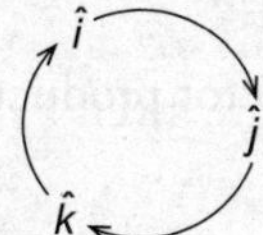

Figure 2.25 Cross product

Therefore, the vector product of two vectors does not obey the commutative law. Also, since $\hat{k}$ is perpendicular to both $\hat{i}$ and $\hat{j}$ and $\hat{i}, \hat{j}$ and $\hat{k}$ form a right-handed system (Fig. 2.25), we have

$$\hat{i} \times \hat{j} = \hat{k} = -\hat{j} \times \hat{i};\ \ \hat{j} \times \hat{k} = \hat{i} = -\hat{k} \times \hat{j} \text{ and } \hat{k} \times \hat{i} = \hat{j} = -\hat{i} \times \hat{k} \quad (2.20)$$

Property 2.3.2

Parallel vectors: Let two proper vectors $\vec{a}$ and $\vec{b}$ be collinear or parallel; then $\theta = 0$ or π. Thus, $\sin\theta = 0$ and $\vec{a} \times \vec{b} = \vec{0}$.

Conversely, if $\vec{a} \times \vec{b} = \vec{0}$, then the vectors $\vec{a}$ and $\vec{b}$ are either parallel or collinear. In particular, $\vec{a} \times \vec{a} = \vec{0}$.

For the three unit vectors $\hat{i}, \hat{j}$ and $\hat{k}$ along three mutually perpendicular axes

$$\hat{i} \times \hat{i} = \hat{j} \times \hat{j} = \hat{k} \times \hat{k} = \vec{0} \tag{2.21}$$

Following are some examples of cross product:

(i) $(3\hat{j}) \times (2\hat{k}) = 6\hat{j} \times \hat{k} = 6\hat{i}$
(ii) $(2\hat{i}) \times (-3\hat{k}) = -6\hat{i} \times \hat{k} = 6\hat{j}$
(iii) $2\hat{j} \times \hat{i} - 3\hat{k} = -2\hat{k} - 3\hat{k} = -5\hat{k}$

Property 2.3.3

Cross product in terms of coordinates: Let $\vec{a} = a_1\hat{i} + a_2\hat{j} + a_3\hat{k}$ and $\vec{b} = b_1\hat{i} + b_2\hat{j} + b_3\hat{k}$ be two given vectors in coordinate form. Then, using the relations (2.20) and (2.21), we get

$$\begin{aligned}
\vec{a} \times \vec{b} &= (a_1\hat{i} + a_2\hat{j} + a_3) \times (b_1\hat{i} + b_2\hat{j} + b_3\hat{k}) \\
&= a_1\hat{i} \times (b_1\hat{i} + b_2\hat{j} + b_3\hat{k}) + a_2\hat{j} \times (b_1\hat{i} + b_2\hat{j} + b_3\hat{k}) \\
&\quad + a_3\hat{k} \times (b_1\hat{i} + b_2\hat{j} + b_3\hat{k}) \\
&= a_1b_1\hat{i} \times \hat{i} + a_1b_2\hat{i} \times \hat{j} + a_1b_3\hat{i} \times \hat{k} \\
&\quad + a_2b_1\hat{j} \times \hat{i} + a_2b_2\hat{j} \times \hat{j} + a_2b_3\hat{j} \times \hat{k} \\
&\quad + a_3b_1\hat{k} \times \hat{i} + a_3b_2\hat{k} \times \hat{j} + a_3b_3\hat{k} \times \hat{k} \\
&= \vec{0} + a_1b_2\hat{k} + a_1b_3(-\hat{j}) + a_2b_1(-\hat{k}) + \vec{0} \\
&\quad + a_2b_3\hat{i} + a_3b_1\hat{j} + a_3b_2(-\hat{i}) + \vec{0} \\
&= (a_2b_3 - a_3b_2)\hat{i} + (a_3b_1 - a_1b_3)\hat{j} + (a_1b_2 - a_2b_1)\hat{k} \\
&= \begin{vmatrix} a_2 & a_3 \\ b_2 & b_3 \end{vmatrix}\hat{i} - \begin{vmatrix} a_1 & a_3 \\ b_1 & b_3 \end{vmatrix}\hat{j} + \begin{vmatrix} a_1 & a_2 \\ b_1 & b_2 \end{vmatrix}\hat{k} = \begin{vmatrix} \hat{i} & \hat{j} & \hat{k} \\ a_1 & a_2 & a_3 \\ b_1 & b_2 & b_3 \end{vmatrix}
\end{aligned} \tag{2.22}$$

This is the vector product in terms of coordinates. In this case, if θ be the angle between $\vec{a}$ and $\vec{b}$, then

$$\sin\theta = \frac{|\vec{a} \times \vec{b}|}{|\vec{a}||\vec{b}|} = \sqrt{\frac{(a_2b_3 - a_3b_2)^2 + (a_3b_1 - a_1b_3)^2 + (a_1b_2 - a_2b_1)^2}{(a_1^2 + a_2^2 + a_3^2)(b_1^2 + b_2^2 + b_3^2)}} \tag{2.23}$$

If $\vec{a}$ and $\vec{b}$ be parallel or collinear, then $\dfrac{a_1}{b_1} = \dfrac{a_2}{b_2} = \dfrac{a_3}{b_3}$.

Example 2.3.1

If $\vec{a} = 2\hat{i} - 3\hat{j} - \hat{k}$ and $\vec{b} = \hat{i} + 4\hat{j} - 2\hat{k}$, find (i) $\vec{a} \times \vec{b}$, (ii) $\vec{b} \times \vec{a}$ and (iii) $(\vec{a} + \vec{b}) \times (\vec{a} - \vec{b})$.

Solution

Given $\vec{a} = 2\hat{i} - 3\hat{j} - \hat{k}$ and $\vec{b} = \hat{i} + 4\hat{j} - 2\hat{k}$.

(i) Using definition (2.22), we get

$$\vec{a} \times \vec{b} = (2\hat{i} - 3\hat{j} - \hat{k}) \times (\hat{i} + 4\hat{j} - 2\hat{k}) = \begin{vmatrix} \hat{i} & \hat{j} & \hat{k} \\ 2 & -3 & -1 \\ 1 & 4 & -2 \end{vmatrix}$$

$$= \begin{vmatrix} -3 & -1 \\ 4 & -2 \end{vmatrix} \hat{i} - \begin{vmatrix} 2 & -1 \\ 1 & -2 \end{vmatrix} \hat{j} + \begin{vmatrix} 2 & -3 \\ 1 & 4 \end{vmatrix} \hat{k}$$

$$= 10\,\hat{i} + 3\,\hat{j} + 11\,\hat{k}$$

(ii) Using definition (2.22), we get

$$\vec{b} \times \vec{a} = (\hat{i} + 4\hat{j} - 2\hat{k}) \times (2\hat{i} - 3\hat{j} + \hat{k}) = -10\,\hat{i} - 3\,\hat{j} - 11\,\hat{k}$$

Comparing with (i), $\vec{a} \times \vec{b} = -\vec{b} \times \vec{a}$.

(iii) $\vec{a} + \vec{b} = (2\hat{i} - 3\hat{j} - \hat{k}) + (\hat{i} + 4\hat{j} - 2\hat{k}) = 3\hat{i} + \hat{j} - 3\hat{k}$, and $\vec{a} - \vec{b} = \hat{i} - 7\hat{j} + \hat{k}$. Using definition (2.22), we get

$$(\vec{a} + \vec{b}) \times (\vec{a} - \vec{b}) = (3\hat{i} + \hat{j} - 3\hat{k}) \times (\hat{i} - 7\hat{j} + \hat{k}) = -20\,\hat{i} - 6\,\hat{j} - 22\,\hat{k}$$

Property 2.3.4

Geometrical significance: In Fig. 2.26, let $\vec{a} = \vec{OA}$ and $\vec{b} = \vec{OB}$ be two given vectors. $\angle AOB = \theta$ is the angle between $\vec{a}$ and $\vec{b}$; $\hat{n}$ is a unit vector in the same direction as the translation of a right-handed screw due to a rotation from $\vec{a}$ to $\vec{b}$ and is perpendicular to the plane of OA and OB (Fig. 2.26). Let $|\vec{OA}| = |\vec{a}| = a$ and $|\vec{OB}| = |\vec{b}| = b$. Now

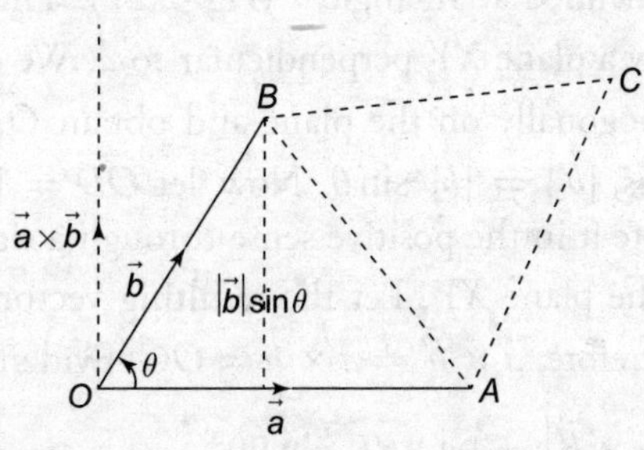

Figure 2.26 Geometrical significance of cross product

$$\left|\vec{a} \times \vec{b}\right| = |\vec{a}|\,|\vec{b}|\,\sin\theta = 2 \times \frac{1}{2}|\vec{a}|\,|\vec{b}|\,\sin\theta$$

$$= 2 \times \frac{1}{2}\text{ base} \times \text{ height of } \triangle OAB$$

$$= 2 \times \text{ area of } \triangle OAB$$

$$= \text{Area of the parallelogram OACB}$$

where OA and OB are the adjacent sides of the parallelogram. Thus, $\vec{a} \times \vec{b}$ and $\frac{1}{2}(\vec{a} \times \vec{b})$ are the vector areas of the parallelogram $OACB$ and triangle OAB, respectively, whose two adjacent sides OA and OB represent $\vec{a}$ and $\vec{b}$, respectively.

Geometrically, the vector product of two vectors represents the vector area of the parallelogram formed by the vectors as adjacent sides.

Property 2.3.5

Associative law: The vector product is associative with scalars, i.e., let $\vec{a}$ and $\vec{b}$ be two given vectors and m and n be scalars. Then

$$mn(\vec{a} \times \vec{b}) = (m\vec{a}) \times (n\vec{b}) = (\vec{a} \times \vec{b})mn = (n\vec{a}) \times (m\vec{b})$$

Property 2.3.6

If either vector be multiplied by a scalar quantity, then the cross product is multiplied by the same, i.e.,

$$(m\vec{a}) \times \vec{b} = m(\vec{a} \times \vec{b}) = \vec{a} \times (m\vec{b}),$$

where m is a scalar quantity. This is the associative property with scalars. The quantities $\vec{a} \times (\vec{b} \times \vec{c})$ and $(\vec{a} \times \vec{b}) \times \vec{c}$ are both defined, but in general, they are not equal. So, associativity with vectors does not hold here.

Property 2.3.7

Distributive law: The vector product is distributive over the addition of vectors, i.e., for any three vectors $\vec{a}$, $\vec{b}$ and $\vec{c}$ we have

$$\vec{a} \times (\vec{b} + \vec{c}) = \vec{a} \times \vec{b} + \vec{a} \times \vec{c}; \quad \text{left distributive}$$

Proof: Let the two vectors $\vec{OA}(= \vec{a})$ and $\vec{OB}(= \vec{b})$ be inclined at an angle θ (Fig. 2.27). Through O, draw a plane XY perpendicular to $\vec{a}$. We project $\vec{b}$ orthogonally on the plane and obtain $\vec{OB'} = \vec{b'}$. Thus, $|\vec{b'}| = |\vec{b}| \sin\theta$. Now, let $\vec{OD} = |\vec{a}|\vec{b'}$ and rotate it in the positive sense through a right angle in the plane XY. Let the resulting vector be $\vec{OC}$. Therefore, $\vec{a} \times \vec{b'} = \vec{a} \times \vec{b} = \vec{OC}$. Evidently

$$\begin{aligned} \left|\vec{a} \times \vec{b'}\right| &= |\vec{a}| \; |\vec{b'}| \; \sin 90^\circ \\ &= |\vec{a}| \; |\vec{b}| \; \sin\theta; \quad \text{as } |\vec{b'}| = |\vec{b}| \; \sin\theta \\ &= |\vec{a} \times \vec{b}|; \qquad \text{by definition} \end{aligned}$$

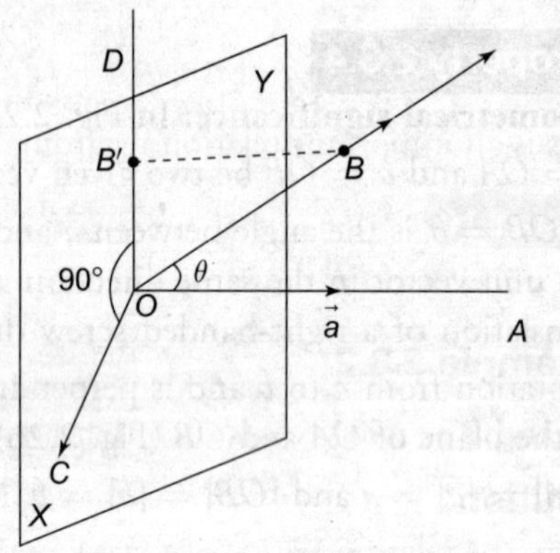

Figure 2.27 Distributive law

Since $\vec{a}$, $\vec{b}$ and $\vec{b'}$ lie on one plane and the sense of rotation from $\vec{a}$ to $\vec{b'}$ is the same as that from $\vec{a}$ to $\vec{b}$, the length and direction of $\vec{a} \times \vec{b'}$ and $\vec{a} \times \vec{b}$ are the same. Thus, to construct the product $\vec{a} \times \vec{b}$, we need to carry out the following three operations:

(i) P: Project the vector $\vec{b}$ on the XY plane, perpendicular to $\vec{a}$ at O to obtain the vector $\vec{b'}$.
(ii) M: Multiply the vector $\vec{b'}$ by $|\vec{a}|$ and obtain $|\vec{a}| \; \vec{b'}$.
(iii) R: Rotate positively through 90° about the vector $\vec{a}$.

Since $|\vec{b}| \sin\theta = |\vec{b'}|$, the resulting vector $\vec{V}$ agrees with $\vec{a} \times \vec{b}$ in magnitude and direction. We may write the sequence of operations as

$$\vec{V} = \vec{a} \times \vec{b} = R\,M\,P\;\vec{b} \tag{2.24}$$

This means that $\vec{b}$ is projected (P), projection multiplied (M) and finally rotated (R). Each of these operations is distributive with respect to addition. Thus

$$\begin{aligned} R\,M\,P(\vec{b}+\vec{c}) &= R\,M\,(P\,\vec{b}+P\,\vec{c}) \\ &= R(M\,P\,\vec{b}+M\,P\,\vec{c}) \\ &= R\,M\,P\,\vec{b}+R\,M\,P\,\vec{c} \end{aligned}$$

Hence, from Eq. (2.24), we get $\vec{a}\times(\vec{b}+\vec{c})=\vec{a}\times\vec{b}+\vec{a}\times\vec{c}$. This is called *left distributive law.*

Similarly, we get the *right distributive law* as $(\vec{a}+\vec{b})\times\vec{c}=\vec{a}\times\vec{c}+\vec{b}\times\vec{c}$.

Property 2.3.8

We know, $|\vec{a}|^2=\vec{a}\cdot\vec{a}=\vec{a}^2$ and $|\vec{b}|^2=\vec{b}\cdot\vec{b}=\vec{b}^2$. Now

$$\begin{aligned} (\vec{a}\times\vec{b})^2 &= (\vec{a}\times\vec{b})\cdot(\vec{a}\times\vec{b}) = |\vec{a}\times\vec{b}|^2 \\ &= |\vec{a}|^2\,|\vec{b}|^2\,\sin^2\theta = |\vec{a}|^2\,|\vec{b}|^2\,(1-\cos^2\theta) \\ &= |\vec{a}|^2\,|\vec{b}|^2 - |\vec{a}|^2\,|\vec{b}|^2\,\cos^2\theta \\ &= |\vec{a}|^2|\vec{b}|^2 - (\vec{a}\cdot\vec{b})^2 \qquad (2.25) \\ \therefore\ (\vec{a}\times\vec{b})^2 + (\vec{a}\cdot\vec{b})^2 &= |\vec{a}|^2|\vec{b}|^2 \end{aligned}$$

This is an important relation regarding the relationship between dot and cross products.

RESULT 10

If $\vec{a}\times\vec{b}=\vec{c}$, then $\vec{c}$ is perpendicular to $\vec{a}$ and $\vec{b}$. Consequently, $\vec{a}\cdot\vec{c}=0$ and $\vec{b}\cdot\vec{c}=0$.

Example 2.3.2

If $\vec{a}$ and $\vec{b}$ be two vectors such that $|\vec{a}|=10$, $|\vec{b}|=1$ and $\vec{a}\cdot\vec{b}=6$, find the value of $|\vec{a}\times\vec{b}|$.

Solution

Using the relation (2.25), we get

$$|\vec{a}\times\vec{b}|^2 = |\vec{a}|^2\,|\vec{b}|^2 - (\vec{a}\cdot\vec{b})^2 = 10^2\cdot 1^2 - 6^2 = 64$$

Therefore, $|\vec{a}\times\vec{b}|=\sqrt{64}=8$.

Example 2.3.3

Prove the law of sines for a plane triangle.

Solution

Let ABC be a triangle on a plane, where $\vec{BC}=\vec{a}$, $\vec{CA}=\vec{b}$ and $\vec{AB}=\vec{c}$ (Fig. 2.9). Then, as in Example (2.1.15)

$$\begin{aligned} &\vec{BC}+\vec{CA}+\vec{AB}=\vec{0};\ i.e., \vec{a}+\vec{b}+\vec{c}=\vec{0} \\ \text{or,}\quad &\vec{a}\times\left(\vec{a}+\vec{b}+\vec{c}\right)=\vec{a}\times\vec{0}\Rightarrow\vec{a}\times\vec{b}=\vec{c}\times\vec{a} \\ &\vec{b}\times\left(\vec{a}+\vec{b}+\vec{c}\right)=\vec{b}\times\vec{0}\Rightarrow\vec{a}\times\vec{b}=\vec{b}\times\vec{c} \end{aligned}$$

$$\vec{c} \times \left(\vec{a} + \vec{b} + \vec{c}\right) = \vec{c} \times \vec{0} \Rightarrow \vec{b} \times \vec{c} = \vec{c} \times \vec{a}$$

Thus, it follows that $\vec{a} \times \vec{b} = \vec{c} \times \vec{a} = \vec{b} \times \vec{c}$, i.e.,

$$|\vec{a}||\vec{b}| \sin(\pi - C) = |\vec{c}||\vec{a}| \sin(\pi - B) = |\vec{b}||\vec{c}| \sin(\pi - A)$$

$$\text{or,} \quad ab \sin C = ca \sin B = bc \sin A$$

$$\text{or,} \quad \frac{\sin C}{c} = \frac{\sin B}{b} = \frac{\sin A}{a}$$

Thus, the sines of the angles of a triangle ABC are proportional to the lengths of their opposite sides.

Example 2.3.4

Prove by the vector method that $\sin(\alpha - \beta) = \sin\alpha \cos\beta - \cos\alpha \sin\beta$.

Solution

Let $\hat{i}$ and $\hat{j}$ denote two unit vectors along two rectangular axes OX and OY, respectively (Fig. 2.28). Let $\vec{OL}$ and $\vec{OM}$ make angles α and β with OX. Let $\overline{OL} = r_1$ and $\overline{OM} = r_2$. Then

$$\vec{OL} = r_1 \cos\alpha \;\; \hat{i} + r_1 \sin\alpha \;\; \hat{j}$$

$$\text{and } \vec{OM} = r_2 \cos\beta \;\; \hat{i} + r_2 \sin\beta \;\; \hat{j}$$

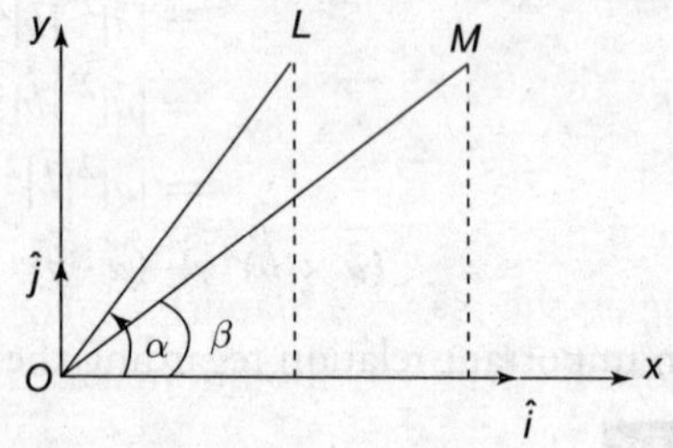

Figure 2.28 $\sin(\alpha - \beta) = \sin\alpha \cos\beta - \cos\alpha \sin\beta$

Now, using the definition of cross products, we have

$$|\vec{OM} \times \vec{OL}| = |r_1 r_2 (\cos\beta \, \hat{i} + \sin\beta \, \hat{j}) \times (\cos\alpha \, \hat{i} + \sin\alpha \, \hat{j})|$$

$$\text{or,} \quad r_1 r_2 \sin(\alpha - \beta) = r_1 r_2 | (\sin\alpha \cos\beta - \cos\alpha \sin\beta) \, \hat{k}|$$

$$\text{or,} \quad \sin(\alpha - \beta) = |\sin\alpha \cos\beta - \cos\alpha \sin\beta|$$

If we choose $\beta = 0$, then both sides $= \sin\alpha$. So, taking the positive sign, we get $\sin(\alpha - \beta) = \sin\alpha \cos\beta - \cos\alpha \sin\beta$.

Example 2.3.5

Given two vectors $\vec{\alpha} = 3\hat{i} - \hat{j}$ and $\vec{\beta} = 2\hat{i} + \hat{j} - 3\hat{k}$, express $\vec{\beta}$ in the form $\vec{\beta} = \vec{\beta}_1 + \vec{\beta}_2$, where $\vec{\beta}_1$ is parallel to $\vec{\alpha}$ and $\vec{\beta}_2$ is perpendicular to $\vec{\alpha}$.

Solution

Let $\vec{\beta}_2 = a_1\hat{i} + a_2\hat{j} + a_3\hat{k}$. Given that $\vec{\beta}_2$ is perpendicular to $\vec{\alpha}$

$$\vec{\beta}_2 \cdot \vec{\alpha} = 0 \Rightarrow 3a_1 - a_2 = 0 \Rightarrow a_2 = 3a_1$$

Now, since $\vec{\beta} = \vec{\beta}_1 + \vec{\beta}_2$, we get

$$\vec{\beta}_1 = \vec{\beta} - \vec{\beta}_2 = (2 - a_1)\hat{i} + (1 - a_2)\hat{j} + (-3 - a_3)\hat{k}$$

Since $\vec{\beta}_1$ is parallel to $\vec{\alpha}$, $\vec{\beta}_1 \times \vec{\alpha} = \vec{0}$. Hence

$$\begin{vmatrix} \hat{i} & \hat{j} & \hat{k} \\ 2-a_1 & 1-a_2 & -3-a_3 \\ 3 & -1 & 0 \end{vmatrix} = \vec{0}$$

$$\Rightarrow (-3-a_3)\hat{i} - (9+3a_3)\hat{j} + (-5+a_1+3a_2)\hat{k} = \vec{0}$$

$$\Rightarrow a_3 = -3, \quad -5+a_1+3a_2 = 0$$

$$\Rightarrow a_3 = -3, \quad -5+10a_1 = 0 \Rightarrow a_3 = -3, a_1 = 1/2, a_2 = 3/2$$

Consequently, $\vec{\beta}_2 = \frac{1}{2}(\hat{i}+3\hat{j}-6\hat{k})$ and $\vec{\beta}_1 = \frac{1}{2}(3\hat{i}-\hat{j})$ and

$$2\hat{i}+\hat{j}-3\hat{k} = \frac{1}{2}(3\hat{i}-\hat{j}) + \frac{1}{2}(\hat{i}+3\hat{j}-6\hat{k})$$

Example 2.3.6

Find the unit vector on the plane of $\vec{a} = \hat{i}+2\hat{j}-\hat{k}$ and $\vec{b} = \hat{i}+\hat{j}-2\hat{k}$ that is perpendicular to the vector $\vec{c} = 2\hat{i}-\hat{j}+\hat{k}$.

Solution

A vector $\vec{d}$ on the plane of $\vec{a}$ and $\vec{b}$ can be written in the form $\vec{d} = x\vec{a}+y\vec{b}$, where x and y are scalars. Now, by the given condition, $\vec{d}$ is perpendicular to $\vec{c}$. Therefore

$$\vec{d}\cdot\vec{c} = 0 \Rightarrow \left(x\vec{\alpha}+y\vec{\beta}\right)\cdot\vec{\gamma} = 0$$

$$\Rightarrow \left\{x\left(\hat{i}+2\hat{j}-\hat{k}\right)+y\left(\hat{i}+\hat{j}-2\hat{k}\right)\right\}\cdot\left(2\hat{i}-\hat{j}+\hat{k}\right) = 0$$

$$\Rightarrow \left\{(x+y)\hat{i}+(2x+y)\hat{j}+(-x-2y)\hat{k}\right\}\cdot\left(2\hat{i}-\hat{j}+\hat{k}\right) = 0$$

$$\Rightarrow 2(x+y)-(2x+y)-x-2y = 0; \quad \text{i.e.,} x+y = 0$$

$$\Rightarrow \vec{d} = x\hat{i}-y\hat{k} = x\left(\hat{i}+\hat{k}\right)$$

$$\Rightarrow \hat{d} = \pm\frac{1}{\sqrt{x^2+y^2}}x\left(\hat{i}+\hat{k}\right) = \frac{1}{\sqrt{2}}\left(\hat{i}+\hat{k}\right)$$

Hence, the required unit vector is $\left(\hat{i}+\hat{k}\right)/\sqrt{2}$

Example 2.3.7

Show by the vector method that the area of the triangle formed by joining the midpoint of one of non-parallel sides of a trapezium to the extremities of the opposite side is half that of the trapezium.

Solution

Let $ABCD$ be a trapezium, where, $AB \parallel CD$ (Fig. 2.29). AD and BC are two non-parallel sides of the trapezium. Let O be the middle point of AD. Now, $\vec{OB} = \vec{OA}+\vec{AB}$, $\vec{OC} = \vec{OD}+\vec{DC}$ and $\vec{AB} = t\,\vec{DC}$, where t is a scalar.

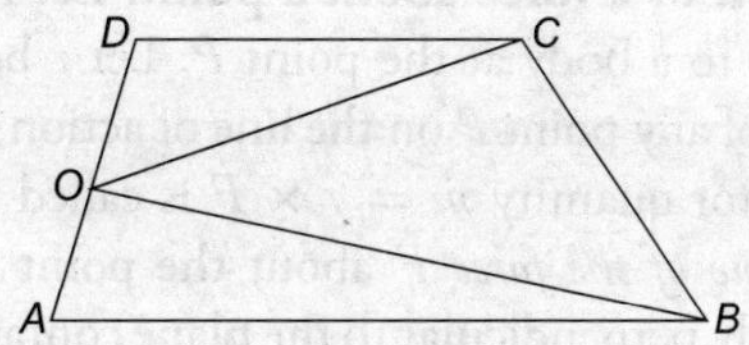

Figure 2.29 Trapezium

Now

$$\begin{aligned}\vec{OB} \times \vec{OC} &= (\vec{OA} + \vec{AB}) \times (\vec{OD} + \vec{DC}) \\ &= \vec{OA} \times \vec{OD} + \vec{OA} \times \vec{DC} + \vec{AB} \times \vec{OD} + \vec{AB} \times \vec{DC} \\ &= \vec{OA} \times \vec{DC} + \vec{AB} \times \vec{OD}; \text{ as } \vec{OA} \times \vec{OD} = \vec{0}, \vec{AB} \times \vec{DC} = \vec{0} \\ &= -\vec{OD} \times \vec{DC} - \vec{AB} \times \vec{OA} \\ &= \vec{DO} \times \vec{DC} + \vec{AB} \times \vec{AO} \\ \Rightarrow \triangle OBC &= \triangle OAB + \triangle OCD\end{aligned}$$

Therefore, trapezium $ABCD = \triangle OBC + \triangle OAB + \triangle OCD = 2\triangle OBC$. Hence, the result follows.

Example 2.3.8

Two non-zero vectors $\vec{r}$ and $\vec{r}'$ are given by $\vec{r} = x\hat{i} + y\hat{j} + z\hat{k}$ and $\vec{r}' = x'\hat{i} + y'\hat{j} + z'\hat{k}$. Prove that the necessary and sufficient condition that $\vec{r}$ and $\vec{r}'$ may be parallel is that $x : y : z = x' : y' : z'$. Choose scalars u and v such that the vectors $(u-1)\hat{i} + 9\hat{j} + 3\hat{k}$ and $2\hat{i} + (v+4)\hat{j} + \hat{k}$ may be co-directional.

Solution

The condition that two non-zero vectors $\vec{r}$ and $\vec{r}'$ are parallel (or co-directional) is $\vec{r} \times \vec{r}' = \vec{0}$. Thus

$$\vec{r} \times \vec{r}' = \begin{vmatrix} \hat{i} & \hat{j} & \hat{k} \\ x & y & z \\ x' & y' & z' \end{vmatrix} = (yz' - y'z)\hat{i} + (x'z - xz')\hat{j} + (xy' - x'y)\hat{k}$$

Since $\hat{i}, \hat{j}$ and $\hat{k}$ are non-coplanar, three scalar coefficients must separately vanish, i.e.,

$$yz' - y'z = 0, \quad x'z - xz' = 0, \quad xy' - x'y = 0$$
$$\text{or,} \quad x : y : z = x' : y' : z'$$

Conversely, let $x : y : z = x' : y' : z'$, then $x = \lambda x'$, $y = \lambda y'$ and $z = \lambda z'$, where λ is a non-zero scalar constant. Using these values, we have

$$\vec{r} \times \vec{r}' = \lambda \begin{vmatrix} \hat{i} & \hat{j} & \hat{k} \\ x & y & z \\ x' & y' & z' \end{vmatrix} = \vec{0}$$

The two given vectors $(u-1)\hat{i} + 9\hat{j} + 3\hat{k}$ and $2\hat{i} + (v+4)\hat{j} + \hat{k}$ are co-directional if

$$(u-1) : 9 : 3 = 2 : (v+4) : 1 \Rightarrow u = 7, \quad v = -1$$

Property 2.3.9

Moment of a force about a point: Let $\vec{F}$ be a force applied to a body at the point P. Let $\vec{r}$ be a position vector of any point P on the line of action of $\vec{F}$. Then, the vector quantity $\vec{m} = \vec{r} \times \vec{F}$ is called the *moment* or *torque of the force* $\vec{F}$ about the point O and it is evidently perpendicular to the plane containing O and the line of $\vec{F}$ (Fig. 2.30).

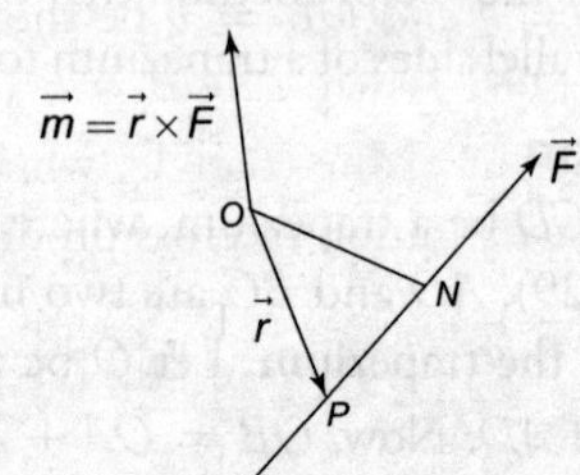

Figure 2.30 Moment of a force

From this definition, we can make the following observations:

(i) Since $\vec{F}$ and $\vec{m}$ are perpendicular to each other, we have $\vec{F} \cdot \vec{m} = 0$.

(ii) The moment vector $\vec{m}$ does not depend on the choice of the point P on the line of $\vec{F}$. Thus, if Q be another point on the line of $\vec{F}$, we have

$$\vec{OQ} \times \vec{F} = (\vec{OP} + \vec{PQ}) \times \vec{F} = \vec{OP} \times \vec{F}; \quad \text{as } \vec{PQ} \times \vec{F} = \vec{0}$$

But $\vec{m}$ does depend on the choice of O.

(iii) Let the applied force $\vec{F}$ be the resultant of a number of concurrent forces $\vec{F_1}, \vec{F_2}, \ldots, \vec{F_n}$, then

$$\begin{aligned}\vec{r} \times \vec{F} &= \vec{r} \times (\vec{F_1} + \vec{F_2} + \cdots + \vec{F_n}) \\ &= \vec{r} \times \vec{F_1} + \vec{r} \times \vec{F_2} + \cdots + \vec{r} \times \vec{F_n}; \quad \text{by distributive law}\end{aligned}$$

Thus, the moment of the resultant of any number of forces about a point is equal to the sum of their separate moments. We shall discuss this further in Chapter 3.

2.4 VECTOR AREA

A plane area bounded by a closed curve without multiple points and traced in a definite sense corresponds to the vector $\vec{A}$, called the *vector area.*

2.4.1 Vector Area of a Parallelogram

Let $\vec{OA} = \vec{a}$ and $\vec{OC} = \vec{b}$ be the adjacent sides of a parallelogram $OABC$ (Fig. 2.31). Let CN be the height of the parallelogram, where $CN = |\vec{b}| \sin\theta$. Then, by the cross product definition

$$\begin{aligned}\vec{a} \times \vec{b} &= |\vec{a}| \; \left|\vec{b}\right| \; \sin\theta \; \hat{n} \\ &= OA.CN \; \hat{n} \qquad (2.26)\end{aligned}$$

Figure 2.31 Vector area of a parallelogram

Now, $OA \cdot CN$ represents the area of the parallelogram $OABC$. Therefore, $OA \cdot CN\hat{n}$ represents the vector area of $OABC$. Thus, $\vec{a} \times \vec{b}$ can be interpreted as vector area $\vec{A}$ of the parallelogram whose adjacent sides are represented by the vectors $\vec{a}$ and $\vec{b}$.

2.4.2 Vector Area of a Triangle

Let $\vec{OA} = \vec{a}$ and $\vec{OB} = \vec{b}$ be the adjacent sides of a triangle OAB, where $\angle AOB = \theta$ (Fig. 2.32). Let BN be the height of the triangle, where $BN = |\vec{b}| \sin\theta$. Using the definition of cross product

$$\begin{aligned}\frac{1}{2}(\vec{a} \times \vec{b}) &= \frac{1}{2}\, |\vec{a}| \; \left|\vec{b}\right| \; \sin\theta \; \hat{n} \\ &= \frac{1}{2} OA \cdot BN \; \hat{n} \qquad (2.27)\end{aligned}$$

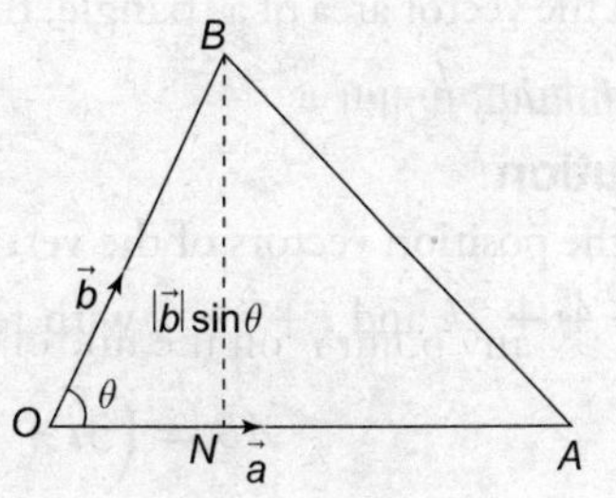

Figure 2.32 Vector area of a triangle

Now, $\frac{1}{2}OA \cdot BN$ represents the area of the triangle OAB. Therefore, $\frac{1}{2}OA \cdot BN\,\hat{n}$ represents the vector area of the triangle OAB. Thus, $\frac{1}{2}\vec{a} \times \vec{b}$ can be interpreted as vector area $\vec{A}$ of the triangle whose adjacent sides are represented by $\vec{a}$ and $\vec{b}$.

DEDUCTION 2.4.1

Let $\vec{a}, \vec{b}$ and $\vec{c}$ be the position vectors of the vertices A, B and C, respectively, of a triangle ABC (Fig. 2.33), with respect to some base point O (not shown in the figure). Then, $\vec{AB} = \vec{b} - \vec{a}$ and $\vec{AC} = \vec{c} - \vec{a}$. Hence, the vector area of the triangle ABC is

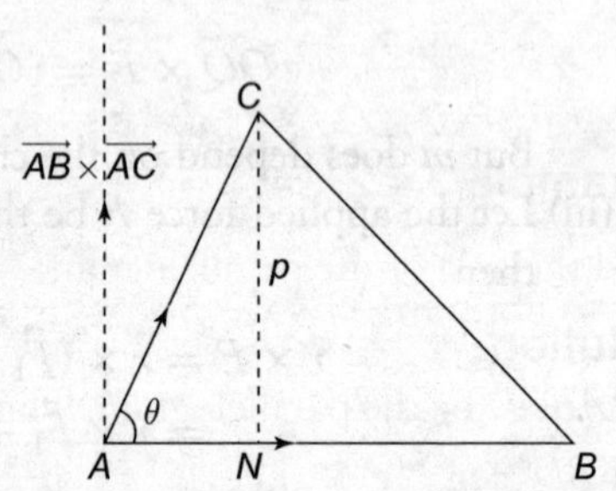

Figure 2.33 Vector area of a triangle

$$\triangle ABC = \frac{1}{2}\vec{AB} \times \vec{AC} = \frac{1}{2}\left[(\vec{b} - \vec{a}) \times (\vec{c} - \vec{a})\right]$$
$$= \frac{1}{2}\left[\vec{b} \times \vec{c} - \vec{b} \times \vec{a} - \vec{a} \times \vec{c} + \vec{a} \times \vec{a}\right]$$
$$= \frac{1}{2}(\vec{a} \times \vec{b} + \vec{b} \times \vec{c} + \vec{c} \times \vec{a}) \qquad (2.28)$$

When the points A, B and C are collinear, the vector area of the triangle vanishes, i.e.,

$$\vec{a} \times \vec{b} + \vec{b} \times \vec{c} + \vec{c} \times \vec{a} = \vec{0} \qquad (2.29)$$

Now, $\vec{AB} \times \vec{AC}$ is a vector which is perpendicular to the plane of the triangle ABC. Thus, we conclude that $\vec{a} \times \vec{b} + \vec{b} \times \vec{c} + \vec{c} \times \vec{a}$ is perpendicular to the plane of the triangle ABC.

DEDUCTION 2.4.2

Let $p = CN$ and the perpendicular distance of C from the straight line AB = the height of the triangle ABC (Fig. 2.33). Then

$$\frac{1}{2}\,|\vec{AB}|\,p = \frac{1}{2}\left|\vec{a} \times \vec{b} + \vec{b} \times \vec{c} + \vec{c} \times \vec{a}\right|$$

$$\text{or,} \qquad p = \frac{\left|\vec{a} \times \vec{b} + \vec{b} \times \vec{c} + \vec{c} \times \vec{a}\right|}{|\vec{b} - \vec{a}|} \qquad (2.30)$$

Example 2.4.1

Find the vector area of a triangle, the position vectors of whose vertices are $2\hat{i}+3\hat{j}-5\hat{k}$, $3\hat{i}+4\hat{j}+2\hat{k}$ and $\hat{i}+\hat{j}-\hat{k}$.

Solution

Let the position vectors of the vertices A, B and C of a triangle ABC be, respectively, $2\hat{i}+3\hat{j}-5\hat{k}$, $3\hat{i}+4\hat{j}+2\hat{k}$ and $\hat{i}+\hat{j}-\hat{k}$ with respect to some base point O (Fig. 2.33). Then

$$\vec{AB} = \left(3\hat{i}+4\hat{j}+2\hat{k}\right) - \left(2\hat{i}+3\hat{j}-5\hat{k}\right) = \hat{i}+\hat{j}+7\hat{k},$$
$$\vec{AC} = \left(\hat{i}+\hat{j}-\hat{k}\right) - \left(2\hat{i}+3\hat{j}-5\hat{k}\right) = -\hat{i}-2\hat{j}+4\hat{k}$$

$$\vec{AB} \times \vec{AC} = \begin{vmatrix} \hat{i} & \hat{j} & \hat{k} \\ 1 & 1 & 7 \\ -1 & -2 & 4 \end{vmatrix} = 18\hat{i} - 11\hat{j} - \hat{k}$$

Hence, the required vector area of the triangle ABC is

$$= \frac{1}{2}(\vec{AB} \times \vec{AC}) = \frac{1}{2}(18\hat{i} - 11\hat{j} - \hat{k})$$

Example 2.4.2

Find the area of the parallelogram, whose diagonals are the vectors $3\hat{i} + \hat{j} - 2\hat{k}$ and $\hat{i} - 3\hat{j} + 4\hat{k}$.

Solution

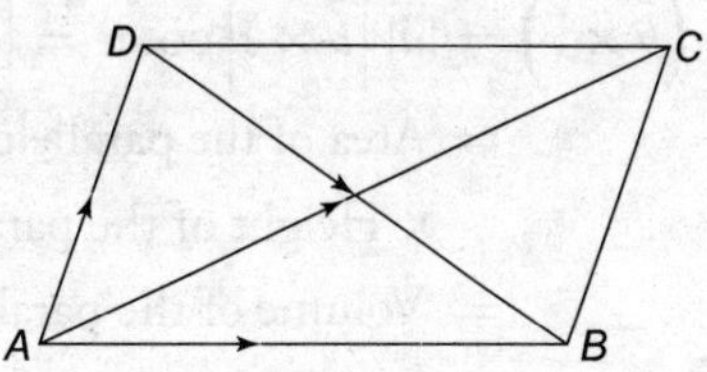

Figure 2.34 Area of a parallelogram

Let $ABCD$ be the parallelogram, whose diagonals are AC and BD (Fig. 2.34). Let $\vec{AC} = 3\hat{i} + \hat{j} - 2\hat{k}$ and $\vec{DB} = \hat{i} - 3\hat{j} + 4\hat{k}$. Now, by the triangle law of addition, $\vec{AB} + \vec{BC} = \vec{AC}$ and $\vec{AD} + \vec{DB} = \vec{AB}$.

$$\therefore\ 2\vec{AB} = \vec{AC} + \vec{BD} = (3\hat{i} + \hat{j} - 2\hat{k}) + (\hat{i} - 3\hat{j} + 4\hat{k})$$

$$\text{or, } \vec{AB} = 2\hat{i} - \hat{j} + \hat{k}$$

$$\text{and } 2\vec{AD} = \vec{AC} - \vec{DB} = 2\hat{i} + 4\hat{j} - 6\hat{k}$$

$$\text{or, } \vec{AD} = \hat{i} + 2\hat{j} - 3\hat{k}$$

Thus, the vector area of the parallelogram $ABCD$ is

$$\vec{AB} \times \vec{AD} = \begin{vmatrix} \hat{i} & \hat{j} & \hat{k} \\ 2 & -1 & 1 \\ 1 & 2 & -3 \end{vmatrix} = \hat{i} + 7\hat{j} + 5\hat{k}$$

Thus, the area of the parallelogram $ABCD$ is

$$\left|\vec{AB} \times \vec{AD}\right| = \sqrt{1^2 + 7^2 + 5^2} = 5\sqrt{3} \text{ sq.units.}$$

2.5 PRODUCT OF THREE VECTORS

As we are already familiar with the scalar and vector products of two vectors, we now extend these ideas for the case of three vectors in the following manner. Here, we shall discuss two types of products of three vectors:

(i) scalar triple product
(ii) vector triple product

2.5.1 Scalar Triple Product

Let $\vec{a}$, $\vec{b}$ and $\vec{c}$ be the three given vectors. Then, the scalar quantity $\vec{a} \cdot (\vec{b} \times \vec{c})$ is known as the *scalar triple product* of the three vectors $\vec{a}$, $\vec{b}$ and $\vec{c}$ and is denoted by $\left[\vec{a}\ \vec{b}\ \vec{c}\right]$. Thus

$$\left[\vec{a}\,\vec{b}\,\vec{c}\right] = \vec{a}\cdot(\vec{b}\times\vec{c}) \tag{2.31}$$

The scalar triple product is also known as the *box product.* Following are some properties of scalar triple product.

Property 2.5.1

Geometrical significance: Let $\vec{a}, \vec{b}$ and $\vec{c}$ be the concurrent edges OA, OB and OC of the parallelepiped (Fig. 2.35). Then, $\vec{b}\times\vec{c}$ is a vector perpendicular to the plane of the face $OBDC$ of the parallelepiped. Let θ be the angle between $\vec{a}$ and $\vec{b}\times\vec{c}$. Then

$$\begin{aligned}\vec{a}\cdot\left(\vec{b}\times\vec{c}\right) &= |\vec{a}|\left|\vec{b}\times\vec{c}\right|\cos\theta = \left|\vec{b}\times\vec{c}\right||\vec{a}|\cos\theta\\ &= \text{Area of the parallelogram } OBDC\\ &\quad\times \text{Height of the parallelepiped}\\ &= \text{Volume of the parallelepiped}\end{aligned}$$

Thus, the scalar triple product $\vec{a}\cdot(\vec{b}\times\vec{c})$ of the three vectors $\vec{a}, \vec{b}$ and $\vec{c}$ is numerically equal to the volume V of the parallelepiped with $\vec{a}, \vec{b}$ and $\vec{c}$ as co-terminus edges. $\vec{a}, \vec{b}$ and $\vec{c}$ form a right-handed system (dextral system), otherwise it would be $-V$ (sinistral system).

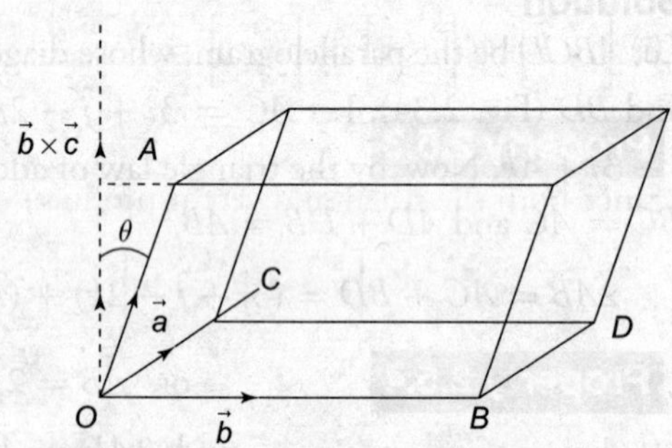

Figure 2.35 Scalar triple product

Therefore, geometrically, the scalar triple product of three vectors represents the volume of the parallelepiped formed by the vectors as the co-terminus edges.

Property 2.5.2

Let $\vec{a} = a_1\hat{i} + a_2\hat{j} + a_3\hat{k}$, $\vec{b} = b_1\hat{i} + b_2\hat{j} + b_3\hat{k}$ and $\vec{c} = c_1\hat{i} + c_2\hat{j} + c_3\hat{k}$ be three given vectors in coordinate form. Then

$$\vec{b}\times\vec{c} = \begin{vmatrix}\hat{i} & \hat{j} & \hat{k}\\ b_1 & b_2 & b_3\\ c_1 & c_2 & c_3\end{vmatrix} = (b_2c_3 - b_3c_2)\hat{i} - (b_1c_3 - b_3c_1)\hat{j} + (b_1c_2 - b_2c_1)\hat{k}$$

$$\begin{aligned}\vec{a}\cdot(\vec{b}\times\vec{c}) &= (b_2c_3 - b_3c_2)a_1 - (b_1c_3 - b_3c_1)a_2 + (b_1c_2 - b_2c_1)a_3\\ &= \begin{vmatrix}a_1 & a_2 & a_3\\ b_1 & b_2 & b_3\\ c_1 & c_2 & c_3\end{vmatrix} = \left[\vec{a}\,\vec{b}\,\vec{c}\right]\end{aligned}$$

This is the scalar triple product in coordinate form. Using the properties of determinants, we get

$$\begin{vmatrix}b_1 & b_2 & b_3\\ c_1 & c_2 & c_3\\ a_1 & a_2 & a_3\end{vmatrix} = \left[\vec{b}\,\vec{c}\,\vec{a}\right], \quad \begin{vmatrix}c_1 & c_2 & c_3\\ a_1 & a_2 & a_3\\ b_1 & b_2 & b_3\end{vmatrix} = \left[\vec{c}\,\vec{a}\,\vec{b}\right]$$

This shows that the vectors can be interchanged in cyclic order without altering the value. Furthermore, we have

$$(\vec{a}\times\vec{b})\cdot\vec{c}=\vec{c}\cdot(\vec{a}\times\vec{b})=\begin{vmatrix} c_1 & c_2 & c_3\\ a_1 & a_2 & a_3\\ b_1 & b_2 & b_3\end{vmatrix}=\begin{vmatrix} a_1 & a_2 & a_3\\ b_1 & b_2 & b_3\\ c_1 & c_2 & c_3\end{vmatrix}=\vec{a}\cdot(\vec{b}\times\vec{c})$$

This shows that the signs of dot and cross product may be interchanged without altering the value. From the above discussion

(i) $\left[\vec{a}\,\vec{b}\,\vec{c}\right]=\left[\vec{b}\,\vec{c}\,\vec{a}\right]=\left[\vec{c}\,\vec{a}\,\vec{b}\right]$ (ii) $\left[\vec{b}\,\vec{c}\,\vec{a}\right]=\left[\vec{c}\,\vec{a}\,\vec{b}\right]=-\left[\vec{b}\,\vec{a}\,\vec{c}\right]=-\left[\vec{a}\,\vec{c}\,\vec{b}\right]$

Also, for the fundamental system, we have

(i) $\left[\hat{i}\,\hat{j}\,\hat{k}\right]=\left[\hat{j}\,\hat{k}\,\hat{i}\right]=\left[\hat{k}\,\hat{i}\,\hat{j}\right]=1$ (ii) $\left[\hat{i}\,\hat{k}\,\hat{j}\right]=\left[\hat{k}\,\hat{j}\,\hat{i}\right]=\left[\hat{j}\,\hat{i}\,\hat{k}\right]=-1$

Property 2.5.3

Since both the scalar and vector products of vectors obey distributive law, we have

$$\left[\vec{a}\;\vec{b}+\vec{d}\;\vec{c}+\vec{r}\right]=\left[\vec{a}\,\vec{b}\,\vec{c}\right]+\left[\vec{a}\,\vec{b}\,\vec{r}\right]+\left[\vec{a}\,\vec{d}\,\vec{c}\right]+\left[\vec{a}\,\vec{d}\,\vec{r}\right]$$

Property 2.5.4

If three vectors $\vec{a}, \vec{b}$ and $\vec{c}$ are non-null and coplanar, then $\left[\vec{a}\,\vec{b}\,\vec{c}\right]=0$, i.e., the volume of a parallelepiped formed by them will vanish. Conversely, if three non-null vectors are such that $\left[\vec{a}\,\vec{b}\,\vec{c}\right]=0$, then $\vec{a}$ is perpendicular to $\vec{b}\times\vec{c}$. Thus, $\vec{a}$ lies on the plane of $\vec{b}$ and $\vec{c}$, i.e., $\vec{a}, \vec{b}$ and $\vec{c}$ are coplanar.

Example 2.5.1

If the points A, B and C, not all lying on the same straight line, have position vectors $\vec{a}, \vec{b}$ and $\vec{c}$ with respect to the given origin, then show that $\vec{a}\times\vec{b}+\vec{b}\times\vec{c}+\vec{c}\times\vec{a}$ is a vector perpendicular to the plane ABC.

Solution

Let P be any point on the plane of A, B and C, whose position vector with respect to the base O as origin is $\vec{p}$. Since the points A, B, C and P are coplanar, the vectors $\vec{AP}=\vec{p}-\vec{a}$, $\vec{AB}=\vec{b}-\vec{a}$ and $\vec{AC}=\vec{c}-\vec{a}$ are also coplanar. The condition for coplanarity gives

$$\left[\vec{AP}\;\vec{AB}\;\vec{AC}\right]=0\Rightarrow\left[\vec{p}-\vec{a}\;\vec{b}-\vec{a}\;\vec{c}-\vec{a}\right]=0$$

$$\Rightarrow(\vec{p}-\vec{a})\cdot\{(\vec{b}-\vec{a})\times(\vec{c}-\vec{a})\}=0$$

$$\Rightarrow(\vec{p}-\vec{a})\cdot(\vec{a}\times\vec{b}+\vec{b}\times\vec{c}+\vec{c}\times\vec{a})=0$$

Since the vector $\vec{p}$ is arbitrary, from the last relation, we conclude that $\vec{a}\times\vec{b}+\vec{b}\times\vec{c}+\vec{c}\times\vec{a}$ is perpendicular to $\vec{p}-\vec{a}$ and is therefore perpendicular to the plane of ABC.

Example 2.5.2

Prove that the points $2\hat{i}-\hat{j}+\hat{k}$, $\hat{i}-3\hat{j}-5\hat{k}$ and $3\hat{i}-4\hat{j}-4\hat{k}$ form the vertices of a right-angled triangle.

Solution

Let $\vec{a}=2\hat{i}-\hat{j}+\hat{k}$, $\vec{b}=\hat{i}-3\hat{j}-5\hat{k}$ and $\vec{c}=3\hat{i}-4\hat{j}-4\hat{k}$, respectively, be the given three vectors. Now

$$\left[\vec{a}\,\vec{b}\,\vec{c}\right]=\begin{vmatrix}2 & -1 & 1\\ 1 & -3 & -5\\ 3 & -4 & -4\end{vmatrix}=0$$

Thus, the given vectors $\vec{a}$, $\vec{b}$ and $\vec{c}$ are coplanar. Again, $\vec{a}\cdot\vec{b}=2+3-5=0$, so $\vec{a}$ is perpendicular to $\vec{b}$. Consequently, the vectors form the sides of a right-angled triangle.

Example 2.5.3

If $\vec{a}\times\vec{b}+\vec{b}\times\vec{c}+\vec{c}\times\vec{a}=\vec{0}$, show that the vectors $\vec{a}$, $\vec{b}$ and $\vec{c}$ are coplanar.

Solution

Taking the dot product with $\vec{a}$ on both sides of the given relation, we get

$$\vec{a}\cdot\left\{\vec{a}\times\vec{b}+\vec{b}\times\vec{c}+\vec{c}\times\vec{a}\right\}=\vec{a}\cdot\vec{0}$$

$$\text{or,}\quad \vec{a}\cdot(\vec{a}\times\vec{b})+\vec{a}\cdot(\vec{b}\times\vec{c})+\vec{a}\cdot(\vec{c}\times\vec{a})=0$$

$$\text{or,}\quad 0+\left[\vec{a}\,\vec{b}\,\vec{c}\right]+0=0\Rightarrow\left[\vec{a}\,\vec{b}\,\vec{c}\right]=0$$

This gives the condition for coplanarity of the three vectors $\vec{a}$, $\vec{b}$ and $\vec{c}$. Hence, the vectors $\vec{a}$, $\vec{b}$ and $\vec{c}$ are coplanar.

Example 2.5.4

If $\vec{a}$, $\vec{b}$ and $\vec{c}$ are three non-coplanar vectors, and $\vec{r}$ is an arbitrary vector, then prove that the vector $\vec{r}$ can be expressed as

$$\vec{r}=\frac{1}{\vec{a}\,\vec{b}\,\vec{c}}\left\{\vec{r}\,\vec{b}\,\vec{a}\vec{a}+\vec{r}\,\vec{c}\,\vec{a}\vec{b}+\vec{r}\,\vec{a}\,\vec{b}\vec{c}\right\}$$

Solution

Since $\vec{a}$, $\vec{b}$ and $\vec{c}$ are three non-coplanar vectors, $\left[\vec{a}\,\vec{b}\,\vec{c}\right]\neq 0$ and the vector $\vec{r}$ can be expressed as a linear combination of the vectors $\vec{a}$, $\vec{b}$ and $\vec{c}$. Thus

$$\vec{r}=l\,\vec{a}+m\,\vec{b}+n\,\vec{c};\quad l,m,n=\ \text{scalars}$$

Taking the dot product on both sides with the vector $\vec{b}\times\vec{c}$, we obtain

$$\vec{r}\cdot(\vec{b}\times\vec{c})=l\,\vec{a}\cdot(\vec{b}\times\vec{c})+m\,\vec{b}\cdot(\vec{b}\times\vec{c})+n\,\vec{c}\cdot(\vec{b}\times\vec{c})$$

$$=l\left[\vec{a}\,\vec{b}\,\vec{c}\right]+0+0=l\left[\vec{a}\,\vec{b}\,\vec{c}\right]$$

$$\Rightarrow l = \frac{[\vec{r}\,\vec{b}\,\vec{c}]}{[\vec{a}\,\vec{b}\,\vec{c}]}; \quad \text{as } \vec{r}\cdot(\vec{b}\times\vec{c}) = [\vec{r}\,\vec{b}\,\vec{c}] \text{ and } [\vec{a}\,\vec{b}\,\vec{c}] \neq 0$$

Similarly, taking the scalar product on both sides, with the vectors $\vec{c}\times\vec{a}$ and $\vec{a}\times\vec{b}$, respectively, we obtain

$$m = \frac{[\vec{r}\,\vec{c}\,\vec{a}]}{[\vec{b}\,\vec{c}\,\vec{a}]} = \frac{[\vec{r}\,\vec{c}\,\vec{a}]}{[\vec{a}\,\vec{b}\,\vec{c}]} \text{ and } n = \frac{[\vec{r}\,\vec{a}\,\vec{b}]}{[\vec{c}\,\vec{a}\,\vec{b}]} = \frac{[\vec{r}\,\vec{a}\,\vec{b}]}{[\vec{a}\,\vec{b}\,\vec{c}]}$$

Using the values of the scalars, l, m and n, we obtain the desired result.

Property 2.5.5

Let us take one of the vertices O of the tetrahedron $OABC$ (Fig. 2.36) as the base point, and let the other vertices A, B and C have the respective position vectors $\vec{a}$, $\vec{b}$ and $\vec{c}$, with respect to the base point O. Therefore, $\vec{OA} = \vec{a}$, $\vec{OB} = \vec{b}$, $\vec{OC} = \vec{c}$. Now, $\frac{1}{2}(\vec{b}\times\vec{c})$ is the vector area of the triangle OBC, and hence, the volume V of the tetrahedron is given by

$$V = \frac{1}{3}\vec{a}\cdot\left\{\frac{1}{2}(\vec{b}\times\vec{c})\right\} = \frac{1}{6}[\vec{a}\,\vec{b}\,\vec{c}] \text{ cu. units} \quad (2.32)$$

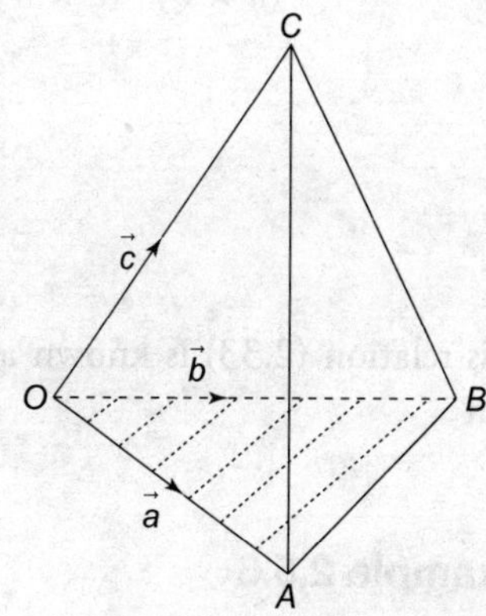

Figure 2.36 Tetrahedron

When the position vector of the four vertices A, B, C and D are, respectively, $\vec{a}, \vec{b}, \vec{c}$ and $\vec{d}$ (Fig. 2.2), then the volume V of the tetrahedron is given by

$$V = \frac{1}{6}\left|(\vec{b}-\vec{a})\times(\vec{c}-\vec{a})\cdot(\vec{d}-\vec{a})\right| \text{ cu. units}$$

In particular, if the four points are coplanar, then the volume of the tetrahedron is zero. We shall discuss this in Chapter 3.

Example 2.5.5

Let $P(1, 3, -1)$, $Q(0, 1, 6)$ and $R(-1, 3, 1)$ be three points in space. Find the coordinates of a point S on the y axis, such that the volume of the tetrahedron $PQRS$ is 10 cubic units.

Solution

Let the position of the fourth vertex be $(0, \lambda, 0)$ on the y axis. Taking P as origin, the position vectors of the other vertices are given by

$$\vec{a} = \vec{PQ} = (0, 1, 6) - (1, 3, -1) = (-1, -2, 7)$$
$$\vec{b} = \vec{PR} = (-1, 3, 1) - (1, 3, -1) = (-2, 0, 2)$$
$$\vec{c} = \vec{PS} = (0, \lambda, 0) - (1, 3, -1) = (-1, \lambda - 3, 1)$$

By the given condition

$$\frac{1}{6}\left[\vec{a}\,\vec{b}\,\vec{c}\right] = \frac{1}{6}\begin{vmatrix} -1 & -2 & 7 \\ -2 & 0 & 2 \\ -1 & \lambda - 3 & 1 \end{vmatrix} = \pm 10$$

$$\text{or,} \quad -12\lambda + 36 = \pm 60 \Rightarrow \lambda = -2, 8$$

Thus, the position vector of S is $(0, -2, 0)$ or $(0, 8, 0)$.

2.5.2 Scalar Product of Four Vectors

Let $\vec{a}, \vec{b}, \vec{c}$ and $\vec{d}$ be four given vectors. Then, $(\vec{a} \times \vec{b}) \cdot (\vec{c} \times \vec{d})$ is called the *scalar product of four vectors* and is given by

$$\begin{aligned}(\vec{a} \times \vec{b}) \cdot (\vec{c} \times \vec{d}) &= \vec{a} \cdot \left\{\vec{b} \times (\vec{c} \times \vec{d})\right\} \\ &= \vec{a} \cdot \left\{(\vec{b} \cdot \vec{d})\vec{c} - (\vec{b} \cdot \vec{c})\vec{d}\right\} \\ &= (\vec{a} \cdot \vec{c})(\vec{b} \cdot \vec{d}) - (\vec{a} \cdot \vec{d})(\vec{b} \cdot \vec{c}) = \begin{vmatrix} \vec{a} \cdot \vec{c} & \vec{a} \cdot \vec{d} \\ \vec{b} \cdot \vec{c} & \vec{b} \cdot \vec{d} \end{vmatrix} \end{aligned} \tag{2.33}$$

This relation (2.33) is known as *Lagrange's identity.* In particular, if we write $\vec{c} = \vec{a}$ and $\vec{d} = \vec{b}$, then

$$(\vec{a} \times \vec{b})^2 = |\vec{a}|^2\,|\vec{b}|^2 - (\vec{a} \cdot \vec{b})^2$$

Example 2.5.6

Let $\vec{a}, \vec{b}, \vec{c}$ and $\vec{d}$ be four given vectors. Prove that

$$(\vec{b} \times \vec{c}) \cdot (\vec{a} \times \vec{d}) + (\vec{c} \times \vec{a}) \cdot (\vec{b} \times \vec{d}) + (\vec{a} \times \vec{b}) \cdot (\vec{c} \times \vec{d}) = 0$$

Hence, deduce that $\sin(\alpha + \beta)\sin(\alpha - \beta) = \sin^2\alpha - \sin^2\beta$.

Solution

The scalar product of four vectors $\vec{a}, \vec{b}, \vec{c}$ and $\vec{d}$ is given by

$$(\vec{a} \times \vec{b}) \cdot (\vec{c} \times \vec{d}) = \begin{vmatrix} \vec{a} \cdot \vec{c} & \vec{a} \cdot \vec{d} \\ \vec{b} \cdot \vec{c} & \vec{b} \cdot \vec{d} \end{vmatrix} \qquad (\vec{c} \times \vec{a}) \cdot (\vec{b} \times \vec{d}) = \begin{vmatrix} \vec{c} \cdot \vec{b} & \vec{c} \cdot \vec{d} \\ \vec{a} \cdot \vec{b} & \vec{a} \cdot \vec{d} \end{vmatrix}$$

$$(\vec{b} \times \vec{c}) \cdot (\vec{a} \times \vec{d}) = \begin{vmatrix} \vec{b} \cdot \vec{a} & \vec{b} \cdot \vec{d} \\ \vec{c} \cdot \vec{a} & \vec{c} \cdot \vec{d} \end{vmatrix}$$

Therefore, for the given four vectors $\vec{a}, \vec{b}, \vec{c}$ and $\vec{d}$, the expression becomes

$$\begin{aligned}&(\vec{b} \times \vec{c}) \cdot (\vec{a} \times \vec{d}) + (\vec{c} \times \vec{a}) \cdot (\vec{b} \times \vec{d}) + (\vec{a} \times \vec{b}) \cdot (\vec{c} \times \vec{d}) \\ &= \begin{vmatrix} \vec{a} \cdot \vec{c} & \vec{a} \cdot \vec{d} \\ \vec{b} \cdot \vec{c} & \vec{b} \cdot \vec{d} \end{vmatrix} + \begin{vmatrix} \vec{c} \cdot \vec{b} & \vec{c} \cdot \vec{d} \\ \vec{a} \cdot \vec{b} & \vec{a} \cdot \vec{d} \end{vmatrix} + \begin{vmatrix} \vec{b} \cdot \vec{a} & \vec{b} \cdot \vec{d} \\ \vec{c} \cdot \vec{a} & \vec{c} \cdot \vec{d} \end{vmatrix}\end{aligned}$$

$$= (\vec{a}\cdot\vec{c})(\vec{b}\cdot\vec{d}) - (\vec{a}\cdot\vec{d})(\vec{b}\cdot\vec{c}) + (\vec{c}\cdot\vec{b})(\vec{a}\cdot\vec{d}) - (\vec{c}\cdot\vec{d})(\vec{a}\cdot\vec{b})$$
$$+(\vec{a}\cdot\vec{b})(\vec{c}\cdot\vec{d}) - (\vec{b}\cdot\vec{d})(\vec{c}\cdot\vec{a}) = 0$$

For the second part, let $\angle AOC = \alpha$, $\angle AOB = \beta$, $\angle COD = \beta$ and $\vec{OA} = \vec{a}$, $\vec{OB} = \vec{b}$, $\vec{OC} = \vec{c}$ and $\vec{OD} = \vec{d}$ (Fig. 2.37). Let us assume that the four vectors $\vec{a}, \vec{b}, \vec{c}$ and $\vec{d}$ are coplanar and $\hat{n}$ is a unit vector perpendicular to this plane of four vectors $\vec{a}, \vec{b}, \vec{c}$ and $\vec{d}$. Now

$$\vec{b}\times\vec{c} = \left|\vec{b}\right| \; |\vec{c}| \sin(\alpha-\beta)\,\hat{n}$$
$$\vec{a}\times\vec{d} = |\vec{a}| \; \left|\vec{d}\right| \sin(\alpha+\beta)\,\hat{n}$$
$$\vec{c}\times\vec{a} = -\vec{a}\times\vec{c} = -\,|\vec{a}| \; |\vec{c}| \sin\alpha\,\hat{n}$$
$$\vec{b}\times\vec{d} = \left|\vec{b}\right| \; \left|\vec{d}\right| \sin\alpha\,\hat{n}$$
$$\vec{a}\times\vec{b} = |\vec{a}| \; \left|\vec{b}\right| \sin\beta\,\hat{n}$$
$$\vec{c}\times\vec{d} = |\vec{c}| \; \left|\vec{d}\right| \sin\beta\,\hat{n}$$

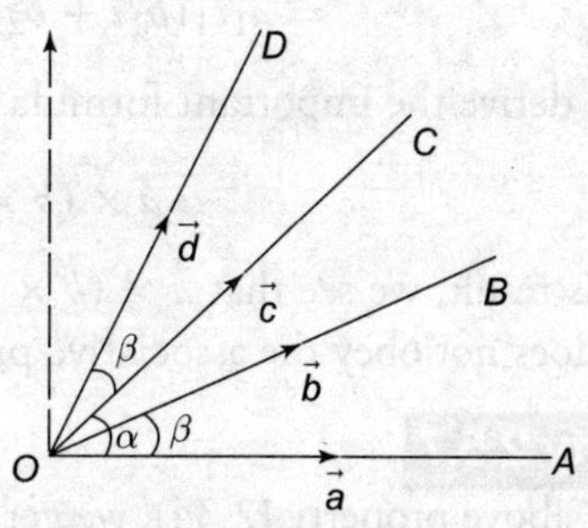

Figure 2.37 $\sin(\alpha+\beta)\sin(\alpha-\beta) = \sin^2\alpha - \sin^2\beta$.

From the first part, we get

$$\left(\vec{b}\times\vec{c}\right)\cdot\left(\vec{a}\times\vec{d}\right) = -(\vec{c}\times\vec{a})\cdot\left(\vec{b}\times\vec{d}\right) - \left(\vec{a}\times\vec{b}\right)\cdot\left(\vec{c}\times\vec{d}\right)$$

$$\text{or,}\quad \left|\vec{b}\right| \; |\vec{c}| \sin(\alpha-\beta)\;\hat{n}.|\vec{a}| \; \left|\vec{d}\right| \sin(\alpha+\beta)\;\hat{n}$$
$$= |\vec{a}| \; |\vec{c}| \sin\alpha\;\hat{n}.\left|\vec{b}\right| \; \left|\vec{d}\right| \sin\alpha\;\hat{n} - |\vec{a}| \; \left|\vec{b}\right| \sin\beta\;\hat{n}.|\vec{c}| \; \left|\vec{d}\right| \sin\beta\;\hat{n}$$

$$\text{or,}\quad |\vec{a}| \; \left|\vec{b}\right| \; |\vec{c}| \; \left|\vec{d}\right| \sin(\alpha+\beta)\sin(\alpha-\beta) = |\vec{a}| \; \left|\vec{b}\right| \; |\vec{c}| \; \left|\vec{d}\right| \left(\sin^2\alpha - \sin^2\beta\right)$$

$$\text{or,}\quad \sin(\alpha+\beta)\sin(\alpha-\beta) = \sin^2\alpha - \sin^2\beta$$

2.5.3 Vector Triple Product

Let $\vec{a}$, $\vec{b}$ and $\vec{c}$ be any three vectors. Then, the quantity $\vec{a}\times(\vec{b}\times\vec{c})$ in the plane of $\vec{b}$ and $\vec{c}$ is called the *vector triple product* of the vectors $\vec{a}$, $\vec{b}$ and $\vec{c}$. Following are some properties of vector triple product.

Property 2.5.6

Let us take three unit vectors $\hat{i}, \hat{j}$ and $\hat{k}$ along the three rectangular axes. Without loss of any generality, we may choose $\vec{a}$ along $\hat{i}$, $\vec{b}$ on the plane of $\hat{i}$ and $\hat{j}$ so that the vectors $\vec{a}$, $\vec{b}$ and $\vec{c}$ can be expressed in the coordinate form $\vec{a} = a_1\,\hat{i}$; $\vec{b} = b_1\,\hat{i} + b_2\,\hat{j}$; $\vec{c} = c_1\,\hat{i} + c_2\,\hat{j} + c_3\,\hat{k}$. Now, $\vec{a}\cdot\vec{c} = a_1\,c_1$, $\vec{a}\cdot\vec{b} = a_1\,b_1$ and

$$\vec{b}\times\vec{c} = \left(b_1\hat{i} + b_2\hat{j}\right)\times\left(c_1\hat{i} + c_2\hat{j} + c_3\hat{k}\right)$$
$$= b_2c_3\hat{i} - b_1c_3\hat{j} + (b_1c_2 - b_2c_1)\hat{k}$$

Therefore, $\vec{a} \times (\vec{b} \times \vec{c})$ is given by

$$\begin{aligned}\vec{a} \times (\vec{b} \times \vec{c}) &= \begin{vmatrix} \hat{i} & \hat{j} & \hat{k} \\ a_1 & 0 & 0 \\ b_2c_3 & -b_1c_3 & b_1c_2 - b_2c_1 \end{vmatrix} = -a_1(b_1c_2 - b_2c_1)\hat{j} - a_1b_1c_3\hat{k} \\ &= (a_1b_1c_1\hat{i} + a_1b_2c_1\hat{j}) - (a_1b_1c_1\hat{i} + a_1b_1c_2\hat{j}a_1b_1c_3\hat{k}) \\ &= a_1c_1(b_1\hat{i} + b_2\hat{j}) - a_1b_1(c_1\hat{i} + c_2\hat{j} + c_3\hat{k}) = (\vec{a} \cdot \vec{c})\vec{b} - (\vec{a} \cdot \vec{b})\vec{c}\end{aligned}$$

Thus, we derive the important formula

$$\vec{a} \times (\vec{b} \times \vec{c}) = (\vec{a} \cdot \vec{c})\vec{b} - (\vec{a} \cdot \vec{b})\vec{c} \tag{2.34}$$

Using this result, we see that $\vec{a} \times (\vec{b} \times \vec{c})$ and $(\vec{a} \times \vec{b}) \times \vec{c}$ are not equal. Thus, the vector triple product does not obey the associative property.

Property 2.5.7

From the above property (2.34), we get

$$\vec{a} \times (\vec{b} \times \vec{c}) = (\vec{a} \cdot \vec{c})\vec{b} - (\vec{a} \cdot \vec{b})\vec{c} = -\left\{(\vec{a} \cdot \vec{b})\vec{c} - (\vec{a} \cdot \vec{c})\vec{b}\right\} = -(\vec{b} \times \vec{c}) \times \vec{a}$$

Similarly, $\vec{a} \times (\vec{b} \times \vec{c}) = -\vec{c} \times (\vec{a} \times \vec{b})$. Using this result, we get

$$\vec{a} \times (\vec{b} \times \vec{c}) + \vec{b} \times (\vec{c} \times \vec{a}) + \vec{c} \times (\vec{a} \times \vec{b}) = \vec{0} \tag{2.35}$$

This relation (2.35) is known as *Jacobi's identity.*

Example 2.5.7

Prove that $\left[\vec{\alpha} + \vec{\beta}\ \vec{\beta} + \vec{\gamma}\ \vec{\gamma} + \vec{\alpha}\right] = 2\left[\vec{\alpha}\ \vec{\beta}\ \vec{\gamma}\right]$.

Solution

We know that the dot product and cross product of vectors are distributive. Now, let $\vec{a} = \vec{\alpha} + \vec{\beta}$, $\vec{b} = \vec{\beta} + \vec{\gamma}$ and $\vec{c} = \vec{\gamma} + \vec{\alpha}$. The scalar product of the three vectors $\vec{a}$, $\vec{b}$ and $\vec{c}$ is given by

$$\begin{aligned}\left[\vec{a}\ \vec{b}\ \vec{c}\right] &= \vec{a} \cdot (\vec{b} \times \vec{c}) = \vec{a} \cdot \left\{(\vec{\beta} + \vec{\gamma}) \times (\vec{\gamma} + \vec{\alpha})\right\} \\ &= \vec{a} \cdot \left\{\vec{\beta} \times \vec{\gamma} + \vec{\beta} \times \vec{\alpha} + \vec{\gamma} \times \vec{\gamma} + \vec{\gamma} \times \vec{\alpha}\right\} \\ &= (\vec{\alpha} + \vec{\beta}) \cdot \left\{\vec{\beta} \times \vec{\gamma} + \vec{\beta} \times \vec{\alpha} + \vec{\gamma} \times \vec{\alpha}\right\}; \text{ as } \vec{\gamma} \times \vec{\gamma} = \vec{0} \\ &= \vec{\alpha} \cdot (\vec{\beta} \times \vec{\gamma}) + \vec{\alpha} \cdot (\vec{\beta} \times \vec{\alpha}) + \vec{\alpha} \cdot (\vec{\gamma} \times \vec{\alpha}) + \vec{\beta} \cdot (\vec{\beta} \times \vec{\gamma}) + \vec{\beta} \cdot (\vec{\beta} \times \vec{\alpha}) + \vec{\beta} \cdot (\vec{\gamma} \times \vec{\alpha}) \\ &= \left[\vec{\alpha}\ \vec{\beta}\ \vec{\gamma}\right] + 0 + 0 + 0 + 0 + \left[\vec{\beta}\ \vec{\gamma}\ \vec{\alpha}\right] = \left[\vec{\alpha}\ \vec{\beta}\ \vec{\gamma}\right] + \left[\vec{\alpha}\ \vec{\beta}\ \vec{\gamma}\right] = 2\left[\vec{\alpha}\ \vec{\beta}\ \vec{\gamma}\right]\end{aligned}$$

which is desired.

Example 2.5.8

Prove that $\left[\vec{a} \times \vec{b}\ \vec{b} \times \vec{c}\ \vec{c} \times \vec{a}\right] = \left[\vec{a}\ \vec{b}\ \vec{c}\right]^2$

Solution

Using the definition of box product, we get

$$\begin{aligned}\left[\vec{a}\times\vec{b}\;\vec{b}\times\vec{c}\;\vec{c}\times\vec{a}\right] &= (\vec{a}\times\vec{b})\cdot\left\{(\vec{b}\times\vec{c})\times(\vec{c}\times\vec{a})\right\}\\ &= (\vec{a}\times\vec{b})\cdot\left\{(\vec{p}\cdot\vec{a})\vec{c}-(\vec{p}\cdot\vec{c})\vec{a}\right\},\ \text{where } \vec{p}=\vec{b}\times\vec{c}\\ &= (\vec{a}\times\vec{b})\cdot\left\{\left[\vec{a}\,\vec{b}\,\vec{c}\right]\vec{c}-\left[\vec{c}\,\vec{b}\,\vec{c}\right]\vec{a}\right\}\\ &= (\vec{a}\times\vec{b})\cdot\vec{c}\left[\vec{a}\,\vec{b}\,\vec{c}\right],\quad \text{as } \left[\vec{c}\,\vec{b}\,\vec{c}\right]=0\\ &= \left[\vec{c}\,\vec{a}\,\vec{b}\right]\left[\vec{a}\,\vec{b}\,\vec{c}\right]=\left[\vec{a}\,\vec{b}\,\vec{c}\right]^2\end{aligned}$$

2.5.4 Vector Product of Four Vectors

Let $\vec{a}, \vec{b}, \vec{c}$ and $\vec{d}$ be four given vectors. Then, $(\vec{a}\times\vec{b})\times(\vec{c}\times\vec{d})$ which is the vector product of two vectors $(\vec{a}\times\vec{b})$ and $(\vec{c}\times\vec{d})$, parallel to the line of intersection of the plane $\vec{a}$ and $\vec{b}$ and that of $\vec{c}$ and $\vec{d}$ is considered a *vector product of four vectors.* Now, we derive certain properties for the vector product of four vectors:

Property 2.5.8

We can look upon $(\vec{a}\times\vec{b})\times(\vec{c}\times\vec{d})$ as a vector triple product. Let $\vec{a}\times\vec{b}=\vec{p}$, then

$$\begin{aligned}(\vec{a}\times\vec{b})\times(\vec{c}\times\vec{d}) &= \vec{p}\times\left(\vec{c}\times\vec{d}\right) = (\vec{p}\cdot\vec{d})\vec{c}-(\vec{p}\cdot\vec{c})\vec{d}\\ &= \left\{(\vec{a}\times\vec{b})\cdot\vec{d}\right\}\vec{c}-\left\{(\vec{a}\times\vec{b})\cdot\vec{c}\right\}\vec{d}\\ &= \left[\vec{a}\,\vec{b}\,\vec{d}\right]\vec{c}-\left[\vec{a}\,\vec{b}\,\vec{c}\right]\vec{d} \qquad (2.36)\end{aligned}$$

Thus, the vector product appears as linear combinations of vectors $\vec{c}$ and $\vec{d}$. Similarly, if we take $\vec{c}\times\vec{d}=\vec{q}$, then,

$$\left(\vec{a}\times\vec{b}\right)\times\left(\vec{c}\times\vec{d}\right)=\left[\vec{a}\,\vec{c}\,\vec{d}\right]\vec{b}-\left[\vec{b}\,\vec{c}\,\vec{d}\right]\vec{a} \qquad (2.37)$$

Thus, the vector product appears as linear combinations of vectors $\vec{a}$ and $\vec{b}$. Any one of the above relations gives the vector product of four vectors. When the vectors $\vec{a}, \vec{b}, \vec{c}$ and $\vec{d}$ are coplanar, then

$$\left[\vec{a}\,\vec{c}\,\vec{d}\right]=0=\left[\vec{b}\,\vec{c}\,\vec{d}\right] \qquad (2.38)$$

Therefore, the condition for coplanarity of the given four vectors is $(\vec{a}\times\vec{b})\times(\vec{c}\times\vec{d})=\vec{0}$.

Example 2.5.9

Prove that $\left[\vec{a}\times\vec{b}\;\vec{c}\times\vec{d}\;\vec{e}\times\vec{f}\right]=\left[\vec{a}\,\vec{b}\,\vec{d}\right]\left[\vec{c}\,\vec{e}\,\vec{f}\right]-\left[\vec{a}\,\vec{b}\,\vec{c}\right]\left[\vec{d}\,\vec{e}\,\vec{f}\right]$.

Solution

Let us put $\vec{p}=\vec{a}\times\vec{b}, \vec{q}=\vec{c}\times\vec{d}, \vec{r}=\vec{e}\times\vec{f}$. Then, we have

$$\vec{q}\times\vec{r}=(\vec{c}\times\vec{d})\times(\vec{e}\times\vec{f})=\left[\vec{c}\,\vec{e}\,\vec{f}\right]\vec{d}-\left[\vec{d}\,\vec{e}\,\vec{f}\right]\vec{c}$$

$$\left[\vec{p}\,\vec{q}\,\vec{r}\right]=\vec{p}\cdot(\vec{q}\times\vec{r})=\left[\vec{c}\,\vec{e}\,\vec{f}\right]\vec{d}\cdot\vec{p}-\left[\vec{d}\,\vec{e}\,\vec{f}\right]\vec{c}\cdot\vec{p}$$

$$=\left[\vec{a}\,\vec{b}\,\vec{d}\right]\left[\vec{c}\,\vec{e}\,\vec{f}\right]-\left[\vec{a}\,\vec{b}\,\vec{c}\right]\left[\vec{d}\,\vec{e}\,\vec{f}\right]$$

Example 2.5.10

Prove that $\left[\vec{a}\times\vec{b}\;\vec{c}\times\vec{d}\;\vec{e}\times\vec{f}\right]=\left[\vec{c}\,\vec{d}\,\vec{a}\right]\left[\vec{b}\,\vec{e}\,\vec{f}\right]-\left[\vec{c}\,\vec{d}\,\vec{b}\right]\left[\vec{a}\,\vec{e}\,\vec{f}\right]$.

Solution

Using the definition

$$\left[\vec{a}\times\vec{b}\;\vec{c}\times\vec{d}\;\vec{e}\times\vec{f}\right]=(\vec{c}\times\vec{d})\cdot\left[(\vec{e}\times\vec{f})\times(\vec{a}\times\vec{b})\right]$$

$$=(\vec{c}\times\vec{d})\cdot\left[\{(\vec{e}\times\vec{f})\cdot\vec{b}\}\vec{a}-\{(\vec{e}\times\vec{f})\cdot\vec{a}\}\vec{b}\right]$$

$$=(\vec{c}\times\vec{d})\cdot\left\{\left[\vec{e}\,\vec{f}\,\vec{b}\right]\vec{a}-\left[\vec{e}\,\vec{f}\,\vec{a}\right]\vec{b}\right\}$$

$$=\left[\vec{c}\,\vec{d}\,\vec{a}\right]\left[\vec{e}\,\vec{f}\,\vec{b}\right]-\left[\vec{c}\,\vec{d}\,\vec{b}\right]\left[\vec{e}\,\vec{f}\,\vec{a}\right]$$

$$=\left[\vec{c}\,\vec{d}\,\vec{a}\right]\left[\vec{b}\,\vec{e}\,\vec{f}\right]-\left[\vec{c}\,\vec{d}\,\vec{b}\right]\left[\vec{a}\,\vec{e}\,\vec{f}\right]$$

Example 2.5.11

Prove that $\left[\vec{a}\times\vec{b}\;\vec{c}\times\vec{d}\;\vec{e}\times\vec{f}\right]=\left[\vec{a}\,\vec{b}\,\vec{e}\right]\left[\vec{f}\,\vec{c}\,\vec{d}\right]-\left[\vec{a}\,\vec{b}\,\vec{f}\right]\left[\vec{e}\,\vec{c}\,\vec{d}\right]$.

Solution

Using the definition

$$\left[\vec{a}\times\vec{b}\;\vec{c}\times\vec{d}\;\vec{e}\times\vec{f}\right]=(\vec{a}\times\vec{b})\cdot\left[(\vec{c}\times\vec{d})\times(\vec{e}\times\vec{f})\right]$$

$$=(\vec{a}\times\vec{b})\cdot\left[\{(\vec{c}\times\vec{d})\cdot\vec{f}\}\vec{e}-\{(\vec{c}\times\vec{d})\cdot\vec{e}\}\vec{f}\right]$$

$$=(\vec{a}\times\vec{b})\cdot\left\{\left[\vec{c}\,\vec{d}\,\vec{f}\right]\vec{e}-\left[\vec{c}\,\vec{d}\,\vec{e}\right]\vec{f}\right\}$$

$$=\left[\vec{a}\,\vec{b}\,\vec{e}\right]\left[\vec{c}\,\vec{d}\,\vec{f}\right]-\left[\vec{a}\,\vec{b}\,\vec{f}\right]\left[\vec{c}\,\vec{d}\,\vec{e}\right]$$

$$=\left[\vec{a}\,\vec{b}\,\vec{e}\right]\left[\vec{f}\,\vec{c}\,\vec{d}\right]-\left[\vec{a}\,\vec{b}\,\vec{f}\right]\left[\vec{e}\,\vec{c}\,\vec{d}\right]$$

Property 2.5.9

Here, we are to determine the linear relation connecting the four given vectors $\vec{a},\vec{b},\vec{c}$ and $\vec{d}$. Considering $(\vec{a}\times\vec{b})\times(\vec{c}\times\vec{d})$ as a vector triple product in two ways, we obtain

$$\left[\vec{a}\,\vec{b}\,\vec{d}\right]\vec{c}-\left[\vec{a}\,\vec{b}\,\vec{c}\right]\vec{d}=\left[\vec{a}\,\vec{c}\,\vec{d}\right]\vec{b}-\left[\vec{b}\,\vec{c}\,\vec{d}\right]\vec{a}$$

$$\text{or,}\quad\left[\vec{a}\,\vec{b}\,\vec{d}\right]\vec{c}-\left[\vec{a}\,\vec{b}\,\vec{c}\right]\vec{d}-\left[\vec{a}\,\vec{c}\,\vec{d}\right]\vec{b}+\left[\vec{b}\,\vec{c}\,\vec{d}\right]\vec{a}=\vec{0}$$

$$\text{or,}\quad \left[\vec{b}\,\vec{c}\,\vec{d}\right]\vec{a}-\left[\vec{a}\,\vec{c}\,\vec{d}\right]\vec{b}+\left[\vec{a}\,\vec{b}\,\vec{d}\right]\vec{c}-\left[\vec{a}\,\vec{b}\,\vec{c}\right]\vec{d}=\vec{0}$$

This is the required relation connecting $\vec{a},\vec{b},\vec{c}$ and $\vec{d}$. If the three vectors $\vec{a},\vec{b}$ and $\vec{c}$ are non-coplanar, then $\vec{a}\,\vec{b}\,\vec{c}\neq 0$. We can rewrite this as

$$\vec{d}=\frac{1}{\left[\vec{a}\,\vec{b}\,\vec{c}\right]}\left\{\left[\vec{b}\,\vec{c}\,\vec{d}\right]\vec{a}-\left[\vec{a}\,\vec{c}\,\vec{d}\right]\vec{b}+\left[\vec{a}\,\vec{b}\,\vec{d}\right]\vec{c}\right\}\tag{2.39}$$

Thus, we have expressed any given vector $\vec{d}$ as a linear combination of any three non-coplanar vectors, so that any four vectors are always linearly dependent.

Example 2.5.12

If the four vectors $\vec{a},\vec{b},\vec{c}$ and $\vec{d}$ are such that $\vec{a}+\vec{b}+\vec{c}+\vec{d}=\vec{0}$, show that

$$\frac{|\vec{a}|}{\left[\vec{\beta}\,\vec{\gamma}\,\vec{\delta}\right]}=\frac{-|\vec{b}|}{\left[\vec{\gamma}\,\vec{\delta}\,\vec{\alpha}\right]}=\frac{|\vec{c}|}{\left[\vec{\delta}\,\vec{\alpha}\,\vec{\beta}\right]}=\frac{-|\vec{d}|}{\left[\vec{\alpha}\,\vec{\beta}\,\vec{\gamma}\right]}$$

where $\vec{\alpha},\vec{\beta},\vec{\gamma}$ and $\vec{\delta}$ are the unit vectors along $\vec{a},\vec{b},\vec{c}$ and $\vec{d}$, respectively.

Solution

Since $\vec{\alpha},\vec{\beta},\vec{\gamma}$ and $\vec{\delta}$ are the unit vectors along the vectors $\vec{a},\vec{b},\vec{c}$ and $\vec{d}$, respectively, we have $\vec{a}=|\vec{a}|\vec{\alpha}$, $\vec{b}=|\vec{b}|\vec{\beta}$, $\vec{c}=|\vec{c}|\vec{\gamma}$ and $\vec{d}=|\vec{d}|\vec{\delta}$. Using the given relation $\vec{a}+\vec{b}+\vec{c}+\vec{d}=\vec{0}$, we get

$$|\vec{a}|\vec{\alpha}+|\vec{b}|\vec{\beta}+|\vec{c}|\vec{\gamma}+|\vec{d}|\vec{\delta}=\vec{0}$$

$$\text{or,}\quad |\vec{a}|\vec{\alpha}\cdot\left(\vec{\beta}\times\vec{\gamma}\right)+|\vec{b}|\vec{\beta}\cdot\left(\vec{\beta}\times\vec{\gamma}\right)+|\vec{c}|\vec{\gamma}\cdot\left(\vec{\beta}\times\vec{\gamma}\right)+|\vec{d}|\vec{\delta}\cdot\left(\vec{\beta}\times\vec{\gamma}\right)=0$$

$$\text{or,}\quad |\vec{a}|\left[\vec{\alpha}\,\vec{\beta}\,\vec{\gamma}\right]+0+0+|\vec{d}|\left[\vec{\delta}\,\vec{\beta}\,\vec{\gamma}\right]=0$$

$$\text{or,}\quad \frac{-|\vec{d}|}{\left[\vec{\alpha}\,\vec{\beta}\,\vec{\gamma}\right]}=\frac{|\vec{a}|}{\left[\vec{\delta}\,\vec{\beta}\,\vec{\gamma}\right]}=\frac{|\vec{a}|}{\left[\vec{\beta}\,\vec{\gamma}\,\vec{\delta}\right]}$$

Similarly, $\dfrac{-|\vec{b}|}{\left[\vec{\gamma}\,\vec{\delta}\,\vec{\alpha}\right]}=\dfrac{|\vec{c}|}{\left[\vec{\delta}\,\vec{\alpha}\,\vec{\beta}\right]}=\dfrac{|\vec{a}|}{\left[\vec{\beta}\,\vec{\gamma}\,\vec{\delta}\right]}$. Hence, the result.

Example 2.5.13

If $\vec{p},\vec{q}$ and $\vec{r}$ and $\vec{a},\vec{b}$ and $\vec{c}$ are two sets of non-coplanar vectors, prove that

$$\left[\vec{p}\,\vec{q}\,\vec{r}\right]\left[\vec{a}\,\vec{b}\,\vec{c}\right]=\begin{vmatrix}\vec{p}\cdot\vec{a} & \vec{p}\cdot\vec{b} & \vec{p}\cdot\vec{c}\\ \vec{q}\cdot\vec{a} & \vec{q}\cdot\vec{b} & \vec{q}\cdot\vec{c}\\ \vec{r}\cdot\vec{a} & \vec{r}\cdot\vec{b} & \vec{r}\cdot\vec{c}\end{vmatrix}$$

Solution

From Eq. (2.39), we see that for four vectors $\vec{p},\vec{q},\vec{r}$ and $\vec{d}$

$$\vec{d}=\frac{1}{\left[\vec{p}\,\vec{q}\,\vec{r}\right]}\left\{\left[\vec{d}\,\vec{q}\,\vec{r}\right]\vec{p}+\left[\vec{d}\,\vec{r}\,\vec{p}\right]\vec{q}+\left[\vec{d}\,\vec{p}\,\vec{q}\right]\vec{r}\right\}\tag{i}$$

where $\vec{p}, \vec{q}$ and $\vec{r}$ are three non-coplanar vectors. Now, using Eq. (2.33), we get

$$(\vec{a} \times \vec{b}) \cdot (\vec{q} \times \vec{r}) = \begin{vmatrix} \vec{q} \cdot \vec{a} & \vec{r} \cdot \vec{a} \\ \vec{q} \cdot \vec{b} & \vec{r} \cdot \vec{b} \end{vmatrix} \tag{ii}$$

Taking $\vec{d} = \vec{a} \times \vec{b}$, we get from result (i)

$$\begin{aligned}
[\vec{p}\,\vec{q}\,\vec{r}]\ \vec{d} &= [\vec{d}\,\vec{q}\,\vec{r}]\vec{p} + [\vec{d}\,\vec{r}\,\vec{p}]\vec{q} + [\vec{d}\,\vec{p}\,\vec{q}]\vec{r} \\
&= \begin{vmatrix} \vec{q} \cdot \vec{a} & \vec{r} \cdot \vec{a} \\ \vec{q} \cdot \vec{b} & \vec{r} \cdot \vec{b} \end{vmatrix}\vec{p} + \begin{vmatrix} \vec{r} \cdot \vec{a} & \vec{p} \cdot \vec{a} \\ \vec{r} \cdot \vec{b} & \vec{p} \cdot \vec{b} \end{vmatrix}\vec{q} + \begin{vmatrix} \vec{p} \cdot \vec{a} & \vec{q} \cdot \vec{a} \\ \vec{p} \cdot \vec{b} & \vec{q} \cdot \vec{b} \end{vmatrix}\vec{r} \\
[\vec{p}\,\vec{q}\,\vec{r}]\,[\vec{a}\,\vec{b}\,\vec{c}] &= \begin{vmatrix} \vec{q} \cdot \vec{a} & \vec{r} \cdot \vec{a} \\ \vec{q} \cdot \vec{b} & \vec{r} \cdot \vec{b} \end{vmatrix}\vec{p} \cdot \vec{c} + \begin{vmatrix} \vec{r} \cdot \vec{a} & \vec{p} \cdot \vec{a} \\ \vec{r} \cdot \vec{b} & \vec{p} \cdot \vec{b} \end{vmatrix}\vec{q} \cdot \vec{c} + \begin{vmatrix} \vec{p} \cdot \vec{a} & \vec{q} \cdot \vec{a} \\ \vec{p} \cdot \vec{b} & \vec{q} \cdot \vec{b} \end{vmatrix}\vec{r} \cdot \vec{c} \\
&= \begin{vmatrix} \vec{q} \cdot \vec{a} & \vec{q} \cdot \vec{b} \\ \vec{r} \cdot \vec{b} & \vec{r} \cdot \vec{b} \end{vmatrix}\vec{p} \cdot \vec{c} - \begin{vmatrix} \vec{p} \cdot \vec{a} & \vec{p} \cdot \vec{b} \\ \vec{r} \cdot \vec{a} & \vec{r} \cdot \vec{b} \end{vmatrix}\vec{q} \cdot \vec{c} + \begin{vmatrix} \vec{p} \cdot \vec{a} & \vec{p} \cdot \vec{b} \\ \vec{q} \cdot \vec{a} & \vec{q} \cdot \vec{b} \end{vmatrix}\vec{r} \cdot \vec{c} \\
&= \begin{vmatrix} \vec{p} \cdot \vec{a} & \vec{p} \cdot \vec{b} & \vec{p} \cdot \vec{c} \\ \vec{q} \cdot \vec{a} & \vec{q} \cdot \vec{b} & \vec{q} \cdot \vec{c} \\ \vec{r} \cdot \vec{a} & \vec{r} \cdot \vec{b} & \vec{r} \cdot \vec{c} \end{vmatrix}
\end{aligned}$$

2.5.5 Reciprocal System of Vectors

Let us consider a system of three non-coplanar vectors $\vec{a}, \vec{b}$ and $\vec{c}$, i.e., $\vec{a}\,\vec{b}\,\vec{c} \neq 0$. Thus, any of these vectors can be expressed as a linear combination of others. Let us define another system of vectors $\vec{a}', \vec{b}'$ and $\vec{c}'$, such that

$$\vec{a}' = \frac{1}{[\vec{a}\,\vec{b}\,\vec{c}]}(\vec{b} \times \vec{c}), \quad \vec{b}' = \frac{1}{[\vec{a}\,\vec{b}\,\vec{c}]}(\vec{c} \times \vec{a}), \quad \vec{c}' = \frac{1}{[\vec{a}\,\vec{b}\,\vec{c}]}(\vec{a} \times \vec{b}) \tag{2.40}$$

Then, $\vec{a}', \vec{b}'$ and $\vec{c}'$ is called the reciprocal system to the set of vectors $\vec{a}, \vec{b}$ and $\vec{c}$. Now, we derive the following important relations:

DEDUCTION 2.5.1

First, let us consider the dot products:

$$\vec{a}' \cdot \vec{a} = \frac{1}{[\vec{a}\,\vec{b}\,\vec{c}]}\left(\vec{b} \times \vec{c}\right) \cdot \vec{a} = \frac{[\vec{a}\,\vec{b}\,\vec{c}]}{[\vec{a}\,\vec{b}\,\vec{c}]} = 1 \quad \text{and} \quad \vec{a}' \cdot \vec{b} = \frac{1}{[\vec{a}\,\vec{b}\,\vec{c}]}\left(\vec{b} \times \vec{c}\right) \cdot \vec{b} = 0$$

Thus, the reciprocal vectors satisfy the following properties:

$$\begin{aligned}
&\vec{a} \cdot \vec{a}' = 1; \quad \vec{a} \cdot \vec{b}' = 0; \quad \vec{a} \cdot \vec{c}' = 0 \\
&\vec{b} \cdot \vec{a}' = 0; \quad \vec{b} \cdot \vec{b}' = 1; \quad \vec{b} \cdot \vec{c}' = 0 \\
&\vec{c} \cdot \vec{a}' = 0; \quad \vec{c} \cdot \vec{b}' = 0; \quad \vec{c} \cdot \vec{c}' = 1
\end{aligned}$$

These relations also imply that the box product $\vec{a}\,\vec{b}\,\vec{c}$ and $\vec{a}'\,\vec{b}'\,\vec{c}'$ must have the same sign, i.e., two sets whose reciprocals to one another are either both right-handed or both left-handed.

DEDUCTION 2.5.2

The expressions of $\vec{a}, \vec{b}$ and $\vec{c}$ can be obtained in terms of $\vec{a}', \vec{b}'$ and $\vec{c}'$ very easily. Since $\vec{a} \cdot \vec{b}' = \vec{a} \cdot \vec{c}' = 0$, $\vec{a}$ is perpendicular to both $\vec{b}'$ and $\vec{c}'$ and hence parallel to $\vec{b}' \times \vec{c}'$. Let $\vec{a} = k(\vec{b}' \times \vec{c}')$, where k is a scalar to be determined. Now, using the property $\vec{a} \cdot \vec{a}' = 1$, we have

$$\vec{a}' \cdot k\left(\vec{b}' \times \vec{c}'\right) = 1$$

$$\text{or,} \quad k\left[\vec{a}'\,\vec{b}'\,\vec{c}'\right] = 1 \Rightarrow k = \frac{1}{\left[\vec{a}'\,\vec{b}'\,\vec{c}'\right]}$$

For this value of k, we have $\vec{a} = \dfrac{1}{\left[\vec{a}'\,\vec{b}'\,\vec{c}'\right]}(\vec{b}' \times \vec{c}')$. Similarly, we get

$$\vec{a} = \frac{1}{\left[\vec{a}'\,\vec{b}'\,\vec{c}'\right]}(\vec{b}' \times \vec{c}'),\ \vec{b} = \frac{1}{\left[\vec{a}'\,\vec{b}'\,\vec{c}'\right]}\left(\vec{c}' \times \vec{a}'\right),\ \vec{c} = \frac{1}{\left[\vec{a}'\,\vec{b}'\,\vec{c}'\right]}(\vec{a}' \times \vec{b}') \tag{2.41}$$

The symmetric relation shows that if $\vec{a}', \vec{b}'$ and $\vec{c}'$ comprise the reciprocal system to $\vec{a}, \vec{b}$ and $\vec{c}$, then the system $\vec{a}, \vec{b}$ and $\vec{c}$ is reciprocal to $\vec{a}', \vec{b}'$ and $\vec{c}'$.

DEDUCTION 2.5.3

Using identity (2.35), we can derive another relation:

$$\vec{a} \times \left(\vec{b} \times \vec{c}\right) + \vec{b} \times (\vec{c} \times \vec{a}) + \vec{c} \times \left(\vec{a} \times \vec{b}\right) = \vec{0}$$

$$\text{or,} \quad \frac{1}{\left[\vec{a}\,\vec{b}\,\vec{c}\right]}\left\{\vec{a} \times \left(\vec{b} \times \vec{c}\right) + \vec{b} \times (\vec{c} \times \vec{a}) + \vec{c} \times \left(\vec{a} \times \vec{b}\right)\right\} = \vec{0}$$

$$\text{or,} \quad \vec{a} \times \vec{a}' + \vec{b} \times \vec{b}' + \vec{c} \times \vec{c}' = \vec{0} \tag{2.42}$$

This is another relation connecting two reciprocal sets.

DEDUCTION 2.5.4

Using Eq. (2.39), any vector $\vec{r}$ on the plane of $\vec{a}, \vec{b}$ and $\vec{c}$ can be expressed in the form

$$\vec{r} = \frac{\left[\vec{r}\,\vec{b}\,\vec{c}\right]}{\left[\vec{a}\,\vec{b}\,\vec{c}\right]}\vec{a} + \frac{\left[\vec{r}\,\vec{c}\,\vec{a}\right]}{\left[\vec{a}\,\vec{b}\,\vec{c}\right]}\vec{b} + \frac{\left[\vec{r}\,\vec{a}\,\vec{b}\right]}{\left[\vec{a}\,\vec{b}\,\vec{c}\right]}\vec{c}$$

$$= (\vec{r} \cdot \vec{a}')\vec{a} + (\vec{r} \cdot \vec{b}')\vec{b} + (\vec{r} \cdot \vec{c}')\vec{c} \tag{2.43}$$

From symmetry, we get

$$\vec{r} = (\vec{r} \cdot \vec{a})\vec{a}' + (\vec{r} \cdot \vec{b})\vec{b}' + (\vec{r}.\vec{c})\vec{c}' \tag{2.44}$$

Example 2.5.14

If $\vec{a} = \hat{i} + \hat{j} - 2\hat{k}$, $\vec{b} = \hat{i} - \hat{j} - 3\hat{k}$ and $\vec{c} = 5\hat{i} - \hat{j} + 8\hat{k}$, then find the reciprocal set of vectors.

Solution

Since $\vec{a} = \hat{\imath} + \hat{\jmath} - 2\hat{k}$, $\vec{b} = \hat{\imath} - \hat{\jmath} - 3\hat{k}$ and $\vec{c} = 5\hat{\imath} - \hat{\jmath} + 8\hat{k}$

$$\left[\vec{a}\,\vec{b}\,\vec{c}\right] = \begin{vmatrix} 1 & 1 & -2 \\ 1 & -1 & -3 \\ 5 & -1 & 8 \end{vmatrix} = -42$$

Also, we have $\vec{b} \times \vec{c} = -11\hat{\imath} - 23\hat{\jmath} + 4\hat{k}$, $\vec{c} \times \vec{a} = -6\hat{\imath} + 18\hat{\jmath} + 6\hat{k}$ and $\vec{a} \times \vec{b} = -5\hat{\imath} + \hat{\jmath} - 2\hat{k}$. Let $\vec{a}'$, $\vec{b}'$ and $\vec{c}'$ be the set of reciprocal vectors. Then, using Eq. (2.40), we get

$$\vec{a}' = \frac{1}{\left[\vec{a}\,\vec{b}\,\vec{c}\right]}(\vec{b} \times \vec{c}) = \frac{-1}{42}(-11\hat{\imath} - 23\hat{\jmath} + 4\hat{k})$$

$$\vec{b}' = \frac{1}{\left[\vec{a}\,\vec{b}\,\vec{c}\right]}(\vec{c} \times \vec{a}) = \frac{-1}{42}(-6\hat{\imath} + 18\hat{\jmath} + 6\hat{k})$$

$$\vec{c}' = \frac{1}{\left[\vec{a}\,\vec{b}\,\vec{c}\right]}(\vec{a} \times \vec{b}) = \frac{-1}{42}(-5\hat{\imath} + \hat{\jmath} - 2\hat{k})$$

EXERCISES

Multiple Choice Questions

1. The angle between $\vec{a}$ and $\vec{b}$ is $\pi/6$, then the angle between $2\vec{a}$ and $3\vec{b}$ is
   (a) $\pi/3$ (b) $\pi/2$ (c) $\pi/6$ (d) none of these
2. The projection of the vector $2\vec{i} + 3\hat{\jmath} - 2\hat{k}$ on the vector $\hat{\imath} - 2\hat{\jmath} + 3\hat{k}$ is
   (a) $2/\sqrt{14}$ (b) $1/\sqrt{14}$ (c) $3/\sqrt{14}$ (d) none of these
3. The points $A(1, 1, 2)$, $B(3, 4, 2)$ and $C(5, 6, 4)$ are the vertices of a triangle. Then the exterior angle of B is
   (a) $\cos^{-1}(-5/\sqrt{39})$ (b) $\cos^{-1}(5/\sqrt{39})$ (c) $\cos^{-1}(5/6)$ (d) none of these
4. A unit vector on the xy plane that makes an angle of $45°$ with the vector $\hat{\imath} + \hat{\jmath}$ and an angle of $60°$ with the vector $3\hat{\imath} - 4\hat{\jmath}$ is
   (a) $\hat{\imath}$ (b) $(\hat{\imath} + \hat{\jmath})/\sqrt{2}$ (c) $(\hat{\imath} - \hat{\jmath})/\sqrt{2}$ (d) none of these
5. The triangle ABC is defined by the vertices $A(1, -2, 2)$, $B(1, 4, 0)$ and $C(-4, 1, 1)$. Then the coordinate of M, the foot of the altitude drawn from the vertex B to the side AC, is
   (a) $(-20/7, -30/7, 10/7)$ (b) $(-20, -30, 10)$ (c) $(2, 3, -1)$ (d) None of these
6. If $\vec{a}$, $\vec{b}$ and $\vec{c}$ are three unit vectors satisfying $\vec{a} + \vec{b} + \vec{c} = \vec{0}$, then the value of $\vec{a} \cdot \vec{b} + \vec{b} \cdot \vec{c} + \vec{c} \cdot \vec{a}$ is
   (a) 0 (b) $-1/2$ (c) $3/2$ (d) $-3/2$
7. If $\vec{a}$ and $\vec{b}$ are two unit vectors and θ is the angle between them, then $|\vec{a} - \vec{b}|/2$ is equal to
   (a) 0 (b) $\pi/2$ (c) $|\sin(\theta/2)|$ (d) $|\cos(\theta/2)|$

8. If $\vec{PO} + \vec{OQ} = \vec{QO} + \vec{OR}$, then P, Q and R are
(a) the vertices of an equilateral triangle (b) the vertices of an isosceles triangle
(c) collinear (d) none of these

9. If $|\vec{a}| = 10$, $|\vec{b}| = 1$ and $\vec{a} \cdot \vec{b} = 6$, then the value of $|\vec{a} \times \vec{b}|$ is
(a) 0 (b) 2 (c) 6 (d) 8

10. If $|\vec{a}| = 2$, $|\vec{b}| = 5$ and $|\vec{a} \times \vec{b}| = 8$, then the value of $\vec{a} \cdot \vec{b}$ is
(a) 0 (b) 2 (c) 6 (d) 8

11. If $\vec{a}$ and $\vec{b}$ be any two vectors such that $|\vec{a} \cdot \vec{b}| = |\vec{a} \times \vec{b}|$, then the angle between them is
(a) $0°$ (b) $15°$ (c) $30°$ (d) $45°$

12. The projection of $\vec{a} = \hat{i} + \hat{j} + \hat{k}$ on the vector $\vec{b} = \hat{i} - 3\hat{j} + 5\hat{k}$ is
(a) $\dfrac{1}{\sqrt{35}}$ (b) $\dfrac{2}{\sqrt{35}}$ (c) $\dfrac{3}{\sqrt{35}}$ (d) $\dfrac{4}{\sqrt{35}}$

13. The projection of the vector $4\hat{i} - 3\hat{j} + \hat{k}$ on the line passing through the points $(2, 3, -1)$ and $(-2, -4, 3)$ is
(a) 1 (b) 2 (c) 3 (d) 4

14. Let $\vec{a}$ and $2\vec{b}$ denote the diagonals of a parallelogram. Then the area of the parallelogram is
(a) $|\vec{a} \times \vec{b}|$ (b) $|\vec{a} \times \vec{b}|/2$ (c) $2|\vec{a} \times \vec{b}|$ (d) none of these

15. The unit vector perpendicular to the plane containing the vectors $\vec{a} = \hat{i} + \hat{j} + \hat{k}$ and $\vec{b} = \hat{i} - \hat{j} - \hat{k}$ is
(a) $\pm\frac{1}{\sqrt{2}}(\hat{i} + \hat{j})$ (b) $\pm\frac{1}{\sqrt{2}}(\hat{i} - \hat{j})$ (c) $\pm\frac{1}{\sqrt{2}}(\hat{j} - \hat{k})$ (d) $\pm\frac{1}{\sqrt{2}}(\hat{j} + \hat{k})$

16. If $\vec{a}$ and $\vec{b}$ are non-zero vectors and $\vec{a} \times \vec{b} = \vec{0}$, then the angle between $\vec{a}$ and $\vec{b}$ must be
(a) $0°$ only (b) $90°$ only (c) $180°$ only (d) $0°$ or $180°$

17. The component of $\vec{a} = \hat{i} - \hat{j} - \hat{k}$ perpendicular to the vector $\vec{b} = 2\hat{i} + \hat{j} - \hat{k}$ is
(a) $(\hat{i} + 2\hat{j} + 2\hat{k})/3$ (b) $(\hat{i} - 4\hat{j} - 2\hat{k})/3$ (c) $(\hat{i} + 4\hat{j} + 2\hat{k})/3$ (d) $(\hat{i} + 2\hat{j} + \hat{k})/3$

18. Two vectors have a magnitude of 3 and 8 units and the angle between them is $60°$. The scalar product of the vectors is
(a) 24 (b) 12 (c) $12\sqrt{3}$ (d) 48

19. The angle between $\vec{a} = 3\hat{i} - 4\hat{j}$ with the x axis is
(a) $\sin^{-1}(\frac{3}{5})$ (b) $67.5°$ (c) $\cos^{-1}(\frac{3}{5})$ (d) $\tan^{-1}(\frac{3}{5})$

20. If the vectors $\vec{a} = \hat{i} - \hat{j} + 2\hat{k}$, $\vec{b} = 2\hat{i} + 4\hat{j} + \hat{k}$ and $\vec{c} = m\hat{i} + \hat{j} + n\hat{k}$ are mutually orthogonal, then $(m, n) =$
(a) $(-3, 2)$ (b) $(2, -3)$ (c) $(-2, 3)$ (d) $(3, -2)$

21. Let $\vec{a}$, $\vec{b}$ and $\vec{c}$ be such that $\vec{a} + \vec{b} = \vec{c}$ and $|\vec{a}| = 3$, $|\vec{b}| = 6$ and $|\vec{c}| = 7$. Then the angle between $\vec{a}$ and $\vec{b}$ is
(a) $\cos^{-1}(\frac{2}{3})$ (b) $\cos^{-1}(\frac{1}{3})$ (c) $\cos^{-1}(\frac{4}{9})$ (d) $\cos^{-1}(\frac{2}{9})$.

22. If $\vec{v}_1$, $\vec{v}_2$ and $\vec{v}_3$ are unit vectors given by $\vec{v}_1 = a\hat{i} + b\hat{j} + c\hat{k}$, $\vec{v}_2 = b\hat{i} + c\hat{j} + a\hat{k}$ and $\vec{v}_3 = c\hat{i} + a\hat{j} + b\hat{k}$, where a, b and c are non-negative real numbers and $\vec{v}_\alpha \cdot \vec{v}_\beta = 0$, for $\alpha \neq \beta$, then
(a) $|\vec{v}_1\ \vec{v}_2\ \vec{v}_3| = 1$ (b) $a + b + c = 1$
(c) $\vec{v}_1 + \vec{v}_2 + \vec{v}_3 = \vec{0}$ (d) $\vec{v}_1\ \vec{v}_2\ \vec{v}_3$ are coplanar

23. The points with position vectors $\vec{v}_1 = a\hat{i} + b\hat{j} + c\hat{k}$, $\vec{v}_2 = b\hat{i} + c\hat{j} + a\hat{k}$ and $\vec{v}_3 = c\hat{i} + a\hat{j} + b\hat{k}$, where a, b and c are non-negative real numbers
(a) are collinear (b) form an equilateral triangle
(c) form an scalene triangle (d) form an right-angled triangle

24. Let A, B, C and D form distinct points on a plane with position vectors $\vec{a}, \vec{b}, \vec{c}$ and $\vec{d}$, respectively. Suppose $(\vec{a} - \vec{d}) \cdot (\vec{b} - \vec{c}) = 0$ and $(\vec{b} - \vec{d}) \cdot (\vec{c} - \vec{a}) = 0$, then
(a) A is the orthocentre of the triangle BCD
(b) B is the orthocentre of the triangle ACD
(c) C is the orthocentre of the triangle ABD
(d) D is the orthocentre of the triangle ABC

25. The value of p, for which the vectors $\vec{a} = p\hat{i} - 4\hat{j} + 3\hat{k}$ and $\vec{b} = 3\hat{i} + p\hat{j} - 2\hat{k}$ are perpendicular, is
(a) 0 (b) 6 (c) -6 (d) -12

26. Let $\vec{\alpha} = \hat{i} + \hat{j} + \hat{k}$ and $\vec{\beta} = 2\hat{i} - \hat{j} + \hat{k}$. The vector of magnitude 3 in the direction $\vec{\alpha} \times \vec{\beta} + \vec{\alpha} + \vec{\beta}$ is
(a) $\vec{0}$ (b) $(2\hat{i} + \hat{j})/\sqrt{5}$ (c) $3(2\hat{i} + \hat{j})/\sqrt{5}$ (d) $(2\hat{i} - \hat{j})/\sqrt{5}$

27. The vectors $\vec{\alpha} = 3\hat{i} - \hat{k}$ and $\vec{\beta} = \hat{i} + 2\hat{j}$ are adjacent sides of a parallelogram. Its area is
(a) $\sqrt{17/2}$ (b) $\sqrt{14}$ (c) $\sqrt{41}$ (d) $\sqrt{7/2}$

28. The projections of $\vec{\alpha} = 2\hat{i} - 3\hat{j} + 6\hat{k}$ and $\vec{\beta} = \lambda\hat{i} + \hat{j} + \hat{k}$ are perpendicular to each other, then the value of λ is
(a) 3/2 (b) 1/2 (c) 5/2 (d) 0

29. The three sides of a parallelepiped are $\vec{a} = 2\hat{i} - 3\hat{j} + 7\hat{k}$, $\vec{b} = \hat{i} + 2\hat{j}$ and $\vec{c} = \hat{j} - \hat{k}$. The volume of this is
(a) 20 units (b) 22 units (c) 25 units (d) zero

30. The angle made by the vector $\vec{a} = 2\hat{i} + 3\hat{j} - \hat{k}$ to the y axis is
(a) $\cos^{-1}\left(\frac{2}{3}\right)$ (b) $\cos^{-1}\left(\frac{3}{\sqrt{14}}\right)$ (c) $\cos^{-1}\left(\frac{3}{2}\right)$ (d) 45°

31. Suppose $\vec{a}, \vec{b}$ and $\vec{c}$ are three vectors of length 1, 1/2 and 1, respectively, such that $\vec{a} \times (\vec{a} \times \vec{c}) + \vec{b} = \vec{0}$. The acute angle between $\vec{a}$ and $\vec{c}$ is
(a) 15° (b) 30° (c) 45° (d) 60°

32. If $\vec{OA}, \vec{OB}$ and $\vec{OC}$ are three distinct vectors lying on the same plane in space, then
(a) $\vec{OA} \cdot (\vec{OB} \times \vec{OC}) = 0$ (b) $\vec{OA} \times (\vec{OB} \times \vec{OC})$ is a vector normal to the plane
(c) There exists three real numbers m, n and p (not all zero) such that $m\vec{OA} + n\vec{OB} + n\vec{OC} = \vec{0}$
(d) $\begin{vmatrix} \vec{OA} \cdot \vec{OA} & \vec{OA} \cdot \vec{OB} & \vec{OA} \cdot \vec{OC} \\ \vec{OB} \cdot \vec{OA} & \vec{OB} \cdot \vec{OB} & \vec{OB} \cdot \vec{OC} \\ \vec{OC} \cdot \vec{OA} & \vec{OC} \cdot \vec{OB} & \vec{OC} \cdot \vec{OC} \end{vmatrix} = 0$

33. A vector whose magnitude is 5 and perpendicular to both $\vec{a} = 2\hat{i} + \hat{j} - 3\hat{k}$ and $\vec{b} = \hat{i} - 2\hat{j} + \hat{k}$ is

(a) $\pm\frac{5\sqrt{3}}{3}(\hat{i}+\hat{j}+\hat{k})$ (b) $\pm\frac{5\sqrt{3}}{3}(\hat{i}+\hat{j}-\hat{k})$

(c) $\pm\frac{5\sqrt{3}}{3}(\hat{i}-\hat{j}+\hat{k})$ (d) $\pm\frac{5\sqrt{3}}{3}(\hat{i}-\hat{j}-\hat{k})$

34. If $\vec{a}=\hat{i}-\hat{j}+\hat{k}$ and $\vec{b}=\hat{i}+\hat{j}+\hat{k}$, then the unit vector perpendicular to the plane containing them is

(a) $\sqrt{2}(\hat{i}+\hat{j})$ (b) $\sqrt{2}(\hat{i}-\hat{j})$ (c) $\frac{1}{\sqrt{2}}(\hat{j}+\hat{k})$ (d) $\frac{1}{\sqrt{2}}(\hat{j}-\hat{k})$

35. If $\vec{b}+\vec{c}, \vec{c}+\vec{a}$ and $\vec{a}+\vec{b}$ are coplanar, then

(a) $\vec{a}, \vec{b}$ and $\vec{c}$ are coplanar (b) $\vec{a}, \vec{b}$ and $\vec{c}$ are non-coplanar

(c) $\vec{a}$ is perpendicular to $\vec{b}+\vec{c}$ (d) $\vec{a}$ is parallel to $\vec{b}+\vec{c}$

36. The value of $\hat{i}\times(\vec{\alpha}\times\hat{i})+\hat{j}\times(\vec{\alpha}\times\hat{j})+\hat{k}\times(\vec{\alpha}\times\hat{k})$ is

(a) $\vec{\alpha}$ (b) $-\vec{\alpha}$ (c) $2\vec{\alpha}$ (d) $-2\vec{\alpha}$

37. The vector $\vec{p}=\vec{a}\times(\vec{b}\times\vec{c})$

(a) lies on the plane of $\vec{a}$ and $\vec{b}$ (b) lies on the plane of $\vec{b}$ and $\vec{c}$

(c) lies on the plane of $\vec{a}$ and $\vec{c}$ (d) lies parallel to $\vec{a}$ and $\vec{b}\times\vec{c}$

38. A vector $\vec{a}$ is along the positive z axis and its vector product with another vector $\vec{b}$ is zero; then $\vec{b}$ could be:

(a) $\hat{i}+\hat{j}$ (b) $4\hat{i}$ (c) $\hat{j}+\hat{k}$ (d) $-7\hat{k}$

39. If $\vec{a}$ and $\vec{b}$ are two arbitrary vectors with magnitudes a and b, respectively, $\left|\vec{a}\times\vec{b}\right|^2$ will equal to

(a) $a^2b^2-(\vec{a}\cdot\vec{b})^2$ (b) $ab-\vec{a}\cdot\vec{b}$ (c) $a^2b^2+(\vec{a}\cdot\vec{b})^2$ (d) $ab+\vec{a}\cdot\vec{b}$

40. The vectors $\vec{a}=\hat{i}+\hat{j}+\hat{k}, \vec{b}=\hat{i}-\hat{j}+\hat{k}$ and $\vec{c}=m\hat{i}+\hat{j}-\hat{k}$ are coplanar or dependent; then the value of m is

(a) 1 (b) 2 (c) −1 (d) 4

41. The value of $(\vec{a}+\vec{b})\cdot\left\{(\vec{b}\times\vec{c})\times(\vec{c}\times\vec{a})\right\}$ equals

(a) $\vec{a}\,\vec{b}\,\vec{c}$ (b) $2\vec{a}\,\vec{b}\,\vec{c}$ (c) $3\vec{a}\,\vec{b}\,\vec{c}$ (d) $\vec{a}\cdot\vec{b}+\vec{b}\cdot\vec{c}+\vec{c}\cdot\vec{a}$

42. Let $\vec{a}, \vec{b}$ and $\vec{c}$ be three non-coplanar vectors in the three-dimensional xyz space. Which of the following vector(s) lies(lie) on the plane containing $\vec{a}$ and $\vec{b}$?

(i) $(\vec{a}\times\vec{b})\times\vec{a}$; (ii) $(\vec{a}\times\vec{b})\times\vec{b}$; (iii) $(\vec{a}\times\vec{b})\times\vec{c}$

(a) (i) only (b) (i) and (ii) only (c) (iii) only (d) (i), (ii) and (iii)

43. Define a binary operation $*$ on the set of all three-dimensional vectors by $\vec{u}*\vec{v}=(\vec{u}\cdot\vec{v})(\vec{u}\times\vec{v})$. Suppose $\vec{a}, \vec{b}$ and $\vec{c}$ are any three non-coplanar vectors satisfying $\vec{a}\cdot\vec{b}\times\vec{c}\neq 0$. If $\vec{a}*(\vec{b}*\vec{c})=(\vec{a}*\vec{b})*\vec{c}$, then

(a) $\vec{a}\cdot\vec{b}=\vec{b}\cdot\vec{c}$ (b) $\vec{a}\times\vec{b}=\vec{b}\times\vec{c}$ (c) $|\vec{a}\cdot\vec{b}|=|\vec{a}\cdot\vec{c}|$ (d) $|\vec{a}\cdot\vec{b}|=|\vec{b}\cdot\vec{c}|$

44. Which of the following vectors is orthogonal to the vector $a\hat{i}+b\hat{j}$, where a and b $(a\neq b)$ are constants, and $\hat{i}$ and $\hat{j}$ are unit orthogonal vectors?

(a) $-b\hat{i}+a\hat{j}$ (b) $-a\hat{i}+b\hat{j}$ (c) $-a\hat{i}-b\hat{j}$ (d) $-b\hat{i}-a\hat{j}$

45. If $\vec{a}, \vec{b}$ and $\vec{c}$ are unit vectors, then $|\vec{a}-\vec{b}|^2+|\vec{b}-\vec{c}|^2+|\vec{c}-\vec{a}|^2$ has its maximum value equal to
(a) 3 (b) 6 (c) 9 (d) 12

46. The angle between the vectors $\vec{a}\times\vec{b}$ and $\vec{a}\times(\vec{a}\times\vec{b})$ is
(a) 60° (b) 0° (c) 90° (d) 30°

47. The area of a triangle formed by the tips of vectors $\vec{a}, \vec{b}$ and $\vec{c}$ is
(a) $\frac{1}{2}(\vec{a}-\vec{b})\cdot(\vec{a}-\vec{c})$ (b) $\frac{1}{2}\left|(\vec{a}-\vec{b})\cdot(\vec{a}-\vec{c})\right|$ (c) $\frac{1}{2}\left|\vec{a}\times\vec{b}\times\vec{c}\right|$ (d) $\frac{1}{2}(\vec{a}\times\vec{b})\cdot\vec{c}$

48. The value of $\left[\vec{a}\times\vec{b}\ \ \vec{b}\times\vec{c}\ \ \vec{c}\times\vec{a}\right]$ is
(a) $\vec{a}\,\vec{b}\,\vec{c}$ (b) $\vec{a}\,\vec{b}\,\vec{c}^2$ (c) $\vec{a}\,\vec{b}\,\vec{c}^3$ (d) 0

49. Let $\vec{a}$ and $\vec{b}$ be two distinct three-dimensional vectors. Then, the component of $\vec{b}$ that is perpendicular to $\vec{a}$ is given by
(a) $\frac{1}{a^2}\{\vec{a}\times(\vec{b}\times\vec{a}\}$ (b) $\frac{1}{b^2}\{\vec{b}\times(\vec{a}\times\vec{b}\}$ (c) $\frac{1}{b^2}\{(\vec{a}\cdot\vec{b})\vec{b}\}$ (d) $\frac{1}{a^2}\{(\vec{b}\cdot\vec{a})\vec{a}\}$

50. The value of $\begin{vmatrix} \vec{a}\cdot\vec{a} & \vec{a}\cdot\vec{b} & \vec{a}\cdot\vec{c} \\ \vec{b}\cdot\vec{a} & \vec{b}\cdot\vec{b} & \vec{b}\cdot\vec{c} \\ \vec{c}\cdot\vec{a} & \vec{c}\cdot\vec{b} & \vec{c}\cdot\vec{c} \end{vmatrix}$ is
(a) $\vec{a}\,\vec{b}\,\vec{c}$ (b) $\vec{a}\,\vec{b}\,\vec{c}^2$ (c) $\vec{a}\,\vec{b}\,\vec{c}^3$ (d) $3\vec{a}\,\vec{b}\,\vec{c}$

51. If the four vectors $\vec{a}, \vec{b}, \vec{c}$ and $\vec{d}$ are coplanar, then
(a) $(\vec{a}\times\vec{b})\cdot(\vec{c}\times\vec{d})=0$ (b) $(\vec{a}\times\vec{c})\cdot(\vec{b}\times\vec{d})=0$
(c) $(\vec{a}\times\vec{b})\times(\vec{c}\times\vec{d})=\vec{0}$ (d) $(\vec{a}\times\vec{c})\times(\vec{b}\times\vec{d})=\vec{0}$

52. The value of $(\vec{a}\times\vec{b})\cdot(\vec{c}\times\vec{d})+(\vec{b}\times\vec{c})\cdot(\vec{a}\times\vec{d})+(\vec{c}\times\vec{a})\cdot(\vec{b}\times\vec{d})$ is
(a) 0 (b) 3 (c) 2 (d) −3

53. If $\vec{a}', \vec{b}'$ and $\vec{c}'$ is the reciprocal system to the vectors $\vec{a}, \vec{b}$ and $\vec{c}$, then the value of $\vec{a}'\,\vec{b}'\,\vec{c}'\vec{a}\,\vec{b}\,\vec{c}$ is
(a) zero (b) 1 (c) 3 (d) none of these

54. If $\vec{a}', \vec{b}'$ and $\vec{c}'$ is the reciprocal system to the three non-coplanar vectors $\vec{a}, \vec{b}$ and $\vec{c}$, then the value of the expression $(\vec{a}'+\vec{b}'+\vec{c}')\cdot(\vec{a}+\vec{b}+\vec{c})$ is
(a) zero (b) 1 (c) 2 (d) 3

55. If $\vec{a}', \vec{b}'$ and $\vec{c}'$ is the reciprocal system to the three non-coplanar vectors $\vec{a}, \vec{b}$ and $\vec{c}$, then the value of the expression $(\vec{a}'+\vec{b}'+\vec{c}')\cdot(\vec{a}+\vec{b})\cdot\vec{a}'+(\vec{b}+\vec{c})\cdot\vec{b}'+(\vec{c}+\vec{a})\cdot\vec{c}'$ is
(a) zero (b) 1 (c) 2 (d) 3

56. If $\vec{a}', \vec{b}'$ and $\vec{c}'$ is the reciprocal system to the vectors $\vec{a}, \vec{b}$ and $\vec{c}$, then the value of $\vec{a}\times\vec{a}'+\vec{b}\times\vec{b}'+\vec{c}\times\vec{c}'$ is
(a) $\vec{0}$ (b) $3\,\vec{a}$ (c) $3\,\vec{b}$ (d) $3\,\vec{c}$

57. If $\vec{a}', \vec{b}'$ and $\vec{c}'$ is the reciprocal system to the three non-coplanar vectors $\vec{a}, \vec{b}$ and $\vec{c}$, then any vector $\vec{r}$ can be expressed as
(a) $\vec{r}=(\vec{r}\times\vec{a})\cdot\vec{a}'+(\vec{r}\times\vec{b})\cdot\vec{b}'+(\vec{r}\times\vec{c})\cdot\vec{c}'$
(b) $\vec{r}=(\vec{r}\times\vec{a}')\cdot\vec{a}+(\vec{r}\times\vec{b}')\cdot\vec{b}+(\vec{r}\times\vec{c}')\cdot\vec{c}$
(c) $\vec{r}=(\vec{r}\cdot\vec{a})\vec{a}'+(\vec{r}\cdot\vec{b})\vec{b}'+(\vec{r}\cdot\vec{c})\vec{c}'$ (d) $\vec{r}=(\vec{r}\cdot\vec{a}')\vec{a}+(\vec{r}\cdot\vec{b}')\vec{b}+(\vec{r}\cdot\vec{c}')\vec{c}$

Answers

1. c	2. a	3. b	4. d	5. a	6. d	7. c	8. c	9. d	10. c	11. d	12. c	13. a
14. a	15. c	16. d	17. b	18. b	19. c	20. a	21. c	22. b	23. b	24. c	25. c	
26. c	27. c	28. a	29. d	30. b	31. b	32. d	33. a	34. c	35. a	36. c	37. b	
38. d	39. a	40. c	41. b	42. d	43. c, d	44. d	45. b	46. c	47. b	48. b	49. a	
50. b	51. c	52. a	53. b	54. d	55. d	56. a	57. d					

Short Answer Questions

1. (a) $\vec{a} = \hat{i} + 3\hat{j} - 2\hat{k}$ and $\vec{b} = 4\hat{i} - 2\hat{j} + 4\hat{k}$, show that $|3\vec{a} + 2\vec{b}| = \sqrt{150}$.
   (b) $\vec{a} = \hat{i} + 3\hat{j} - 2\hat{k}$ and $\vec{b} = 4\hat{i} - 2\hat{j} + 4\hat{k}$, show that $(2\vec{a} + \vec{b}) \cdot (\vec{a} - 2\vec{b}) = -14$.
2. If $\vec{a} = (a_1, a_2, a_3)$, $\vec{b} = (b_1, b_2, b_3)$ and $\vec{c} = (c_1, c_2, c_3)$, compute $\vec{a} \cdot \vec{b}$, $\vec{b} \cdot \vec{c}$ and $\vec{c} \cdot \vec{a}$.
3. For any vector $\vec{a} = a_1\hat{i} + a_2\hat{j} + a_3\hat{k}$, show that
   (a) $a = \sqrt{\vec{a} \cdot \vec{a}} = \sqrt{a_1^2 + a_2^2 + a_3^2}$ (b) $\vec{a} = (\vec{a} \cdot \hat{i})\hat{i} + (\vec{a} \cdot \hat{j})\hat{j} + (\vec{a} \cdot \hat{k})\hat{k}$
4. Prove that
   (a) the dot product of two unit vectors is the cosine of the angle between them.
   (b) the dot product of two parallel vectors is the product of their modulli.
   (c) the dot product of a vector by itself is the square of its length.
   (d) the scalar product of two non-zero vectors is zero if and only if they are at right angles to each other.
5. If the scalar product of a vector $\vec{r}$ with each of three non-coplanar vectors is zero, prove that $\vec{r}$ must be a zero vector.
6. Find the projection of the vector $\vec{a} = 4\hat{i} - 3\hat{j} + \hat{k}$ on the line passing through the points $(2, 3, -1)$ and $(-2, -4, 3)$. *Ans*: 1
7. If $\vec{a} = -2\hat{i} + 2\hat{j} - \hat{k}$ and $\vec{b} = 3\hat{i} + 6\hat{j} + 2\hat{k}$, find
   (a) projection of $\vec{a}$ along $\vec{b}$ (b) projection of $\vec{b}$ along $\vec{a}$ (c) angle between $\vec{a}$ and $\vec{b}$
   Ans: (a) 4/7, (b) 4/3, (c) 79°
8. Find the projection of the vector $\vec{a}$ on the vector $\vec{b}$, where
   (a) $\vec{a} = 2\hat{i} - 3\hat{j} + 6\hat{k}$ and $\vec{b} = \hat{i} + 2\hat{j} + 2\hat{k}$ *Ans*: 8/3
   (b) $\vec{a} = 2\hat{i} + \hat{j} - \hat{k}$ and $\vec{b} = -6\hat{i} + 2\hat{j} - 3\hat{k}$ *Ans*: −1
9. Deduce a vector on the *xy* plane perpendicular to $6\hat{i} + 4\hat{j} + 3\hat{k}$ and equal in length.
   Ans: $\sqrt{61/13}(2\hat{i} - 3\hat{j})$
10. Find out the unit vector which is perpendicular to vector $l\hat{i} + m\hat{j} + n\hat{k}$ and lies on the
    (a) *xy* plane (b) *yz* plane (c) *zx* plane
    Ans: (a) $\pm(m\hat{i} - l\hat{j})/\sqrt{l^2 + m^2}$, (b) $\pm(n\hat{j} - m\hat{k})/\sqrt{n^2 + m^2}$, (c) $\pm(n\hat{i} - l\hat{k})/\sqrt{l^2 + n^2}$
11. Find a vector $\vec{a}$ such that $\vec{a} \times \hat{j} = \hat{i} - \vec{a}$. *Ans*: $\frac{1}{2}(\hat{i} - \hat{k})$
12. The position vectors of A, B, C and D are, respectively, $\vec{a} = \hat{i} + \hat{j} + \hat{k}$, $\vec{b} = 2\hat{i} + 3\hat{j}$, $\vec{c} = 3\hat{i} + 5\hat{j} - 2\hat{k}$ and $\vec{d} = -\hat{j} + \hat{k}$. Evaluate $\vec{AB} \cdot \vec{CD}$ and $\vec{AB} \times \vec{CD}$. *Ans*: $-18, \vec{0}$
13. Given the vectors $\vec{a} = 3\hat{i} + \hat{j} - 5\hat{k}$ and $\vec{b} = -\hat{i} + 2\hat{j} + 6\hat{k}$, calculate
    (a) the length of each *Ans*: $\sqrt{50}, \sqrt{41}$
    (b) $\vec{a} \cdot \vec{b}$ and $\vec{a} \times \vec{b}$ *Ans*: -25, $34\hat{i} - 13\hat{j} + 10\hat{k}$

(c) $(\vec{a} \times (\vec{a} \times \vec{b})$ (d) a unit vector in the direction of $\vec{b}$

14. If D is the midpoint of the side BC of a triangle ABC, show that $\vec{DA} \cdot \vec{DB} + \vec{DA} \cdot \vec{DC} = 0$.
15. Given two points A and B, what is the locus of the point P such that (i) $\vec{PA} \cdot \vec{PB} < 0$ (ii) $\vec{PA} \cdot \vec{PB} = 0$ (iii) $\vec{PA} \cdot \vec{PB} > 0$
16. Is it true that
(a) $\vec{a} \cdot \vec{b} = \vec{a} \cdot \vec{c} \Rightarrow \vec{b} = \vec{c}$ (b) $\vec{a} \cdot \vec{b} = 0 \Rightarrow \vec{a}$ is perpendicular to $\vec{c}$
17. Give two points O and A identify the locus of the point P in each of the following cases:
(a) $\vec{OP} \cdot \vec{OA} > 0$ (b) $\vec{OP} \cdot \vec{OA} = 0$ (c) $\vec{OP} \cdot \vec{OA} < 0$
18. If $\vec{a} \cdot \vec{b} = \vec{a} \cdot \vec{c}$, prove that the line joining the end points of $\vec{b}$ and $\vec{c}$ is perpendicular to $\vec{a}$.
19. Given that $\vec{a}$ and $\vec{b}$ are two vectors of lengths 1 and 2, respectively, and $\vec{a} \cdot \vec{b} = \sqrt{3}$. What is the angle between the vectors $\vec{a}$ and $\vec{b}$?
20. Express $\vec{b} = \vec{\beta}_1 + \vec{\beta}_2$, where $\vec{\beta}_1$ is parallel to $\vec{a}$ and $\vec{\beta}_2$ is perpendicular to $\vec{a}$, given
(a) $\vec{a} = \hat{\imath} - 2\hat{\jmath} + \hat{k}$ and $\vec{b} = \hat{\imath} + \hat{\jmath} - \hat{k}$ *Ans:* $-\frac{1}{3}(\hat{\imath} - 2\hat{\jmath} + \hat{k}), \frac{1}{3}(4\hat{\imath} + \hat{\jmath} - 2\hat{k})$
(b) $\vec{a} = 3\hat{\imath} - \hat{\jmath}$ and $\vec{b} = 2\hat{\imath} + \hat{\jmath} - 3\hat{k}$ *Ans:* $(\frac{3}{2}, -\frac{1}{2}, 0), (\frac{1}{2}, \frac{3}{2}, -3)$
21. (a) Show that the vectors $9\hat{\imath} + \hat{\jmath} - 6\hat{k}$ and $4\hat{\imath} - 6\hat{\jmath} + 5\hat{k}$ are perpendicular to each other.
(b) Find the vector $\vec{a}$ which is orthogonal to $\vec{b} = 4\hat{\imath} - \hat{\jmath} - 2\hat{k}$ and $\vec{c} = 5\hat{\imath} + \hat{\jmath} + \hat{k}$ and $\vec{a} \cdot \vec{d} = 24$, where $\vec{d} = \hat{\imath} - \hat{\jmath} + \hat{k}$. *Ans:* $\hat{\imath} - 14\hat{\jmath} + 9\hat{k}$
(c) Show that each of the vectors $(\vec{a} \cdot \vec{c})\vec{b} - (\vec{a} \cdot \vec{b})\vec{c}$ and $\frac{1}{|\vec{a}|^2}\{\vec{b} - (\vec{a} \cdot \vec{b})\vec{a}\}$ is perpendicular to the vector $\vec{a}$.
(d) If $\vec{a}$, $\vec{b}$ and $\vec{c}$ are three mutually perpendicular vectors of equal magnitude, then show that $\vec{a} + \vec{b} + \vec{c}$ is equally inclined to $\vec{a}$, $\vec{b}$ and $\vec{c}$.
(e) Show that the vectors $\hat{\imath} + 2\hat{\jmath} + \hat{k}$, $\hat{\imath} + \hat{\jmath} - 3\hat{k}$ and $7\hat{\imath} - 4\hat{\jmath} + \hat{k}$ are mutually perpendicular.
22. Show by the vector method that
(a) the diagonals of a rhombus are at right angles
(b) the altitudes of a triangles are concurrent
(c) the midpoint of the hypotenuse of a right-angled triangle is equidistant from its vertices
(d) the right bisectors of the sides of a triangle are concurrent
(e) a parallelogram whose diagonals are equal is a rectangle
(f) a quadrilateral whose diagonals are equal is a rectangle
(g) a quadrilateral whose diagonals bisect each other at right angles is a rhombus
(h) the median to the base of an isoceles triangle is perpendicular to the base
23. Find the angle between the following pair of vectors:
(a) $\vec{a} = 3\hat{\imath} + 2\hat{\jmath} - 6\hat{k}$ and $\vec{b} = 4\hat{\imath} - 3\hat{\jmath} + \hat{k}$ *Ans:* 90°
(b) $\vec{a} = 4\hat{\imath} - 2\hat{\jmath} + 4\hat{k}$ and $\vec{b} = 3\hat{\imath} - 6\hat{\jmath} - 2\hat{k}$ *Ans:* $\cos^{-1} 8/21$
24. Find the acute angles that the line joining the points $(1, -3, 2)$ and $(3, -5, 1)$ makes with the coordinate axes. *Ans:* $\cos^{-1} 2/3, \cos^{-1} 2/3, \cos^{-1} 1/3$
25. Determine the value of the constant p, for which the vectors
(a) $\vec{a} = p\hat{\imath} - 4\hat{\jmath} + 3\hat{k}$ and $\vec{b} = 3\hat{\imath} + p\hat{\jmath} - 2\hat{k}$ are perpendicular *Ans:* -6
(b) $\vec{a} = p\hat{\imath} - 2\hat{\jmath} + \hat{k}$ and $\vec{b} = 2p\hat{\imath} + p\hat{\jmath} - 4\hat{k}$ are perpendicular *Ans:* $2, -1$
(c) $\vec{a} = 2\hat{\imath} + \hat{\jmath} + p\hat{k}$ and $\vec{b} = 2\hat{\imath} + p\hat{\jmath} + \hat{k}$ are perpendicular *Ans:* 2

(d) $\vec{a}+p\vec{b}$ and $\vec{a}-p\vec{b}$ will be perpendicular to one another, where $|\vec{a}|=3$ and $|\vec{b}|=4$ *Ans*: $\pm 3/4$

26. If $\hat{e}_1$ and $\hat{e}_2$ be two unit vectors and θ be the angle between them, then show that
(a) $2\sin\frac{1}{2}\theta=|\hat{e}_1-\hat{e}_2|$ (b) $2\cos\frac{1}{2}\theta=|\hat{e}_1+\hat{e}_2|$

27. (a) If $\vec{a}$, $\vec{b}$ and $\vec{c}$ be three non-coplanar vectors and $\vec{d}\cdot\vec{a}=\vec{d}\cdot\vec{b}=\vec{d}\cdot\vec{c}=0$, then show that $\vec{d}$ is a zero vector.
(b) If $\vec{a}$, $\vec{b}$ and $\vec{c}$ be three vectors such that $\vec{a}\times\vec{b}=\vec{c}$ and $\vec{b}\times\vec{c}=\vec{a}$, then show that the three vectors $\vec{a}$, $\vec{b}$ and $\vec{c}$ are orthogonal in pairs and $|\vec{b}|=1$, $|\vec{c}|=|\vec{a}|$.

28. Find the projection of the vector
(a) $\vec{a}=2\hat{\imath}+\hat{\jmath}-3\hat{k}$ on the vector $\vec{b}=\hat{\imath}-2\hat{\jmath}+\hat{k}$. *Ans*: $-3/\sqrt{14}$
(b) $\vec{a}+\vec{b}$ in the direction of $\vec{c}$, where $\vec{a}=2\hat{\imath}+\hat{\jmath}+\hat{k}$, $\vec{b}=\hat{\imath}-2\hat{\jmath}+3\hat{k}$ and $\vec{c}=3\hat{\imath}-4\hat{\jmath}+2\hat{k}$. *Ans*: $21/\sqrt{29}$
(c) $\vec{a}+\vec{c}$ in the direction of $\vec{b}$, where $\vec{a}=2\hat{\imath}+\hat{\jmath}+\hat{k}$, $\vec{b}=\hat{\imath}-2\hat{\jmath}+2\hat{k}$ and $\vec{c}=3\hat{\imath}-4\hat{\jmath}+2\hat{k}$. *Ans*: $17/3$
(d) $\vec{a}=2\hat{\imath}-\frac{3}{2}\hat{\jmath}+\frac{4}{5}\hat{k}$ on the vector $\vec{b}=\hat{\imath}+\sqrt{2}\hat{\jmath}+\sqrt{6}\hat{k}$.

29. If $\vec{a}=\hat{\imath}-2\hat{\jmath}$, $\vec{b}=\hat{\jmath}+\hat{k}$, find the component of $\vec{a}$ along $\vec{b}$. *Ans*: $-(\hat{\jmath}+\hat{k})$

30. Find the values of p and q such that
(a) the vector $\vec{c}-p\,\vec{a}-q\,\vec{b}$ is perpendicular to both $\vec{a}$ and $\vec{b}$, where $\vec{a}=\hat{\imath}+\hat{\jmath}-2\hat{k}$, $\vec{b}=-\hat{\imath}+2\hat{\jmath}+3\hat{k}$ and $\vec{c}=5\hat{\imath}+8\hat{k}$. *Ans*: $-1, 1$
(b) $|\vec{a}|=|\vec{b}|$ and the vectors $\vec{a}=2\hat{\imath}+\hat{\jmath}+p\hat{k}$ and $\vec{b}=3\hat{\imath}+q\hat{\jmath}-\hat{k}$ are mutually perpendicular. *Ans*: $41/12, -31/12$
(c) the vector $\vec{c}-p\vec{b}-q\vec{a}$ is orthogonal to both $\vec{a}$ and $\vec{b}$, where $\vec{a}=\hat{\imath}+\hat{\jmath}-2\hat{k}$, $\vec{b}=\hat{\imath}-2\hat{\jmath}-3\hat{k}$ and $\vec{c}=5\hat{\imath}+8\hat{k}$. *Ans*: $-1, -1$

31. (a) Given three vectors $\vec{a}=\frac{1}{7}(2,3,6)$, $\vec{b}=\frac{1}{7}(3,-6,2)$ and $\vec{c}=\frac{1}{7}(6,2,-3)$. Show that the vectors $\vec{a}$, $\vec{b}$ and $\vec{c}$ are each of unit length and that they are mutually perpendicular.
(b) Prove that the three vectors $\vec{a}=\hat{\imath}+2\hat{\jmath}+\hat{k}$ and $\vec{b}=\hat{\imath}+\hat{\jmath}-3\hat{k}$, $\vec{c}=7\hat{\imath}-4\hat{\jmath}+\hat{k}$ are at right angles to each other.

32. (a) If $\vec{a}\times\vec{b}=\vec{c}$, $\vec{b}\times\vec{c}=\vec{a}$ and $\vec{c}\times\vec{a}=\vec{b}$, then show that the vectors $\vec{a}$, $\vec{b}$ and $\vec{c}$ are mutually perpendicular.
(b) If $\vec{a}$, $\vec{b}$ and $\vec{c}$ are three mutually perpendicular unit vectors such that $\vec{a}\times\vec{b}=\vec{c}$, then prove that $\vec{b}\times\vec{c}=\vec{a}$ and $\vec{c}\times\vec{a}=\vec{b}$.

33. Let the position vectors of the vertices of a triangle be $\hat{\imath}+\hat{\jmath}+2\hat{k}$, $2\hat{\imath}+2\hat{\jmath}+3\hat{k}$ and $3\hat{\imath}-\hat{\jmath}-\hat{k}$. Find the vector and scalar area of the triangle. *Ans*: $\frac{1}{2}\sqrt{42}$ sq. units

34. (a) If a vector $\vec{a}$ be perpendicular to two non-collinear vectors $\vec{b}$ and $\vec{c}$, then show that $\vec{a}$ is perpendicular to every vector on the plane of $\vec{b}$ and $\vec{c}$.
(b) Prove that the necessary and sufficient condition for parallelism of two proper vectors $\vec{a}$ and $\vec{b}$ is $\vec{a}\times\vec{b}=\vec{0}$.

35. Find in each case a unit vector $\vec{c}$ perpendicular to $\vec{a}$ and coplanar with $\vec{a}$ and $\vec{b}$.

36. (a) If $\vec{a}$ and $\vec{b}$ be two vectors such that $|\vec{a}| = 4$, $|\vec{b}| = 5$ and the angle between $\vec{a}$ and $\vec{b}$ be $60°$, find the value of $|\vec{a} \times \vec{b}|$. *Ans*: $10\sqrt{3}$
(b) If $\vec{a}$ and $\vec{b}$ be two mutually perpendicular vectors, prove that $|\vec{a} \times \vec{b}| = |\vec{a}|\ |\vec{b}|$.
37. Find two mutually perpendicular vectors of equal magnitude, each having an x component equal to 5 and each perpendicular to the vector $3\hat{j} + 4\hat{k}$. *Ans*: $5\hat{i} \pm 4\hat{j} \mp 3\hat{k}$
38. Find the scalar area of the square, of which $\vec{r} = a_1\hat{i} + a_2\hat{j}$ is a semi-diagonal. *Ans*: $2(a_1^2 + a_2^2)$ sq. units
39. (a) Show that $(\vec{a} - \vec{b}) \times (\vec{a} + \vec{b}) = 2\vec{a} \times \vec{b}$. Interpret the result for two non-collinear vectors $\vec{a}$ and $\vec{b}$.
(b) Verify $(\vec{a} - \vec{b}) \times (\vec{a} + \vec{b}) = 2\vec{a} \times \vec{b}$, when $\vec{a} = 2\hat{i} + 3\hat{j} + \hat{k}$ and $\vec{b} = 3\hat{i} - \hat{j} + 2\hat{k}$.
40. If $\vec{a} = 3\hat{i} - \hat{j} + 2\hat{k}$, $\vec{b} = 2\hat{i} + \hat{j} - \hat{k}$ and $\vec{c} = \hat{i} - 2\hat{j} + 2\hat{k}$, then show that
(a) $(\vec{a} \times \vec{b}) \times \vec{c} \neq \vec{a} \times (\vec{b} \times \vec{c})$ (b) $(\vec{a} + \vec{b}) \times \vec{c} = \vec{a} \times \vec{c} + \vec{b} \times \vec{c}$
(c) $\vec{a} \times (\vec{b} + \vec{c}) = \vec{a} \times \vec{b} + \vec{a} \times \vec{c}$
41. If $\vec{a}, \vec{b}, \vec{c}$ and $\vec{d}$ be four distinct vectors such that $\vec{a} \times \vec{b} = \vec{c} \times \vec{d}$ and $\vec{a} \times \vec{c} = \vec{b} \times \vec{d}$, then show that the vectors $\vec{a} - \vec{d}$ and $\vec{b} - \vec{c}$ are collinear.
42. The vertices of a quadrilateral are, respectively, given by $3\hat{i} - 4\hat{j} + \hat{k}$, $4\hat{i} + 3\hat{j} - \hat{k}$, $2\hat{i} + 3\hat{j} - \hat{k}$ and $3\hat{i} + 2\hat{j} - 4\hat{k}$. Show that the diagonals are at right angles.
43. Evaluate $\left[\vec{a}\,\vec{b}\,\vec{c}\right]$ for the following cases:
(a) $\vec{a} = 2\hat{i} - 3\hat{j} + \hat{k}$, $\vec{b} = \hat{i} - \hat{j} + 2\hat{k}$ and $\vec{c} = 2\hat{i} + \hat{j} + \hat{k}$ *Ans*: -12
(b) $\vec{a} = 2\hat{i} - 4\hat{j} + 5\hat{k}$, $\vec{b} = \hat{i} - \hat{j} + \hat{k}$ and $\vec{c} = 3\hat{i} - 5\hat{j} + 2\hat{k}$ *Ans*: -8
Verify in each case whether $\vec{a} \cdot (\vec{b} \times \vec{c}) = (\vec{a} \times \vec{b}) \cdot \vec{c}$
44. Show that $\left[x\vec{a} + y\vec{b}\ \vec{c}\ \vec{d}\right] = x\left[\vec{a}\,\vec{c}\,\vec{d}\right] + y\left[\vec{b}\,\vec{c}\,\vec{d}\right]$.
45. Determine the unit vectors which make an angle of $60°$ with $\hat{i} - \hat{j}$ and an angle $60°$ with $\hat{i} + \hat{k}$. *Ans*: $(0, -1/\sqrt{2}, 1/\sqrt{2})$
46. Evaluate $\vec{a} \times (\vec{b} \times \vec{c})$ for $2\hat{i} - 10\hat{j} + 2\hat{k}$, $3\hat{i} + \hat{j} + 2\hat{k}$ and $2\hat{i} + \hat{j} + 3\hat{k}$ and interpret your result geometrically.
47. For any three vectors $\vec{a}, \vec{b}$ and $\vec{c}$, prove that
(a) $\vec{a} \times (\vec{b} \times \vec{c})$, $\vec{b} \times (\vec{c} \times \vec{a})$ and $\vec{c} \times (\vec{a} \times \vec{b})$ are coplanar.
(b) $\vec{a} \times (\vec{b} \times \vec{c}) + \vec{b} \times (\vec{c} \times \vec{a}) + \vec{c} \times (\vec{a} \times \vec{b}) = \vec{0}$.
48. (a) If $\vec{a}, \vec{b}, \vec{c}$ and $\vec{d}$ are coplanar vectors, show that $(\vec{a} \times \vec{b}) \times (\vec{c} \times \vec{d}) = \vec{0}$.
(b) If $\vec{a}$ and $\vec{b}$ lie in a plane normal to the plane containing $\vec{c}$ and $\vec{d}$, then show that $(\vec{a} \times \vec{b}) \cdot (\vec{c} \times \vec{d}) = 0$.
(c) If $\vec{a}, \vec{b}$ and $\vec{c}$ be three mutually perpendicular vectors, show that $\vec{a}\,\vec{b}\,\vec{c}^2 = |\vec{a}|^2\ |\vec{b}|^2\ |\vec{c}|^2$.
(d) If $\vec{a}, \vec{b}$ and $\vec{c}$ be three unit vectors such that $\vec{b}$ and $\vec{c}$ are non-parallel and $2\vec{a} \times (\vec{b} \times \vec{c}) = \vec{b}$, find the angles which $\vec{a}$ makes with $\vec{b}$ and $\vec{c}$. *Ans*: $\pi/2, \pi/3$
(e) If $\vec{a}$ and $\vec{c}$ are perpendicular, then show that the vectors $\vec{a} \times (\vec{b} \times \vec{c})$ and $(\vec{a} \times \vec{b}) \times \vec{c}$ are also perpendicular.
49. (a) Show that the vector $\vec{a} + \vec{b} + \vec{c}$ is perpendicular to the plane through the extremities of the vectors $\vec{b} \times \vec{c}$, $\vec{c} \times \vec{a}$ and $\vec{a} \times \vec{b}$.

(b) If the three vectors $\vec{a}, \vec{b}$ and $\vec{c}$ are non-coplanar, then prove that $\vec{b} \times \vec{c}, \vec{c} \times \vec{a}$ and $\vec{a} \times \vec{b}$ are also non-coplanar.

50. If $\vec{a}, \vec{b}$ and $\vec{c}$ and $\vec{a}', \vec{b}'$ and $\vec{c}'$ be two reciprocal system of vectors, prove that
(a) $\vec{a} \cdot \vec{a}' + \vec{b} \cdot \vec{b}' + \vec{c} \cdot \vec{c}' = 3$ (b) $\vec{a} \times \vec{a}' + \vec{b} \times \vec{b}' + \vec{c} \times \vec{c}' = \vec{0}$
(c) $\left[\vec{a}'\, \vec{b}'\, \vec{c}'\right]\left[\vec{a}\, \vec{b}\, \vec{c}\right] = 1$ (d) $\vec{a}' \times \vec{b}' + \vec{b}' \times \vec{c}' + \vec{c}' \times \vec{a}' = \dfrac{1}{\left[\vec{a}\, \vec{b}\, \vec{c}\right]}\{\vec{a} + \vec{b} + \vec{c}\}$, $\left[\vec{a}\, \vec{b}\, \vec{c}\right] \neq 0$
(e) $\left[\vec{b} \times \vec{c}\, \vec{c} \times \vec{a}\, \vec{a} \times \vec{b}\right] = \left[\vec{a}\, \vec{b}\, \vec{c}\right]^3 \left[\vec{a}'\, \vec{b}'\, \vec{c}'\right]$.
(f) $\left[\vec{a} \times \vec{a}'\, \vec{b} \times \vec{b}'\, \vec{c} \times \vec{c}'\right] + \left[\vec{a} \times \vec{b}'\, \vec{b} \times \vec{c}'\, \vec{c} \times \vec{a}'\right] + \left[\vec{a} \times \vec{c}'\, \vec{b} \times \vec{a}'\, \vec{c} \times \vec{b}'\right] = 0$.

51. Find the system of vectors reciprocal to $\vec{a}$ and $\vec{b}$ and $\vec{a} \times \vec{b}$.

52. Show that the orthogonal vector triad $\hat{i}, \hat{j}$ and $\hat{k}$ form a self-reciprocal system.

53. Prove that
(a) the only right-handed self-reciprocal set of vectors comprises the unit vectors $\hat{i}, \hat{j}$ and $\hat{k}$.
(b) there is one and only one set of vectors reciprocal to the given set of non-coplanar vectors $\vec{a}, \vec{b}$ and $\vec{c}$.

Long Answer Questions

1. Find the vector of
(a) magnitude 5 perpendicular to both the vectors $2\hat{i} + \hat{j} - 3\hat{k}$ and $\hat{i} - 2\hat{j} + \hat{k}$.
Ans: $\pm\frac{5\sqrt{3}}{3}(\hat{i} + \hat{j} + \hat{k})$
(b) magnitude 11 perpendicular to the plane of vectors $2\hat{i} - \hat{j} + \hat{k}$ and $\hat{i} - 2\hat{j} - \hat{k}$.
Ans: $\pm\frac{11\sqrt{3}}{3}(\hat{i} + \hat{j} - \hat{k})$

2. If $\vec{a} + \vec{b} + \vec{c} = \vec{0}$ and $|\vec{a}| = 7$, $|\vec{b}| = 3$, $|\vec{c}| = 5$, find the angle between the directions of $\vec{b}$ and $\vec{c}$.
Ans: 60°

3. Let $\vec{a}, \vec{b}$ and $\vec{c}$ be three vectors whose lengths are 3, 4 and 5 units, respectively. If $\vec{a}$ is perpendicular to $\vec{b} + \vec{c}$, $\vec{b}$ is perpendicular to $\vec{c} + \vec{a}$ and $\vec{c}$ is perpendicular to $\vec{a} + \vec{b}$, find $|\vec{a} + \vec{b} + \vec{c}|$.
Ans: $5\sqrt{2}$

4. Find the unit vector $\vec{c}$ which is perpendicular to each of the unit vectors $\vec{a} = \frac{1}{\sqrt{3}}\left(\hat{i} + \hat{j} + \hat{k}\right)$ and $\vec{b} = \frac{1}{\sqrt{14}}\left(\hat{i} - 3\hat{j} + 2\hat{k}\right)$. Verify that $\vec{b} \times \vec{c} = \vec{a}$ and $\vec{c} \times \vec{a} = \vec{b}$.

5. Show that the angle between
(a) two vectors $\hat{i} - 8\hat{j} - 6\hat{k}$ and $-2\hat{i} + 3\hat{j} + \hat{k}$ is $\theta = \cos^{-1}\left(\frac{-32}{\sqrt{1414}}\right)$.
(b) OP and OQ, where the position vectors of P and Q are, respectively, $3\hat{i} + 7\hat{j} - 4\hat{k}$ and $6\hat{i} - 2\hat{j} + 12\hat{k}$ with respect to the origin O is $\theta = \cos^{-1}\left(\frac{-11}{\sqrt{851}}\right)$.

6. Find the perpendicular distance of a point whose position vector is $\vec{a}$ from the plane through three points with position vectors $\vec{b}, \vec{c}$ and $\vec{d}$.

7. Show that the unit vector
(a) parallel to the xy plane and perpendicular to the vector $4\hat{i} - 3\hat{j} + \hat{k}$ is $\dfrac{1}{5}(3\hat{i} + 4\hat{j})$.
(b) parallel to the xy plane and perpendicular to the vector $2\hat{i} + 3\hat{j} + 4\hat{k}$ is $\pm\dfrac{1}{\sqrt{13}}(2\hat{i} - 3\hat{j})$.

(c) coplanar with of the vectors $\vec{\alpha} = \hat{\imath} + \hat{\jmath} + 2\hat{k}$ and $\vec{\beta} = \hat{\imath} + 2\hat{\jmath} + \hat{k}$ and perpendicular to the vector $\vec{\gamma} = \hat{\imath} + \hat{\jmath} + \hat{k}$ is $\frac{1}{\sqrt{2}}\left(\hat{\jmath} - \hat{k}\right)$.

8. (a) The vectors join the origin to the points $A(2, -1, 1)$, $B(3, 0, 1)$ and $C(1, -2, 3)$. Find a unit vector perpendicular to the plane. *Ans:* $\pm\frac{\sqrt{2}}{2}(\hat{\imath} - \hat{\jmath})$

(b) If the position vectors of the three points A, B and C be $\hat{\imath} + \hat{\jmath} + \hat{k}$, $2\hat{\imath} + 3\hat{\jmath} + \hat{k}$ and $3\hat{\imath} - \hat{\jmath} + 4\hat{k}$, respectively, find a vector perpendicular to the plane ABC. *Ans:* $\pm 3(2\hat{\imath} - \hat{\jmath} - 2\hat{k})$

9. Consider a vector $\vec{a} = 5\hat{\imath} + 6\hat{\jmath} + 5\hat{k}$. Deduce a vector $\vec{b}$ which lies on the xy plane and is perpendicular to $\vec{a}$. Deduce also the unit vector which is normal to both $\vec{a}$ and $\vec{b}$.
Ans: $\pm(6\hat{\imath} - 5\hat{\jmath})/\sqrt{61}$; $\pm(25\hat{\imath} + 30\hat{\jmath} - 61\hat{k})/\sqrt{5246}$

10. Deduce a vector $\vec{b}$ which lies on the xy plane and is perpendicular to $\vec{a}$. Deduce also the unit vector which is normal to both $\vec{a}$ and $\vec{b}$.

11. Prove that a necessary and sufficient condition that $\vec{a} \times (\vec{b} \times \vec{c}) = (\vec{a} \times \vec{b}) \times \vec{c}$ is $(\vec{a} \times \vec{c}) \times \vec{b} = \vec{0}$. Discuss the cases where $\vec{a} \cdot \vec{b} = 0$ or $\vec{b} \cdot \vec{c} = 0$.

12. (a) A vector $\vec{\delta}$ is perpendicular to both $\vec{\alpha} = 4\hat{\imath} + 5\hat{\jmath} - \hat{k}$ and $\vec{\beta} = \hat{\imath} - 4\hat{\jmath} + 5\hat{k}$ and satisfies $\vec{\delta} \cdot \vec{\gamma} = 2$, where $\vec{\gamma} = 3\hat{\imath} + \hat{\jmath} - \hat{k}$. Find $\vec{\delta}$. *Ans:* $\frac{2}{3}(\hat{\imath} - \hat{\jmath} - \hat{k})$

(b) If $\vec{a} = \hat{\imath} + \hat{\jmath} + \hat{k}$ and $\vec{c} = \hat{\jmath} - \hat{k}$ be two vectors and the vector $\vec{b}$ satisfies $\vec{a} \times \vec{b} = \vec{c}$ and $\vec{a} \cdot \vec{b} = 3$, find $\vec{b}$. *Ans:* $\frac{1}{3}(5\hat{\imath} + 2\hat{\jmath} + 2\hat{k})$

(c) If $\vec{a} = \hat{\imath} + 3\hat{\jmath} + 3\hat{k}$, $\vec{b} = 4\hat{\imath} + 5\hat{\jmath} + \hat{k}$ and $\vec{c} = 3\hat{\imath} + 3\hat{\jmath} + \hat{k}$ be three vectors and the vector $\vec{d}$ satisfies $\vec{a} \cdot \vec{d} = 5$ and $\vec{d} \times \vec{b} = \vec{c}$, find $\vec{d}$. *Ans:* $\frac{1}{22}(14\hat{\imath} + 33\hat{\jmath} - \hat{k})$

13. (a) If $\vec{a}$ and $\vec{b}$ are two non-collinear vectors such that $\vec{a} = \vec{c} + \vec{d}$, where $\vec{c}$ is a vector parallel to $\vec{b}$ and $\vec{d}$ is a vector perpendicular to $\vec{b}$. Show that the expression for $\vec{c}$ and $\vec{d}$ in terms of $\vec{a}$ and $\vec{b}$ are $\vec{c} = \frac{\vec{a} \cdot \vec{b}}{\left|\vec{b}\right|^2}\vec{b}$ and $\vec{d} = \vec{a} - \frac{\vec{a} \cdot \vec{b}}{\left|\vec{b}\right|^2}\vec{b}$.

(b) Express $\vec{b}$ as a sum of two vectors such that one part is parallel to $\vec{a}$ and another part is perpendicular to $\vec{a}$. *Ans:* $\frac{\vec{b} \cdot \vec{a}}{\vec{a} \cdot \vec{a}}\vec{a}$, $\frac{1}{\vec{a} \cdot \vec{a}}\vec{a} \times (\vec{b} \times \vec{a})$

(c) If $\vec{r}$ be resolved into two components, one in the direction of unit vector $\hat{e}$ and the other perpendicular to $\hat{e}$, then find the components. *Ans:* $(\vec{r} \cdot \hat{e})\hat{e}$, $\vec{r} - (\vec{r} \cdot \hat{e})\hat{e}$,

14. Find the area of a triangle ABC by the vector method
(a) where $\vec{a}$, $\vec{b}$ and $\vec{c}$ are the position vectors of A, B and C, respectively.
(b) whose three vertices are $A(3, -1, 2)$, $B(1, -1, -3)$ and $C(4, -3, 1)$. *Ans:* $\sqrt{165}/2$

15. Show that the vector area of the plane quadrilateral $ABCD$ is $\frac{1}{2}(\vec{AC} \times \vec{BD})$. Show further that if $\vec{AB} = \vec{b}$, $\vec{AD} = \vec{d}$ and $\vec{AC} = m\vec{b} + n\vec{d}$, then the vector area of $ABCD$ is $\frac{1}{2}(m + n)(\vec{b} + \vec{d})$.

16. Find the area of a parallelogram by the vector method
(a) having $\vec{\alpha} = \hat{\imath} + 2\hat{\jmath} + 3\hat{k}$ and $\vec{\beta} = 3\hat{\imath} - 2\hat{\jmath} + \hat{k}$ as its diagonals. *Ans:* $-(\hat{\imath} + \hat{\jmath} - \hat{k})$
(b) whose three vertices are $(1, 2, 3)$, $(-2, 1, -4)$ and $(3, 4, -2)$.

17. Show that the parallelogram, whose diagonals are $\vec{\alpha} = 2\hat{\imath} + 3\hat{\jmath} - 6\hat{k}$ and $\vec{\beta} = 3\hat{\imath} - 4\hat{\jmath} - \hat{k}$ is a rhombus, each side being length $5\sqrt{3}/2$ units.
18. Prove by the vector method that in any triangle ABC,
(a) $\dfrac{\sin A}{a} = \dfrac{\sin B}{b} = \dfrac{\sin C}{c}$ (b) $a\cos B - b\cos A = \dfrac{a^2 - b^2}{c}$ (c) $\cos C = \dfrac{a^2 + b^2 - c^2}{2ab}$
(d) If D is the midpoint of BC, then $AB^2 + AC^2 = 2(AD^2 + BD^2)$.
19. Find the volume of the parallelepiped,
(a) whose edges are represented by the vectors $3\hat{\imath} + 2\hat{\jmath} - 4\hat{k}$, $3\hat{\imath} + \hat{\jmath} + 3\hat{k}$ and $\hat{\imath} - 2\hat{\jmath} + \hat{k}$.
Ans: 49 cu. units
(b) whose edges are represented by the vectors $-3\hat{\imath} + 7\hat{\jmath} + 5\hat{k}$, $5\hat{\imath} + 7\hat{\jmath} - 3\hat{k}$ and $7\hat{\imath} - 5\hat{\jmath} - 3\hat{k}$.
Ans: 304 cu. units
20. (a) Using the vector method, find the volume of the tetrahedron with vertices at $(0, 0, 0)$, $(-1, 1, 1)$, $(1, -1, 1)$ and $(1, 1, -1)$.
Ans: 2/3 cu. units
(b) Given $A(1, 3, -1), B(0, 1, 6)$ and $C(-1, 3, 1)$ are three points in space. Find the coordinates of a point D on the y axis such that volume of the tetrahedron $ABCD$ is 8 units.
Ans: $(0, 7, 0)$ or $(0, -1, 0)$.
(c) Given $A(3, 2, 1), B(-2, 0, -3)$ and $C(0, 0, -2)$ are three points. Find the locus of the variable point D, if the volume of the tetrahedron $ABCD$ is 5 units.
Ans: $2x + 3y - 4z = 38$, $2x + 3y - 4z + 22 = 0$.
(d) $PQRS$ is a tetrahedron with $(-5, -4, 8)$, $(2, 3, 1)$, $(4, 1, 2)$, $(6, 3, 7)$ as the coordinates of P, Q, R and S, respectively. Show that the distance of the point P from the plane determined by Q, R and S is $49/\sqrt{17}$.
21. Show by the vector method that
(a) the area of the triangle formed by joining the midpoint of one of the non-parallel sides of a trapezium to the extremities of the opposite side is half that of the trapezium.
(b) if D is the midpoint of the sides BC of the triangle ABC, then $AB^2 + AC^2 = 2(AD^2 + BD^2)$.
(c) the circumcentres are centroid and the orthocentre of the triangle are collinear and the centroid divides the join of the circumcentre and the orthocentre in the ratio 1 : 2.
22. If two pairs of opposite edges of a tetrahedron are at right angles, then show that the third pair is also at a right angle. Further, show that for such a tetrahedron, the sum of the squares of each pair of opposite edges is the same.
23. If $\vec{a}$ and $\vec{b}$ be any two vectors with $a = |\vec{a}|$ and $b = |\vec{b}|$, show that
(a) $\left(\vec{a} \times \vec{b}\right)^2 = |\vec{a}|^2|\vec{b}|^2 - \left(\vec{a} \cdot \vec{b}\right)^2$
(b) $\left|\vec{a} \times \vec{b}\right|^2 \; |\vec{a} \times \vec{c}|^2 - \left\{(\vec{a} \times \vec{b}) \cdot (\vec{a} \times \vec{c})\right\}^2 = |\vec{a}|^2 \left[\vec{a}\,\vec{b}\,\vec{c}\right]^2$
(c) $\left[\vec{b} \times \vec{c} \;\; \vec{c} \times \vec{a} \;\; \vec{a} \times \vec{b}\right] = \left[\vec{a} \;\; \vec{b} \;\; \vec{c}\right]^2$
(d) $(1 + a^2)(1 + b^2) = |1 - \vec{a} \cdot \vec{b}|^2 + |\vec{a} + \vec{b} + \vec{a} \times \vec{b}|^2$.
24. Show that the three vector products
(a) $\vec{\alpha} \times \vec{\beta}$, $\vec{\beta} \times \vec{\gamma}$ and $\vec{\gamma} \times \vec{\alpha}$ are equal when $\vec{\alpha} + \vec{\beta} + \vec{\gamma} = \vec{0}$.
(b) $\vec{\alpha} \times \vec{\beta}$, $\vec{\beta} \times \vec{\gamma}$ and $\vec{\gamma} \times \vec{\alpha}$ are coplanar, $\vec{\alpha}, \vec{\beta}$ and $\vec{\gamma}$ are coplanar.
25. Establish the following identities:

(a) $2\vec{\alpha} = \hat{i} \times (\vec{\alpha} \times \hat{i}) + \hat{j} \times (\vec{\alpha} \times \hat{j}) + \hat{k} \times \left(\vec{\alpha} \times \hat{k}\right)$, where $\vec{\alpha}$ = any vector.

(b) $\vec{a} \times \vec{b} = \{(\hat{i} \times \vec{a}) \cdot \vec{b}\}\hat{i} + \{(\hat{j} \times \vec{a}) \cdot \vec{b}\}\hat{j} + \{(\hat{k} \times \vec{a}) \cdot \vec{b}\}\hat{k}$.

(c) $\left[\vec{p}\,\vec{q}\,\vec{r}\right]\left[\vec{a}\,\vec{b}\,\vec{c}\right] = \begin{vmatrix} \vec{p}\cdot\vec{a} & \vec{p}\cdot\vec{b} & \vec{p}\cdot\vec{c} \\ \vec{q}\cdot\vec{a} & \vec{q}\cdot\vec{b} & \vec{q}\cdot\vec{c} \\ \vec{r}\cdot\vec{a} & \vec{r}\cdot\vec{b} & \vec{r}\cdot\vec{c} \end{vmatrix}$.

26. Prove by vector method that
(a) $\cos(\alpha - \beta) = \cos\alpha\ \cos\beta + \sin\alpha\ \sin\beta$
(b) $\sin(\alpha - \beta) = \sin\alpha\ \cos\beta - \cos\alpha\ \sin\beta$
(c) $\cos(\alpha - \beta)\cos(\alpha + \beta) = \cos^2\alpha - \sin^2\beta$

27. Prove that the distributive law $\vec{a} \times \left(\vec{b} + \vec{c}\right) = \vec{a} \times \vec{b} + \vec{a} \times \vec{c}$ for a vector product. Does it follow from $\vec{a} \times \vec{b} = \vec{a} \times \vec{c}$ that $\vec{b} = \vec{c}$? Give reasons for your answer.

28. Reduce the expression $(\vec{b} + \vec{c}) \cdot \{(\vec{c} + \vec{a}) \times (\vec{a} + \vec{b})\}$ to its simplest form and prove that it vanishes when $\vec{a}, \vec{b}$ and $\vec{c}$ are also non-coplanar.
Conversely, if the three vectors $\vec{a}, \vec{b}$ and $\vec{c}$ be non-coplanar, then prove that $(\vec{b} + \vec{c})$, $(\vec{c} + \vec{a})$, $(\vec{a} + \vec{b})$ are also non-coplanar.

29. If $\vec{\alpha}, \vec{\beta}$ and $\vec{\gamma}$ are the position vectors of the three distinct points A, B and C, respectively, and satisfy the relation $\vec{\alpha} \times \vec{\beta} + \vec{\beta} \times \vec{\gamma} + \vec{\gamma} \times \vec{\alpha} = \vec{0}$, show that A, B and C lie on the same straight line.

30. Prove that
(a) $\left[\vec{b} \times \vec{c}\ \ \vec{c} \times \vec{a}\ \ \vec{a} \times \vec{b}\right] = \left[\vec{a}\,\vec{b}\,\vec{c}\right]^2$
(b) $\left[\vec{a} \times \vec{b}\ \ \vec{c} \times \vec{d}\ \ \vec{e} \times \vec{f}\right] = \left[\vec{a}\,\vec{b}\,\vec{d}\right]\left[\vec{c}\,\vec{e}\,\vec{f}\right] - \left[\vec{a}\,\vec{b}\,\vec{c}\right]\left[\vec{d}\,\vec{e}\,\vec{f}\right]$
(c) $\left[\vec{p}\,\vec{q}\,\vec{r}\right](\vec{a} \times \vec{b}) = \begin{vmatrix} \vec{p}\cdot\vec{a} & \vec{p}\cdot\vec{b} & \vec{p} \\ \vec{q}\cdot\vec{a} & \vec{q}\cdot\vec{b} & \vec{q} \\ \vec{r}\cdot\vec{a} & \vec{r}\cdot\vec{b} & \vec{r} \end{vmatrix}$, where $\vec{p}, \vec{q}$ and $\vec{r}$ and $\vec{a}, \vec{b}$ and $\vec{c}$ are two sets of non-coplanar vectors
(d) For any three vectors $\vec{a}, \vec{b}$ and $\vec{c}$, prove that

$$\left[\vec{a}\,\vec{b}\,\vec{c}\right]^2 = \left\{\vec{a} \cdot \left(\vec{b} \times \vec{c}\right)\right\}^2 = \begin{vmatrix} \vec{a}\cdot\vec{a} & \vec{a}\cdot\vec{b} & \vec{a}\cdot\vec{c} \\ \vec{a}\cdot\vec{b} & \vec{b}\cdot\vec{b} & \vec{b}\cdot\vec{c} \\ \vec{a}\cdot\vec{c} & \vec{b}\cdot\vec{c} & \vec{c}\cdot\vec{c} \end{vmatrix}$$

(e) $(\vec{a} \times \vec{b}) \times (\vec{c} \times \vec{d}) + (\vec{a} \times \vec{d}) \times (\vec{b} \times \vec{c}) + (\vec{a} \times \vec{c}) \times (\vec{d} \times \vec{b}) = -2\left[\vec{b}\,\vec{c}\,\vec{d}\right]\vec{a}$
(f) $\vec{a} \times \{\vec{b} \times (\vec{c} \times \vec{d})\} = (\vec{b} \cdot \vec{d})\vec{a} \times \vec{c} - (\vec{b} \cdot \vec{c})\vec{a} \times \vec{d}$
(g) $(\vec{a} \times \vec{b}) \times (\vec{a} \times \vec{c}) \cdot \vec{d} = (\vec{a} \cdot \vec{d})\ \left[\vec{a}\,\vec{b}\,\vec{c}\right]$ (h) $\vec{a} \times \{\vec{a} \times (\vec{a} \times \vec{b})\} = (\vec{a} \cdot \vec{a})\vec{b} \times \vec{a}$

31. Show that the four points with position vectors $\vec{a}, \vec{b}, \vec{c}$ and $\vec{d}$ are coplanar if and only if

$$\left[\vec{a}\,\vec{b}\,\vec{c}\right] = \left[\vec{b}\,\vec{c}\,\vec{d}\right] + \left[\vec{c}\,\vec{a}\,\vec{d}\right] + \left[\vec{a}\,\vec{b}\,\vec{d}\right]$$

32. Show that if $\vec{a} = a_1\hat{l} + a_2\hat{m} + a_3\hat{n}$, $\vec{b} = b_1\hat{l} + b_2\hat{m} + b_3\hat{n}$ and $\vec{c} = c_1\hat{l} + c_2\hat{m} + c_3\hat{n}$, where $\hat{l}, \hat{m}$ and $\hat{n}$ are three non-coplanar vectors, then

$$\left[\vec{a}\,\vec{b}\,\vec{c}\right] = \begin{vmatrix} a_1 & a_2 & a_3 \\ b_1 & b_2 & b_3 \\ c_1 & c_2 & c_3 \end{vmatrix} \left[\hat{l}\,\hat{m}\,\hat{n}\right]$$

33. Show that if $\vec{A} = x_1\vec{a} + y_1\vec{b} + z_1\vec{c}$, $\vec{B} = x_2\vec{a} + y_2\vec{b} + z_2\vec{c}$ and $\vec{C} = x_3\vec{a} + y_3\vec{b} + z_3\vec{c}$, prove that

$$\left[\vec{A}\,\vec{B}\,\vec{C}\right] = \vec{A}\cdot\vec{B}\times\vec{C} \begin{vmatrix} x_1 & y_1 & z_1 \\ x_2 & y_2 & z_2 \\ x_3 & y_3 & z_3 \end{vmatrix} \left(\vec{a}\cdot\vec{b}\times\vec{c}\right)$$

34. Show that four vectors $\vec{a}, \vec{b}, \vec{c}$ and $\vec{d}$

$$\left(\vec{a}\times\vec{b}\right)\times\left(\vec{c}\times\vec{d}\right) = \left[\vec{a}\,\vec{b}\,\vec{d}\right]\vec{c} - \left[\vec{a}\,\vec{b}\,\vec{c}\right]\vec{d} = \left[\vec{a}\,\vec{c}\,\vec{d}\right]\vec{b} - \left[\vec{b}\,\vec{c}\,\vec{d}\right]\vec{a}.$$

Hence deduce that $\left[\vec{a}\,\vec{b}\,\vec{c}\right]\vec{d} = \left[\vec{b}\,\vec{c}\,\vec{d}\right]\vec{a} + \left[\vec{c}\,\vec{a}\,\vec{d}\right]\vec{b} + \left[\vec{a}\,\vec{b}\,\vec{d}\right]\vec{c}$.

35. (a) If $\vec{a}, \vec{b}$ and $\vec{c}$ are three non-coplanar vectors, any vector $\vec{d}$ can be expressed as

i. $\vec{d} = \dfrac{\vec{a}\cdot\vec{d}}{\left[\vec{a}\,\vec{b}\,\vec{c}\right]}(\vec{b}\times\vec{c}) + \dfrac{\vec{b}\cdot\vec{d}}{\left[\vec{a}\,\vec{b}\,\vec{c}\right]}(\vec{c}\times\vec{a}) + \dfrac{\vec{c}\cdot\vec{d}}{\left[\vec{a}\,\vec{b}\,\vec{c}\right]}(\vec{a}\times\vec{b})$

ii. $\vec{d} = \dfrac{1}{\left[\vec{a}\,\vec{b}\,\vec{c}\right]}\left\{\left[\vec{b}\,\vec{c}\,\vec{d}\right]\vec{a} - \left[\vec{a}\,\vec{c}\,\vec{d}\right]\vec{b} + \left[\vec{a}\,\vec{b}\,\vec{d}\right]\vec{c}\right\}$

(b) Show that three non-coplanar vectors $\vec{a}, \vec{b}$ and $\vec{c}$ can be expressed as

i. $\vec{a} = \dfrac{\vec{a}\cdot\vec{a}}{\left[\vec{a}\,\vec{b}\,\vec{c}\right]}(\vec{b}\times\vec{c}) + \dfrac{\vec{a}\cdot\vec{b}}{\left[\vec{a}\,\vec{b}\,\vec{c}\right]}(\vec{c}\times\vec{a}) + \dfrac{\vec{a}\cdot\vec{c}}{\left[\vec{a}\,\vec{b}\,\vec{c}\right]}(\vec{a}\times\vec{b})$

ii. $\vec{b} = \dfrac{\vec{b}\cdot\vec{b}}{\left[\vec{b}\,\vec{a}\,\vec{c}\right]}(\vec{a}\times\vec{c}) + \dfrac{\vec{b}\cdot\vec{a}}{\left[\vec{b}\,\vec{a}\,\vec{c}\right]}(\vec{c}\times\vec{b}) + \dfrac{\vec{b}\cdot\vec{c}}{\left[\vec{b}\,\vec{a}\,\vec{c}\right]}(\vec{b}\times\vec{a})$

iii. $\vec{c} = \dfrac{\vec{c}\cdot\vec{c}}{\left[\vec{c}\,\vec{b}\,\vec{a}\right]}(\vec{b}\times\vec{a}) + \dfrac{\vec{c}\cdot\vec{b}}{\left[\vec{c}\,\vec{b}\,\vec{a}\right]}(\vec{a}\times\vec{c}) + \dfrac{\vec{c}\cdot\vec{a}}{\left[\vec{c}\,\vec{b}\,\vec{a}\right]}(\vec{c}\times\vec{b})$

(c) Show that three non-coplanar vectors $\vec{a}\times\vec{b}$, $\vec{b}\times\vec{c}$ and $\vec{c}\times\vec{a}$ can be expressed as

i. $\vec{b}\times\vec{c} = \dfrac{(\vec{b}\times\vec{c})\cdot(\vec{b}\times\vec{c})}{\left[\vec{a}\,\vec{b}\,\vec{c}\right]}\vec{a} + \dfrac{(\vec{b}\times\vec{c})\cdot(\vec{c}\times\vec{a})}{\left[\vec{a}\,\vec{b}\,\vec{c}\right]}\vec{b} + \dfrac{(\vec{b}\times\vec{c})\cdot(\vec{a}\times\vec{b})}{\left[\vec{a}\,\vec{b}\,\vec{c}\right]}\vec{c}$

ii. $\vec{c}\times\vec{a} = \dfrac{(\vec{c}\times\vec{a})\cdot(\vec{c}\times\vec{a})}{\left[\vec{a}\,\vec{b}\,\vec{c}\right]}\vec{a} + \dfrac{(\vec{b}\times\vec{c})\cdot(\vec{c}\times\vec{a})}{\left[\vec{a}\,\vec{b}\,\vec{c}\right]}\vec{b} + \dfrac{(\vec{c}\times\vec{a})\cdot(\vec{a}\times\vec{b})}{\left[\vec{a}\,\vec{b}\,\vec{c}\right]}\vec{c}$

iii. $\vec{a}\times\vec{b} = \dfrac{(\vec{a}\times\vec{b})\cdot(\vec{a}\times\vec{b})}{\left[\vec{a}\,\vec{b}\,\vec{c}\right]}\vec{a} + \dfrac{(\vec{a}\times\vec{b})\cdot(\vec{c}\times\vec{a})}{\left[\vec{a}\,\vec{b}\,\vec{c}\right]}\vec{b} + \dfrac{(\vec{b}\times\vec{c})\cdot(\vec{a}\times\vec{b})}{\left[\vec{a}\,\vec{b}\,\vec{c}\right]}\vec{c}$

36. Given that $\vec{a} = \frac{1}{\left[\vec{a}'\,\vec{b}'\,\vec{c}'\right]}(\vec{b}' \times \vec{c}'),\ \vec{b} = \frac{1}{\left[\vec{a}'\,\vec{b}'\,\vec{c}'\right]}(\vec{c}' \times \vec{a}')$ and $\vec{c} = \frac{1}{\left[\vec{a}'\,\vec{b}'\,\vec{c}'\right]}(\vec{a}' \times \vec{b}')$, where $\vec{a}', \vec{b}'$ and $\vec{c}'$ are three non-coplanar vectors. Show that if $\left[\vec{a}'\,\vec{b}'\,\vec{c}'\right] = V$, then $\left[\vec{a}\,\vec{b}\,\vec{c}\right] = 1/V$.

37. Suppose $\vec{a}' = \frac{1}{\left[\vec{a}\,\vec{b}\,\vec{c}\right]}(\vec{b} \times \vec{c}),\ \vec{b}' = \frac{1}{\left[\vec{a}\,\vec{b}\,\vec{c}\right]}(\vec{c} \times \vec{a})$ and $\vec{c}' = \frac{1}{\left[\vec{a}\,\vec{b}\,\vec{c}\right]}(\vec{a} \times \vec{b})$, prove that $\vec{a} = \frac{1}{\left[\vec{a}'\,\vec{b}'\,\vec{c}'\right]}(\vec{b}' \times \vec{c}'),\ \vec{b} = \frac{1}{\left[\vec{a}'\,\vec{b}'\,\vec{c}'\right]}(\vec{c}' \times \vec{a}')$ and $\vec{c} = \frac{1}{\left[\vec{a}'\,\vec{b}'\,\vec{c}'\right]}(\vec{a}' \times \vec{b}')$.

38. Find a set of vectors reciprocal to the set

(a) $-\hat{i} + \hat{j} + \hat{k},\ \hat{i} - \hat{j} + \hat{k}$ and $\hat{i} + \hat{j} - \hat{k}$ *Ans*: $\frac{1}{2}(\hat{j} + \hat{k}),\ \frac{1}{2}(\hat{i} + \hat{k}),\ \frac{1}{2}(\hat{i} + \hat{j})$,

(b) $2\hat{i} + 3\hat{j} - \hat{k},\ \hat{i} - \hat{j} - 2\hat{k}$ and $-\hat{i} + 2\hat{j} + 2\hat{k}$

Ans: $\frac{1}{3}(2\hat{i} + \hat{k}),\ \frac{1}{3}(-8\hat{i} + 3\hat{j} - 7\hat{k}),\ \frac{1}{3}(-7\hat{i} + 3\hat{j} - 5\hat{k})$,

(c) $\hat{i} + 2\hat{j} + 3\hat{k},\ 5\hat{i} - \hat{j} - \hat{k}$ and $\hat{i} + \hat{j} - \hat{k}$

Ans: $\frac{1}{28}(2\hat{i} + 4\hat{j} + 6\hat{k}),\ \frac{1}{28}(5\hat{i} - 4\hat{j} + \hat{k}),\ \frac{1}{28}(\hat{i} + 9\hat{j} - 11\hat{k})$,

39. Show that the system reciprocal to $\vec{a}, \vec{b}$ and $\vec{a} \times \vec{b}$ is

$$\frac{1}{(\vec{a} \times \vec{b})^2}\{\vec{b} \times (\vec{a} \times \vec{b})\},\ \frac{1}{(\vec{a} \times \vec{b})^2}\{(\vec{a} \times \vec{b}) \times \vec{a}\},\ \frac{1}{(\vec{a} \times \vec{b})^2}\{\vec{a} \times \vec{b}\}$$

40. Show that the orthogonal vector triad $\hat{i}, \hat{j}$ and $\hat{k}$ form a self-reciprocal system. Hence, show that $\vec{r} = (\vec{r} \cdot \hat{i})\hat{i} + (\vec{r} \cdot \hat{j})\hat{j} + (\vec{r} \cdot \hat{k})\hat{k}$.

41. Show that any vector $\vec{r}$ can be expressed as

(a) $\vec{r} = (\vec{r} \cdot \vec{a})\vec{a}' + (\vec{r} \cdot \vec{b})\vec{b}' + (\vec{r} \cdot \vec{c})\vec{c}'$ (b) $\vec{r} = (\vec{r} \cdot \vec{a}')\vec{a} + (\vec{r} \cdot \vec{b}')\vec{b} + (\vec{r} \cdot \vec{c}')\vec{c}$

if $\vec{a}, \vec{b}$ and $\vec{c}$ and $\vec{a}', \vec{b}'$ and $\vec{c}'$ be two reciprocal systems of vectors.

42. Prove that $\left[\vec{b} \times \vec{c}\right]\left[\vec{c} \times \vec{a}\right]\left[\vec{a} \times \vec{b}\right] = \left[\vec{a}'\,\vec{b}'\,\vec{c}'\right]\left[\vec{a}\,\vec{b}\,\vec{c}\right]^3$, where $\left[\vec{a}\,\vec{b}\,\vec{c}\right]$ and $\left[\vec{a}'\,\vec{b}'\,\vec{c}'\right]$ are reciprocal sets of vectors.

Applications of Vector Algebra 3

In Chapters 1 and 2, we learnt about the various operations involving vectors like vector addition and subtraction, scalar multiplication of a vector and scalar and vector products of two or more vectors. In this chapter, we shall make use of such operations to solve vector equations and apply it to the various problems of metric geometry (derive the equations of lines, planes and spheres) and mechanics.

3.1 VECTOR EQUATIONS

An equation involving the addition, subtraction, dot product or cross product of known and unknown scalar and vector quantities is called a *vector equation*. For example, $\vec{r}+\vec{a}=\vec{b}$, where $\vec{a}$ and $\vec{b}$ are known vectors, is a vector equation.

A *solution of a vector equation* is a vector which, when substituted in this equation, satisfies the equation. For example, the vector $\vec{b}-\vec{a}$ is the solution of the vector equation $\vec{r}+\vec{a}=\vec{b}$. A vector equation with unknown scalars and known vector coefficients may frequently be solved by the use of scalar and vector products.

Theorem 3.1

The vector equation $\lambda\vec{r}+\mu\vec{a}=\nu\vec{b}$, where $\lambda \neq 0, \mu$ and ν are constant scalars and $\vec{a}$ and $\vec{b}$ are constant vectors, has a unique solution $\vec{r}=\frac{1}{\lambda}(\nu\vec{b}-\mu\vec{a})$.

Proof

Existence: Firstly, we suppose that the given equation admits a solution. Therefore, we must have

$$\begin{aligned}
&(\lambda\vec{r}+\mu\vec{a})-\mu\vec{a}=\nu\vec{b}-\mu\vec{a}\\
\Rightarrow\ &\lambda\vec{r}+(\mu\vec{a}-\mu\vec{a})=\nu\vec{b}-\mu\vec{a}, \text{ by associative law}\\
\Rightarrow\ &\lambda\vec{r}+\vec{0}=\nu\vec{b}-\mu\vec{a}\Rightarrow\vec{r}=\frac{1}{\lambda}(\nu\vec{b}-\mu\vec{a}) \qquad (3.1)
\end{aligned}$$

Substituting the value of $\vec{r}$ as in Eq. (3.1) in the given equation, we obtain

$$\begin{aligned}
\lambda\vec{r}+\mu\vec{a}&=\lambda\left\{\frac{1}{\lambda}(\nu\vec{b}-\mu\vec{a})\right\}+\mu\vec{a}=(\nu\vec{b}-\mu\vec{a})+\mu\vec{a}\\
&=\nu\vec{b}+(-\mu\vec{a}+\mu\vec{a}), \text{ by associative law}\\
&=\nu\,\vec{b}+\vec{0}=\nu\,\vec{b}
\end{aligned}$$

Thus, it is seen that (3.1) satisfies the given equation $\lambda\vec{r}+\mu\vec{a}=\nu\vec{b}$. Therefore, (3.1) gives the solution of the given equation.

Uniqueness: If possible, let $\vec{r}_1$ and $\vec{r}_2$ be two solutions of the given equation. Then, $\lambda\vec{r}_1 + \mu\vec{a} = \nu\vec{b}$ and $\lambda\vec{r}_2 + \mu\vec{a} = \nu\vec{b}$. Comparing the two equations, we get

$$\begin{aligned}
&\lambda\vec{r}_1 + \mu\vec{a} = \lambda\vec{r}_2 + \mu\vec{a} \\
&\Rightarrow \lambda\vec{r}_1 + \mu\vec{a} - (\lambda\vec{r}_2 + \mu\vec{a}) = \lambda\vec{r}_2 + \mu\vec{a} - (\lambda\vec{r}_2 + \mu\vec{a}) \\
&\Rightarrow \lambda(\vec{r}_1 - \vec{r}_2) + \mu\vec{a} - \mu\,\vec{a} = \vec{0} \\
&\Rightarrow \lambda(\vec{r}_1 - \vec{r}_2) = \vec{0}, \text{ by distributive law} \\
&\Rightarrow \vec{r}_1 - \vec{r}_2 = \vec{0} \Rightarrow \vec{r}_1 = \vec{r}_2, \text{ as } \lambda \neq 0
\end{aligned}$$

This shows that the solution is unique.

Example 3.1.1

Solve for $\vec{r}$, where $\vec{r} \cdot \vec{a} = p$, and $\vec{a}$ and p are the given vector and scalar.

Solution

We rewrite the given equation as

$$\vec{r} \cdot \vec{a} = p\,\frac{\vec{a} \cdot \vec{a}}{|\vec{a}|^2} \Rightarrow \left(\vec{r} - \frac{p}{|\vec{a}|^2}\,\vec{a}\right) \cdot \vec{a} = 0$$

This shows that $\vec{r} - \dfrac{p}{|\vec{a}|^2}\,\vec{a}$ is perpendicular to $\vec{a}$, i.e., parallel to $\vec{a} \times \vec{b}$, where $\vec{b}$ is an arbitrary vector. So the solution of the given equation can be written as

$$\vec{r} - \frac{p}{|\vec{a}|^2}\,\vec{a} = t(\vec{a} \times \vec{b}) \Rightarrow \vec{r} = \frac{p}{|\vec{a}|^2}\,\vec{a} + t(\vec{a} \times \vec{b}), t = \text{ scalar} \tag{3.2}$$

Also, using (3.2), we get

$$\begin{aligned}
\vec{r} \cdot \vec{a} &= \left\{\frac{p}{|\vec{a}|^2}\,\vec{a} + t(\vec{a} \times \vec{b})\right\} \cdot \vec{a} \\
&= \frac{p}{|\vec{a}|^2}\,\vec{a} \cdot \vec{a} + t(\vec{a} \times \vec{b}) \cdot \vec{a} = \frac{p}{|\vec{a}|^2}\,|\vec{a}|^2 + 0 = p
\end{aligned}$$

Thus, Eq. (3.2) satisfies $\vec{r} \cdot \vec{a} = p$ for every value of the vector $\vec{b}$ and scalar t. Therefore, (3.2) gives the general solution of the given vector equation and the solution is unique.

Example 3.1.2

Solve for $\vec{r}$, $\vec{r} \times \vec{b} = \vec{a} \times \vec{b}$, where $\vec{a}$, $\vec{b}$ and $\vec{c}$ are the three given vectors.

Solution

Let $\vec{r}$ be any solution of the given equation. The equation can be rewritten as

$$\vec{r} \times \vec{b} = \vec{a} \times \vec{b} \Rightarrow (\vec{r} - \vec{a}) \times \vec{b} = \vec{0}$$

Thus, we see that $\vec{r} - \vec{a}$ is parallel to $\vec{b}$. Therefore

$$\vec{r} - \vec{a} = t\,\vec{b} \Rightarrow \vec{r} = \vec{a} + t\,\vec{b} \tag{3.3}$$

where t is a scalar. Also, using Eq. (3.3), we get

$$\begin{aligned}\vec{r}\times\vec{b} &= (\vec{a}+t\vec{b})\times\vec{b} = \vec{a}\times\vec{b}+t\vec{b}\times\vec{b}\\ &= \vec{a}\times\vec{b}+\vec{0} = \vec{a}\times\vec{b};\ \text{as } \vec{b}\times\vec{b}=\vec{0}\end{aligned}$$

Thus, Eq. (3.3) satisfies $\vec{r}\times\vec{b} = \vec{a}\times\vec{b}$ for every value of the scalar. Therefore, (3.3) gives the general solution of the given vector equation.

Geometrically, we know that the points whose position vectors are given by Eq. (3.3) for different values of the scalar t lie on the straight line which passes through the point, with position vector $\vec{a}$, and is parallel to the vector $\vec{b}$.

Theorem 3.2

The vector equation $y\vec{a}+z\vec{b}=\vec{c}$, where y and z are unknown constant scalars and $\vec{a}$, $\vec{b}$ and $\vec{c}$ are the given non-zero vectors, admits solutions for the unknown scalars.

Proof: Taking the cross product of the given equation $y\vec{a}+z\vec{b}=\vec{c}$ with $\vec{b}$, we get

$$\begin{aligned}&\vec{c}\times\vec{b} = y(\vec{a}\times\vec{b})+z(\vec{b}\times\vec{b}) = y(\vec{a}\times\vec{b}) \quad \text{as } \vec{b}\times\vec{b}=\vec{0}\\ \Rightarrow\ &(\vec{c}\times\vec{b})\cdot(\vec{a}\times\vec{b}) = y(\vec{a}\times\vec{b})\cdot(\vec{a}\times\vec{b}) = y|\vec{a}\times\vec{b}|^2\\ \Rightarrow\ &y=\{(\vec{c}\times\vec{b})\cdot(\vec{a}\times\vec{b})\}/|\vec{a}\times\vec{b}|^2 \qquad (3.4)\end{aligned}$$

This is correct, provided that the vectors $\vec{a}$ and $\vec{b}$ are not parallel to each other. Proceeding further, we obtain $z=\{(\vec{c}\times\vec{a})\cdot(\vec{b}\times\vec{a})\}/|\vec{b}\times\vec{a}|^2$.

Example 3.1.3

Solve for $\vec{r}$, where $\vec{r}\cdot\vec{a}=t$, $\vec{r}\times\vec{b}=\vec{c}$ and $\vec{a}\times\vec{b}\neq\vec{0}$, t is a given scalar and $\vec{a}$, $\vec{b}$ and $\vec{c}$ are the given vectors.

Solution

The cross multiplication of $\vec{a}$ with the given relation $\vec{r}\times\vec{b}=\vec{c}$ gives

$$\begin{aligned}&\vec{a}\times(\vec{r}\times\vec{b}) = \vec{a}\times\vec{c} \Rightarrow (\vec{a}\cdot\vec{b})\vec{r}-(\vec{a}\cdot\vec{r})\vec{b} = \vec{a}\times\vec{c}\\ \Rightarrow\ &(\vec{a}\cdot\vec{b})\vec{r} = \vec{a}\times\vec{c}+(\vec{r}\cdot\vec{a})\vec{b} = \vec{a}\times\vec{c}+t\vec{b}\\ \Rightarrow\ &\vec{r} = \frac{(\vec{a}\times\vec{c})+(\vec{a}\cdot\vec{r})\vec{b}}{\vec{a}\cdot\vec{b}} = \frac{(\vec{a}\times\vec{c})+t\vec{b}}{\vec{a}\cdot\vec{b}} \qquad (3.5)\end{aligned}$$

Hence, the solution of $\vec{r}\times\vec{b}=\vec{c}$, if it exists, is given by Eq. (3.5). Direct substitution of (3.5) in $\vec{r}\times\vec{b}=\vec{c}$ with $\vec{r}\cdot\vec{a}=t$ verifies the solution. Hence, the solution is unique. Thus, a vector is uniquely determined when its scalar and vector products with two known non-perpendicular vectors are given.

Example 3.1.4

Solve for $\vec{r}$, where $\vec{r}\cdot\vec{a}=5$, $\vec{r}\times\vec{b}=\vec{c}$, $\vec{a}=\hat{\imath}+3\hat{\jmath}+2\hat{k}$, $\vec{b}=4\hat{\imath}+5\hat{\jmath}+\hat{k}$ and $\vec{c}=3\hat{\imath}+3\hat{\jmath}+\hat{k}$.

Solution

Since $\vec{a} = \hat{\imath} + 3\hat{\jmath} + 2\hat{k}$ and $\vec{b} = 4\hat{\imath} + 5\hat{\jmath} + \hat{k}$, $\vec{a} \cdot \vec{b} = 1 \cdot 4 + 3 \cdot 5 + 2 \cdot 1 = 21$. Since $\vec{a} = \hat{\imath} + 3\hat{\jmath} + 2\hat{k}$ and $\vec{c} = 3\hat{\imath} + 3\hat{\jmath} + \hat{k}$

$$\vec{a} \times \vec{c} = \begin{vmatrix} \hat{\imath} & \hat{\jmath} & \hat{k} \\ 1 & 3 & 2 \\ 3 & 3 & 1 \end{vmatrix} = -3\hat{\imath} + 5\hat{\jmath} - 6\hat{k}$$

Here, given $t = 5$, using Eq. (3.5), the solution for $\vec{r}$ is given by

$$\vec{r} = \frac{1}{21}\left[(\vec{a} \times \vec{c}) + 5\vec{b}\right] = \frac{1}{21}\left[(-3\hat{\imath} + 5\hat{\jmath} - 6\hat{k}) + 5(4\hat{\imath} + 5\hat{\jmath} + \hat{k})\right]$$

$$= \frac{17}{21}\hat{\imath} + \frac{30}{21}\hat{\jmath} - \frac{1}{21}\hat{k}$$

Theorem 3.3

A necessary and sufficient condition that the vector equation $\vec{a} \times \vec{r} = \vec{b}$; $\vec{a}$ and $\vec{b}$ are the given vectors with $\vec{a} \neq \vec{0}$ for a solution that $\vec{a} \cdot \vec{b} = 0$.

Proof

Condition necessary: Firstly, we consider that the vector equation $\vec{a} \times \vec{r} = \vec{b}$ admits a solution. Taking the dot product with $\vec{a}$, we get

$$\vec{a} \cdot (\vec{a} \times \vec{r}) = \vec{a} \cdot \vec{b} \Rightarrow [\vec{a}\ \vec{a}\ \vec{r}] = \vec{a} \cdot \vec{b} \Rightarrow 0 = \vec{a} \cdot \vec{b}$$

Condition sufficient: Consider three non-coplanar vectors $\vec{a}$, $\vec{b}$ and $\vec{b} \times \vec{a}$. So, every vector $\vec{r}$ can be expressed as a linear combination of the vectors $\vec{a}$, $\vec{b}$ and $\vec{b} \times \vec{a}$ as

$$\vec{r} = x\vec{a} + y\vec{b} + z(\vec{b} \times \vec{a})$$

where x, y and z are scalars. From the given vector equation $\vec{a} \times \vec{r} = \vec{b}$, we get

$$\vec{a} \times \left[x\vec{a} + y\vec{b} + z(\vec{b} \times \vec{a})\right] = \vec{b}$$

$$\Rightarrow\ x(\vec{a} \times \vec{a}) + y(\vec{a} \times \vec{b}) + z\vec{a} \times (\vec{b} \times \vec{a}) = \vec{b}$$

$$\Rightarrow\ y(\vec{a} \times \vec{b}) + z\left[(\vec{a} \cdot \vec{a})\vec{b} - (\vec{a} \cdot \vec{b})\vec{a}\right] = \vec{b}$$

$$\Rightarrow\ y(\vec{a} \times \vec{b}) + (za^2 - 1)\vec{b} = \vec{0};\ \text{as } |\vec{a}| = a \text{ and } \vec{a} \cdot \vec{b} = 0$$

Since $\vec{a} \times \vec{b}$ and $\vec{b}$ are non-collinear, we get $y = 0$ and $z = a^{-2}$. Therefore

$$\vec{r} = x\,\vec{a} + \frac{1}{\vec{a} \cdot \vec{a}}\,(\vec{b} \times \vec{a}) = x\,\vec{a} + a^{-2}\,(\vec{b} \times \vec{a}) \qquad (3.6)$$

By direct substitution in the given equation, this value of $\vec{r}$ is a solution, whatever scalar x may be. Hence, Eq. (3.6) is the general solution of the given equation, x being the parameter.

Geometrically speaking, the points whose position vectors satisfy the given equation lie on the straight line which passes through the point with position vector $a^{-2}\,(\vec{b} \times \vec{a})$ and is parallel to $\vec{a}$.

RESULT 11

A vector can be uniquely determined when its dot and cross products with two given non-orthogonal vectors are known.

Proof: Let $\vec{r}$ be any vector whose dot and cross products with two given non-orthogonal vectors are known. We take two given non-orthogonal vectors $\vec{a}$ and $\vec{b}$ and assume that $\vec{r} \times \vec{a} = \vec{p}$ and $\vec{r} \cdot \vec{b} = k$. Hence, we have

$$\vec{b} \times (\vec{r} \times \vec{a}) = \vec{b} \times \vec{p}$$

$$\text{or,} \quad (\vec{b} \cdot \vec{a})\vec{r} - (\vec{b} \cdot \vec{r})\vec{a} = \vec{b} \times \vec{p}$$

$$\text{or,} \quad (\vec{b} \cdot \vec{a})\vec{r} - k\,\vec{a} = \vec{b} \times \vec{p}, \text{ as } \vec{r} \cdot \vec{b} = k$$

$$\text{or,} \quad \vec{r} = \frac{1}{\vec{b} \cdot \vec{a}}\left[k\,\vec{a} + \vec{b} \times \vec{p}\right] \tag{3.7}$$

This satisfies the given equation. Let the given equations $\vec{r} \times \vec{a} = \vec{p}$ and $\vec{r} \cdot \vec{b} = k$ have two different solutions $\vec{r}_1$ and $\vec{r}_2$. Then, $\vec{r}_1 \times \vec{a} = \vec{p}, \vec{r}_1 \cdot \vec{b} = k$ and $\vec{r}_2 \times \vec{a} = \vec{p}, \vec{r}_2 \cdot \vec{b} = k$. Comparing, we get

$$(\vec{r}_1 - \vec{r}_2) \times \vec{a} = \vec{0}; \quad (\vec{r}_1 - \vec{r}_2) \cdot \vec{b} = 0$$

Hence, the vector $\vec{r}_1 - \vec{r}_2$ is parallel and perpendicular to $\vec{a}$ and $\vec{b}$, respectively, which contradicts the fact that $\vec{a}$ and $\vec{b}$ are non-orthogonal. Thus, $\vec{r}_1 = \vec{r}_2$ and the solution is unique.

RESULT 12

A vector can be uniquely determined when its dot products with three given non-coplanar vectors are known.

Proof: Let $\vec{r}$ be any vector whose dot products with three given non-coplanar vectors are known. We take the three given non-coplanar vectors as $\vec{a}, \vec{b}$ and $\vec{c}$. Let $\vec{l}, \vec{m}$ and $\vec{m}$ be the set of vectors reciprocal to the vectors $\vec{a}, \vec{b}$ and $\vec{c}$, respectively. We also assume that

$$\vec{r} \cdot \vec{a} = \lambda, \vec{r} \cdot \vec{b} = \mu \text{ and } \vec{r} \cdot \vec{c} = \nu.$$

Hence, using Eq.(2.44), we get

$$\vec{r} = (\vec{r} \cdot \vec{a})\vec{l} + (\vec{r} \cdot \vec{b})\vec{m} + (\vec{r} \cdot \vec{c})\vec{n} = \lambda\vec{l} + \mu\vec{m} + \nu\vec{n}$$

Example 3.1.5

Let $\vec{a}$ and $\vec{c}$ be the given vectors and let k be a given scalar. Determine all vectors $\vec{r}$ such that $\vec{a} \times \vec{r} = \vec{c}$ and $\vec{a} \cdot \vec{r} = k$.

Solution

Case 1: If $\vec{a} \cdot \vec{c} = 0$, there is no solution for $\vec{a} \cdot (\vec{a} \times \vec{r}) = 0$ and, hence, the equation $\vec{a} \times \vec{r} = \vec{c}$ becomes inconsistent.

Case 2: Let $\vec{a} \cdot \vec{c} = 0$ and $\vec{a} \neq \vec{0}$. Now, $\vec{a} \times \vec{r} = \vec{c}$ leads to

$$(\vec{a} \times \vec{r}) \times \vec{a} = \vec{c} \times \vec{a} \Rightarrow (\vec{a} \cdot \vec{a})\vec{r} - (\vec{a} \cdot \vec{r})\vec{a} = \vec{c} \times \vec{a}$$

$$\Rightarrow a^2\vec{r} - k\vec{a} = \vec{c} \times \vec{a} \Rightarrow \vec{r} = \frac{1}{a^2}[\vec{c} \times \vec{a} + k\vec{a}] \tag{3.8}$$

Since $\vec{a} \cdot \vec{r} = k$ and $\vec{a} \neq \vec{0}$, this value of $\vec{r}$, as in Eq. (3.8), satisfies both the equations.

Case 3: If $\vec{a} = \vec{0}$, there is no solution unless $\vec{c} = \vec{0}$ and $k = 0$. In this case, $\vec{b}$ can be any vector.

Example 3.1.6

Find the scalar λ and the vector $\vec{r}$, where $\vec{r}\cdot\vec{a} = 1$, $\lambda\vec{a} + \vec{r}\times\vec{a} = \vec{b}$, $\vec{a} = \hat{i} + 2\hat{j} + 2\hat{k}$ and $\vec{b} = \hat{i} + 2\hat{j} - 2\hat{k}$.

Solution

Since $\vec{a} = \hat{i} + 2\hat{j} + 2\hat{k}$ and $\vec{b} = \hat{i} + 2\hat{j} - 2\hat{k}$, $\vec{a}\cdot\vec{b} = 1\cdot 1 + 2\cdot 2 + 2\cdot(-2) = 1$ and $|\vec{a}| = \sqrt{1^2 + 2^2 + 2^2} = 3$. Also

$$\vec{a}\times\vec{b} = \begin{vmatrix} \hat{i} & \hat{j} & \hat{k} \\ 1 & 2 & 2 \\ 1 & 2 & -2 \end{vmatrix} = -8\hat{i} + 4\hat{j}$$

Now, taking the dot product of $\lambda\vec{a} + \vec{r}\times\vec{a} = \vec{b}$ with $\vec{a}$, we obtain

$$\lambda\vec{a}\cdot\vec{a} + (\vec{r}\times\vec{a})\cdot\vec{a} = \vec{b}\cdot\vec{a}$$

$$\Rightarrow \lambda\,|\vec{a}|^2 + 0 = \vec{b}\cdot\vec{a} \Rightarrow \lambda\times 9 = 1 \Rightarrow \lambda = 1/9$$

Again, taking the cross product of $\lambda\vec{a} + \vec{r}\times\vec{a} = \vec{b}$ with $\vec{a}$, we obtain

$$\lambda\vec{a}\times\vec{a} + (\vec{r}\times\vec{a})\times\vec{a} = \vec{b}\times\vec{a}$$

$$\Rightarrow (\vec{r}\cdot\vec{a})\vec{a} - (\vec{a}\cdot\vec{a})\vec{r} = \vec{b}\times\vec{a}$$

$$\Rightarrow \vec{a} - 9\,\vec{r} = \vec{b}\times\vec{a} = 8\hat{i} - 4\hat{j}$$

$$\Rightarrow \vec{r} = \frac{1}{9}\left[\hat{i} + 2\hat{j} + 2\hat{k} - (8\hat{i} - 4\hat{j})\right] = \frac{1}{9}(-7\hat{i} + 6\hat{j} + 2\hat{k})$$

Example 3.1.7

Solve for $\vec{r}$ from the simultaneous equations $\vec{a}\times\vec{r} = \vec{a}\times\vec{b}$ and $\vec{a}\cdot(\vec{r} - \vec{b}) = 2|\vec{a}|^2$.

Solution

From the given first relation $\vec{a}\times\vec{r} = \vec{a}\times\vec{b}$, we get $\vec{a}\times(\vec{r} - \vec{b}) = \vec{0}$, i.e., $\vec{a}$ and $\vec{r} - \vec{b}$ are collinear, and $\vec{r} - \vec{b} = t\,\vec{a}$, where t is a scalar. Using the second equation $\vec{a}\cdot(\vec{r} - \vec{b}) = 2|\vec{a}|^2$, we get

$$\vec{a}\cdot t\,\vec{a} = 2|\vec{a}|^2 \Rightarrow t = 2$$

Therefore, the unique solution is $\vec{r} = 2\vec{a} + \vec{b}$, which satisfies both the equations.

Example 3.1.8

Solve for $\vec{r}$ from the simultaneous equations $\vec{r}\times\vec{b} = \vec{c}\times\vec{b}$ and $\vec{r}\cdot\vec{a} = 0$, provided that $\vec{a}$ is not perpendicular to $\vec{b}$.

Solution

Suppose that $\vec{r}$ is a solution of the given equations. Rewriting the first equation, we get

$$\vec{r}\times\vec{b} = \vec{c}\times\vec{b} \Rightarrow (\vec{r} - \vec{c})\times\vec{b} = \vec{0}$$

We see that $\vec{r} - \vec{c}$ and $\vec{b}$ are collinear, so

$$\vec{r} - \vec{c} = t\,\vec{b} \Rightarrow \vec{r} = \vec{c} + t\,\vec{b}$$

where t is a scalar. Substituting in $\vec{r}\cdot\vec{a}=0$, we obtain

$$(\vec{c}+t\,\vec{b})\cdot\vec{a}=0 \Rightarrow t=-\frac{\vec{c}\cdot\vec{a}}{\vec{b}\cdot\vec{a}};\quad \text{for } \vec{b}\cdot\vec{a}\neq 0$$

Thus, the solution is given by

$$\vec{r}=\vec{c}+t\,\vec{b}=\vec{c}-\frac{\vec{c}\cdot\vec{a}}{\vec{b}\cdot\vec{a}}\,\vec{b} \tag{3.9}$$

We may also easily verify that $\vec{r}$ given by Eq. (3.9) does satisfy the given equations. Hence, $\vec{r}$ given by Eq. (3.9) is a *unique solution* of the given equations.

It may be seen that $\vec{r}\cdot\vec{a}=0$ represents the plane which passes through the origin and is normal to the vector $\vec{a}$. Thus, the solution in question represents the point of intersection of a line and a plane.

Theorem 3.4

Solve the vector equation $t\vec{r}+\vec{r}\times\vec{a}=\vec{b}$, where $\vec{a}$ and $\vec{b}$ are the given vectors and t is a given non-zero scalar.

Proof: Given $t\vec{r}+\vec{r}\times\vec{a}=\vec{b}$. Taking the dot product with $\vec{a}$, we get

$$t(\vec{r}\cdot\vec{a})+(\vec{r}\times\vec{a})\cdot\vec{a}=\vec{b}\cdot\vec{a}$$

$$\Rightarrow \vec{r}\cdot\vec{a}=\frac{1}{t}(\vec{b}\cdot\vec{a});\ \text{as } [\vec{r}\,\vec{a}\,\vec{a}]=0 \tag{3.10}$$

Taking the cross product with $\vec{a}$ on both sides of the given relation, we get

$$\begin{aligned}
& t(\vec{r}\times\vec{a})+(\vec{r}\times\vec{a})\times\vec{a}=\vec{b}\times\vec{a}\\
\Rightarrow\ & t(\vec{r}\times\vec{a})+\{(\vec{a}\cdot\vec{r})\vec{a}-(\vec{a}\cdot\vec{a})\vec{r}\}=\vec{b}\times\vec{a}\\
\Rightarrow\ & t(\vec{b}-t\vec{r})-a^2\vec{r}=\vec{b}\times\vec{a}-(\vec{a}\cdot\vec{r})\vec{a};\ \text{as } \vec{r}\times\vec{a}=\vec{b}-t\vec{r}\\
\Rightarrow\ & -(t^2+a^2)\vec{r}=\vec{b}\times\vec{a}-t\vec{b}-(\vec{r}\cdot\vec{a})\vec{a}\\
\Rightarrow\ & (t^2+a^2)\vec{r}=-\vec{b}\times\vec{a}+t\vec{b}+\frac{1}{t}(\vec{b}\cdot\vec{a})\vec{a};\ \text{from Eq. (3.10)}\\
\Rightarrow\ & \vec{r}=\frac{1}{t^2+a^2}\left[\vec{a}\times\vec{b}+\frac{1}{t}(\vec{a}\cdot\vec{b})\vec{a}+t\vec{b}\right]
\end{aligned} \tag{3.11}$$

Example 3.1.9

Solve the equation for $\vec{r}$, where $\vec{r}\cdot\vec{a}=t$, $\vec{a}$ is the given vector and t is the given scalar.

Solution

Given $\vec{r}\cdot\vec{a}=t$. This equation can be written as

$$\vec{r}\cdot\vec{a}=t=t\,\frac{\vec{a}\cdot\vec{a}}{a^2}\Rightarrow\left(\vec{r}-\frac{t}{a^2}\,\vec{a}\right)\cdot\vec{a}=0$$

To satisfy this equation, the general solution can be written in the form

$$\vec{r}-\frac{t}{a^2}\,\vec{a}=\vec{h}\times\vec{a}\Rightarrow\vec{r}=\frac{t}{a^2}\,\vec{a}+\vec{h}\times\vec{a}$$

where $\vec{h}$ is an arbitrary vector.

Example 3.1.10

Solve the vector equation for $\vec{r}$, where $t\vec{r} + (\vec{r}\cdot\vec{b})\vec{a} = \vec{c}$; $t \neq 0$ is the given scalar and $\vec{a}, \vec{b}$ and $\vec{c}$ are the given vectors.

Solution

Given $t\vec{r} + (\vec{r}\cdot\vec{b})\vec{a} = \vec{c}$. Multiplying both sides scalarly by $\vec{b}$, we get

$$t(\vec{r}\cdot\vec{b}) + (\vec{r}\cdot\vec{b})(\vec{a}\cdot\vec{b}) = \vec{c}\cdot\vec{b}$$

$$\Rightarrow\ (\vec{r}\cdot\vec{b})\left[t + (\vec{a}\cdot\vec{b})\right] = \vec{c}\cdot\vec{b} \Rightarrow \vec{r}\cdot\vec{b} = \frac{1}{t + (\vec{a}\cdot\vec{b})}(\vec{c}\cdot\vec{b})$$

Also, from the given relation, we have

$$(\vec{r}\cdot\vec{b})\vec{a} = \vec{c} - t\vec{r} \Rightarrow \vec{r} = \frac{1}{t}\left[\vec{c} - \frac{\vec{c}\cdot\vec{b}}{t + (\vec{a}\cdot\vec{b})}\vec{a}\right]$$

Example 3.1.11

Solve: $p\vec{x} + (\vec{x}\cdot\vec{b})\vec{x} = \vec{a}\times\vec{b} + \vec{c}$, where p is a known scalar.

Solution

Taking the scalar product of both sides of the given equation with $\vec{b}$, we have

$$p\vec{x}\cdot\vec{b} + (\vec{x}\cdot\vec{b})\vec{x}\cdot\vec{b} = (\vec{a}\times\vec{b})\cdot\vec{b} + \vec{c}\cdot\vec{b}$$

$$\Rightarrow\ (\vec{x}\cdot\vec{b})^2 + p\vec{x}\cdot\vec{b} - \vec{c}\cdot\vec{b} = 0$$

$$\Rightarrow\ \vec{x}\cdot\vec{b} = \frac{1}{2}\left[-p \pm \sqrt{p^2 + 4\vec{c}\cdot\vec{b}}\right]$$

Since p is a known scalar and $\vec{b}$ is a known vector, $\vec{x}\cdot\vec{b}$ is a fixed scalar. Using this value of $\vec{x}\cdot\vec{b}$, we have

$$p\vec{x} + (\vec{x}\cdot\vec{b})\vec{x} = \vec{a}\times\vec{b} + \vec{c}$$

$$p\vec{x} + \frac{1}{2}\left[-p \pm \sqrt{p^2 + 4\,\vec{c}\cdot\vec{b}}\right]\vec{x} = \vec{a}\times\vec{b} + \vec{c}$$

$$\Rightarrow\ \vec{x} = \frac{1}{p \pm \sqrt{p^2 + 4\vec{b}\cdot\vec{c}}}2(\vec{a}\times\vec{b} + \vec{c})$$

The locus of $\vec{x}$ is a plane passing through the point with position vector $\frac{1}{p + \vec{a}\cdot\vec{b}}\vec{c}$ and is perpendicular to the vector $\vec{b}$, where $p + \vec{a}\cdot\vec{b} \neq 0$. When $p + \vec{a}\cdot\vec{b} = 0$, but $p \neq 0$, the locus of $\vec{x}$ is a straight line passing through the point with position vector $\frac{1}{\vec{b}\cdot\vec{a}}2(\vec{a}\times\vec{b} + \vec{c})$ and is parallel to the vector $\vec{a}$. When $p = 0$, the locus of $\vec{x}$ is the plane $\vec{x}\cdot\vec{b} = \frac{\vec{c}\cdot\vec{a}}{|\vec{a}|^2}$, when $\vec{a} \neq \vec{0}$. If $\vec{a} = \vec{0}$ with $p = 0$, then $\vec{c} = \vec{0}$ and $\vec{x}$ is an arbitrary vector.

3.2 APPLICATIONS IN GEOMETRY

In this section, concepts of product of vectors is applied to find equations of straight lines and planes in non-parametric forms. Equations of planes in normal form, distance between two planes, distance of a point from a plane measured in a given direction, angle between a line and a plane, perpendicular distance of a point from a straight line, shortest distance between two skew lines and related properties are discussed. Finally, some examples are solved to illustrate the applications of product of vectors in space geometry.

3.2.1 Vector Equation of a Straight Line

Vector algebra supplies several indispensable tools which we can use to study straight lines under various conditions.

A straight line passing through a given point and parallel to a given vector

Let A be a given point, with position vector $\vec{\alpha}$ relative to the origin O, i.e., $\vec{OA} = \vec{\alpha}$ (Fig. 3.1). Let $\vec{r}$ be the position vector of any point P on the required straight line, through A and parallel to a given vector $\vec{\beta}$. Hence, we can take $\vec{AP} = t\vec{\beta}$, where t is a real number. t is positive for points on one side of A and negative for points on the other side. From Fig. 3.1

$$\vec{r} = \vec{OP} = \vec{OA} + \vec{AP} = \vec{\alpha} + t\ \vec{\beta} \quad (3.12)$$

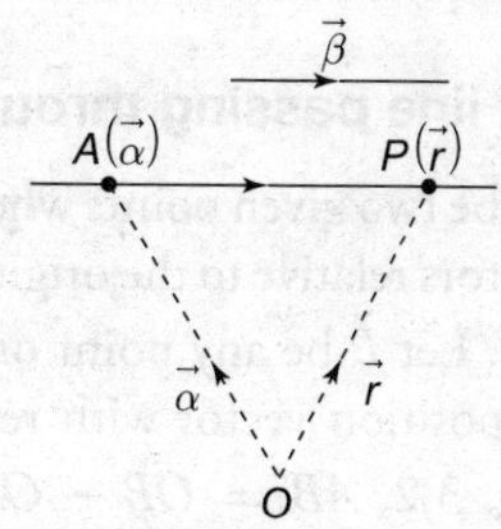

Figure 3.1 Vector equation of a straight line

Each point P of the line arises for some value of the scalar t and, conversely, for each value of scalar t, $\vec{\alpha} + t\vec{\beta}$ is the position vector of a point on the line. Hence, Eq. (3.12) is the required *parametric equation of the line* in vector form.

(i) If the straight line passes through the origin, then the equation of the straight line (3.12) becomes the parametric form $\vec{r} = t\,\vec{\beta}$, as $\vec{\alpha} = \vec{0}$. The non-parametric form of the equation is $\vec{r} \times \vec{\beta} = \vec{0}$.

(ii) If we take $\vec{r} = (x, y, z)$; $\vec{\alpha} = (\alpha_1, \alpha_2, \alpha_3)$ and $\vec{\beta} = (\beta_1, \beta_2, \beta_3)$, then Eq. (3.12) gives

$$\frac{x - \alpha_1}{\beta_1} = \frac{y - \alpha_2}{\beta_2} = \frac{z - \alpha_3}{\beta_3} = t$$

This is the three-dimensional Cartesian equation of a straight line passing through the point $(\alpha_1, \alpha_2, \alpha_3)$ having direction ratios β_1, β_2 and β_3.

(iii) **Non-parametric form**: From the above discussion (Fig. 3.1), we see that $\vec{AP} = \vec{r} - \vec{\alpha}$ and $\vec{\beta}$ are parallel to each other, and hence, we obtain

$$(\vec{r} - \vec{\alpha}) \times \vec{\beta} = \vec{0} \Rightarrow \vec{r} \times \vec{\beta} = \vec{\alpha} \times \vec{\beta} \quad (3.13)$$

This is the equation of the straight line in non-parametric form.

Example 3.2.1

Find the vector equation of a straight line passing through the point $\hat{i}+\hat{j}+2\hat{k}$ and parallel to the vector $\hat{i}+\hat{j}-\hat{k}$.

Solution

The required straight line passes through the point $\vec{\alpha}=\hat{i}+\hat{j}+2\hat{k}$ and is parallel to the vector $\vec{\beta}=\hat{i}+\hat{j}-\hat{k}$. Let $\vec{r}$ be any point on the straight line. Using Eq. (3.12), we have

$$\vec{r}=\vec{\alpha}+t\,\vec{\beta}=\hat{i}+\hat{j}+2\hat{k}+t[\hat{i}+\hat{j}-\hat{k}]$$
$$=(1+t)\hat{i}+(1+t)\hat{j}+(2-t)\hat{k}$$

This is the parametric equation of the required straight line. Using Eq. (3.13), we have

$$\vec{r}\times(\hat{i}+\hat{j}-\hat{k})=(\hat{i}+\hat{j}+2\hat{k})\times(\hat{i}+\hat{j}-\hat{k})=-3(\hat{i}-\hat{j})$$

This is the required non-parametric equation of the straight line.

A straight line passing through two given points

Let A and B be two given points whose respective position vectors relative to the origin O are $\vec{\alpha}$ and $\vec{\beta}$ (Fig. 3.2). Let P be any point on the straight line whose position vector with respect to O is $\vec{r}$. From Fig. 3.2, $\vec{AB}=\vec{OB}-\vec{OA}=\vec{\beta}-\vec{\alpha}$. So, the required line passes through the point A whose position vector is $\vec{\alpha}$ and is parallel to the vector $\vec{\beta}-\vec{\alpha}$. Hence, the vector equation of the straight line is

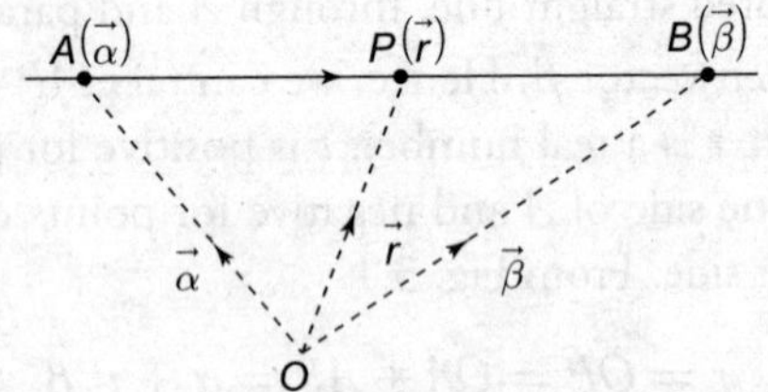

Figure 3.2 Vector equation of a straight line

$$\vec{r}=\vec{\alpha}+t(\vec{\beta}-\vec{\alpha})=(1-t)\vec{\alpha}+t\vec{\beta} \tag{3.14}$$

where t is a scalar quantity. Every point P on the line AB has a one-to-one correspondence with a vector $\vec{r}$ satisfying Eq. (3.14). Hence, Eq. (3.14) is the *vector equation of the straight line AB in parametric form*. Writting $1-t=s$, Eq. (3.14) can be written in the form

$$\vec{r}=s\,\vec{\alpha}+(1-s)\,\vec{\beta} \tag{3.15}$$

(i) If we take $\vec{r}=(x,y,z)$, $\vec{\alpha}=(\alpha_1,\alpha_2,\alpha_3)$ and $\vec{\beta}=(\beta_1,\beta_2,\beta_3)$, then Eq. (3.14) gives

$$\frac{x-\alpha_1}{\beta_1-\alpha_1}=\frac{y-\alpha_2}{\beta_2-\alpha_2}=\frac{z-\alpha_3}{\beta_3-\alpha_3}=t \tag{3.16}$$

This is the Cartesian equation of a straight line passing through the points $(\alpha_1,\alpha_2,\alpha_3)$ and $(\beta_1,\beta_2,\beta_3)$.

(ii) **Non-parametric form**: From the above discussion (Fig. 3.2), we see that $\vec{AP}=\vec{r}-\vec{\alpha}$ and $\vec{\beta}-\vec{\alpha}$ are parallel to each other, and hence, we obtain

$$(\vec{r}-\vec{\alpha})\times(\vec{\beta}-\vec{\alpha})=\vec{0}\Rightarrow\vec{r}\times(\vec{\beta}-\vec{\alpha})=\vec{\alpha}\times\vec{\beta} \tag{3.17}$$

This is the equation of the straight line in non-parametric form.

(iii) Equation (3.14) can be written as $(1-t)\vec{\alpha}+t\vec{\beta}-\vec{r}=\vec{0}$. Let $x=(1-t), y=t$ and $z=-1$. Then, the above equation can be written as

$$x\vec{\alpha}+y\vec{\beta}+z\vec{r}=\vec{0},\ \text{with } x+y+z=1-t+t-1=0$$

Thus, Eq. (3.14) verifies the necessary and sufficient conditions for collinearity of the points A, B and P.

Example 3.2.2

Find the vector equation of the straight line passing through the points $\hat{i}-\hat{j}+\hat{k}$ and $3\hat{i}-\hat{k}$.

Solution

The required straight line passes through the points $\vec{\alpha}=\hat{i}-\hat{j}+\hat{k}$ and $\vec{\beta}=3\hat{i}-\hat{k}$. Let $\vec{r}$ be any point on the straight line. Using Eq. (3.14), we have

$$\begin{aligned}\vec{r}&=(1-t)\vec{\alpha}+t\,\vec{\beta}=(1-t)\left[\hat{i}-\hat{j}+\hat{k}\right]+t\left[3\hat{i}-\hat{k}\right]\\&=(1+2t)\hat{i}-(1-t)\hat{j}+(1-2t)\hat{k};t=\text{ scalar}\end{aligned}$$

This is the parametric equation of the required straight line. Using Eq.(3.17), we have

$$\vec{r}\times(2\hat{i}+\hat{j}-2\hat{k})=(\hat{i}-\hat{j}+\hat{k})\times(2\hat{i}+\hat{j}-2\hat{k})=\hat{i}+4\hat{j}+3\hat{k}$$

This is the required non-parametric equation of the straight line. Using Eq. (3.16), we get

$$\frac{x-1}{3-1}=\frac{y+1}{0-(-1)}=\frac{z-1}{-1-1}\Rightarrow\frac{x-1}{2}=\frac{y+1}{1}=\frac{z-1}{-2}$$

This is the Cartesian equation of a straight line passing through the points.

A straight line passing through a point and perpendicular to two given vectors

We consider that A is a given point whose position vector is $\vec{\alpha}$ with respect to some base point O, i.e., $\vec{OA}=\vec{\alpha}$ (Fig. 3.3). Let $\vec{\beta}$ and $\vec{\gamma}$ be the two given vectors. Let P be any point, with position vector $\vec{r}$, on the line. Then, $\vec{AP}=\vec{OP}-\vec{OA}=\vec{r}-\vec{\alpha}$. Now, since the line AP is perpendicular to both the vectors $\vec{\beta}$ and $\vec{\gamma}$, the line is parallel to the vector $\vec{\beta}\times\vec{\gamma}$. Thus, the equation of the straight line AP that passes through $A(\vec{\alpha})$ and is parallel to $\vec{\beta}\times\vec{\gamma}$ is obtained as

$$(\vec{r}-\vec{\alpha})\times(\vec{\beta}\times\vec{\gamma})=\vec{0} \quad (3.18)$$

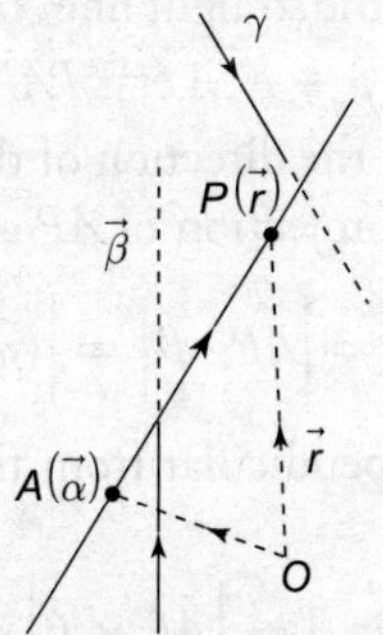

Figure 3.3 Vector equation of a straight line

Example 3.2.3

Find the equation of the straight line passing through a point $A(\vec{p})$ and which is perpendicular to the line $\vec{r}=\vec{\alpha}+t\vec{\beta}$.

Solution

Let the required line be parallel to the vector $\vec{\delta}$ (Fig. 3.4). It passes through the point $A(\vec{p})$. Then, its equation is $\vec{r} = \vec{p}+t\vec{\beta}$, where $\vec{\delta}\cdot\vec{\beta} = 0$.

At the point of intersection

$$\vec{p} + k\vec{\delta} = \vec{\alpha} + t\vec{\beta}$$

$$\Rightarrow \vec{\beta} \times (\vec{p} + k\vec{\delta} - \vec{\alpha}) = \vec{\beta} \times t\vec{\beta} = \vec{0}$$

$$\Rightarrow k\vec{\delta} \times \vec{\beta} = (\vec{\alpha} - \vec{p}) \times \vec{\beta}$$

$$\Rightarrow k\vec{\beta} \times (\vec{\delta} \times \vec{\beta}) = \vec{\beta} \times \{(\vec{\alpha} - \vec{p}) \times \vec{\beta}\}$$

$$\Rightarrow k|\vec{\beta}|^2\vec{\delta} = \vec{\beta} \times \{(\vec{\alpha} - \vec{p}) \times \vec{\beta}\}$$

$$\Rightarrow k\vec{\delta} = \left[\vec{\beta} \times \{(\vec{\alpha} - \vec{p}) \times \vec{\beta}\}\right]/|\vec{\beta}|^2$$

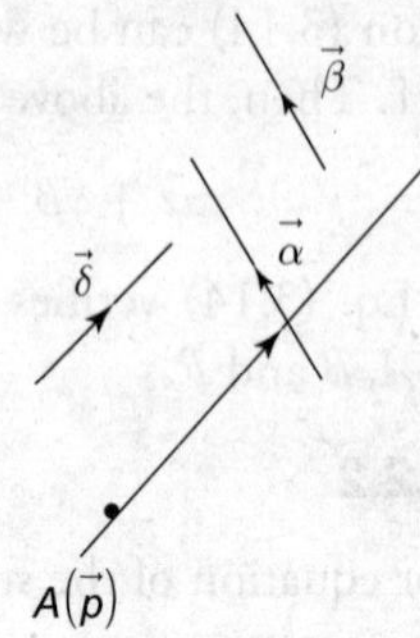

Figure 3.4 Equation of a straight line

Hence, the required equation of the straight line is

$$\vec{r} = \vec{p} + \frac{1}{|\vec{\beta}|^2}\left[\vec{\beta} \times \{(\vec{\alpha} - \vec{p}) \times \vec{\beta}\}\right]$$

$$= \vec{p} + s\vec{\beta} \times \left\{(\vec{\alpha} - \vec{p}) \times \vec{\beta}\right\}, \quad s = 1/|\vec{\beta}|^2$$

DEDUCTION 3.2.1

Distance of a point from a straight line: The equation of the straight line AB passing through the point $A(\vec{\alpha})$ and parallel to the vector $\vec{\beta}$ is of the form (3.12) as $\vec{r} = \vec{\alpha} + t\,\vec{\beta}$ (Fig. 3.5).

Let P be a point whose position vector is $\vec{p}$, whose distance from the straight line AN is to be determined. Thus, $\vec{AP} = \vec{p} - \vec{\alpha}$. Let $\angle PAN = \theta$ and $\hat{\beta}$ be the unit vectors in the direction of the vector $\vec{\beta}$. AN is the length of the projection of AP along the line AN. So

$$AN = \left|\vec{AP}\cdot\hat{\beta}\right| = \left|(\vec{p} - \vec{\alpha})\cdot\hat{\beta}\right|$$

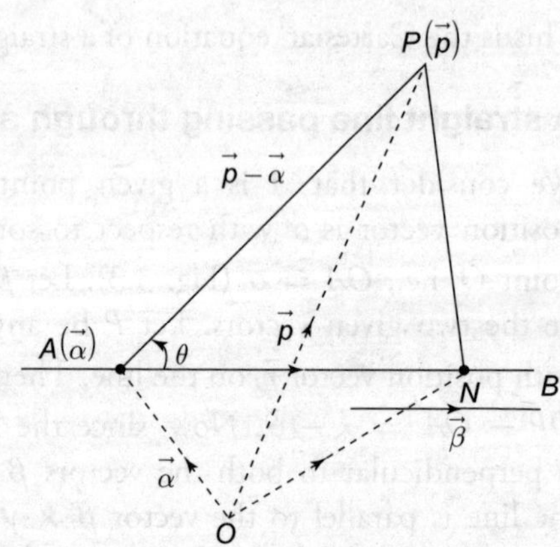

Figure 3.5 Perpendicular distance from a point to a straight line

PN is the perpendicular from the point P to the line AN. Then

$$PN = AP\ \sin\theta = \left|\vec{AP} \times \hat{\beta}\right|$$

$$= \left|(\vec{p} - \vec{\alpha}) \times \hat{\beta}\right| = \frac{1}{|\vec{\beta}|}\left[\left|(\vec{p} - \vec{\alpha}) \times \vec{\beta}\right|\right] \quad (3.19)$$

The projection vector of the foot of the perpendicular N relative to the origin O is

$$\vec{ON} = \vec{OA} + \vec{AN} = \vec{\alpha} + \left|(\vec{p} - \vec{\alpha}) \times \hat{\beta}\right|\,\hat{\beta} \quad (3.20)$$

The length of the perpendicular drawn from P on AN is

$$|PN| = \left|\vec{\alpha} + \left|(\vec{p} - \vec{\alpha}) \times \hat{\beta}\right| \hat{\beta} - \vec{p}\right| \quad (3.21)$$

Example 3.2.4

Find the distance of the point $P(\vec{p})$ from the straight line $\vec{r} = \vec{\alpha} + t\,\vec{\beta}$, where t is scalar, measured in the direction of the vector $\vec{\gamma}$.

Solution

The unit vector in the direction of $\vec{\gamma}$ is $\hat{\gamma} = \dfrac{1}{|\vec{\gamma}|}\vec{\gamma}$. PN is drawn (Fig. 3.6) parallel to vector $\vec{\gamma}$ and meets the given line $\vec{r} = \vec{\alpha} + t\,\vec{\beta}$ at N. Since $\vec{PN}$ and $\hat{\gamma}$ are collinear, $\vec{PN} = PN\,\hat{\gamma}$. Then

$$\vec{ON} = \vec{OP} + \vec{PN} = \vec{p} + PN\,\hat{\gamma}$$

This gives the position vector of N relative to the origin O. Since N is a point on the line $\vec{r} = \vec{\alpha} + t\,\vec{\beta}$, we have, for some value of t, say t_0

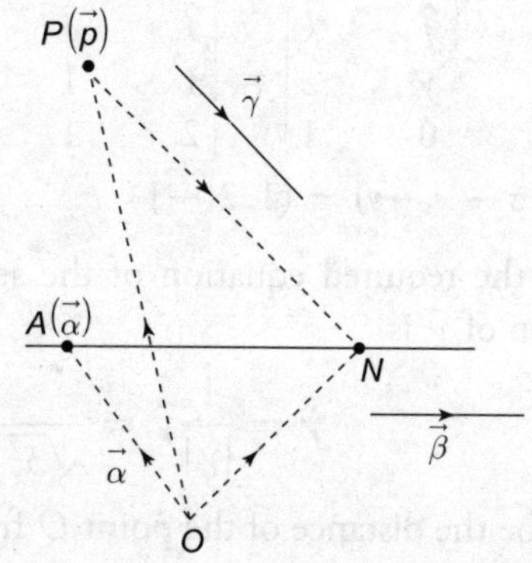

Figure 3.6 Vector equation of a straight line

$$\vec{ON} = \vec{\alpha} + t_0\,\vec{\beta} = \vec{p} + PN\,\hat{\gamma}$$

$$\Rightarrow (\vec{p} - \vec{\alpha}) \times \vec{\beta} + PN\,\hat{\gamma} \times \vec{\beta} = \vec{0}$$

$$\Rightarrow \left|\vec{\beta} \times \hat{\gamma}\right| PN = \left|(\vec{p} - \vec{\alpha}) \times \vec{\beta}\right|$$

$$\Rightarrow PN = \frac{1}{|\vec{\beta} \times \hat{\gamma}|}|(\vec{p} - \vec{\alpha}) \times \vec{\beta}|$$

This is the distance of $P(\vec{p})$ from the line $\vec{r} = \vec{\alpha} + t\,\vec{\beta}$, where t is scalar, measured in the direction of the vector $\vec{\gamma}$.

Deduction 3.2.2

If we want to find the perpendicular distance of the point $P(\vec{p})$ from the line $\vec{r} = \vec{\alpha} + t\,\vec{\beta}$, then taking $\vec{\gamma}$ as a vector perpendicular to the line $\vec{r} = \vec{\alpha} + t\,\vec{\beta}$, we have

$$\vec{\beta} \cdot \vec{\gamma} = 0 \Rightarrow |\vec{\beta} \times \hat{\gamma}| = |\vec{\beta}|\,|\hat{\gamma}|\,\sin 90^\circ = |\vec{\beta}|$$

Then, the perpendicular distance of $P(\vec{p})$ from the given line $\vec{r} = \vec{\alpha} + t\,\vec{\beta}$ is

$$PN = \frac{1}{|\vec{\beta}|}\left|(\vec{p} - \vec{\alpha}) \times \vec{\beta}\right| = \left|(\vec{p} - \vec{\alpha}) \times \hat{\beta}\right|$$

Example 3.2.5

Find the vector equation of the straight line passing through the points $A(1, 1, 3)$ and $B(2, 1, 4)$. Hence, find the distance of the point $Q(1, -1, 2)$ from the line AB measured parallel to the vector $\vec{\gamma} = (3, 2, 4)$.

Solution

Let $P(\vec{r})$ be any point on the line AB (Fig. 3.7). Let $\vec{\alpha} = \vec{OA} = (1,1,3)$, $\vec{\beta} = \vec{OB} = (2,1,4)$ and $\vec{p} = \vec{OQ} = (1,-1,2)$. Therefore, $\vec{\beta} - \vec{\alpha} = \vec{OB} - \vec{OA} = (2,1,4) - (1,1,3) = (1,0,1)$. From Fig. 3.7, we see that $\vec{PA}$ and $\vec{AB}$ are collinear. So, $\vec{PA} \times \vec{AB} = \vec{0}$, i.e.

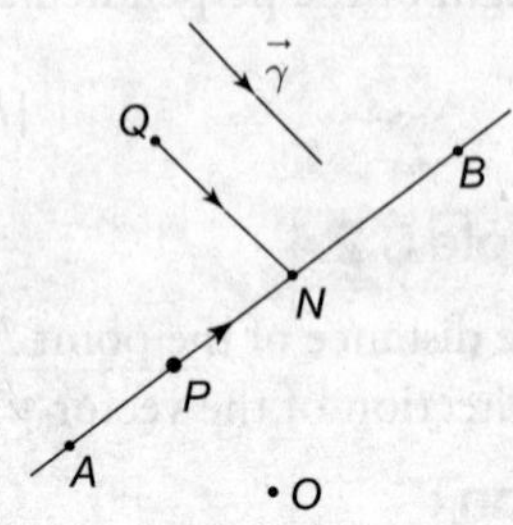

Figure 3.7 Distance from a point to a straight line

$$(\vec{r} - \vec{\alpha}) \times (\vec{\beta} - \vec{\alpha}) = \vec{0} \Rightarrow \vec{r} \times (\vec{\beta} - \vec{\alpha}) = \vec{\alpha} \times \vec{\beta}$$

$$\Rightarrow \begin{vmatrix} \hat{i} & \hat{j} & \hat{k} \\ x & y & z \\ 1 & 0 & 1 \end{vmatrix} = \begin{vmatrix} \hat{i} & \hat{j} & \hat{k} \\ 1 & 1 & 3 \\ 2 & 1 & 4 \end{vmatrix}$$

$$\Rightarrow (y, z - x, -y) = (1, 2, -1)$$

This is the required equation of the straight line in Cartesian form. The unit vector $\hat{\gamma}$ in the direction of $\vec{\gamma}$ is

$$\hat{\gamma} = \frac{1}{|\vec{\gamma}|}\vec{\gamma} = \frac{1}{\sqrt{3^2 + 2^2 + 4^2}}(3,2,4) = \frac{1}{\sqrt{29}}(3,2,4)$$

If QN be the distance of the point Q from the line AB measured parallel to $\vec{\gamma}$, then

$$\vec{QN} = QN\,\hat{\gamma} = QN \cdot \frac{1}{\sqrt{29}}(3,2,4)$$

From Fig. 3.7, we get

$$\vec{ON} = \vec{OQ} + \vec{QN} = (1,-1,2) + \frac{QN}{\sqrt{29}}(3,2,4)$$

Since N is a point on the line AB, $\vec{ON} \times (\vec{b} - \vec{a}) = \vec{a} \times \vec{b}$. Thus

$$\begin{vmatrix} \hat{i} & \hat{j} & \hat{k} \\ 1 & -1 & 2 \\ 1 & 0 & 1 \end{vmatrix} + \frac{QN}{\sqrt{29}}\begin{vmatrix} \hat{i} & \hat{j} & \hat{k} \\ 3 & 2 & 4 \\ 1 & 0 & 1 \end{vmatrix} = (1,2,-1)$$

$$\Rightarrow (-1,1,1) + \frac{QN}{\sqrt{29}}(2,1,-2) = (1,2,-1)$$

$$\Rightarrow \frac{QN}{\sqrt{29}} = 1 \Rightarrow QN = \sqrt{29} \text{ units}$$

Example 3.2.6

Find the perpendicular distance of the point $P(\vec{p})$ from the straight line joining the points $A(\vec{\alpha})$ and $B(\vec{\beta})$.

Solution

The equation of the straight line AB passing through the points $A(\vec{\alpha})$ and $B(\vec{\beta})$ is of the form (3.14), as $\vec{r} = (1-t)\vec{\alpha} + t\,\vec{\beta}$ (Fig. 3.8). Let P be a point whose position vector is $\vec{p}$, whose distance from the straight line AN is to be determined. Thus, $\vec{AP} = \vec{p} - \vec{\alpha}$. Let $\angle PAN = \theta$ and $\hat{\gamma}$ be the unit vectors in the direction of the vector $\vec{\gamma} = \vec{\beta} - \vec{\alpha}$. AN is the length projection of AP

along the line AN. So

$$AN = |\vec{AP} \cdot \hat{\gamma}| = |(\vec{p} - \vec{\alpha}) \cdot \hat{\gamma}|$$

PN is the perpendicular from the point P to the line AN. Then

$$PN = AP \sin\theta = \left|\vec{AP} \times \hat{\gamma}\right|$$

$$= \left|(\vec{p} - \vec{\alpha}) \times \hat{\gamma}\right|$$

$$= \frac{1}{|\vec{\gamma}|}\left[|(\vec{p} - \vec{\alpha}) \times \vec{\gamma}|\right]$$

$$= \frac{1}{|\vec{\beta} - \vec{\alpha}|}\left[|(\vec{p} - \vec{\alpha}) \times (\vec{\beta} - \vec{\alpha})|\right]$$

$$= \frac{1}{|\vec{\beta} - \vec{\alpha}|}\left[|\vec{p} \times \vec{\beta} - \vec{p} \times \vec{\alpha} - \vec{\alpha} \times \vec{\beta} + \vec{\alpha} \times \vec{\alpha})|\right]$$

$$\therefore PN = \frac{1}{|\vec{\beta} - \vec{\alpha}|}|\vec{p} \times \vec{\beta} + \vec{\alpha} \times \vec{p} - \vec{\alpha} \times \vec{\beta}|$$

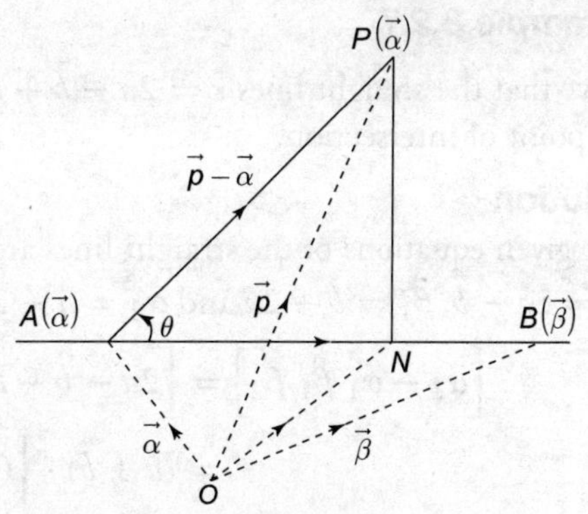

Figure 3.8 Perpendicular distance from a point to a straight line

The projection vector of the foot of the perpendicular N relative to the origin O is

$$\vec{ON} = \vec{OA} + \vec{AN} = \vec{\alpha} + \frac{1}{|\vec{\beta} - \vec{\alpha}|}\left|\vec{p} \times \vec{\beta} + \vec{\alpha} \times \vec{p} - \vec{\alpha} \times \vec{\beta}\right| \hat{\gamma} \tag{3.22}$$

Deduction 3.2.3

Condition for intersection of two straight lines:
Let us consider the following two straight lines

$L_1 : \vec{r} = \vec{\alpha}_1 + t\,\vec{\beta}_1$ and $L_2 : \vec{r} = \vec{\alpha}_2 + s\,\vec{\beta}_2$ (3.23)

where t and s are scalar quantities. The first straight line L_1 passes through the point $\vec{\alpha}_1$ and is parallel to the vector $\vec{\beta}_1$ (Fig. 3.9). The second line L_2 passes through the point $\vec{\alpha}_2$ and is parallel to the vector $\vec{\beta}_2$. Thus, in order to meet the line L_1 with L_2 at a point, they must lie on a plane, i.e., the vectors $\vec{\beta}_1$, $\vec{\beta}_2$ and $\vec{\alpha}_2 - \vec{\alpha}_1$ will be coplanar. Therefore, the condition for coplanarity gives

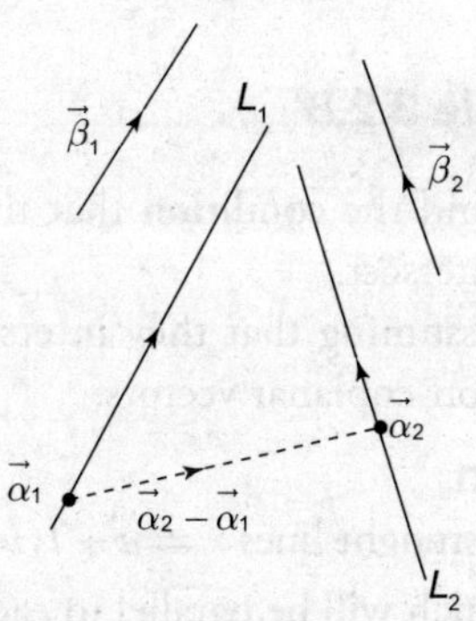

Figure 3.9 Vector equation of a straight line

$$(\vec{\alpha}_2 - \vec{\alpha}_1) \times \{\vec{\beta}_1 \cdot \vec{\beta}_2\} = 0$$

$$\Rightarrow \left[\vec{\alpha}_2 - \vec{\alpha}_1 \;\; \vec{\beta}_1 \;\; \vec{\beta}_2\right] = 0$$

$$\Rightarrow \left[\vec{\alpha}_2 \;\; \vec{\beta}_1 \;\; \vec{\beta}_2\right] = \left[\vec{\alpha}_1 \;\; \vec{\beta}_1 \;\; \vec{\beta}_2\right] \tag{3.24}$$

This is the condition for the intersection of two lines. At the point of intersection of the two straight lines, the values of $\vec{r}$ will be the same.

Example 3.2.7

Prove that the straight lines $\vec{r} = 2\vec{a} - \vec{b} + t(\vec{b} + 2\vec{a})$ and $\vec{r} = \vec{a} - 2\vec{b} + s(2\vec{b} + \vec{a})$ intersect. Find the point of intersection.

Solution

The given equations of the straight lines are of the form $\vec{r} = \vec{\alpha}_1 + t\,\vec{\beta}_1$ and $\vec{r} = \vec{\alpha}_2 + s\,\vec{\beta}_2$, where $\vec{\alpha}_1 = 2\vec{a} - \vec{b}, \vec{\beta}_1 = \vec{b} + 2\vec{a}$ and $\vec{\alpha}_2 = \vec{a} - 2\vec{b}, \vec{\beta}_2 = 2\vec{b} + \vec{a}$. Now

$$\left[\vec{\alpha}_2 - \vec{\alpha}_1\ \vec{\beta}_1\ \vec{\beta}_2\right] = \left[2\vec{a} - \vec{b} - \vec{a} + 2\vec{b}\ \ \vec{b} + 2\vec{a}\ \ 2\vec{b} + \vec{a}\right]$$
$$= (\vec{a} + \vec{b}) \cdot \left\{(\vec{b} + 2\vec{a}) \times (2\vec{b} + \vec{a})\right\}$$
$$= (\vec{a} + \vec{b}) \cdot \left\{(\vec{b} \times \vec{a}) + 4(\vec{a} \times \vec{b})\right\}$$
$$= 3(\vec{a} + \vec{b}) \cdot (\vec{a} \times \vec{b}) = 3\left\{\vec{a} \cdot (\vec{a} \times \vec{b}) + \vec{b} \cdot (\vec{a} \times \vec{b})\right\} = 0$$

Thus, the straight lines $\vec{r} = 2\vec{a} - \vec{b} + t(\vec{b} + 2\vec{a})$ and $\vec{r} = \vec{a} - 2\vec{b} + s(2\vec{b} + \vec{a})$ intersect. At the point of intersection of the lines, we have

$$2\vec{a} - \vec{b} + t(\vec{b} + 2\vec{a}) = \vec{a} - 2\vec{b} + s(2\vec{b} + \vec{a})$$
$$\Rightarrow\ (1 + 2t - s)\,\vec{a} + (1 + t - 2s)\,\vec{b} = \vec{0}$$

Since $\vec{a}$ and $\vec{b}$ are non-collinear, we have $1 + 2t - s = 0$ and $1 + t - 2s = 0$. Solving, we get $t = -1/3$ and $s = 1/3$. Thus, the position vector of the point of intersection is obtained as

$$\vec{r} = 2\vec{a} - \vec{b} + t(\vec{b} + 2\vec{a}) = 2\vec{a} - \vec{b} - \frac{1}{3}(\vec{b} + 2\vec{a}) = \frac{4}{3}(\vec{a} - \vec{b})$$

Example 3.2.8

(i) Find the condition that the straight lines $\vec{r} = \vec{a} + t(\vec{b} \times \vec{c})$ and $\vec{r} = \vec{b} + s(\vec{c} \times \vec{a})$ may intersect.

(ii) Assuming that they intersect, find the point of intersection, where $\vec{a}$, $\vec{b}$ and $\vec{c}$ are three non-coplanar vectors.

Solution

(i) If the straight lines $\vec{r} = \vec{a} + t(\vec{b} \times \vec{c})$ and $\vec{r} = \vec{b} + s(\vec{c} \times \vec{a})$ intersect, then they lie on a common plane, which will be parallel to each of the vectors $\vec{b} \times \vec{c}$, $\vec{c} \times \vec{a}$ and $\vec{a} - \vec{b}$. Thus, the three vectors $\vec{b} \times \vec{c}$, $\vec{c} \times \vec{a}$ and $\vec{a} - \vec{b}$ are coplanar. Therefore, their scalar triple product vanishes, i.e.

$$\left\{(\vec{b} \times \vec{c}) \times (\vec{c} \times \vec{a})\right\} \cdot (\vec{a} - \vec{b}) = 0$$
$$\Rightarrow\ \left(\left[\vec{b}\,\vec{c}\,\vec{a}\right]\vec{c} - \left[\vec{c}\,\vec{c}\,\vec{a}\right]\vec{b}\right) \cdot (\vec{a} - \vec{b}) = 0$$
$$\Rightarrow\ \left[\vec{a}\,\vec{b}\,\vec{c}\right]\left(\vec{c} \cdot \vec{a} - \vec{c} \cdot \vec{b}\right) = 0; \text{ as } \left[\vec{c}\,\vec{c}\,\vec{a}\right] = 0$$
$$\Rightarrow\ \vec{a} \cdot \vec{c} = \vec{b} \cdot \vec{c}; \text{ as } \left[\vec{a}\,\vec{b}\,\vec{c}\right] \neq 0$$

This is the condition for the intersection of the two straight lines $\vec{r} = \vec{a} + t(\vec{b} \times \vec{c})$ and $\vec{r} = \vec{b} + s(\vec{c} \times \vec{a})$.

(ii) Now, we assume that $\vec{a} \cdot \vec{c} = \vec{b} \cdot \vec{c}$, i.e., the two lines intersect. To find the point of intersection of the two lines, we use the fact that $\vec{b} \times \vec{c}$, $\vec{c} \times \vec{a}$ and $\vec{a} \times \vec{b}$ are non-coplanar vectors, since $\vec{a}$, $\vec{b}$ and $\vec{c}$ are three non-coplanar vectors. Hence, we can express $\vec{a}$ and $\vec{b}$ as a linear combination of the non-coplanar vectors $\vec{b} \times \vec{c}$, $\vec{c} \times \vec{a}$ and $\vec{a} \times \vec{b}$. Let x, y and z be scalars, then

$$\begin{aligned}
&\vec{a} = x(\vec{b} \times \vec{c}) + y(\vec{c} \times \vec{a}) + z(\vec{a} \times \vec{b}) \\
\Rightarrow\ &\vec{a} \cdot \vec{a} = x(\vec{b} \times \vec{c}) \cdot \vec{a} + y(\vec{c} \times \vec{a}) \cdot \vec{a} + z(\vec{a} \times \vec{b}) \cdot \vec{a} \\
\Rightarrow\ &\vec{a} \cdot \vec{a} = x\left[\vec{a}\,\vec{b}\,\vec{c}\right] + y\left[\vec{a}\,\vec{c}\,\vec{a}\right] + z\left[\vec{a}\,\vec{a}\,\vec{b}\right] \\
\Rightarrow\ &\vec{a} \cdot \vec{a} = x\left[\vec{a}\,\vec{b}\,\vec{c}\right] \text{ as } \left[\vec{a}\,\vec{c}\,\vec{a}\right] = 0 = \left[\vec{a}\,\vec{a}\,\vec{b}\right] \\
\Rightarrow\ &x = \frac{1}{\left[\vec{a}\,\vec{b}\,\vec{c}\right]}(\vec{a} \cdot \vec{a}); \quad \text{as } \left[\vec{a}\,\vec{b}\,\vec{c}\right] \neq 0
\end{aligned}$$

Similarly, $y = \dfrac{1}{\left[\vec{a}\,\vec{b}\,\vec{c}\right]}(\vec{a} \cdot \vec{b})$ and $z = \dfrac{1}{\left[\vec{a}\,\vec{b}\,\vec{c}\right]}(\vec{a} \cdot \vec{c})$. Using these values, the vector $\vec{a}$ can be written as

$$\vec{a} = \frac{\vec{a} \cdot \vec{a}}{\left[\vec{a}\,\vec{b}\,\vec{c}\right]}(\vec{b} \times \vec{c}) + \frac{\vec{a} \cdot \vec{b}}{\left[\vec{a}\,\vec{b}\,\vec{c}\right]}(\vec{c} \times \vec{a}) + \frac{\vec{a} \cdot \vec{c}}{\left[\vec{a}\,\vec{b}\,\vec{c}\right]}(\vec{a} \times \vec{b})$$

Similarly, we have

$$\vec{b} = \frac{\vec{b} \cdot \vec{a}}{\left[\vec{a}\,\vec{b}\,\vec{c}\right]}(\vec{b} \times \vec{c}) + \frac{\vec{b} \cdot \vec{b}}{\left[\vec{a}\,\vec{b}\,\vec{c}\right]}(\vec{c} \times \vec{a}) + \frac{\vec{b} \cdot \vec{c}}{\left[\vec{a}\,\vec{b}\,\vec{c}\right]}(\vec{a} \times \vec{b})$$

Thus, the equations of the two straight lines can be written as

$$\vec{r} = \frac{\vec{a} \cdot \vec{a}}{\left[\vec{a}\,\vec{b}\,\vec{c}\right]}(\vec{b} \times \vec{c}) + \frac{\vec{a} \cdot \vec{b}}{\left[\vec{a}\,\vec{b}\,\vec{c}\right]}(\vec{c} \times \vec{a}) + \frac{\vec{a} \cdot \vec{c}}{\left[\vec{a}\,\vec{b}\,\vec{c}\right]}(\vec{a} \times \vec{b}) + t(\vec{b} \times \vec{c})$$

$$\vec{r} = \frac{\vec{b} \cdot \vec{a}}{\left[\vec{a}\,\vec{b}\,\vec{c}\right]}(\vec{b} \times \vec{c}) + \frac{\vec{b} \cdot \vec{b}}{\left[\vec{a}\,\vec{b}\,\vec{c}\right]}(\vec{c} \times \vec{a}) + \frac{\vec{b} \cdot \vec{c}}{\left[\vec{a}\,\vec{b}\,\vec{c}\right]}(\vec{a} \times \vec{b}) + s(\vec{c} \times \vec{a})$$

Since these two lines intersect, we have a value for t (and also s) for which $\vec{r}$ given by the two equations is equal. Equating the coefficient of $\vec{b} \times \vec{c}$ from both the equations, we have

$$\frac{\vec{a} \cdot \vec{a}}{\left[\vec{a}\,\vec{b}\,\vec{c}\right]} + t = \frac{\vec{b} \cdot \vec{a}}{\left[\vec{a}\,\vec{b}\,\vec{c}\right]} \Rightarrow t = \frac{\vec{b} \cdot \vec{a}}{\left[\vec{a}\,\vec{b}\,\vec{c}\right]} - \frac{\vec{a} \cdot \vec{a}}{\left[\vec{a}\,\vec{b}\,\vec{c}\right]}$$

Then, for this value of t, $\vec{r}$ from the first equation becomes

$$\vec{r} = \frac{1}{\left[\vec{a}\,\vec{b}\,\vec{c}\right]}\left\{(\vec{b}\cdot\vec{a})(\vec{b}\times\vec{c}) + (\vec{a}\cdot\vec{b})(\vec{c}\times\vec{a}) + (\vec{a}\cdot\vec{c})(\vec{a}\times\vec{b})\right\}$$

This gives the point of intersection of the two straight lines.

3.2.2 Angle Bisector of Two Intersecting Straight Lines

Let $\vec{\alpha}$ and $\vec{\beta}$ be the unit vectors along the straight lines OA and OB intersecting at the point O, taken as origin (Fig. 3.10). Let $\vec{r}$ be the position vector of any point P on the internal bisector. Through P, draw PM parallel to OB which meets OA at M. Since $\angle OPM = \angle BOP = \angle MOP$, we get $OM = PM$. The unit vector in the direction of OA is $\hat{\alpha} = \frac{1}{|\vec{\alpha}|}\vec{\alpha}$ and that in the direction of OB is $\hat{\beta} = \frac{1}{|\vec{\beta}|}\vec{\beta}$. Hence

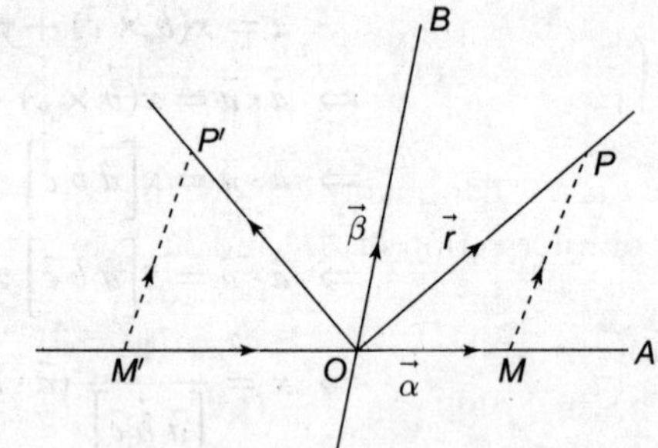

Figure 3.10 Bisector of an angle

$$\vec{OM} = t\,\hat{\alpha} = \frac{t}{|\vec{\alpha}|}\vec{\alpha}, \quad \vec{MP} = t\,\hat{\beta} = \frac{t}{|\vec{\beta}|}\vec{\beta}$$

where t is a scalar quantity. From Fig. 3.10

$$\vec{r} = \vec{OP} = \vec{OM} + \vec{MP} = t\left(\frac{1}{|\vec{\alpha}|}\vec{\alpha} + \frac{1}{|\vec{\beta}|}\vec{\beta}\right) \tag{3.25}$$

This is the vector equation of the internal bisector OP of the angle $\angle AOB$ between the lines OA and OB. Again, if we take a point P' on the external bisector whose position vector is $\vec{r}$, and draw $P'M'$ parallel to OB, as shown in Fig. 3.10, then

$$\vec{r} = \vec{OP'} = \vec{OM'} + \vec{M'P'} = t\left(\frac{1}{|\vec{\alpha}|}\vec{\alpha} - \frac{1}{|\vec{\beta}|}\vec{\beta}\right) \tag{3.26}$$

This is the vector equation of the external bisector OP' of the angle $\angle AOB$ between the lines OA and OB.

DEDUCTION 3.2.4

Let the equations of the straight lines OA and OB be $\vec{r} = \vec{\gamma} + t\vec{\alpha}$ and $\vec{r} = \vec{\gamma} + t\vec{\beta}$ so that their point of intersection is the point O, whose position vector is $\vec{\gamma}$, and they are parallel to the vectors $\vec{\alpha}$ and $\vec{\beta}$, respectively. Then, the equation of the bisector becomes

$$\vec{r} = \vec{\gamma} + t\left(\frac{1}{|\vec{\alpha}|}\vec{\alpha} \pm \frac{1}{|\vec{\beta}|}\vec{\beta}\right) \tag{3.27}$$

Example 3.2.9

Show that the internal bisector of an angle of a triangle divides the opposite side internally in the ratio of the sides containing the angle.

Solution

Let $\vec{c}$ and $\vec{b}$ be the respective position vectors of B and C relative to A (Fig. 3.11). Then, $\vec{AB} = \vec{c}$ and $\vec{AC} = \vec{b}$. Let $\vec{r}$ be the position vector of any point on the bisector AD. With A as origin, the equation of the internal bisector of the angle $\angle A$ is given by

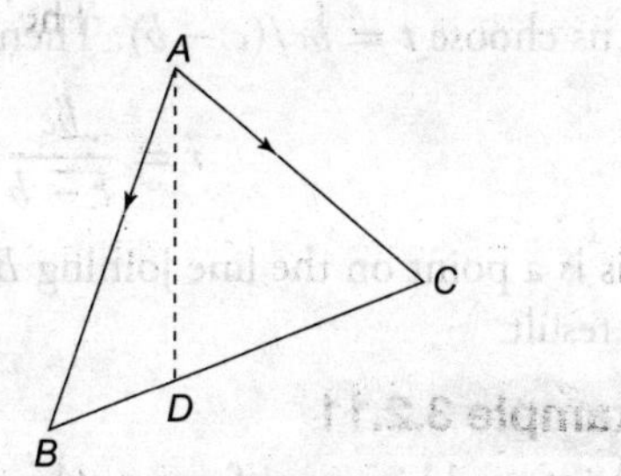

Figure 3.11 Internal bisector of a triangle

$$\vec{r} = t\left(\frac{1}{|\vec{b}|}\vec{b} + \frac{1}{|\vec{c}|}\vec{c}\right) = t\,\frac{1}{|\vec{b}|\,|\vec{c}|}(|\vec{b}|\,\vec{c} + |\vec{c}|\,\vec{b})$$

where t is a scalar quantity. Now, the equation of the straight line BC is $\vec{r} = \vec{b} + s(\vec{c} - \vec{b})$. At the point of intersection of AD and BC, we have

$$\vec{b} + s(\vec{c} - \vec{b}) = \frac{t}{|\vec{b}|\,|\vec{c}|}(|\vec{b}|\,\vec{c} + |\vec{c}|\,\vec{b})$$

$$\Rightarrow \left[1 - s - \frac{t}{|\vec{b}|}\right]\vec{b} + \left[s - \frac{t}{|\vec{c}|}\right]\vec{c} = \vec{0}$$

$$\Rightarrow 1 - s - \frac{t}{|\vec{b}|} = 0, s - \frac{t}{|\vec{c}|}, \quad \text{as } \vec{b} \text{ and } \vec{c} \text{ are non-collinear}$$

$$\Rightarrow t = \frac{|\vec{b}|\,|\vec{c}|}{|\vec{b}| + |\vec{c}|} \text{ and } s = \frac{|\vec{b}|}{|\vec{b}| + |\vec{c}|}$$

Substituting t in the equation of AD, we get the position vector of the point D as

$$\vec{r} = \frac{t}{|\vec{b}|\,|\vec{c}|}(|\vec{b}|\vec{c} + |\vec{c}|\vec{b}) = \frac{|\vec{b}|\,|\vec{c}|}{|\vec{b}| + |\vec{c}|} \cdot \frac{1}{|\vec{b}|\,|\vec{c}|}(|\vec{b}|\,\vec{c} + |\vec{c}|\,\vec{b})$$

$$= \frac{1}{|\vec{b}|\,|\vec{c}|}(|\vec{b}|\,\vec{c} + |\vec{c}|\,\vec{b})$$

This shows that the position vector of the point D divides the side BC in the ratio $|\vec{b}| : |\vec{c}|$, i.e., in the ratio $AB : AC$.

Example 3.2.10

Show that the external bisector of an angle of a triangle divides the opposite side externally in the ratio of the other two sides.

Solution

Let us consider an external bisector AE of the angle $\angle BAC$ of a triangle ABC (Fig. 3.12). With A as the vector origin, $\vec{AB} = \vec{b}$, where $|\vec{b}| = c$, $\vec{AC} = \vec{c}$, and $|\vec{c}| = b$. Then, the vector equation of the bisector AE is

$$\vec{r} = t\left(\frac{1}{b}\vec{c} - \frac{1}{c}\vec{b}\right) = \frac{t}{bc}(c\,\vec{c} - b\,\vec{b}); \quad t = \text{ scalar}$$

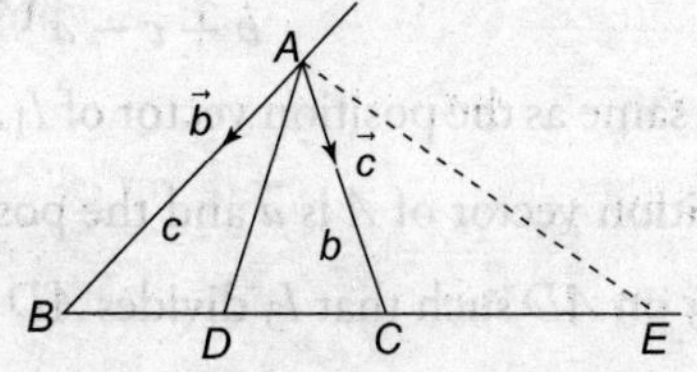

Figure 3.12 External bisector AE of a triangle ABC

Let us choose $t = bc/(c - b)$. Then, the position vector of any point on AE is given by

$$\vec{r} = \frac{bc}{c-b}\,\frac{1}{bc}(c\,\vec{c} - b\,\vec{b}) = \frac{1}{c-b}(c\,\vec{c} - b\,\vec{b})$$

This is a point on the line joining B and C dividing BC externally in the ratio $c : b$. This proves the result.

Example 3.2.11

The internal bisector of one angle of a triangle and the external bisectors of the other two angles of a triangle are concurrent.

Solution

Let ABC be a triangle. Suppose $\vec{a}, \vec{b}$ and $\vec{c}$ are the position vectors of the vertices A, B and C, respectively, with respect to a certain vector origin O (Fig. 3.13) and AD is the internal bisector of $\angle BAC$, BE is the external bisector of $\angle ABC$ and CF is the external bisector of $\angle ACB$. If the external bisector of $\angle ABC$ meets AC (produced) at E, then E divides AC externally in the ratio $c : -b$, and hence, the position vector of E is $\frac{1}{c-a}(c\,\vec{c} - a\,\vec{a})$. We take a point I_1 on BE such that $BI_1 : I_1E = (c - a) : b$. Then, the position vector of I_1 is

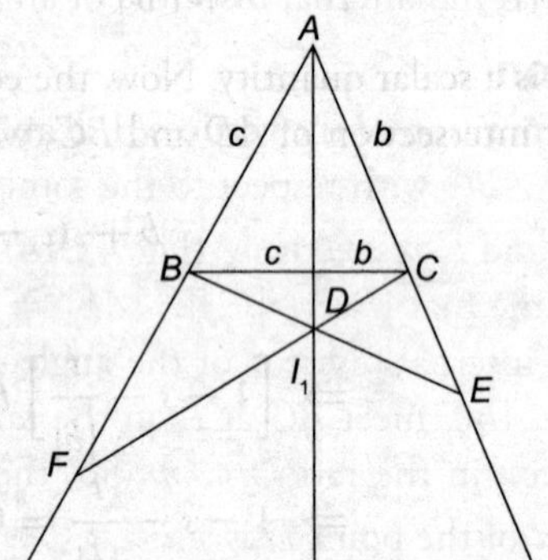

Figure 3.13 Concurrence of one internal and two external bisectors

$$= \frac{1}{(c-a)+b}\left[(c-a)\,\frac{1}{c-a}(c\,\vec{c} - a\,\vec{a}) + b\,\vec{b}\right]$$

$$= \frac{1}{b+c-a}\left(b\,\vec{b} + c\,\vec{c} - a\,\vec{a}\right)$$

If the external bisector of $\angle ACB$ meets AB (produced) externally at F, then the position vector of F is $\frac{1}{b-a}(b\,\vec{c} - a\,\vec{a})$. We take a point I_2 on CF such that $CI_2 : I_2F = (b - a) : c$. Then, the position vector of I_2 is

$$= \frac{1}{(b-a)+c}\left[(b-a)\,\frac{1}{b-a}(b\,\vec{b} - a\,\vec{a}) + c\,\vec{c}\right]$$

$$= \frac{1}{b+c-a}\left(b\,\vec{b} + c\,\vec{c} - a\,\vec{a}\right)$$

i.e., the same as the position vector of I_1. I_1 and I_2 coincide. On the internal bisector AD of $\angle BAC$, the position vector of A is $\vec{a}$ and the position vector of D is $\frac{1}{b+c}(b\,\vec{b} + c\,\vec{c})$. We take an external point I_3 on AD such that I_3 divides AD in the ratio $(b + c) : -a$. Thus, the position vector of I_3 is

$$= \frac{1}{(b+c)-a}\left[(b+c)\,\frac{1}{b+c}(b\,\vec{b} - c\,\vec{c}) + (-a)\,\vec{a}\right]$$

$$= \frac{1}{b+c-a}\left(b\,\vec{b}+c\,\vec{c}-a\,\vec{a}\right)$$

This is the same as the position vector of I_1 or I_2, i.e., the three points I_1, I_2 and I_3 coincide. In other words, the three bisectors meet at a point such that the position vector of the point of concurrence is $\frac{1}{b+c-a}(b\,\vec{b}+c\,\vec{c}-a\,\vec{a})$. The point of concurrence is the centroid of the points $\vec{a}, \vec{b}$ and $\vec{c}$ with associated numbers $-a, b$ and c, respectively.

Example 3.2.12

Show that the internal bisectors of an angle of a triangle are concurrent.

Solution

Let the position of the vertices A, B and C of the triangle ABC with respect to the some origin O be $\vec{a}, \vec{b}$ and $\vec{c}$, respectively (Fig. 3.14). Further, we assume that $|BC| : |CA| : |AB| = l : m : n$. Let the internal bisector of the angle $\angle A$ of the triangle ABC meet BC at D in the ratio $|AB| : |AC|$, i.e., in the ratio $n : m$. So, the position vector $\vec{d}$ of the point D is

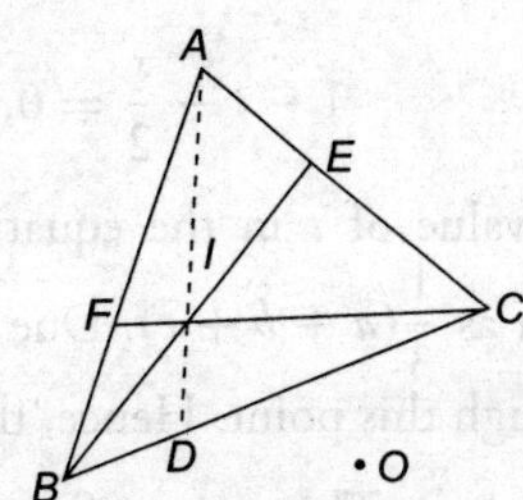

Figure 3.14 Internal bisectors of a triangle

$$\vec{d} = \frac{1}{m+n}(m\vec{b}+n\vec{c})$$

This is relative to the base point O. Now, the position vector $\vec{I}$ of the point I, which divides AD in the ratio $(m+n) : l$, is

$$\begin{aligned}\vec{I} &= \frac{1}{(m+n)+l}\left\{l\vec{a}+(m+n)\vec{d}\right\}\\ &= \frac{1}{l+m+n}\left\{l\vec{a}+(m+n)\frac{1}{m+n}(m\vec{b}+n\vec{c})\right\}\\ &= \frac{1}{l+m+n}\left\{l\vec{a}+m\vec{b}+n\vec{c}\right\}\end{aligned}$$

The symmetry of the above result shows that the other two internal bisectors also pass through the point I. Hence, the internal bisectors of the angles of a triangle meet at a point, and therefore they are concurrent.

Example 3.2.13

Prove that the medians of a triangle are concurrent and find the point of concurrence.

Solution

Let the position of the vertices A, B and C of the triangle ABC with respect to the some origin O be $\vec{a}, \vec{b}$ and $\vec{c}$, respectively (Fig. 3.14). If D, E and F are the middle points of the sides CB, AC and BA, then the position vectors are $\frac{1}{2}(\vec{b}+\vec{c})$, $\frac{1}{2}(\vec{c}+\vec{a})$ and $\frac{1}{2}(\vec{a}+\vec{b})$, respectively. So, using

Eq. (3.14), the vector equations of the straight lines AD and BE are

$$\vec{r} = (1-t)\vec{a} + t\frac{1}{2}(\vec{b}+\vec{c}) \text{ and } \vec{r} = (1-s)\vec{b} + s\frac{1}{2}(\vec{c}+\vec{a})$$

If the two straight lines AD and BE intersect, at the point of intersection, we have

$$(1-t)\vec{a} + \frac{t}{2}(\vec{b}+\vec{c}) = (1-s)\vec{b} + \frac{s}{2}(\vec{c}+\vec{a})$$

$$\Rightarrow \left(1-t-\frac{s}{2}\right)\vec{a} + \left(\frac{t}{2}-1+s\right)\vec{b} + \frac{t-s}{2}\vec{c} = \vec{0}$$

Since $\vec{a}$, $\vec{b}$ and $\vec{c}$ are non-collinear, we have

$$1-t-\frac{s}{2} = 0, \frac{t}{2}-1+s = 0, t-s = 0 \Rightarrow t = s = \frac{2}{3}$$

Using the value of t in the equation of AD, we obtain the position vector of their point of intersection as $\frac{1}{3}(\vec{a}+\vec{b}+\vec{c})$. Due to symmetry in $\vec{a}$, $\vec{b}$ and $\vec{c}$, we see that the third median also passes through this point. Hence, the three medians are concurrent and the point of concurrence is $\frac{1}{3}(\vec{a}+\vec{b}+\vec{c})$. This point of concurrence is the centroid of the triangle. This shows that the centroid trisects the medians of a triangle.

Example 3.2.14

Show that the middle points of the three diagonals of a complete quadrilateral are collinear.

Solution

A complete quadrilateral is composed of four straight lines, no three of which are concurrent. Let $ABCD$ be a complete quadrilateral and P, Q and R be the respective middle points of the diagonals BD, CA and EF (Fig. 3.15). Let $\vec{a}, \vec{b}, \vec{c}, \vec{d}, \vec{e}, \vec{f}, \vec{p}, \vec{q}$ and $\vec{r}$ be the respective position vectors of the points A, B, C, D, E, F, P, Q and R relative to some origin O. Then

$$\vec{p} = \frac{1}{2}(\vec{b}+\vec{d}), \vec{q} = \frac{1}{2}(\vec{c}+\vec{a}) \text{ and } \vec{r} = \frac{1}{2}(\vec{e}+\vec{f})$$

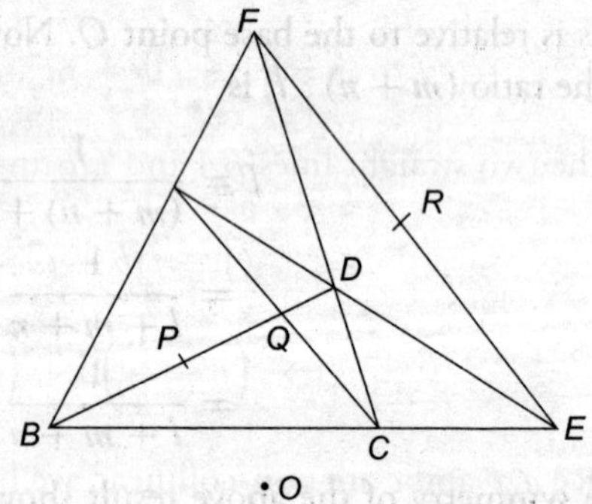

Figure 3.15 Complete quadrilateral

Since E lies on the two lines BC and AD and F lies on the two lines BA and CD separately, we have

$$(1-s)\vec{b} + s\vec{c} - \vec{e} = \vec{0}, (1-t)\vec{a} + t\vec{d} - \vec{e} = \vec{0}$$

$$(1-u)\vec{b} + u\vec{a} - \vec{f} = \vec{0}, (1-v)\vec{c} + v\vec{d} - \vec{f} = \vec{0}$$

where s, t, u and v are scalars. Adding all these relations, we get

$$(1-t+u)\vec{a} + (2-s-u)\vec{b} + (1+s-v)\vec{c} + (t+v)\vec{d} - 2(\vec{e}+\vec{f}) = \vec{0}$$

Setting $1-t+u=1+s-v$ and $2-s-u=t+v$, so that $s+t=1=u+v$, the above equation becomes

$$(2-s-u)(\vec{b}+\vec{d})+(1-t+u)(\vec{c}+\vec{a})-2(\vec{e}+\vec{f})=\vec{0}$$

$$\Rightarrow x\vec{p}+y\vec{q}-z\vec{r}=\vec{0}$$

where $x+y+z=(2-s-u)+(1-t+u)-2=1-(s+t)=0$. This shows that the points P, Q and R are collinear.

Example 3.2.15

In a triangle ABC, D divides BC in the ratio $2:1$ and E divides CA in the ratio $3:1$. The line DE meets the line AB in F. Find the ratio in which F divides AB.

Solution

Let the position of the vertices A, B and C of the triangle ABC with respect to the some origin O be $\vec{a}$, $\vec{b}$ and $\vec{c}$, respectively (Fig. 3.16). Then, the position vectors $\vec{d}$ and $\vec{e}$ of the points D and E are, respectively

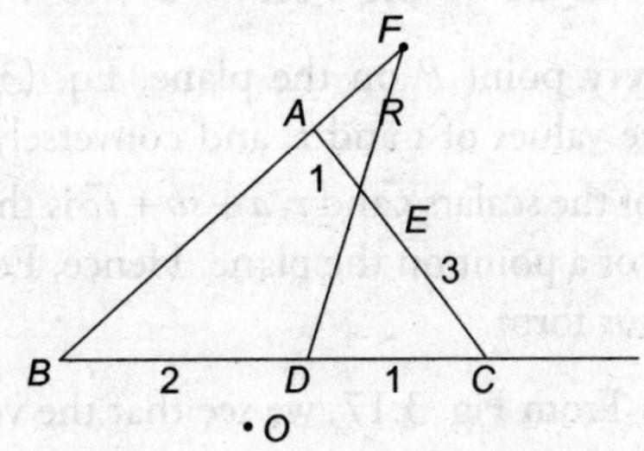

Figure 3.16 Triangle ABC

$$\vec{d}=\vec{OD}=\frac{1}{3}(2\vec{c}+\vec{b}),\quad \vec{e}=\vec{OE}=\frac{1}{3}(3\vec{a}+\vec{c})$$

Using Eq. (3.14), the vector equations of the straight lines BA and DE are

$$\vec{r}=(1-t)\vec{b}+t\vec{a} \text{ and } \vec{r}=(1-s)\frac{1}{3}(2\vec{c}+\vec{b})+s\frac{1}{4}(3\vec{a}+\vec{c})$$

If the two straight lines BA and DE intersect, at the point of intersection, we have

$$(1-t)\vec{b}+t\vec{a}=(1-s)\frac{1}{3}(2\vec{c}+\vec{b})+s\frac{1}{4}(3\vec{a}+\vec{c})$$

$$\Rightarrow \left(t-\frac{3s}{4}\right)\vec{a}+\left(1-t-\frac{1}{3}+\frac{s}{3}\right)\vec{b}+\left(\frac{5s}{12}-\frac{2}{3}\right)\vec{c}=\vec{0}$$

Since $\vec{a}$, $\vec{b}$ and $\vec{c}$ are non-collinear, we have

$$t-\frac{3s}{4}=0,\ 1-t-\frac{1}{3}+\frac{s}{3}=0,\ \frac{5s}{12}-\frac{2}{3}=0 \Rightarrow t=\frac{6}{5}, s=\frac{8}{5}$$

Inserting the value of t in the equation of BA, we get the position vector of F as $\frac{1}{5}(6\vec{a}-\vec{b})$, which shows that F divides the line AB externally in the ratio $1:6$.

3.3 PLANE

A plane is a surface which is purely flat, such that if any two points are considered on the surface and joined by a straight line, the straight line lies wholly on the surface. The surface is known as the *plane surface* or *flat surface*. Now, we study the plane under various conditions using the tools of vector algebra.

Vector equation of the plane passing through a given point and parallel to two given non-collinear vectors

Let P be any point on the plane whose position vector relative to the origin O is $\vec{r}$ (Fig. 3.17). Let A be a point whose position vector is $\vec{a}$ relative to the origin O. If the required plane is parallel to the two straight lines given by the vectors $\vec{b}$ and $\vec{c}$, then the vector $\vec{AP}$ is coplanar with the vectors $\vec{b}$ and $\vec{c}$. Thus, we can take $\vec{AP} = s\vec{b} + t\vec{c}$, where s and t are scalars. Now, from Fig. 3.17

$$\vec{r} = \vec{OP} = \vec{OA} + \vec{AP} = \vec{a} + s\vec{b} + t\vec{c} \quad (3.28)$$

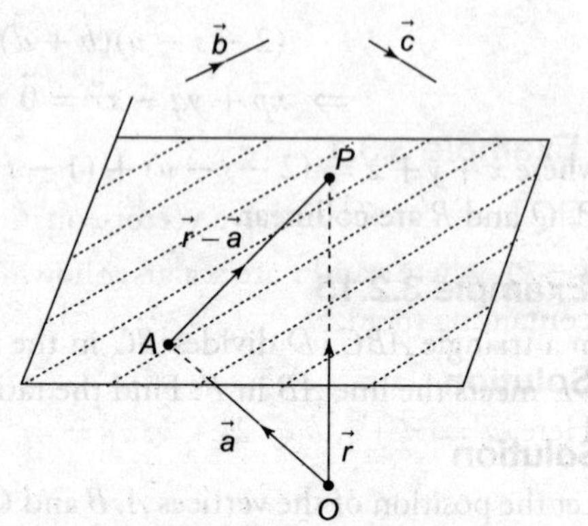

Figure 3.17 Plane

For every point P on the plane, Eq. (3.28) gives definite values of s and t, and conversely, for each value of the scalars s and t, $\vec{a}+s\vec{b}+t\vec{c}$ is the position vector of a point on the plane. Hence, Eq. (3.28) is the required parametric equation of the plane in vector form.

(i) From Fig. 3.17, we see that the vector $\vec{AP} = \vec{r} - \vec{a}$ is coplanar with $\vec{b}$ and $\vec{c}$, so

$$(\vec{r} - \vec{a}) \cdot \vec{b} \times \vec{c} = 0 \Rightarrow [\vec{r}\,\vec{b}\,\vec{c}] = [\vec{a}\,\vec{b}\,\vec{c}] \quad (3.29)$$

This is the equation of the required plane in non-parametric form. This result can also be obtained from Eq. (3.28) by multiplying scalarly by $\vec{b} \times \vec{c}$. From the above form of the plane, we obtain

$$|\vec{r}| \cos\theta \, |\vec{b} \times \vec{c}| = [\vec{a}\,\vec{b}\,\vec{c}]$$

So, the length of the perpendicular from the origin to the plane is

$$d = |r| \cos\theta = \frac{[\vec{a}\,\vec{b}\,\vec{c}]}{|\vec{b} \times \vec{c}|} \quad (3.30)$$

(ii) The equation of the plane through the origin and parallel to the vectors $\vec{b}$ and $\vec{c}$ is $[\vec{r}\,\vec{b}\,\vec{c}] = 0$.

(iii) The equation of a plane containing the two straight lines $\vec{r} = \vec{a} + t\,\vec{b}$ and $\vec{r} = \vec{a}_1 + s\,\vec{c}$ (t and s are scalars) is given by Eq. (3.29), as the plane passes through the point $\vec{a}$ and is parallel to the vectors $\vec{b}$ and $\vec{c}$.

Again, the above plane also passes through the vector $\vec{a}_1$. So, we have $[\vec{a}_1\,\vec{b}\,\vec{c}] = [\vec{a}\,\vec{b}\,\vec{c}]$. This is the condition for coplanarity of the above lines.

Since the plane is parallel to both the vectors $\vec{b}$ and $\vec{c}$, $\vec{b} \times \vec{c}$ is perpendicular to the plane. Hence, a unit vector normal to the plane is $\frac{1}{|\vec{b} \times \vec{c}|}(\vec{b} \times \vec{c})$.

(iv) Let $\vec{r} = (x, y, z)$, $\vec{a} = (x_1, y_1, z_1)$, $\vec{b} = (l_1, m_1, n_1)$ and $\vec{c} = (l_2, m_2, n_2)$, then the Cartesian form of the equation of the plane is

$$\begin{vmatrix} x - x_1 & y - y_1 & z - z_1 \\ l_1 & m_1 & n_1 \\ l_2 & m_2 & n_2 \end{vmatrix} = 0 \tag{3.31}$$

Example 3.3.1

The lines L_1 and L_2 have vector equations given by $\vec{r} = 3\hat{i} + \hat{j} + t(2\hat{j} + \hat{k})$ and $\vec{r} = 4\hat{k} + s(\hat{i} + \hat{j} - \hat{k})$, where s and t are scalars. Show that L_1 and L_2 intersect and find the equation of the plane containing them.

Solution

Here, $\vec{a} = 3\hat{i} + \hat{j}$, $\vec{b} = 2\hat{j} + \hat{k}$, $\vec{a}_1 = 4\hat{k}$ and $\vec{c} = \hat{i} + \hat{j} - \hat{k}$. Now

$$\left[\vec{a}_1\,\vec{b}\,\vec{c}\right] = \begin{vmatrix} 0 & 0 & 4 \\ 0 & 2 & 1 \\ 1 & 1 & -1 \end{vmatrix} = -8$$

and similarly $\left[\vec{a}\,\vec{b}\,\vec{c}\right] = -8$. Therefore, $\left[\vec{a}_1\,\vec{b}\,\vec{c}\right] = \left[\vec{a}\,\vec{b}\,\vec{c}\right]$. Thus, the two straight lines L_1 and L_2 are coplanar. Using Eq. (3.29), the equation of the plane containing them is given by

$$\left[\vec{r}\,\vec{b}\,\vec{c}\right] = \left[\vec{a}\,\vec{b}\,\vec{c}\right] \Rightarrow \vec{r}\cdot(-3\hat{i} + \hat{j} - 2\hat{k}) = -8$$

Example 3.3.2

Find the vector equation of a plane containing the two parallel straight lines $\vec{r} = \vec{a} + t\,\vec{c}$ and $\vec{r} = \vec{b} + s\,\vec{c}$ (t and s are scalars).

Solution

Let $\vec{r}$ be the position vector of any point P on the required plane, relative to the origin O. Since the plane contains two parallel straight lines $\vec{r} = \vec{a} + t\,\vec{c}$ and $\vec{r} = \vec{b} + s\,\vec{c}$, the straight lines as well as the plane passes through the points $A(\vec{a})$ and $B(\vec{b})$. Now, the straight lines as well as the plane are parallel to the vector $\vec{c}$. Thus, $\overrightarrow{BP}$, $\overrightarrow{BA}$ and $\vec{c}$ are coplanar vectors and the condition for coplanarity gives

$$\overrightarrow{BP}\cdot\overrightarrow{BA}\times\vec{c} = 0 \Rightarrow (\vec{r} - \vec{b})\cdot(\vec{a} - \vec{b})\times\vec{c} = 0$$

$$\Rightarrow \vec{r}\cdot(\vec{a} - \vec{b})\times\vec{c} = \vec{b}\cdot(\vec{a} - \vec{b})\times\vec{c}$$

$$\Rightarrow \vec{r}\cdot(\vec{a} - \vec{b})\times\vec{c} = \left[\vec{b}\,\vec{a}\,\vec{c}\right], \text{ as } \left[\vec{b}\,\vec{b}\,\vec{c}\right] = 0$$

This equation is satisfied by every point of the plane, and hence, is the equation of the required plane.

Vector equation of a plane passing through two given points and parallel to a given straight line

Let A and B be two given points whose position vectors are $\vec{a}$ and $\vec{b}$, with reference to some base point O (Fig. 3.18), i.e., $\overrightarrow{OA} = \vec{a}$ and $\overrightarrow{OB} = \vec{b}$. Let P be any point on the plane whose position vector relative to the origin O is $\vec{r}$. Let the given straight line be parallel to the vector $\vec{c}$. Then, the required plane will pass through the point whose position vector is $\vec{a}$ and lie parallel to the vectors $\vec{c}$ and $\vec{b}-\vec{a}$. Hence, the parametric equation of the required plane is

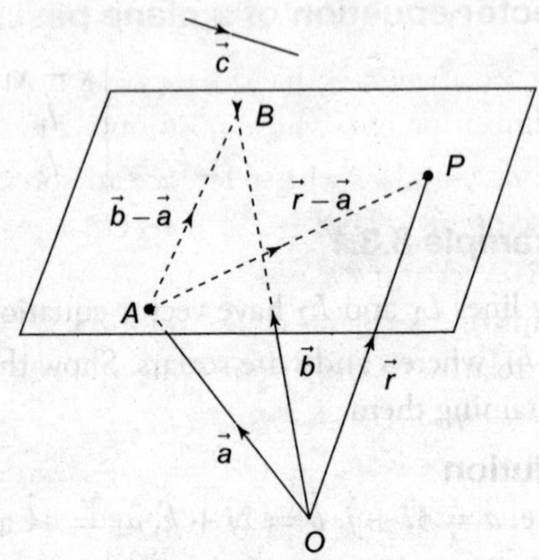

Figure 3.18 Plane

$$\vec{r}=\vec{a}+t(\vec{b}-\vec{a})+s\vec{c}=(1-t)\vec{a}+t\vec{b}+s\vec{c} \quad (3.32)$$

where t and s are scalar constants.

(i) From Fig. 3.18, we see that $\vec{AP}=\vec{r}-\vec{a}$, $\vec{AB}=\vec{b}-\vec{a}$ and $\vec{c}$ are coplanar vectors. Hence

$$\vec{AP}\cdot\vec{AB}\times\vec{c}=0\Rightarrow(\vec{r}-\vec{a})\cdot(\vec{b}-\vec{a})\times\vec{c}=0$$

$$\Rightarrow \vec{r}\cdot(\vec{b}-\vec{a})\times\vec{c}=\vec{a}\cdot\vec{b}\times\vec{c}$$

$$\Rightarrow \left[\vec{r}\ \vec{b}-\vec{a}\ \vec{c}\right]=\left[\vec{a}\,\vec{b}\,\vec{c}\right],\text{ as }\left[\vec{a}\,\vec{a}\,\vec{c}\right]=0 \quad (3.33)$$

Because of the one-to-one correspondence between point P on the plane and position vector $\vec{r}$ satisfying Eq.(3.33), this is the equation of the required plane in non-parametric form. $(\vec{b}-\vec{a})\times\vec{c}$ is the normal to the plane.

(i) Let $\vec{r}=(x,y,z)$, $\vec{a}=(x_1,y_1,z_1)$, $\vec{b}=(x_2,y_2,z_2)$ and $\vec{c}=(l_1,m_1,n_1)$. Then, the Cartesian form of the equation of the plane is

$$\begin{vmatrix} x-x_1 & y-y_1 & z-z_1 \\ x_2-x_1 & y_2-y_1 & z_2-z_1 \\ l_1 & m_1 & n_1 \end{vmatrix}=0 \quad (3.34)$$

Example 3.3.3

Find the position vector of the point of intersection between the straight line $\vec{r}=\vec{a}+t\,\vec{c}$ and the plane $(\vec{r}-\vec{b})\cdot\vec{d}=0$.

Solution

For the intersection of the straight line $\vec{r}=\vec{a}+t\,\vec{c}$ and the plane $(\vec{r}-\vec{b})\cdot\vec{d}=0$, we have

$$(\vec{a}+t\,\vec{c}-\vec{b})\cdot\vec{d}=0$$

$$\text{or, } t\,\vec{c}\cdot\vec{d}=(\vec{b}-\vec{a})\cdot\vec{d}\Rightarrow t=\frac{(\vec{b}-\vec{a})\cdot\vec{d}}{\vec{c}\cdot\vec{d}}$$

Hence, the required vector is

$$\vec{r}=\vec{a}+t\,\vec{c}=\vec{a}+\frac{(\vec{b}-\vec{a})\cdot\vec{d}}{\vec{c}\cdot\vec{d}}\vec{c}$$

Vector equation of a plane passing through three given non-collinear points

Let A, B and C be three given non-collinear points, whose position vectors are $\vec{a}$, $\vec{b}$ and $\vec{c}$, with reference to some base point O (Fig. 3.19), i.e., $\vec{OA} = \vec{a}$, $\vec{OB} = \vec{b}$ and $\vec{OC} = \vec{c}$. Let P be any point on the plane whose position vector relative to the origin O be $\vec{r}$. Then, the vectors $\vec{AB} = \vec{OB} - \vec{OA} = \vec{b} - \vec{a}$ and $\vec{AC} = \vec{OC} - \vec{OA} = \vec{c} - \vec{a}$ are parallel to the plane. The required plane thus passes through the point A and is parallel to the vectors $\vec{b} - \vec{a}$ and $\vec{c} - \vec{a}$. Hence, the parametric equation of the required plane is

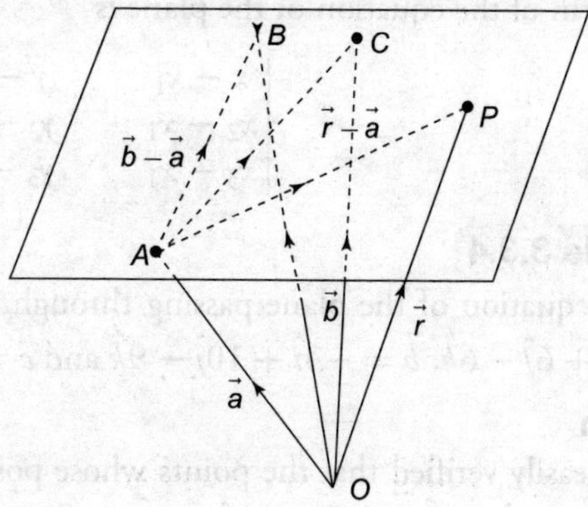

Figure 3.19 Plane

$$\vec{r} - \vec{a} = s(\vec{b} - \vec{a}) + t(\vec{c} - \vec{a})$$

$$\Rightarrow \vec{r} = \vec{a} + s(\vec{b} - \vec{a}) + t(\vec{c} - \vec{a}) = (1 - s - t)\vec{a} + s\vec{b} + t\vec{c}, \tag{3.35}$$

where t and s are scalar constants.

(i) The equation of the plane passing through three given non-collinear points can also be obtained from the condition that the vectors $\vec{r} - \vec{a}$, $\vec{a} - \vec{b}$ and $\vec{b} - \vec{c}$ are coplanar, i.e.

$$(\vec{r} - \vec{a}) \cdot (\vec{a} - \vec{b}) \times (\vec{b} - \vec{c}) = 0$$

$$\Rightarrow (\vec{r} - \vec{a}) \cdot \left(\vec{b} \times \vec{c} + \vec{c} \times \vec{a} + \vec{a} \times \vec{b}\right) = 0$$

$$\Rightarrow \vec{r} \cdot \left(\vec{b} \times \vec{c} + \vec{c} \times \vec{a} + \vec{a} \times \vec{b}\right) = \left[\vec{a}\,\vec{b}\,\vec{c}\right]. \tag{3.36}$$

Because of the one-to-one correspondence between point P on the plane and position vector $\vec{r}$ satisfying Eq.(3.36), this is the equation of the required plane in non-parametric form. $\vec{b} \times \vec{c} + \vec{c} \times \vec{a} + \vec{a} \times \vec{b}$ is the normal to the plane. Geometrically, this normal represents twice the vector area of the triangle ABC.

The length of the perpendicular p from the origin to the plane is

$$p = \frac{\left[\vec{a}\,\vec{b}\,\vec{c}\right]}{|\vec{b} \times \vec{c} + \vec{c} \times \vec{a} + \vec{a} \times \vec{b}|}$$

(ii) Eq. (3.35) can be rewritten as

$$(1 - s - t)\vec{a} + s\vec{b} + t\vec{c} - \vec{r} = \vec{0}$$

$$\Rightarrow x\vec{a} + y\vec{b} + z\vec{c} + p\vec{r} = \vec{0}$$

where $x + y + z + p = (1 - s - t) + s + t - 1 = 0$. Thus, the four position vectors $\vec{a}$, $\vec{b}$, $\vec{c}$ and $\vec{r}$ are coplanar.

(iii) Let $\vec{r} = (x, y, z)$, $\vec{a} = (x_1, y_1, z_1)$, $\vec{b} = (x_2, y_2, z_2)$ and $\vec{c} = (x_3, y_3, z_3)$. Then, the Cartesian form of the equation of the plane is

$$\begin{vmatrix} x - x_1 & y - y_1 & z - z_1 \\ x_2 - x_1 & y_2 - y_1 & z_2 - z_1 \\ x_3 - x_1 & y_3 - y_1 & z_3 - z_1 \end{vmatrix} = 0 \tag{3.37}$$

Example 3.3.4

Find the equation of the plane passing through the points whose position vectors are given by $\vec{a} = -2\hat{\imath} + 6\hat{\jmath} - 6\hat{k}$, $\vec{b} = -3\hat{\imath} + 10\hat{\jmath} - 9\hat{k}$ and $\vec{c} = -5\hat{\imath} - 6\hat{k}$.

Solution

It can be easily verified that the points whose position vectors are given by $\vec{a} = -2\hat{\imath} + 6\hat{\jmath} - 6\hat{k}$, $\vec{b} = -3\hat{\imath} + 10\hat{\jmath} - 9\hat{k}$ and $\vec{c} = -5\hat{\imath} - 6\hat{k}$ are non-collinear. Now

$$\vec{a} \times \vec{b} = (-2\hat{\imath} + 6\hat{\jmath} - 6\hat{k}) \times (-3\hat{\imath} + 10\hat{\jmath} - 9\hat{k}) = 6\hat{\imath} - 2\hat{k}$$

$$\vec{b} \times \vec{c} = (-3\hat{\imath} + 10\hat{\jmath} - 9\hat{k}) \times (-5\hat{\imath} - 6\hat{k}) = -60\hat{\imath} + 27\hat{\jmath} + 50\hat{k}$$

$$\vec{c} \times \vec{a} = (-5\hat{\imath} - 6\hat{k}) \times (-2\hat{\imath} + 6\hat{\jmath} - 6\hat{k}) = 36\hat{\imath} - 18\hat{\jmath} - 30\hat{k}$$

$$\left[\vec{a}\,\vec{b}\,\vec{c}\right] = (-2\hat{\imath} + 6\hat{\jmath} - 6\hat{k}) \cdot (-60\hat{\imath} + 27\hat{\jmath} + 50\hat{k}) = -18$$

$$\vec{a} \times \vec{b} + \vec{b} \times \vec{c} + \vec{c} \times \vec{a} = -18\hat{\imath} + 9\hat{\jmath} + 18\hat{k}$$

Using Eq. (3.36), we obtain the required equation of the plane as

$$\vec{r} \cdot \left(-18\hat{\imath} + 9\hat{\jmath} + 18\hat{k}\right) = -18 \Rightarrow \vec{r} \cdot \left(2\hat{\imath} - \hat{\jmath} - 2\hat{k}\right) = 2$$

Example 3.3.5

Find an equation for the plane perpendicular to the vector $\vec{a} = 2\hat{\imath} + 3\hat{\jmath} + 6\hat{k}$ and passing through the terminal point of the vector $\vec{b} = \hat{\imath} + 5\hat{\jmath} + 3\hat{k}$. Hence, find the distance from the origin to the plane.

Solution

Let $\vec{r}$ be the position vector of the point P, and Q the terminal point of B. Since $\vec{PQ} = \vec{b} - \vec{r}$ is perpendicular to $\vec{a}$

$$(\vec{b} - \vec{r}) \cdot \vec{a} = 0 \Rightarrow \vec{r} \cdot \vec{a} = \vec{b} \cdot \vec{a}$$

This is the required equation of the plane in vector form. In rectangular form, this equation becomes

$$(x\hat{\imath} + y\hat{\jmath} + z\hat{k}) \cdot (2\hat{\imath} + 3\hat{\jmath} + 6\hat{k}) = (\hat{\imath} + 5\hat{\jmath} + 3\hat{k}) \cdot (2\hat{\imath} + 3\hat{\jmath} + 6\hat{k})$$

$$\Rightarrow 2x + 3y + 6z = 1 \cdot 2 + 5 \cdot 3 + 3 \cdot 6 = 35$$

A unit vector in the direction of $\vec{a}$ is

$$\hat{a} = \frac{1}{|\vec{a}|}\vec{a} = \frac{1}{\sqrt{2^2 + 3^2 + 6^2}}(2\hat{\imath} + 3\hat{\jmath} + 6\hat{k}) = \frac{1}{7}(2\hat{\imath} + 3\hat{\jmath} + 6\hat{k})$$

The distance from the origin to the plane is the projection of $\vec{b}$ on $\hat{a}$. Thus

$$p = \vec{b} \cdot \hat{a} = (\hat{i} + 5\hat{j} + 3\hat{k}) \cdot (2\hat{i} + 3\hat{j} + 6\hat{k})/7 = 35/7 = 5$$

Vector equation of a plane in normal form

Let the required plane pass through a point A with position vector $\vec{OA} = \vec{a}$ (Fig. 3.20). Let $\hat{n}$ be a unit vector perpendicular to the plane from the origin to the plane. Let $ON = p$ be the perpendicular distance of the origin from the plane. Since $ON = OA \cos\theta = \vec{a} \cdot \hat{n}$, $p = \vec{a} \cdot \hat{n}$ (Fig. 3.20). Let P be any point on the plane with position vector $\vec{OP} = \vec{r}$. Then, $\vec{AP} = \vec{r} - \vec{a}$ is a vector on the plane, and hence

$$\vec{AP} \cdot \hat{n} = 0 \Rightarrow (\vec{r} - \vec{a}) \cdot \hat{n} = 0$$

$$\Rightarrow \vec{r} \cdot \hat{n} = \vec{a} \cdot \hat{n} = p \qquad (3.38)$$

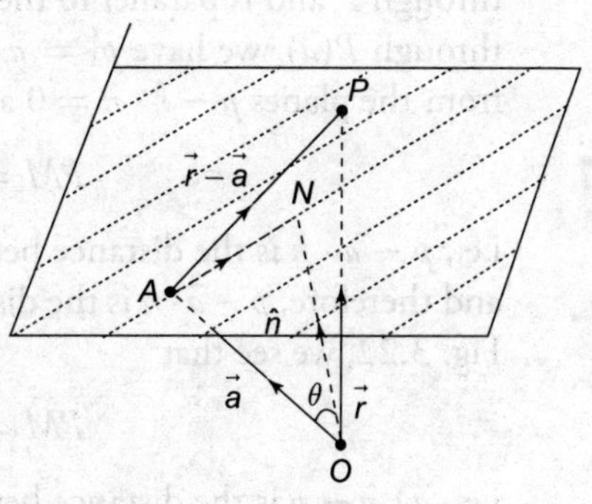

Figure 3.20 Plane in normal form

Because of the one-to-one correspondence between a point on the plane and a vector $\vec{r}$ satisfying Eq. (3.38), this is the vector equation of the plane.

(i) If $\vec{n}$ is not a unit vector, but is the normal to the plane, then $p \neq \vec{a} \cdot \vec{n}$. Eq. (3.38) then takes the form $\vec{r} \cdot \vec{n} = \vec{a} \cdot \vec{n}$, which is the equation of the plane not in normal form. In this case, the normal form of the equation is

$$\vec{r} \cdot \hat{n} = \vec{a} \cdot \hat{n} = p/|\vec{n}| = q. \qquad (3.39)$$

(ii) Let the plane $\vec{r} \cdot \hat{n} = p$, given by Eq. (3.38), intersect the x, y and z axes at A, B and C, respectively. Further, let the lengths of the intercepts made by the given plane with the x, y and z axes be α, β and γ, respectively. Then, the position vectors of A, B and C are $\alpha\,\hat{i}$, $\beta\,\hat{j}$ and $\gamma\,\hat{k}$, respectively. Since the points A, B and C lie on the given plane

$$\alpha\,\hat{i} \cdot \hat{n} = p, \beta\,\hat{j} \cdot \hat{n} = p \text{ and } \gamma\,\hat{k} \cdot \hat{n} = p$$

$$\Rightarrow \alpha = \frac{p}{\hat{i} \cdot \hat{n}}, \quad \beta = \frac{p}{\hat{j} \cdot \hat{n}}, \quad \gamma = \frac{p}{\hat{k} \cdot \hat{n}}$$

These are the lengths of the intercepts made by the plane.

(iii) Let $\vec{r} = (x, y, z)$ and $\hat{n} = (l, m, n)$. Then, from Eq. (3.38), we get $lx + my + nz = p$, which is the Cartesian equation of the plane in normal form.

(iv) **Angle between two intersecting planes**: Let θ be the angle of inclination of the intersecting planes $\vec{r} \cdot \hat{n}_1 = p_1$ and $\vec{r} \cdot \hat{n}_2 = p_2$. Then, θ is the angle between the normals $\hat{n}_1$ and $\hat{n}_2$. Hence

$$\hat{n}_1 \cdot \hat{n}_2 = \cos\theta \Rightarrow \theta = \cos^{-1}(\hat{n}_1 \cdot \hat{n}_2)$$

Similarly, the angle of intersection between two intersecting planes $\vec{r} \cdot \vec{n}_1 = p_1$ and $\vec{r} \cdot \vec{n}_2 = p_2$ is given by

$$\vec{n}_1 \cdot \vec{n}_2 = |\vec{n}_1|\,|\vec{n}_2| \cos\theta \Rightarrow \theta = \cos^{-1}\left(\frac{\vec{n}_1 \cdot \vec{n}_2}{|\vec{n}_1|\,|\vec{n}_2|}\right)$$

From this, we conclude that the planes are parallel to each other if $\hat{n}_1 = \pm\hat{n}_2$. The equation of any plane parallel to the plane $\vec{r} \cdot \hat{n} = p$ is $\vec{r} \cdot \hat{n} = p'$, where p' is any scalar.

(v) **Distance of a point from a plane**: The distance of a point $P(\vec{a})$ from the plane $p - \vec{r} \cdot \hat{n} = 0$ is the distance between the plane $p - \vec{r} \cdot \hat{n} = 0$ and the plane $p' - \vec{r} \cdot \hat{n} = 0$, which passes through P and is parallel to the plane $p - \vec{r} \cdot \hat{n} = 0$. Since the plane $p' - \vec{r} \cdot \hat{n} = 0$ passes through $P(\vec{a})$, we have $p' = \vec{a} \cdot \hat{n}$. p and p' are, respectively, the distance of the origin O from the planes $p - \vec{r} \cdot \hat{n} = 0$ and $p' - \vec{r} \cdot \hat{n} = 0$. Hence (Fig. 3.21)

$$PM = ON - ON' = p - p'$$

i.e., $p - \vec{a} \cdot \hat{n}$ is the distance between the parallel planes $p - \vec{r} \cdot \hat{n} = 0$ and $p' - \vec{r} \cdot \hat{n} = 0$, and therefore, $p - \vec{a} \cdot \hat{n}$ is the distance of the point $P(\vec{a})$ from the plane $p - \vec{r} \cdot \hat{n} = 0$. From Fig. 3.22, we see that

$$PM = ON' - ON = p' - p$$

i.e., $\vec{a} \cdot \hat{n} - p$ is the distance between the parallel planes $p - \vec{r} \cdot \hat{n} = 0$ and $p' - \vec{r} \cdot \hat{n} = 0$, and therefore, $\vec{a} \cdot \hat{n} - p$ is the distance of the point $P(\vec{a})$ from the plane $p - \vec{r} \cdot \hat{n} = 0$. Thus, $|p - \vec{a} \cdot \hat{n}|$ is the magnitude (distance) of the point $P(\vec{a})$ from the plane $p - \vec{r} \cdot \hat{n} = 0$.

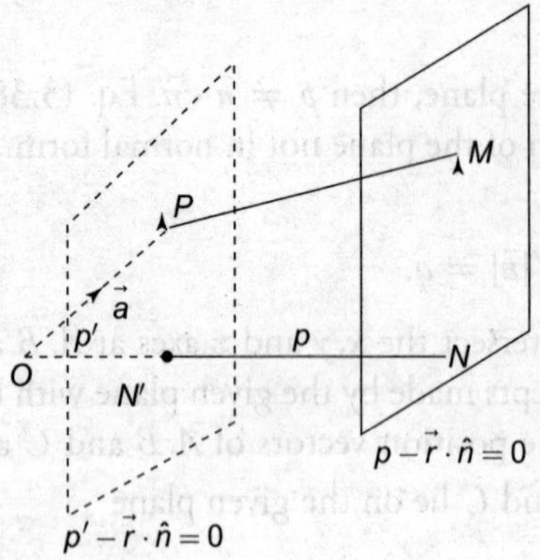

Figure 3.21 P, O lie on same side of the plane

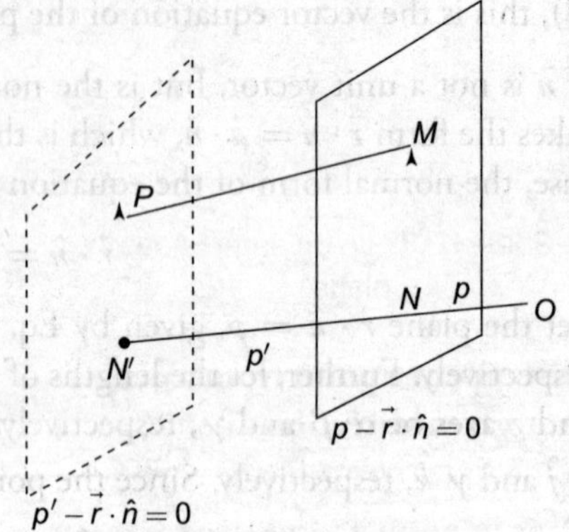

Figure 3.22 P, O lie on opposite sides of the plane

(vi) **Position of two points with respect to a plane**: Let $A(\vec{a})$ and $B(\vec{b})$ be two points; the line joining A and B meets the plane $p - \vec{r} \cdot \hat{n} = 0$ at the point P in the ratio $AP : PB = \lambda : \mu$, $\lambda \neq 0$ and $\mu \neq 0$ (Figs 3.23 and 3.24). Then, the position vector of P relative to O as origin is given by

$$\vec{OP} = \frac{1}{\lambda + \mu}\left[\lambda\vec{b} + \mu\vec{a}\right]; \lambda + \mu \neq 0, \lambda \neq 0, \mu \neq 0$$

It satisfies the equation $p - \vec{r} \cdot \hat{n} = 0$. Hence

$$p\,(\lambda + \mu) = (\lambda\vec{b} + \mu\vec{a}) \cdot \hat{n}$$

$$\Rightarrow \lambda(p - \vec{b} \cdot \hat{n}) = -\mu(p - \vec{a} \cdot \hat{n})$$

$$\Rightarrow \frac{\lambda}{\mu} = -\frac{p - \vec{a}\cdot\hat{n}}{p - \vec{b}\cdot\hat{n}}$$

Now, A and B lie on the opposite sides of the plane $p - \vec{r}\cdot\hat{n} = 0$ (Fig. 3.23) if P is the internal point of division of the line AB, i.e., if $\frac{\lambda}{\mu}$ is positive or equivalently if $p - \vec{a}\cdot\hat{n}$ and $p - \vec{b}\cdot\hat{n}$ are positive signs. A and B lie on the same side of the plane $p - \vec{r}\cdot\hat{n} = 0$ (Fig. 3.24) if P is the external point of the division of the line AB (produced), i.e., if $\frac{\lambda}{\mu}$ is negative or equivalently if $p - \vec{a}\cdot\hat{n}$ and $p - \vec{b}\cdot\hat{n}$ have the same sign. Thus, two points A and B lie on the same side or on opposite sides of the plane $p - \vec{r}\cdot\hat{n} = 0$ depending on whether $p - \vec{a}\cdot\hat{n}$ and $p - \vec{b}\cdot\hat{n}$ have the same sign or opposite signs.

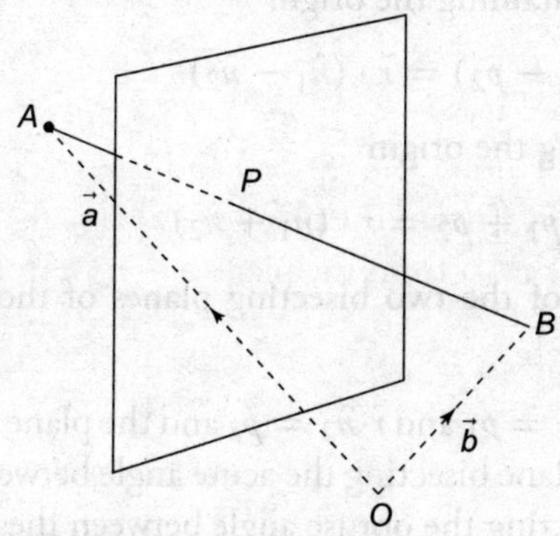

Figure 3.23 *P*, *O* lie on same side of the plane

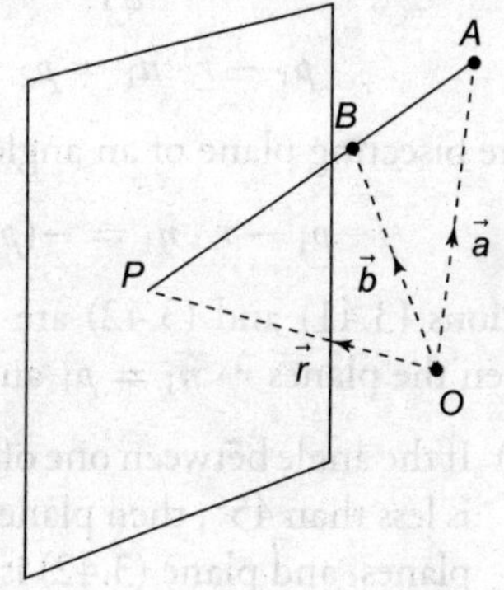

Figure 3.24 Position of points with respect to a plane

(vii) **Distance of a point from a plane measured in a given direction**: Let $P(\vec{a})$ be a given point and $\vec{r}\cdot\hat{n} = p$ be the equation of a given plane in normal form (Fig. 3.25). Suppose $\hat{b}$ be is given unit vector. PN is drawn parallel to $\hat{b}$, which meets the plane $\vec{r}\cdot\hat{n} = p$ at N. Then, $\vec{PN} = PN\,\hat{b}$ and

$$\vec{ON} = \vec{OP} + \vec{PN} = \vec{a} + PN\,\hat{b}$$

Since N lies on the plane, we have

$$\vec{ON}\cdot\hat{n} = p \Rightarrow \left[\vec{a} + PN\,\hat{b}\right]\cdot\hat{n} = p$$

$$\Rightarrow PN\,\hat{b}\cdot\hat{n} = p - \hat{a}\cdot\hat{n}$$

$$\Rightarrow PN = \frac{p - \hat{a}\cdot\hat{n}}{\hat{b}\cdot\hat{n}}$$

$$\Rightarrow |PN| = \frac{|p - \hat{a}\cdot\hat{n}|}{|\hat{b}\cdot\hat{n}|} \qquad (3.40)$$

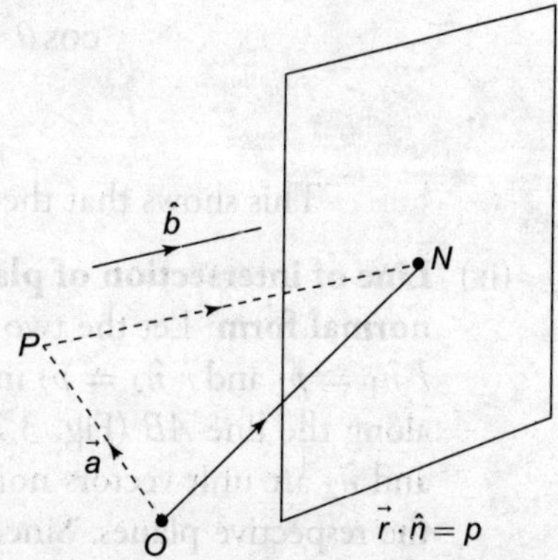

Figure 3.25 Distance of a point from a plane measured in a given direction

This is the required distance of the point P from the plane $\vec{r} \cdot \hat{n} = p$, measured parallel to $\vec{b}$. If $\hat{b}$ is perpendicular to the plane $\vec{r} \cdot \hat{n} = p$, then $\hat{b} \cdot \hat{n} = 1$ and $|PN| = |p - \hat{a} \cdot \hat{n}|$ gives the distance of the point P from the plane.

(viii) **Planes bisecting the angles between two planes**: Let the equations (in normal form) of the two intersecting planes be $\vec{r} \cdot \hat{n}_1 = p_1$ and $\vec{r} \cdot \hat{n}_2 = p_2$, where $\hat{n}_1$ and $\hat{n}_2$ are the unit vectors normal to the respective planes such that $\hat{n}_1 \times \hat{n}_2 \neq \vec{0}$. Any point on either of the two bisecting planes (internal or external) will be equidistant from the respective planes. The two perpendicular distances will have the same sign for points on the bisecting plane of the angle (between the two planes) which contains the origin, and opposite signs for points on the other bisecting plane.

Thus, for the bisecting plane of an angle containing the origin

$$p_1 - \vec{r} \cdot \hat{n}_1 = p_2 - \vec{r} \cdot \hat{n}_2 \Rightarrow (p_1 - p_2) = \vec{r} \cdot (\hat{n}_1 - \hat{n}_2) \tag{3.41}$$

For the bisecting plane of an angle not containing the origin

$$p_1 - \vec{r} \cdot \hat{n}_1 = -(p_2 - \vec{r} \cdot \hat{n}_2) \Rightarrow p_1 + p_2 = \vec{r} \cdot (\hat{n}_1 + \hat{n}_2) \tag{3.42}$$

Equations (3.41) and (3.42) are the equations of the two bisecting planes of the angle between the planes $\vec{r} \cdot \hat{n}_1 = p_1$ and $\vec{r} \cdot \hat{n}_2 = p_2$.

(a) If the angle between one of the planes $\vec{r} \cdot \hat{n}_1 = p_1$ and $\vec{r} \cdot \hat{n}_2 = p_2$ and the plane (3.41) is less than 45°, then plane (3.41) is the plane bisecting the acute angle between the planes, and plane (3.42) is the plane bisecting the obtuse angle between the planes $\vec{r} \cdot \hat{n}_1 = p_1$ and $\vec{r} \cdot \hat{n}_2 = p_2$.

(b) If θ be the angle between the bisecting planes $\vec{r} \cdot \hat{n}_1 = p_1$ and $\vec{r} \cdot \hat{n}_2 = p_2$, then

$$\begin{aligned} \cos\theta &= \frac{(\hat{n}_1 + \hat{n}_2) \cdot (\hat{n}_1 - \hat{n}_2)}{|\hat{n}_1 + \hat{n}_2|\,|\hat{n}_1 - \hat{n}_2|} = \frac{|\hat{n}_1|^2 - |\hat{n}_2|^2}{|\hat{n}_1 + \hat{n}_2|\,|\hat{n}_1 - \hat{n}_2|} \\ &= 0; \quad \text{as } |\hat{n}_1| = |\hat{n}_2| = 1 \end{aligned}$$

This shows that the bisecting planes are at right angles.

(ix) **Line of intersection of planes in normal form**: Let the two planes $\vec{r} \cdot \hat{n}_1 = p_1$ and $\vec{r} \cdot \hat{n}_2 = p_2$ intersect along the line AB (Fig. 3.26). $\hat{n}_1$ and $\hat{n}_2$ are unit vectors normal to the respective planes. Since AB is a line common to both the planes $\vec{r} \cdot \hat{n}_1 = p_1$ and $\vec{r} \cdot \hat{n}_2 = p_2$, $\hat{n}_1$ and $\hat{n}_2$ are both perpendicular to the line AB. Therefore, the line AB is parallel to the vector $\hat{n}_1 \times \hat{n}_2$.

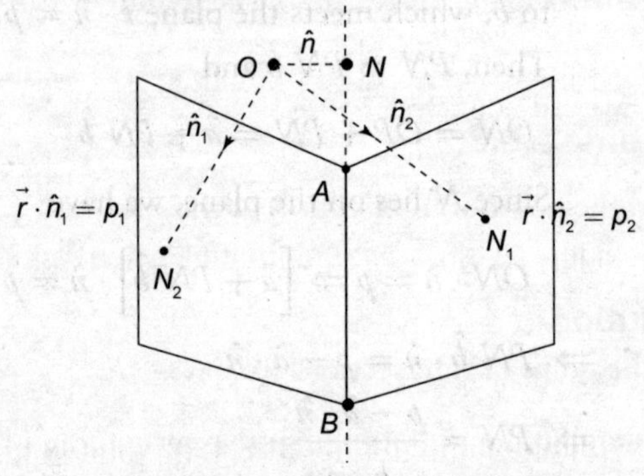

Figure 3.26 Line of intersection

Let ON be the perpendicular to the line AB, drawn from the origin O. Then, $\vec{ON}$, $\hat{n}_1$ and $\hat{n}_2$

are coplanar, and hence

$$\vec{n} = \vec{ON} = a\,\hat{n}_1 + b\,\hat{n}_2$$

for scalars a and b. Since the point N lies on both the planes, the position vector of N, $\vec{n}$ satisfies both the equations $\vec{r}\cdot\hat{n}_1 = p_1$ and $\vec{r}\cdot\hat{n}_2 = p_2$. Therefore

$$p_1 - (a\,\hat{n}_1 + b\,\hat{n}_2)\cdot\hat{n}_1 = 0,\ \ p_2 - (a\,\hat{n}_1 + b\,\hat{n}_2)\cdot\hat{n}_2 = 0$$
$$\Rightarrow p_1 = a + (\hat{n}_1\cdot\hat{n}_2)b \text{ and } p_2 = b + (\hat{n}_1\cdot\hat{n}_2)a$$
$$\Rightarrow a = \frac{p_1 - (\hat{n}_1\cdot\hat{n}_2)p_2}{1-(\hat{n}_1\cdot\hat{n}_2)^2},\quad b = \frac{p_2 - (\hat{n}_1\cdot\hat{n}_2)p_1}{1-(\hat{n}_1\cdot\hat{n}_2)^2} \tag{3.43}$$

as $\hat{n}_1\cdot\hat{n}_1 = 1 = \hat{n}_2\cdot\hat{n}_2$. Thus, $\vec{n}$ is completely determined. Hence, the equation of the line of intersection AB is

$$\left.\begin{array}{ll}(\vec{r}-\vec{n})\times(\hat{n}_1\times\hat{n}_2) = \vec{0}; & \text{non-parametric form}\\ \vec{r} = \vec{n} + t(\hat{n}_1\times\hat{n}_2); & \text{parametric form, } t \text{ parameter}\end{array}\right\} \tag{3.44}$$

Example 3.3.6

What is the equation of the plane which passes through the line of intersection of the planes $\vec{r}\cdot\hat{n}_1 = q_1$ and $\vec{r}\cdot\hat{n}_2 = q_2$ and is parallel to the line of intersection of the planes $\vec{r}\cdot\hat{n}_3 = q_3$ and $\vec{r}\cdot\hat{n}_4 = q_4$?

Solution

Any plane passing through the line of intersection of the two planes $\vec{r}\cdot\hat{n}_1 = q_1$ and $\vec{r}\cdot\hat{n}_2 = q_2$ is of the form

$$\vec{r}\cdot(\hat{n}_1 - \lambda\hat{n}_2) = q_1 - \lambda q_2;\ \ \lambda = \text{ a parameter}$$

But the required plane is parallel to $\hat{n}_3\times\hat{n}_4$, and hence, perpendicular to the normal $\hat{n}_1 - \vec{n}_2$. Therefore

$$(\hat{n}_1 - \lambda\hat{n}_2)\cdot(\hat{n}_3\times\hat{n}_4) = 0 \Rightarrow \lambda = \frac{\hat{n}_1\cdot\hat{n}_3\times\hat{n}_4}{\hat{n}_2\cdot\hat{n}_3\times\hat{n}_4} = \frac{[\hat{n}_1\ \hat{n}_3\ \hat{n}_4]}{[\hat{n}_2\ \hat{n}_3\ \hat{n}_4]}$$

For this value of λ, we obtain the required equation in the form

$$[\hat{n}_2\ \hat{n}_3\ \hat{n}_4](\vec{r}\cdot\hat{n}_1 - q_1) = [\hat{n}_1\ \hat{n}_3\ \hat{n}_4](\vec{r}\cdot\hat{n}_2 - q_2)$$

Example 3.3.7

Find the point of intersection of the the line $\vec{r} = 2\vec{a} + \vec{b} + m(\vec{b}+\vec{c})$ and the plane $\vec{r} = \vec{a} + u(\vec{b} - \vec{c}) + v(\vec{a}+\vec{b}-\vec{c})$.

Solution

At the point of intersection, we have

$$2\vec{a} + \vec{b} + m(\vec{b}+\vec{c}) = \vec{a} + u(\vec{b}-\vec{c}) + v(\vec{a}+\vec{b}-\vec{c})$$
$$\Rightarrow\ (1-v)\vec{a} + (1+m-u-v)\vec{b} + (m+u+v)\vec{c} = \vec{0}$$

Since $\vec{a}$, $\vec{b}$ and $\vec{c}$ are non-coplanar, we have

$$1 - v = 0, 1 + m - u - v = 0 \text{ and } m + u + v = 0$$

Solving, we get $v = 1, m = u = -1/2$. Hence, the position vector of the point of intersection is $\vec{r} = 2\vec{a} + \vec{b} - \frac{1}{2}(\vec{b} + \vec{c})$.

Example 3.3.8

Find the vector equation of the plane through the point $8\hat{i} + 2\hat{j} + 2\hat{k}$ and perpendicular to each of the planes $\vec{r} \cdot (2\hat{i} - \hat{j} + 2\hat{k}) = 0$ and $\vec{r} \cdot (\hat{i} + 3\hat{j} - 5\hat{k}) = 0$.

Solution

The required plane will be parallel to each of the vectors $2\hat{i} - \hat{j} + 2\hat{k}$ and $\hat{i} + 3\hat{j} - 5\hat{k}$ as it is perpendicular to the given planes. Now, the vector

$$\vec{p} = (2\hat{i} - \hat{j} + 2\hat{k}) \times (\hat{i} + 3\hat{j} - 5\hat{k}) = -\hat{i} + 12\hat{j} + 7\hat{k}$$

This is perpendicular to each of the vectors $2\hat{i} - \hat{j} + 2\hat{k}$ and $\hat{i} + 3\hat{j} - 5\hat{k}$. Hence, $\vec{p}$ is a vector normal to the required plane. Hence, the equation of the required plane passing through the point $8\hat{i} + 2\hat{j} + 2\hat{k}$ is given by

$$\left\{\vec{r} - (8\hat{i} + 2\hat{j} + 2\hat{k})\right\} \cdot (-\hat{i} + 12\hat{j} + 7\hat{k}) = 0$$

$$\Rightarrow \vec{r} \cdot (-\hat{i} + 12\hat{j} + 7\hat{k}) = (8\hat{i} + 2\hat{j} + 2\hat{k}) \cdot (-\hat{i} + 12\hat{j} + 7\hat{k})$$

$$\Rightarrow \vec{r} \cdot (-\hat{i} + 12\hat{j} + 7\hat{k}) = -8 + 24 + 14 = 30$$

Example 3.3.9

Find the position vector of the point of intersection of the straight line joining the points $\hat{i} + \hat{j} + \hat{k}$ and $3\hat{i} + 2\hat{j} - \hat{k}$ with the plane $\vec{r} \cdot (\hat{k} - \hat{j}) = 5$.

Solution

Let $\vec{r}$ be the position vector of any point on the line passing through the points $\hat{i} + \hat{j} + \hat{k}$ and $3\hat{i} + 2\hat{j} - \hat{k}$. Then, the equation of the line is given by

$$\begin{aligned}\vec{r} &= \hat{i} + \hat{j} + \hat{k} + t(3\hat{i} + 2\hat{j} - \hat{k} - \hat{i} - \hat{j} - \hat{k}) \\ &= (\hat{i} + \hat{j} + \hat{k}) + t(2\hat{i} + \hat{j} - 2\hat{k})\end{aligned}$$

where t is a scalar. If this intersects the given plane $\vec{r} \cdot (\hat{k} - \hat{j}) = 5$, then we have

$$\left[(\hat{i} + \hat{j} + \hat{k}) + t(2\hat{i} + \hat{j} - 2\hat{k})\right] \cdot (\hat{k} - \hat{j}) = 5$$

$$\Rightarrow (1 + 2t) \cdot 0 + (1 + t) \cdot (-1) + (1 - 2t) \cdot (1) = 5 \Rightarrow t = -\frac{5}{3}$$

Therefore, the required point of intersection obtained by inserting the value of t is

$$\vec{r} = \hat{i} + \hat{j} + \hat{k} - \frac{5}{3}(2\hat{i} + \hat{j} - 2\hat{k}) = -\frac{7}{3}\hat{i} - \frac{2}{3}\hat{j} + \frac{13}{3}\hat{k}$$

Example 3.3.10

Show that the equation of the plane which contains the line $\vec{r} = t\,\vec{\alpha}$ and is perpendicular to the plane containing the lines $\vec{r} = t_1\,\vec{\beta}$ and $\vec{r} = t_2\,\vec{\gamma}$ is $\left[\vec{r}\,\vec{\alpha}\,\vec{\beta}\times\vec{\gamma}\right] = 0$.

Solution

The required plane is perpendicular to the plane which contain the lines $\vec{r} = t_1\vec{\beta}$ and $\vec{r} = t_2\,\vec{\gamma}$. Hence, the required plane is parallel to $\vec{\beta}\times\vec{\gamma}$. By the given condition, it passes through the origin and is parallel to $\vec{\alpha}$. Hence, $\vec{\alpha}\times(\vec{\beta}\times\vec{\gamma})$ is the normal to the required plane.

Let $\vec{r}$ be the position vector of any point on the plane with reference to some vector origin O. Since the required plane contains the origin O, we have

$$(\vec{r}-\vec{0})\cdot\left\{\vec{\alpha}\times(\vec{\beta}\times\vec{\gamma})\right\} = 0$$

$$\vec{r}\cdot\left\{\vec{\alpha}\times(\vec{\beta}\times\vec{\gamma})\right\} = 0 \Rightarrow \left[\vec{r}\,\vec{\alpha}\,\vec{\beta}\times\vec{\gamma}\right] = 0$$

The angle between a straight line and a plane

Let the straight line $\vec{r} = \vec{a} + t\,\hat{b}$ intersect the plane $p - \vec{r}\cdot\hat{n} = 0$ at an angle θ (Fig. 3.27). Then, $90^\circ - \theta$ is the angle made by the line $\vec{r} = \vec{a} + t\,\hat{b}$ with the normal $\hat{n}$ to the plane $p - \vec{r}\cdot\hat{n} = 0$. $\hat{b}$ is the unit vector parallel to the given line, hence $90^\circ - \theta$ is the angle between $\hat{b}$ and $\hat{n}$ and $\cos(90^\circ - \theta) = \hat{b}\cdot\hat{n} \Rightarrow$ $\theta = \sin^{-1}(\hat{b}\cdot\hat{n})$ This determines the angle between the straight line $\vec{r} = \vec{a} + t\,\hat{b}$ and the plane $p - \vec{r}\cdot\hat{n} = 0$.

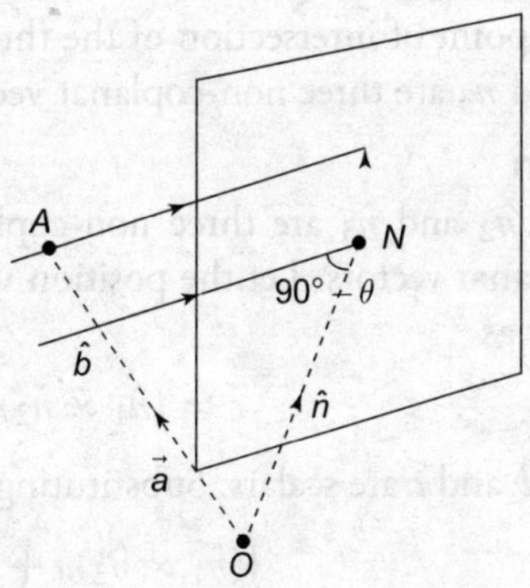

Figure 3.27 Angle between a straight line and a plane

Example 3.3.11

Find the equation of the plane which contains the origin and the line of intersection of the planes $\vec{r}\cdot\hat{n}_1 = p_1$ and $\vec{r}\cdot\hat{n}_2 = p_2$.

Solution

The equation of the plane through the intersection of the planes $\vec{r}\cdot\hat{n}_1 = p_1$ and $\vec{r}\cdot\hat{n}_2 = p_2$ is

$$(\vec{r}\cdot\hat{n}_1 - p_1) + \lambda(\vec{r}\cdot\hat{n}_2 - p_2) = 0,\ \ \lambda = \text{ constant}$$

Since this plane passes through the origin, we have

$$(\vec{0}\cdot\hat{n}_1 - p_1) + \lambda(\vec{0}\cdot\hat{n}_2 - p_2) = 0 \Rightarrow \lambda = -\frac{p_1}{p_2}$$

Thus, the required equation of the plane is

$$\left(\vec{r}\cdot\hat{n}_1 - p_1\right) - \frac{p_1}{p_2}\left(\vec{r}\cdot\hat{n}_2 - p_2\right) = 0$$

$$\Rightarrow \vec{r}\cdot\left(p_2\,\hat{n}_1 - p_1\,\hat{n}_2\right) = 0$$

Example 3.3.12

Find the equation of the plane which contains the line $\vec{r} = t_1\,\vec{a}$ and is perpendicular to the plane containing the straight lines $\vec{r} = t_2\,\vec{b}$ and $\vec{r} = t_3\,\vec{c}$.

Solution

We know that any vector that is perpendicular to both the vectors $\vec{b}$ and $\vec{c}$ is parallel to the vector $\vec{b}\times\vec{c}$. Thus, according to the given condition, the required plane is parallel to the vector $\vec{b}\times\vec{c}$. Also, the required plane contains the line $\vec{r} = t_1\,\vec{a}$, and hence, it is parallel to the vector $\vec{a}$. Therefore, the required plane is perpendicular to the vector $\vec{a}\times(\vec{b}\times\vec{c})$ and also passes through the origin. Thus

$$(\vec{r}-\vec{0})\cdot\vec{a}\times(\vec{b}\times\vec{c}) = 0 \Rightarrow \vec{r}\cdot\vec{a}\times(\vec{b}\times\vec{c}) = 0$$

This is the required equation of the plane.

Example 3.3.13

Find the point of intersection of the three planes $\vec{r}\cdot\hat{n}_1 = p_1$, $\vec{r}\cdot\hat{n}_2 = p_2$ and $\vec{r}\cdot\hat{n}_3 = p_3$, where $\hat{n}_1, \hat{n}_2$ and $\hat{n}_3$ are three non-coplanar vectors, not necessarily unit vectors.

Solution

Since $\hat{n}_1, \hat{n}_2$ and $\hat{n}_3$ are three non-coplanar vectors, $\hat{n}_1\times\hat{n}_2$, $\hat{n}_2\times\hat{n}_3$ and $\hat{n}_3\times\hat{n}_1$ are three non-coplanar vectors. Let the position vector $\vec{r}$ of the point of intersection of the three planes be expressed as

$$\vec{r} = (\hat{n}_1\times\hat{n}_2)a + (\hat{n}_2\times\hat{n}_3)b + (\hat{n}_3\times\hat{n}_1)c$$

where a, b and c are scalars. Substituting this value in $\vec{r}\cdot\hat{n}_1 = p_1$, we get

$$\{(\hat{n}_1\times\hat{n}_2)a + (\hat{n}_2\times\hat{n}_3)b + (\hat{n}_3\times\hat{n}_1)c\}\cdot\hat{n}_1 = p_1$$

$$\Rightarrow a\left[\hat{n}_1\,\hat{n}_2\,\hat{n}_3\right] = p_1 \Rightarrow a = \frac{p_1}{\left[\hat{n}_1\,\hat{n}_2\,\hat{n}_3\right]}$$

Similarly, $b = \dfrac{p_2}{\left[\hat{n}_1\,\hat{n}_2\,\hat{n}_3\right]}$ and $c = \dfrac{p_3}{\left[\hat{n}_1\,\hat{n}_2\,\hat{n}_3\right]}$. Using these values of a, b and c, we get

$$\vec{r} = \frac{1}{\left[\hat{n}_1\,\hat{n}_2\,\hat{n}_3\right]}\left\{p_1\,(\hat{n}_1\times\hat{n}_2) + p_2\,(\hat{n}_2\times\hat{n}_3) + p_3\,(\hat{n}_3\times\hat{n}_1)\right\}$$

Example 3.3.14

Find the equation of the plane through the point $(4, 3, -1)$ and perpendicular to the vector $3\hat{i} - 4\hat{j} + 2\hat{k}$. Find also the distance of the point $(1, 1, 1)$ from the plane.

Solution

Let P be any point on the plane whose position vector is $\vec{r} = x\hat{i} + y\hat{j} + z\hat{k}$. Since the plane passes through the point A whose position vector is $4\hat{i} + 3\hat{j} - \hat{k}$, the vector $\vec{PA} = \vec{r} - (4\hat{i} + 3\hat{j} - \hat{k})$ is parallel to the plane, i.e., $\vec{PA}$ is perpendicular to the vector $3\hat{i} - 4\hat{j} + 2\hat{k}$. Hence

$$\left[\vec{r} - (4\hat{i} + 3\hat{j} - \hat{k})\right]\cdot(3\hat{i} - 4\hat{j} + 2\hat{k}) = 0$$

$$\Rightarrow \vec{r}\cdot(3\hat{i}-4\hat{j}+2\hat{k}) = (4\hat{i}+3\hat{j}-\hat{k})\cdot(3\hat{i}-4\hat{j}+2\hat{k}) = -2$$

$$\Rightarrow \vec{r}\cdot\frac{1}{\sqrt{3^2+(-4)^2+2^2}}(3\hat{i}-4\hat{j}+2\hat{k}) = -2\,\frac{1}{\sqrt{3^2+(-4)^2+2^2}}$$

$$\Rightarrow \vec{r}\cdot\hat{n} = p, \quad \text{where } \hat{n} = \frac{1}{\sqrt{29}}(3\hat{i}-4\hat{j}+2\hat{k}), p = -\frac{2}{\sqrt{29}}$$

The perpendicular distance of the point $(1,1,1)$ from the plane is

$$|\vec{p}-\vec{a}\cdot\hat{n}| = \left|-\frac{2}{\sqrt{29}} - (\hat{i}+\hat{j}+\hat{k})\cdot\frac{1}{\sqrt{29}}(3\hat{i}-4\hat{j}+2\hat{k})\right| = \frac{3}{\sqrt{29}} \text{ units}$$

Example 3.3.15

Find the projection of the line $\vec{r} = \vec{a}+t\vec{b}$ on the plane $\vec{r}\cdot\vec{n} = q$.

Solution

The equation of the line through any point $\vec{a}+t\vec{b}$ of the given line normal to the given plane is $\vec{r} = \vec{a}+t\vec{b}+s\vec{n}$, so the projection of the point $\vec{a}+t\,\vec{b}$ on the given plane is given for the value of s, satisfying

$$(\vec{a}+t\vec{b}+s\vec{n})\cdot\vec{n} = q \Rightarrow s = \frac{1}{|\vec{n}|^2}\left\{q-(\vec{a}+t\vec{b})\cdot\vec{n}\right\}$$

Thus, the projection of the point $\vec{a}+t\vec{b}$ is the point

$$\vec{a}+t\vec{b}+\frac{1}{|\vec{n}|^2}\left\{q-(\vec{a}+t\vec{b})\cdot\vec{n}\right\}\vec{n} = \left[\vec{a}+\frac{(q-\vec{a}\cdot\vec{n})}{|\vec{n}|^2}\vec{n}\right]+t\left[\vec{b}-\frac{\vec{b}\cdot\vec{n}}{|\vec{n}|^2}\vec{n}\right]$$

Hence, the required projection is the line

$$\vec{r} = \left[\vec{a}+\frac{(q-\vec{a}\cdot\vec{n})}{|\vec{n}|^2}\vec{n}\right]+t\left[\vec{b}-\frac{\vec{b}\cdot\vec{n}}{|\vec{n}|^2}\vec{n}\right]$$

3.3.1 Shortest Distance Between Two Skew Lines

Skew lines are two non-intersecting, non-parallel lines. The line of shortest distance between two skew lines is the line which intersects both the skew lines at right angles. The length of the intercept between the two skew lines is called the *shortest distance between the lines.*

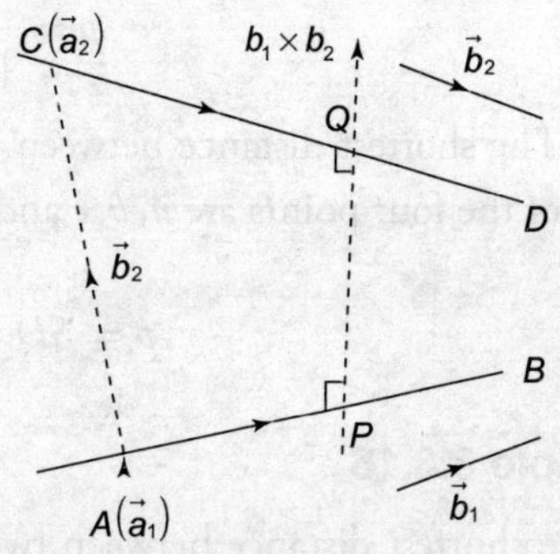

Figure 3.28 Shortest distance between two skew lines

Let the equations of the two skew lines AB and CD be, respectively, $\vec{r} = \vec{a}_1+t\,\vec{b}_1$ and $\vec{r} = \vec{a}_2+s\,\vec{b}_2$, where t and s are scalars (Fig. 3.28). The line AB passes through the point $A(\vec{a}_1)$ and is parallel to $\vec{b}_1$, and the second line CD passes through the point $C(\vec{a}_2)$ and is parallel to $\vec{b}_2$. Let PQ be the line of shortest distance. It is perpendicular to both the lines AB and CD and so parallel to $\vec{b} = \vec{b}_1 \times \vec{b}_2$. Since the given lines are not parallel, $\vec{b} = \vec{b}_1 \times \vec{b}_2 \neq \vec{0}$. Thus, the shortest distance between the lines AB and CD is given by

$$\begin{aligned} |PQ| &= \text{Projection of } AC \text{ on the line } PQ \\ &= \text{Projection of } |\vec{a}_1 - \vec{a}_2| \text{ on } \hat{b} \\ &= (\vec{a}_1 - \vec{a}_2) \cdot \hat{b} = \left| \frac{1}{|\vec{b}_1 \times \vec{b}_2|} (\vec{a}_1 - \vec{a}_2) \cdot (\vec{b}_1 \times \vec{b}_2) \right| \\ &= \frac{\left|(\vec{a}_1 - \vec{a}_2) \cdot (\vec{b}_1 \times \vec{b}_2)\right|}{|\vec{b}_1 \times \vec{b}_2|}; \quad \vec{b}_1 \times \vec{b}_2 \neq \vec{0} \end{aligned} \tag{3.45}$$

(i) If the given two straight lines $\vec{r} = \vec{a}_1 + t\,\vec{b}_1$ and $\vec{r} = \vec{a}_2 + s\,\vec{b}_2$ intersect, then they are coplanar and the shortest distance between them is zero, i.e.

$$(\vec{a}_1 - \vec{a}_2) \cdot (\vec{b}_1 \times \vec{b}_2) = 0 \Rightarrow \left[\vec{a}_1 - \vec{a}_2 \ \ \vec{b}_1 \ \ \vec{b}_2\right] = 0$$

(ii) The shortest distance is the line of intersection of the two planes drawn through the two skew lines. The equation of the plane through AB and PQ is

$$(\vec{r} - \vec{a}_1) \cdot \{\vec{b}_1 \times (\vec{b}_1 \times \vec{b}_2)\} = 0 \tag{3.46}$$

The equation of the plane through CD and PQ is

$$(\vec{r} - \vec{a}_2) \cdot \{\vec{b}_2 \times (\vec{b}_1 \times \vec{b}_2)\} = 0 \tag{3.47}$$

The planes (3.46) and (3.47) determine the equation of the straight line PQ.

(iii) Let O, Q, R and S be four points, no three of which are collinear. $\vec{b}, \vec{c}$ and $\vec{d}$ are the respective position vectors with O as origin. The equation of the line OQ is $\vec{r} = t\,\vec{b}$ and the equation of the line RS is $\vec{r} = \vec{c} + s(\vec{d} - \vec{c})$. Hence, the shortest distance between the lines OQ and RS is given by

$$p = \frac{1}{|\vec{b} \times (\vec{d} - \vec{c})|} \, |\vec{c} \cdot \vec{b} \times (\vec{d} - \vec{c})|$$

(iv) The shortest distance between two straight lines AB and CD, where the position vectors of the four points are $\vec{a}, \vec{b}, \vec{c}$ and $\vec{d}$, is

$$p = SD = \frac{|(\vec{c} - \vec{a}) \cdot \{(\vec{b} - \vec{a}) \times (\vec{d} - \vec{c})\}|}{|(\vec{b} - \vec{a}) \times (\vec{d} - \vec{c})|}$$

Example 3.3.16

Find the shortest distance between two skew lines $\vec{r} = \vec{a}_1 + t\,\vec{b}_1$ and $\vec{r} = \vec{a}_2 + t\,\vec{b}_2$, where t is a scalar and $\vec{a}_1, \vec{b}_1, \vec{a}_2$ and $\vec{b}_2$ are vectors with coordinates $(1, -2, 3), (2, 1, 1), (-2, 2, -1)$ and $(-3, 1, 2)$, respectively. Find out where the line of shortest distance meets the given lines.

Solution

The first line $\vec{r} = \vec{a}_1 + t\,\vec{b}_1$ passes through the point $\vec{a}_1 = (1, -2, 3)$ and is parallel to the vector $\vec{b}_1 = (2, 1, 1)$. Similarly, for the second, $\vec{a}_2 = (-2, 2, -1)$ and $\vec{b}_2 = (-3, 1, 2)$. The line of shortest distance meets the lines perpendicularly. Hence, $\vec{b}_1 \times \vec{b}_2$ is parallel to the line of shortest distance. Now

$$\vec{a}_2 - \vec{a}_1 = (-2, 2, -1) - (1, -2, 3) = (-3, 4, -4)$$
$$\vec{b}_1 \times \vec{b}_2 = (2, 1, 1) \times (3, 1, 2) = (1, -7, 5)$$
$$|\vec{b}_1 \times \vec{b}_2| = \sqrt{1^2 + 7^2 + 5^2} = \sqrt{75} = 5\sqrt{3}$$

The required shortest distance between the skew lines $\vec{r} = \vec{a}_1 + t\,\vec{b}_1$ and $\vec{r} = \vec{a}_2 + t\,\vec{b}_2$ is the length of projection of AB (refer to Fig. 3.28) along the line of shortest distance. Using Eq.(3.45)

$$p = \frac{|(-3\hat{\imath} + 4\hat{\jmath} - 4\hat{k}) \cdot (\hat{\imath} - 7\hat{\jmath} + 5\hat{k})|}{5\sqrt{3}} = \frac{51}{5\sqrt{3}} = \frac{17\sqrt{3}}{5} \text{ units}$$

This is the required shortest distance. Now

$$\vec{b}_1 \times (\vec{b}_1 \times \vec{b}_2) = \begin{vmatrix} \hat{\imath} & \hat{\jmath} & \hat{k} \\ 2 & 1 & 1 \\ 1 & -7 & 5 \end{vmatrix} = 12\hat{\imath} - 9\hat{\jmath} - 15\hat{k}$$

Similarly, $\vec{b}_2 \times (\vec{b}_1 \times \vec{b}_2) = 19\hat{\imath} + 17\hat{\jmath} + 20\hat{k}$. $\vec{b}_1 \times (\vec{b}_1 \times \vec{b}_2)$ is the normal to the plane containing the line $\vec{r} = \vec{a}_1 + t\,\vec{b}_1$ and the line of shortest distance. $\vec{b}_2 \times (\vec{b}_1 \times \vec{b}_2)$ is the normal to the plane containing the line $\vec{r} = \vec{a}_2 + t\,\vec{b}_2$ and the line of shortest distance. Hence, the equation of the plane containing the line $\vec{r} = \vec{a}_1 + t\,\vec{b}_1$ and the line of shortest distance (Eq. 3.46) is

$$\vec{r} \cdot \{\vec{b}_1 \times (\vec{b}_1 \times \vec{b}_2)\} = \vec{a}_1 \cdot \{\vec{b}_1 \times (\vec{b}_1 \times \vec{b}_2)\}$$
$$\vec{r} \cdot (12, -9, -15) = (1, -2, 3) \cdot (12, -9, -15) = -15$$

The equation of the plane containing the line $\vec{r} = \vec{a}_2 + t\,\vec{b}_2$ and the line of shortest distance (Eq. 3.47) is

$$\vec{r} \cdot \{\vec{b}_2 \times (\vec{b}_1 \times \vec{b}_2)\} = \vec{a}_2 \cdot \{\vec{b}_2 \times (\vec{b}_1 \times \vec{b}_2)\}$$
$$\vec{r} \cdot (19, 17, 20) = (-2, 2, 1) \cdot (19, 17, 20) = 16$$

The points where the line of shortest distance meets the lines $\vec{r} = \vec{a}_1 + t\,\vec{b}_1$ and $\vec{r} = \vec{a}_2 + t\,\vec{b}_2$ are the points P and Q, where the second line meets the plane $\vec{r} \cdot (12, -9, -15) = -15$ and the first line meets the plane $\vec{r} \cdot (19, 17, 20) = 16$, respectively. To find the coordinates of P, we have

$$(\vec{a}_2 + t\,\vec{b}_2) \cdot (12, -9, -15) = 75$$
$$\Rightarrow \Big[(-2, 2, 1) + t(-3, 1, 2)\Big] \cdot (12, -9, -15) = -15$$
$$\Rightarrow -75t = 42 \Rightarrow t = -\frac{14}{25}$$

Therefore, using $\vec{r} = \vec{a}_2 + t\,\vec{b}_2$, we get

$$\vec{r} = (-2, 2, 1) - \frac{14}{25}(-3, 1, 2) = \left(-\frac{8}{25}, \frac{36}{25}, \frac{3}{25}\right)$$

This gives the point P. Similarly, solving the first line $\vec{r} = \vec{a}_1 + t\,\vec{b}_1$ and the plane $\vec{r}\cdot(19, 17, 20) = 16$, we have another point Q.

Example 3.3.17

A straight line intersects two non-coplanar straight lines and moves parallel to a fixed plane. Prove that the locus of a point which divides the intercept in the constant ratio is a straight line.

Solution

Take O, the midpoint of the line of shortest distance PQ, as origin (Fig. 3.28). Suppose $\vec{OP} = \vec{a}$, then $\vec{OQ} = -\vec{a}$. Let $\vec{b}_1$ and $\vec{b}_2$ be two vectors parallel to the given lines AB and CD. Then, the equations of the lines will be

$$\vec{r} = \vec{a} + t\vec{b}_1, \quad \vec{r} = -\vec{a} + s\vec{b}_2$$

Suppose the fixed plane has the equation $\vec{r}\cdot\hat{n} = p$. Two points A and C, one on each line, have the position vectors $\vec{a} + t\vec{b}_1$ and $-\vec{a} + s\vec{b}_2$ (t and s are fixed constants). The join of A and C parallel to the fixed plane of

$$\left\{\vec{a} + t\vec{b}_1 - \left(-\vec{a} + s\vec{b}_2\right)\right\}\cdot\hat{n} = 0$$

$$\Rightarrow 2\vec{a}\cdot\hat{n} + t\vec{b}_1\cdot\hat{n} - s\vec{b}_2\cdot\hat{n} = 0$$

$$\Rightarrow s = 2\frac{\vec{a}\cdot\hat{n}}{\vec{b}_1\cdot\hat{n}} + t\frac{\vec{b}_1\cdot\hat{n}}{\vec{b}_2\cdot\hat{n}}$$

Take a point D on AC such that $AD : DC = x : y$, Then, the position vector of D is given by

$$\vec{r} = \frac{1}{x+y}\left\{y(\vec{a} + t\vec{b}_1) + x\left(-\vec{a} + s\vec{b}_2\right)\right\}$$

$$= \frac{1}{x+y}\left[y(\vec{a} + t\vec{b}_1) + x\left(-\vec{a} + 2\frac{\vec{a}\cdot\hat{n}}{\vec{b}_2\cdot\hat{n}}\vec{b}_2 + t\frac{\vec{b}_1.\hat{n}}{\vec{b}_2\cdot\hat{n}}\vec{b}_2\right)\right]$$

$$= \frac{y-x}{x+y}\vec{a} + \left\{\frac{t}{x+y}\left(y\vec{b}_1 + x\frac{\vec{b}_1\cdot\hat{n}}{\vec{b}_2\cdot\hat{n}}\vec{b}_2\right) + \frac{2x}{x+y}\frac{\vec{a}\cdot\hat{n}}{\vec{b}_2\cdot\hat{n}}\vec{b}_2\right\}$$

$$= \vec{c} + \lambda\vec{d}$$

where $\vec{c}$ and $\vec{d}$ are vectors involving $x, y, t, \vec{a}, \vec{b}_1$ and $\vec{b}_2$. This will imply that the line is a straight line.

Example 3.3.18

Show that the distance of a corner of a unit cube from the diagonal not passing through it is $\sqrt{6}/3$ units.

Solution

We consider that $OABCDEFG$ is a unit cube, where O is considered as origin and A, C and E are on the x, y and z axes, respectively (Fig. 3.29). Let $\hat{i}, \hat{j}$ and $\hat{k}$ be the unit vectors along the x, y and z axes, respectively. Then, the position vectors of B and E are $\hat{i}+\hat{j}$ and $\hat{k}$, respectively. Therefore, the equation of the straight line BE is

$$\vec{r} = \hat{k} + t(\hat{i}+\hat{j}-\hat{k}), t = \text{parameter}$$

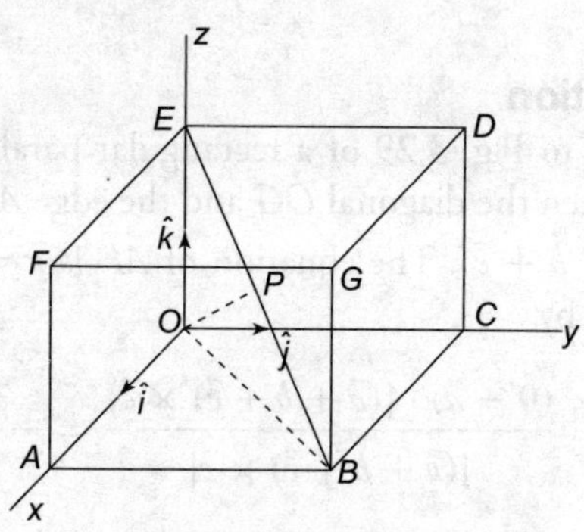

Figure 3.29 Unit cube

Now, the unit vector corresponding to the vector $\hat{i}+\hat{j}-\hat{k}$ is $\frac{1}{\sqrt{3}}(\hat{i}+\hat{j}-\hat{k})$. Thus, the perpendicular distance OP of the straight line BE from O (the diagonal BE does not pass through the corner O) is

$$OP = \left|(\vec{0}-\hat{k}) \times \frac{1}{\sqrt{3}}(\hat{i}+\hat{j}-\hat{k})\right| = \left|\frac{1}{\sqrt{3}}(\hat{i}-\hat{j})\right| = \frac{\sqrt{6}}{3} \text{ units}$$

Example 3.3.19

Show that the vector equation of a plane through the point $\hat{i}+2\hat{j}-\hat{k}$ and perpendicular to the line of intersection of the planes $\vec{r}\cdot(3\hat{i}-\hat{j}+\hat{k}) = 1$ and $\vec{r}\cdot(\hat{i}+4\hat{j}-2\hat{k}) = 2$ is $\vec{r}\cdot(-2\hat{i}+7\hat{j}+13\hat{k})+1 = 0$.

Solution

The two equations of the planes can be written as $\vec{r}\cdot\vec{n}_1 = p_1$, where $\vec{n}_1 = 3\hat{i}-\hat{j}+\hat{k}, p_1 = 1$. $n_2 = \hat{i}+4\hat{j}-2\hat{k}$ and $p_2 = 2$. Thus, the unit normals are given by

$$\hat{n}_1 = \frac{1}{\sqrt{11}}(3,-1,1), \hat{n}_2 = \frac{1}{\sqrt{21}}(1,4,-2)$$

$$\Rightarrow \hat{n} = \hat{n}_1 \times \hat{n}_2 = \frac{1}{\sqrt{11}}\frac{1}{\sqrt{21}}(3,-1,1)\times(1,4,-2) = \frac{1}{\sqrt{11\times 21}}(-2,7,13)$$

This vector $\hat{n}$ is perpendicular to the required plane. Let $P(\vec{r})$ be the position vector of any point on the plane, which passes through the point $\vec{\alpha} = \hat{i}+2\hat{j}-\hat{k}$. Then, the required equation of the plane is

$$(\vec{r}-\vec{\alpha})\cdot\hat{n} = 0 \Rightarrow \vec{r}\cdot\hat{n} = \vec{\alpha}\cdot\hat{n}$$

$$\Rightarrow \vec{r}\cdot\frac{1}{\sqrt{11\times 21}}(-2,7,13) = (1,2,-1)\cdot\frac{1}{\sqrt{11\times 21}}(-2,7,13)$$

$$\Rightarrow \vec{r}\cdot(-2,7,13) = -1$$

Example 3.3.20

Prove that the shortest distances between the diagonals of a rectangular parallelepiped, whose sides are a, b and c and the edges that do not meet it are $\frac{bc}{\sqrt{b^2+c^2}}, \frac{ca}{\sqrt{c^2+a^2}}$ and $\frac{ab}{\sqrt{a^2+b^2}}$.

Solution

Refer to Fig. 3.29 of a rectangular parallelopiped. Suppose we are to find the shortest distance between the diagonal OG and the edge AF, which does not meet OG. The equation of OG is $\vec{r} = t(\vec{a} + \vec{b} + \vec{c})$. The equation of AF is $\vec{r} = \vec{a} + s\vec{c}$. Hence, the shortest distance between them is given by

$$\frac{(\vec{0} - \vec{a}) \cdot \{(\vec{a} + \vec{b} + \vec{c}) \times \vec{c}\}}{|(\vec{a} + \vec{b} + \vec{c}) \times \vec{c}|} = -\frac{\vec{a} \cdot \vec{b} \times \vec{c}}{|(\vec{a} + \vec{b} + \vec{c}) \times \vec{c}|} = -\frac{abc}{c\sqrt{a^2 + b^2}} = -\frac{ab}{\sqrt{a^2 + b^2}}$$

Therefore, the numerical value of the shortest distance $= \dfrac{ab}{\sqrt{a^2 + b^2}}$. Similarly, the shortest distance between OG and the other edges that do not meet it (i.e., CB and DO) are $\dfrac{bc}{\sqrt{b^2 + c^2}}$ and $\dfrac{ca}{\sqrt{c^2 + a^2}}$.

3.4 SPHERE

A sphere is the locus of a variable point in space, whose distance from a fixed point in the space is always a constant. The fixed point is known as the centre of the sphere and the constant length of any point on the sphere from the centre is known as its radius.

3.4.1 Sphere with Given Radius and Centre

We consider a sphere of radius ρ units (Fig. 3.30), whose centre is at C, with a position vector $\vec{\alpha}$ relative to a fixed point O chosen as origin. Let P be any point with position vector $\vec{r}$ relative to the origin O, on the surface of the sphere. Then, $\vec{CP} = \vec{OP} - \vec{OC} = \vec{r} - \vec{\alpha}$ and

$$|\vec{CP}| = |\vec{r} - \vec{\alpha}| = \rho \tag{3.48}$$

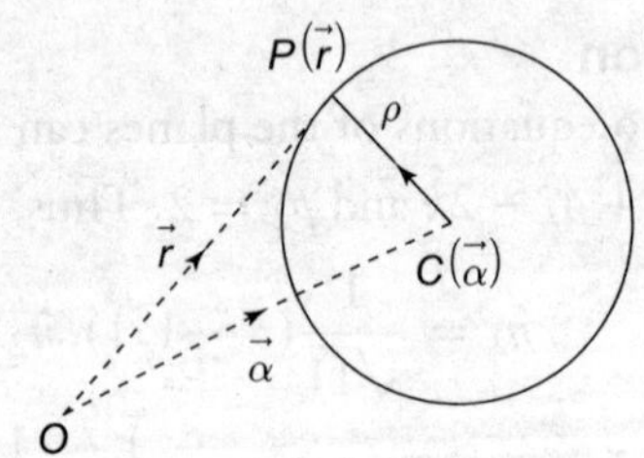

Figure 3.30 Sphere with given radius and centre

This is the equation of the circle. Also

$$\begin{aligned}
|\vec{CP}|^2 = \rho^2 &\Rightarrow \vec{CP} \cdot \vec{CP} = \rho^2 \\
&\Rightarrow (\vec{r} - \vec{\alpha}) \cdot (\vec{r} - \vec{\alpha}) = \rho^2 \\
&\Rightarrow \vec{r} \cdot \vec{r} - 2\vec{r} \cdot \vec{\alpha} + \vec{\alpha} \cdot \vec{\alpha} = \rho^2 \\
&\Rightarrow |\vec{r}|^2 - 2\vec{r} \cdot \vec{\alpha} + k = 0; \quad \text{where } k = |\vec{\alpha}|^2 - \rho^2
\end{aligned} \tag{3.49}$$

Equation (3.49) is the vector equation of the sphere, whose centre is at C and whose radius is ρ.

(i) If the centre of the sphere is at origin, i.e., O coincides with C, then $\vec{CP} = \vec{r}$ and the equation of the sphere reduces to $|\vec{r}|^2 = \rho^2$, i.e., $|\vec{r}| = \rho$.

(ii) When the sphere passes through origin, i.e., O is on the sphere, $|\vec{OC}| = |\vec{\alpha}| = \rho$, and hence, the equation of the sphere reduces to $|\vec{r}|^2 - 2\vec{r} \cdot \vec{\alpha} = 0$.

Example 3.4.1

Find the condition that the two spheres $|\vec{r}-\vec{\alpha}_1| = \rho_1$ and $|\vec{r}-\vec{\alpha}_2| = \rho_2$ may intersect orthogonally.

Solution

The two spheres are $\Gamma_1 : |\vec{r} - \vec{\alpha}_1| = \rho_1$ and $\Gamma_2 : |\vec{r} - \vec{\alpha}_2| = \rho_2$ (Fig. 3.31).

Let C_1 and C_2 with position vectors $\vec{\alpha}_1$ and $\vec{\alpha}_2$ (with respect to the origin O) be the centres of the spheres. Then, $\vec{C_1C_2} = \vec{OC_1} - \vec{OC_2} = \vec{\alpha}_2 - \vec{\alpha}_1$. If the spheres intersect at right angles at P, then the tangent plane at P to the first sphere Γ_1 will pass through C_2 and that to the second sphere Γ_2 will pass through C_1. Here, $\triangle C_1PC_2$ is a right-angled triangle. Therefore

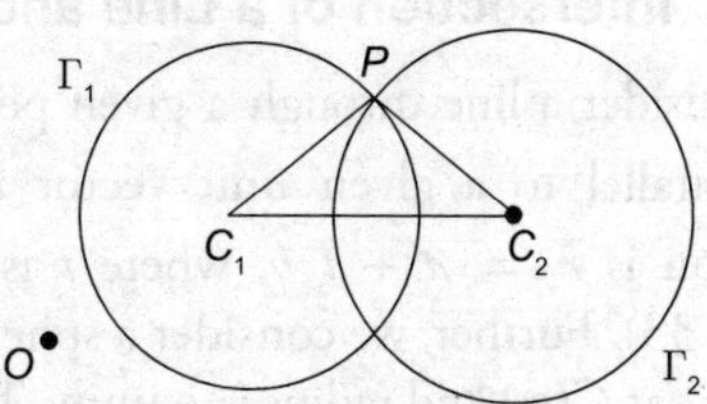

Figure 3.31 Spheres intersect orthogonally

$$C_1C_2^2 = C_1P^2 + C_2P^2 \Rightarrow |\vec{\alpha}_1 - \vec{\alpha}_2|^2 \rho_1^2 + \rho_2^2$$

This is the required condition.

Example 3.4.2

O is any fixed point and P is any point on a given sphere. Q is a point taken on OP such that $OP : OQ$ is a fixed ratio. Show that the locus of Q is a sphere.

Solution

We take O as origin. Let the equation of the fixed sphere be $|\vec{r}|^2 - 2\vec{r} \cdot \vec{\alpha} + k = 0$. Let $\vec{OQ} = \vec{\rho}$ and $OP : OQ = \lambda (> 0)$. Then, $\vec{OP} = \lambda\ \vec{OQ}$, i.e., $\vec{r} = \lambda\ \vec{\rho}$. Since the point P lies on the fixed sphere, we have

$$\lambda^2\ |\vec{\rho}|^2 - 2\lambda\ \vec{\rho} \cdot \vec{\alpha} + k = 0$$

$$\text{or,}\quad |\vec{\rho}|^2 - 2\ \vec{\rho}\ \frac{1}{\lambda} \cdot \vec{\alpha} + \frac{k}{\lambda^2} = 0$$

Hence, the locus of the point Q is $|\vec{r}|^2 - 2\ \vec{r}\ \dfrac{1}{\lambda} \cdot \vec{\alpha} + \dfrac{k}{\lambda^2} = 0$, and it represents a sphere.

3.4.2 Sphere with Given Extremities of a Diameter

Let the two given points A and B be the extremities of a diameter of a sphere with position vectors $\vec{\alpha}$ and $\vec{\beta}$, respectively (Fig. 3.32), with respect to the base point O. Let P be any point on the sphere having position vector $\vec{r}$ with respect to the origin O. Then, $\vec{AP} = \vec{r} - \vec{\alpha}$ and $\vec{BP} = \vec{r} - \vec{\beta}$. Now, for the sphere, $\vec{AP}$ is perpendicular to $\vec{BP}$, so $\angle APB = 90°$. As this is a semi-circle, we have

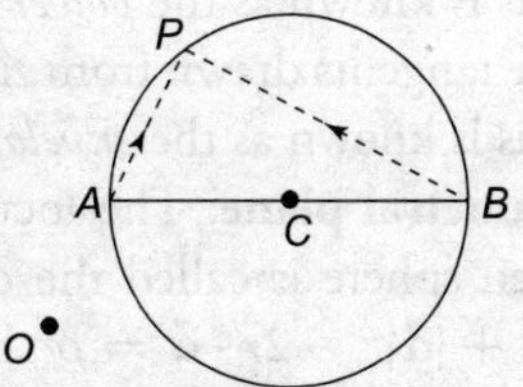

Figure 3.32 Sphere with given extremities of a diameter

$$(\vec{r} - \vec{\alpha}) \cdot (\vec{r} - \vec{\beta}) = 0 \tag{3.50}$$

Thus, this is the vector equation as it is satisfied by the position vector of any point on the sphere.

3.4.3 Intersection of a Line and a Sphere

We consider a line through a given point $A(\vec{a})$ and parallel to a given unit vector $\hat{b}$, so its equation is $\vec{r} = \vec{a} + t\,\hat{b}$, where t is a scalar (Fig. 3.33). Further, we consider a sphere whose centre is at $C(\vec{\alpha})$ and radius is ρ units. Then, the equation of the sphere is

$$|\vec{r}|^2 + |\vec{\alpha}|^2 - 2\vec{r}\cdot\vec{\alpha} = \rho^2$$

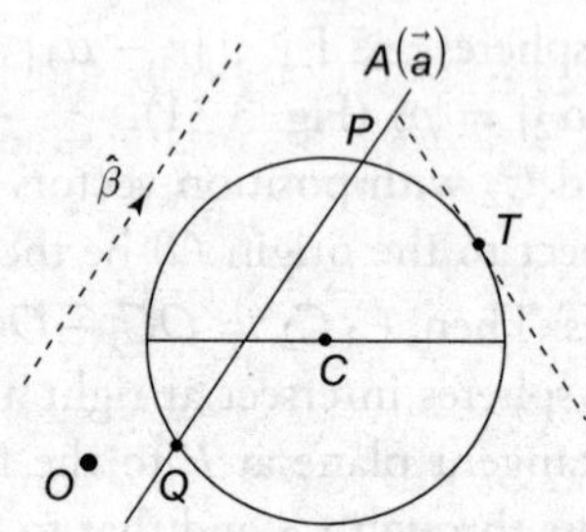

Figure 3.33 Intersection of a line and a sphere

For the points of intersection, eliminating $\vec{r}$ between these equations and using the fact that $|\hat{\beta}| = 1$, we obtain the required condition as

$$|\vec{a} + t\,\hat{b}|^2 + |\vec{\alpha}|^2 - 2(\vec{a} + t\,\hat{b})\cdot\vec{\alpha} = \rho^2$$

$$\text{or,}\quad (\vec{a} + t\,\hat{b})\cdot(\vec{a} + t\,\hat{b}) + |\vec{\alpha}|^2 - 2\vec{a}\cdot\vec{\alpha} + 2t\,\hat{b}\cdot\vec{\alpha} = \rho^2$$

$$\text{or,}\quad t^2 + 2t\hat{b}\cdot(\vec{a} - \vec{\alpha}) + |\vec{a}|^2 - 2\vec{a}\cdot\vec{\alpha} + |\vec{\alpha}|^2 = \rho^2 \tag{3.51}$$

This yields two values t_1 and t_2 of t, corresponding to each of which, we have points of intersection as P and Q. This shows that a straight line cannot meet a sphere at more than two points. Thus, the position vectors of the points P and Q are $\vec{a} + t_1\,\hat{b}$ and $\vec{a} + t_2\,\hat{b}$, respectively, where $t_1 = AP$ and $t_2 = AQ$.

(i) If P and Q coincide at a point T on the sphere, then the two roots of the quadratic in t are equal and the line becomes the tangent and the condition is

$$4\left\{\hat{\beta}\cdot(\vec{a} - \vec{\alpha})\right\}^2 - 4\left[|\vec{a}|^2 - 2\vec{a}\cdot\vec{\alpha} + |\vec{\alpha}|^2 - \rho^2\right] = 0$$

(ii) The square of the length of the tangent from any point to the sphere, i.e., the quantity AT^2, is given by

$$AT^2 = AP\cdot AQ = |\vec{a}|^2 - 2\vec{a}\cdot\vec{\alpha} + |\vec{\alpha}|^2 - \rho^2$$

This is known as the *power of point* A with respect to the sphere.

(iii) The tangents drawn from A generate a cone having its vertex A and enveloping the sphere. This is known as the *enveloping cone*.

(iv) **Diametral plane**: The locus of the points which bisect a system of parallel chords of a given sphere is called the diametral plane, all chords being parallel to a given line. Let $|\vec{r}|^2 + |\vec{\alpha}|^2 - 2\vec{r}\cdot\vec{\alpha} = \rho^2$ be a given sphere and let a chord parallel to a given unit vector $\hat{b}$ and passing through the point $\vec{a}$ be $\vec{r} = \vec{a} + t\hat{b}$. The points of intersection of the sphere and the plane are given by Eq.(3.51).

If $\vec{a}$ is the middle point of the chord $\vec{r} = \vec{a} + t\hat{b}$, then the two values of t must be equal and opposite in sign, i.e.

$$t_1 + t_2 = 0 \Rightarrow \hat{b}\cdot(\vec{a} - \vec{\alpha}) = 0$$

Thus, the locus of the point $\vec{a}$ is $\hat{b} \cdot (\vec{r} - \vec{\alpha}) = 0$, which is the required diametral plane.

3.4.4 Tangent Plane at a Given Point

The point of intersection of the sphere $|\vec{r}|^2 + |\vec{\alpha}|^2 - 2\vec{r} \cdot \vec{\alpha} = \rho^2$ and the line $\vec{r} = \vec{a} + t\hat{b}$ is given by Eq.(3.51). Since $\vec{a}$ lies on the sphere, $|\vec{a}|^2 + |\vec{\alpha}|^2 - 2\vec{a} \cdot \vec{\alpha} = \rho^2$. Therefore, Eq.(3.51) reduces to

$$t^2 + 2t\hat{b} \cdot (\vec{a} - \vec{\alpha}) = 0 \Rightarrow t = 0 \text{ and } t = -2\hat{b} \cdot (\vec{a} - \vec{\alpha})$$

Obviously, one value of t is zero. If the line touches the circle, both the values of t must be identical, i.e., $\hat{b} \cdot (\vec{a} - \vec{\alpha}) = 0$. From the equation of the line $\vec{r} = \vec{a} + t\hat{b}$, we get $\hat{b} = (\vec{r} - \vec{a})/t$. Therefore

$$\hat{b} \cdot (\vec{a} - \vec{\alpha}) = 0 \Rightarrow \frac{1}{t}(\vec{r} - \vec{a}) \cdot (\vec{a} - \vec{\alpha}) = 0$$
$$\Rightarrow \vec{r} \cdot (\vec{a} - \vec{\alpha}) = \vec{a} \cdot (\vec{a} - \vec{\alpha}) \tag{3.52}$$

This is an equation of the form $\vec{r} \cdot \hat{n} = q$, so Eq.(3.52) represents a plane that is perpendicular to the radius through the point $\vec{a}$. Also, from Eq.(3.52), we get

$$\vec{r} \cdot \vec{a} - \vec{r} \cdot \vec{\alpha} = \vec{a} \cdot \vec{a} - \vec{a} \cdot \vec{\alpha}$$
$$\Rightarrow \vec{r} \cdot \vec{a} - \vec{r} \cdot \vec{\alpha} - \vec{a} \cdot \vec{a} + \vec{a} \cdot \vec{\alpha} + (|\vec{r}|^2 + |\vec{\alpha}|^2 - 2\vec{r} \cdot \vec{\alpha} - \rho^2) = 0$$
$$\Rightarrow \vec{r} \cdot \vec{a} - \vec{\alpha} \cdot (\vec{r} + \vec{a}) + |\vec{\alpha}|^2 - \rho^2 = 0 \tag{3.53}$$

This is the required equation of the tangent plane to the given sphere at the given point $\vec{a}$.

(i) **Condition for tangency of any plane to the sphere**: Let $|\vec{r}|^2 + |\vec{\alpha}|^2 - 2\vec{r} \cdot \vec{\alpha} = \rho^2$ be a given sphere whose centre is $\vec{\alpha}$ and radius is ρ. Let the given plane be $\vec{r} \cdot \hat{n} = q$. Since the tangent plane at any point is perpendicular to the radius through that point, the square of the perpendicular from the centre to the tangent plane must be equal to the square of the radius. Hence, the required condition will be

$$\left(\frac{q - \vec{\alpha} \cdot \hat{n}}{|\hat{n}|}\right)^2 = \rho^2$$

(ii) **Condition for orthogonality**: Let the equations of two spheres be $|\vec{r}|^2 + |\vec{\alpha}_1|^2 - 2\vec{r} \cdot \vec{\alpha}_1 = \rho_1^2$ and $|\vec{r}|^2 + |\vec{\alpha}_2|^2 - 2\vec{r} \cdot \vec{\alpha}_2 = \rho_2^2$. The spheres will intersect orthogonally if the tangent plane to any of them at the point of intersection passes through the centre of the other. So the square of the distance between the centres must be equal to the sum of the squares at their radii. Thus

$$(\vec{\alpha}_1 - \vec{\alpha}_2)^2 = \rho_1^2 + \rho_2^2$$

(iii) **Polar plane**: The polar plane of the given point with respect to a sphere is the locus of the points, the tangent plane at which passes through the given point. Let $|\vec{r}|^2 + |\vec{\alpha}|^2 - 2\vec{r} \cdot \vec{\alpha} = \rho^2$ be a given sphere whose centre is $\vec{\alpha}$ and radius is ρ and $\vec{d}$ be any given point. Let $\vec{a}$ be any point on the sphere, the tangent plane at which passes through $\vec{d}$. The tangent at the point $\vec{a}$ to the sphere is

$$\vec{r}\cdot\vec{a}-\vec{\alpha}\cdot(\vec{r}+\vec{a})+|\vec{\alpha}|^2-\rho^2=0$$

It passes through the point $\vec{d}$, so

$$\vec{d}\cdot\vec{a}-\vec{\alpha}\cdot(\vec{d}+\vec{a})+|\vec{\alpha}|^2-\rho^2=0$$

Therefore, the locus of $\vec{a}$ is

$$\vec{d}\cdot\vec{r}-\vec{\alpha}\cdot(\vec{d}+\vec{r})+|\vec{\alpha}|^2-\rho^2=0$$
$$\Rightarrow \vec{r}\cdot\vec{d}-\vec{\alpha}\cdot(\vec{r}+\vec{d})+|\vec{\alpha}|^2-\rho^2=0$$

(iv) **Radical plane**: The locus of the point which moves so that the squares of the lengths of the tangents drawn from it to the spheres are equal (or powers of the point with respect to two spheres are equal) is called the radical plane of the two spheres. Let $|\vec{r}|^2+|\vec{\alpha}_1|^2-2\vec{r}\cdot\vec{\alpha}_1=\rho_1^2$ and $|\vec{r}|^2+|\vec{\alpha}_2|^2-2\vec{r}\cdot\vec{\alpha}_2=\rho_2^2$ be two given spheres. Let $\vec{a}$ be the position vector of a point which moves so that the square of the tangents drawn from it to the spheres are equal. Then

$$|\vec{a}|^2+|\vec{\alpha}_1|^2-2\vec{a}\cdot\vec{\alpha}_1=\rho_1^2 \text{ and } |\vec{a}|^2+|\vec{\alpha}_2|^2-2\vec{a}\cdot\vec{\alpha}_2=\rho_2^2$$
$$2\vec{a}\cdot(\vec{\alpha}_1-\vec{\alpha}_2)=\rho_1^2-\rho_2^2$$

Thus, the locus of $\vec{a}$ is

$$2\vec{r}\cdot(\vec{\alpha}_1-\vec{\alpha}_2)=\rho_1^2-\rho_2^2$$

This equation shows that the radial plane is perpendicular to $\vec{\alpha}_1-\vec{\alpha}_2$. Thus, the radical plane of two spheres is perpendicular to the line joining the centres of the given sphere.

Example 3.4.3

If any tangent plane to the sphere $|\vec{r}|^2-d^2=0$ makes intercepts a, b and c on the axes, prove by the vector method that $a^{-2}+b^{-2}+c^{-2}=d^{-2}$.

Solution

Let $\vec{r}\cdot\vec{n}=p$ be a plane which touches the given sphere at a point. The perpendicular distance from the centre of the sphere is equal to its radius, i.e., $\frac{p}{|\vec{n}|}=d$. Now, we assume that $\vec{n}=n_1\hat{i}+n_2\hat{j}+n_3\hat{k}$. If the plane $\vec{r}\cdot\vec{n}=p$ makes intercepts a, b and c on the axes, then the position vectors of the intercepting points are $a\hat{i}$, $b\hat{j}$ and $c\hat{k}$, respectively. Thus

$$a=\frac{p}{\hat{i}\cdot\vec{n}}=\frac{p}{n_1}, b=\frac{p}{\hat{j}\cdot\vec{n}}=\frac{p}{n_2} \text{ and } c=\frac{p}{\hat{k}\cdot\vec{n}}=\frac{p}{n_3}$$
$$\Rightarrow \frac{1}{d^2}=\frac{|\vec{n}|^2}{p^2}=\frac{n_1^2+n_2^2+n_3^2}{p^2}=\frac{1}{a^2}+\frac{1}{b^2}+\frac{1}{c^2}$$

3.5 VOLUME OF A TETRAHEDRON

Let $ABCD$ be a tetrahedron and ΔABC be the base, with D as the fourth vertex (Fig. 3.34). Let the position vector of the vertices A, B, C and D of the tetrahedron, with respect to some vector origin O, be $\vec{\alpha}, \vec{\beta}, \vec{\gamma}$ and $\vec{\delta}$, respectively. DL is perpendicular to the plane of ΔABC, which is the *altitude* or *height* of the tetrahedron. Now, $\vec{AB} \times \vec{AC} = (\vec{\beta} - \vec{\alpha}) \times (\vec{\gamma} - \vec{\alpha})$ is perpendicular to the plane ABC and represents a vector parallel to LD. Its magnitude is equal to twice the area of ΔABC. Also LD is the projection of AD on LD. Thus, the volume of the tetrahedron is

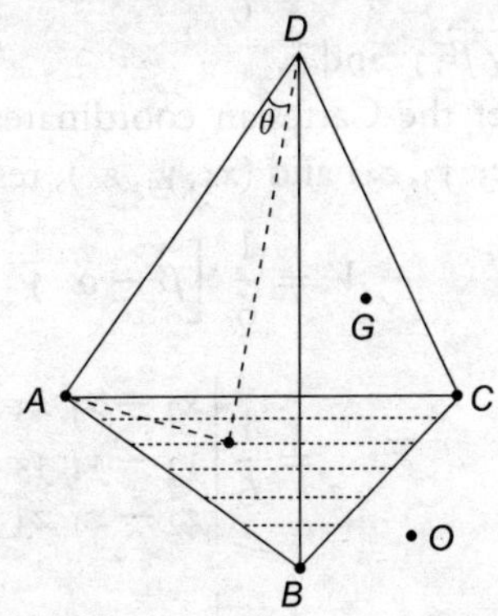

Figure 3.34 Volume of a tetrahedron

$$V = \frac{1}{3}\ \Delta\, ABC \cdot LD = \frac{1}{3} \cdot \frac{1}{2} |\vec{AB} \times \vec{AC}|\ (AD\ \cos\theta)$$

$$= \frac{1}{6}\ |\vec{AB} \times \vec{AC}|\ |\vec{AD}|\ \cos\theta = \frac{1}{6}\ (\vec{AB} \times \vec{AC}) \cdot \vec{AD}$$

$$= \frac{1}{6}\left\{(\vec{\beta} - \vec{\alpha}) \times (\vec{\gamma} - \vec{\alpha})\right\} \cdot (\vec{\delta} - \vec{\alpha}) \text{ cubic units}$$

$$= \frac{1}{6}\left[\vec{\beta} - \vec{\alpha}\ \ \vec{\gamma} - \vec{\alpha}\ \ \vec{\delta} - \vec{\alpha}\right] \text{ cubic units} \tag{3.54}$$

$$= \frac{1}{6}\left\{\left[\vec{\beta}\ \vec{\gamma}\ \vec{\delta}\right] + \left[\vec{\gamma}\ \vec{\alpha}\ \vec{\delta}\right] + \left[\vec{\delta}\ \vec{\alpha}\ \vec{\beta}\right] - \left[\vec{\alpha}\ \vec{\beta}\ \vec{\gamma}\right]\right\} \text{ cubic units}$$

(i) When the four given vectors $\vec{\alpha}, \vec{\beta}, \vec{\gamma}$ and $\vec{\delta}$ are coplanar, then the volume of the tetrahedron vanishes, i.e.

$$\left[\vec{\beta}\ \vec{\gamma}\ \vec{\delta}\right] + \left[\vec{\gamma}\ \vec{\alpha}\ \vec{\delta}\right] + \left[\vec{\delta}\ \vec{\alpha}\ \vec{\beta}\right] - \left[\vec{\alpha}\ \vec{\beta}\ \vec{\gamma}\right] = 0$$

This is the condition for the coplanarity of four points A, B, C and D.

(ii) The shortest distance between two opposite edges AC and BD of the tetrahedron $ABCD$ is

$$p = \frac{(\vec{\delta} - \vec{\alpha}) \cdot \{(\vec{\gamma} - \vec{\alpha}) \times (\vec{\delta} - \vec{\beta})\}}{|(\vec{\gamma} - \vec{\alpha}) \times (\vec{\delta} - \vec{\beta})|} = \frac{\left[\vec{\delta} - \vec{\alpha}\ \ \vec{\gamma} - \vec{\alpha}\ \ \vec{\delta} - \vec{\beta}\right]}{|(\vec{\gamma} - \vec{\alpha}) \times (\vec{\delta} - \vec{\beta})|} \text{ cubic units}$$

The edges AC and BD are, respectively, parallel to $\vec{\gamma} - \vec{\alpha}$ and $\vec{\delta} - \vec{\beta}$ and the points lie on each line.

(iii) If one of the vertices, say A, is taken as vector origin, then the volume of the tetrahedron $ABCD$ is

$$V = \frac{1}{6}\left[\vec{\beta}\ \ \vec{\gamma}\ \ \vec{\delta}\right] = \frac{1}{6}\begin{vmatrix} \vec{\beta}\cdot\vec{\beta} & \vec{\beta}\cdot\vec{\gamma} & \vec{\beta}\cdot\vec{\delta} \\ \vec{\gamma}\cdot\vec{\beta} & \vec{\gamma}\cdot\vec{\gamma} & \vec{\gamma}\cdot\vec{\delta} \\ \vec{\delta}\cdot\vec{\beta} & \vec{\delta}\cdot\vec{\gamma} & \vec{\delta}\cdot\vec{\delta} \end{vmatrix}^{1/2} \text{ cubic units} \tag{3.55}$$

This is equal to $\frac{1}{6}$ (Volume of the parallelepiped) whose three conterminous edges are given by $\vec{\beta}, \vec{\gamma}$ and $\vec{\delta}$.

(iv) Let the Cartesian coordinates of the vertices A, B, C and D be (x_1, y_1, z_1), (x_2, y_2, z_2), (x_3, y_3, z_3) and (x_4, y_4, z_4), respectively. Using Eq. (3.55), we get

$$V = \frac{1}{6}\left[\vec{\beta} - \vec{\alpha} \;\; \vec{\gamma} - \vec{\alpha} \;\; \vec{\delta} - \vec{\alpha}\right]$$

$$= \frac{1}{6}\begin{vmatrix} x_2 - x_1 & x_3 - x_1 & x_4 - x_1 \\ y_2 - y_1 & y_3 - y_1 & y_4 - y_1 \\ z_2 - z_1 & z_3 - z_1 & z_4 - z_1 \end{vmatrix} = \frac{1}{6}\begin{vmatrix} x_2 - x_1 & x_3 - x_1 & x_4 - x_1 & x_1 \\ y_2 - y_1 & y_3 - y_1 & y_4 - y_1 & y_1 \\ z_2 - z_1 & z_3 - z_1 & z_4 - z_1 & z_1 \\ 1 & 1 & 1 & 1 \end{vmatrix}$$

$$= \frac{1}{6}\begin{vmatrix} x_1 & x_2 & x_3 & 1 \\ y_1 & y_2 & y_3 & 1 \\ z_1 & z_2 & z_3 & 1 \\ 1 & 1 & 1 & 1 \end{vmatrix}. \tag{3.56}$$

Example 3.5.1

Find the volume of the tetrahedron whose vertices are $A(2, -1, -3)$, $B(4, 1, 3)$, $C(3, 2, -1)$ and $D(1, 4, 2)$.

Solution

Refer to Fig. 3.34, of a tetrahedron. Here $\vec{AB} = (2, 2, 6)$, $\vec{AC} = (1, 3, 2)$ and $\vec{AD} = (-1, 5, 5)$. Now

$$\vec{AB} \times \vec{AC} = (2\hat{i} + 2\hat{j} + 6\hat{k}) \times (\hat{i} + 3\hat{j} + 2\hat{k}) = -14\hat{i} + 2\hat{j} + 4\hat{k}$$

Therefore, the volume of the tetrahedron is

$$V = \frac{1}{6}\,(\vec{AB} \times \vec{AC}) \cdot \vec{AD} = \frac{1}{6}(-14, 2, 4) \cdot (-1, 5, 5) = \frac{1}{6} \times 44 = \frac{22}{3} = 7\frac{1}{3} \text{ cubic units}$$

Example 3.5.2

In a tetrahedron, if two pairs of opposite edges are perpendicular, the third pair is also perpendicular to each other and the sum of the squares of two opposite edges is the same for each other.

Solution

Refer to Fig. 3.34. Let the position vector of the vertices A, B, C and D of the tetrahedron with respect to the some vector origin O be $\vec{\alpha}, \vec{\beta}, \vec{\gamma}$ and $\vec{\delta}$, respectively. Thus, $\vec{AB} = \vec{\beta} - \vec{\alpha}$, $\vec{AC} = \vec{\gamma} - \vec{\alpha}$ and $\vec{AD} = \vec{\delta} - \vec{\alpha}$. Thus, $\vec{BD} = \vec{\delta} - \vec{\beta}$. Suppose the pair of opposite edges (BD, AC) are perpendicular and another pair of opposite edges (AD, BC) is perpendicular. Therefore

$$\vec{BD} \cdot \vec{AC} = 0 \Rightarrow (\vec{\delta} - \vec{\beta}) \cdot (\vec{\gamma} - \vec{\alpha}) = 0$$

$$\vec{AD} \cdot \vec{BC} = 0 \Rightarrow (\vec{\delta} - \vec{\alpha}) \cdot (\vec{\gamma} - \vec{\beta}) = 0$$

$$\Rightarrow (\vec{\delta} - \vec{\beta}) \cdot (\vec{\gamma} - \vec{\alpha}) = (\vec{\delta} - \vec{\alpha}) \cdot (\vec{\gamma} - \vec{\beta})$$

$$\Rightarrow -\vec{\delta} \cdot \vec{\alpha} - \vec{\beta} \cdot \vec{\gamma} = -\vec{\delta} \cdot \vec{\beta} - \vec{\alpha} \cdot \vec{\gamma} \Rightarrow (\vec{\beta} - \vec{\alpha}) \cdot (\vec{\delta} - \vec{\gamma}) = 0$$

i.e., $\vec{AB}$ is perpendicular to $\vec{CD}$. The sum of the squares on BD and AC is

$$\begin{aligned} BD^2 + AC^2 &= \vec{BD}\cdot\vec{BD} + \vec{AC}\cdot\vec{AC} \\ &= (\vec{\delta}-\vec{\beta})\cdot(\vec{\delta}-\vec{\beta}) + (\vec{\gamma}-\vec{\alpha})\cdot(\vec{\gamma}-\vec{\alpha}) \\ &= \vec{\delta}\cdot\vec{\delta} - 2\vec{\delta}\cdot\vec{\beta} + \vec{\beta}\cdot\vec{\beta} + \vec{\gamma}\cdot\vec{\gamma} - 2\vec{\gamma}\cdot\vec{\alpha} + \vec{\alpha}\cdot\vec{\alpha} \\ &= \vec{\delta}\cdot\vec{\delta} + \vec{\beta}\cdot\vec{\beta} + \vec{\gamma}\cdot\vec{\gamma} + \vec{\alpha}\cdot\vec{\alpha} - 2[\vec{\beta}\cdot\vec{\alpha} + \vec{\delta}\cdot\vec{\gamma}] \\ &= (\vec{\beta}-\vec{\alpha})\cdot(\vec{\beta}-\vec{\alpha}) + (\vec{\delta}-\vec{\gamma})\cdot(\vec{\delta}-\vec{\gamma}) \\ &= \vec{AB}\cdot\vec{AB} + \vec{CD}\cdot\vec{CD} = AB^2 + CD^2 \end{aligned}$$

Example 3.5.3

Find the volume of a tetrahedron in terms of the lengths of the three edges which meet in a point and the angles which they make with each other.

Solution

In Fig. 3.34, $ABCD$ is a tetrahedron with the three edges AB, AC and AD concurrent at A. Taking A as origin, let $\vec{AB} = \vec{b}, \vec{AC} = \vec{c}, \vec{AD} = \vec{d}$; $\angle CAD = \alpha$, $\angle DAB = \beta$ and $\angle BAC = \gamma$. Now

$$[\vec{b}\,\vec{c}\,\vec{d}]^2 = \begin{vmatrix} \vec{b}\cdot\vec{b} & \vec{b}\cdot\vec{c} & \vec{b}\cdot\vec{d} \\ \vec{c}\cdot\vec{b} & \vec{c}\cdot\vec{c} & \vec{c}\cdot\vec{d} \\ \vec{d}\cdot\vec{b} & \vec{d}\cdot\vec{c} & \vec{d}\cdot\vec{d} \end{vmatrix} = \begin{vmatrix} b^2 & bc\cos\gamma & bd\cos\beta \\ bc\cos\gamma & c^2 & cd\cos\alpha \\ bd\cos\beta & cd\cos\alpha & d^2 \end{vmatrix}$$

$$= b^2c^2d^2 \begin{vmatrix} 1 & \cos\gamma & \cos\beta \\ \cos\gamma & 1 & \cos\alpha \\ \cos\beta & \cos\alpha & 1 \end{vmatrix}$$

Thus, the required volume of the tetrahedron is

$$V = \frac{1}{6}[\vec{b}\,\vec{c}\,\vec{d}] = \frac{bcd}{6}\begin{vmatrix} 1 & \cos\gamma & \cos\beta \\ \cos\gamma & 1 & \cos\alpha \\ \cos\beta & \cos\alpha & 1 \end{vmatrix}^{1/2}$$

Example 3.5.4

Show that six planes, each passing through one edge of a tetrahedron and bisecting the opposite edge, meet in a point.

Solution

Refer to Fig. 3.34. Let the position vectors of the vertices A, B, C and D of the tetrahedron with respect to the some vector origin O be $\vec{\alpha}, \vec{\beta}, \vec{\gamma}$ and $\vec{\delta}$, respectively. The plane containing AB and bisecting DC contains the points $\vec{\alpha}, \vec{\beta}$ and $\frac{1}{2}(\vec{\gamma}+\vec{\delta})$. Let $\vec{r}$ be the position vector of any point on this plane. Then, the vectors $\vec{r}-\vec{\alpha}, \vec{r}-\vec{\beta}$ and $\vec{r}-\frac{1}{2}(\vec{\gamma}+\vec{\delta})$ are coplanar, and hence, the condition for coplanarity gives

$$\left[\vec{r}-\vec{\alpha}\;\; \vec{r}-\vec{\beta}\;\; \vec{r}-\frac{1}{2}(\vec{\gamma}+\vec{\delta})\right] = 0$$

This gives the equation of the plane. Now, we see that $\vec{g} = \frac{1}{4}(\vec{\alpha}+\vec{\beta}+\vec{\gamma}+\vec{\delta})$ lies on the plane. It follows that the six planes, each passing through one edge and bisecting the opposite edge, meet at $G(\vec{g})$ (by symmetry).

Example 3.5.5

Find, by the vector method, the volume of the tetrahedron $ABCD$ with vertices $A(1,1,-1)$, $B(3,-2,-2)$ $C(4,3,2)$ and $D(5,5,3)$. Hence, find the distance of the point D from the plane ABC.

Solution

We consider the face ABC of the tetrahedron $ABCD$ as its base and D as its fourth vertex (Fig. 3.34). Let the position vector of the vertices A, B, C and D of the tetrahedron with respect to the some vector origin O be $\vec{\alpha}$, $\vec{\beta}$, $\vec{\gamma}$ and $\vec{\delta}$, respectively. Then

$$\vec{\beta}-\vec{\alpha}=-3\hat{\imath}-2\hat{\jmath}-3\hat{k}, \vec{\gamma}-\vec{\alpha}=-\hat{\imath}-5\hat{\jmath}-4\hat{k}, \vec{\delta}-\vec{\alpha}=\hat{\imath}+2\hat{\jmath}+\hat{k}$$

Using Eq. (3.56), we obtain the required volume of the tetrahedron as

$$V=\frac{1}{6}\left[\vec{\beta}-\vec{\alpha}\ \ \vec{\gamma}-\vec{\alpha}\ \ \vec{\delta}-\vec{\alpha}\right]$$
$$=\frac{1}{6}\begin{vmatrix}-3 & -2 & -3\\ -1 & -5 & -4\\ 1 & 2 & 1\end{vmatrix}=2$$

Now, $\vec{AB}=-3\hat{\imath}-2\hat{\jmath}-3\hat{k}$ and $\vec{AC}=-\hat{\imath}-5\hat{\jmath}-4\hat{k}$, so

$$\vec{AB}\times\vec{AC}=-(-2\hat{\imath}-2\hat{\jmath}-3\hat{k})(-\hat{\imath}-5\hat{\jmath}-4\hat{k})=-7\hat{\imath}-9\hat{\jmath}+13\hat{k}$$
$$\Rightarrow \left|\vec{AB}\times\vec{AC}\right|=\sqrt{(-7)^2+(-9)^2+13^2}=\sqrt{299}\text{ units}$$
$$\Rightarrow \text{ Area of the triangle } ABC=\frac{1}{2}\left|\vec{AB}\times\vec{AC}\right|=\frac{1}{2}\sqrt{299}\text{ units}$$

Thus, the volume V of the tetrahedron is given by

$$V=\frac{1}{3}\times\text{ area of the triangle } ABC\times\text{ Height}$$
$$\Rightarrow \text{ Height }=\frac{3V}{\text{Area of the triangle } ABC}=\frac{6}{\frac{1}{2}\sqrt{299}}=\frac{12}{\sqrt{299}}\text{ units}$$

This is the required distance of the point D from the plane ABC.

Example 3.5.6

Show that the volume of a tetrahedron bounded by the four planes $\vec{r}\cdot(l\hat{\imath}+m\hat{\jmath})=0, \vec{r}\cdot(m\hat{\jmath}+n\hat{k})=0, \vec{r}\cdot(l\hat{\imath}+n\hat{k})=0$ and $\vec{r}\cdot(l\hat{\imath}+m\hat{\jmath}+n\hat{k})=p$ is $2p^3/3lmn$.

Solution

The first three planes $\vec{r}\cdot(l\hat{\imath}+m\hat{\jmath})=0, \vec{r}\cdot(m\hat{\jmath}+n\hat{k})=0$ and $\vec{r}\cdot(l\hat{\imath}+n\hat{k})=0$ pass through the point $A(0,0,0)$, and hence, they meet at A. Suppose the planes $\vec{r}\cdot(l\hat{\imath}+m\hat{\jmath})=0, \vec{r}\cdot(m\hat{\jmath}+n\hat{k})=0$ and $\vec{r}\cdot(l\hat{\imath}+m\hat{\jmath}+n\hat{k})=p$ meet at the point $B(\vec{\beta})$, where $\vec{\beta}=\beta_1\hat{\imath}+\beta_2\hat{\jmath}+\beta_3\hat{k}$. Using these equations of planes, we have

$$(\beta_1\hat{\imath}+\beta_2\hat{\jmath}+\beta_3\hat{k})\cdot(l\hat{\imath}+m\hat{\jmath})=0\Rightarrow \quad l\beta_1+m\beta_2=0$$
$$(\beta_1\hat{\imath}+\beta_2\hat{\jmath}+\beta_3\hat{k})\cdot(m\hat{\jmath}+n\hat{k})=0\Rightarrow \quad m\beta_2+n\beta_3=0$$

$$(\beta_1\hat{i}+\beta_2\hat{j}+\beta_3\hat{k})\cdot(l\hat{i}+m\hat{j}+n\hat{k})=p\Rightarrow l\beta_1+m\beta_2+n\beta_3=p$$

Solving these equations, we have $\beta_1=\frac{p}{l}$, $\beta_2=-\frac{p}{m}$ and $\beta_3=\frac{p}{n}$. Therefore, the position vector of the point B is $\vec{\beta}=\frac{p}{l}\hat{i}-\frac{p}{m}\hat{j}+\frac{p}{n}\hat{k}$. Similarly, if $\vec{r}\cdot(m\hat{j}+n\hat{k})=0$, $\vec{r}\cdot(l\hat{i}+n\hat{k})=0$ and $\vec{r}\cdot(l\hat{i}+m\hat{j}+n\hat{k})=p$ meet at $C(\vec{\gamma})$ and the planes $\vec{r}\cdot(l\hat{i}+m\hat{j})=0$, $\vec{r}\cdot(l\hat{i}+n\hat{k})=0$ and $\vec{r}\cdot(l\hat{i}+m\hat{j}+n\hat{k})=p$ meet at $D(\vec{\delta})$, then as before we have

$$\vec{\gamma}=\frac{p}{l}\hat{i}+\frac{p}{m}\hat{j}-\frac{p}{n}\hat{k},\quad \vec{\delta}=-\frac{p}{l}\hat{i}+\frac{p}{m}\hat{j}+\frac{p}{n}\hat{k}$$

Hence, the volume of the tetrahedron $OABC$ bounded by the given four planes is

$$V_2=\frac{1}{6}[\vec{\beta}\;\vec{\gamma}\;\vec{\delta}]=\frac{1}{6}\begin{vmatrix}\frac{p}{l} & -\frac{p}{m} & \frac{p}{n}\\ \frac{p}{l} & \frac{p}{m} & -\frac{p}{n}\\ -\frac{p}{l} & \frac{p}{m} & \frac{p}{n}\end{vmatrix}$$

$$=\frac{1}{6}\,\frac{p^3}{lmn}\begin{vmatrix}1 & -1 & 1\\ 1 & 1 & -1\\ -1 & 1 & 1\end{vmatrix}=\frac{2}{3}\,\frac{p^3}{lmn}$$

Example 3.5.7

Find the equation of the sphere which is inscribed in the tetrahedron bounded by the planes $\vec{r}\cdot\hat{i}=0$, $\vec{r}\cdot\hat{j}=0$, $\vec{r}\cdot\hat{k}=0$ and $\vec{r}\cdot(\hat{i}+\hat{j}+\hat{k})=a$.

Solution

Since the sphere is inscribed in the tetrahedron, the given planes are the tangent planes to the sphere. Let the position vector of the centre P (Fig. 3.35) of the sphere be $\vec{\alpha}=\alpha_1\hat{i}+\alpha_2\hat{j}+\alpha_3\hat{k}$ with respect to some base point O, considered as origin. Consider the plane $\vec{r}\cdot\hat{i}=0$, which can be written as $\vec{r}\cdot\vec{n}_1=p_1$, where $\vec{n}_1=\hat{i}$ and $p_1=0$. Let PN_1 be the perpendicular distance from the centre P to this plane. The line PN_1 passes through $\vec{\alpha}$ and is parallel to $\vec{n}_1$, so its equation is $\vec{r}=\vec{\alpha}+t_1\,\vec{n}_1$. For the common point N_1 of intersection

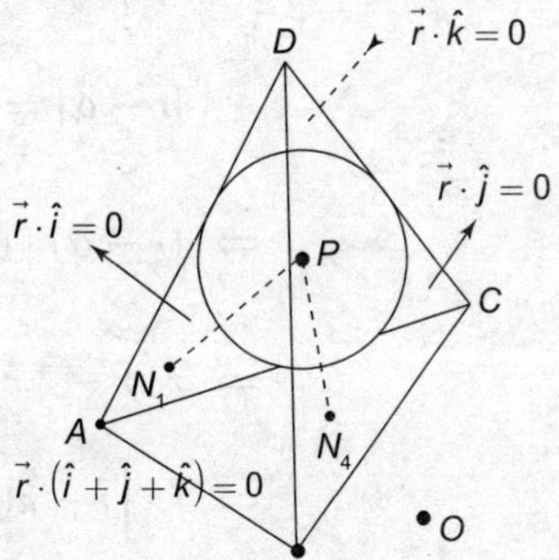

Figure 3.35 A sphere inscribed in a tetrahedron

$$(\vec{\alpha}+t_1\,\vec{n}_1)\cdot\vec{n}_1=0\Rightarrow\vec{\alpha}\cdot\vec{n}_1+t_1\,\vec{n}_1\cdot\vec{n}_1=0$$
$$\Rightarrow\vec{\alpha}\cdot\hat{i}+t_1\,|\hat{i}|^2=0\Rightarrow t_1=-\vec{\alpha}\cdot\hat{i}=-\alpha_1$$

Thus, the position vector of N_1 is $\vec{r}=\vec{\alpha}-\alpha_1\,\hat{i}$. Therefore

$$\vec{PN_1}=\vec{\alpha}-\alpha_1\,\hat{i}-\vec{\alpha}=-\alpha_1\,\hat{i}\Rightarrow|\vec{PN_1}|=\alpha_1$$

Similarly, $|\vec{PN_2}|=\alpha_2$ and $|\vec{PN_3}|=\alpha_3$. Let PN_4 be the perpendicular distance from the point P to the plane $\vec{r}\cdot(\hat{i}+\hat{j}+\hat{k})=a$. The equation of the line PN_4, which passes through the point $\vec{\alpha}$ and is parallel to $\hat{i}+\hat{j}+\hat{k}$, is $\vec{r}=\vec{\alpha}+t_4(\hat{i}+\hat{j}+\hat{k})$. The point of intersection N_4 of the plane and the straight line is given by

$$\left[\vec{\alpha} + t_4(\hat{\imath}+\hat{\jmath}+\hat{k})\right]\cdot(\hat{\imath}+\hat{\jmath}+\hat{k}) = a$$

$$\Rightarrow \alpha_1+\alpha_2+\alpha_3+t_4\cdot 3 = a \Rightarrow t_4 = \frac{a-(\alpha_1+\alpha_2+\alpha_3)}{3}$$

Thus, the position vector of N_4 is $\vec{r} = \vec{\alpha} + \left[\frac{a-(\alpha_1+\alpha_2+\alpha_3)}{3}\right](\hat{\imath}+\hat{\jmath}+\hat{k})$. Therefore

$$\vec{PN_4} = \vec{\alpha} + \left[\frac{a-(\alpha_1+\alpha_2+\alpha_3)}{3}\right](\hat{\imath}+\hat{\jmath}+\hat{k}) - \vec{\alpha}$$

$$\Rightarrow |\vec{PN_4}| = \left[\frac{a-(\alpha_1+\alpha_2+\alpha_3)}{3}\right]\left|\hat{\imath}+\hat{\jmath}+\hat{k}\right| = \{a-(\alpha_1+\alpha_2+\alpha_3)\}/\sqrt{3}$$

The perpendiculars from the centre to the given planes are equal to the radius of the sphere. Thus

$$\alpha_1 = \alpha_2 = \alpha_3 = \frac{a-(\alpha_1+\alpha_2+\alpha_3)}{\sqrt{3}}$$

$$\Rightarrow \alpha_1 = \alpha_2 = \alpha_3 = \frac{a}{3+\sqrt{3}}, \text{ solving}$$

$$\Rightarrow \vec{\alpha} = \text{Centre of the sphere} = \frac{a}{3+\sqrt{3}}(\hat{\imath}+\hat{\jmath}+\hat{k})$$

The radius of the sphere is $\frac{a}{3+\sqrt{3}}$. Thus, the equation of the sphere is

$$|\vec{r}-\vec{\alpha}| = \frac{a}{3+\sqrt{3}} \Rightarrow |\vec{r}-\vec{\alpha}|^2 = \frac{a^2}{(3+\sqrt{3})^2}$$

$$\Rightarrow (\vec{r}-\vec{\alpha})\cdot(\vec{r}-\vec{\alpha}) = \frac{a^2}{(3+\sqrt{3})^2}$$

$$\Rightarrow \vec{r}\cdot\vec{r} - 2\,\vec{r}\cdot\vec{\alpha} + |\vec{\alpha}|^2 = \frac{a^2}{(3+\sqrt{3})^2}$$

$$\Rightarrow \vec{r}\cdot\left[\vec{r} - a(\hat{\imath}+\hat{\jmath}+\hat{k})\right] = 0$$

3.5.1 Regular Tetrahedron

A tetrahedron whose edges are all equal is called a *regular tetrahedron.* The faces of a regular tetrahedron are equilateral triangles, and hence, the angle between any two of its concurrent edges is 60° (Fig. 3.36). Consider a regular tetrahedron $DABC$ of side length a (say) and take A as origin. With reference to A, suppose the position vectors of B, C and D are $\vec{\beta}, \vec{\gamma}$ and $\vec{\delta}$, respectively. Since the tetrahedron is regular, we have

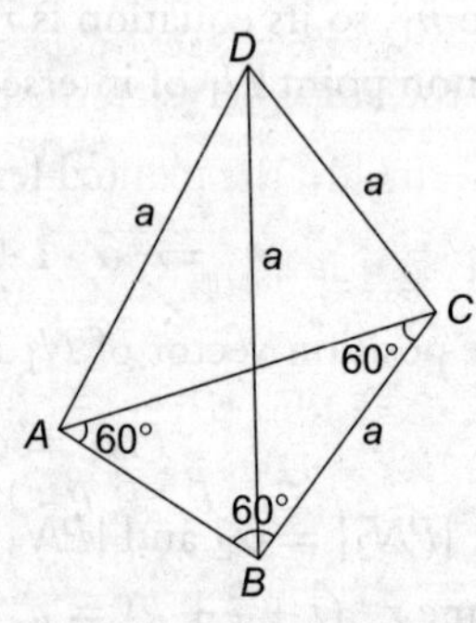

Figure 3.36 Regular tetrahedron

(i) $|\vec{\beta}| = |\vec{\gamma}| = |\vec{\delta}| = a$ (say).

(ii) $|\vec{\beta} \cdot \vec{\gamma}| = |\vec{\beta}|\,|\vec{\gamma}| \cos 60^\circ = a^2 \cdot \frac{1}{2}$. Similarly, $|\vec{\beta} \cdot \vec{\delta}| = |\vec{\delta} \cdot \vec{\gamma}| = \frac{1}{2}a^2$.

(iii) $\vec{\beta} \cdot \vec{\beta} = \vec{\gamma} \cdot \vec{\gamma} = \vec{\delta} \cdot \vec{\delta} = a^2$. (iv) $|\vec{\beta} - \vec{\gamma}| = |\vec{\gamma} - \vec{\delta}| = |\vec{\beta} - \vec{\delta}| = a$.

Property 3.5.1

The angle between any two plane faces of a regular tetrahedron is $\cos^{-1} 1/3$.

Proof: Suppose we are required to find the angle θ between two plane faces ABC and ABD (Fig. 3.37).

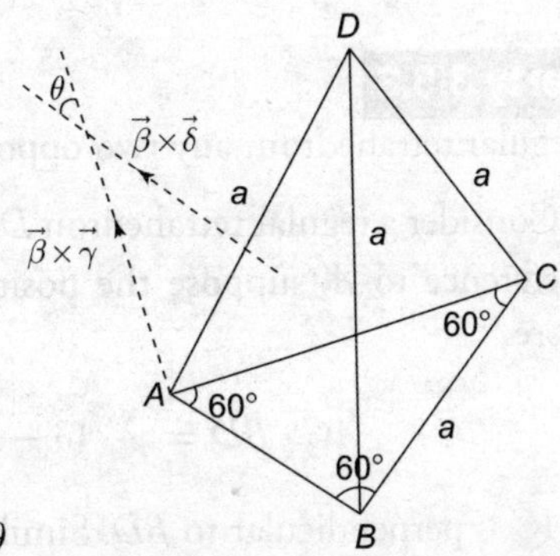

Figure 3.37 Regular tetrahedron

Evidently, $\vec{\beta} \times \vec{\gamma}$ and $\vec{\beta} \times \vec{\delta}$ are the directions of the normals to the two planes. Since the angle between two planes is the angle between their normals, we have

$$(\vec{\beta} \times \vec{\gamma}) \cdot (\vec{\beta} \times \vec{\delta}) = |\vec{\beta} \times \vec{\gamma}|\,|\vec{\beta} \times \vec{\delta}| \cos\theta$$

or, $$\begin{vmatrix} \vec{\beta} \cdot \vec{\beta} & \vec{\beta} \cdot \vec{\delta} \\ \vec{\gamma} \cdot \vec{\beta} & \vec{\gamma} \cdot \vec{\delta} \end{vmatrix} = |\vec{\beta}|\,|\vec{\gamma}| \sin 60^\circ |\vec{\beta}|\,|\vec{\delta}| \sin 60^\circ \cos\theta$$

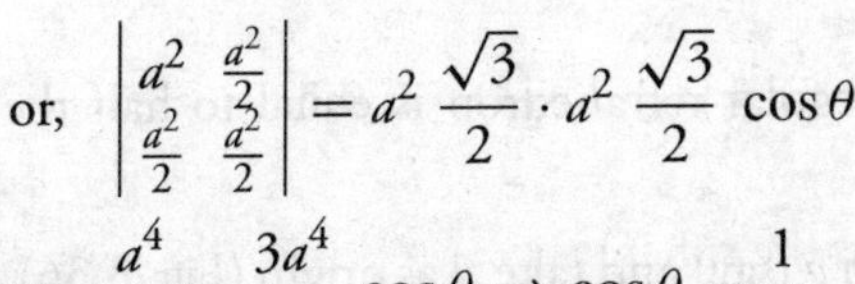

or, $$\begin{vmatrix} a^2 & \frac{a^2}{2} \\ \frac{a^2}{2} & \frac{a^2}{2} \end{vmatrix} = a^2 \frac{\sqrt{3}}{2} \cdot a^2 \frac{\sqrt{3}}{2} \cos\theta$$

or, $$\frac{a^4}{4} = \frac{3a^4}{4} \cos\theta \Rightarrow \cos\theta = \frac{1}{3}$$

Similarly, we can find the angle between other pairs of plane faces.

Property 3.5.2

For a regular tetrahedron, the angle θ between any edge and the face not containing that edge is given by $\cos\theta = 1/\sqrt{3}$.

Proof: Let θ be the angle between any plane face ABC and the edge BD (which is not on the plane ABC). The direction of the normal to the plane ABC is $\vec{\beta} \times \vec{\gamma}$. The equation of the straight line BD, which passes through $\vec{\beta}$ and is parallel to $\vec{\delta} - \vec{\beta}$ is

$$\vec{r} = \vec{\beta} + t(\vec{\delta} - \vec{\beta}), \quad t = \text{ scalar}$$

Since θ is the angle between the plane ABC and the edge BD, $\frac{\pi}{2} - \theta$ is the angle between the line and the normal to the plane. Hence

$$(\vec{\beta} \times \vec{\gamma}) \cdot (\vec{\delta} - \vec{\beta}) = |\vec{\beta} \times \vec{\gamma}|\,|\vec{\delta} - \vec{\beta}| \cos\left(\frac{\pi}{2} - \theta\right)$$

or, $$(\vec{\beta} \times \vec{\gamma}) \cdot \vec{\delta} = |\vec{\beta}|\,|\vec{\gamma}| \sin 60^\circ \; a \; \sin\theta, \quad \text{as } (\vec{\beta} \times \vec{\gamma}) \cdot \vec{\beta} = 0$$

or, $$\begin{vmatrix} \vec{\beta} \cdot \vec{\beta} & \vec{\beta} \cdot \vec{\gamma} & \vec{\beta} \cdot \vec{\delta} \\ \vec{\gamma} \cdot \vec{\beta} & \vec{\gamma} \cdot \vec{\gamma} & \vec{\gamma} \cdot \vec{\delta} \\ \vec{\delta} \cdot \vec{\beta} & \vec{\delta} \cdot \vec{\gamma} & \vec{\delta} \cdot \vec{\delta} \end{vmatrix}^{1/2} = a^2 \cdot \frac{\sqrt{3}}{2} \; a \sin\theta$$

$$\text{or,}\quad \begin{vmatrix} a^2 & \frac{a^2}{2} & \frac{a^2}{2} \\ \frac{a^2}{2} & a^2 & \frac{a^2}{2} \\ \frac{a^2}{2} & \frac{a^2}{2} & a^2 \end{vmatrix}^{1/2} = a^3\,\frac{\sqrt{3}}{2}\,\sin\theta$$

$$\text{or,}\quad \sqrt{\frac{a^6}{2}} = \frac{\sqrt{3}\,a^3}{2}\sin\theta \Rightarrow \sin\theta = \sqrt{\frac{2}{3}}$$

Property 3.5.3

For a regular tetrahedron, any two opposite edges are perpendicular to each other.

Proof: Consider a regular tetrahedron *DABC* of side length *a* (say) and take *A* as origin (Fig. 3.36). With reference to *A*, suppose the position vectors of *B*, *C* and *D* are $\vec{\beta}, \vec{\gamma}$ and $\vec{\delta}$, respectively. Therefore

$$\vec{AC}\cdot\vec{BD} = \vec{\gamma}\cdot(\vec{\delta}-\vec{\beta}) = \vec{\gamma}\cdot\vec{\delta} - \vec{\gamma}\cdot\vec{\beta} = \frac{a^2}{2} - \frac{a^2}{2} = 0$$

Thus, *AC* is perpendicular to *BD*. Similarly, we proceed for the opposite edges.

Property 3.5.4

The shortest distance between two opposite edges of a regular tetrahedron is equal to half the diagonal of the square described on the edge.

Proof: Consider a regular tetrahedron *DABC* of side length *a* (say) and take *A* as origin (Fig. 3.36). With reference to *A*, suppose the position vectors of *B*, *C* and *D* are $\vec{\beta}, \vec{\gamma}$ and $\vec{\delta}$, respectively. The edge *AC* passes through $A(\vec{0})$ and is parallel to $\vec{\gamma}$, so the equation of the straight line *AC* is $\vec{r} = \vec{0} + t\,\vec{\gamma}$. The edge *BD* passes through $\vec{\beta}$ and is parallel to $\vec{\delta}-\vec{\beta}$, so the equation is $\vec{r} = \vec{\beta} + s(\vec{\delta}-\vec{\beta})$. Therefore, the shortest distance between the skew lines is given by

$$p = \vec{\beta}\cdot\frac{1}{|\vec{\gamma}\times(\vec{\delta}-\vec{\beta})|}\left[\vec{\gamma}\times(\vec{\delta}-\vec{\beta})\right]$$

$$= \frac{\vec{\beta}\cdot(\vec{\gamma}\times\vec{\delta}) - \vec{\beta}\cdot(\vec{\gamma}\times\vec{\beta})}{|\vec{\gamma}|\,|\vec{\delta}-\vec{\beta}|\sin 90^\circ} = \frac{1}{a\cdot a\cdot 1}\left[\vec{\beta}\;\vec{\gamma}\;\vec{\delta}\right];\ \text{as } \vec{\beta}\cdot(\vec{\gamma}\times\vec{\beta}) = 0$$

$$= \frac{1}{a^2}\begin{vmatrix} \vec{\beta}\cdot\vec{\beta} & \vec{\beta}\cdot\vec{\gamma} & \vec{\beta}\cdot\vec{\delta} \\ \vec{\gamma}\cdot\vec{\beta} & \vec{\gamma}\cdot\vec{\gamma} & \vec{\gamma}\cdot\vec{\delta} \\ \vec{\delta}\cdot\vec{\beta} & \vec{\delta}\cdot\vec{\gamma} & \vec{\delta}\cdot\vec{\delta} \end{vmatrix}^{1/2} = \frac{1}{a^2}\begin{vmatrix} a^2 & \frac{a^2}{2} & \frac{a^2}{2} \\ \frac{a^2}{2} & a^2 & \frac{a^2}{2} \\ \frac{a^2}{2} & \frac{a^2}{2} & a^2 \end{vmatrix}^{1/2}$$

$$= = \frac{1}{2}\sqrt{2}\,a = \frac{1}{2}\times \text{ Diagonal of the square described on the edge}$$

Property 3.5.5

For a regular tetrahedron, the distance of any vertex from the opposite face is $\sqrt{2/3}\,a$, *a* being the length of any edge.

Proof: Consider a regular tetrahedron *DABC* of side length *a* (say) and take *A* as origin (Fig. 3.38).

With reference to A, suppose the position vectors of B, C and D are $\vec{\beta}, \vec{\gamma}$ and $\vec{\delta}$, respectively. The plane ABC passes through $A(\vec{0})$ and is perpendicular to $\vec{\beta} \times \vec{\gamma}$, so the equation of the plane ABC is $\vec{r} \cdot (\vec{\beta} \times \vec{\gamma}) = 0$. The equation of the plane ABC can be written as $\vec{r} \cdot \vec{n} = p$, where $\vec{n} = \vec{\beta} \times \vec{\gamma}$ and $p = 0$. Now, if DL is the perpendicular from D to the plane ABC, then DL is parallel to $\vec{n}$ and passes through $\vec{\delta}$. The equation of the line DL is $\vec{r} = \vec{\delta} + t\,\vec{n}$. The position vector of the point L, which is the point of intersection of the line DL with the plane, is

Figure 3.38 Regular tetrahedron

$$(\vec{\delta} + t\,\vec{n}) \cdot \vec{n} = 0 \Rightarrow t = -\vec{\delta} \cdot \vec{n}/|\vec{n}|^2$$

Thus, the position vector of L is $\vec{AL} = \vec{\delta} - \{\vec{\delta} \cdot \vec{n}/|\vec{n}|^2\}\,\vec{n}$. Therefore

$$\vec{DL} = \vec{AL} - \vec{AD} = \vec{\delta} - \frac{\vec{\delta} \cdot \vec{n}}{|\vec{n}|^2}\,\vec{n} - \vec{\delta} = -\frac{\vec{\delta} \cdot \vec{n}}{|\vec{n}|^2}\,\vec{n}$$

Thus, the distance of $D(\vec{\delta})$ from the plane ABC is given by

$$|\vec{DL}| = \left| -\frac{\vec{\delta} \cdot \vec{n}}{|\vec{n}|^2}\,\vec{n} \right| = \frac{|\vec{\delta} \cdot \vec{n}|}{|\vec{n}|} = \frac{|\vec{\delta} \cdot (\vec{\beta} \times \vec{\gamma})|}{|\vec{\beta} \times \vec{\gamma}|}$$

$$= \frac{\left[\vec{\beta}\ \vec{\gamma}\ \vec{\delta}\right]}{|\vec{\beta}|\,|\vec{\gamma}| \sin 60^\circ} = \frac{2}{a^2\sqrt{3}}\sqrt{\frac{a^6}{2}} = \sqrt{\frac{2}{3}}\,a$$

3.6 APPLICATIONS IN MECHANICS

We have noted that several entities in mechanics are vector quantities, and thus, the results of vector algebra are readily applied to mechanics.

3.6.1 Concurrent Forces

A force has magnitude and direction as well as a definite line of action. To represent a force, we need a localised vector.

An n number of several forces $\vec{F}_1, \vec{F}_2, \cdots, \vec{F}_n$ acting on a body at a point O is equivalent to a single force $\vec{F}$, called their *resultant*. This is represented by

$$\vec{F} = \vec{F}_1 + \vec{F}_2 + \cdots + \vec{F}_n = \sum_{k=1}^{n} \vec{F}_k \qquad (3.57)$$

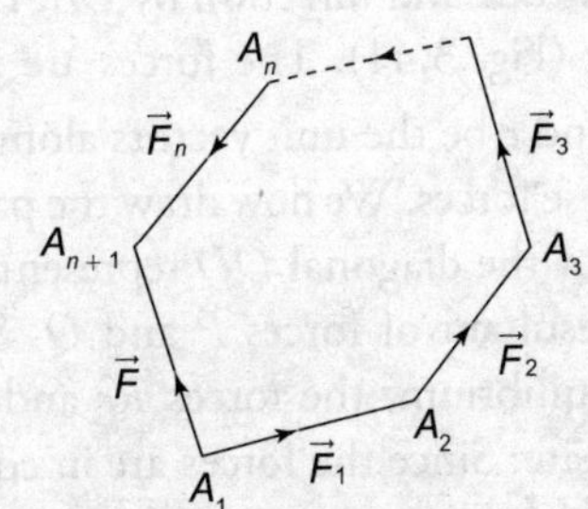

Figure 3.39 Resultant of several concurrent forces

This acts through the point O. In order to obtain the resultant $\vec{F}$, we construct the vector polygon (Fig. 3.39), i.e., a polygon whose lengths and directions of the sides are those of vectors $\vec{F}_1, \vec{F}_2, \cdots, \vec{F}_n$. Thus, suppose $A_1A_2A_3 \ldots A_nA_{n+1}$, where $\vec{A_1A_2} = \vec{F_1}$ $\vec{A_2A_3} = \vec{F_2}, \ldots, \vec{A_nA_{n+1}} = \vec{F_n}$. The closing side A_1A_{n+1} of the polygon taken in the reverse order in which the forces have been drawn represents $\vec{F}$. If the vector polygon is closed, then the forces will be in equilibrium, and the resultant $\vec{F} = \vec{0}$. Further, the vector polygon is not necessarily on the plane unless the forces are coplanar.

Theorem 3.5

Triangle law of forces: If three coplanar concurrent forces acting at a point on a body can be represented in magnitude and direction but not in position by the three sides of a triangle taken in order, then the forces are in equilibrium, and conversely.

Proof: We consider that the forces $\vec{P}, \vec{Q}$ and $\vec{R}$ acting at the point O (Fig. 3.40) are represented in magnitude and direction by the sides AB, BC and CA of a triangle ABC, taken in order.
Then, from Fig. 3.40, it is found that

$$\begin{aligned}\vec{P} + \vec{Q} + \vec{R} &= \vec{AB} + \vec{BC} + \vec{CA} \\ &= \vec{AB} + (\vec{BC} + \vec{CA}) \\ &= \vec{AB} + \vec{BA} = \vec{0}\end{aligned}$$

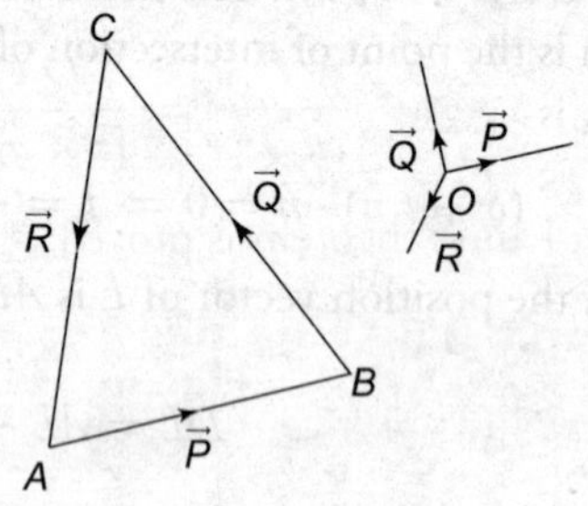

Figure 3.40 Triangle law of forces

Thus, the forces are in equilibrium. Now, we prove the converse part as follows: The resultant of $\vec{AB}$ and $\vec{BC}$ is $\vec{AC}$. Since the forces $\vec{P}, \vec{Q}$ and $\vec{R}$ are in equilibrium, the force $\vec{R}$ is equal and opposite to that of the resultant of the forces $\vec{P}$ and $\vec{Q}$. Hence, $\vec{R}$ must be represented in magnitude and direction by $\vec{CA}$. Therefore, the sides of a triangle considered in order represent the forces $\vec{P}, \vec{Q}$ and $\vec{R}$. Hence, the theorem follows.

Theorem 3.6

Lami's theorem: If three coplanar forces acting at a point are in equilibrium, then the magnitude of each force is proportional to the sine of the angle between the other two.

Proof: Let P, Q and R be the magnitude of three concurrent forces acting at O and be represented in magnitude and direction by OA, OB and OC, respectively (Fig. 3.41). The forces are in equilibrium. Let $\vec{a}, \vec{b}$ and $\vec{c}$ be the unit vectors along the lines of action of these forces. We now draw the parallelogram $OADB$ so that the diagonal OD represents the magnitude of the resultant of forces P and Q. Since the forces are in equilibrium, the forces $R\hat{c}$ and $\vec{OD}$ are equal and opposite. Since the forces are in equilibrium, we have

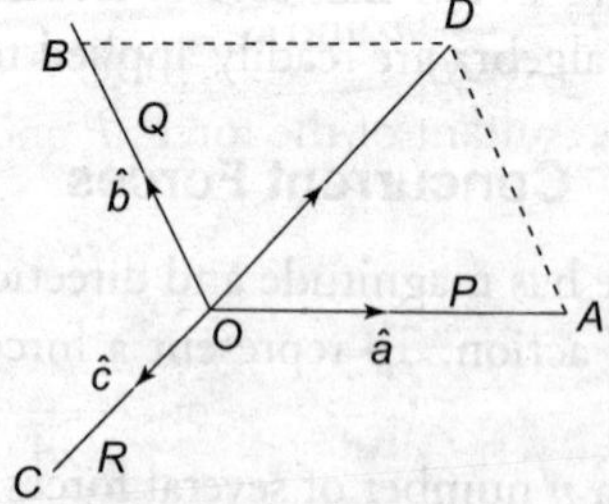

Figure 3.41 Lami's theorem

$$P\,\vec{a} + Q\,\vec{b} + R\,\vec{c} = \vec{0}$$

On taking the cross product of the above relation with $\vec{a}$, we have

$$\begin{aligned}P\,(\vec{a} \times \vec{a}) + Q\,(\vec{b} \times \vec{a}) + R\,(\vec{c} \times \vec{a}) &= \vec{0} \times \vec{a} \\ \Rightarrow Q\,(\vec{b} \times \vec{a}) + R\,(\vec{c} \times \vec{a}) &= \vec{0}; \text{ as } \vec{a} \times \vec{a} = \vec{0}\end{aligned}$$

$$\Rightarrow Q\,(\vec{a}\times\vec{b}) = R\,(\vec{c}\times\vec{a}) \Rightarrow \frac{Q}{|\vec{c}\times\vec{a}|} = \frac{R}{|\vec{a}\times\vec{b}|}$$

Similarly, taking the cross product of the above relation with $\vec{b}$, we have

$$P\,(\vec{a}\times\vec{b}) + Q\,(\vec{b}\times\vec{b}) + R\,(\vec{c}\times\vec{b}) = \vec{0}\times\vec{b}$$

$$\Rightarrow P\,(\vec{a}\times\vec{b}) = R\,(\vec{b}\times\vec{c}) \Rightarrow \frac{P}{|\vec{b}\times\vec{c}|} = \frac{R}{|\vec{a}\times\vec{b}|}$$

Combining the above results, we get

$$\frac{P}{|\vec{b}\times\vec{c}|} = \frac{Q}{|\vec{c}\times\vec{a}|} = \frac{R}{|\vec{a}\times\vec{b}|} \Rightarrow \frac{P}{\sin\widehat{bc}} = \frac{Q}{\sin\widehat{ca}} = \frac{R}{\sin\widehat{ab}}$$

Hence, Lami's theorem is proved.

Example 3.6.1

Forces $\vec{P}$ and $\vec{Q}$ whose resultant is $\vec{R}$ act at a point O. If any transversal cuts the lines of action of the forces $\vec{P}$, $\vec{Q}$ and $\vec{R}$ at the points A, B and C, respectively, show that $\dfrac{P}{OA} + \dfrac{Q}{OB} = \dfrac{R}{OC}$, where P, Q and R are the magnitudes of $\vec{P}$, $\vec{Q}$ and $\vec{R}$, respectively.

Solution

Let the position vectors of the points A, B and C with respect to the point O be $\vec{a}$, $\vec{b}$ and $\vec{c}$, respectively, (Fig. 3.42), i.e., $\vec{OA} = \vec{a}$, $\vec{OB} = \vec{b}$ and $\vec{OC} = \vec{c}$. Therefore, the unit vectors along the directions $\vec{OA}$, $\vec{OB}$ and $\vec{OC}$ are $\dfrac{1}{OA}\vec{a}$, $\dfrac{1}{OB}\vec{b}$ and $\dfrac{1}{OC}\vec{c}$, respectively. Then the vectors representing the given forces are $\vec{P} = \dfrac{P}{OA}\vec{a}$, $\vec{Q} = \dfrac{Q}{OB}\vec{b}$ and $\vec{R} = \dfrac{R}{OC}\vec{c}$, respectively. Since $\vec{R}$ is the resultant of the forces $\vec{P}$ and $\vec{Q}$, we have

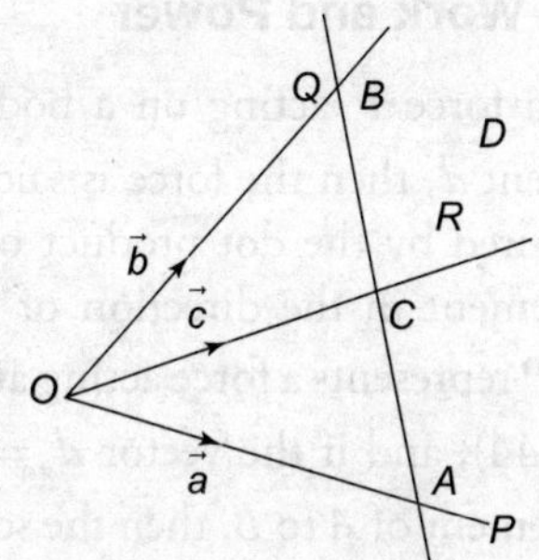

Figure 3.42 Transverse cut of the lines of action

$$\vec{P} + \vec{Q} = \vec{R} \Rightarrow \frac{P}{OA}\vec{a} + \frac{Q}{OB}\vec{b} = \frac{R}{OC}\vec{c}$$

$$\text{or,}\quad \frac{P}{OA}\vec{a} + \frac{Q}{OB}\vec{b} - \frac{R}{OC}\vec{c} = \vec{0}$$

Now, as the points A, B and C are collinear, the sum of the coefficients of $\vec{a}$, $\vec{b}$ and $\vec{c}$ is zero, and hence, we have

$$\frac{P}{OA} + \frac{Q}{OB} - \frac{R}{OC} = 0 \Rightarrow \frac{P}{OA} + \frac{Q}{OB} = \frac{R}{OC}$$

Example 3.6.2

Prove by the vector method that three concurrent forces represented in magnitude and direction by the medians of a triangle are in equilibrium.

Solution

Let the forces $\vec{P}, \vec{Q}$ and $\vec{R}$ be represented by the vectors $\vec{AD}, \vec{BE}$ and $\vec{CF}$, respectively, where AD, BE and CF are three medians of the triangle ABC (Fig. 3.43). Then, $\vec{P} = \vec{AD}$, $\vec{Q} = \vec{BE}$ and $\vec{R} = \vec{CF}$. Let these forces meet at the point G, which is the centroid of the triangle. Now

Figure 3.43 Medians of a triangle are in equilibrium

$$\vec{AD} = \vec{AG} + \vec{GD} = \vec{GD} - \vec{GA},$$

$$\vec{BE} = \vec{GE} - \vec{GB}; \quad \vec{CF} = \vec{GF} - \vec{GC}$$

$$\vec{GD} = \frac{1}{2}(\vec{GB} + \vec{GC}); \quad \vec{GE} = \frac{1}{2}(\vec{GC} + \vec{GA}); \quad \vec{GF} = \frac{1}{2}(\vec{GA} + \vec{GB})$$

Also, $\vec{P} + \vec{Q} + \vec{R} = \vec{AD} + \vec{BE} + \vec{CF}$. So

$$\vec{P} + \vec{Q} + \vec{R} = \frac{1}{2}(\vec{GB} + \vec{GC}) - \vec{GA} + \frac{1}{2}(\vec{GC} + \vec{GA}) - \vec{GB} + \frac{1}{2}(\vec{GA} + \vec{GB}) - \vec{GC} = \vec{0}$$

Hence, the forces $\vec{P}, \vec{Q}$ and $\vec{R}$ are in equilibrium.

3.6.2 Work and Power

When a force $\vec{F}$ acting on a body undergoes a displacement $\vec{d}$, then the force is said to do work which is measured by the dot product of the force and the displacement in the direction of force. If the vector $\vec{F} = \vec{AP}$ represents a force acting at a point A of a body (Fig. 3.44), and if the vector $\vec{d} = \vec{AB}$ represents the displacement of A to B, then the scalar product $\vec{F} \cdot \vec{d}$ is the work done by the force during the displacement.

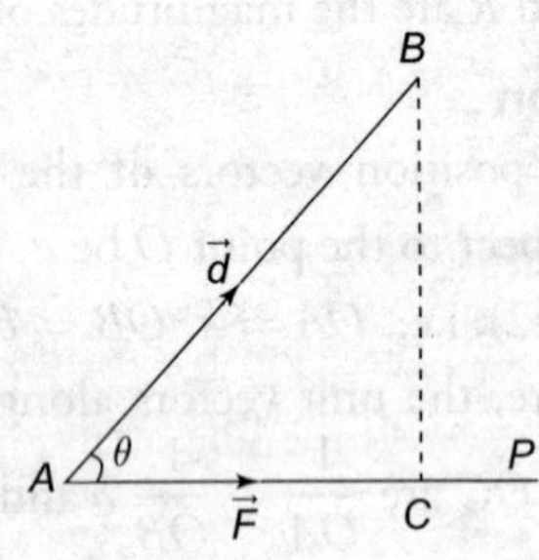

Figure 3.44 Work done by a force

(i) Let $F = |\vec{F}|$, $d = |\vec{d}|$ and θ be the angle between the force vector $\vec{F}$ and displacement vector $\vec{d}$ (Fig. 3.44). Now, we draw the perpendicular BC on AP. So the projection of AB on AP is AC. Thus, if W is the work done by the force during displacement, then by the definition of work, we have

$$W = F \cdot AC = F \cdot AB \cos\theta = Fd \cos\theta = \vec{F} \cdot \vec{d} \tag{3.58}$$

From the above, it follows that if $\vec{d}$ is perpendicular to $\vec{F}$, then $W = 0$. Thus, if the point of application of the force is displaced in a direction perpendicular to the direction of the force, then the work done by the force is zero.

(ii) Now, we consider that a system of n forces $\vec{F}_1, \vec{F}_2, \cdots, \vec{F}_n$ acting on a body at a point A is equivalent to a resultant single force $\vec{F}$, as given in Eq. (3.57). If $\vec{d}$ is the displacement produced after the application of force, then the total work done W is

$$\begin{aligned} W &= \vec{F} \cdot \vec{d} = \left[\vec{F}_1 + \vec{F}_2 + \cdots + \vec{F}_n\right] \vec{d} \\ &= \vec{F}_1 \cdot \vec{d} + \vec{F}_2 \cdot \vec{d} + \cdots + \vec{F}_n \cdot \vec{d}. \end{aligned} \tag{3.59}$$

Power is defined as the rate of doing work. If the vector $\vec{F}$ represents a force acting on a body, and if the vector $\vec{v}$ represents the velocity of the particle, then the scalar product $\vec{F} \cdot \vec{v}$ represents the power.

Example 3.6.3

Show that the work is done by a force $\vec{F} = 5\hat{\imath} + 10\hat{\jmath} + 15\hat{k}$ when it produces a displacement from the point $3\hat{\imath} - \hat{\jmath} - 6\hat{k}$ to $\hat{\imath} + 3\hat{k}$.

Solution

Let us consider a fixed point O, chosen as origin (Fig. 3.45).
Relative to O, let A and B be the points whose position vectors are, respectively, $3\hat{\imath} - \hat{\jmath} - 6\hat{k}$ and $\hat{\imath} + 3\hat{k}$. Then, the displacement vector is

$$\vec{d} = \vec{AB} = \vec{OB} - \vec{OA} = (\hat{\imath} + 3\hat{k}) - (3\hat{\imath} - \hat{\jmath} - 6\hat{k})$$
$$= -2\hat{\imath} + \hat{\jmath} + 9\hat{k}$$

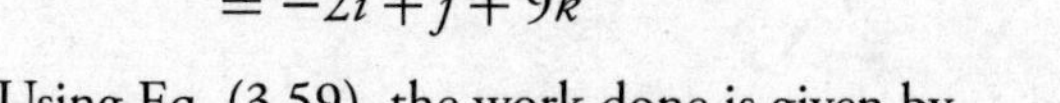

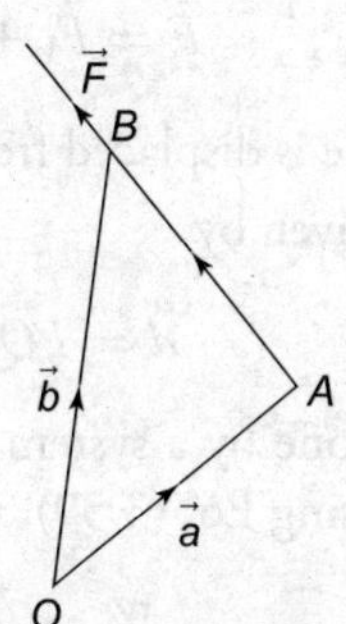

Figure 3.45 Work done by a force

Using Eq. (3.59), the work done is given by

$$W = \vec{F} \cdot \vec{d} = (5\hat{\imath} + 10\hat{\jmath} + 15\hat{k}) \cdot (-2\hat{\imath} + \hat{\jmath} + 9\hat{k})$$
$$= 5 \cdot (-2) + 10 \cdot 1 + 15 \cdot 9 = -10 + 10 + 135 = 135 \text{ units}$$

Example 3.6.4

A particle, acted on by a force of 15 units, is displaced from the point $\hat{\imath} + \hat{\jmath} + \hat{k}$ to the point $2\hat{\imath} + \hat{\jmath} + 3\hat{k}$. If the line of action of the force is the vector $\hat{\imath} + 2\hat{\jmath} + 2k$, find the work done by the force.

Solution

Let us consider a fixed point O, chosen as origin (Fig. 3.45). Relative to O, let A and B be the points whose position vectors are, respectively, $\hat{\imath} + \hat{\jmath} + \hat{k}$ and $2\hat{\imath} + \hat{\jmath} + 3\hat{k}$. Then, the displacement vector is

$$\vec{d} = \vec{AB} = \vec{OB} - \vec{OA} = (2\hat{\imath} + \hat{\jmath} + 3\hat{k}) - (\hat{\imath} + \hat{\jmath} + \hat{k}) = \hat{\imath} + 2\hat{k}$$

The unit vector along the direction of force is

$$= \frac{1}{\sqrt{1^2 + 2^2 + 2^2}}(\hat{\imath} + 2\hat{\jmath} + 2\hat{k})$$
$$= \frac{1}{3}(\hat{\imath} + 2\hat{\jmath} + 2\hat{k})$$

Using Eq. (3.59), the work done is given by

$$W = \vec{F} \cdot \vec{d} = \frac{15}{3}(\hat{\imath} + 2\hat{\jmath} + 2\hat{k}) \cdot (\hat{\imath} + 2\hat{k})$$
$$= \frac{15}{3}\Big[1 \cdot 1 + 2 \cdot 0 + 2 \cdot 2\Big] = 5 \times (1 + 0 + 4) = 25 \text{ units}$$

Example 3.6.5

A particle, acted on by two constant forces $\hat{\imath}+\hat{\jmath}-3\hat{k}$ and $2\hat{\imath}+\hat{\jmath}-\hat{k}$, is displaced from the point $\hat{\imath}+\hat{\jmath}-\hat{k}$ to the point $2\hat{\imath}+4\hat{\jmath}+\hat{k}$. Find the work done by the forces.

Solution

Let $\vec{F}$ be the resultant of the forces $\vec{F}_1=\hat{\imath}+\hat{\jmath}-3\hat{k}$ and $\vec{F}_2=2\hat{\imath}+\hat{\jmath}-\hat{k}$. Then, we get

$$\vec{F}=\vec{F}_1+\vec{F}_2=\hat{\imath}+\hat{\jmath}-3\hat{k}+2\hat{\imath}+\hat{\jmath}-\hat{k}=3\hat{\imath}+2\hat{\jmath}-4\hat{k}$$

If the particle is displaced from the point $P(\hat{\imath}+\hat{\jmath}-\hat{k})$ to $Q(2\hat{\imath}+4\hat{\jmath}+\hat{k})$, then the displacement vector $\vec{d}$ is given by

$$\vec{d}=\vec{PQ}=(2\hat{\imath}+4\hat{\jmath}+\hat{k})-(\hat{\imath}+\hat{\jmath}-\hat{k})=\hat{\imath}+3\hat{\jmath}+2\hat{k}$$

The work done by a system of forces acting on a particle is the same as the work done by their resultant. Using Eq. (3.59), the work done is given by

$$W=\vec{F}\cdot\vec{d}=(3\hat{\imath}+2\hat{\jmath}-4\hat{k})\cdot(\hat{\imath}+3\hat{\jmath}+2\hat{k})=1 \text{ unit}$$

Example 3.6.6

Forces of 6, 7 and 2 pounds of weight act on a particle along the vectors with coordinates $(6,2,3)$, $(3,-2,6)$ and $(2,-3,-6)$, respectively. If the particle is displaced from the point $(2,-1,-3)$ to the point $(5,-1,1)$, then show that the work done is $53\frac{4}{7}$ feet pound weight, the unit of length being one foot.

Solution

The unit vectors in the direction of the vectors $(6,2,3)$, $(3,-2,6)$ and $(2,-3,-6)$ are $\frac{1}{7}(6,2,3)$, $\frac{1}{7}(3,-2,6)$ and $\frac{1}{7}(2,-3,-6)$, respectively. If $\vec{F}$ be the resultant force, then it is given by

$$\vec{F}=\frac{6}{7}(6,2,3)+\frac{7}{7}(3,-2,6)+\frac{2}{7}(2,-3,-6)=\frac{1}{7}(61,-8,48)$$

If the particle is displaced from the point $P(2,-1,-3)$ to $Q(5,-1,1)$, then the displacement vector $\vec{d}$ is given by

$$\vec{d}=\vec{PQ}=(5,-1,1)-(2,-1,-3)=(3,0,4)$$

The work done by a system of forces acting on a particle is the same as the work done by their resultant. Using Eq. (3.59), the work done is given by

$$\begin{aligned}W=\vec{F}\cdot\vec{d}&=\frac{1}{7}(61,-8,48)\cdot(3,0,4)\\&=53\frac{4}{7}\text{ feet pound weight}\end{aligned}$$

Example 3.6.7

A force of magnitude 5 units acts on a particle at the point $A(1,-1,3)$ in the direction $2\hat{\imath}+\hat{\jmath}+3\hat{k}$, causing three successive displacements from A to $B(2,-1,7)$, B to $C(1,3,5)$ and finally C to

$D(2,7,4)$. Find, by the vector method, the work done by the force in bringing the particle from A to D.

Solution

Let $\vec{F}$ be the given force, so $|\vec{F}| = 5$ units (given). The unit vector in the direction of $\vec{F}$ is $(2\hat{i}+\hat{j}+3\hat{k})/\sqrt{14}$. Thus, $\vec{F} = 5(2\hat{i}+\hat{j}+3\hat{k})/\sqrt{14}$. Successive displacements are

$$\vec{d_1} = \vec{AB} = (2,-1,7) - (1,-1,3) = (1,0,4)$$
$$\vec{d_2} = \vec{BC} = (1,3,5) - (2,-1,7) = (-1,4,-2)$$
$$\vec{d_3} = \vec{CD} = (2,7,4) - (1,3,5) = (1,4,-1)$$

The resultant displacement is

$$\vec{d} = \vec{d_1} + \vec{d_2} + \vec{d_3} = \hat{i} + 8\hat{j} + \hat{k}$$

Hence, the required work done by the force $\vec{F}$ is

$$W = \vec{F}\cdot\vec{d} = \frac{5}{\sqrt{14}}(2\hat{i}+\hat{j}+3\hat{k})\cdot(\hat{i}+8\hat{j}+\hat{k})$$
$$= \frac{65}{14}\sqrt{14}\text{ units}$$

Example 3.6.8

The forces acting on a particle have magnitudes of 4, 3 and 2 kg weight and act in the direction of the vectors $3\hat{i}+2\hat{j}+\hat{k}$, $\hat{i}+3\hat{j}+2\hat{k}$ and $2\hat{i}+3\hat{j}-\hat{k}$, respectively. These forces remain constant while the particle is displaced from the point $A(3\hat{i}+\hat{j}-5\hat{k})$ to $B(4\hat{i}-\hat{j}+\hat{k})$. Find the work done by the forces, the unit of length being 1 metre.

Solution

Let $\vec{F_1}, \vec{F_2}$ and $\vec{F_3}$ be the three given forces whose magnitudes are $|\vec{F_1}| = 4$ kg weight, $|\vec{F_2}| = 3$ kg weight and $F_3 = 2$ kg weight. Then

$$\vec{F_1} = \frac{4}{\sqrt{14}}(3\hat{i}+2\hat{j}+\hat{k}), \vec{F_2} = \frac{3}{\sqrt{14}}(\hat{i}+3\hat{j}+2\hat{k}) \text{ and } \vec{F_3} = \frac{2}{\sqrt{14}}(2\hat{i}+3\hat{j}-\hat{k})$$

If $\vec{R}$ be the resultant of the forces $\vec{F_1}, \vec{F_2}$ and $\vec{F_3}$, it is given by

$$\vec{R} = \vec{F_1} + \vec{F_2} + \vec{F_3} = \frac{1}{\sqrt{14}}(19\hat{i}+23\hat{j}+8\hat{k})$$

This acts at the point A. If $\vec{d}$ be the displacement from A to B, then

$$\vec{d} = \vec{AB} = (4\hat{i}-\hat{j}+\hat{k}) - (3\hat{i}+\hat{j}-5\hat{k}) = \hat{i}-2\hat{j}+6\hat{k}$$

Thus, the work done is given by

$$W = \vec{R}\cdot\vec{d} = \frac{1}{\sqrt{14}}(19\hat{i}+23\hat{j}+8\hat{k})\cdot(\hat{i}-2\hat{j}+6\hat{k})$$
$$= \frac{21}{\sqrt{14}}\text{ kg weight metre}$$

3.6.3 Rotation of a Rigid Body

Let us consider the motion of a rigid body about a fixed axis ON (Fig. 3.46). Let ϕ be the angle between two planes passing through ON, one fixed in space and the other fixed in the body. Then, $\omega = \dfrac{d\phi}{dt}$ is the angular velocity of the rigid body about ON.

If $\hat{n}$ be the unit vector along ON, the axis of rotation, then $\vec{\omega} = \omega\,\hat{n}$ is the angular velocity vector of the rigid body about ON.

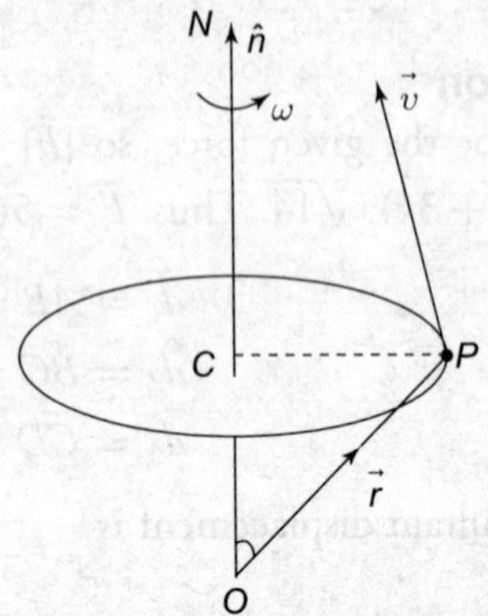

Figure 3.46 Rotation of a rigid body about a fixed axis

Let P be any point on the rigid body and $\vec{OP} = \vec{r}$, where O is any fixed point in space. Let $\vec{v}$ be the velocity of the point P with reference to O as origin. Let PC be drawn perpendicular to the axis ON. Then, P describes a circle of radius PC with the centre at C. The plane of rotation of P is perpendicular to ON. Let $\angle PON = \theta$.

Then, $PC = OP\sin\theta = |\vec{r}|\sin\theta$. $\vec{v}$ is along the tangent to the circular path at P, and hence, $\vec{v}$ is perpendicular to the plane OPC and its magnitude is $\omega(PC) = \omega|\vec{r}|\sin\theta$. The sense of $\vec{v}$ is such that $\vec{\omega}, \vec{r}$ and $\vec{v}$ form a right-handed system. Hence, $\vec{v} = \vec{\omega} \times \vec{r}$.

Example 3.6.9

A rigid body spins with an angular velocity of 4 radians per second about an axis through the point $\hat{i}+\hat{j}-\hat{k}$ and parallel to $3\hat{i}+2\hat{j}+\hat{k}$. Find the velocity of the particle at the point $4\hat{i}+\hat{j}-\hat{k}$ of the body and also find the speed of the particle at this point.

Solution

Let the axis of the spinning rigid body be along ON and $\hat{n}$ be the unit vector along $\vec{ON}$ (Fig. 3.46). Then, $\hat{n} = \frac{1}{\sqrt{14}}(3\hat{i}+2\hat{j}+\hat{k})$. The angular velocity of the rigid body is $\omega = 4$ radian/sec. Therefore, the angular velocity vector of the rigid body is

$$\vec{\omega} = \omega\,\hat{n} = \frac{4}{\sqrt{14}}(3\hat{i}+2\hat{j}+\hat{k})$$

The velocity of the particle at the point $P(4\hat{i}+\hat{j}-\hat{k})$ is

$$\vec{v} = \vec{\omega}\times\vec{r} = \vec{\omega}\times\vec{OP} = \frac{4}{\sqrt{14}}(3\hat{i}+2\hat{j}+\hat{k}) \times \left\{(4\hat{i}+\hat{j}-\hat{k}) - (\hat{i}+\hat{j}-\hat{k})\right\}$$

$$= \frac{4}{\sqrt{14}}(3\hat{i}+2\hat{j}+\hat{k}) \times (3\hat{i}) = \frac{12}{\sqrt{14}}(\hat{j}-2\hat{k})$$

The speed of the particle at P is $v = |\vec{v}| = \frac{6}{7}\sqrt{70}$ metres/sec.

Example 3.6.10

A rigid body rotates with a uniform angular velocity of 2 radians per second about an axis passing through the points $(0,1,2)$ and $(1,3,-2)$. Find by the vector method, the velocity of the point $(3,6,4)$ of the rigid body.

Solution

Let $A(0,1,2)$ and $B(1,3,-2)$ be two given points on the axis. The unit vector along the axis of rotation of the rigid body is

$$\hat{n} = \frac{1}{|\vec{AB}|}\vec{AB} = \frac{1}{\sqrt{21}}(1,2,-4)$$

$P(3,6,4)$ is a point on the body. The angular velocity vector of the rigid body is

$$\vec{\omega} = \omega\,\hat{n} = 2\hat{n} = \frac{2}{\sqrt{21}}(\hat{\imath} + 2\hat{\jmath} - 4\hat{k})$$

The linear velocity of the particle at the point $P(3\hat{\imath} + 6\hat{\jmath} + 4\hat{k})$ is

$$\vec{v} = \vec{\omega} \times \vec{r} = \vec{\omega} \times \vec{AP} = \frac{2}{\sqrt{21}}(\hat{\imath} + 2\hat{\jmath} - 4\hat{k}) \times \left\{(3\hat{\imath} + 6\hat{\jmath} + 4\hat{k}) - (\hat{\jmath} + 2\hat{k})\right\}$$

$$= \frac{2}{\sqrt{21}}(\hat{\imath} + 2\hat{\jmath} - 4\hat{k}) \times (3\hat{\imath} + 5\hat{\jmath} + 2\hat{k}) = \frac{2}{\sqrt{21}}(24\hat{\imath} - 14\hat{\jmath} - \hat{k})$$

Hence, the magnitude of the linear velocity is $|\frac{2}{\sqrt{21}}(24\hat{\imath} - 14\hat{\jmath} - \hat{k})| = \frac{2\sqrt{773}}{\sqrt{21}}$

3.6.4 Moment of a Force

Moment of a force and torque are two important quantities in mechanics. Let $\vec{F}$ be a force acting at a point P whose position vector is $\vec{r}$ with reference to a fixed point O, chosen as base. Consider a particle P of mass m, whose position vector at any any time t is $\vec{r}$ (Fig. 3.47), with respect to the base point O.

The moment of force $\vec{F}$ or torque $\vec{\tau}$ about O is defined as

$$\vec{\tau} = \vec{r} \times \vec{F} \qquad (3.60)$$

Figure 3.47 Moment of a force or torque

about an axis through O perpendicular to the plane of $\vec{r}$ and $\vec{F}$. A force causes linear acceleration, whereas a torque causes angular acceleration.

Example 3.6.11

A force of 15 units acts in the direction of the vector $\hat{\imath} - 2\hat{\jmath} + 2\hat{k}$ and passes through a point $2\hat{\imath} - 2\hat{\jmath} + 2\hat{k}$. Find the moment of the force $\hat{\imath} + 2\hat{k}$ about the point $\hat{\imath} + \hat{\jmath} + \hat{k}$.

Solution

Let us consider a fixed point O, chosen as origin. Relative to O, let A and B be the points whose position vectors are, respectively, $\hat{\imath} + \hat{\jmath} + \hat{k}$ and $2\hat{\imath} - 2\hat{\jmath} + 2\hat{k}$. Then

$$\vec{r} = \vec{AB} = \vec{OB} - \vec{OA}$$

$$= (2\hat{\imath} - 2\hat{\jmath} + 2\hat{k}) - (\hat{\imath} + \hat{\jmath} + \hat{k}) = \hat{\imath} - 3\hat{\jmath} + \hat{k}$$

The unit vector along the direction of the vector $\hat{\imath} - 2\hat{\jmath} + 2\hat{k}$ is obtained as

$$= \frac{1}{\sqrt{1^2 + (-2)^2 + 2^2}}(\hat{\imath} - 2\hat{\jmath} + 2\hat{k}) = \frac{1}{3}(\hat{\imath} - 2\hat{\jmath} + 2\hat{k})$$

Let $\vec{F}$ be the given force whose magnitude is of 15 units. Then

$$\vec{F} = \frac{15}{3}(\hat{\imath} - 2\hat{\jmath} + 2\hat{k}) = 5(\hat{\imath} - 2\hat{\jmath} + 2\hat{k})$$

If $\vec{M}$ be the moment of the force about the point P, we obtain

$$\vec{M} = \vec{r} \times \vec{F} = (\hat{\imath} - 3\hat{\jmath} + \hat{k}) \times 5(\hat{\imath} - 2\hat{\jmath} + 2\hat{k}) = 5(-4\hat{\imath} - \hat{\jmath} + \hat{k}) \text{ units}$$

Therefore, the magnitude of the torque is

$$|\vec{M}| = 5|-4\hat{\imath} - \hat{\jmath} + \hat{k}| = 5\sqrt{(-4)^2 + (-1)^2 + 1^2} = 15\sqrt{2} \text{ units}$$

Theorem 3.7

Varignon's theorem: The algebraic sum of the vector moments of any number of coplanar concurrent forces about a point O on their plane is equal to the vector moment of their resultant about the same point O.

Proof: Let $\vec{F}_1, \vec{F}_2, \cdots, \vec{F}_n$ be coplanar concurrent forces acting at a point O_1 (say) and $\vec{R}$ be their resultant (Fig. 3.48). Then, $\vec{R} = \vec{F}_1 + \vec{F}_2 + \cdots + \vec{F}_n$. Let O be any given point on the plane of the forces $\vec{F}_1, \vec{F}_2, \cdots, \vec{F}_n$, where $\vec{OO_1} = \vec{r}$. Then, the vector moment of the force $\vec{F}_k$ about O is

$$\vec{M}_k = \vec{r} \times \vec{F}_k; \quad k = 1, 2, \cdots, n$$

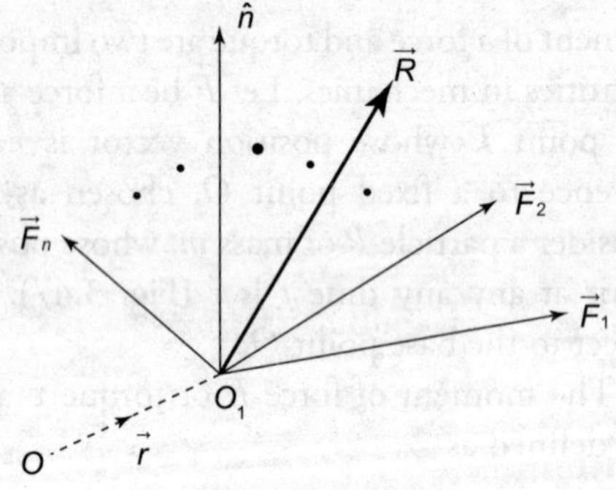

Figure 3.48 Varignon's theorem

If $\hat{n}$ be the unit vector perpendicular to the plane of $\vec{OO_1}$, which is also the plane of the forces $\vec{F}_1, \vec{F}_2, \cdots, \vec{F}_n$, then $\hat{n}$ and $\vec{M}_k$, for each $k = 1, 2, \cdots, n$, are collinear vectors. Therefore, the algebraic sum of the vector moments of the forces $\vec{F}_1, \vec{F}_2, \cdots, \vec{F}_n$ about O is

$$\begin{aligned}\vec{M} &= \vec{M}_1 + \vec{M}_2 + \cdots + \vec{M}_n = \vec{r} \times \vec{F}_1 + \vec{r} \times \vec{F}_2 + \cdots + \vec{r} \times \vec{F}_n \\ &= \vec{r} \times \left[\vec{F}_1 + \vec{F}_2 + \cdots + \vec{F}_n\right]; \text{ distributive property} \\ &= \vec{r} \times \vec{R} = \text{ Vector moment of their resultant } \vec{R} \text{ about } O\end{aligned}$$

Example 3.6.12

Forces of magnitude 5, 3 and 1 units act at the point $B(3\hat{\imath} + 4\hat{\jmath} + 2\hat{k})$ along the directions of the vectors $2\hat{\imath} + \hat{\jmath} - 2\hat{k}$, $\hat{\imath} + 2\hat{\jmath} - \hat{k}$ and $2\hat{\imath} + 2\hat{\jmath} - \hat{k}$, respectively. Find the torque of each force about the point $A(\hat{\imath} - \hat{\jmath} - \hat{k})$. Also verify Varignon's theorem.

Solution

The given forces acting at the point $B(3\hat{\imath}+4\hat{\jmath}+2\hat{k})$ are $\vec{F}_1=\frac{5}{3}(2\hat{\imath}+\hat{\jmath}-2\hat{k})$, $\vec{F}_2=\hat{\imath}+2\hat{\jmath}-\hat{k}$ and $\vec{F}_3=\frac{1}{3}(2\hat{\imath}+2\hat{\jmath}-\hat{k})$. The torque of the forces $\vec{F}_1, \vec{F}_2$ and $\vec{F}_3$ about the point $A(\hat{\imath}-\hat{\jmath}-\hat{k})$ are

$$\vec{M}_1=\vec{AB}\times\vec{F}_1=(2\hat{\imath}+5\hat{\jmath}+3\hat{k})\times\frac{5}{3}(2\hat{\imath}+\hat{\jmath}-2\hat{k})=\frac{5}{3}(-13\hat{\imath}+10\hat{\jmath}-8\hat{k})$$

$$\vec{M}_2=\vec{AB}\times\vec{F}_2=(2\hat{\imath}+5\hat{\jmath}+3\hat{k})\times(\hat{\imath}+2\hat{\jmath}-2\hat{k})=-16\hat{\imath}+7\hat{\jmath}-\hat{k}$$

$$\vec{M}_3=\vec{AB}\times\vec{F}_3=(2\hat{\imath}+5\hat{\jmath}+3\hat{k})\times\frac{1}{3}(2\hat{\imath}+2\hat{\jmath}-\hat{k})=\frac{1}{3}(-11\hat{\imath}+8\hat{\jmath}-6\hat{k})$$

To verify Varignon's theorem, we have the algebraic sum of the moments of the forces $\vec{F}_1, \vec{F}_2$ and $\vec{F}_3$ about the point A as

$$\vec{M}=\vec{M}_1+\vec{M}_2+\vec{M}_3=\frac{1}{3}(-124\hat{\imath}+79\hat{\jmath}-49\hat{k})$$

The resultant of the forces $\vec{F}_1, \vec{F}_2$ and $\vec{F}_3$ is given by

$$\vec{R}=\vec{F}_1+\vec{F}_2+\vec{F}_3=\frac{1}{3}(15\hat{\imath}+13\hat{\jmath}-17\hat{k})$$

Hence, the moment of $\vec{r}$ about the point A is

$$\vec{AB}\times\vec{R}=(2\hat{\imath}+5\hat{\jmath}+3\hat{k})\times\frac{1}{3}(15\hat{\imath}+13\hat{\jmath}-17\hat{k})=\frac{1}{3}(-124\hat{\imath}+79\hat{\jmath}-49\hat{k})$$

Thus, we have $\vec{M}=\vec{AB}\times\vec{R}$, and hence, Varignon's theorem is verified.

Example 3.6.13

A force $\vec{F}$ of magnitude 10 units acts along the line $\frac{x-2}{5}=\frac{y-1}{4}=\frac{z-3}{3}$. Find the moment of the force $\vec{F}$ about the z axis.

Solution

Given that $|\vec{F}|=10$ units. The line of action of the force $\vec{F}$ is $\frac{x-2}{5}=\frac{y-1}{4}=\frac{z-3}{3}$ and $P(2,1,3)$ is a point on this line. An unit vector along the line of action of force $\vec{F}$ is $\frac{1}{5\sqrt{2}}(5\hat{\imath}+4\hat{\jmath}+3\hat{k})$. Therefore

$$\vec{F}=10\times\frac{1}{5\sqrt{2}}(5\hat{\imath}+4\hat{\jmath}+3\hat{k})=\sqrt{2}(5\hat{\imath}+4\hat{\jmath}+3\hat{k})$$

Origin O is a point on the z axis and $\hat{n}=(0,0,1)$ is a unit vector along the z axis. Therefore, the moment of the force $\vec{F}$ about the z axis is

$$\left[\vec{OP}\ \vec{F}\ \hat{n}\right]=\sqrt{2}\begin{vmatrix}2 & 1 & 3\\ 5 & 4 & 3\\ 0 & 0 & 1\end{vmatrix}=3\sqrt{2}\text{ units}$$

Example 3.6.14

A force $\vec{F} = F_1\hat{i} + F_2\hat{j} + F_3\hat{j}$ acts at the point $\vec{a} = a_1\hat{i} + a_2\hat{j} + a_3\hat{k}$. Find the moment of this force about the line $\vec{r} = \vec{b} + t\hat{c}$, where t is a scalar, and $\vec{b} = b_1\hat{i} + b_2\hat{j} + b_3\hat{k}$ and $\hat{c} = l\hat{i} + m\hat{j} + n\hat{k}$ are unit vector.

Solution

The given line $\vec{r} = \vec{b} + t\hat{c}$ passes through the point $B(\vec{b})$ and is parallel to the vector $\hat{c}$ (Fig. 3.49). $A(\vec{A})$ is a point on the line of action of the force $\vec{F}$. Now

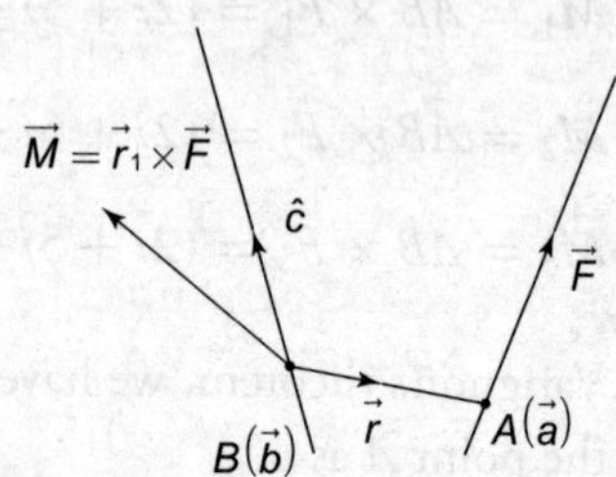

Figure 3.49 Moment of a force

$$\vec{r}_1 = \vec{BA} = \vec{a} - \vec{b}$$
$$= (a_1 - b_1)\hat{i} + (a_2 - b_2)\hat{j} + (a_3 - b_3)\hat{k}$$

The vector moment of force $\vec{F}$ about the point B is

$$\vec{M} = \vec{r}_1 \times \vec{F} = \begin{vmatrix} \hat{i} & \hat{j} & \hat{k} \\ a_1 - b_1 & a_2 - b_2 & a_3 - b_3 \\ F_1 & F_2 & F_3 \end{vmatrix}$$

Hence, the moment of $\vec{F}$ about the line $\vec{r} = \vec{b} + t\hat{c}$ is

$$\vec{M} \cdot \hat{c} = \left[\vec{r}_1\, \vec{F}\, \hat{c}\right] = \begin{vmatrix} l & m & n \\ a_1 - b_1 & a_2 - b_2 & a_3 - b_3 \\ F_1 & F_2 & F_3 \end{vmatrix}$$

$$= l\left\{(a_2 - b_2)F_3 - (a_3 - b_3)F_2\right\} + m\left\{(a_3 - b_3)F_1 - (a_1 - b_1)F_3\right\} + n\left\{(a_1 - b_1)F_2 - (a_2 - b_2)F_1\right\}$$

EXERCISES

Multiple Choice Questions

1. If the lines $\vec{r} = \vec{a} + t(\vec{b} \times \vec{c})$ and $\vec{r} = \vec{b} + s(\vec{c} \times \vec{a})$ intersect, then
   (a) $\vec{a} \cdot \vec{b} = \vec{b} \cdot \vec{c}$ (b) $\vec{a} \cdot \vec{c} = \vec{b} \cdot \vec{c}$ (c) $\vec{a} \cdot \vec{b} = \vec{a} \cdot \vec{c}$ (d) $\vec{b} \cdot \vec{a} = \vec{c} \cdot \vec{a}$
2. The vector equation of the plane which contains the line $\vec{r} = \vec{a} + t\,\vec{b}$ and is perpendicular to the plane $\vec{r} \cdot \vec{d} = q$ is
   (a) $[\vec{r}\,\vec{a}\,\vec{d}] = [\vec{a}\,\vec{b}\,\vec{d}]$ (b) $[\vec{r}\,\vec{a}\,\vec{b}] = [\vec{a}\,\vec{b}\,\vec{d}]$
   (c) $[\vec{r}\,\vec{b}\,\vec{d}] = [\vec{r}\,\vec{a}\,\vec{d}]$ (d) $[\vec{r}\,\vec{b}\,\vec{d}] = [\vec{a}\,\vec{b}\,\vec{d}]$
3. The vector lying on a plane of $\vec{a}$ and $\vec{b}$ and perpendicular to $\vec{a}$ is given by
   (a) $\vec{a} \times (\vec{a} \times \vec{b})$ (b) $\vec{b} \times (\vec{a} \times \vec{b})$ (c) $\vec{a} \times \vec{b}$ (d) $\vec{b} \times (\vec{b} \times \vec{a})$

4. The vector $\vec{a} \times (\vec{b} \times \vec{c})$ lies on a plane of
(a) $\vec{a}$ and $\vec{c}$ (b) $\vec{a}$ and $\vec{b}$ (c) $\vec{b}$ and $\vec{c}$ (d) $\vec{a} \times \vec{b}$

5. The vector equation of the line of intersection of the planes $\vec{r} = \vec{b} + t_1(\vec{b} - \vec{a}) + s_1(\vec{a} + \vec{c})$ and $\vec{r} = \vec{c} + t_2(\vec{b} - \vec{c}) + s_2(\vec{a} + \vec{b})$, $\vec{a}$, $\vec{b}$ and $\vec{c}$ being coplanar, is
(a) $\vec{r} = \vec{a} + t_1(\vec{b} + \vec{a})$ (b) $\vec{r} = \vec{b} + t_1(\vec{a} + \vec{c})$
(c) $\vec{r} = \vec{c} + t_1(\vec{a} + \vec{b})$ (d) $\vec{r} = \vec{c} + t_1(\vec{a} \times \vec{b})$

6. The vector equation of the straight line passing through the point $\vec{c}$ and parallel to the plane $\vec{r} \cdot \vec{n} = p$ and perpendicular to the line $\vec{r} = \vec{a} + t\,\vec{b}$ is
(a) $\vec{r} = \vec{c} + s(\vec{b} \times \vec{n})$ (b) $\vec{r} = \vec{c} + s(\vec{a} \times \vec{n})$
(c) $\vec{r} = \vec{a} + s(\vec{b} \times \vec{n})$ (d) $\vec{r} = \vec{a} + s(\vec{c} \times \vec{n})$

7. The vector equation of the straight line passing through the point $\vec{a}$ and parallel to the line of intersection of the planes $\vec{r} \cdot \vec{n}_1 = p_1$ and $\vec{r} \cdot \vec{n}_2 = p_2$ is
(a) $\vec{r} = \vec{a} + s(\vec{n}_1 \times \vec{n}_2)$ (b) $\vec{r} = \vec{a} + s(\vec{a} \times \vec{n}_1)$
(c) $\vec{r} = \vec{a} + s(\vec{a} \times \vec{n}_2)$ (d) $\vec{r} = \vec{a} + s\{\vec{a} \times (\vec{n}_1 \times \vec{n}_2)\}$

8. The point of intersection where the line $\vec{r} = \vec{a} + t\,\vec{b}$ meets the plane $\vec{r} \cdot \vec{n} = p$ is
(a) $\vec{0}$ (b) $\vec{a} + \dfrac{p - \vec{a} \cdot \vec{n}}{\vec{b} \cdot \vec{n}}$ (c) $\vec{a} + \dfrac{p - \vec{b} \cdot \vec{n}}{\vec{a} \cdot \vec{n}}$ (d) does not intersect

9. The shortest distance between two skew lines $\vec{r} = \vec{a} + t\,\vec{b}$ and $\vec{r} = \vec{c} + s\,\vec{d}$ is
(a) $\dfrac{1}{|\vec{a} \times \vec{c}|}(\vec{b} - \vec{d}) \cdot (\vec{a} \times \vec{c})$ (b) $\dfrac{1}{|\vec{b} \times \vec{d}|}(\vec{a} - \vec{c}) \cdot (\vec{b} \times \vec{d})$
(c) $\dfrac{1}{|\vec{a} \times \vec{d}|}(\vec{b} - \vec{c}) \cdot (\vec{a} \times \vec{d})$ (d) $\dfrac{1}{|\vec{a} \times \vec{b}|}(\vec{b} - \vec{d}) \cdot (\vec{a} \times \vec{b})$

10. The vector equation of the plane which contains the line $\vec{r} = t\vec{a}$ and is perpendicular to the plane containing the lines $\vec{r} = u\vec{b}$ and $\vec{r} = v\vec{c}$ is
(a) $\vec{r} \cdot \vec{a} \times (\vec{b} \times \vec{c}) = 0$ (b) $[\vec{r} - \vec{a}\;\vec{b}\;\vec{c}] = 0$
(c) $[\vec{r} - \vec{b}\;\vec{a}\;\vec{c}] = 0$ (d) $[\vec{r} - \vec{c}\;\vec{a}\;\vec{b}] = 0$

11. The volume of the tetrahedron whose vertices are $\hat{j} + 2\hat{k}$, $3\hat{i} + \hat{k}$, $4\hat{i} + 3\hat{j} + 6\hat{k}$ and $2\hat{i} + 3\hat{j} + 2\hat{k}$ is
(a) 2 units (b) 3 units (c) 6 units (d) 9 units

12. The equation of the sphere whose centre is the origin and radius $\vec{a}$ is
(a) $r^2 - 2\vec{r} \cdot \vec{a} = 0$ (b) $\vec{r} = \vec{a}$ (c) $r^2 = a^2$ (d) $\vec{r} \cdot \vec{a} = k$

Answers

1. b 2. d 3. a 4. c 5. d 6. a 7. a 8. b 9. b 10. a 11. c 12. c

Short Answer Questions

1. Solve the following vector equations:

(a) $\vec{r} = \vec{a} \times \vec{b}$, where $\vec{a} = \hat{i} + \hat{j} + \hat{k}$ and $\vec{b} = 3\hat{i} + 4\hat{j} + 5\hat{k}$. *Ans:* $\vec{r} = \hat{i} - 2\hat{j} + \hat{k}$
(b) $\vec{a} \times \vec{r} = \vec{a} \times \vec{b}$ and $\vec{a} \cdot (\vec{r} - \vec{b}) = 2|\vec{a}|^2$. *Ans:* $2\vec{a} + \vec{b}$

(c) $\vec{r} = \vec{a} \times (\vec{b} \times \vec{c})$, where $\vec{a} = \hat{\imath} + \hat{\jmath} + \hat{k}$, $\vec{b} = \hat{\imath} - \hat{\jmath} - \hat{k}$ and $\vec{c} = 3\hat{\imath} + 4\hat{\jmath} + 5\hat{k}$.

Ans: $\vec{r} = 15\hat{\imath} - 8\hat{\jmath} - 7\hat{k}$

(d) $\vec{r} \times \vec{b} = \vec{a} \times \vec{b}$ and $\vec{r} \cdot \vec{c} = 0$, where $\vec{a} = \hat{\imath} - \hat{\jmath} - \hat{k}$, $\vec{b} = 2\hat{\imath} + 3\hat{\jmath} + 4\hat{k}$ and $\vec{c} = \frac{1}{2}\hat{\imath} + \hat{\jmath} + \hat{k}$.

Ans: $\vec{r} = \frac{1}{16}(22\hat{\imath} - 7\hat{\jmath} - 4\hat{k})$

(e) $\vec{r} \times \vec{b} = \vec{c} \times \vec{b}$ and $\vec{r} \cdot \vec{a} = 0$, where $\vec{a} = 2\hat{\imath} + 3\hat{\jmath} - \hat{k}$, $\vec{b} = 3\hat{\imath} - \hat{\jmath} + \hat{k}$ and $\vec{c} = \hat{\imath} + \hat{\jmath} + 3\hat{k}$.

Ans: $\vec{r} = 2(-\hat{\imath} + \hat{\jmath} + \hat{k})$

(f) $\vec{r} + (\vec{r} \cdot \vec{b})\vec{a} = \vec{c}$, where $\vec{a} = \hat{\imath} + 2\hat{\jmath} + \hat{k}$, $\vec{b} = 2\hat{\imath} + \hat{\jmath} + \hat{k}$ and $\vec{c} = \hat{\imath} + \hat{\jmath} + 2\hat{k}$.

Ans: $\vec{r} = \frac{1}{6}(\hat{\imath} - 4\hat{\jmath} + 7\hat{k})$

(g) $\vec{r} + 2\,\vec{r} \times \vec{a} = \vec{b}$, where $\vec{a} = \hat{\imath} + \hat{\jmath}$ and $\vec{b} = \hat{\imath} + \hat{\jmath} - \hat{k}$. *Ans:* $\vec{r} = \frac{1}{9}(-\hat{\imath} + 5\hat{\jmath} + \hat{k})$

(h) $\vec{a} \times \vec{r} = \vec{b} + \vec{a}$, and $\vec{r} \cdot \vec{a} = 3$ where $\vec{a} = 2\hat{\imath} + 3\hat{\jmath} - \hat{k}$ and $\vec{b} = -\hat{\imath} - 2\hat{\jmath} + 2\hat{k}$.

Ans: $\vec{r} = \hat{\imath} + \hat{\jmath}$

2. If $\vec{a}$ be the position vector of a given point in space and $\vec{r}$ be the position vector of any arbitrary point, find the locus of $\vec{r}$, if

   (a) $|\vec{r} - \vec{a}| = 5$ (b) $(\vec{r} - \vec{a}) \cdot \vec{a} = 0$

3. Prove that the straight lines $\vec{r} = \vec{a} + t(\vec{b} + \vec{c})$ and $\vec{r} = \vec{b} + t(\vec{a} + \vec{c})$ intersect. Find the point of intersection.

   Ans: $\vec{a} + \vec{b} + \vec{c}$

4. Prove that the straight lines $\vec{r} = -3\vec{a} + 6\vec{b} + t(-4\vec{a} + 3\vec{b} + \vec{c})$ and $\vec{r} = -2\vec{a} + 7\vec{c} + s(-4\vec{a} + \vec{b} + \vec{c})$ intersect. Find the point of intersection.

5. Prove that

   (a) the external bisector of the angle $\angle A$ divides the side BC externally in the ratio $AB : AC$.
   (b) the centroid trisects the medians of a triangle.

6. Find the vector equation of a line which passes through the

   (a) origin and parallel to the vector $3\hat{\imath} - 2\hat{\jmath} + 4\hat{k}$.
   (b) point $3\hat{\imath} - \hat{\jmath} + \hat{k}$ and parallel to the vector $\hat{\imath} + \hat{\jmath} + \hat{k}$.
   (c) point $2\hat{\imath} + \hat{\jmath} - 3\hat{k}$ and parallel to the vector $\hat{\imath} + 2\hat{\jmath} + \hat{k}$.
   (d) points whose position vectors are $3\hat{\imath} + \hat{\jmath} + \hat{k}$ and $2\hat{\imath} + 2\hat{\jmath} - \hat{k}$.

7. Find the equation of the plane passing through the points whose position vectors are given by

   (a) $\vec{a} = 2\hat{\imath} + 3\hat{\jmath} - \hat{k}$, $\vec{b} = 3\hat{\imath} + 4\hat{\jmath} - 2\hat{k}$ and $\vec{c} = \hat{\imath} - 3\hat{\jmath} + 2\hat{k}$.
   (b) $\vec{a} = 2\hat{\imath} + \hat{\jmath} + 3\hat{k}$, $\vec{b} = -3\hat{\imath} - \hat{\jmath} + 2\hat{k}$ and $\vec{c} = 2\hat{\jmath} - \hat{k}$.

8. Find the equation of the plane containing two parallel lines $\vec{r} = \vec{a} + t\,\vec{b}$ and $\vec{r} = \vec{c} + s\,\vec{b}$.

   Ans: $\vec{r} \cdot (\vec{a} - \vec{c}) \times \vec{b} = [\vec{a}\,\vec{c}\,\vec{b}]$

9. Find the equation of the plane

   (a) passing through the origin and parallel to the vectors $\vec{a} = \hat{\imath} + 2\hat{\jmath} + 3\hat{k}$ and $\vec{b} = 2\hat{\imath} - \hat{\jmath} - 4\hat{k}$.

(b) passing through the points $\vec{a} = \hat{i} + \hat{j} + \hat{k}$, $\vec{b} = 2\hat{i} - \hat{j} + 3\hat{k}$ and parallel to the vector $3\hat{i} - 2\hat{j} + 4\hat{k}$.

10. Find the reflection of the

(a) point $\vec{a}$, in the plane $\vec{r} \cdot \vec{n} = q$. *Ans*: $\vec{a} + \dfrac{2}{|\vec{n}|^2}(\vec{q} - \vec{a} \cdot \vec{n})\vec{n}$

(b) line $\vec{r} = \vec{a} + t\vec{b}$ in the plane $\vec{r} \cdot \vec{n} = q$. *Ans*: $\vec{r} = \vec{a} + \dfrac{2}{|\vec{n}|^2}(\vec{q} - \vec{a} \cdot \vec{n})\vec{n} + t\left(\vec{b} - 2\dfrac{\vec{b} \cdot \vec{n}}{|\vec{n}|^2}\vec{n}\right)$

(c) point $\vec{a}$, in the line $\vec{r} = \vec{b} + t\vec{c}$. *Ans*: $2\vec{b} - \vec{a} + \dfrac{2}{|\vec{c}|^2}\{(\vec{a} - \vec{b}) \cdot \vec{c}\}\vec{c}$

11. Find the position vector of the point of intersection between the straight line $\vec{r} = \vec{a} + t\,\vec{c}$ and the plane $(\vec{r} - \vec{b}) \cdot \vec{d} = 0$. *Ans*: $\vec{a} + \frac{(\vec{b}-\vec{a})\cdot\vec{d}}{\vec{c}\cdot\vec{d}}\,\vec{c}$

12. Find the lengths of the intercepts made by the plane $\vec{r} \cdot \vec{n} = p$ on the coordinate axes.

13. Find the angle between the planes $\vec{r} \cdot (2\hat{i} + 3\hat{j} + \hat{k}) = 7$ and $\vec{r} \cdot (3\hat{i} - 2\hat{j} + 5\hat{k}) = 5$. *Ans*: $\cos^{-1} 5/2\sqrt{132}$

14. Find the equation of a plane perpendicular to a given vector $\vec{a}$ and at distance p from the origin. Express this equation in rectangular coordinates. *Ans*: $\vec{r} \cdot \hat{n} = p, \hat{n} = \vec{a}/|\vec{a}|$

15. Show that the points $(1, -1, 3)$ and $(3, 3, 3)$ are equidistant from the plane $\vec{r} \cdot (4\hat{i} - 3\hat{j} + \hat{k}) = 18$.

16. Find the equation of the plane which contains the origin and the line of intersection of the planes $\vec{r} \cdot \hat{n}_1 = p_1$ and $\vec{r} \cdot \hat{n}_2 = p_2$. *Ans*: $\vec{r} \cdot (p_2\,\hat{n}_1 - p_1\,\hat{n}_2) = 0$

17. If $\vec{\alpha}$ be the position vector of a given point, and $\vec{r}$ the position vector of any variable point P, find the locus of P if $(\vec{r} - \vec{\alpha}) \cdot \vec{r} = 0$. *Ans*: Sphere, centre $\frac{1}{2}\vec{\alpha}$

18. Find the distance of the point $\vec{a}$ from the plane $\vec{r} \cdot \vec{n} = q$ measured parallel to the line $\vec{r} = \vec{b} + t\vec{c}$. *Ans*: $|\vec{q} - \vec{a} \cdot \vec{n}|\ |\vec{c}|/|\vec{c} \cdot \vec{n}|$

19. Find the condition that the two spheres $|\vec{r}|^2 - 2\vec{r} \cdot \vec{\alpha}_1 + k_1 = 0$ and $|\vec{r}|^2 - 2\vec{r} \cdot \vec{\alpha}_2 + k_2 = 0$ may intersect orthogonally. *Ans*: $k_1 + k_2 = 2\vec{\alpha}_1 \cdot \vec{\alpha}_2$

20. Prove that equation of the sphere passing through the origin O and the points $A(a\,\hat{i})$, $B(b\,\hat{j})$ and $C(c\,\hat{k})$ is $\vec{r} \cdot \{\vec{r} - (a\hat{i} + b\hat{j} + c\hat{k}) = 0$.

21. Find the equation of the straight line passing through the point $\vec{c}$ and parallel to the plane $\vec{r} \cdot \vec{n} = p$ and perpendicular to the line $\vec{r} = \vec{a} + t\,\vec{b}$. *Ans*: $\vec{r} = \vec{c} + s(\vec{b} \times \vec{n})$

22. Prove that the diameter subtends a right angle at any point on the surface of the sphere.

23. A straight line is drawn from a point O to meet a fixed sphere in P. A point Q is taken in OP such that $OP : OQ$ is a fixed ratio. Show that the locus of Q is a sphere.

24. Prove that the locus of a point, the sum of the squares of whose distances from n given points is constant, is a sphere.

25. If the plane of the point P passes through Q, then show that the polar plane of Q passes through P.

26. Find the locus of a point whose powers with respect to two spheres are equal.

27. Prove by the vector method that the triangle inscribed in a semi-circle is right angled.
28. If the line joining the centre O of a sphere to any point P meets the polar plane of P in Q, prove that $OP \cdot OQ = a^2$, where a is the radius of the sphere.
29. Show that the locus of the straight lines which intersect the sphere $r^2 - 2\vec{r} \cdot \vec{a} + k = 0$ and bisected at a given point $\vec{p}$ is $(\vec{r} - \vec{p}) \cdot (\vec{p} - \vec{a}) = 0$.
30. Find the centre and radius of the sphere $r^2 - 2\vec{r} \cdot \vec{a} + k = 0$.
31. Let $OABC$ be a tetrahedron such that $\vec{OA} = \vec{a}$, $\vec{OB} = \vec{b}$ and $\vec{OC} = \vec{c}$. Find the equations of the six faces.
32. Let $OABCDEFG$ be a parallelepiped such that $\vec{OA} = \vec{a}$, $\vec{OB} = \vec{b}$ and $\vec{OC} = \vec{c}$. Find the equation of the six faces.

Long Answer Questions

1. Solve the following vector equations:

   (a) $p\vec{x} + (\vec{x} \cdot \vec{b})\vec{x} = \vec{a} \times \vec{b} + \vec{c}$. *Ans:* $\vec{x} = \dfrac{1}{p \pm \sqrt{p^2 + 4\vec{b} \cdot \vec{c}}} 2(\vec{a} \times \vec{b} + \vec{c})$

   (b) $p\vec{x} + (\vec{x} \cdot \vec{b})\vec{x} = \vec{b}$, where p is a known scalar and $\vec{b}$ is a known vector. Interpret the result. *Ans:* $\vec{x} \cdot \vec{b} = \dfrac{1}{2}\left[-p \pm \sqrt{p^2 + 4|\vec{b}|^2}\right]$

   (c) $\vec{x} \cdot \vec{a} = 0, \vec{x} \times \vec{b} = \vec{c} \times \vec{b}, \vec{a} \cdot \vec{b} \neq 0$. *Ans:* $\vec{x} = \vec{c} - \dfrac{\vec{a} \cdot \vec{c}}{\vec{a} \cdot \vec{b}} \vec{b}$

   (d) $\vec{r} \cdot \vec{n}_1 = 1$ and $\vec{r} \cdot \vec{n}_2 = 1$.

   Ans: $\vec{r} = \dfrac{|\vec{n}_2|^2 - \vec{n}_2 \cdot \vec{n}_1}{|\vec{n}_2 \times \vec{n}_1|^2} \vec{n}_1 + \dfrac{|\vec{n}_1|^2 - \vec{n}_1 \cdot \vec{n}_2}{|\vec{n}_1 \times \vec{n}_2|^2} \vec{n}_2 + t\, \vec{n}_1 \times \vec{n}_2$, t scalar

   (e) $\vec{r} \times \vec{a} = \vec{b}$ and $\vec{r} \times \vec{c} = \vec{d}$, where $\vec{a}$, $\vec{b}$ and $\vec{c}$ are the given non-coplanar vectors, with $\vec{b} \cdot \vec{c} + \vec{a} \cdot \vec{d} = 0$. *Ans:* $\vec{r} = \dfrac{1}{[\vec{a}\, \vec{b}\, \vec{c}]}\left\{(\vec{d} \cdot \vec{a})\vec{b} + (\vec{b} \cdot \vec{b})\vec{c} - (\vec{d} \cdot \vec{b})\vec{a}\right\}$

2. Solve the simultaneous vector equations for $\vec{x}$ and $\vec{y}$, where

   (a) $\vec{x} + \vec{c} \times \vec{y} = \vec{a}$ and $\vec{y} + \vec{c} \times \vec{x} = \vec{b}$, where $\vec{c} \neq \vec{0}$.

   Ans: $\vec{x} = \dfrac{1}{1 + |\vec{c}|^2}\left[\vec{a} + \vec{b} \times \vec{c} + (\vec{c} \cdot \vec{a})\vec{c}\right], \vec{y} = \dfrac{1}{1 + |\vec{c}|^2}\left[\vec{b} + \vec{a} \times \vec{c} + (\vec{c} \cdot \vec{b})\vec{c}\right]$

   (b) $\vec{x} + \vec{y} = \vec{a}, \vec{x} \times \vec{y} = \vec{b}$ and $\vec{x} \cdot \vec{y} = 1$.

   Ans: $\vec{x} = \dfrac{1}{|\vec{a}|^2}(\vec{a} + \vec{a} \times \vec{b})\ \vec{y} = \vec{a} - \dfrac{1}{|\vec{a}|^2}(\vec{a} + \vec{a} \times \vec{b})$

3. Show that the vector equation $\vec{x} \times \vec{a} + (\vec{x} \cdot \vec{b})\vec{c} = \vec{d}$ is satisfied, if

$$\vec{x} = t\,\vec{a} + \frac{1}{(\vec{a} \cdot \vec{c})\ |\vec{a}|^2}\left[\vec{a} \times \{\vec{a} \times (\vec{d} \times \vec{c})\}\right], \quad t = \text{ scalar.}$$

4. Show that the solution of the equation $t\vec{r} + \vec{r} \times \vec{a} = \vec{b}$, where t is a non-zero scalar and $\vec{a}$ and $\vec{b}$ are two vectors, can be expressed as

$$\vec{r} = \frac{1}{t^2 + a^2}\left[\vec{a} \times \vec{b} + \frac{1}{t}(\vec{a} \cdot \vec{b})\vec{a} + t\vec{b}\right].$$

5. Find the condition for the equations $\vec{r} \times \vec{a} = \vec{b}$ and $\vec{r} \times \vec{c} = \vec{d}$ to be consistent. Assuming the condition for consistency to be satisfied, solve the equations.

6. Decompose a vector $\vec{r}$ as a linear combination of a vector $\vec{a}$ and another vector perpendicular to $\vec{a}$ and coplanar with $\vec{r}$ and $\vec{a}$.

7. Express a vector $\vec{a}$ as the sum of two component vectors, one parallel and the other perpendicular to the vector $\vec{b}$ in the form

$$\vec{a} = \frac{1}{b^2}\left[(\vec{a} \cdot \vec{b})\vec{b} + \vec{b} \times (\vec{a} \times \vec{b})\right]$$

8. Solve the vector equation for $\vec{r}$ if $k\vec{r} + \vec{r} \times \vec{a} = \vec{b}$, where t is a non-zero given scalar and $\vec{a}$ and $\vec{b}$ are the two given vectors.

9. Find $\vec{r}$ and scalar λ, where $\vec{a} \times \vec{r} = \vec{b} + \lambda\vec{a}$; $\vec{a} \cdot \vec{r} = 2$ given that $\vec{a} = \hat{i} + 2\hat{j} - \hat{k}$ and $\vec{b} = 2\hat{i} - \hat{j} + \hat{k}$. *Ans*: $\lambda = \frac{1}{6}, \vec{r} = \frac{1}{6}(\hat{i} + \hat{j} + 3\hat{k})$

10. Let $\vec{a}$ and $\vec{c}$ be two given vectors and let k be a given scalar. Determine all vectors $\vec{r}$ such that $\vec{a} \times \vec{r} = \vec{c}$ and $\vec{a} \cdot \vec{b} = k$.

11. Find the vector equation of the plane through the point $(2, 3, -1)$ and perpendicular to the vector $3\hat{i} - 4\hat{j} + 2\hat{k}$. Find also the distance of the point $(0, 0, 0)$ from the plane.

Ans: $13/\sqrt{74}$

12. Find the vector equation of the plane through the point $(4, 3, -1)$ and perpendicular to the vector $3\hat{i} - 4\hat{j} + 2\hat{k}$. Find also the distance of the point $(1, 2, 3)$ from the plane.

Ans: $1/\sqrt{26}$

13. Show that the perpendicular distance of a point having position vector $\vec{p}$ from the plane passing through three points $\vec{a}, \vec{b}$ and $\vec{c}$ is

$$p = \frac{\left[\vec{a}\,\vec{b}\,\vec{c}\right] + \left[\vec{b}\,\vec{p}\,\vec{c}\right] + \left[\vec{p}\,\vec{a}\,\vec{c}\right] + \left[\vec{a}\,\vec{p}\,\vec{b}\right]}{|\vec{a} \times \vec{b} + \vec{b} \times \vec{c} + \vec{c} \times \vec{a}|}$$

14. Find the equation of a plane containing the two straight lines $\vec{r} = \vec{a} + t\,\vec{b}$ and $\vec{r} = \vec{c} + s\,\vec{d}$ (t and s are scalars). Find also a unit vector normal to the plane.

Ans: $\left[\vec{r}\,\vec{b}\,\vec{d}\right] = \left[\vec{a}\,\vec{b}\,\vec{d}\right], \frac{1}{|\vec{b} \times \vec{d}|}(\vec{b} \times \vec{d})$

15. Show that the vector equation of a plane containing the two parallel straight lines $\vec{r} = \vec{a} + t\,\vec{b}$ and $\vec{r} = \vec{c} + s\,\vec{b}$ (t and s are scalars) is $\vec{r} \cdot (\vec{a} - \vec{c}) \times \vec{b} = \left[\vec{c}\,\vec{a}\,\vec{b}\right]$.

16. Let O, Q, R and S be given four points, no three of which are collinear. If $\vec{b}, \vec{c}$ and $\vec{d}$ be the respective position vectors with O as origin, find the shortest distance between the lines OQ and RS. *Ans*: $\dfrac{|\vec{c}\cdot\vec{b}\times(\vec{d}-\vec{c})|}{|\vec{b}\times(\vec{d}-\vec{c})|}$

17. Prove that the shortest distance between two straight lines AB and CD, where the position vectors of the four points are $\vec{a}, \vec{b}, \vec{c}$ and $\vec{d}$, is

$$p = SD = \frac{|(\vec{c}-\vec{a})\cdot\{(\vec{b}-\vec{a})\times(\vec{d}-\vec{c})\}|}{|(\vec{b}-\vec{a})\times(\vec{d}-\vec{c})|}$$

18. Find the shortest distance between two skew lines $\vec{r} = \vec{a}_1 + t\,\vec{b}_1$ and $\vec{r} = \vec{a}_2 + s\,\vec{b}_2$, where $\vec{a}_1, \vec{b}_1, \vec{a}_2$ and $\vec{b}_2$ are vectors with coordinates $(6,2,2), (-4,0,-1), (1,-2,2)$ and $(3,-2,-2)$, respectively. Find where the line of shortest distance meets the given lines. *Ans*: 9 units, $(-1,-2,-3), (5,4,0)$

19. Find in terms of λ, the shortest distance between the lines passing through the points $A(2,1,3)$, $B(3,1,\lambda)$, $C(1,1,5)$ and $D(2,-1,4)$. For what value of λ are the lines coplanar? *Ans*: $\lambda = 1$

20. Find the equation of the plane through the line of intersection of the planes $\vec{r}\cdot\vec{n}_1 = 1$ and $\vec{r}\cdot\vec{n}_2 = 1$ and perpendicular to the plane $\vec{r}\cdot\vec{n}_3 = 1$.
Ans: $(\vec{r}\cdot\vec{n}_1 - 1)(\vec{n}_2\cdot\vec{n}_3) = (\vec{r}\cdot\vec{n}_2 - 1)(\vec{n}_1\cdot\vec{n}_3)$

21. Forces $\vec{P}, \vec{Q}$ and $\vec{R}$ act at a point O and are in equilibrium. If any transversal cuts their lines of action of the forces $\vec{P}, \vec{Q}$ and $\vec{R}$ at the points A, B and C, respectively, show that $\dfrac{P}{OA} + \dfrac{Q}{OB} + \dfrac{R}{OC} = 0$, where P, Q and R are the magnitudes of $\vec{P}, \vec{Q}$ and $\vec{R}$, respectively.

22. If a force given by $\vec{F} = 2\hat{i} + 4\hat{j} - \hat{k}$ displaces a particle from the position $A(1,1,2)$ to $B(3,-1,-1)$, find the work done by the force. *Ans*: -1 units

23. (a) A particle acted on by two constant forces $4\hat{i} + \hat{j} - 3\hat{k}$ and $3\hat{i} + \hat{j} - \hat{k}$ is displaced from the point $\hat{i} + 2\hat{j} + 3\hat{k}$ to the point $5\hat{i} + 4\hat{j} - \hat{k}$. Find the total work done by the forces. *Ans*: 48 units

(b) A particle acted on by two constant forces $5\hat{i} + 2\hat{j} + \hat{k}$ and $2\hat{i} - \hat{j} - 3\hat{k}$ is displaced from origin to the point $4\hat{i} + \hat{j} - 3\hat{k}$. Find the total work done by the forces. *Ans*: 35 units

(c) A particle acted on by two constant forces $2\hat{i} + 3\hat{j} - \hat{k}$ and $2\hat{i} - \hat{j} - 3\hat{k}$ is displaced from the point $\hat{i} + 3\hat{j} + 2\hat{k}$ to the point $4\hat{i} + 5\hat{j} - \hat{k}$. Find the total work done by the forces. *Ans*: 7 units

(d) A particle acted on by two constant force $4\hat{i} + \hat{j} - 3\hat{k}$ and $3\hat{i} + \hat{j} - \hat{k}$ is displaced from the point $\hat{i} + 2\hat{j} + 3\hat{k}$ to the point $5\hat{i} + 4\hat{j} + \hat{k}$. Find the total work done by the forces. *Ans*: 40 units

24. Show that the square of any straight line is equal to

(a) the sum of the squares of its projections on three mutually perpendicular straight lines.
(b) half the sum of the squares of its projections on three mutually perpendicular planes.

25. Prove that, for a regular tetrahedron, the perpendiculars from the vertices to the opposite faces meet the faces at their centroids.
26. Show that the six planes through the middle point of each edge of a tetrahedron and perpendicular to the opposite edge meet in a point.
27. Find the equation of the sphere which is inscribed in the tetrahedron bounded by the planes $\vec{r}\cdot\hat{i}=0, \vec{r}\cdot\hat{j}=0, \vec{r}\cdot\hat{k}=0$ and $\vec{r}\cdot(\hat{i}+\hat{j}+\hat{k})=a(3-\sqrt{3})$. *Ans*: $|\vec{r}-a(\hat{i}+\hat{j}+\hat{k})|=a$
28. Find the shortest distance between the diagonals of a rectangular parallelepiped whose sides are a, b and c and the edges not meeting it.
29. Prove that the intersections of the bisectors of the exterior angles of any triangle with opposite sides are collinear.
30. In a triangle ABC, D divides BC in the ratio $3:2$ and E divides CA in the ratio $1:3$. The lines AD and BE meet at H and CH in F. Show that F divides AB in the ratio $2:1$.
31. A rigid body rotates with an angular velocity 2 radians per second about an axis passing through the point $\hat{i}-2\hat{j}+\hat{k}$ and parallel to the vector $\hat{i}-\hat{j}-\hat{k}$. Find the magnitude of the linear velocity of the point whose position vector is $\hat{i}+\hat{j}+\hat{k}$. *Ans*: $2\sqrt{26}$ units
32. (a) A force $F_1\hat{i}+F_2\hat{j}+F_3\hat{k}$ acts at the point $a\hat{i}+b\hat{j}+c\hat{k}$. Find its torque about the origin. Find also the moments of the force about the three coordinate axes and the line $\frac{x}{l}=\frac{y}{m}=\frac{z}{n}$, where $l^2+m^2+n^2=1$.

    Ans: $(bF_3-cF_2)\hat{i}+(cF_1-aF_3)\hat{j}+(aF_2-bF_1)\hat{k}$; $(bF_3-cF_2), (cF_1-aF_3), (aF_2-bF_1)$; $l(bF_3-cF_2)+m(cF_1-aF_3)+n(aF_2-bF_1)$

    (b) Find the magnitude of the torque about the point $\hat{i}+\hat{j}+3\hat{k}$ of the force represented by $\hat{i}+2\hat{j}+2\hat{k}$ acting through the point $2\hat{i}+\hat{k}$. *Ans*: $\sqrt{29}$ units

    (c) A force $\vec{F}=7\hat{i}-3\hat{j}-4\hat{k}$ acts along a line $\vec{r}=2\hat{i}-3\hat{j}-\hat{k}+t(3\hat{i}+\hat{j}+2\hat{k})$, t is a parameter. Find the torque of the force $\vec{F}$ about the point $A(5\hat{i}+3\hat{j}+\hat{k})$. Find also the moment of the force $\vec{F}$ about the line. *Ans*: $18\hat{i}-26\hat{j}+51\hat{k}$; $130/\sqrt{14}$ units

    (d) A force $\vec{F}$ of magnitude 12 units acts along the line joining the points $(2,1,-1)$ and $(3,2,1)$. Evaluate the torque of the force $\vec{F}$ about the point $(3,-2,-5)$, and hence, find the moment of the force $\vec{F}$ about the through the point $(3,-2,-5)$ having direction ratios 5, 3 and 4. Ans: $25\sqrt{5}/\sqrt{3}$ units

    (e) A force of magnitude 5 units acts along the line joining the points $A(3,-1,-6)$ and $B(8,4,-1)$. Find the torque of the force about the point $(1,2,-3)$. *Ans*: $\frac{25}{3}(-\hat{j}+\hat{k})$
33. The angular velocity of a rotating body about an axis of rotation is given by $\vec{\omega}=4\hat{i}+\hat{j}-2\hat{k}$. Find the linear velocity of a point on the body whose position vector relative to a point on the axis of rotation is $2\hat{i}-3\hat{j}+\hat{k}$. *Ans*: $-5\hat{i}-8\hat{j}-14\hat{j}$
34. A rigid body is rotating about an axis through the point $(3,1,2)$. If the particle at the point $(2,1,3)$ has the velocity $-5\hat{i}+8\hat{j}+\hat{k}$ and that at the point $(4,1,2)$ has the velocity $2\hat{j}-\hat{k}$, then find the magnitude and direction of the angular velocity of the body.

    Ans: 3 units, $\frac{1}{3}(2\hat{i}+\hat{j}+2\hat{k})$

35. A rigid body is spinning with angular velocity 5 radians per second about an axis parallel to $\hat{i}+\hat{j}+\hat{k}$ and passing through the point $\hat{i}+2\hat{j}-\hat{k}$. Find the velocity of the particle at the point $2\hat{i}+\hat{j}-\hat{k}$. *Ans:* $\frac{5}{\sqrt{3}}(\hat{i}+\hat{j}-2\hat{k})$

36. A rigid body is spinning with angular speed 4 radians per second about an axis through $\hat{i}+3\hat{j}-\hat{k}$ in the direction of the vector $3\hat{j}-\hat{k}$. Find the velocity of the particle at the point $4\hat{i}-2\hat{j}+\hat{k}$. *Ans:* $\frac{4}{\sqrt{10}}(\hat{i}-3\hat{j}-9\hat{k})$

37. (a) A force of 5 units acts in the direction of the vector $\hat{i}-\hat{j}-\hat{k}$ and passes through a point $\hat{i}-2\hat{j}+2\hat{k}$. Find the moment of the force about the point $\hat{i}-\hat{j}+\hat{k}$. *Ans:* $-\frac{5}{3}\sqrt{3}(\hat{j}+\hat{k})$

    (b) A force of 15 units acts in the direction $\hat{i}-\hat{j}-\hat{k}$ and passes through a point $\hat{i}+\hat{j}+\hat{k}$. Find the moment of the force about the point $2(\hat{i}-\hat{j}+\hat{k})$. *Ans:* $\hat{i}-\hat{k}$

    (c) A force of $15\sqrt{3}$ units acts in the direction $\hat{i}-\hat{j}+\hat{k}$ and passes through a point $\hat{i}-\hat{j}+5\hat{k}$. Find the moment of the force about the point $\hat{i}+\hat{j}+\hat{k}$).

    Ans: $-10\sqrt{3}(\hat{i}+2\hat{j}+\hat{k})$

38. Prove that the moment of a force $\hat{b}$ acting along a line $\vec{r}=\vec{a}+t\hat{b}$ about the line $\vec{r}=\vec{c}+s\hat{d}$ is equal to the moment of a force $\hat{d}$ acting along the line $\vec{r}=\vec{c}+s\hat{d}$ about the line $\vec{r}=\vec{a}+t\hat{b}$, where $\hat{b}$ and $\hat{d}$ are unit vectors.

Vector Differentiation 4

In this chapter, we shall study the concept of scalar and vector point functions and scalar and vector fields: limit, continuity, differentiability and directional derivative of these functions at a point in any direction. We shall also discuss the concepts of differentiability, partial differentiation and total differential of vector functions and their properties. A careful comparison between the concepts of scalar calculus and vector calculus will reveal that the definitions of various properties in these two concepts are quite analogous. The use of vector calculus to study the differential geometry of space curves and the mechanics of particles will be discussed.

4.1 SCALAR AND VECTOR FIELDS

In physical problems, we often come across quantities such as temperature of a liquid, distance between two points, density of a gas, velocity and acceleration of a particle or a body, tangent to a curve and normal to a surface. In the previous chapter, we learnt that physical quantities can be categorised either as scalar or as vector. You may have noticed that some physical quantities, whether scalars or vectors, are variable, i.e., their values are not constant or static but change with change in the variable. For example, the density of a gas, which is a scalar quantity, changes from place to place and is different at different times. Similarly, the tangent to a curve may have different directions at different points on the curve. This variable character of scalars and vectors gives rise to scalar functions and vector functions. Further, you may have noticed that at different positions, the temperature or velocity of a body does not remain the same. The distribution of temperature or velocity is therefore defined at each point of a given domain in space, which leads to the idea of scalar fields and vector fields.

4.1.1 Scalar Fields

Definition 4.1.1

Let $\mathcal{D}$ be any region (domain) in space and P be any point of region $\mathcal{D}$. If, for each point P of $\mathcal{D}$, there exists a scalar, denoted by $f(P)$, then f is said to be a *scalar point function* for the region $\mathcal{D}$.

(i) We consider the Cartesian coordinate system in space and take coordinates of a fixed point P_0 as (x_0, y_0, z_0) and a variable point P as (x, y, z). Then, the scalar point function f is given by the well-known formula

$$f(P) = f(x, y, z) = \sqrt{(x - x_0)^2 + (y - y_0)^2 + (z - z_0)^2}$$

(ii) Temperature at any point P of a body occupying a region $\mathcal{D}$ is an example of a scalar point function.

In most applications, the domain of a scalar function is a curve, a surface or a three-dimensional region in space. The function f associates with each point in domain $\mathcal{D}$ as a scalar (a real number) and we say that a *scalar field* is obtained in $\mathcal{D}$.

Definition 4.1.2

If ϕ is a scalar point function defined in a region $\mathcal{D}$ of space, then the set of points P of the region together with functional values $\phi(P)$ is called the *scalar field function* or simply a *scalar field* over $\mathcal{D}$.

In a plane, for instance, the equation of a curve is given by $f(x, y) =$ a constant, i.e., these are curves along which f has a constant value for all points in the xy plane. Similarly, a surface may be given by $\phi(x, y, z) =$ a constant, i.e., these are surfaces for which ϕ has a constant value for all points in space. In these examples, scalar functions f and ϕ, having plane and space as their respective domains, are scalar fields. Some more examples of scalar fields are the density of the air of the earth's atmosphere; the pressure within a region through which a compressible fluid flows; the electrostatic potential of a system of charges; temperature distribution in a medium; and mass distribution in a body.

We can represent a scalar field by a formula as well as pictorially. For a pictorial representation, we may use curves and surfaces. The pictorial representation of scalar fields takes the form of maps showing the physical geography of a region (indicating hills, lakes, land above or below sea level, etc.).

4.1.2 Vector Fields

Definition 4.1.3

If, to each point P of a certain region $\mathcal{G}$ in space, a vector $\vec{F}(P)$ is assigned, then $\vec{F}(P)$ is said to be a *vector point function* for the region $\mathcal{G}$.

There are many examples of vector functions in physics and mathematics:

(i) Velocity, acceleration and force are all vector functions.
(ii) The gravitational force exerted by the sun on a unit mass is also a variable vector, depending on the position of the mass, and thus represents a vector function.

A vector point function of $\vec{r}$ and/or t may be of the following forms:

(i) $F = F(\vec{r})$, where F is a scalar (ii) $\vec{f} = \vec{f}(\vec{r})$, where $\vec{f}$ is a vector

(iii) $F = F(\vec{r}, t)$; where F and t are scalars (iv) $\vec{f} = \vec{f}(\vec{r}, t)$; where $\vec{f}$ is a vector and t is a scalar

Definition 4.1.4

The collection of all vector functions $\vec{F}(P)$ is called a *vector field on* $\mathcal{G}$.

Let $\left[a, b\right]$ be a bounded closed interval. If $\vec{F}$ is a function which assigns to each point $t \in \left[a, b\right]$ a unique vector, $\vec{F}(t)$ is called a *vector field function* or a *vector field.* The interval $\left[a, b\right]$ is the domain of the function $\vec{F}$ and for a fixed $t_0 \in \left[a, b\right]$ we have a fixed vector $\vec{F}_0 = \vec{F}(t_0)$.

(i) A simple example of a vector field is the field defined by the vector $\vec{r} = x\hat{i} + y\hat{j} + z\hat{k}$.
(ii) A physical example of a vector field is given by the particles of a fluid under flow.

(iii) Electrostatic and magnetostatic are examples of fluid fields.
(iv) At any instant, the velocity vector $\vec{v}(P)$ of a rotating body constitutes a vector field, called the velocity field of rotation.
(v) A particle A of mass M attracts particle B of mass m. According to Newton's law of attraction/gravitation

$$\left|\vec{F}\right| = \frac{GMm}{d^2}; \quad G = \text{Gravitational constant}$$

where d is the distance between A and B. Hence, $\vec{F}$ defines a vector field in space.

A vector function $\vec{F}$ of a scalar variable t can be expressed analytically as

$$\vec{F}(t) = F_1(t)\,\hat{i} + F_2(t)\,\hat{j} + F_3(t)\,\hat{k}$$

where $F_1(t), F_2(t)$ and $F_3(t)$ are scalar functions of t, each having the same domain as that of $\vec{F}(t)$.

4.2 VECTOR CALCULUS

You have already learnt about the basic concepts of limit, continuity, differentiability and partial differentiation of scalar functions in calculus. We now begin by introducing the idea of limit and continuity of a vector function. We know that the limit concept is a basic and useful tool for the analysis of functions. It enables us to study derivatives, improper integrals and other important features of functions. We shall introduce the concepts of limit and continuity in an informal manner without going into the intricacies of the delta and epsilon method.

4.2.1 Limit and Continuity of Vector Functions

Definition 4.2.1

A vector function $\vec{f}(t)$ of scalar variable t is said to tend to a *limit* (constant vector) $\vec{a}$ as t tends to t_0 (which is a limit point of $\mathcal{D}$), if, corresponding to a pre-assigned $\varepsilon > 0$, there exists $\delta = \delta(\varepsilon) > 0$, such that

$$|\vec{f}(t) - \vec{a}| < \varepsilon, \text{ whenever } |t - t_0| < \delta$$

In notation, $\lim_{t \to t_0} \vec{f}(t) = \vec{a}$ or $\vec{f}(t) \to \vec{a}$ as $t \to t_0$. Here, $|\vec{f}(t) - \vec{a}|$ is the *length of the vector* $\vec{f}(t) - \vec{a}$. If we take $\vec{a} = a_1\hat{i} + a_2\hat{j} + a_3\hat{k}$, then the above limit exists if and only if the following limits exist:

$$\lim_{t \to t_0} f_1(t) = a_1; \; \lim_{t \to t_0} f_2(t) = a_2; \; \lim_{t \to t_0} f_3(t) = a_3$$

Further, if $\vec{a}$ is a non-null vector and t is closed to t_0, then $\vec{f}(t)$ and $\vec{a}$ have the same magnitude and direction; otherwise the direction of $\vec{f}(t)$ may vary arbitrarily.

If $\vec{f}(t)$ and $\vec{g}(t)$ are two vector functions and $h(t)$ is a scalar function, then

(i) $\lim_{t \to t_0} \left[\vec{f}(t) \pm \vec{g}(t)\right] = \lim_{t \to t_0} \vec{f}(t) \pm \lim_{t \to t_0} \vec{g}(t)$

(ii) $\lim_{t \to t_0} \left[\vec{f}(t) \cdot \vec{g}(t)\right] = \lim_{t \to t_0} \vec{f}(t) \cdot \lim_{t \to t_0} \vec{g}(t)$

(iii) $\lim_{t \to t_0} \left[\vec{f}(t) \times \vec{g}(t)\right] = \lim_{t \to t_0} \vec{f}(t) \times \lim_{t \to t_0} \vec{g}(t)$

(iv) $\lim_{t \to t_0} \left[h(t)\, \vec{g}(t)\right] = \lim_{t \to t_0} h(t) \ \lim_{t \to t_0} \vec{g}(t)$

provided all the limits exist. Note that t_0 is the limit point of the domains of all the vector functions and scalar functions under consideration.

Definition 4.2.2

The vector point function $\vec{f}(t)$ defined on $[a, b]$ is said to be *continuous* at a point $t_0 \in [a, b]$, if, for any pre-assigned $\varepsilon > 0$, there exists $\delta = \delta(\varepsilon) > 0$, such that

$$|\vec{f}(t) - \vec{f}(t_0)| < \varepsilon, \text{ whenever } |t - t_0| < \delta$$

In notation, $\lim_{t \to t_0} \vec{f}(t) = \vec{f}(t_0)$. If $\vec{f}(t)$ is continuous at all points in $[a, b]$, then we say $\vec{f}(t)$ is continuous in $[a, b]$ and if we write $\vec{f}(t) = f_1(t)\hat{i} + f_2(t)\hat{j} + f_3(t)\hat{k}$, then $\vec{f}(t)$ is continuous if $f_1(t), f_2(t)$ and $f_3(t)$ are continuous functions, and conversely.

Using the results of the limits given above, it can be shown that if $\vec{f}(t)$ and $\vec{g}(t)$ are two continuous vector point functions, then their sum, difference, scalar product, vector product and scalar multiple of any one are also continuous functions.

4.2.2 Ordinary Derivatives

Let $\vec{F}(t)$ be a vector single-valued function depending on a single scalar variable t. Relative to a fixed origin O, let $\vec{OP}$ be the value $\vec{F}(t)$ corresponding to some fixed value of t of the scalar variable (Fig. 4.1). Now, suppose the value $t + \triangle t$ of t corresponds to the value $\vec{F}(t + \triangle t)$ at Q. The increment in $\vec{F}(t)$ when t changes to $t + \triangle t$ is

$$\vec{F}(t + \triangle t) - \vec{F}(t) = \vec{OQ} - \vec{OP} = \vec{PQ}$$
$$= \triangle\vec{F}(t), \text{ say} \qquad (4.1)$$

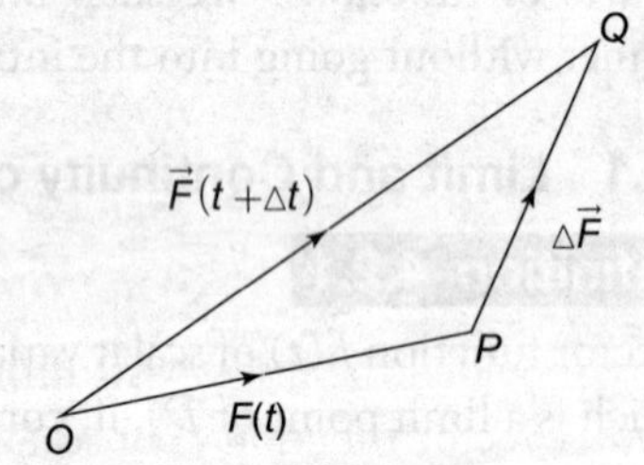

Figure 4.1 Increment of a vector

Now, $\triangle t$ is a scalar. Consider the ratio

$$\frac{\triangle\vec{F}}{\triangle t} = \frac{\vec{F}(t + \triangle t) - \vec{F}(t)}{\triangle t} = \frac{\vec{PQ}}{\triangle t}$$

This is itself a vector extending along PQ and having the same direction as that of $\triangle\vec{F}$. The derivative of the vector function $\vec{F}(t)$ with respect to the scalar variable t is given by

$$\vec{F}'(t) = \frac{d\vec{F}}{dt} = \lim_{\triangle t \to 0} \frac{\triangle\vec{F}}{\triangle t} = \lim_{\triangle t \to 0} \frac{\vec{F}(t + \triangle t) - \vec{F}(t)}{\triangle t} \qquad (4.2)$$

provided the limit exists finitely. Note that the differential coefficient $\frac{d\vec{F}}{dt}$ is itself a vector depending on the scalar t, and is called the *derivative* or *differential coefficient* of $\vec{F}$ with respect to the scalar t.

(i) Let $\vec{F}$ be a vector function of scalar variable t, whose derivative $\vec{F}'$ is itself a differentiable function of t in its domain. Then, the second order derivative of $\vec{F}$ is given by

$$\vec{F}''(t) = \frac{d^2\vec{F}}{dt^2} = \lim_{\Delta t \to 0} \frac{\vec{F}'(t+\Delta t) - \vec{F}'(t)}{\Delta t} \tag{4.3}$$

If the limit exists, then $\vec{F}''(t)$ is called the *second order derivative* of $\vec{F}$ with respect to the scalar variable t. Similarly, we can define more higher order derivatives of vector function $\vec{F}$ such as $\frac{d^3\vec{F}}{dt^3}, \frac{d^4\vec{F}}{dt^4}$, etc.

(ii) In the explicit form, let $\vec{F}(t) = F_1(t)\hat{i} + F_2(t)\hat{j} + F_3(t)\hat{k}$. Then

$$\begin{aligned} F'(t) &= \frac{d\vec{F}(t)}{dt} = \lim_{\Delta t \to 0} \frac{\vec{F}(t+\Delta t) - \vec{F}(t)}{\Delta t} \\ &= \lim_{\Delta t \to 0} \left[\frac{F_1(t+\Delta t) - F_1(t)}{\Delta t}\hat{i} + \frac{F_2(t+\Delta t) - F_2(t)}{\Delta t}\hat{j} + \frac{F_3(t+\Delta t) - F_3(t)}{\Delta t}\hat{k} \right] \\ &= \frac{dF_1(t)}{dt}\hat{i} + \frac{dF_2(t)}{dt}\hat{j} + \frac{dF_3(t)}{dt}\hat{k} \end{aligned} \tag{4.4}$$

provided the derivatives of the component scalar functions $F_1(t)$, $F_2(t)$ and $F_3(t)$ with respect to t exist. Higher order derivatives $\frac{d^2\vec{F}}{dt^2}, \frac{d^3\vec{F}}{dt^3}, \cdots$ are similarly defined as in scalar calculus.

(iii) We may also write $\vec{F}'(t_0)$ as

$$\vec{F}'(t_0) = \frac{d\vec{F}}{dt} = \lim_{t \to t_0} \frac{\vec{F}(t) - \vec{F}(t_0)}{t - t_0} = \lim_{\Delta t \to 0} \frac{\vec{F}(t_0+\Delta t) - \vec{F}(t_0)}{\Delta t} \tag{4.5}$$

provided the limit exists finitely.

(iv) A vector function $\vec{F}$ of scalar variable t is said to be *differentiable* in its domain if it has finite derivatives for all values of t in the domain of definition.

(v) Since the difference of two vectors in a vector and division of a vector by a scalar does not change its nature, it follows that $\vec{F}'$ itself is a vector function, whose domain will be a subset of the domain of vector function $\vec{F}$.

(vi) Every derivable vector function is continuous, but a continuous function $\vec{F}$ may not be differentiable.

Rules of Differentiation

Let $\vec{F}(t)$ and $\vec{G}(t)$ be two differentiable vector functions of a scalar variable t. $\phi(t)$ is a differentiable scalar function of t and $\vec{C}$ is a constant vector. Then

(i) $\frac{d\vec{C}}{dt} = \vec{0}$ (ii) $\frac{d}{dt}\left[\vec{F}(t) \pm \vec{G}(t)\right] = \frac{d\vec{F}}{dt} \pm \frac{d\vec{G}}{dt}$ (iii) $\frac{d}{dt}(\phi\vec{F}) = \phi\frac{d\vec{F}}{dt} + \frac{d\phi}{dt}\vec{F}$

(iv) $\frac{d}{dt}(\vec{F}\cdot\vec{G}) = \vec{F}\cdot\frac{d\vec{G}}{dt} + \frac{d\vec{F}}{dt}\cdot\vec{G}$ (v) $\frac{d}{dt}(\vec{F}\times\vec{G}) = \vec{F}\times\frac{d\vec{G}}{dt} + \frac{d\vec{F}}{dt}\times\vec{G}$

Following are the derivatives of triple products. Let $\vec{a}$, $\vec{b}$ and $\vec{c}$ be differentiable vector functions of a scalar variable t. Then

(i) $\frac{d}{dt}\left\{\vec{a}\cdot(\vec{b}\times\vec{c})\right\} = \frac{d\vec{a}}{dt}\cdot(\vec{b}\times\vec{c}) + \vec{a}\cdot\left(\frac{d\vec{b}}{dt}\times\vec{c}\right) + \vec{a}\cdot\left(\vec{b}\times\frac{d\vec{c}}{dt}\right)$

(ii) $\frac{d}{dt}\left\{\vec{a}\times(\vec{b}\times\vec{c})\right\} = \frac{d\vec{a}}{dt}\times(\vec{b}\times\vec{c}) + \vec{a}\times\left(\frac{d\vec{b}}{dt}\times\vec{c}\right) + \vec{a}\times\left(\vec{b}\times\frac{d\vec{c}}{dt}\right)$

If $\vec{F}$ is a differentiable vector function of a scalar variable u and u itself is a scalar function of a scalar variable t, then $\vec{F}$ is a differentiable vector function of scalar variable t and

$$\frac{d\vec{F}}{dt} = \frac{d\vec{F}}{du}\frac{du}{dt}$$

Example 4.2.1

If $\vec{F}(t) = e^{-t}\,\hat{i} + \log(t^2+1)\hat{j} - \tan t\,\hat{k}$, evaluate

$(i)\ \frac{d\vec{F}}{dt}$, $(ii)\ \frac{d^2\vec{F}}{dt^2}$, $(iii)\ |\frac{d\vec{F}}{dt}|$, $(iv)\ |\frac{d^2\vec{F}}{dt^2}|$ at $t=0$.

Solution

Here, the vector function $\vec{F}(t)$ is given in explicit form. Thus, using Eq. (4.4), we get

(i)
$$\frac{d\vec{F}}{dt} = \frac{d}{dt}(e^{-t})\hat{i} + \frac{d}{dt}\{\log(t^2+1)\}\hat{j} - \frac{d}{dt}(\tan t)\hat{k}$$
$$= -e^{-t}\hat{i} + \frac{2t}{t^2+1}\hat{j} - \sec^2 t\hat{k} = -\hat{i} - \hat{k} \quad \text{at } t=0$$

(ii)
$$\frac{d^2\vec{F}}{dt^2} = \frac{d}{dt}\left(\frac{d\vec{F}}{dt}\right) = -\frac{d}{dt}(e^{-t})\hat{i} + \frac{d}{dt}\left(\frac{2t}{t^2+1}\right)\hat{j} - \frac{d}{dt}(\sec^2 t)\hat{k}$$
$$= e^{-t}\hat{i} + \frac{2(1-t^2)}{(t^2+1)^2}\hat{j} - 2\sec^2 t\tan t\hat{k} = \hat{i} + 2\hat{j}; \text{ at } t=0$$

(iii)
$$\left|\frac{d\vec{F}}{dt}\right| = \left|-\hat{i} - \hat{k}\right| = \sqrt{(-1)^2 + (-1)^2} = \sqrt{2}$$

(iv)
$$\left|\frac{d^2\vec{F}}{dt^2}\right| = |\hat{i} + 2\hat{j}| = \sqrt{(1)^2 + (2)^2} = \sqrt{5}$$

Example 4.2.2

Prove that $\frac{d}{dt}\left[\vec{p}\,\vec{q}\,\vec{r}\right] = \left[\frac{d\vec{p}}{dt}\,\vec{q}\,\vec{r}\right] + \left[\vec{p}\,\frac{d\vec{q}}{dt}\,\vec{r}\right] + \left[\vec{p}\,\vec{q}\,\frac{d\vec{r}}{dt}\right]$, where $\vec{p}, \vec{q}$ and $\vec{r}$ are functions of t.

Hence, find $\frac{d}{dt}\left[\vec{r}\,\frac{d\vec{r}}{dt}\,\frac{d^2\vec{r}}{dt^2}\right]$, where $\vec{r}$ is a function of t.

Solution

Since $[\vec{p}\,\vec{q}\,\vec{r}] = \vec{p}\cdot(\vec{q}\times\vec{r})$, we have

$$\frac{d}{dt}\left[\vec{p}\,\vec{q}\,\vec{r}\right] = \frac{d\vec{p}}{dt}\cdot(\vec{q}\times\vec{r}) + \vec{p}\cdot\left(\frac{d\vec{q}}{dt}\times\vec{r}\right) + \vec{p}\cdot\left(\vec{q}\times\frac{d\vec{r}}{dt}\right)$$

$$= \left[\frac{d\vec{p}}{dt}\,\vec{q}\,\vec{r}\right] + \left[\vec{p}\,\frac{d\vec{q}}{dt}\,\vec{r}\right] + \left[\vec{p}\,\vec{q}\,\frac{d\vec{r}}{dt}\right]$$

Substituting $\vec{r}$ for $\vec{p}$, $\dfrac{d\vec{r}}{dt}$ for $\vec{q}$ and $\dfrac{d^2\vec{r}}{dt^2}$ for $\vec{r}$, we get

$$\frac{d}{dt}\left[\vec{r}\,\frac{d\vec{r}}{dt}\,\frac{d^2\vec{r}}{dt^2}\right] = \left[\frac{d\vec{r}}{dt}\,\frac{d\vec{r}}{dt}\,\frac{d^2\vec{r}}{dt^2}\right] + \left[\vec{r}\,\frac{d^2\vec{r}}{dt^2}\,\frac{d^2\vec{r}}{dt^2}\right] + \left[\vec{r}\,\frac{d\vec{r}}{dt}\,\frac{d^3\vec{r}}{dt^3}\right]$$

$$= \left[\vec{r}\,\frac{d\vec{r}}{dt}\,\frac{d^3\vec{r}}{dt^3}\right];\text{ as } \left[\frac{d\vec{r}}{dt}\,\frac{d\vec{r}}{dt}\,\frac{d^2\vec{r}}{dt^2}\right] = 0, \left[\vec{r}\,\frac{d^2\vec{r}}{dt^2}\,\frac{d^2\vec{r}}{dt^2}\right] = 0$$

Example 4.2.3

If $\vec{r} = a\cos t\hat{i} + a\sin t\hat{j} + at\tan\alpha\hat{k}$, then find the values of $\left|\dfrac{d\vec{r}}{dt} \times \dfrac{d^2\vec{r}}{dt^2}\right|$ and $\left[\dfrac{d\vec{r}}{dt}\,\dfrac{d^2\vec{r}}{dt^2}\,\dfrac{d^3\vec{r}}{dt^3}\right]$.

Solution

Since $\vec{r} = a\cos t\hat{i} + a\sin t\hat{j} + at\tan\alpha\hat{k}$

$$\frac{d\vec{r}}{dt} = -a\sin t\hat{i} + a\cos t\hat{j} + a\tan\alpha\hat{k}$$

$$\frac{d^2\vec{r}}{dt^2} = -a\cos t\hat{i} - a\sin t\hat{j};\quad \frac{d^3\vec{r}}{dt^3} = a\sin t\hat{i} - a\cos t\hat{j}$$

Therefore, using the definition of cross product, we get

$$\frac{d\vec{r}}{dt} \times \frac{d^2\vec{r}}{dt^2} = \begin{vmatrix} \hat{i} & \hat{j} & \hat{k} \\ -a\sin t & a\cos t & a\tan\alpha \\ -a\cos t & -a\sin t & 0 \end{vmatrix}$$

$$= a^2\sin t\tan\alpha\,\hat{i} - a^2\cos t\tan\alpha\,\hat{j} + a^2\,\hat{k}$$

$$\Rightarrow \left|\frac{d\vec{r}}{dt} \times \frac{d^2\vec{r}}{dt^2}\right| = \sqrt{a^4\sin^2 t\tan^2\alpha + a^4\cos^2 t\tan^2\alpha + a^4}$$

$$= \sqrt{a^4\tan^2\alpha + a^4} = a^2\sec\alpha$$

Using the definition of scalar triple product, we get

$$\left[\frac{d\vec{r}}{dt}\,\frac{d^2\vec{r}}{dt^2}\,\frac{d^3\vec{r}}{dt^3}\right] = \left(\frac{d\vec{r}}{dt} \times \frac{d^2\vec{r}}{dt^2}\right)\cdot\frac{d^3\vec{r}}{dt^3}$$

$$= \left[a^2\sin t\tan\alpha\,\hat{i} - a^2\cos t\tan\alpha\,\hat{j} + a^2\,\hat{k}\right]\cdot(a\sin t\hat{i} - a\cos t\hat{j})$$

$$= a^3(\sin^2 t + \cos^2 t)\tan\alpha = a^3\tan\alpha$$

Example 4.2.4

If $\vec{r} = \vec{a}\cos nt + \vec{b}\sin nt$, where $\vec{a}$, $\vec{b}$ and n are constants, prove that

$$\frac{d^2\vec{r}}{dt^2} + n^2\vec{r} = \vec{0} \text{ and } \vec{r} \times \frac{d\vec{r}}{dt} = n(\vec{a} \times \vec{b}).$$

Solution

Since $\vec{r} = \vec{a}\cos nt + \vec{b}\sin nt$

$$\frac{d\vec{r}}{dt} = \frac{d}{dt}\left[\vec{a}\cos nt + \vec{b}\sin nt\right] = -n\vec{a}\sin nt + n\vec{b}\cos nt$$

$$\frac{d^2\vec{r}}{dt^2} = n\frac{d}{dt}\left[-\vec{a}\sin nt + \vec{b}\cos nt\right] = -n^2\left[\vec{a}\cos nt + \vec{b}\sin nt\right] = -n^2\vec{r}$$

$$\vec{r}\times\frac{d\vec{r}}{dt} = \left[\vec{a}\cos nt + \vec{b}\sin nt\right]\times\left[-n\vec{a}\sin nt + n\vec{b}\cos nt\right]$$

$$= (\vec{a}\times\vec{b})n(\cos^2 nt + \sin^2 nt) = n(\vec{a}\times\vec{b})$$

Example 4.2.5

If $r' = \frac{d\vec{r}}{dt}$ and $r'' = \frac{d^2\vec{r}}{dt^2}$, evaluate $\frac{d}{dt}\left[\vec{r}\times(\vec{r}'\times\vec{r}'')\right]$ and $\frac{d^2}{dt^2}\left[\vec{r}\times(\vec{r}'\times\vec{r}'')\right]$, where $\vec{r}$ is a function of t.

Solution

Using the definition of derivative, we get

$$\frac{d}{dt}\left[\vec{r}\times(\vec{r}'\times\vec{r}'')\right] = \frac{d}{dt}\vec{r}\times(\vec{r}'\times\vec{r}'') + \vec{r}\times\left(\frac{d}{dt}\vec{r}'\times\vec{r}''\right) + \vec{r}\times\left(\vec{r}'\times\frac{d}{dt}\vec{r}''\right)$$

$$= \vec{r}'\times(\vec{r}'\times\vec{r}'') + \vec{r}\times(\vec{r}''\times\vec{r}'') + \vec{r}\times(\vec{r}'\times\vec{r}''')$$

$$= \vec{r}'\times(\vec{r}'\times\vec{r}'') + \vec{r}\times(\vec{r}'\times\vec{r}'''), \text{ as } \vec{r}''\times\vec{r}'' = \vec{0}$$

Again, differentiating with respect to t, we get

$$\frac{d^2}{dt^2}\left[\vec{r}\times(\vec{r}'\times\vec{r}'')\right] = \frac{d}{dt}\left[\vec{r}'\times(\vec{r}'\times\vec{r}'') + \vec{r}\times(\vec{r}'\times\vec{r}''')\right]$$

$$= \frac{d}{dt}\vec{r}'\times(\vec{r}'\times\vec{r}'') + \vec{r}'\times\left(\frac{d}{dt}\vec{r}'\times\vec{r}''\right) + \vec{r}'\times\left(\vec{r}'\times\frac{d}{dt}\vec{r}''\right)$$

$$+\frac{d}{dt}\vec{r}\times(\vec{r}'\times\vec{r}''') + \vec{r}\times\left(\frac{d}{dt}\vec{r}'\times\vec{r}'''\right) + \vec{r}\times\left(\vec{r}'\times\frac{d}{dt}\vec{r}'''\right)$$

$$= \vec{r}''\times(\vec{r}'\times\vec{r}'') + \vec{r}'\times\left(\vec{r}''\times\vec{r}''\right) + \vec{r}'\times\left(\vec{r}'\times\vec{r}'''\right) + \vec{r}'\times(\vec{r}'\times\vec{r}''')$$

$$+\vec{r}\times\left(\vec{r}''\times\vec{r}'''\right) + \vec{r}\times\left(\vec{r}'\times\vec{r}''''\right)$$

$$= \vec{r}''\times(\vec{r}'\times\vec{r}'') + 2\vec{r}'\times\left(\vec{r}'\times\vec{r}'''\right) + \vec{r}\times\left(\vec{r}''\times\vec{r}'''\right) + \vec{r}\times\left(\vec{r}'\times\vec{r}''''\right)$$

Example 4.2.6

Evaluate $\frac{d}{dt}\left[\frac{1}{r^2}\vec{r} + \frac{r}{\vec{a}\cdot\vec{r}}\vec{b}\right]$, where $\vec{r}$ is a vector function of t, r is its length and $\vec{a}$ is a constant vector.

Solution

Using the rules of differentiation, we get

$$\frac{d}{dt}\left[\frac{1}{r^2}\vec{r} + \frac{r}{\vec{a}\cdot\vec{r}}\vec{b}\right] = \frac{d}{dt}\left[\frac{1}{r^2}\vec{r}\right] + \frac{d}{dt}\left[\frac{r}{\vec{a}\cdot\vec{r}}\vec{b}\right],\ r' = \frac{dr}{dt}$$

$$= \frac{1}{r^4}\left[r^2\,\vec{r}' - 2\,r\,r'\,\vec{r}\right] + \frac{1}{(\vec{a}\cdot\vec{r})^2}\left[(\vec{a}\cdot\vec{r})r'\,\vec{b} - (\vec{a}\cdot\vec{r}')\,r\,\vec{b}\right]$$

$$= \frac{1}{r^2}\,\vec{r}' - \frac{2r'}{r^3}\,\vec{r} + \frac{r'}{\vec{a}\cdot\vec{r}}\,\vec{b} - \frac{r(\vec{a}\cdot\vec{r}')}{(\vec{a}\cdot\vec{r})^2}\,\vec{b}$$

Result 13

Geometrical interpretation of vector derivative:
Let $\vec{r} = \vec{r}(t)$ be a continuous and single-valued vector function of a scalar variable t. With respect to O as origin, $\vec{r}$ and $\vec{r}+\Delta\vec{r}$ are the position vectors of two neighbouring points P and Q, respectively, on the continuous curve $\vec{r} = \vec{r}(t)$ corresponding to the values of t and $t + \Delta t$ of the scalar variable (Fig. 4.2). Then

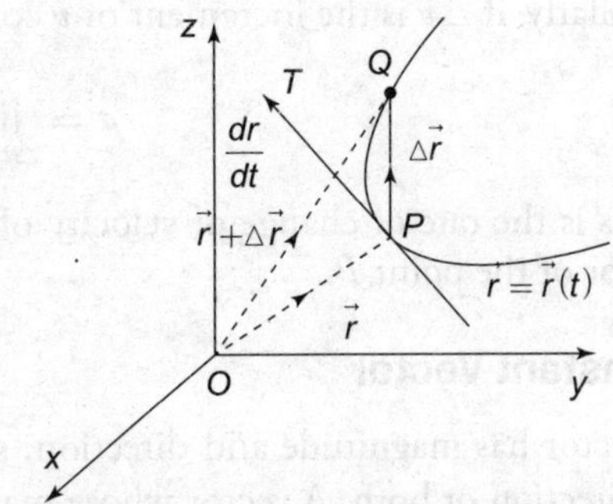

Figure 4.2 Geometrical interpretation of vector derivative

$$\vec{PQ} = \vec{OQ} - \vec{OP}$$

$$= \vec{r} + \Delta\vec{r} - \vec{r} = \Delta\vec{r} \Rightarrow \frac{\Delta\vec{r}}{\Delta t} = \frac{\vec{PQ}}{\Delta t}$$

When $Q \to P$ and $\Delta t \to 0$, the chord PQ tends to the tangent PT to the curve at P. Hence, $\lim\limits_{\Delta t \to 0} \frac{\Delta\vec{r}}{\Delta t} = \frac{d\vec{r}}{dt}$, assuming the existence of the limit. Thus, geometrically, $\frac{d\vec{r}}{dt}$ represents a vector along the tangent to the curve $\vec{r} = \vec{r}(t)$ at P in the sense of increasing t and whose modulus $\left|\frac{d\vec{r}}{dt}\right|$ represents the absolute value of the slope of the tangent at P.

Example 4.2.7

Interpret the result $\vec{a} \times \frac{d\vec{b}}{dt} = \vec{b} \times \frac{d\vec{a}}{dt}$ geometrically.

Solution

From the given relation, we get

$$\vec{a} \times \frac{d\vec{b}}{dt} + \frac{d\vec{a}}{dt} \times \vec{b} = \vec{0} \Rightarrow \frac{d}{dt}(\vec{a} \times \vec{b}) = \vec{0}$$

$$\Rightarrow \vec{a} \times \vec{b} \text{ is a constant vector } = \vec{c}, \text{ say}$$

Then, $\vec{a}$ and $\vec{b}$ are parallel to a fixed plane whose normal is $\vec{c}$, and $\vec{a}$ and $\vec{b}$ represent the adjacent sides of a parallelogram whose vector area is a constant vector and whose scalar area is a constant scalar.

Result 14

Physical interpretation of vector derivative: In Fig. 4.2, we have seen that

$$\frac{d\vec{r}}{dt} = \lim_{\Delta t \to 0} \frac{\Delta\vec{r}}{\Delta t} = \lim_{Q \to P} \frac{\vec{PQ}}{\Delta t}$$

If t denotes time, then $\triangle\vec{r}$ represents displacement during time $\triangle t$, and hence, $\dfrac{d\vec{r}}{dt}$ is the rate of change of displacement of the point P with respect to time t, which is the velocity vector $\vec{v}$ of the point P at any time t. Thus

$$\vec{v} = \lim_{\triangle t \to 0} \frac{\triangle\vec{r}}{\triangle t} = \frac{d\vec{r}}{dt}$$

Similarly, if $\triangle\vec{v}$ is the increment of $\vec{v}$ corresponding to the increment $\triangle t$ of t, then

$$\vec{a} = \lim_{\triangle t \to 0} \frac{\triangle\vec{v}}{\triangle t} = \frac{d\vec{v}}{dt} = \frac{d^2\vec{r}}{dt^2}$$

This is the rate of change of velocity of the point P with respect to time t. $\vec{a}$ is the acceleration vector of the point P.

Constant Vector

A vector has magnitude and direction, so if it changes, there may be a change in its magnitude or direction or both. A vector whose magnitude and direction are fixed is called a *constant vector*. It can easily be shown that differentiation of a constant vector is a null vector. Below are two important theorems on vectors with constant magnitude and vectors with fixed direction.

Theorem 4.1

A differentiable vector function is constant if and only if its derivative vanishes.

Proof: Consider a differentiable vector function $\vec{f}(t)$ of scalar variable t, which is constant. Then, we have $\vec{f}(t+\triangle t) = \vec{f}(t)$. Thus

$$\frac{d\vec{f}}{dt} = \lim_{\triangle t \to 0} \frac{\vec{f}(t+\triangle t) - \vec{f}(t)}{\triangle t} = \lim_{\triangle t \to 0} \frac{\vec{0}}{\triangle t} = \vec{0}; \qquad \text{as } \triangle t \neq 0$$

So the condition is necessary. Conversely, let us assume that $\dfrac{d\vec{f}}{dt} = \vec{0}$ and $\vec{f}(t) = f_1(t)\hat{i} + f_2(t)\hat{j} + f_3(t)\hat{k}$. Therefore

$$\frac{d\vec{f}}{dt} = \frac{df_1}{dt}\hat{i} + \frac{df_2}{dt}\hat{j} + \frac{df_3}{dt}\hat{k} = \vec{0}$$

$$\Rightarrow \quad \frac{df_1}{dt} = 0, \quad \frac{df_2}{dt} = 0, \quad \frac{df_3}{dt} = 0$$

as $\hat{i}, \hat{j}$ and $\hat{k}$ are non-coplanar vectors. This implies that $f_1(t)$, $f_2(t)$ and $f_3(t)$ are constant scalar functions, and consequently, $\vec{f}(t)$ is a constant vector.

Theorem 4.2

The derivative of a non-zero and non-constant differentiable vector function with constant magnitude is orthogonal to itself.

Proof: We consider a differentiable vector function $\vec{f}(t)$ of scalar variable t, whose magnitude is constant. Thus, $f = |\vec{f}(t)| =$ a constant, and hence, we obtain $\dfrac{df}{dt} = 0$. Now

$$\frac{d}{dt}(\vec{f}\cdot\vec{f}) = \frac{d\vec{f}}{dt}\cdot\vec{f} + \vec{f}\cdot\frac{d\vec{f}}{dt}$$

$$\Rightarrow \frac{d}{dt}(|\vec{f}|^2) = 2\vec{f}\cdot\frac{d\vec{f}}{dt} \Rightarrow \frac{d}{dt}(f^2) = 2\vec{f}\cdot\frac{d\vec{f}}{dt}$$

$$\Rightarrow 2f\frac{df}{dt} = 2\vec{f}\cdot\frac{d\vec{f}}{dt} \Rightarrow \vec{f}\cdot\frac{d\vec{f}}{dt} = f\frac{df}{dt} = 0;\ \text{as}\ \frac{df}{dt} = 0$$

Since $\vec{f}(t)$ is non-zero and non-constant, it is orthogonal to its derivative.

Example 4.2.8

For the vector function $\vec{f}(t) = \sin t\hat{i} + \cos t\hat{j} + \hat{k}$ of scalar variable t, verify Theorem 4.2.

Solution

Since $\vec{f}(t) = \sin t\hat{i} + \cos t\hat{j} + \hat{k}$, $|\vec{f}(t)| = \sqrt{\sin^2 t + \cos^2 t + 1} = \sqrt{2} =$ constant. Now

$$\frac{d\vec{f}}{dt} = \frac{d}{dt}(\sin t\hat{i} + \cos t\hat{j} + \hat{k}) = \cos t\hat{i} - \sin t\hat{j}$$

$$\Rightarrow \vec{f}\cdot\frac{d\vec{f}}{dt} = (\sin t\hat{i} + \cos t\hat{j} + \hat{k})\cdot(\cos t\hat{i} - \sin t\hat{j}) = 0$$

This shows that $\frac{d\vec{f}}{dt}$ is orthogonal to $\vec{f}$. Hence, Theorem 4.2 is verified.

Theorem 4.3

Vectors with constant magnitude: A necessary and sufficient condition that a proper vector $\vec{f}(t)$ has a constant magnitude (length) is that $\vec{f}\cdot\frac{d\vec{f}}{dt} = 0$.

Proof: First, let us assume that the vector differentiable function $\vec{f}(t)$ has a constant magnitude. Then, $|\vec{f}(t)| =$ constant, i.e., $|\vec{f}(t)|^2 =$ constant, i.e., $\vec{f}(t)\cdot\vec{f}(t) =$ constant. Differentiating, we get

$$\frac{d}{dt}\left[\vec{f}(t)\cdot\vec{f}(t)\right] = 0$$

$$\Rightarrow \vec{f}\cdot\frac{d\vec{f}}{dt} + \vec{f}\cdot\frac{d\vec{f}}{dt} = 0$$

$$\Rightarrow 2\vec{f}\cdot\frac{d\vec{f}}{dt} = 0 \Rightarrow \vec{f}\cdot\frac{d\vec{f}}{dt} = 0 \tag{4.6}$$

Thus, the condition is necessary. Conversely, if the given condition $\vec{f}\cdot\frac{d\vec{f}}{dt} = 0$ holds, then

$$\vec{f}\cdot\frac{d\vec{f}}{dt} = 0 \Rightarrow \frac{d}{dt}(\vec{f}\cdot\vec{f}) = 0 \Rightarrow \vec{f}\cdot\vec{f} = |\vec{f}(t)|^2 = \text{constant}$$

Hence, the theorem. This theorem is related to vectors of constant magnitude. Also, from this theorem, the derivative of a vector of constant magnitude is perpendicular to that vector.

Theorem 4.4

Vectors with constant direction: The necessary and sufficient condition for a proper vector $\vec{F}(t)$ to have a constant direction is $\vec{F} \times \frac{d\vec{F}}{dt} = \vec{0}$.

Proof: Any vector $\vec{F}(t)$ can be written as $\vec{F}(t) = \phi(t)\vec{f}(t)$, where $\phi(t) = |\vec{F}(t)|$, the magnitude of $\vec{F}(t)$ and accordingly $\vec{f}(t)$ is a vector function with modulus unity for every value of t. Therefore

$$\frac{d\vec{F}}{dt} = \phi(t)\frac{d\vec{f}}{dt} + \frac{d\vec{\phi}}{dt}\vec{f}$$

Multiplying vectorically by $\vec{F}(t)$, we get

$$\begin{aligned} \vec{F} \times \frac{d\vec{F}}{dt} &= \phi(t)\vec{f}(t) \times \left[\phi(t)\frac{d\vec{f}}{dt} + \frac{d\vec{\phi}}{dt}\vec{f}\right] \\ &= \phi^2(t)\vec{f}(t) \times \frac{d\vec{f}}{dt} + \phi(t)\frac{d\vec{\phi}}{dt}\vec{f}(t) \times \vec{f}(t) \\ &= \phi^2\vec{f} \times \frac{d\vec{f}}{dt};\ \text{as } \vec{f}(t) \times \vec{f}(t) = \vec{0} \end{aligned} \tag{4.7}$$

When $\vec{F}(t)$ remains parallel to a fixed line, $\vec{f}(t)$ is a constant vector, and hence, $\frac{d\vec{f}}{dt} = \vec{0}$ and as such the condition follows necessarily. Conversely, since $\vec{F}(t) \neq \vec{0}$, using Eq. (4.7), the given condition leads to

$$\vec{F} \times \frac{d\vec{F}}{dt} = \vec{0} = \phi^2\vec{f} \times \frac{d\vec{f}}{dt} \Rightarrow \vec{f} \times \frac{d\vec{f}}{dt} = \vec{0}$$

However, $\vec{f}$ has a constant length unity. Hence, by Theorem 4.3, we get $\vec{f} \cdot \frac{d\vec{f}}{dt} = 0$. These two equations are contradictory unless $\frac{d\vec{f}}{dt} = \vec{0}$; that is, the unit vector $\vec{f}$ is constant, which means that $\vec{F}$ is parallel to a fixed line.

Example 4.2.9

For the vector function $\vec{F}(t) = e^t(\hat{i} + \hat{j} + \hat{k})$ of scalar variable t, verify Theorem 4.4.

Solution

We see that the given vector $\vec{F}(t) = e^t(\hat{i} + \hat{j} + \hat{k})$ is parallel to the constant vector $\hat{i} + \hat{j} + \hat{k}$, and hence, the vector function has constant direction. Now

$$\begin{aligned} \frac{d\vec{F}}{dt} &= \frac{d}{dt}\{e^t(\hat{i} + \hat{j} + \hat{k})\} = e^t(\hat{i} + \hat{j} + \hat{k}) \\ \Rightarrow \vec{F} \times \frac{d\vec{F}}{dt} &= e^t(\hat{i} + \hat{j} + \hat{k}) \times e^t(\hat{i} + \hat{j} + \hat{k}) = \vec{0} \end{aligned}$$

Hence, Theorem 4.4 is verified.

4.2.3 Partial Derivatives

In the case of a vector function which depends on several independent scalar variables, we can define the partial derivatives. Let us consider a vector function $\vec{F} = \vec{F}(x, y, z)$, which depends on three independent scalar variables x, y and z. Then, the *first order partial derivative* of $\vec{F}(x, y, z)$ with respect to x is defined as

$$\vec{F}_x = \frac{\partial \vec{F}}{\partial x} = \lim_{\triangle x \to 0} \frac{\vec{F}(x + \triangle x, y, z) - \vec{F}(x, y, z)}{\triangle x}$$

provided the limit exists finitely. If $\vec{F}(x, y, z) = F_1(x, y, z)\hat{i} + F_2(x, y, z)\hat{j} + F_3(x, y, z)\hat{k}$, then

$$\frac{\partial \vec{F}}{\partial x} = \frac{\partial F_1}{\partial x}\,\hat{i} + \frac{\partial F_2}{\partial x}\,\hat{j} + \frac{\partial F_1}{\partial z}\,\hat{k}$$

Similarly, the partial derivatives of $\vec{F}$ with respect to y and z are, respectively, defined as

$$\frac{\partial \vec{F}}{\partial y} = \lim_{\triangle y \to 0} \frac{\vec{F}(x, y + \triangle y, z) - \vec{F}(x, y, z)}{\triangle y}; \frac{\partial \vec{F}}{\partial z} = \lim_{\triangle z \to 0} \frac{\vec{F}(x, y, z + \triangle z) - \vec{F}(x, y, z)}{\triangle z}$$

provided the limit exists finitely. Similarly, we can define higher order partial derivatives of $\vec{F}$, such as

$$\frac{\partial^2 \vec{F}}{\partial x^2} = \frac{\partial}{\partial x}\left(\frac{\partial \vec{F}}{\partial x}\right) = \lim_{\triangle x \to 0} \frac{\vec{F}_x(x + \triangle x, y, z) - \vec{F}_x(x, y, z)}{\triangle x}$$

$$\frac{\partial^2 \vec{F}}{\partial x \partial y} = \frac{\partial}{\partial x}\left(\frac{\partial \vec{F}}{\partial y}\right) = \lim_{\triangle x \to 0} \frac{\vec{F}_y(x + \triangle x, y, z) - \vec{F}_y(x, y, z)}{\triangle x}$$

$$\frac{\partial^2 \vec{F}}{\partial y \partial x} = \frac{\partial}{\partial y}\left(\frac{\partial \vec{F}}{\partial x}\right) = \lim_{\triangle y \to 0} \frac{\vec{F}_x(x, y + \triangle y, z) - \vec{F}_x(x, y, z)}{\triangle y}$$

provided the limit exists finitely. In a similar manner, we can define partial derivatives of third and higher orders. If $\vec{F}$ is a vector function of two variables x and y such that $\dfrac{\partial^2 \vec{F}}{\partial x \partial y}$ and $\dfrac{\partial^2 \vec{F}}{\partial y \partial x}$ are both continuous functions, then $\dfrac{\partial^2 \vec{F}}{\partial x \partial y} = \dfrac{\partial^2 \vec{F}}{\partial y \partial x}$. If $\vec{F}$, $\vec{G}$ and $\vec{H}$ are vector functions of two scalar variables x and y, and ϕ is a scalar function of x and y, then

(i) $\dfrac{d}{dt}(\phi\,\vec{F}) = \dfrac{d\phi}{dt}\,\vec{F} + \phi\,\dfrac{d\vec{F}}{dt}$

(ii) $\dfrac{\partial}{\partial x}(\vec{F} \pm \vec{G}) = \dfrac{\partial \vec{F}}{\partial x} + \dfrac{\partial \vec{G}}{\partial x}$ and $\dfrac{\partial}{\partial y}(\vec{F} \pm \vec{G}) = \dfrac{\partial \vec{F}}{\partial y} + \dfrac{\partial \vec{G}}{\partial y}$

(iii) $\dfrac{\partial}{\partial x}(\vec{F} \cdot \vec{G}) = \dfrac{\partial \vec{F}}{\partial x} \cdot \vec{G} + \vec{F} \cdot \dfrac{\partial \vec{G}}{\partial x}$ and $\dfrac{\partial}{\partial y}(\vec{F} \cdot \vec{G}) = \dfrac{\partial \vec{F}}{\partial y} \cdot \vec{G} + \vec{F} \cdot \dfrac{\partial \vec{G}}{\partial y}$

(iv) $\dfrac{\partial}{\partial x}(\vec{F} \times \vec{G}) = \dfrac{\partial \vec{F}}{\partial x} \times \vec{G} + \vec{F} \times \dfrac{\partial \vec{G}}{\partial x}$ and $\dfrac{\partial}{\partial y}(\vec{F} \times \vec{G}) = \dfrac{\partial \vec{F}}{\partial y} \times \vec{G} + \vec{F} \times \dfrac{\partial \vec{G}}{\partial y}$

(v) $\dfrac{d}{dt}\left[\vec{F}\,\vec{G}\,\vec{H}\right] = \left[\dfrac{d\vec{F}}{dt}\,\vec{G}\,\vec{H}\right] + \left[\vec{F}\,\dfrac{d\vec{G}}{dt}\,\vec{H}\right] + \left[\vec{F}\,\vec{G}\,\dfrac{d\vec{H}}{dt}\right]$

(vi) $\dfrac{d}{dt}\left\{\vec{F} \times (\vec{G} \times \vec{H})\right\} = \dfrac{d\vec{F}}{dt} \times (\vec{G} \times \vec{H}) + \vec{F} \times \left(\dfrac{d\vec{G}}{dt} \times \vec{H}\right) + \vec{F} \times \left(\vec{G} \times \dfrac{d\vec{H}}{dt}\right)$

Similarly, we can define the total differential of $\vec{F}$ as

$$d\vec{F} = \frac{\partial \vec{F}}{\partial x}\,dx + \frac{\partial \vec{F}}{\partial y}\,dy + \frac{\partial \vec{F}}{\partial z}\,dz$$

Let $\vec{F} = F_1(x,y,z)\hat{i} + F_2(x,y,z)\hat{j} + F_3(x,y,z)\hat{k}$ and $x = x(t), y = y(t), z = z(t)$ be scalar functions of t having continuous derivatives. Then, the total differential coefficient is given by

$$\frac{d\vec{F}}{dt} = \frac{\partial \vec{F}}{\partial x}\frac{dx}{dt} + \frac{\partial \vec{F}}{\partial y}\frac{dy}{dt} + \frac{\partial \vec{F}}{\partial z}\frac{dz}{dt}$$

Example 4.2.10

If $\phi(x,y,z) = xy^2 z$ and $\vec{F} = xz\hat{i} - xy^2\hat{j} + yz^2\hat{k}$, evaluate $\dfrac{\partial^3}{\partial x^2\,\partial z}(\phi\,\vec{F})$ at $(2,-1,1)$.

Solution

For the given ϕ and $\vec{F}$, we have

$$\frac{\partial}{\partial z}(\phi\vec{F}) = \frac{\partial}{\partial z}\left[x^2y^2z^2\hat{i} - x^2y^4z\hat{j} + xy^3z^3\hat{k}\right]$$

$$= 2x^2y^2z\hat{i} - x^2y^4\hat{j} + 3xy^3z^2\hat{k}$$

$$\frac{\partial^2}{\partial x\partial z}(\phi\vec{F}) = \frac{\partial}{\partial x}\left[2x^2y^2z\hat{i} - x^2y^4\hat{j} + 3xy^3z^2\hat{k}\right]$$

$$= 4xy^2z\,\hat{i} - 2xy^4\,\hat{j} + 3y^3z^2\,\hat{k}$$

$$\frac{\partial^3}{\partial x^2\partial z}(\phi\vec{F}) = \frac{\partial}{\partial x}\left[4xy^2z\,\hat{i} - 2xy^4\,\hat{j} + 3y^3z^2\,\hat{k}\right]$$

$$= 4y^2z\,\hat{i} - 2y^4\,\hat{j} = 4\hat{i} - 2\hat{j}, \text{ at } (2,-1,1)$$

Example 4.2.11

Show that the function $\vec{u}(x,y) = e^{-\lambda x}(\vec{c}_1 \sin\lambda y + \vec{c}_2 \cos\lambda y)$ satisfies the partial differential equation $\dfrac{\partial^2\vec{u}}{\partial x^2} + \dfrac{\partial^2\vec{u}}{\partial y^2} = \vec{0}$.

Solution

Since $\vec{u}(x,y) = e^{-\lambda x}(\vec{c}_1 \sin\lambda y + \vec{c}_2 \cos\lambda y)$, we get

$$\frac{\partial\vec{u}}{\partial x} = -\lambda e^{-\lambda x}(\vec{c}_1 \sin\lambda y + \vec{c}_2 \cos\lambda y) = -\lambda\vec{u}; \quad \frac{\partial^2\vec{u}}{\partial x^2} = -\lambda\frac{\partial\vec{u}}{\partial x} = -\lambda(-\lambda\vec{u}) = \lambda^2\vec{u}$$

$$\frac{\partial\vec{u}}{\partial y} = \lambda e^{-\lambda x}(\vec{c}_1 \cos\lambda y - \vec{c}_2 \sin\lambda y); \quad \frac{\partial^2\vec{u}}{\partial y^2} = \lambda^2 e^{-\lambda x}(-\vec{c}_1 \sin\lambda y - \vec{c}_2 \cos\lambda y) = -\lambda^2\vec{u}$$

Hence, $\vec{u}(x,y)$ satisfies the Laplace equation $\dfrac{\partial^2 \vec{u}}{\partial x^2} + \dfrac{\partial^2 \vec{u}}{\partial y^2} = \vec{0}$.

Example 4.2.12

If $\vec{F} = xyz\hat{i} + xz^2\hat{j} - y^3\hat{k}$ and $\vec{G} = x^3\hat{i} - xyz\hat{j} + x^2z\hat{k}$, evaluate

(i) $\dfrac{\partial^2 \vec{F}}{\partial x \partial y}, \dfrac{\partial^2 \vec{G}}{\partial x \partial y}$ at the origin (ii) $\dfrac{\partial^2 \vec{F}}{\partial y^2} \times \dfrac{\partial^2 \vec{G}}{\partial x^2}$ at $(1,1,0)$

Solution

Since $\vec{F} = xyz\hat{i} + xz^2\hat{j} - y^3\hat{k}$, we have

$$\frac{\partial \vec{F}}{\partial x} = \frac{\partial}{\partial x}\left[xyz\hat{i} + xz^2\hat{j} - y^3\hat{k}\right] = yz\hat{i} + z^2\hat{j}, \quad \frac{\partial^2 \vec{F}}{\partial x^2} = \frac{\partial}{\partial x}\left[yz\hat{i} + z^2\hat{j}\right] = \vec{0}$$

$$\frac{\partial \vec{F}}{\partial y} = \frac{\partial}{\partial y}\left[xyz\hat{i} + xz^2\hat{j} - y^3\hat{k}\right] = xz\hat{i} - 3y^2\hat{k}, \quad \frac{\partial^2 \vec{F}}{\partial y^2} = \frac{\partial}{\partial y}\left[xz\hat{i} - 3y^2\hat{k}\right] = -6y\hat{k}$$

$$\frac{\partial^2 \vec{F}}{\partial x \partial y} = \frac{\partial}{\partial x}\left[xz\hat{i} - 3y^2\hat{k}\right] = z\hat{i} = \vec{0}, \text{ at origin}$$

$$\frac{\partial^2 \vec{F}}{\partial y \partial x} = \frac{\partial}{\partial y}\left[yz\hat{i} + z^2\hat{j}\right] = \vec{0}, \text{ at origin}$$

Since $\vec{G} = x^3\hat{i} - xyz\hat{j} + x^2z\hat{k}$, we have

$$\frac{\partial \vec{G}}{\partial x} = \frac{\partial}{\partial x}\left[x^3\hat{i} - xyz\hat{j} + x^2z\hat{k}\right] = 3x^2\hat{i} - yz\hat{j} + 2xz\hat{k}$$

$$\frac{\partial^2 \vec{G}}{\partial x^2} = \frac{\partial}{\partial x}\left[3x^2\hat{i} - yz\hat{j} + 2xz\hat{k}\right] = 6x\hat{i} + 2z\hat{k}$$

$$\frac{\partial \vec{G}}{\partial y} = \frac{\partial}{\partial y}\left[x^3\hat{i} - xyz\hat{j} + x^2z\hat{k}\right] = -xz\hat{j}, \quad \frac{\partial^2 \vec{G}}{\partial y^2} = \frac{\partial}{\partial y}\left[-xz\hat{j}\right] = \vec{0}$$

$$\frac{\partial^2 \vec{G}}{\partial x \partial y} = \frac{\partial}{\partial x}\left[-xz\hat{j}\right] = -z\hat{j} = \vec{0}, \text{ at origin}$$

$$\frac{\partial^2 \vec{G}}{\partial y \partial x} = \frac{\partial}{\partial y}\left[3x^2\hat{i} - yz\hat{j} + 2xz\hat{k}\right] = -z\hat{j} + 2x\hat{k} = \vec{0}, \text{ at origin}$$

Therefore

$$\frac{\partial^2 \vec{F}}{\partial y^2} \times \frac{\partial^2 \vec{G}}{\partial x^2} = -6y\,\hat{k} \times (6x\hat{i} + 2z\hat{k}) = -36xy\hat{k} \times \hat{i} - 12yz\hat{k} \times \hat{k}$$

$$= -36xy\hat{j} = -36\hat{j}, \text{ at } (1,1,0)$$

4.2.4 Directional Derivative

The directional derivative of a function at a point P is the rate of variation of the function in a particular direction at P. If P is any point in the region of definition of the scalar point function

$f(P)$ and PQ is the line segment drawn from P on one side of P only, then the directional derivative of $f(P)$ at P in the direction PQ is defined by the scalar quantity

$$\lim_{Q \to P} \frac{f(Q) - f(P)}{PQ}$$

provided the limit exists. The directional derivative of a vector point function, which is also a vector quantity, may be similarly defined.

In a three-dimensional space, consider a system of rectangular Cartesian axes ox, oy and oz through the point O. We may take the line segment PQ joining $P(x, y, z)$ and $Q(x + \triangle x, y + \triangle y, z + \triangle z)$ at a distance $\triangle s$ apart. This makes angles α, β and γ with the x, y and z axes, respectively, whose direction cosines are l, m and n. Then

$$\frac{\triangle x}{\triangle s} = \cos\alpha = l, \quad \frac{\triangle y}{\triangle s} = \cos\beta = m, \quad \frac{\triangle z}{\triangle s} = \cos\gamma = n$$

where $\triangle s = \sqrt{(\triangle x)^2 + (\triangle y)^2 + (\triangle z)^2}$. Thus, the directional derivative of the scalar point function $f(P)$ at a given point P in the direction of PQ is

$$\begin{aligned}
\frac{df}{ds} &= \lim_{Q \to P} \frac{f(Q) - f(P)}{PQ} = \lim_{\triangle s \to 0} \frac{\triangle f}{\triangle s} \\
&= \lim_{\triangle s \to 0} \frac{\triangle f(x + \triangle x, y + \triangle y, z + \triangle z) - f(x, y, z)}{\triangle s} \\
&= \lim_{\triangle s \to 0} \frac{1}{\triangle s}\left[\frac{\partial f}{\partial x}\triangle x + \frac{\partial f}{\partial y}\triangle y + \frac{\partial f}{\partial z}\triangle z + \varepsilon_1 \triangle x + \varepsilon_2 \triangle y + \varepsilon_3 \triangle z\right] \\
&= \frac{\partial f}{\partial x}\lim_{\triangle s \to 0}\frac{\triangle x}{\triangle s} + \frac{\partial f}{\partial y}\lim_{\triangle s \to 0}\frac{\triangle y}{\triangle s} + \frac{\partial f}{\partial z}\lim_{\triangle s \to 0}\frac{\triangle z}{\triangle s} \\
&= \frac{\partial f}{\partial x}\frac{dx}{ds} + \frac{\partial f}{\partial y}\frac{dy}{ds} + \frac{\partial f}{\partial z}\frac{dz}{ds} \\
&= l\, f_x(x, y, z) + m\, f_y(x, y, z) + n\, f_z(x, y, z) \qquad (4.8)
\end{aligned}$$

The expression (4.8) represents the directional derivative of a scalar point function at a given point $P(x, y, z)$ and in the direction of the unit vector $\hat{\alpha} = (l, m, n)$.

(i) If the line PQ is parallel to the x axis, then $\hat{\alpha} = (1, 0, 0)$ and the directional derivative in the direction parallel to the x axis is $f_x(x, y, z)$ and those parallel to the y and z axes are, respectively, $f_y(x, y, z)$ and $f_z(x, y, z)$.

(ii) The expression of the directional derivative of a vector point function $\vec{F}(P)$ at the point P in the direction of the vector $\vec{\alpha}$ may similarly derived.

(iii) Let $\phi(x, y, z)$ be the temperature at a point P in a certain region and Q in the neighbouring point. Then, the directional derivative $\frac{d\phi}{ds}$ represents the rate of change of temperature with respect to distance at P in a direction towards Q.

Example 4.2.13

Let $f = x^3 + y^3 + z^3$; find the directional derivative of f at $(1, -1, 2)$ in the direction of the vector $\hat{\imath} + 2\hat{\jmath} + \hat{k}$.

Solution

The unit vector of $\hat{\imath} + 2\hat{\jmath} + \hat{k}$ is $(\hat{\imath} + 2\hat{\jmath} + \hat{k})/\sqrt{6} = (l, m, n)$. From Eq. (4.8), the directional derivative is given by

$$\frac{df}{ds} = (l\hat{\imath} + m\hat{\jmath} + n\hat{k}) \cdot (f_x\hat{\imath} + f_y\hat{\jmath} + f_z\hat{k})$$
$$= \frac{1}{\sqrt{6}}(\hat{\imath} + 2\hat{\jmath} + \hat{k}) \cdot (3x^2\hat{\imath} + 3y^2\hat{\jmath} + 3z^2\hat{k})$$
$$= \frac{3}{\sqrt{6}}\left[1 \cdot 1^2 + 2 \cdot (-1)^2 + 1 \cdot 2^2\right] = \frac{7}{2}\sqrt{6}$$

4.3 APPLICATIONS IN DIFFERENTIAL GEOMETRY

Differential geometry involves the study of space curves and surfaces defined on a space. In this section, we shall discuss a few elementary concepts of differential geometry of curves.

4.3.1 Space Curve

A *space curve* is an aggregate of points whose coordinates are functions of a single variable t, called a *parameter*, and is represented by $\vec{r} = \vec{r}(t)$. A space curve can also be described in terms of arc length s, measured from the fixed point on the curve up to a variable point on it. If $\vec{r}(s)$ is the position vector of a point $P(x, y, z)$ with respect to the origin O, then as s changes, P describes a space curve whose position vector is given by $x = x(s), y = y(s)$ and $z = z(s)$. Thus, a space curve can be represented by the locus of the point P, whose position vector is given by

$$\vec{r}(s) = x(s)\,\hat{\imath} + y(s)\,\hat{\jmath} + z(s)\,\hat{k}; \quad s_0 \le s \le s_1 \tag{4.9}$$

$x(s), y(s)$ and $z(s)$ have continuous derivatives of all orders. For example,

(i) The straight line through the point $(\alpha_1, \alpha_2, \alpha_3)$ with proportional direction cosines (l, m, n) has the parametric representation $x = lt + \alpha_1, y = mt + \alpha_2$ and $z = nt + \alpha_3$.

(ii) The circle in the xy plane with the centre at the point $(\alpha_1, \alpha_2, 0)$ and radius a has the parametric equation $x = \alpha_1 + a\cos t,\ \ y = \alpha_2 + a\sin t$ and $z = 0$.

(iii) The parametric equations $x = a\cos t, y = a\sin t$ and $z = bt (b \neq 0)$ represents a curve, known as a *circular helix*. This curve is an example of a space curve which is *twisted*, that is, which does not lie on a plane.

(iv) The parametric equations $x = at, y = bt^2, z = ct^3 (abc \neq 0)$ represent a curve, known as a *twisted cubic*. This curve is also an example of a space curve which is *twisted*, that is, which does not lie on a plane.

Tangent at a point

Let $P(x, y, z)$ be any point on the space curve $\vec{r} = \vec{r}(s)$ and Q be a neighbouring point of P on the curve (Fig. 4.3), where $\widehat{AP} = s$ and $\widehat{AQ} = s + \Delta s$, A being the fixed point on the curve; then $\widehat{PQ} = \Delta s$ and $\vec{PQ} = \Delta\vec{r}$. The tangent vector to the space curve at P is $\lim\limits_{\Delta s \to 0} \dfrac{\Delta\vec{r}}{\Delta s} = \dfrac{d\vec{r}}{ds}$, assuming the existence of the limit. Differentiating $\vec{r}(s) = x(s)\,\hat{\imath} + y(s)\,\hat{\jmath} + z(s)\,\hat{k}$ with respect to the parameter s, we get

$$\frac{d\vec{r}}{ds} = \frac{dx}{ds}\hat{i} + \frac{dy}{ds}\hat{j} + \frac{dz}{ds}\hat{k} = \vec{T}, \text{ say} \quad (4.10)$$

This is a vector in the direction of the tangent at P. Further

$$\left|\frac{d\vec{r}}{ds}\right| = \sqrt{\left(\frac{dx}{ds}\right)^2 + \left(\frac{dy}{ds}\right)^2 + \left(\frac{dz}{ds}\right)^2}$$

$$= \sqrt{\frac{(ds)^2}{(ds)^2}} = 1$$

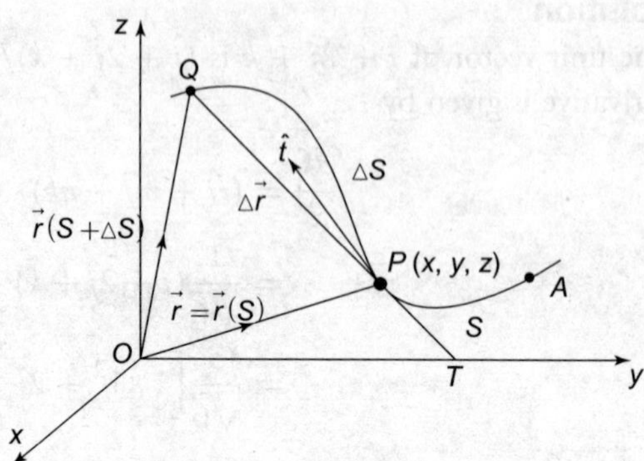

Figure 4.3 Tangent vector

Hence, $\frac{d\vec{r}}{ds}$ is a unit tangent vector in the direction of the tangent at P and is denoted by $\hat{t}$. The direction cosines of the tangent are $\frac{dx}{ds}, \frac{dy}{ds}$ and $\frac{dz}{ds}$. If $\frac{d\vec{r}}{ds}$ is not a unit vector, then the unit tangent vector is given by

$$\hat{t} = \frac{1}{\left|\frac{d\vec{r}}{ds}\right|}\frac{d\vec{r}}{ds} = \frac{1}{\sqrt{\left(\frac{dx}{ds}\right)^2 + \left(\frac{dy}{ds}\right)^2 + \left(\frac{dz}{ds}\right)^2}}\left[\frac{dx}{ds}\hat{i} + \frac{dy}{ds}\hat{j} + \frac{dz}{ds}\hat{k}\right]$$

Let the equation of the curve be represented in terms of a parameter t of the form $\vec{r} = \vec{f}(t)$, other than the arc length s. Then, the unit tangent vector is given by

$$\hat{t} = \frac{d\vec{r}}{ds} = \frac{d\vec{r}}{dt}\Big/\frac{ds}{dt}$$

Let $\vec{R}$ be any point on the tangent line. Then, the equation of the tangent line at P to the curve is given by

$$(\vec{R} - \vec{r}) \times \hat{t} = \vec{0}$$

$$\Rightarrow \vec{R} - \vec{r} = \lambda\,\hat{t} \Rightarrow \vec{R} = \vec{r} + \lambda\,\hat{t}, \; \lambda = \text{ scalar} \quad (4.11)$$

Example 4.3.1

Given an equation of space curve $x = t, y = t^2, z = \frac{2}{3}t^3$. Determine

(i) the tangent vector to any point on the space curve.
(ii) the unit tangent vector to the space curve at any point on the curve.
(iii) the unit tangent vector to the space curve at the point on the curve where $t = 1$.
(iv) the equation of the tangent at the point on the curve where $t = 1$.

Solution

The position vector at any point on the space curve is $\vec{r} = t\hat{i} + t^2\hat{j} + \frac{2}{3}t^3\hat{k}$. The position vector of the point where $t = 1$ on the curve is $\vec{r}_0 = \hat{i} + \hat{j} + \frac{2}{3}\hat{k}$.

(i) A tangent vector to the curve at any point is

$$\vec{T} = \frac{d\vec{r}}{dt} = \frac{d}{dt}\left[t\hat{i} + t^2\hat{j} + \frac{2}{3}t^3\hat{k}\right] = \hat{i} + 2t\hat{j} + 2t^2\hat{k}$$

(ii) The magnitude of the vector is

$$|\vec{T}| = \left|\frac{d\vec{r}}{dt}\right| = \sqrt{1^2 + (2t)^2 + (2t^2)^2} = 1 + 2t^2$$

However, $\frac{d\vec{r}}{dt} = \frac{d\vec{r}}{ds}\frac{ds}{dt} = \hat{t}\,\frac{ds}{dt}$, so

$$\left|\frac{d\vec{r}}{dt}\right| = |\hat{t}|\,\frac{ds}{dt} \Rightarrow \frac{ds}{dt} = \left|\frac{d\vec{r}}{dt}\right| = 1 + 2t^2$$

Then, the required unit tangent vector is given by

$$\hat{t} = \frac{d\vec{r}}{dt}\bigg/\frac{ds}{dt} = \frac{1}{1+2t^2}\left[\hat{i} + 2t\hat{j} + 2t^2\hat{k}\right]$$

(iii) The unit tangent vector at the point on the curve where $t = 1$ is given by

$$\hat{t} = \frac{1}{1 + 2\cdot 1^2}\left[\hat{i} + 2\cdot 1\hat{j} + 2\cdot 1^2\hat{k}\right] = \frac{1}{3}\left[\hat{i} + 2\hat{j} + 2\hat{k}\right]$$

(iv) Let $\vec{R}$ be any point on the tangent at the point $\vec{r}_0$ on the space curve. Then

$$(\vec{R} - \vec{r}_0) \times \hat{t} = \vec{0} \Rightarrow \left\{\vec{R} - \left(\hat{i} + \hat{j} + \frac{2}{3}\hat{k}\right)\right\} \times \frac{1}{3}\left[\hat{i} + 2\hat{j} + 2\hat{k}\right] = \vec{0}$$

$$\Rightarrow \vec{R} = \left(\hat{i} + \hat{j} + \frac{2}{3}\hat{k}\right) + \lambda\left(\hat{i} + 2\hat{j} + 2\hat{k}\right);\ \lambda = \text{ scalar}$$

Normal at a Point

For the space curve $\vec{r} = \vec{r}(s)$, we know the unit tangent $\hat{t}$ is a vector of constant length (Fig. 4.4). Then

$$\hat{t}\cdot\hat{t} = 1 \Rightarrow \frac{d}{ds}(\hat{t}\cdot\hat{t}) = 0 \Rightarrow \hat{t}\cdot\frac{d\hat{t}}{ds} = 0$$

Hence, the rate at which $\hat{t}$ changes with s, i.e., $\frac{d\hat{t}}{ds}$, which is measured as curvature of the curve at the point P, is perpendicular to the unit tangent vector $\hat{t}$ and is thus the direction of the normal to the curve $\vec{r} = \vec{r}(s)$ at the point P. If $\hat{n}$ is the unit normal vector, called the *principal normal*, at the point P, then we can write

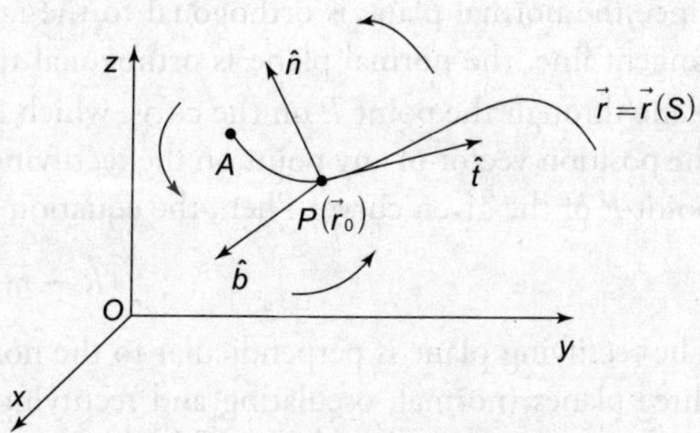

Figure 4.4 Principal normal and binormal

$$\frac{d\hat{t}}{ds} \times \hat{n} = \vec{0} \Rightarrow \frac{d\hat{t}}{ds} = \frac{d^2\vec{r}}{ds^2} = \kappa\,\hat{n},\ \kappa = \text{ positive scalar} \qquad (4.12)$$

The quantity κ is called the *curvature* and its reciprocal ρ is called the *radius of curvature* of the curve at P. This derivative reveals how fast the unit tangent vector is changing direction as a point moves along the curve. Let $\vec{R}$ be the position vector of any point on the normal. Then, the equation of the normal at $P(\vec{r}_0)$ to the curve is given by

$$(\vec{R}-\vec{r}_0)\times\hat{n}=\vec{0}$$
$$\Rightarrow \vec{R}-\vec{r}_0=\mu\,\hat{n}\Rightarrow\vec{R}=\vec{r}_0+\mu\,\hat{n},\quad \mu=\text{ scalar} \tag{4.13}$$

The *osculating plane* or *the plane of curvature* to a curve at a point P (Fig. 4.5) is the plane containing the tangent line and principal normal at P. Let $\vec{R}$ be the position vector of any point on the osculating plane and $\vec{r}_0$ be the position vector of a specified point P of the given curve. The equation of the *osculating plane* is given by

$$\left[\vec{R}-\vec{r}_0\;\hat{t}\;\hat{n}\right]=0\Rightarrow\left[\vec{R}\;\hat{t}\;\hat{n}\right]=\left[\vec{r}_0\;\hat{t}\;\hat{n}\right] \tag{4.14}$$

The plane through a point P on the curve, perpendicular to the tangent at P, is called the *normal plane* at that point (Fig. 4.5). From Eq. (4.14), we see that the normal plane is orthogonal to the osculating plane at any point. Let $\vec{R}$ be the position vector of any point on the normal plane and $\vec{r}_0$ be the position vector of a specified point P on the given curve. Then, the equation of the normal plane is given by

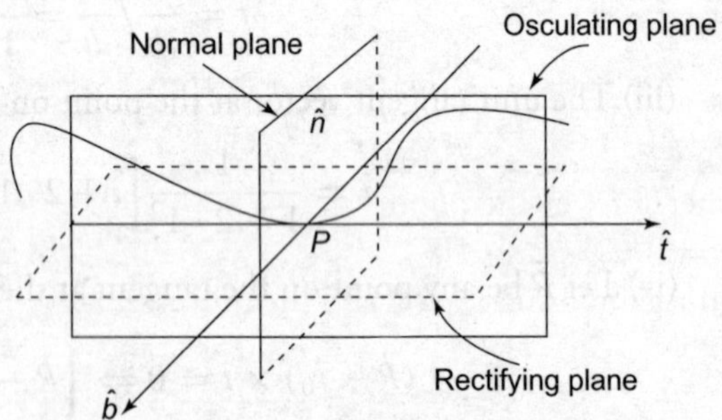

Figure 4.5 Three fundamental planes

$$(\vec{R}-\vec{r}_0)\cdot\hat{t}=(\vec{R}-\vec{r}_0)\cdot\frac{d\vec{r}}{ds}=0 \tag{4.15}$$

Since the normal plane is orthogonal to the tangent line and the osculating plane contains the tangent line, the normal plane is orthogonal to the osculating plane. The *rectifying plane* is the plane through the point P on the curve which is perpendicular to the principal normal. Let $\vec{R}$ be the position vector of any point on the rectifying plane and $\vec{r}_0$ be the position vector of a specified point P of the given curve. Then, the equation of the plane is given by

$$(\vec{R}-\vec{r}_0)\cdot\hat{n}=0 \tag{4.16}$$

The rectifying plane is perpendicular to the normal plane as well as to the osculating plane. The three planes (normal, osculating and rectifying) are orthogonal to each other. These planes are known as the *fundamental planes* for a space curve at any point.

Example 4.3.2

Given an equation of space curve as in Example (4.3.1). Determine
(i) the curvature at the point where $t=1$ on the space curve.
(ii) the unit normal vector to the space curve at the point where $t=1$ on the curve.
(iii) the equation of the normal at the point on the curve where $t=1$.
(iv) the equations of the osculating, normal and rectifying planes.

Solution

The position vector at any point on the space curve is $\vec{r}=t\hat{i}+t^2\hat{j}+\frac{2}{3}t^3\hat{k}$.

(i) From Example (4.3.1), we get $\hat{t}=\frac{1}{1+2t^2}\left[\hat{i}+2t\hat{j}+2t^2\hat{k}\right]$ and so

$$\frac{d\hat{t}}{dt} = \frac{1}{(1+2t^2)^2}\Big[(1+2t^2)(2\hat{j}+4t\hat{k}) - (\hat{i}+2t\hat{j}+2t^2\hat{k})(4t)\Big]$$
$$= \frac{1}{(1+2t^2)^2}\Big[-4t\hat{i}+(2-4t^2)\hat{j}+4t\hat{k}\Big]$$
$$\Rightarrow \frac{d\hat{t}}{ds} = \frac{d\hat{t}}{dt}\Big/\frac{ds}{dt} = \frac{1}{(1+2t^2)^3}\Big[-4t\hat{i}+(2-4t^2)\hat{j}+4t\hat{k}\Big]$$

Using Eq. (4.12), we get

$$\left|\kappa\,\hat{n}\right| = \left|\frac{d\hat{t}}{ds}\right| = \frac{\sqrt{(-4t)^2+(2-4t^2)^2+(4t)^2}}{(1+2t^2)^3}$$

$$\Rightarrow \kappa = \frac{\sqrt{16t^2+4+16t^4-16t^2+16t^2}}{(1+2t^2)^3} = \frac{2(1+2t^2)}{(1+2t^2)^3} = \frac{2}{(1+2t^2)^2}$$

Therefore, at the point where $t=1$, the curvature is $\kappa = 2/9$ and the radius of curvature is $\rho = 9/2$.

(ii) Again, using Eq. (4.12), the unit normal at any point on the curve is given by

$$\hat{n} = \frac{1}{\kappa}\frac{d\hat{t}}{ds} = \frac{(1+2t^2)^2}{2}\cdot\frac{1}{(1+2t^2)^3}\Big[-4t\hat{i}+(2-4t^2)\hat{j}+4t\hat{k}\Big]$$
$$= \frac{1}{1+2t^2}\Big[-2t\hat{i}+(1-2t^2)\hat{j}+2t\hat{k}\Big]$$

At the point on the curve where $t=1$, the unit normal is given by

$$\hat{n} = \frac{1}{1+2\cdot 1^2}\Big[-2\cdot 1\hat{i}+(1-2\cdot 1^2)\hat{j}+2\cdot 1\hat{k}\Big] = \frac{1}{3}(-2\hat{i}-\hat{j}+2\hat{k})$$

(iii) Let $\vec{R}$ be the position vector of any point on the normal to the space curve at $\vec{r}_0$. Then, from Eq. (4.13), we get

$$(\vec{R}-\vec{r}_0)\times\hat{n} = \vec{0} \Rightarrow \left\{\vec{R}-\left(\hat{i}+\hat{j}+\frac{2}{3}\hat{k}\right)\right\}\times\frac{1}{3}(-2\hat{i}-\hat{j}+2\hat{k}) = \vec{0}$$
$$\Rightarrow \vec{R} = \left(\hat{i}+\hat{j}+\frac{2}{3}\hat{k}\right)+\mu(-2\hat{i}-\hat{j}+2\hat{k});\ \mu = \text{scalar}$$

(iv) Using Eq. (4.14), the equation of the osculating plane is $\left[\vec{R}\ \hat{t}\ \hat{n}\right] = \left[\vec{r}_0\ \hat{t}\ \hat{n}\right]$, i.e.

$$\begin{vmatrix} x & y & z \\ \frac{1}{3} & \frac{2}{3} & \frac{2}{3} \\ -\frac{2}{3} & \frac{-1}{3} & \frac{2}{3} \end{vmatrix} = \begin{vmatrix} 1 & 1 & \frac{2}{3} \\ \frac{1}{3} & \frac{2}{3} & \frac{2}{3} \\ -\frac{2}{3} & \frac{-1}{3} & \frac{2}{3} \end{vmatrix}$$

Using Eq. (4.15), the equation of the normal plane is $(\vec{R}-\vec{r}_0)\cdot\frac{d\vec{r}}{ds} = 0$, i.e.

$$\left\{\vec{R}-\left(\hat{i}+\hat{j}+\frac{2}{3}\hat{k}\right)\right\}\cdot\frac{1}{3}\Big[\hat{i}+2\hat{j}+2\hat{k}\Big] = 0$$

Using Eq. (4.16), the equation of the rectifying plane is $(\vec{R}-\vec{r}_0)\cdot\hat{n}=0$, i.e.

$$\left\{\vec{R}-\left(\hat{\imath}+\hat{\jmath}+\frac{2}{3}\hat{k}\right)\right\}\cdot\frac{1}{3}(-2\hat{\imath}-\hat{\jmath}+2\hat{k})=0$$

Binormal

We introduce another unit vector $\hat{b}$, defined by

$$\hat{b}=\hat{t}\times\hat{n} \tag{4.17}$$

This is along a straight line perpendicular to the osculating plane and is *binormal* to the curve at P. The vector equation of the binormal at the point P is

$$(\vec{R}-\vec{r})\times\hat{b}=\vec{0}\Rightarrow\vec{R}=\vec{r}+\lambda\,\hat{b} \tag{4.18}$$

From the definition, we see that the osculating plane is perpendicular to the binormal at the point P on the curve. Thus, if $\vec{R}$ is the position vector of any point on the osculating plane and $\vec{r}_0$ is the position vector of a specified point P of the given curve, then the equation of the plane is

$$(\vec{R}-\vec{r}_0)\cdot\hat{b}=0 \tag{4.19}$$

Example 4.3.3

Given an equation of space curve as in Example (4.3.1). Determine

(i) the unit binormal vector to the space curve at the point where $t=1$ on the curve.

(ii) the equation of the osculating plane.

Solution

(i) From Examples (4.3.1) and (4.3.2), we get

$$\hat{t}=\frac{1}{1+2t^2}\left[\hat{\imath}+2t\hat{\jmath}+2t^2\hat{k}\right]\text{ and }\hat{n}=\frac{1}{1+2t^2}\left[-2t\hat{\imath}+(1-2t^2)\hat{\jmath}+2t\hat{k}\right]$$

Therefore, from Eq. (4.17), we get

$$\hat{b}=\hat{t}\times\hat{n}=\begin{vmatrix}\hat{\imath}&\hat{\jmath}&\hat{k}\\ \frac{1}{1+2t^2}&\frac{2t}{1+2t^2}&\frac{2t^2}{1+2t^2}\\ \frac{-2t}{1+2t^2}&\frac{1-2t^2}{1+2t^2}&\frac{2t}{1+2t^2}\end{vmatrix}=\frac{1}{(1+2t^2)^2}\begin{vmatrix}\hat{\imath}&\hat{\jmath}&\hat{k}\\ 1&2t&2t^2\\ -2t&1-2t^2&2t\end{vmatrix}$$

$$=\frac{1}{(1+2t^2)^2}\Big[\{2t\cdot 2t-2t^2(1-2t^2)\}\hat{\imath}+\{2t^2\cdot(-2t)-1\cdot 2t\}\hat{\jmath}$$
$$+\{1\cdot(1-2t^2)-2t(-2t)\}\hat{k}\Big]$$

$$=\frac{1}{(1+2t^2)^2}(1+2t^2)\left[2t^2\hat{\imath}-2t\hat{\jmath}+\hat{k}\right]=\frac{1}{1+2t^2}\left[2t^2\hat{\imath}-2t\hat{\jmath}+\hat{k}\right]$$

At the point where $t=1$, the unit binormal is $\hat{b}=\frac{1}{3}\left[2\hat{\imath}-2\hat{\jmath}+\hat{k}\right]$.

(ii) Using Eq. (4.19), the equation of the osculating plane is $(\vec{R}-\vec{r}_0)\cdot\hat{b}=\vec{0}$, i.e.

$$\left\{\vec{R}-\left(\hat{\imath}+\hat{\jmath}+\frac{2}{3}\hat{k}\right)\right\}\cdot\frac{1}{3}\left[2\hat{\imath}-2\hat{\jmath}+\hat{k}\right]=0$$

Frenet–Serret Formulae

The unit tangent vector $\hat{t}$, the unit normal vector $\hat{n}$ and the unit binormal vector $\hat{b}$ form a right-handed system of mutually orthogonal unit vectors. Accordingly

$$\hat{b} = \hat{t} \times \hat{n}, \quad \hat{n} = \hat{b} \times \hat{t}, \quad \hat{t} = \hat{n} \times \hat{b}$$

The triad formed by these vectors is called the *principal triad.* All vectors associated with the curve at the point P can be expressed as a linear combination of this principal triad. Differentiating the relation (4.17), we have

$$\frac{d\hat{b}}{ds} = \frac{d\hat{t}}{ds} \times \hat{n} + \hat{t} \times \frac{d\hat{n}}{ds} = \frac{1}{\rho}\hat{n} \times \hat{n} + \hat{t} \times \frac{d\hat{n}}{ds} = \hat{t} \times \frac{d\hat{n}}{ds} \tag{4.20}$$

Using the relation (4.20), we get

$$\hat{t} \cdot \frac{d\hat{b}}{ds} = \hat{t} \cdot \left(\hat{t} \times \frac{d\hat{n}}{ds}\right) = \left[\hat{t}\ \hat{t}\ \frac{d\hat{n}}{ds}\right] = 0$$

$$\Rightarrow \hat{t} \text{ and } \frac{d\hat{n}}{ds} \text{ are perpendicular}$$

The unit vector $\hat{b}$ is of constant length. Then

$$\hat{b} \cdot \hat{b} = 1 \Rightarrow \frac{d}{ds}(\hat{b} \cdot \hat{b}) = 0 \Rightarrow \hat{b} \cdot \frac{d\hat{b}}{ds} = 0$$

$$\Rightarrow \hat{b} \text{ and } \frac{d\hat{b}}{ds} \text{ are perpendicular}$$

From relation (4.20), we see that $\frac{d\hat{b}}{ds}$ is perpendicular to $\hat{t}$ and also to $\hat{b}$, and hence, is parallel to $\hat{n}$. Thus, we may write

$$\frac{d\hat{b}}{ds} = -\tau\, \hat{n} \tag{4.21}$$

As in the formula (4.21), it can be shown that τ is the arc rate of turning of binormal and is known as the *torsion* of the curve at P. This is also the arc rate of rotation of the osculating plane. Torsion is regarded as positive when, with an increase in s, the binormal rotates right-handed relative to $\hat{t}$. The quantity $1/\tau$ is called the *radius of torsion.*

Torsion measures the twisting of the curve at the point P on the curve. Again, from the result $\hat{n} = \hat{b} \times \hat{t}$, we obtain

$$\frac{d\hat{n}}{ds} = \frac{d\hat{b}}{ds} \times \hat{t} + \hat{b} \times \frac{d\hat{t}}{ds} = \hat{b} \times \kappa\hat{n} - \tau\hat{n} \times \hat{t} = \tau\hat{b} - \kappa\hat{t} \tag{4.22}$$

A set of relations involving the derivatives of the fundamental vectors $\hat{t}$, $\hat{n}$ and $\hat{b}$ given by (4.12), (4.21) and (4.22) are together called the *Frenet–Serret formulae,* given by

$$\frac{d\hat{t}}{ds} = \kappa\, \hat{n}, \quad \frac{d\hat{b}}{ds} = -\tau\, \hat{n}, \quad \frac{d\hat{n}}{ds} = \tau\hat{b} - \kappa\hat{t} \tag{4.23}$$

In matrix notation, the *Frenet–Serret formulae* (4.23) can be written as

$$\frac{d}{ds}\begin{pmatrix}\hat{t}\\ \hat{n}\\ \hat{b}\end{pmatrix}=\begin{pmatrix}0 & \kappa & 0\\ -\kappa & 0 & \tau\\ 0 & -\tau & 0\end{pmatrix}\begin{pmatrix}\hat{t}\\ \hat{n}\\ \hat{b}\end{pmatrix} \tag{4.24}$$

Example 4.3.4

Given an equation of space curve as in Example (4.3.1). Determine the torsion to the space curve at the point where $t = 1$ on the curve.

Solution

From Example (4.3.3), we get $\hat{b}=\frac{1}{1+2t^2}\left[2t^2\hat{i}-2t\hat{j}+\hat{k}\right]$, and so

$$\begin{aligned}\frac{d\hat{b}}{dt}&=\frac{1}{(1+2t^2)^2}\left[(1+2t^2)(4t\hat{i}-2\hat{j})-(2t^2\hat{i}-2t\hat{j}+\hat{k})(4t)\right]\\ &=\frac{1}{(1+2t^2)^2}\left[4t\hat{i}-(2-4t^2)\hat{j}-4t\hat{k}\right]\end{aligned}$$

$$\Rightarrow\quad \frac{d\hat{b}}{ds}=\frac{d\hat{b}}{dt}\Big/\frac{ds}{dt}=\frac{1}{(1+2t^2)^3}\left[4t\hat{i}-(2-4t^2)\hat{j}-4t\hat{k}\right]$$

Using Eq. (4.21), we get

$$\left|\tau\,\hat{n}\right|=\left|\frac{d\hat{b}}{ds}\right|=\frac{\sqrt{(-4t)^2+(2-4t^2)^2+(4t)^2}}{(1+2t^2)^3}$$

$$\Rightarrow \tau=\frac{\sqrt{16t^2+4+16t^4-16t^2+16t^2}}{(1+2t^2)^3}=\frac{2(1+2t^2)}{(1+2t^2)^3}=\frac{2}{(1+2t^2)^2}$$

Therefore, at the point where $t = 1$, the torsion is $\tau = 2/9$ and radius of torsion is $\sigma = 9/2$.

Example 4.3.5

Show that the Frenet–Serret formulae can be expressed in the form

$$\frac{d\hat{t}}{ds}=\vec{\omega}\times\hat{t},\quad \frac{d\hat{n}}{ds}=\vec{\omega}\times\hat{n},\quad \frac{d\hat{b}}{ds}=\vec{\omega}\times\hat{b}$$

where $\vec{\omega}$ is to be found by you.

Solution

We know that $\hat{b}=\hat{t}\times\hat{n}$, $\hat{n}=\hat{b}\times\hat{t}$ and $\hat{t}=\hat{n}\times\hat{b}$. Using these relations in the Frenet–Serret formulae (4.23), we find that

$$\frac{d\hat{t}}{ds}=\kappa\,\hat{n}=\kappa(\hat{b}\times\hat{t})=\kappa\hat{b}\times\hat{t}+\tau\hat{t}\times\hat{t}=(\kappa\hat{b}+\tau\hat{t})\times\hat{t}$$

$$\frac{d\hat{b}}{ds}=-\tau\,\hat{n}=\tau\hat{t}\times\hat{b}+\kappa\hat{b}\times\hat{b}=(\kappa\hat{b}+\tau\hat{t})\times\hat{b}$$

$$\frac{d\hat{n}}{ds}=\tau\hat{b}-\kappa\hat{t}=\tau\hat{t}\times\hat{n}+\kappa\hat{b}\times\hat{n}=(\kappa\hat{b}+\tau\hat{t})\times\hat{n}$$

Thus, the Frenet–Serret formulae can be expressed in the form $\frac{d\hat{t}}{ds} = \vec{\omega} \times \hat{t}$, $\frac{d\hat{n}}{ds} = \vec{\omega} \times \hat{n}$, $\frac{d\hat{b}}{ds} = \vec{\omega} \times \hat{b}$, where $\vec{\omega} = \kappa\hat{b} + \tau\hat{t}$, $\tau\hat{t}$, known as the *Darboux vector.*

Example 4.3.6

Given an equation of space curve $x = 3\cos t, y = 3\sin t, z = 4t$. Determine

(i) the tangent and unit tangent vector to any point on the space curve.

(ii) the principal normal vector, curvature and radius of curvature to any point on the space curve.

(iii) the binormal vector, torsion and radius of torsion to any point on the space curve.

Solution

The position vector at any point on the space curve is $\vec{r} = 3\cos t\hat{i} + 3\sin t\hat{j} + 4t\hat{k}$.

(i) A tangent vector to the curve at any point is

$$\vec{T} = \frac{d\vec{r}}{dt} = \frac{d}{dt}\left[3\cos t\hat{i} + 3\sin t\hat{j} + 4t\hat{k}\right] = -3\sin t\hat{i} + 3\cos t\hat{j} + 4\hat{k}$$

The magnitude of the tangent vector is

$$|\vec{T}| = \left|\frac{d\vec{r}}{dt}\right| = \sqrt{(-3)^2\sin^2 t + (3)^2\cos^2 t + (4)^2} = 5$$

However, $\frac{d\vec{r}}{dt} = \frac{d\vec{r}}{ds}\frac{ds}{dt} = \hat{t}\frac{ds}{dt}$, so

$$\left|\frac{d\vec{r}}{dt}\right| = |\hat{t}|\frac{ds}{dt} \Rightarrow \frac{ds}{dt} = \left|\frac{d\vec{r}}{dt}\right| = 5 \text{ as } |\hat{t}| = 1$$

Then, the required unit tangent vector is given by

$$\hat{t} = \frac{d\vec{r}}{dt}\bigg/\frac{ds}{dt} = \frac{1}{5}\left[-3\sin t\hat{i} + 3\cos t\hat{j} + 4\hat{k}\right]$$

(ii) Using the expression for $\hat{t}$, we get

$$\frac{d\hat{t}}{dt} = \frac{1}{5}\frac{d}{dt}\left[-3\sin t\hat{i} + 3\cos t\hat{j} + 4\hat{k}\right] = \frac{1}{5}\left[-3\cos t\hat{i} - 3\sin t\hat{j}\right]$$

$$\Rightarrow \frac{d\hat{t}}{ds} = \frac{d\hat{t}}{dt}\bigg/\frac{ds}{dt} = \frac{1}{25}\left[-3\cos t\hat{i} - 3\sin t\hat{j}\right]$$

Using Eq. (4.12), we get

$$\left|\kappa\,\hat{n}\right| = \left|\frac{d\hat{t}}{ds}\right| = \frac{\sqrt{(-3)^2\cos^2 t + (3)^2\sin^2 t}}{25} = \frac{3}{25} = \kappa$$

Therefore, the curvature is $\kappa = 3/25$ and radius of curvature is $\rho = 25/3$. Thus, the required unit normal vector is given by

$$\hat{n} = \frac{1}{\kappa}\frac{d\vec{t}}{ds} = \frac{25}{3}\cdot\frac{1}{25}\left[-3\cos t\hat{i} + 3\sin t\hat{j}\right] = \frac{1}{3}\left[-3\cos t\hat{i} + 3\sin t\hat{j}\right]$$

(iii) Using the expressions for $\hat{t}$ and $\hat{n}$, we get

$$\hat{b} = \hat{t} \times \hat{n} = \begin{vmatrix} \hat{i} & \hat{j} & \hat{k} \\ -\frac{3}{5}\sin t & \frac{3}{5}\cos t & \frac{4}{5} \\ -\cos t & -\sin t & 0 \end{vmatrix}$$

$$= \frac{1}{5}\left[\{0 + 4\sin t\}\hat{i} + \{-4\cos t + 0\}\hat{j} + (3\sin^2 t + 3\cos^2 t)\hat{k}\right]$$

$$= \frac{1}{5}\left[4\sin t\hat{i} - 4\cos t\hat{j} + 3\hat{k}\right]$$

This is the required unit binormal vector. Now, using the expression for $\hat{b}$, we get

$$\frac{d\hat{b}}{dt} = \frac{1}{5}\left[4\cos t\,\hat{i} + 4\sin t\,\hat{j}\right] \Rightarrow \frac{d\hat{b}}{ds} = \frac{d\hat{b}}{dt} / \frac{ds}{dt} = \frac{1}{25}\left[4\cos t\,\hat{i} + 4\sin t\,\hat{j}\right]$$

Using Eq. (4.21), we get

$$\left|\tau\,\hat{n}\right| = \left|\frac{d\hat{b}}{ds}\right| = \frac{\sqrt{(4\cos t)^2 + (4\sin t)^2}}{25} = \frac{4}{25} \Rightarrow \tau = \frac{4}{25}$$

Therefore, the torsion is $\tau = 4/25$ and radius of torsion is $\sigma = 25/4$.

4.3.2 Curvature and Torsion of Space Curve $\vec{r} = \vec{f}(u)$

Consider the space curve $\vec{r} = \vec{f}(u)$, where u is a variable parameter. Differentiating the equation of the given curve with respect to u, we get

$$\dot{\vec{r}} = \frac{d\vec{r}}{du} = \frac{d\vec{r}}{ds}\frac{ds}{du} = \hat{t}\,\dot{s} \tag{4.25}$$

$$\Rightarrow \left|\dot{\vec{r}}\right| = \left|\hat{t}\,\dot{s}\right| = |\hat{t}|\,|\dot{s}| = |\dot{s}|$$

where s is the arc length of the given curve, measured from a fixed point on the given curve. Now, differentiating with respect to u, we obtain

$$\ddot{\vec{r}} = \frac{d\hat{t}}{du}\dot{s} + \hat{t}\frac{d\dot{s}}{du} = \frac{d\hat{t}}{ds}\frac{ds}{du}\dot{s} + \hat{t}\frac{d\dot{s}}{du}$$

$$= \frac{d\hat{t}}{ds}\dot{s}^2 + \hat{t}\,\ddot{s} = \kappa\,\hat{n}\dot{s}^2 + \hat{t}\,\ddot{s}, \text{ using Eq. (4.23)}$$

By using the Frenet–Serret formula (Eq. (4.23)), we get

$$\dddot{\vec{r}} = 2\dot{s}\ddot{s}\kappa\,\hat{n} + \dot{s}^2\dot{\kappa}\,\hat{n} + \dot{s}^2\kappa\frac{d\hat{n}}{du} + \frac{d\hat{t}}{du}\ddot{s} + \hat{t}\dddot{s}$$

$$= 2\dot{s}\ddot{s}\kappa\,\hat{n} + \dot{s}^2\dot{\kappa}\,\hat{n} + \dot{s}^3\kappa\frac{d\hat{n}}{ds} + \frac{d\hat{t}}{ds}\dot{s}\ddot{s} + \hat{t}\dddot{s}$$

$$= 2\dot{s}\ddot{s}\kappa\,\hat{n} + \dot{s}^2\dot{\kappa}\,\hat{n} + \dot{s}^3\kappa(\tau\,\hat{b} - \kappa\,\hat{t}) + \kappa\,\hat{n}\dot{s}\ddot{s} + \hat{t}\dddot{s}$$

$$= (\dddot{s} - \kappa^2\dot{s}^3)\hat{t} + (3\dot{s}\ddot{s}\kappa + \dot{s}^2\dot{\kappa})\hat{n} + \tau\dot{s}^3\kappa\,\hat{b} \tag{4.26}$$

Using the definition of cross product, we get

$$\dot{\vec{r}} \times \ddot{\vec{r}} = \hat{t}\,\dot{s} \times \left[\kappa\,\hat{n}\dot{s}^2 + \hat{t}\,\ddot{s}\right] = \kappa\,\dot{s}^3\hat{t} \times \hat{n} + \dot{s}\,\ddot{s}\,\hat{t} \times \hat{t}$$
$$= \kappa\,\dot{s}^3\hat{t} \times \hat{n} + \vec{0} = \kappa\,\dot{s}^3\hat{b}, \text{ using Eq. (4.23)}$$

This implies

$$|\dot{\vec{r}} \times \ddot{\vec{r}}| = |\kappa\,\dot{s}^3\hat{b}| = |\dot{\vec{r}}|^3\kappa, \text{ as } |\dot{s}| = |\dot{\vec{r}}|$$
$$\Rightarrow \kappa = \frac{1}{|\dot{\vec{r}}|^3}|\dot{\vec{r}} \times \ddot{\vec{r}}| = \frac{1}{|\dot{\vec{f}}|^3}|\dot{\vec{f}} \times \ddot{\vec{f}}| \qquad (4.27)$$

Using the definition of scalar triple product, we get

$$\left[\dot{\vec{r}}\;\ddot{\vec{r}}\;\dddot{\vec{r}}\right] = \dot{\vec{r}} \times \ddot{\vec{r}} \cdot \dddot{\vec{r}} = \left\{(\dddot{s} - \kappa^2\dot{s}^3)\hat{t} + (3\dot{s}\ddot{s}\kappa + \dot{s}^2\dot{\kappa})\hat{n} + \tau\dot{s}^3\kappa\hat{b}\right\} \cdot \dot{s}^3\kappa\hat{b}$$
$$= \dot{s}^3\kappa(\dddot{s} - \kappa^2\dot{s}^3)\hat{t} \cdot \hat{b} + (3\dot{s}\ddot{s}\kappa + \dot{s}^2\dot{\kappa})\dot{s}^3\kappa\hat{n} \cdot \hat{b} + \tau\dot{s}^6\kappa^2\hat{b} \cdot \hat{b}$$
$$= \dot{s}^3\kappa(\dddot{s} - \kappa^2\dot{s}^3) \cdot 0 + (3\dot{s}\ddot{s}\kappa + \dot{s}^2\dot{\kappa})\dot{s}^3\kappa \cdot 0 + \tau\dot{s}^6\kappa^2 \cdot 1 = \tau\dot{s}^6\kappa^2$$
$$\Rightarrow \tau = \frac{1}{\dot{s}^6\kappa^2}\left[\dot{\vec{r}}\;\ddot{\vec{r}}\;\dddot{\vec{r}}\right] = \frac{1}{|\dot{\vec{r}} \times \ddot{\vec{r}}|^2}\left[\dot{\vec{f}}\;\ddot{\vec{f}}\;\dddot{\vec{f}}\right] \qquad (4.28)$$

Equations (4.27) and (4.28) are, respectively, the formulae for evaluating the curvature and torsion of the space curve $\vec{r} = \vec{f}(u)$, where u is the parameter different from arc length s.

Example 4.3.7

Show that the curvature is the same at all points on the curve whose vector equation is $\vec{r} = 4\cos u\hat{i} + 4\sin u\hat{j} + 2u\hat{k}$.

Solution

The given vector equation is $\vec{r} = 4\cos u\hat{i} + 4\sin u\hat{j} + 2u\hat{k}$. Differentiating, we get

$$\dot{\vec{r}} = -4\sin u\hat{i} + 4\cos u\hat{j} + 2\hat{k};\ \ddot{\vec{r}} = -4\cos u\hat{i} - 4\sin u\hat{j}$$
$$\Rightarrow \dot{\vec{r}} \times \ddot{\vec{r}} = \begin{vmatrix} \hat{i} & \hat{j} & \hat{k} \\ -4\sin u & 4\cos u & 2 \\ -4\cos u & -4\sin u & 0 \end{vmatrix} = 8(\sin u\hat{i} - \cos u\hat{j} + 2\hat{k})$$
$$\Rightarrow |\dot{\vec{r}} \times \ddot{\vec{r}}| = 8\sqrt{\sin^2 u + \cos^2 u + 4} = 8\sqrt{5}$$
$$\text{and } |\dot{\vec{r}}|^3 = \left\{16(\sin^2 u + \cos^2 u) + 4\right\}^{3/2} = (20)^{3/2} = 40\sqrt{5}$$

Using Eq. (4.27) for curvature, we get

$$\kappa = \frac{1}{|\dot{\vec{r}}|^3}|\dot{\vec{r}} \times \ddot{\vec{r}}| = \frac{8\sqrt{5}}{40\sqrt{5}} = \frac{1}{5}$$

This is independent of the parameter u, and hence, is the same for all points of the given space curve.

Example 4.3.8

Find the curvature and torsion of the curve $\vec{r} = ae^u\hat{i} + ae^{-u}\hat{j} + \sqrt{2}au\hat{k}$.

Solution

The given vector equation is $\vec{r} = ae^u\hat{i} + ae^{-u}\hat{j} + \sqrt{2}au\hat{k}$. Differentiating, we get

$$\dot{\vec{r}} = ae^u\hat{i} - ae^{-u}\hat{j} + \sqrt{2}a\hat{k};\ \ddot{\vec{r}} = ae^u\hat{i} + ae^{-u}\hat{j};\ \dddot{\vec{r}} = ae^u\hat{i} - ae^{-u}\hat{j}$$

$$\Rightarrow \dot{\vec{r}} \times \ddot{\vec{r}} = \begin{vmatrix} \hat{i} & \hat{j} & \hat{k} \\ ae^u & ae^{-u} & \sqrt{2}a \\ ae^u & -ae^{-u} & 0 \end{vmatrix} = \sqrt{2}a^2(-e^{-u}\hat{i} + e^u\hat{j} - \sqrt{2}\hat{k})$$

$$\Rightarrow |\dot{\vec{r}} \times \ddot{\vec{r}}| = \sqrt{2}\sqrt{(-\sqrt{2}a^2e^{-u})^2 + (\sqrt{2}a^2e^u)^2 + (-2a^2)^2} = \sqrt{2}a^2(e^u + e^{-u})$$

$$\text{and}\quad |\dot{\vec{r}}| = \sqrt{(ae^u)^2 + (-ae^{-u})^2 + (\sqrt{2}a)^2} = a(e^u + e^{-u})$$

Using Eq. (4.27) for curvature, we get

$$\kappa = \frac{1}{|\dot{\vec{r}}|^3}|\dot{\vec{r}} \times \ddot{\vec{r}}| = \frac{\sqrt{2}a^2(e^u + e^{-u})}{a^3(e^u + e^{-u})^3} = \frac{\sqrt{2}}{a(e^u + e^{-u})^2}$$

Using the definition of scalar triple product, we get

$$\left[\dot{\vec{r}}\,\ddot{\vec{r}}\,\dddot{\vec{r}}\right] = a(e^u\hat{i} - e^{-u}\hat{j}) \cdot \sqrt{2}a^2(-e^{-u}\hat{i} + e^u\hat{j} - \sqrt{2}\hat{k}) = -2\sqrt{2}a^3$$

Using Eq. (4.28) for torsion, we get

$$\tau = \frac{\left[\dot{\vec{r}}\,\ddot{\vec{r}}\,\dddot{\vec{r}}\right]}{|\dot{\vec{r}} \times \ddot{\vec{r}}|^2} = \frac{-2\sqrt{2}a^3}{2a^4(e^u + e^{-u})^2} = \frac{-\sqrt{2}}{a(e^u + e^{-u})^2}$$

4.3.3 Tangent Plane and Normal to a Surface $\vec{r} = \vec{r}(u, v)$

Let us consider the vector equation $\vec{r} = \vec{r}(u, v)$, where u and v are two scalar variables, both restricted to lie at bounded intervals. If we give a fixed value u_0 to u, then $\vec{r} = \vec{r}(u_0, v)$ represents a curve which can be denoted by $u = u_0$. Similarly, $u = u_1$ defines a curve $\vec{r} = \vec{r}(u_1, v)$. As u varies, $\vec{r} = \vec{r}(u, v)$ represents a curve moving in a space and this moving curve generates a surface. Thus, $\vec{r} = \vec{r}(u, v)$ represents the surface (Fig. 4.6).

By assigning definite values to u and v, say u_0 and v_0, we obtain a point (u_0, v_0) on the surface $\vec{r} = \vec{r}(u, v)$. The curves $u = u_0$ and $v = v_0$ intersect at a point (u_0, v_0). The pair of numbers (u, v) defines the curvilinear coordinates on the surface (discussed in Section 4.4).

Let $P(u_0, v_0)$ be a point on the surface S represented by a vector equation $\vec{r} = \vec{r}(u, v)$ (Fig. 4.6). $\dfrac{\partial \vec{r}}{\partial u}$ is the partial derivative of $\vec{r}$ with respect to u when $v = v_0$, and so $\dfrac{\partial \vec{r}}{\partial u}$ at P represents a vector tangent to the space curve $v = v_0$ at P. Similarly, $\dfrac{\partial \vec{r}}{\partial v}$ at P represents a vector tangent to the space curve $u = u_0$ at P.

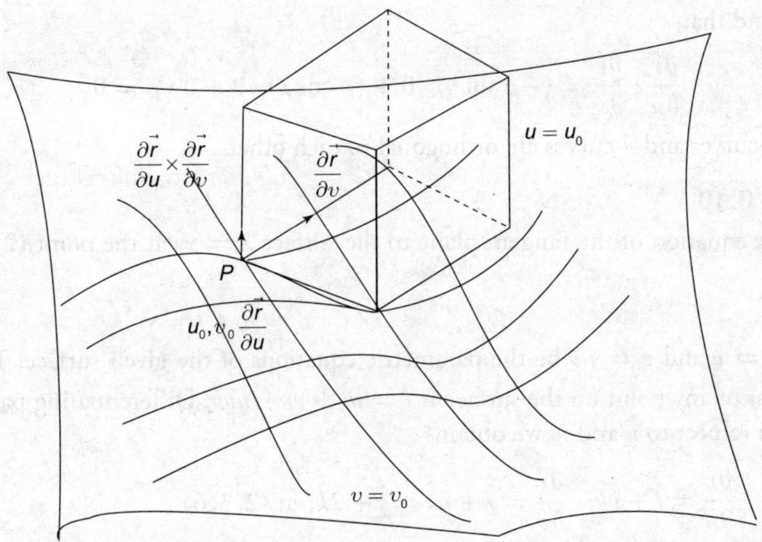

Figure 4.6 Tangent plane and normal to a surface

$\frac{\partial \vec{r}}{\partial u}$ and $\frac{\partial \vec{r}}{\partial v}$ represent vectors at P, which is tangent to the space curves $v = v_0$ and $u = u_0$ which lie on the surface $\mathcal{S}$ at P. Hence, $\frac{\partial \vec{r}}{\partial u}$ and $\frac{\partial \vec{r}}{\partial v}$ are vectors which are tangents to the surface at P and so we can conclude that $\frac{\partial \vec{r}}{\partial u} \times \frac{\partial \vec{r}}{\partial v}$ is a vector normal to the surface $\vec{r} = \vec{r}(u, v)$ at P. We define the unit normal to the considered surface as

$$\hat{n} = \frac{1}{\left|\frac{\partial \vec{r}}{\partial u} \times \frac{\partial \vec{r}}{\partial v}\right|}\left(\frac{\partial \vec{r}}{\partial u} \times \frac{\partial \vec{r}}{\partial v}\right) \tag{4.29}$$

If $\vec{r}$ is the position vector of any point on the tangent plane to the surface $\vec{r} = \vec{r}(u, v)$ at $P(u_0, v_0)$, then the equation of the tangent plane to the surface at P is

$$\left(\vec{r} - \vec{r}_0\right) \cdot \left(\frac{\partial \vec{r}}{\partial u} \times \frac{\partial \vec{r}}{\partial v}\right) = 0 \tag{4.30}$$

where $\vec{r}_0 = (u_0, v_0)$ is the position vector of P.

Example 4.3.9

Show that the u-curves and v-curves are orthogonal for the surface $\vec{r} = \rho \cos u\hat{i} + \rho \sin u\hat{j} + \rho v\hat{k}$, where $\rho =$ constant.

Solution

Differentiating partially the equation $\vec{r} = \rho \cos u\hat{i} + \rho \sin u\hat{j} + \rho v\hat{k}$ of the given surface with respect to u and v, we obtain

$$\frac{\partial \vec{r}}{\partial u} = -\rho \sin u\hat{i} + \rho \cos u\hat{j}; \quad \frac{\partial \vec{r}}{\partial v} = \rho\hat{k}$$

Further, we find that

$$\frac{\partial \vec{r}}{\partial u} \cdot \frac{\partial \vec{r}}{\partial v} = (-\rho \sin u) \cdot 0 + (\rho \cos u) \cdot 0 + 0 \cdot \rho = 0$$

Hence, the u-curves and v-curves are orthogonal to each other.

Example 4.3.10

Show that the equation of the tangent plane to the surface $z = xy$ at the point $(2, 3, 6)$ is $3x + 2y - z = 0$.

Solution

Let $x = u, y = v$ and $z = uv$ be the parametric equations of the given surface. Therefore, the position vector of any point on the surface is $\vec{r} = u\hat{\imath} + v\hat{\jmath} + uv\hat{k}$. Differentiating partially the last equation with respect to u and v, we obtain

$$\frac{\partial \vec{r}}{\partial u} = \hat{\imath} + v\hat{k}; \quad \frac{\partial \vec{r}}{\partial v} = \hat{\jmath} + u\hat{k} = \hat{\jmath} + 2\hat{k}, \text{ at } (2, 3, 6)$$

$$\Rightarrow \frac{\partial \vec{r}}{\partial u} \times \frac{\partial \vec{r}}{\partial v} = \begin{vmatrix} \hat{\imath} & \hat{\jmath} & \hat{k} \\ 1 & 0 & 3 \\ 0 & 1 & 2 \end{vmatrix} = -3\hat{\imath} - 2\hat{\jmath} + \hat{k}$$

$$\Rightarrow \hat{n} = \frac{1}{\sqrt{(-3)^2 + (-2)^2 + 1^2}}(-3\hat{\imath} - 2\hat{\jmath} + \hat{k}) = \frac{1}{\sqrt{14}}(-3\hat{\imath} - 2\hat{\jmath} + \hat{k})$$

This is the unit normal to the surface at the point $(2, 3, 6)$. Using Eq. (4.30), we obtain the required equation of the tangent plane as

$$\left[\vec{r} - (2\hat{\imath} + 3\hat{\jmath} + 6\hat{k})\right] \cdot (-3\hat{\imath} - 2\hat{\jmath} + \hat{k}) = 0$$

$$\Rightarrow \left[(x-2)\hat{\imath} + (y-3)\hat{\jmath} + (z-6)\hat{k}\right] \cdot (-3\hat{\imath} - 2\hat{\jmath} + \hat{k}) = 0$$

$$\Rightarrow 3(x-2) + 2(y-3) - (z-6) = 0 \Rightarrow 3x + 2y - z = 6$$

Example 4.3.11

Find the unit tangent vector, the unit normal vector, the unit binormal vector, the curvature and the torsion at any point of the circular helix $\vec{r} = a(\cos u\hat{\imath} + \sin u\hat{\jmath} + u \cot \alpha \hat{k})$, a being a constant.

Solution

Let s be the arc length of any point of the given curve measured from a fixed point on the curve. Differentiating the equation of the given curve with respect to s, we obtain

$$\hat{t} = \frac{d\vec{r}}{ds} = a(-\sin u\hat{\imath} + \sin u\hat{\jmath} + u \cot \alpha \hat{k})\frac{du}{ds}$$

$$\Rightarrow |\hat{t}| = |a(-\sin u\hat{\imath} + \sin u\hat{\jmath} + u \cot \alpha \hat{k})\frac{du}{ds}|$$

$$\Rightarrow 1 = \frac{du}{ds}\sqrt{(-a \sin u)^2 + (a \cos u)^2 + (a \cot \alpha)^2}$$

$$\Rightarrow 1 = \frac{du}{ds} a \operatorname{cosec} \alpha \Rightarrow \frac{du}{ds} = \frac{\sin \alpha}{a}$$

Therefore, using the value of $\frac{du}{ds}$, we obtain the unit tangent vector as

$$\hat{t} = \sin\alpha(-\sin u\hat{i} + \cos u\hat{j} + \cot\alpha\hat{k})$$

Further differentiating with respect to s, we obtain

$$\frac{d\hat{t}}{ds} = -\sin\alpha(\cos u\hat{i} + \sin u\hat{j})\frac{du}{ds} = -\frac{\sin^2\alpha}{a}(\cos u\hat{i} + \sin u\hat{j})$$

$$\text{or, } \kappa\hat{n} = -\frac{\sin^2\alpha}{a}(\cos u\hat{i} + \sin u\hat{j})$$

$$\text{or, } \kappa = |\kappa\hat{n}| = \frac{\sin^2\alpha}{a}\sqrt{(\cos u\hat{i} + \sin u\hat{j}} = \frac{\sin^2\alpha}{a}$$

This is the required curvature. Hence, the unit normal vector is $\hat{n} = -(\cos u\hat{i} + \sin u\hat{j})$. Taking the cross product, we find that

$$\hat{b} = \hat{t} \times \hat{n} = \begin{vmatrix} \hat{i} & \hat{j} & \hat{k} \\ -\sin\alpha\sin u & \sin\alpha\cos u & \cos\alpha \\ -\cos u & -\sin u & 0 \end{vmatrix}$$

$$= \cos\alpha\sin u\hat{i} - \cos\alpha\cos u\hat{j} + \sin\alpha\hat{k}$$

This is the required unit binormal vector. Differentiating with respect to s, we get

$$\frac{d\hat{b}}{ds} = \cos\alpha(\cos u\hat{i} + \sin u\hat{j})\frac{du}{ds} = \frac{\sin\alpha\cos\alpha}{a}(\cos u\hat{i} + \sin u\hat{j})$$

$$\Rightarrow \tau\hat{n} = \frac{\sin\alpha\cos\alpha}{a}(\cos u\hat{i} + \sin u\hat{j})$$

$$\Rightarrow \tau = \frac{\sin\alpha\cos\alpha}{a}\sqrt{(\cos u)^2 + (\sin u)^2)} = \frac{\sin\alpha\cos\alpha}{a}$$

This is the required torsion.

Example 4.3.12

Find the equations of the tangent plane and normal line to the surface $4z = x^2 - y^2$ at the point $(3, 1, 2)$.

Let $x = u, y = v$ and $z = (u^2 - v^2)/4$ be the parametric equation of the surface. The position vector of any point on the given surface $\vec{r} = \vec{r}(u, v)$ is $\vec{r} = u\hat{i} + v\hat{j} + 1/4\,(u^2 - v^2)\hat{k}$. We have

$$\frac{\partial\vec{r}}{\partial u} = \hat{i} + \frac{1}{2}u\hat{k} = \hat{i} + \frac{3}{2}\hat{k}; \frac{\partial\vec{r}}{\partial v} = \hat{j} - \frac{1}{2}v\hat{k} = \hat{j} - \frac{1}{2}\hat{k} \text{ at } (3, 1, 2)$$

$$\Rightarrow \vec{n} = \frac{\partial\vec{r}}{\partial u} \times \frac{\partial\vec{r}}{\partial v} = \begin{vmatrix} \hat{i} & \hat{j} & \hat{k} \\ 1 & 0 & \frac{3}{2} \\ 0 & 1 & -\frac{1}{2} \end{vmatrix} = -\frac{3}{2}\hat{i} + \frac{1}{2}\hat{j} + \hat{k}$$

The position vector of the point $(3, 1, 2)$ is $r_0 = 3\hat{i} + \hat{j} + 2\hat{k}$. Thus, for a point $\vec{r} = x\hat{i} + y\hat{j} + z\hat{k}$ on the tangent plane to the given surface at $(3, 1, 2)$, we have

$$(\vec{r} - \vec{r}_0) \cdot \vec{n} = 0 \Rightarrow 3x - y - 2z = 4$$

This is the equation of the tangent plane to the given surface at (3, 1, 2). The equation of the normal to the given surface at (3, 1, 2) is

$$(\vec{r} - \vec{r}_0) \times \vec{n} = \vec{0} \Rightarrow x = 3 - 3\lambda, y = 1 + \lambda, z = 2 + \lambda$$

when λ is a parameter.

4.4 CURVILINEAR COORDINATES

The Cartesian coordinates of a point (x, y, z) are determined by following a family of straight paths from the origin: first along the x axis, then parallel to the y axis, then parallel to the z axis, as in Fig. 1.23. In curvilinear coordinate systems, these paths can be curved. The two types of curvilinear coordinates which we will consider are cylindrical and spherical coordinates. Instead of referencing a point in terms of sides of a rectangular parallelepiped, as with Cartesian coordinates, we will think of the point as lying on a cylinder or sphere. Cylindrical coordinates are often used when there is symmetry around the z axis, while spherical coordinates are useful when there is symmetry about the origin.

4.4.1 Transformation of Coordinates

Let (x, y, z) denote the rectangular Cartesian coordinates of any point P in a three-dimensional space $\mathcal{R}$. Consider three continuously differentiable scalar point functions of three independent variable u, v and w as

$$x = x(u, v, w), \quad y = (u, v, w), \quad z = z(u, v, w) \tag{4.31}$$

defined in any region $\mathcal{R}$ of a three-dimensional space. Suppose that (4.31) can be solved for u, v and w in terms of x, y and z uniquely, i.e.,

$$u = u(x, y, z), \quad v = v(x, y, z), \quad w = w(x, y, z) \tag{4.32}$$

provided the Jacobian J of the transformation

$$J = \begin{vmatrix} \dfrac{\partial u}{\partial x} & \dfrac{\partial u}{\partial y} & \dfrac{\partial u}{\partial z} \\ \dfrac{\partial v}{\partial x} & \dfrac{\partial v}{\partial y} & \dfrac{\partial v}{\partial z} \\ \dfrac{\partial w}{\partial x} & \dfrac{\partial w}{\partial y} & \dfrac{\partial w}{\partial z} \end{vmatrix} \neq 0 \tag{4.33}$$

The functions in (4.31) and (4.32) are assumed to be single-valued and to have continuous derivations so that the corresponds between the triads (x, y, z) and (u, v, w) is unique. The sets of equations (4.31) or (4.32) define a transformation of coordinates.

Given a point P with rectangular Cartesian coordinates (x, y, z) we can, from (4.32), associate a unique set of coordinates (u, v, w), called the curvilinear coordinates of P.

A point (x_0, y_0, z_0) is called the *singular point of the transformation* (4.32) if any one or all of u, v and w is indeterminate at that point.

4.4.2 Coordinate Surface and Curves

Let (x, y, z) be the rectangular Cartesian coordinates of a point P and (u, v, w) be the corresponding curvilinear coordinates. The loci of the equations $u = c_1, v = c_2, w = c_3$, c_1, c_2 and c_3 being constants, are called *coordinate surfaces* and they are called u_1-surface, u_2-surface and u_3-surface, respectively (Fig. 4.7). Through each point, there passes a system of three coordinate surfaces. The three curves of intersection of the three coordinate surfaces through any point, taken in pairs, are called the *coordinate curves* or *lines* through the point. The three curves through any point (c_1, c_2, c_3) are

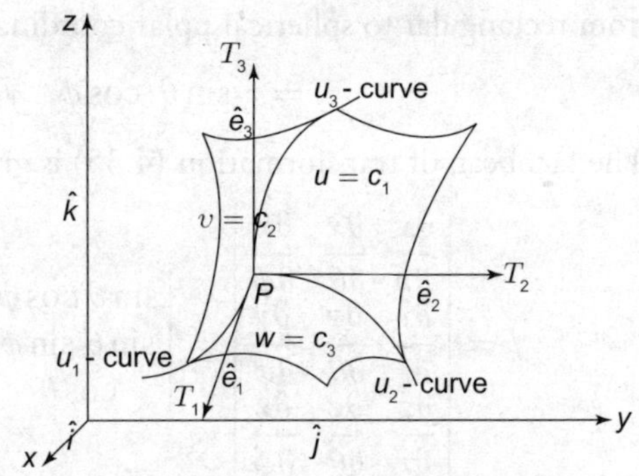

Figure 4.7 Reciprocal system

$$v = c_2, w = c_3; \quad w = c_3, u = c_1; \quad u = c_1, v = c_2 \tag{4.34}$$

If the three coordinate surfaces through every point are mutually orthogonal, i.e., if the three tangent planes, and accordingly the three normals to the three surfaces, through every point being the line of intersection of the tangent planes to the two coordinate surfaces through the curve, lies along the normal to the third surface.

PT_1, PT_2 and PT_3 are tangents to the coordinate curves $u = c_1$, $v = c_2$ and $w = c_3$, respectively, and these tangents are taken as coordinate axes of the curvilinear system. The directions of the axes depend on the choice of the point P, and the unit vectors associated with them are not necessarily constant.

If, at every point P, the coordinate axes are mutually perpendicular, i.e., if the coordinate surfaces intersect at right angles at P, then this coordinate system is called an *orthogonal curvilinear coordinate system* and u, v and w are called the *orthogonal curvilinear coordinates* of P.

Example 4.4.1

Determine the transformation equations from rectangular to spherical polar coordinates, and hence, find its inverse transformation.

Solution

Let $P(x, y, z)$ be any point in a rectangular Cartesian coordinate system (Fig. 4.8). Let $OP = r$; θ and ϕ be, respectively, the angle which OP makes with the z axis and the angle which OM makes with the x axis, where $PM(= z)$ is perpendicular to the xy plane. Hence, the spherical polar coordinates of P are (r, θ, ϕ). Here, $u = r, v = \theta$ and $w = \phi$. From Fig. 4.8, we get

$$x = OA = OM \cos\phi = (OP \sin\theta)\cos\phi$$
$$= r\sin\theta \cos\phi$$
$$y = OB = OM \sin\phi = r\sin\theta \sin\phi$$
$$z = OC = OP \cos\theta = r \cos\theta$$

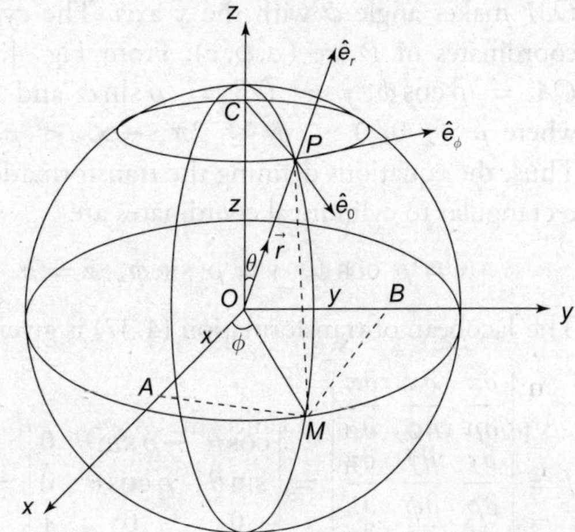

Figure 4.8 Spherical polar coordinates

where $r \geq 0, 0 \leq \theta \leq \pi$ and $0 \leq \phi \leq 2\pi$. Thus, the equations defining the transformation from rectangular to spherical polar coordinates are

$$x = r\sin\theta\cos\phi,\quad y = r\sin\theta\sin\phi,\quad z = r\cos\theta \tag{4.35}$$

The Jacobean of transformation (4.35) is given by

$$J = \begin{vmatrix} \frac{\partial x}{\partial r} & \frac{\partial x}{\partial \theta} & \frac{\partial x}{\partial \phi} \\ \frac{\partial y}{\partial r} & \frac{\partial y}{\partial \theta} & \frac{\partial y}{\partial \phi} \\ \frac{\partial z}{\partial r} & \frac{\partial z}{\partial \theta} & \frac{\partial z}{\partial \phi} \end{vmatrix} = \begin{vmatrix} \sin\theta\cos\phi & r\cos\theta\cos\phi & -r\sin\theta\sin\phi \\ \sin\theta\sin\phi & r\cos\theta\sin\phi & r\sin\theta\cos\phi \\ \cos\theta & -r\sin\theta & 0 \end{vmatrix}$$

$$= \cos\theta \begin{vmatrix} r\cos\theta\cos\phi & -r\sin\theta\sin\phi \\ r\cos\theta\sin\phi & r\sin\theta\cos\phi \end{vmatrix} + r\sin\theta \begin{vmatrix} \sin\theta\cos\phi & -r\sin\theta\sin\phi \\ \sin\theta\sin\phi & r\sin\theta\cos\phi \end{vmatrix}$$

$$= r^2\sin\theta\cos^2\theta(\sin^2\phi + \cos^2\phi) + r^2\sin^3\theta(\sin^2\phi + \cos^2\phi)$$

$$= r^2\sin\theta \neq 0$$

Thus, the inverse of the transformation (4.35) is given by

$$r = \sqrt{x^2+y^2+z^2},\quad \theta = \sin^{-1}\frac{z}{\sqrt{x^2+y^2+z^2}},\quad \phi = \tan^{-1}\frac{y}{x} \tag{4.36}$$

Example 4.4.2

Determine the transformation equations from rectangular to cylindrical coordinates, and hence, find its inverse transformation.

Solution

Let $P(x, y, z)$ be any point in the rectangular Cartesian coordinate system (Fig. 4.9). Let $OP = \rho$. PM is perpendicular to the xy plane, MA is perpendicular to the x axis and MB is perpendicular to the y axis. OM makes angle ϕ with the x axis. The cylindrical coordinates of P are (ρ, ϕ, z). From Fig. 4.9, $x = OA = \rho\cos\phi$, $y = OB = \rho\sin\phi$ and $z = z$, where $\rho \geq 0, 0 \leq \phi \leq 2\pi, -\infty < z < \infty$. Thus, the equations defining the transformation from rectangular to cylindrical coordinates are

$$x = \rho\cos\phi,\quad y = \rho\sin\phi,\quad z = z \tag{4.37}$$

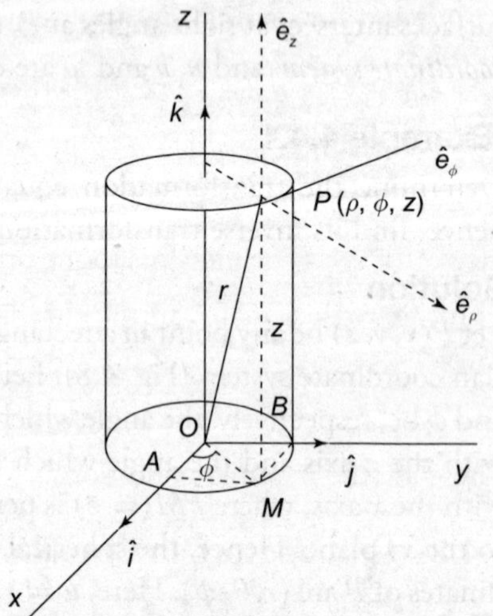

Figure 4.9 Cylindrical coordinates

The Jacobean of transformation (4.37) is given by

$$J = \begin{vmatrix} \frac{\partial x}{\partial \rho} & \frac{\partial x}{\partial \phi} & \frac{\partial x}{\partial z} \\ \frac{\partial y}{\partial \rho} & \frac{\partial y}{\partial \phi} & \frac{\partial y}{\partial z} \\ \frac{\partial z}{\partial \rho} & \frac{\partial z}{\partial \phi} & \frac{\partial z}{\partial z} \end{vmatrix} = \begin{vmatrix} \cos\theta & -\rho\sin\theta & 0 \\ \sin\theta & \rho\cos\theta & 0 \\ 0 & 0 & 1 \end{vmatrix} = \rho \neq 0$$

Therefore, the inverse transformation of (4.37) is

$$\rho = \sqrt{x^2 + y^2}, \ \phi = \tan^{-1}\frac{y}{x}, \quad z = z \tag{4.38}$$

For points on the z axis $(x = 0, y = 0)$, note that ϕ is indeterminate. Such points are called the *singular points* of the transformation.

Example 4.4.3

Describe the coordinate surfaces and coordinate curves for (a) cylindrical coordinates and (b) spherical coordinates.

Solution

(a) Refer to Fig. 4.9. The coordinate surfaces (or level surfaces) are:

(i) $\rho = c_1$, i.e., $x^2 + y^2 = c_1^2$ which are cylinders co-axial with the z axis.

(ii) $\phi = c_2$, i.e., $y = \tan c_2$ which are planes through the z axis.

(iii) $z = c_3$ which are planes perpendicular to the z axis.

These surfaces meet at the point $P(\rho, \phi, z)$.

(b) Refer to Fig. 4.8. The coordinate surfaces (or level surfaces) are:

(i) $r = c_1$, i.e., $x^2 + y^2 + z^2 = c_1^2$ which are concentric spheres with the centre at O.

(ii) $\theta = c_2$, i.e., $x^2 + y^2 = z^2 \tan^2 c_2$ which are right circular cones with the z axis and vertex origin.

(iii) $\phi = c_3$, i.e., $y = x \tan c_3$, i.e., planes through the z axis.

These surfaces meet at the point $P(r, \theta, \phi)$.

DEDUCTION 4.4.1

Condition for orthogonality: Using Eq. (4.31), we get

$$\vec{r} = x\,\hat{i} + y\,\hat{j} + z\,\hat{k} = x(u, v, w)\hat{i} + y(u, v, w)\hat{j} + z(u, v, w)\hat{k} = \vec{r}(u, v, w), \text{ say} \tag{4.39}$$

The coordinate curve $v = c_2, w = c_3$ through (c_1, c_2, c_3) is the same as $\vec{r} = \vec{r}(u, c_2, c_3)$; $u =$ parameter. The tangent to this coordinate u-curve (for which v and w are constant) is parallel to the vector $\dfrac{\partial \vec{r}}{\partial u}$. The tangents to the other coordinates curves are parallel to the vectors $\dfrac{\partial \vec{r}}{\partial v}$ and $\dfrac{\partial \vec{r}}{\partial w}$. Thus, for orthogonality of the curvilinear coordinate system, we have

$$\frac{\partial \vec{r}}{\partial u} \cdot \frac{\partial \vec{r}}{\partial v} = 0, \quad \frac{\partial \vec{r}}{\partial v} \cdot \frac{\partial \vec{r}}{\partial w} = 0, \quad \frac{\partial \vec{r}}{\partial w} \cdot \frac{\partial \vec{r}}{\partial u} = 0 \tag{4.40}$$

These are the required conditions.

4.4.3 Unit Vectors in Curvilinear Systems

Let $\vec{r} = x\hat{i} + y\hat{j} + z\hat{k}$ be the position vector of a point P in the rectangular Cartesian coordinate system. Then, the transformed equation (4.31) can be written as (4.39). Then, the unit tangent vector to the u-curve at P (for which v and w are constants) is

$$\hat{e}_1 = \frac{1}{\left|\frac{\partial \vec{r}}{\partial u}\right|}\frac{\partial \vec{r}}{\partial u} \Rightarrow \frac{\partial \vec{r}}{\partial u} = h_1\ \hat{e}_1; \ h_1 = \left|\frac{\partial \vec{r}}{\partial u}\right| \tag{4.41}$$

Similarly, if $\hat{e}_2$ and $\hat{e}_3$ are unit tangent vectors to the v and w curves at P, then

$$\frac{\partial \vec{r}}{\partial v} = h_2\,\hat{e}_2, \quad \frac{\partial \vec{r}}{\partial w} = h_3\,\hat{e}_3; \quad h_2 = \left|\frac{\partial \vec{r}}{\partial v}\right|, h_3 = \left|\frac{\partial \vec{r}}{\partial w}\right| \tag{4.42}$$

The quantities h_1, h_2 and h_3 are called *scalar factors.* The unit vectors $\hat{e}_1, \hat{e}_2$ and $\hat{e}_3$ are the directions of increasing u_1, u_2 and u_3, respectively.

We know the vectors $\vec{\nabla} u$, $\vec{\nabla} v$ and $\vec{\nabla} w$ lie along the normals to the coordinate surfaces given by (4.34), which are the level surfaces of the functions u, v and w. Let us take three mutually orthogonal unit vectors

$$\hat{E}_1 = \frac{1}{|\vec{\nabla} u|}\,\vec{\nabla} u, \quad \hat{E}_2 = \frac{1}{|\vec{\nabla} v|}\,\vec{\nabla} v, \quad \hat{E}_3 = \frac{1}{|\vec{\nabla} w|}\,\vec{\nabla} w \tag{4.43}$$

at P that have directions normal to the respective level surfaces, given by Eq. (4.34). We know $|\vec{\nabla} u|$ is the directional derivative of u along the direction of the normal to the surface $u = c_1$, i.e., along the tangent to the curve $v = c_2$ and $w = c_3$. If ds_1 denotes the differential to length along this level curve, then, using Eq. (4.42), we have

$$|\vec{\nabla} u| = \frac{du}{ds_1} = \left|\frac{\partial \vec{r}}{\partial u}\right|^{-1} = \frac{1}{h_1}$$

where h_1 is given in Eq. (4.41). Similarly, we obtain the expressions for h_2 and h_3. Thus, we see from Eq. (4.43) that

$$\hat{E}_1 = h_1\,\vec{\nabla} u, \quad \hat{E}_2 = h_2\,\vec{\nabla} v, \quad \hat{E}_3 = h_3\,\vec{\nabla} w \tag{4.44}$$

This is a right-handed system of mutually orthogonal unit vectors. These systems of vectors are different for different points. Now

$$du = \vec{\nabla} u \cdot d\vec{r} = \left(\vec{\nabla} u \cdot \frac{\partial \vec{r}}{\partial u}\right) du + \left(\vec{\nabla} v \cdot \frac{\partial \vec{r}}{\partial v}\right) dv + \left(\vec{\nabla} u \cdot \frac{\partial \vec{r}}{\partial w}\right) dw$$

$$\Rightarrow \vec{\nabla} u \cdot \frac{\partial \vec{r}}{\partial u} = 1, \; \vec{\nabla} u \cdot \frac{\partial \vec{r}}{\partial v} = 0, \; \vec{\nabla} u \cdot \frac{\partial \vec{r}}{\partial w} = 0$$

Similarly, we have

$$\vec{\nabla} v \cdot \frac{\partial \vec{r}}{\partial u} = 0, \; \vec{\nabla} v \cdot \frac{\partial \vec{r}}{\partial v} = 1, \; \vec{\nabla} v \cdot \frac{\partial \vec{r}}{\partial w} = 0$$

$$\vec{\nabla} w \cdot \frac{\partial \vec{r}}{\partial u} = 0, \; \vec{\nabla} w \cdot \frac{\partial \vec{r}}{\partial w} = 0, \; \vec{\nabla} w \cdot \frac{\partial \vec{r}}{\partial w} = 1$$

Using these results, we get

$$\hat{e}_p \cdot \hat{E}_q = \delta^p_q = \begin{cases} 1, & p = q \\ 0, & p \neq q \end{cases}$$

Thus, the sets $\hat{e}_1, \hat{e}_2$ and $\hat{e}_3$ and $\hat{E}_1, \hat{E}_2$ and $\hat{E}_3$ constitute a reciprocal system of vectors. Thus, at each point P of a curvilinear system, there exists in general, two sets of unit vectors, $\hat{e}_1, \hat{e}_2$ and $\hat{e}_3$ tangent to the coordinate curves and $\hat{E}_1, \hat{E}_2$ and $\hat{E}_3$ normal to the coordinate surfaces (Fig. 4.10).

Let the vector $\vec{A}$ have the respective components A_1, A_2 and A_3 and a_1, a_2 and a_3 in the two coordinate systems. Then, $\vec{A}$ can be represented as

$$\vec{A} = A_1\hat{e}_1 + A_2\hat{e}_2 + A_3\hat{e}_3 = a_1\hat{E}_1 + a_2\hat{E}_2 + a_3\hat{E}_3$$

in terms of unit base vectors $\hat{e}_1, \hat{e}_2$ and $\hat{e}_3$ and $\hat{E}_1, \hat{E}_2$ and $\hat{E}_3$. Also, in terms of unitary base vectors $\frac{\partial \vec{r}}{\partial u}, \frac{\partial \vec{r}}{\partial v}$ and $\frac{\partial \vec{r}}{\partial w}$ and $\vec{\nabla} u$, $\vec{\nabla} v$ and $\vec{\nabla} w$ the vector $\vec{A}$ can be represented as

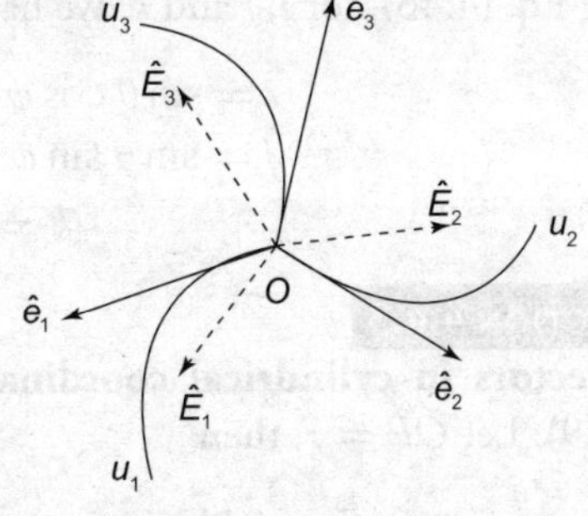

Figure 4.10 Reciprocal system

$$\vec{A} = C_1\frac{\partial \vec{r}}{\partial u} + C_2\frac{\partial \vec{r}}{\partial v} + C_3\frac{\partial \vec{r}}{\partial w} = C_1\vec{\alpha}_1 + C_2\vec{\alpha}_2 + C_3\vec{\alpha}_3$$

$$\vec{A} = c_1\vec{\nabla} u + c_2\vec{\nabla} v + c_3\vec{\nabla} w = c_1\vec{\beta}_1 + c_2\vec{\beta}_2 + c_3\vec{\beta}_3$$

where C_1, C_2 and C_3 are called the *contravariant components* of $\vec{A}$ and c_1, c_2 and c_3 are called the *covariant component* of $\vec{A}$. Also

$$\vec{\alpha}_1 = \frac{\partial \vec{r}}{\partial u},\quad \vec{\alpha}_2 = \frac{\partial \vec{r}}{\partial v},\quad \vec{\alpha}_3 = \frac{\partial \vec{r}}{\partial w},\quad \vec{\beta}_1 = \vec{\nabla} u,\quad \vec{\beta}_2 = \vec{\nabla} v,\quad \vec{\beta}_3 = \vec{\nabla} w$$

DEDUCTION 4.4.2

Unit vectors in spherical polar coordinates: Let the unit vectors be, respectively, $\hat{e}_\rho, \hat{e}_\theta$ and $\hat{e}_\phi$ (Fig. 4.8). Let $OP = r$, then

$$\vec{r} = \vec{OP} = x\,\hat{\imath} + y\,\hat{\jmath} + z\,\hat{k} = \rho\sin\theta\,\cos\phi\,\hat{\imath} + \rho\,\sin\theta\,\sin\phi\,\hat{\jmath} + \rho\cos\theta\,\hat{k} \quad (4.45)$$

The tangent vectors to ρ, θ and ϕ are given by

$$\frac{\partial \vec{r}}{\partial \rho} = \sin\theta\cos\phi\,\hat{\imath} + \sin\theta\sin\phi\,\hat{\jmath} + \cos\theta\,\hat{k}$$

$$\frac{\partial \vec{r}}{\partial \theta} = \rho\cos\theta\cos\phi\,\hat{\imath} + \rho\cos\theta\sin\phi\,\hat{\jmath} - \rho\sin\theta\,\hat{k}$$

$$\frac{\partial \vec{r}}{\partial \phi} = -\rho\sin\theta\sin\phi\,\hat{\imath} - \rho\sin\theta\sin\phi\,\hat{\jmath}$$

Therefore, $h_1 = \left|\frac{\partial \vec{r}}{\partial \rho}\right| = 1$, $h_2 = \left|\frac{\partial \vec{r}}{\partial \theta}\right| = \rho$ and $h_3 = \left|\frac{\partial \vec{r}}{\partial \phi}\right| = \rho\sin\theta$, then

$$\left.\begin{aligned}\hat{e}_\rho &= \frac{\partial \vec{r}}{\partial \rho} = \sin\theta\cos\phi\,\hat{\imath} + \sin\theta\sin\phi\,\hat{\jmath} + \cos\theta\,\hat{k}\\ \hat{e}_\theta &= \frac{1}{\rho}\frac{\partial \vec{r}}{\partial \theta} = \cos\theta\cos\phi\,\hat{\imath} + \cos\theta\sin\phi\,\hat{\jmath} - \sin\theta\,\hat{k}\\ \hat{e}_\theta &= \frac{1}{\rho\sin\theta}\frac{\partial \vec{r}}{\partial \phi} = -\sin\phi\,\hat{\imath} - \sin\phi\,\hat{\jmath}\end{aligned}\right\} \quad (4.46)$$

Solving Eq. (4.46), for $\hat{i}, \hat{j}$ and $\hat{k}$, we have

$$\left.\begin{aligned}\hat{i} &= \sin\theta\cos\phi\,\hat{e}_\rho + \cos\theta\cos\phi\,\hat{e}_\theta - \sin\phi\,\hat{e}_\phi \\ \hat{j} &= \sin\theta\sin\phi\,\hat{e}_\rho + \cos\theta\sin\phi\,\hat{e}_\theta + \cos\phi\,\hat{e}_\phi \\ \hat{k} &= \cos\theta\,\hat{e}_\rho - \sin\theta\,\hat{e}_\theta\end{aligned}\right\} \tag{4.47}$$

DEDUCTION 4.4.3

Unit vectors in cylindrical coordinates: Let the unit vectors be, respectively, $\hat{e}_\rho, \hat{e}_\phi$ and $\hat{e}_z$ (Fig. 4.9). Let $OP = r$, then

$$\vec{r} = \vec{OP} = x\,\hat{i} + y\,\hat{j} + z\,\hat{k} = \rho\cos\phi\,\hat{i} + \rho\sin\phi\,\hat{j} + z\,\hat{k} \tag{4.48}$$

The tangent vectors to the ρ, ϕ and z are given by

$$\frac{\partial\vec{r}}{\partial\rho} = \cos\phi\,\hat{i} + \sin\phi\,\hat{j}, \quad \frac{\partial\vec{r}}{\partial\phi} = -\rho\sin\phi\,\hat{i} + \rho\cos\phi\,\hat{j}, \quad \frac{\partial\vec{r}}{\partial z} = \hat{k}$$

Therefore, $h_1 = \left|\frac{\partial\vec{r}}{\partial\rho}\right| = 1$, $h_2 = \left|\frac{\partial\vec{r}}{\partial\phi}\right| = \rho$ and $h_3 = \left|\frac{\partial\vec{r}}{\partial z}\right| = 1$, then

$$\hat{e}_\rho = \frac{\partial\vec{r}}{\partial\rho} = \cos\phi\,\hat{i} + \sin\phi\,\hat{j}, \hat{e}_\phi = \frac{1}{\rho}\frac{\partial\vec{r}}{\partial\phi} = -\sin\phi\,\hat{i} + \cos\phi\,\hat{j}, \hat{e}_z = \frac{\partial\vec{r}}{\partial z} = \hat{k} \tag{4.49}$$

Solving Eq. (4.49), for $\hat{i}, \hat{j}$ and $\hat{k}$, we have

$$\hat{i} = \cos\phi\hat{e}_\rho - \sin\phi\hat{e}_\phi, \quad \hat{j} = \sin\phi\hat{e}_\rho + \cos\phi\hat{e}_\phi, \quad \hat{k} = \hat{e}_z \tag{4.50}$$

Example 4.4.4

Prove that the spherical coordinate system is orthogonal.

Solution

Let the unit vectors be, respectively, $\hat{e}_\rho, \hat{e}_\theta$ and $\hat{e}_\phi$ (Fig. 4.8) and given by Eq. (4.46). It is quite clear that

$$\begin{aligned}\hat{e}_\rho \cdot \hat{e}_\theta &= \left[\sin\theta\cos\phi\hat{i} + \sin\theta\sin\phi\hat{j} + \cos\theta\hat{k}\right] \cdot \left[\cos\theta\cos\phi\hat{i} + \cos\theta\sin\phi\hat{j} - \sin\theta\hat{k}\right] \\ &= \sin\theta\cos\theta\cos^2\phi + \sin\theta\cos\theta\sin^2\phi - \sin\theta\cos\theta \\ &= \sin\theta\cos\theta - \sin\theta\cos\theta = 0\end{aligned}$$

Similarly, $\hat{e}_\rho \cdot \hat{e}_\phi = 0 = \hat{e}_\phi \cdot \hat{e}_\theta$. Thus, the tangent vectors are mutually perpendicular, and hence, the spherical polar coordinate system is orthogonal.

Example 4.4.5

Prove that the cylindrical coordinate system is orthogonal.

Solution

Let the unit vectors be, respectively, $\hat{e}_\rho, \hat{e}_\theta$ and $\hat{e}_z$ (Fig. 4.8) and given by Eq. (4.49). It is quite clear that

$$\begin{aligned}\hat{e}_\rho \cdot \hat{e}_\phi &= \left[\cos\phi\,\hat{i} + \sin\phi\,\hat{j}\right] \cdot \left[-\sin\phi\,\hat{i} + \cos\phi\,\hat{j}\right] \\ &= \sin\phi\cos\phi - \sin\phi\cos\phi = 0\end{aligned}$$

Similarly, $\hat{e}_\rho \cdot \hat{e}_z = 0 = \hat{e}_\phi \cdot \hat{e}_z$. Thus, the tangent vectors are mutually perpendicular, and hence, the cylindrical coordinate system is orthogonal.

Example 4.4.6

If u, v and w are orthogonal curvilinear coordinates, show that the Jacobian of x, y and z with respect to u, v and w is $J = h_1 h_2 h_3$, where, h_1, h_2 and h_3 are the scale factors.

Solution

We have $\vec{r} = x\hat{i} + y\hat{j} + z\hat{k}$, where $x = x(u, v, w)$, $y = y(u, v, w)$ and $z = z(u, v, w)$. Now

$$\frac{\partial \vec{r}}{\partial u} = \frac{\partial x}{\partial u}\hat{i} + \frac{\partial y}{\partial u}\hat{j} + \frac{\partial z}{\partial u}\hat{k}, \quad \frac{\partial \vec{r}}{\partial v} = \frac{\partial x}{\partial v}\hat{i} + \frac{\partial y}{\partial v}\hat{j} + \frac{\partial z}{\partial v}\hat{k}, \quad \frac{\partial \vec{r}}{\partial w} = \frac{\partial x}{\partial w}\hat{i} + \frac{\partial y}{\partial w}\hat{j} + \frac{\partial z}{\partial w}\hat{k}$$

Then, the unit tangent vectors $\hat{e}_1$, $\hat{e}_2$ and $\hat{e}_3$ to the u, v and w curves at P are given by Eq. (4.41) and (4.42) as

$$\frac{\partial \vec{r}}{\partial u} = h_1 \hat{e}_1; \quad \frac{\partial \vec{r}}{\partial v} = h_2 \hat{e}_2, \quad \frac{\partial \vec{r}}{\partial w} = h_3 \hat{e}_3$$

where $h_1 = \left|\frac{\partial \vec{r}}{\partial u}\right|$, $h_2 = \left|\frac{\partial \vec{r}}{\partial v}\right|$ and $h_3 = \left|\frac{\partial \vec{r}}{\partial w}\right|$. Using these results, the Jacobian transformation is given by

$$J = \begin{vmatrix} \frac{\partial x}{\partial u} & \frac{\partial y}{\partial u} & \frac{\partial z}{\partial u} \\ \frac{\partial x}{\partial v} & \frac{\partial y}{\partial v} & \frac{\partial z}{\partial v} \\ \frac{\partial x}{\partial w} & \frac{\partial y}{\partial w} & \frac{\partial z}{\partial w} \end{vmatrix} = \frac{\partial \vec{r}}{\partial u} \cdot \frac{\partial \vec{r}}{\partial v} \times \frac{\partial \vec{r}}{\partial w} = h_1 \hat{e}_1 \cdot h_2 \hat{e}_2 \times h_3 \hat{e}_3$$
$$= h_1\, h_2\, h_3 \hat{e}_1 \cdot \hat{e}_2 \times \hat{e}_3 = h_1 h_2 h_3$$

4.4.4 Arc Length, Surface Area and Volume Element

We consider an orthogonal curvilinear coordinate system. From Eq. (4.39), $\vec{r} = \vec{r}(u, v, w)$, so

$$d\vec{r} = \frac{\partial \vec{r}}{\partial u}\, du + \frac{\partial \vec{r}}{\partial v}\, dv + \frac{\partial \vec{r}}{\partial w}\, dw$$
$$= h_1\, du\, \hat{e}_1 + h_2\, dv\, \hat{e}_2 + h_3\, dw\, \hat{e}_3 \tag{4.51}$$

Because of the assumed orthogonality of the curvilinear coordinate system, we have $\hat{e}_1 \cdot \hat{e}_2 = \hat{e}_2 \cdot \hat{e}_3 = \hat{e}_3 \cdot \hat{e}_1 = 0$ and $\hat{e}_1 \cdot \hat{e}_1 = \hat{e}_2 \cdot \hat{e}_2 = \hat{e}_3 \cdot \hat{e}_3 = 1$. For such an orthogonal curvilinear system, the differential of length ds is given by

$$ds^2 = d\vec{r} \cdot d\vec{r} = \left[h_1 du\hat{e}_1 + h_2 dv\hat{e}_2 + h_3 dw\hat{e}_3\right] \cdot \left[h_1 du\hat{e}_1 + h_2 dv\hat{e}_2 + h_3 dw\hat{e}_3\right]$$
$$= h_1^2 (du)^2 + h_2^2 (dv)^2 + h_2^2 (dw)^2$$
$$= \left|\frac{\partial \vec{r}}{\partial u}\right|^2 (du)^2 + \left|\frac{\partial \vec{r}}{\partial v}\right|^2 (dv)^2 + \left|\frac{\partial \vec{r}}{\partial w}\right|^2 (dw)^2 \tag{4.52}$$

$(ds)^2$ expressed in this form is known as the quadratic differential form. To interpret h_1, h_2 and h_3 geometrically, we note that along the u-curve, v and w are constants so that $dv = dw = 0$. Therefore, from Eq. (4.51), $d\vec{r} = h_1\, du\, \hat{e}_1$. If ds_1 denotes the differential of length along the coordinate curves $v = c_2$ and $w = c_3$, so that $dv = 0$ and $dw = 0$, then

$$(ds_1)^2 = d\vec{r} \cdot d\vec{r} = h_1^2\,(du)^2 = \left|\frac{\partial \vec{r}}{\partial u}\right|^2 (du)^2 \tag{4.53}$$

Similarly, $ds_2 = h_2\,dv$ and $ds_3 = h_3\,dw$. Now, $\hat{e}_1 \times \hat{e}_2 = \hat{e}_3, \hat{e}_2 \times \hat{e}_3 = \hat{e}_1$ and $\hat{e}_3 \times \hat{e}_1 = \hat{e}_2$. We consider an infinitesimal rectangular parallelepiped with one vertex at P and three concurrent edges along the three tangents to the coordinate curves. The lengths of the edges of the parallelepiped are ds_1, ds_2 and ds_3. The areas of the three intersecting faces are

$$dA_1 = \left|(h_1\,du\,\hat{e}_1) \times (h_2\,dv\,\hat{e}_2)\right| = h_1\,h_2\,du\,dv$$

$$dA_2 = \left|(h_2\,dv\,\hat{e}_2) \times (h_3\,dw\,\hat{e}_3)\right| = h_2\,h_3\,dv\,dw$$

$$dA_3 = \left|(h_3\,dw\,\hat{e}_3) \times (h_1\,du\,\hat{e}_1)\right| = h_3\,h_1\,dw\,du$$

The volume of the parallelopiped is

$$dV = \left|(h_1\,du\,\hat{e}_1) \cdot (h_2\,dv\,\hat{e}_2) \times (h_3\,dw\,\hat{e}_3)\right| = h_1\,h_2\,h_3\,du\,dv\,dw \tag{4.54}$$

DEDUCTION 4.4.4

Arc length, scale factor and volume element in cylindrical coordinates: The position vector is given by $\vec{r} = \rho\cos\phi\,\hat{i} + \rho\sin\phi\,\hat{j} + z\,\hat{k}$, so

$$\begin{aligned} d\vec{r} &= \frac{\partial \vec{r}}{\partial \rho}\,d\rho + \frac{\partial \vec{r}}{\partial \phi}\,d\phi + \frac{\partial \vec{r}}{\partial z}\,dz \\ &= (\cos\phi\,\hat{i} + \sin\phi\,\hat{j})d\rho + (-\rho\sin\phi\,\hat{i} + \rho\cos\phi\,\hat{j})d\phi + \hat{k}\,dz \\ &= \left[\cos\phi\,d\rho - \rho\sin\phi\,d\phi\right]\hat{i} + \left[\sin\phi\,d\rho + \rho\cos\phi\,d\phi\right]\hat{j} + \hat{k}\,dz \end{aligned}$$

Thus, the differential of arc length in cylindrical coordinates (ρ, ϕ, z) is given by

$$\begin{aligned} ds^2 = d\vec{r} \cdot d\vec{r} &= \left[\cos\phi\,d\rho - \rho\sin\phi\,d\phi\right]^2 + \left[\sin\phi\,d\rho + \rho\cos\phi\,d\phi\right]^2 + (dz)^2 \\ &= (d\rho)^2 + \rho^2\,(d\phi)^2 + (dz)^2 \end{aligned} \tag{4.55}$$

Comparing with Eq. (4.52), we obtain the scale factors as $h_1 = 1$, $h_2 = \rho$ and $h_3 = 1$. The edges of the volume element in cylindrical coordinates have magnitudes $\rho d\phi$, $d\rho$ and dz (Fig. 4.11).

Thus, the edges are given by $ds_1 = h_1 du_1 = (1)(d\rho) = d\rho$, $ds_2 = h_2 du_2 = \rho\ d\phi$ and $ds_3 = (1)(dz) = dz$. Using Eq. (4.54), the volume element in cylindrical coordinates is given by

$$\begin{aligned} V &= h_1\,h_2\,h_3\,d\rho\,d\phi\,dz = ds_1\,ds_2\,ds_3 \\ &= d\rho\,\rho\,d\phi\,dz = \rho\,d\rho\,d\phi\,dz \end{aligned}$$

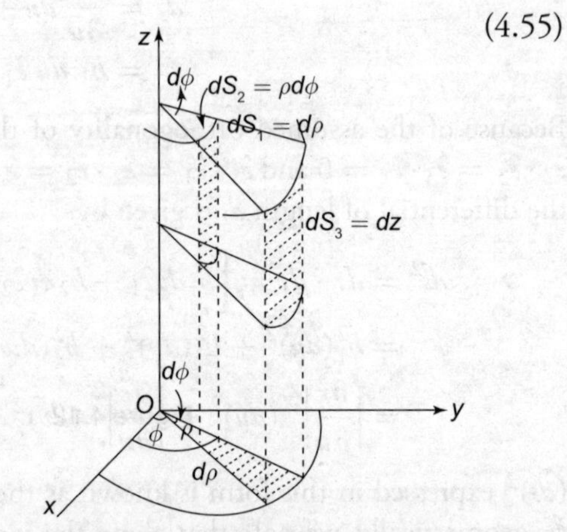

Figure 4.11 Line and volume element

DEDUCTION 4.4.5

Arc length, scale factor and volume element in spherical polar coordinates: The position vector is given by $\vec{r} = \rho \sin\theta \cos\phi \, \hat{\imath} + \rho \sin\theta \sin\phi \, \hat{\jmath} + \cos\theta \, \hat{k}$, so

$$
\begin{aligned}
d\vec{r} &= \frac{\partial \vec{r}}{\partial \rho} \, d\rho + \frac{\partial \vec{r}}{\partial \theta} \, d\theta + \frac{\partial \vec{r}}{\partial \phi} \, d\phi \\
&= (\sin\theta \cos\phi \, \hat{\imath} + \sin\theta \sin\phi \, \hat{\jmath} + \cos\theta \hat{k}) d\rho + \rho(\cos\theta \cos\phi \, \hat{\imath} + \cos\theta \sin\phi \, \hat{\jmath} \\
&\quad - \sin\theta \hat{k}) d\theta + \rho \sin\theta(-\sin\phi \hat{\imath} + \cos\phi \hat{\jmath}) \, d\phi \\
&= \Big[\sin\theta \cos\phi \, d\rho + \rho \cos\theta \cos\phi \, d\theta - \rho \sin\theta \sin\phi \Big] \hat{\imath} + \Big[\sin\theta \sin\phi \, d\rho \\
&\quad + \rho \cos\theta \sin\phi \, d\theta + \rho \sin\theta \cos\phi d\phi \Big] \hat{\jmath} + \Big[\cos\theta \, d\rho - \rho \sin\theta d\theta \Big] \hat{k}
\end{aligned}
$$

Thus, the differential of arc length in spherical polar coordinates (ρ, θ, ϕ) is given by

$$
\begin{aligned}
ds^2 = d\vec{r} \cdot d\vec{r} &= \Big[\sin\theta \cos\phi \, d\rho + \rho \cos\theta \cos\phi \, d\theta - \rho \sin\theta \sin\phi \Big]^2 + \Big[\sin\theta \sin\phi \, d\rho \\
&\quad + \rho \cos\theta \sin\phi \, d\theta + \rho \sin\theta \cos\phi d\phi \Big]^2 + \Big[\cos\theta \, d\rho - \rho \sin\theta d\theta \Big]^2 \\
&= (d\rho)^2 + \rho^2 \, (d\theta)^2 + \rho^2 \sin^2\theta (d\phi)^2 \qquad (4.56)
\end{aligned}
$$

Comparing with Eq. (4.52), we obtain the scale factors as $h_1 = 1$, $h_2 = \rho$ and $h_3 = \rho \sin\theta$. The edges of the volume element in spherical polar coordinates have magnitudes $\rho d\phi$, $d\rho$ and dz (Fig. 4.12).

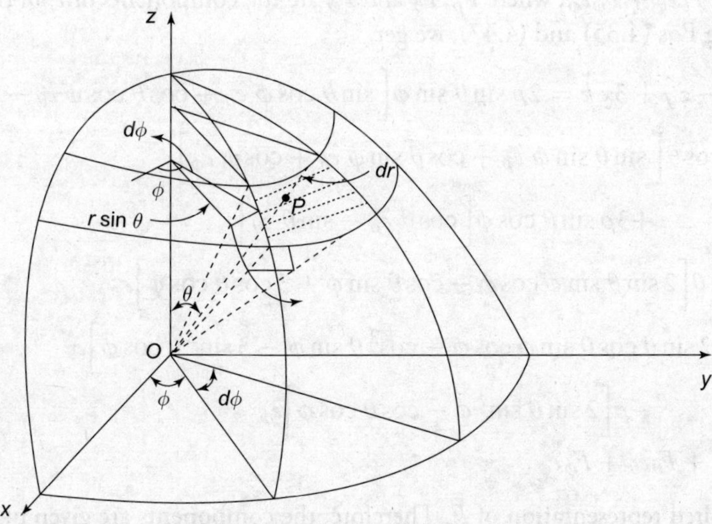

Figure 4.12 Line and volume element

Thus, the edges are given by $ds_1 = h_1 du_1 = (1)(d\rho) = d\rho$, $ds_2 = h_2 du_2 = \rho \, d\phi$ and $ds_3 = (1)(dz) = dz$. Using Eq. (4.54), the volume element in spherical polar coordinates is given by

$$V = h_1\, h_2\, h_3\, d\rho\, d\theta\, d\phi = ds_1\, ds_2\, ds_3$$
$$= d\rho\, \rho\, d\theta\, \rho\, \sin\theta\, d\phi = \rho^2\, \sin\theta\, d\rho\, d\theta\, d\phi$$

Example 4.4.7

Represent the vector $\vec{F} = z\hat{i} - 2x\hat{j} + y\hat{k}$ in cylindrical coordinates and determine F_ρ, F_ϕ and F_z.

Solution

Let $\vec{F} = F_\rho \hat{e}_\rho + F_\phi \hat{e}_\phi + F_z \hat{e}_z$, where F_ρ, F_ϕ and F_z are the components of $\vec{F}$ in the $\hat{e}_\rho$, $\hat{e}_\phi$ and $\hat{e}_z$ directions. Using Eqs. (4.35) and (4.50), we get

$$\begin{aligned}\vec{F} &= z(\cos\phi\hat{e}_\rho - \sin\phi\hat{e}_\phi) - 2x\hat{j} + y\hat{k} = z(\cos\phi\hat{e}_\rho - \sin\phi\hat{e}_\phi)\\ &\quad -2\rho\cos\phi(\sin\phi\hat{e}_\rho + \cos\phi\hat{e}_\phi) + \rho\sin\phi\hat{e}_z\\ &= \left[z\cos\phi - 2\rho\cos\phi\sin\phi\right]\hat{e}_\rho - \left[z\sin\phi + 2\rho\cos^2\phi\right]\hat{e}_\phi + \rho\sin\phi\hat{e}_z\end{aligned}$$

This is the required representation of $\vec{F}$. Therefore, the components are given by

$$F_\rho = z\cos\phi - 2\rho\cos\phi\sin\phi, F_\phi = z\sin\phi + 2\rho\cos^2\phi \text{ and } F_z = \rho\sin\phi$$

Example 4.4.8

Represent the vector $\vec{F} = 2y\hat{i} - z\hat{j} + 3x\hat{k}$ in spherical polar coordinates and determine F_ρ, F_θ and F_ϕ.

Solution

Let $\vec{F} = F_\rho \hat{e}_\rho + F_\theta \hat{e}_\theta + F_\phi \hat{e}_\phi$, where F_ρ, F_θ and F_ϕ are the components of $\vec{F}$ in the $\hat{e}_\rho$, $\hat{e}_\theta$ and $\hat{e}_\phi$ directions. Using Eqs (4.35) and (4.47), we get

$$\begin{aligned}\vec{F} &= 2y\,\hat{i} - z\,\hat{j} + 3x\,\hat{k} = 2\rho\sin\theta\sin\phi\Big[\sin\theta\cos\phi\,\hat{e}_\rho + \cos\theta\cos\phi\,\hat{e}_\theta - \sin\phi\,\hat{e}_\phi\Big]\\ &\quad -\rho\cos\theta\Big[\sin\theta\sin\phi\,\hat{e}_\rho + \cos\theta\sin\phi\,\hat{e}_\theta + \cos\phi\,\hat{e}_\phi\Big]\\ &\quad +3\rho\sin\theta\cos\phi\Big[\cos\theta\,\hat{e}_\rho - \sin\theta\,\hat{e}_\theta\Big]\\ &= \rho\sin\theta\Big[2\sin\theta\sin\phi\cos\phi - \cos\theta\sin\phi + 3\cos\theta\cos\phi\Big]\hat{e}_\rho\\ &\quad +\rho\Big[2\sin\theta\cos\theta\sin\phi\cos\phi - \cos^2\theta\sin\phi - 3\sin^2\theta\cos\phi\Big]\hat{e}_\theta\\ &\quad -\rho\Big[2\sin\theta\sin^2\phi - \cos\theta\cos\phi\Big]\hat{e}_\phi\\ &= F_\rho\hat{e}_\rho + F_\theta\hat{e}_\theta + F_\phi\hat{e}_\phi\end{aligned}$$

This is the required representation of $\vec{F}$. Therefore, the components are given by

$$\begin{aligned}F_\rho &= \rho\sin\theta\Big[2\sin\theta\sin\phi\cos\phi - \cos\theta\sin\phi + 3\cos\theta\cos\phi\Big]\\ F_\theta &= \rho\Big[2\sin\theta\cos\theta\sin\phi\cos\phi - \cos^2\theta\sin\phi - 3\sin^2\theta\cos\phi\Big]\\ F_\phi &= -\rho\Big[2\sin\theta\sin^2\phi - \cos\theta\cos\phi\Big]\end{aligned}$$

4.5 APPLICATIONS IN MECHANICS

A body is said to be in motion when it changes its position with respect to other bodies. Since the motion of a body of finite size is quite complicated, in that it can rotate about an axis passing through itself in addition to displacement, we begin the study of mechanics by considering the motion of a particle of matter called *material point* and then extend this study to systems of particles and bodies of finite size.

Let P be the position of the particle whose position vector is $\vec{r}$ with respect to some convenient origin O of a frame of reference (Fig. 4.13). Since P is in motion, $\vec{r}$ is a function $\vec{r}(t)$ of time t measured from a suitable initial moment to which we assign the value $t = 0$. Then, the velocity $\vec{v}$ is the time rate of change of the displacement of a particle, given by

$$\vec{v} = \lim_{\Delta t \to 0} \frac{\Delta \vec{r}}{\Delta t} = \lim_{\Delta t \to 0} \frac{\Delta \vec{r}}{\Delta s}\frac{\Delta s}{\Delta t} = \frac{d\vec{r}}{dt} = \dot{\vec{r}} \quad (4.57)$$

Figure 4.13 Frame of reference

where the dot on a symbol denotes, as usual, the derivative with respect to time. As is evident from Fig. 4.13, in the limit $\Delta t \to 0$, $\left|\frac{\Delta \vec{r}}{\Delta s}\right| \to 1$ and the direction of $\Delta\vec{r}$ tend to that of the tangent to the trajectory at P, and therefore, we see that

$$\frac{d\vec{r}}{ds} = \lim_{\Delta t \to 0} \frac{\Delta \vec{r}}{\Delta s} = \hat{t}$$

where $\hat{t}$ is the unit tangent vector to the trajectory at P and is directed in the sense in which the variable t increases along the path. The acceleration vector $\vec{a}$ is similarly defined as the time rate of change of velocity vector

$$\vec{a} = \frac{d\vec{v}}{dt} = \frac{d^2\vec{r}}{dt^2} = \ddot{\vec{r}} \quad (4.58)$$

Newton's second law of motion is the basic law of mechanics, and it states that

$$\vec{F}^{(e)} = \frac{d}{dt}(m\vec{v}) = m\frac{d\vec{v}}{dt} = m\vec{a} \quad (4.59)$$

where $\vec{a} = \ddot{\vec{r}}$ is acceleration produced in the motion of the particle on applying the external force $\vec{F}^{(e)}$. The reference frame in which Eq. (4.59) is valid is called the *inertial* or *Galilean system.*

Example 4.5.1

A particle of mass 2 units moves along the space curve defined by $\vec{r} = (4t^2 - t^3)\hat{\imath} - 5t\hat{\jmath} + (t^4 - 2)\hat{k}$. Find at any time t,(i) the velocity, (ii) the momentum, (iii) the acceleration and (iv) the force field.

Solution

Given $\vec{r} = (4t^2 - t^3)\hat{\imath} - 5t\hat{\jmath} + (t^4 - 2)\hat{k}$.

(i) Using Eq. (4.57), the velocity at any time t is given by

$$\vec{v} = \frac{d\vec{r}}{dt} = (8t - 3t^2)\hat{\imath} - 5\hat{\jmath} + 4t^3\hat{k}$$

(ii) The momentum $\vec{p}$ is given by

$$\vec{p} = m\vec{v} = 2\left[(8t - 3t^2)\hat{i} - 5\hat{j} + 4t^3\hat{k}\right]$$

(iii) Using Eq. (4.58), the acceleration at any time t is given by

$$\vec{a} = \frac{d\vec{v}}{dt} = \frac{d^2\vec{r}}{dt^2} = (8 - 6t)\hat{i} + 12t^2\hat{k}$$

(iv) The force field at any time t is given by

$$\vec{F} = m\frac{d\vec{v}}{dt} = \frac{d\vec{p}}{dt} = 2\left[(8 - 6t)\hat{i} + 12t^2\hat{k}\right]$$

Example 4.5.2

A particle moves along the curve $x = 1 + t^3, y = t^2$ and $z = 5 + 2t$, t being the time variable. Find the magnitude of the components of its velocity and acceleration at $t = 1$ in the direction $\hat{i} + \hat{j} + 3\hat{k}$.

Solution

The position vector of any point on the curve is $\vec{r} = (1 + t^3)\hat{i} + t^2\hat{j} + (5 + 2t)\hat{k}$. Differentiating with respect to t, we get

$$\vec{v} = \frac{d\vec{r}}{dt} = 3t^2\,\hat{i} + 2t\,\hat{j} + 2\,\hat{k} = 3\hat{i} + 2\hat{j} + 2\hat{k}, \text{ at } t = 1$$

Now, the unit vector in the direction of $\vec{\alpha} = \hat{i} + \hat{j} + 3\hat{k}$ is $\hat{\alpha} = (\hat{i} + \hat{j} + 3\hat{k})/\sqrt{11}$. Thus, the magnitude of $\vec{v}$ in the direction of the unit vector $\hat{\alpha}$ is given by

$$\vec{v} \cdot \hat{\alpha} = (3\hat{i} + 2\hat{j} + 2\hat{k}) \cdot \frac{1}{\sqrt{11}}(\hat{i} + \hat{j} + 3\hat{k}) = \frac{11}{\sqrt{11}} = \sqrt{11}$$

Further differentiation gives

$$\vec{a} = \frac{d\vec{v}}{dt} = 6t\,\hat{i} + 2\,\hat{j} = 6\hat{i} + 2\hat{j}, \text{ at } t = 1$$

Thus, the magnitude of the acceleration $\vec{a}$ in the direction of the unit vector $\hat{\alpha}$ is given by

$$\vec{a} \cdot \hat{\alpha} = (6\hat{i} + 2\hat{j}) \cdot \frac{1}{\sqrt{11}}(\hat{i} + \hat{j} + 3\hat{k}) = \frac{8}{\sqrt{11}} = 8\sqrt{11}/11$$

DEDUCTION 4.5.1

Rectangular components: In rectangular components (Fig. 4.14), the position vector is given by $\vec{r} = x\hat{i} + y\hat{j} + z\hat{k}$. Therefore, from Eqs (4.57) and (4.58), $\vec{v}$ and $\vec{a}$ are

$$\vec{v} = \dot{x}\hat{i} + \dot{y}\hat{j} + \dot{z}\hat{k}; \quad \vec{a} = \ddot{x}\hat{i} + \ddot{y}\hat{j} + \ddot{z}\hat{k} \quad (4.60)$$

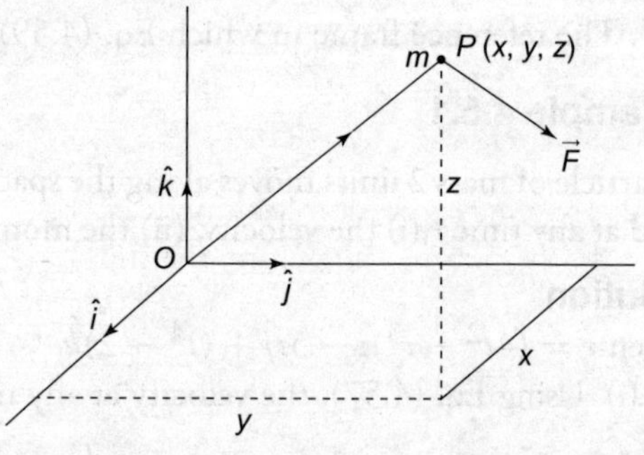

Figure 4.14 Rectangular coordinates

We call (x, y, z) the rectangular components of position (or the rectangular coordinates) of the point P. Similarly, $(\dot{x}, \dot{y}, \dot{z})$ and

$(\ddot{x}, \ddot{y}, \ddot{z})$ are the rectangular components of velocity and acceleration, respectively. The distance P from the origin and the speed of P are

$$r = |\vec{r}| = \sqrt{x^2 + y^2 + z^2}; \quad v = |\vec{v}| = \sqrt{\dot{x}^2 + \dot{y}^2 + \dot{z}^2}$$

Expressing the resultant force on the mass m at point P in rectangular components $\vec{F} = F_x\hat{i} + F_y\hat{j} + F_z\hat{k}$ and Eq. (4.59) gives three scalar equations of motion

$$F_x = m\,\ddot{x}, \quad F_y = m\,\ddot{y}, \quad F_z = m\,\ddot{z} \tag{4.61}$$

Example 4.5.3

Linear harmonic oscillator: A block of mass m moves along a frictionless table, towards a fixed point O on it by a force proportional to the distance from O. Initially, the block is at a distance x_0 from O and released. Determine the kinetic energy and position at any time.

Solution

A block slides on a frictionless horizontal table (Fig. 4.15) indicated by the x axis. When the body is displaced to the right, the direction of the force exerted by the spring on the body points to the left, and when the body is displaced to the left, the direction of the force points to the right (restoring force). When the mass is displaced through the distance x, the restoring force set up in the stretched spring is $\vec{F} = -kx\,\hat{i}$, where k is the constant of proportionality, often called the *spring constant, elastic constant, stiffness factor* or *modulus of elasticity*, and $\hat{i}$ is the unit vector in the positive x direction. The equation of motion can be written as

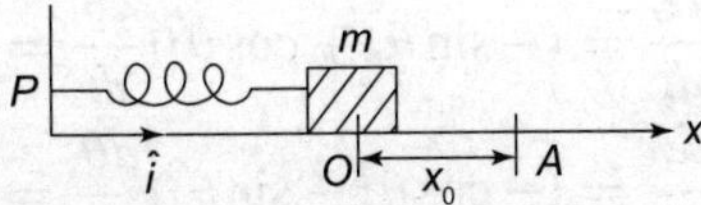

Figure 4.15 Linear harmonic oscillator

$$m\frac{d^2x\hat{i}}{dt^2} = m\frac{d\vec{v}}{dt} = \vec{F} = -kx\,\hat{i} \Rightarrow m\ddot{x} + kx = 0$$

This vibrating system is called a *simple harmonic oscillator* or *linear harmonic oscillator*. Now

$$\int_{x_0}^{x} F(x)dx = m\int_{x_0}^{x} \frac{dv}{dt}dx = m\int_{t_0}^{t} \frac{dv}{dt}vdt$$

$$\Rightarrow \int_{x_0}^{x} (-kx)dx = m\int_{t_0}^{t} \frac{d}{dt}(\frac{1}{2}v^2)dt$$

$$\Rightarrow -\frac{1}{2}kx^2 + \frac{1}{2}kx_0^2 = \frac{1}{2}mv^2 - \frac{1}{2}mv_0^2$$

$$\Rightarrow \frac{1}{2}mv^2 = \frac{1}{2}mv_0^2 + \frac{1}{2}kx_0^2 - \frac{1}{2}kx^2$$

This gives the expression for kinetic energy. Let the mass be released initially from rest at a distance x_0 from the origin. Therefore, the initial conditions are $x = x_0, v = v_0 = 0$ at $t = 0$. Thus

$$v = \frac{dx}{dt} = \sqrt{\frac{k}{m}}\sqrt{x_0^2 - x^2} \Rightarrow \int_{x_0}^{x} \frac{dx}{\sqrt{x_0^2 - x^2}} = \sqrt{\frac{k}{m}}\int_0^t dt$$

$$\Rightarrow x = x_0 \sin(\omega t + \frac{\pi}{2}) = x_0 \cos \omega t; \quad \omega = \sqrt{k/m}$$

Many problems of oscillations can be discussed using this single mathematical model.

DEDUCTION 4.5.2

Radial and transverse components: Consider a particle P moving on a plane, its position being described in polar coordinates (r, θ), as in Fig. 4.16. Let $\hat{e}_r$ be the unit vector along OP and $\hat{e}_\theta$ the unit vector perpendicular to $\hat{e}_r$. Then, we can see from Fig. 4.16 that $\hat{e}_r = \cos\theta \hat{i} + \sin\theta \hat{j}$ and $\hat{e}_\theta = -\sin\theta \hat{i} + \cos\theta \hat{j}$. Since $\hat{e}_r$ and $\hat{e}_\theta$ are time dependent as the particle moves along the curve Γ, we have

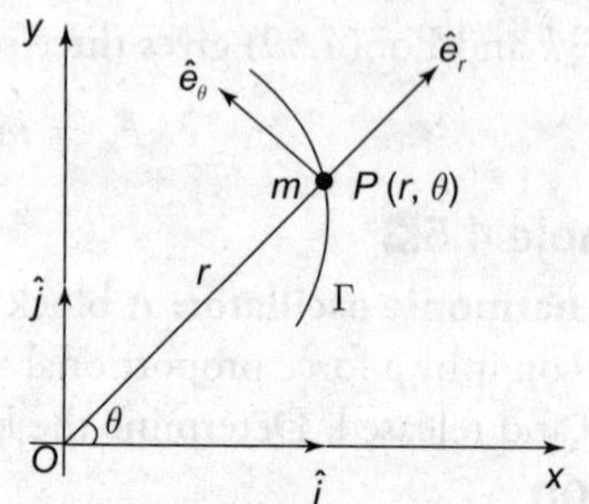

Figure 4.16 Polar coordinate system

$$\frac{d\hat{e}_r}{dt} = (-\sin\theta \hat{i} + \cos\theta \hat{j})\frac{d\theta}{dt} = \dot{\theta}\,\hat{e}_\theta$$

$$\frac{d\hat{e}_\theta}{dt} = (-\cos\theta \hat{i} - \sin\theta \hat{j})\frac{d\theta}{dt} = -\dot{\theta}\,\hat{e}_r$$

Since $\vec{r} = r\,\hat{e}_r$, from Eq. (4.57), the velocity vector is

$$\vec{v} = \dot{\vec{r}} = \dot{r}\,\hat{e}_r + r\,\dot{\hat{e}}_r = \dot{r}\,\hat{e}_r + r\dot{\theta}\,\hat{e}_\theta \tag{4.62}$$

The components $v_r = \dot{r}$ and $v_\theta = r\dot{\theta}$ of the velocity $\vec{v}$ are, respectively, called the *radial* and *transverse velocities* of the particle P. Again differentiating with respect to time, from Eq. (4.58) we obtain the acceleration as

$$\begin{aligned}\vec{a} = \dot{\vec{v}} &= \ddot{r}\,\hat{e}_r + \dot{r}\,\dot{\hat{e}}_r + \dot{r}\dot{\theta}\,\hat{e}_\theta + r\ddot{\theta}\,\hat{e}_\theta + r\dot{\theta}\,\dot{\hat{e}}_\theta \\ &= \ddot{r}\,\hat{e}_r + \dot{r}\,\dot{\theta}\,\hat{e}_\theta + \dot{r}\dot{\theta}\,\hat{e}_\theta + r\ddot{\theta}\,\hat{e}_\theta + r\dot{\theta}\,(-\dot{\theta}\hat{e}_r) \\ &= (\ddot{r} - r\dot{\theta}^2)\hat{e}_r + (r\ddot{\theta} + 2\dot{r}\dot{\theta})\hat{e}_\theta\end{aligned} \tag{4.63}$$

The components $a_r = \ddot{r} - r\dot{\theta}^2$ and $a_\theta = r\ddot{\theta} + 2\dot{r}\dot{\theta}$ of the acceleration $\vec{v}$ are, respectively, called the *radial* and *transverse accelerations* of the particle P. With the help of Eq. (4.59), the equations of motion of a particle in the *radial* and *transverse* directions can be obtained as

$$F_r = m(\ddot{r} - r\dot{\theta}^2), \quad F_\theta = m(r\ddot{\theta} + 2\dot{r}\dot{\theta}) = m\frac{1}{r}\frac{d}{dt}(r^2\dot{\theta}) \tag{4.64}$$

Example 4.5.4

Write down the cross radial equation of motion for a simple pendulum in terms of the angle of deflection θ, and hence, obtain the corresponding equation for simple harmonic motion, when θ is small.

Solution

Consider a point mass m attached to an extensible string of length l and suspended from S (Fig. 4.17).

Let P be the position of the particle at any time t such that $<PSO = \theta$ is the angular displacement from the vertical SO at any time t, where $\widehat{AP} = s = l\theta$. Now, the cross radial equation of motion of a particle is

$$m\frac{1}{r}\frac{d}{dt}(r^2\dot{\theta}) = -mg\sin\theta$$

Now, $r = l$ and when θ is small, then $\sin\theta \approx \theta$, so

$$\frac{1}{l}\frac{d}{dt}(l^2\dot{\theta}) = -g\sin\theta \approx -g\,\theta$$

$$\Rightarrow \ddot{\theta} = \frac{d^2\theta}{dt^2} = -\frac{g}{l}\,\theta$$

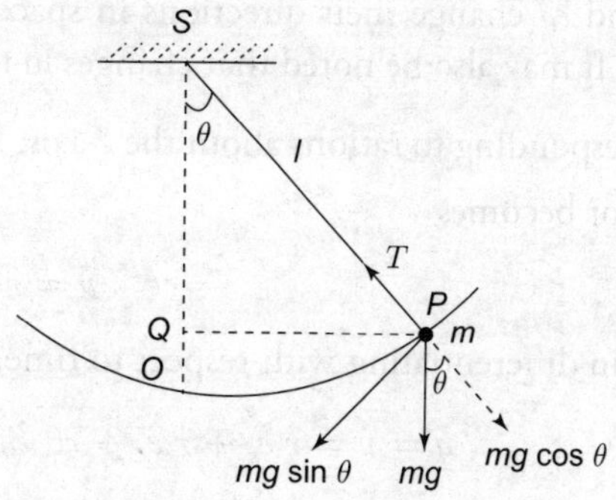

Figure 4.17 Simple pendulum

This is the required equation of simple harmonic motion.

Example 4.5.5

Find the radial acceleration of a particle moving along the curve $r = ae^{\theta\cot\alpha}$; a and α being constant and *theta* $= \omega =$ a constant.

Solution

For the given equiangular spiral $r = ae^{\theta\cot\alpha}$, we have

$$\dot{r} = a\cot\alpha\, e^{\theta\cot\alpha}\,\dot{\theta} = \omega\cot\alpha\, r;\quad \ddot{r} = \omega\cot\alpha\,\dot{r} = \omega^2\cot^2\alpha\, r$$

Using Eq. (4.63), the radial acceleration is given by

$$a_\theta = \ddot{r} - r\dot{\theta}^2 = \omega^2\cot^2\alpha\, r - r\,\omega^2 = (\cot^2\alpha - 1)\omega^2\, r$$

DEDUCTION 4.5.3

Cylindrical components: The cylindrical coordinates (r, θ, z) are shown in Fig. 4.18. From the geometry, the cylindrical and rectangular coordinates are related by $x = r\cos\theta, y = r\sin\theta$ and $z = z$. As shown in Fig. 4.18, the position vector of P is $\vec{r} = \overrightarrow{OP} = r\hat{e}_r + z\hat{e}_z$. It may be noted that the unit vectors $\hat{e}_r, \hat{e}_\theta$ and $\hat{e}_z$ form a mutually orthogonal triad, whose directions are given by the directions in which P moves for small increases in r, θ and z, respectively. Differentiating with respect to time, from Eq. (4.57) we obtain the velocity as

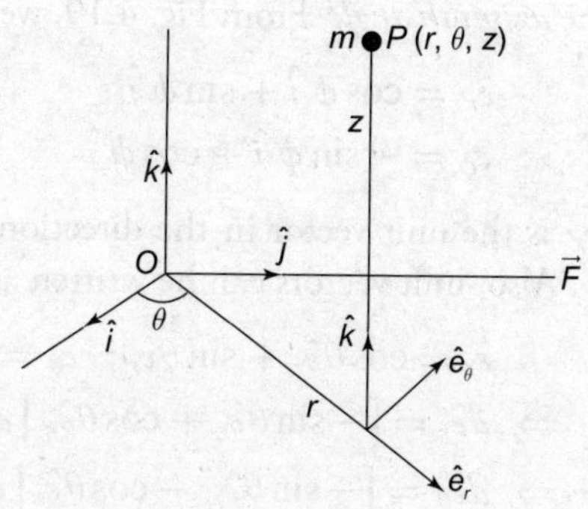

Figure 4.18 Cylindrical coordinates

$$\vec{v} = \dot{\vec{r}} = \dot{r}\,\hat{e}_r + r\,\dot{\hat{e}}_r + \dot{z}\,\hat{e}_z$$

$\hat{e}_r$ and $\hat{e}_\theta$ change their directions in space as P moves, while e_z always remains parallel to the z axis. It may also be noted that changes in the directions of $\hat{e}_r$ and $\hat{e}_\theta$ are solely due to changes in θ, corresponding to rations about the z axis. Now, $\frac{d\hat{e}_r}{dt} = \dot{\theta}\,\hat{e}_\theta$; $\frac{d\hat{e}_\theta}{dt} = -\dot{\theta}\,\hat{e}_r$. Therefore, the velocity vector becomes

$$\vec{v} = \dot{r}\,\hat{e}_r + r\,\dot{\theta}\,\hat{e}_\theta + \dot{z}\,\hat{e}_z$$

Again differentiating with respect to time, from Eq. (4.58) we obtain the acceleration as

$$\begin{aligned}\vec{a} = \dot{\vec{v}} &= \ddot{r}\,\hat{e}_r + \dot{r}\,\dot{\hat{e}}_r + \dot{r}\dot{\theta}\,\hat{e}_\theta + r\ddot{\theta}\,\hat{e}_\theta + r\dot{\theta}\,\dot{\hat{e}}_\theta + \dot{z}\,\dot{\hat{e}}_z + \ddot{z}\,\hat{e}_z;\ \text{as } \dot{\hat{e}}_z = 0\\ &= \ddot{r}\,\hat{e}_r + \dot{r}\,\dot{\theta}\,\hat{e}_\theta + \dot{r}\dot{\theta}\,\hat{e}_\theta + r\ddot{\theta}\,\hat{e}_\theta + r\dot{\theta}\,(-\dot{\theta}\hat{e}_r) + \ddot{z}\,\hat{e}_z\\ &= (\ddot{r} - r\dot{\theta}^2)\hat{e}_r + (r\ddot{\theta} + 2\dot{r}\dot{\theta})\hat{e}_\theta + \ddot{z}\,\hat{e}_z\\ &= (\ddot{r} - r\dot{\theta}^2)\hat{e}_r + \frac{1}{r}\frac{d}{dt}(r^2\dot{\theta})\hat{e}_\theta + \ddot{z}\,\hat{e}_z \end{aligned} \tag{4.65}$$

With the help of Eq. (4.59), the equations of motion of a particle in the cylindrical coordinate system can be obtained as

$$F_r = m(\ddot{r} - r\dot{\theta}^2),\quad F_\theta = m\frac{1}{r}\frac{d}{dt}(r^2\dot{\theta}),\quad F_z = m\ddot{z} \tag{4.66}$$

DEDUCTION 4.5.4

Spherical polar components: From Fig. 4.19, the spherical polar (r, θ, ϕ) and rectangular coordinates are related by $x = r\sin\theta\cos\phi, y = r\sin\theta\sin\phi$ and $z = r\cos\theta$. Here, θ is the angle between the radius vector and the z axis which is called *zenith angle*, while ϕ is the angle between the projection of radius vector OP in the xy pane and the x axis measured in the direction shown in Fig.4.19 and is called the *azimuth angle.* From Fig. 4.19, we see that

$$\hat{e}_{r'} = \cos\phi\,\hat{i} + \sin\phi\,\hat{j};$$
$$\hat{e}_\phi = -\sin\phi\,\hat{i} + \cos\phi\,\hat{j}$$

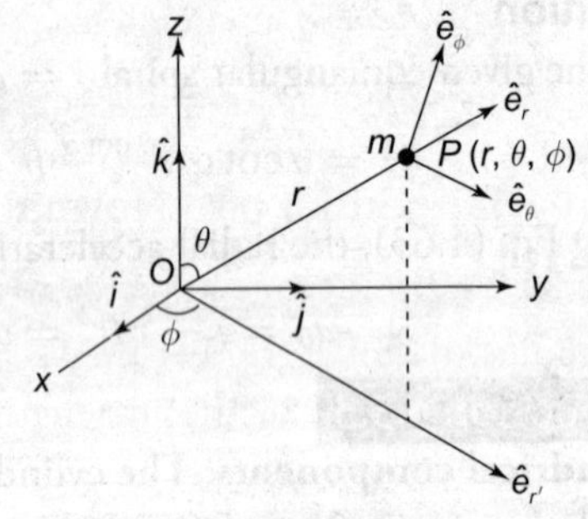

Figure 4.19 Spherical polar coordinates

where $\hat{e}_{r'}$ is the unit vector in the direction r' in the xy plane. Thus, $d\hat{e}_{r'} = d\phi\,\hat{e}_\phi$ and $d\hat{e}_\phi = -d\phi\,\hat{e}_{r'}$. Also, unit vectors can be written as

$$\hat{e}_r = \cos\theta\hat{e}_z + \sin\theta\hat{e}_{r'};\quad \hat{e}_\theta = \cos\theta\hat{e}_{r'} - \sin\theta\hat{e}_z;\quad \hat{e}_{r'} = \sin\theta\hat{e}_r + \cos\theta\hat{e}_\theta$$
$$\Rightarrow d\hat{e}_r = \left[-\sin\theta\hat{e}_z + \cos\theta\hat{e}_{r'}\right]d\theta + \sin\theta\,d\hat{e}_{r'} = d\theta\,\hat{e}_\theta + \sin\theta\,d\phi\,\hat{e}_\phi$$
$$\Rightarrow d\hat{e}_\theta = \left[-\sin\theta\hat{e}_{r'} - \cos\theta\hat{e}_z\right]d\theta + \cos\theta\,d\hat{e}_{r'} = -d\theta\,\hat{e}_r + \cos\theta\,d\phi\,\hat{e}_\phi$$
$$\Rightarrow d\hat{e}_\phi = \left[-\cos\phi\,\hat{e}_{r'} - \sin\phi\,\hat{e}_y\right]d\phi = -d\phi\,\hat{e}_{r'}$$

The position vector for the point P can be written as $\vec{r} = r\,\hat{e}_r$. Thus, in spherical coordinates, we obtain the expression for the velocity as

$$\frac{d\vec{r}}{dt} = \frac{dr}{dt}\,\hat{e}_\phi + r\,\frac{d\hat{e}_r}{dt} = \frac{dr}{dt}\,\hat{e}_r + r\left[\frac{d\theta}{dt}\hat{e}_\theta + \sin\theta\,\frac{d\phi}{dt}\,\hat{e}_\phi\right]$$

$$= \frac{dr}{dt}\,\hat{e}_r + r\,\frac{d\theta}{dt}\,\hat{e}_\theta + r\sin\theta\,\frac{d\phi}{dt}\,\hat{e}_\phi$$

$$= \dot{r}\,\hat{e}_r + r\dot{\theta}\,\hat{e}_\theta + r\dot{\phi}\sin\theta\,\hat{e}_\phi \tag{4.67}$$

In spherical coordinates, we obtain the expression for the acceleration as

$$\vec{a} = \frac{d^2\vec{r}}{dt^2} = \frac{d}{dt}\Big[\dot{r}\,\hat{e}_r + r\dot{\theta}\,\hat{e}_\theta + r\dot{\phi}\sin\theta\,\hat{e}_\phi\Big]$$

$$= \ddot{r}\,\hat{e}_r + \dot{r}\,\frac{d}{dt}(\hat{e}_r) + \dot{r}\dot{\theta}\hat{e}_\theta + r\ddot{\theta}\,\hat{e}_\theta + r\dot{\theta}\frac{d\hat{e}_\theta}{dt} + \dot{r}\dot{\phi}\sin\theta\,\hat{e}_\phi$$

$$+ r\ddot{\phi}\sin\theta\,\hat{e}_\phi + r\dot{\phi}\cos\theta\,\dot{\theta}\hat{e}_\phi + r\dot{\phi}\sin\theta\frac{d\hat{e}_\phi}{dt}$$

$$= \ddot{r}\,\hat{e}_r + \dot{r}\Big[\dot{\theta}\,\hat{e}_\theta + \sin\theta\,d\phi\,\hat{e}_\phi\Big] + 2\dot{r}\,\dot{\theta}\,\hat{e}_\theta + r\dot{\theta}\Big[-\dot{\theta}\,\hat{e}_r + \cos\theta\,d\phi\,\hat{e}_\phi\Big] + r\,\dot{\phi}\,\dot{\theta}\sin\theta\,\dot{\theta}\hat{e}_\theta$$

$$+ \sin\theta\,d\phi\,\hat{e}_\phi - r\,\dot{\phi}\sin\theta(\sin\theta\,\dot{\theta}\hat{e}_\theta + \sin\theta\,d\phi\,\hat{e}_r + \cos\theta\,\dot{\theta}\hat{e}_\theta + \sin\theta\,d\phi\,\hat{e}_\theta)\dot{\phi}$$

$$= (\ddot{r} - r\dot{\theta}^2 - r\dot{\phi}^2\sin^2\theta)\hat{e}_r + (r\ddot{\theta} + 2\dot{r}\dot{\theta} - r\dot{\phi}^2\sin\theta\cos\theta)\hat{e}_\theta$$

$$+ (r\ddot{\phi}\sin\theta + 2\dot{r}\dot{\phi}\sin\theta + 2r\dot{\theta}\dot{\phi}\cos\theta)\hat{e}_\phi \tag{4.68}$$

With the help of Eq. (4.59), the equations of motion of a particle in the spherical polar coordinate system can be obtained as

$$F_r = m(\ddot{r} - r\dot{\theta}^2 - r\dot{\phi}^2\sin^2\theta), \quad F_\theta = m(r\ddot{\theta} + 2\dot{r}\dot{\theta} - r\dot{\phi}^2\sin\theta\cos\theta)$$

$$F_\phi = m(r\ddot{\phi}\sin\theta + 2\dot{r}\dot{\phi}\sin\theta + 2r\dot{\theta}\dot{\phi}\cos\theta) \tag{4.69}$$

4.5.1 Tangential and Normal Acceleration

Let $\hat{t}$ be the unit tangent vector to the space curve, $\hat{n}$ the unit principal normal and $\rho = 1/k$ the radius of curvature. Let the vector equation of the curve be $\vec{r} = \vec{f}(t)$, t being regarded as a parameter. Now, if s be the arc length of the curve, measured from a fixed point on the curve, then

$$\vec{v} = \frac{d\vec{r}}{dt} = \frac{d\vec{r}}{ds}\frac{ds}{dt} = \hat{t}\frac{ds}{dt} = \hat{t}v$$

where $v = |\vec{v}| = \dfrac{ds}{dt}$ is called the speed of the particle along the curve. From this, it follows that the velocity has the direction of the unit tangent vector $\hat{t}$ and the magnitude of the speed. So there is no normal component of velocity for this motion. If the unit vector $\hat{t}$ may be regarded as the function of the arc length s of the trajectory, then

$$\frac{d\hat{t}}{dt} = \frac{d\hat{t}}{ds}\frac{ds}{dt} = (k\hat{n})\frac{ds}{dt} = \frac{v}{\rho}\hat{n} : v = \frac{ds}{dt}$$

where we use Frenet–Serret's first formula (4.23). Further

$$\vec{a} = \frac{d\vec{v}}{dt} = \frac{d}{dt}(v\hat{t}) = \frac{dv}{dt}\hat{t} + v\frac{d\hat{t}}{dt}$$

$$= \frac{dv}{dt}\hat{t} + v\frac{v}{\rho}\hat{n} = \frac{dv}{dt}\hat{t} + \frac{v^2}{\rho}\hat{n}$$

This shows that the component of acceleration is $\frac{dv}{dt}$, the rate of increase of speed in the direction tangent to the path, but independent of the shape of the curve and $v^2/\rho = \kappa v^2$ in the direction of the principal normal to the path. Thus, the acceleration vector $\vec{a}$ lies on the oscillating plane.

Example 4.5.6

A particle moves according to the law $\vec{r} = \cos t\hat{i} + \sin t\hat{j} + t^2\hat{k}$. Find the tangential and normal components of acceleration of the particle.

Solution

The position vector of any point on the curve is $\vec{r} = \cos t\hat{i} + \sin t\hat{j} + t^2\hat{k}$. Then

$$\vec{v} = \frac{d\vec{r}}{dt} = -\sin t\hat{i} + \cos t\hat{j} + 2t\hat{k}; \frac{d\vec{v}}{dt} = -\cos t\hat{i} - \sin t\hat{j} + 2\hat{k}$$

This is the linear velocity (vector) of the particle at any time t. Let v be the magnitude of $\vec{v}$; then it is given by

$$v = |\vec{v}| = \sqrt{\sin^2 t + \cos^2 t + 4t^2} = \sqrt{4t^2+1}$$

$$\Rightarrow \frac{dv}{dt} = \frac{1}{2}\frac{8t}{\sqrt{4t^2+1}} = \frac{4t}{\sqrt{4t^2+1}}$$

Therefore, the tangential component of acceleration of the particle at any time t is $\frac{4t}{\sqrt{4t^2+1}}$.

Now, let $\hat{t}$ be the unit tangent vector in the direction of $\vec{v}$, so $\vec{v} = v\hat{t}$, i.e., $\hat{t} = \frac{1}{v}\vec{v}$. Differentiating it with respect to s and using the Frenet–Serret formula, we get

$$\frac{d\hat{t}}{ds} = \frac{dt}{ds}\frac{d\hat{t}}{dt} = \frac{1}{v}\frac{d}{dt}\left(\frac{1}{v}\vec{v}\right) = \frac{1}{v^2}\frac{d\vec{v}}{dt} - \frac{1}{v^3}\frac{dv}{dt}\vec{v}$$

$$\Rightarrow \kappa\,\hat{n} = \frac{1}{4t^2+1}(-\cos t\hat{i} - \sin t\hat{j} + 2\hat{k})$$

$$-\frac{1}{(4t^2+1)^{3/2}}\frac{4t}{\sqrt{4t^2+1}}(-\sin t\hat{i} + \cos t\hat{j} + 2t\hat{k})$$

$$= \frac{1}{(4t^2+1)^2}\left[\left\{-(4t^2+1)\cos t + 4t\sin t\right\}\hat{i} - \left\{(4t^2+1)\sin t + 4t\cos t\right\}\hat{j} + \left\{2(4t^2+1) - 8t^2\right\}\hat{k}\right]$$

$$\Rightarrow \kappa = \frac{1}{(4t^2+1)^2}\sqrt{(4t^2+1)^2 + 16t^2 + 5}$$

$$\Rightarrow \kappa = \frac{1}{(4t^2+1)^2}\sqrt{(4t^2+1)(4t^2+5)} = \frac{\sqrt{4t^2+5}}{(4t^2+1)^{3/2}}$$

Hence, the magnitude of normal acceleration is

$$\frac{v^2}{\rho} = v^2\kappa = (4t^2+1)\cdot\frac{\sqrt{4t^2+5}}{(4t^2+1)^{3/2}} = \frac{\sqrt{4t^2+5}}{\sqrt{4t^2+1}}$$

4.5.2 Uniform Motion on a Circle

We consider that a particle moves with constant angular velocity ω along the circle of radius ρ (Fig. 4.20). Let us take the centre O of the circle as the origin. Let P be the position of the particle at any time t, whose position vector is $\vec{r}$ with respect to O. Then

$$\vec{r} = \rho\cos\theta\hat{i} + \rho\sin\theta\hat{j} = \rho(\cos\theta\hat{i} + \sin\theta\hat{j}),$$

where $\theta = \angle XOP$ and $\hat{i}$ and $\hat{j}$ are the constant unit vectors along the x and y axes, respectively. Therefore

$$\vec{v} = \frac{d\vec{r}}{dt} = \rho(-\sin\theta\hat{i} + \cos\theta\hat{j})\frac{d\theta}{dt}$$

$$= \rho\omega(-\sin\theta\hat{i} + \cos\theta\hat{j}); \text{ as } \frac{d\theta}{dt} = \omega$$

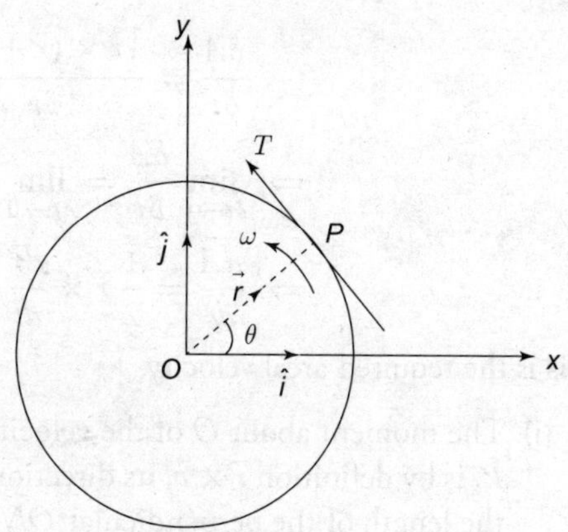

Figure 4.20 Uniform motion on a circle

This gives the velocity vector. Its magnitude is

$$v = |\vec{v}| = \sqrt{(-\rho\omega\sin\theta)^2 + (\rho\omega\cos\theta)^2} = \rho\omega$$

and direction along the tangent PT to the circle. Thus, for uniform motion on a circle, the tangential component of velocity is $\rho\omega$, but there is no component of velocity along the normal. Now, the acceleration vector is given by

$$\vec{a} = \frac{d\vec{v}}{dt} = \rho\omega(-\cos\theta\hat{i} - \sin\theta\hat{j})\frac{d\theta}{dt} = -\omega^2\vec{r}$$

This shows that the direction of acceleration is the opposite of that of $\vec{r}$, i.e., towards O, the centre of the circle, and magnitude $|\vec{a}| = \omega^2|\vec{r}| = \rho\omega^2$.

Thus, for uniform motion on a circle, there is no tangential component of acceleration, but the magnitude of the normal component is $\rho\omega^2$.

4.5.3 Areal Velocity

The areal velocity of a moving point P about the origin O is the rate of description of the vector area swept out by the line segment OP.

We consider that a particle moves in space and thereby its locus will be a space curve. We take two neighbouring points P and Q as the positions of the moving particle at times t and $t + \delta t$, respectively. Let the position vectors of P and Q be, respectively, $\vec{r}$ and $\vec{r} + \delta\vec{r}$, relative to a fixed point O, chosen as origin (Fig. 4.21). We draw a perpendicular ON of length p from the

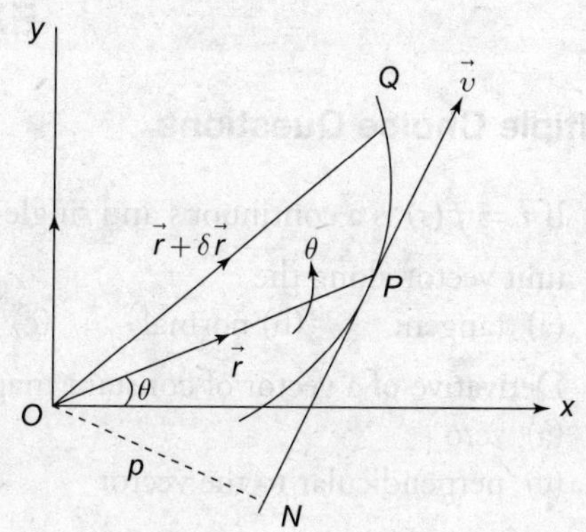

Figure 4.21 Areal velocity

origin to the tangent NP at P. Now, if $\delta\vec{A}$ is the vector area of the triangle OPQ, then

$$\frac{\delta\vec{A}}{\delta t} = \frac{\frac{1}{2}\vec{r}\times(\vec{r}+\delta\vec{r})}{\delta t} = \frac{1}{2}\vec{r}\times\frac{\delta\vec{r}}{\delta t}, \text{ as } \vec{r}\times\vec{r}=\vec{0}$$

$$\Rightarrow \lim_{\delta t\to 0}\frac{\delta\vec{A}}{\delta t} = \lim_{\delta t\to 0}\frac{1}{2}\vec{r}\times\frac{\delta\vec{r}}{\delta t}, \text{ proceeding the limit}$$

$$\Rightarrow \frac{d\vec{A}}{dt} = \frac{1}{2}\vec{r}\times\frac{d\vec{r}}{dt} = \frac{1}{2}\vec{r}\times\vec{v} \tag{4.70}$$

This is the required areal velocity.

(i) The moment about O of the velocity vector $\vec{v}$, considered localised along the tangent line P, is by definition $\vec{r}\times\vec{v}$, its direction being at right angles to the plane of $\vec{r}$ and $\vec{v}$. Let p be the length of the perpendicular ON to the tangent at P. Then

$$\vec{r}\times\vec{v} = OP\cdot v\ \sin\angle OPN\ \hat{\xi} = pv\,\hat{\xi} \tag{4.71}$$

where v is the speed of P and $\hat{\xi}$ is the unit vector perpendicular (normal) to the plane containing $\vec{r}$ and $\vec{v}$. Therefore

$$2\times\ \text{Areal velocity} = \text{Moment of } \vec{v} \text{ about } O = pv\hat{\xi} \tag{4.72}$$

(ii) If the particle moves on the plane OPN (plane determined by $\vec{r}$ and $\vec{v}$), then using Eq. 4.62, we get

$$\vec{r}\times\vec{v} = \vec{r}\times\left(\dot{r}\,\hat{e}_r + r\dot{\theta}\,\hat{e}_\theta\right) = r\,\dot{\theta}\,(\vec{r}\times\hat{e}_\theta), \text{ as } \vec{r}\times\hat{e}_r = \vec{0}$$

$$= r^2\,\dot{\theta}\,\hat{\xi} \Rightarrow pv = r^2\,\dot{\theta} = r^2\omega \tag{4.73}$$

where $\dot{\theta} = \omega$ is the rate of turning of OP. Hence, if we denote the magnitude of the areal velocity by $h/2$, then

$$h = |\vec{r}\times\vec{v}| = pv = r^2\,\dot{\theta} = r^2\omega \tag{4.74}$$

EXERCISES

Multiple Choice Questions

1. If $\vec{r} = \vec{f}(s)$ is a continuous and single-valued function of the scalar variable s, then $\frac{d\vec{r}}{ds}$ is a unit vector along the
   (a) tangent (b) normal (c) binormal (d) radius vector
2. Derivative of a vector of constant magnitude is
   (a) zero (b) parallel to the vector
   (c) perpendicular to the vector (d) none of these

3. The necessary and sufficient condition for the vector $\vec{f}$ to have a constant direction is
(a) $\vec{f} \cdot \dfrac{d\vec{f}}{dt} = 0$ (b) $\vec{f} \times \dfrac{d\vec{f}}{dt} = \vec{0}$ (c) $\vec{f} = \vec{0}$ (d) $\dfrac{d\vec{f}}{dt} = \vec{0}$

4. The necessary and sufficient condition for the vector $\vec{f}$ to have a constant magnitude is
(a) $\vec{f} \cdot \dfrac{d\vec{f}}{dt} = 0$ (b) $\vec{f} \times \dfrac{d\vec{f}}{dt} = \vec{0}$ (c) $\vec{f} = \vec{0}$ (d) $\dfrac{d\vec{f}}{dt} = \vec{0}$

5. If $\vec{A}(t) = 2t\hat{i} + \frac{1}{3}t^3\hat{j} + \frac{1}{5}t^5\hat{k}$, then $\left(\dfrac{d\vec{r}}{dt} \times \dfrac{d^2\vec{r}}{dt^2}\right) \cdot \dfrac{d^3\vec{r}}{dt^3}$ at $t = 1$ is
(a) 0 (b) 12 (c) 22 (d) 32

6. If $\phi(x, y, z) = xy^2z$ and $\vec{F} = xz\hat{i} - xy^2\hat{j} + yz^2\hat{k}$, then at $(2, -1, 1)$ $\dfrac{\partial^2}{\partial x\, \partial z}(\phi\, \vec{F}) =$
(a) $8\hat{i} - 4\hat{j} - 3\hat{k}$ (b) $-8\hat{i} - 4\hat{j} - 3\hat{k}$ (c) $8\hat{i} + 4\hat{j} - 3\hat{k}$ (d) $8\hat{i} - 4\hat{j} + 3\hat{k}$

7. If $\vec{r} = 3t\hat{i} + 3t^2\hat{j} + 2t^3\hat{k}$, then $\left[\dfrac{d\vec{r}}{dt}\ \dfrac{d^2\vec{r}}{dt^2}\ \dfrac{d^3\vec{r}}{dt^3}\right] =$
(a) 0 (b) 116 (c) 216 (d) 512

8. The unit tangent vector to the curve $\vec{r} = 3\cos t\hat{i} + 3\sin t\hat{j} + 4t\hat{k}$ at any time t is
(a) $\hat{t} = -3\sin t\hat{i} + 3\cos t\hat{j} + 4\hat{k}$ (b) $\hat{t} = \dfrac{1}{5}(-3\sin t\hat{i} + 3\cos t\hat{j} + 4\hat{k})$
(c) $\hat{t} = \dfrac{1}{\sqrt{9 + 16t^2}}(3\cos t\hat{i} + 3\sin t\hat{j} + 4t\hat{k})$ (d) $\hat{t} = (\cos t\hat{i} + \sin t\hat{j}$

9. If $\hat{r}$ and $\hat{\theta}$ are unit vectors along the direction of $\vec{r}$ and the direction of increasing θ, respectively, then
(a) $\hat{r} = \dot{\theta}\hat{\theta}$ and $\hat{\theta} = \dot{\theta}\hat{r}$ (b) $\hat{r} = \dot{\theta}\hat{\theta}$ and $\hat{\theta} = -\dot{\theta}\hat{r}$
(c) $\hat{r} = \dot{r}\hat{\theta}$ and $\hat{\theta} = \dot{\theta}\hat{r}$ (d) $\hat{r} = \dot{r}\hat{r}$ and $\hat{\theta} = \dot{\theta}\hat{\theta}$

10. If $\hat{r}, \hat{\theta}$ and $\hat{\phi}$ are the unit vectors in spherical coordinates, then $\hat{\theta}$ may be expressed in terms of $\hat{i}, \hat{j}$ and $\hat{k}$ as
(a) $\sin\theta\cos\theta\hat{i} + \sin\theta\sin\phi\hat{j} + \cos\theta\hat{k}$ (b) $\cos\theta\cos\phi\hat{i}$
(c) $-\sin\phi\hat{i} + \cos\phi\hat{j}$ (d) $\cos\phi\hat{i} + \sin\phi\hat{j}$

11. A particle moves on a given line with a constant speed v. At a certain time t, it is at point P on its straight line path. O is a fixed point. Then, $\vec{OP} \times \vec{v}$ is
(a) dependent on the position of the particle
(b) depends only on the x coordinate of the motion
(c) independent of position
(d) perpendicular to the velocity $\vec{v}$

Answers

1. a 2. c 3. b 4. a 5. d 6. a 7. c 8. b 9. b 10. a 11. c

Short Answer Questions

1. Find the first and second derivatives of the following, with respect to t, where $\vec{a}$ and $\vec{b}$ are the given vector functions of the scalar variable t.

(a) $a^2 + \dfrac{1}{a^2}$ *Ans*: $2\vec{a}\dot{\vec{a}}(1 - \dfrac{1}{a^4})$; $\dfrac{2}{a^3}\vec{a}'''(a^4 - 1) + \dfrac{2}{a^4}\vec{a}'^2(4a^8 + a^4 - 1)$

(b) $\vec{a} \times (\vec{b} \times \vec{c})$ *Ans*: $\vec{a}' \times (\vec{b} \times \vec{c}) + \vec{a} \times (\vec{b}' \times \vec{c}) + \vec{a} \times (\vec{b} \times \vec{c}')$;

$$\vec{a}'' \times (\vec{b} \times \vec{c}) + \vec{a} \times (\vec{b}'' \times \vec{c}) + \vec{a} \times (\vec{b} \times \vec{c}'')$$
$$+2\vec{a}' \times (\vec{b}' \times \vec{c}) + 2\vec{a}' \times (\vec{b} \times \vec{c}') + 2\vec{a} \times (\vec{b}' \times \vec{c}')$$

(c) $\left[\vec{a}\ \dfrac{d\vec{a}}{dt}\ \dfrac{d^2\vec{a}}{dt^2}\right]$ *Ans*: $\left[\vec{a}\ \dfrac{d\vec{a}}{dt}\ \dfrac{d^3\vec{a}}{dt^3}\right]$; $\left[\vec{a}\ \dfrac{d^2\vec{a}}{dt^2}\ \dfrac{d^3\vec{a}}{dt^3}\right] + \left[\vec{a}\ \dfrac{d\vec{a}}{dt}\ \dfrac{d^4\vec{a}}{dt^4}\right]$.

(d) If $\vec{f}(t)$ is a differentiable vector function and $f = |\vec{f}(t)|$, prove that $\vec{f} \cdot \dfrac{d\vec{f}}{dt} = f\dfrac{df}{dt}$.

(e) Prove that the derivative of a vector of constant magnitude is perpendicular to that vector.

(f) $\vec{a} \times \left(\dfrac{d\vec{a}}{dt} \times \dfrac{d^2\vec{a}}{dt^2}\right)$ *Ans*: $\dfrac{d\vec{a}}{dt} \times \left(\dfrac{d\vec{a}}{dt} \times \dfrac{d^2\vec{a}}{dt^2}\right) + \vec{a} \times \left(\dfrac{d\vec{a}}{dt} \times \dfrac{d^3\vec{a}}{dt^3}\right)$;

$$\frac{d^2\vec{a}}{dt^2} \times \left(\frac{d\vec{a}}{dt} \times \frac{d^2\vec{a}}{dt^2}\right) + 2\frac{d\vec{a}}{dt} \times \left(\frac{d\vec{a}}{dt} \times \frac{d^3\vec{a}}{dt^3}\right)$$
$$+\vec{a} \times \left(\frac{d^2\vec{a}}{dt^2} \times \frac{d^3\vec{a}}{dt^3}\right) + \vec{a} \times \left(\frac{d\vec{a}}{dt} \times \frac{d^4\vec{a}}{dt^4}\right).$$

(g) $\dfrac{1}{a^2 + b^2}(\vec{a} + \vec{b})$ *Ans*: $\dfrac{\vec{a}' + \vec{b}'}{a^2 + b^2} - \dfrac{2\{a^2\vec{a}' + b^2\vec{b}' + ab(\vec{a} + \vec{b})\}}{(a^2 + b^2)^2}$; $\dfrac{\vec{a}'' + \vec{b}''}{a^2 + b^2}$

$$-\frac{-2\{a\vec{a}^2 + b\vec{b}^2\} + (\vec{a} + \vec{b})\{\vec{a}'(\vec{b}' - \vec{b}) + \vec{b}'(\vec{a}' - \vec{a})\} - a^2\vec{a}'' - b^2\vec{b}'' - ab(\vec{a}' + \vec{b}')}{(a^2 + b^2)^2}$$
$$-\frac{4\{a^2\vec{a}' + b^2\vec{b}' + ab(\vec{a}' + \vec{b}')\}(2a\vec{a}' + 2b\vec{b}')}{(a^2 + b^2)^3}.$$

(h) $\dfrac{1}{\vec{a} \cdot \vec{b}}(\vec{a} \times \vec{b})$ *Ans*: $\dfrac{1}{\vec{a} \cdot \vec{b}}\left\{\left(\dfrac{d\vec{a}}{dt} \times \vec{b}\right) + \left(\vec{a} \times \dfrac{d\vec{b}}{dt}\right)\right\}$;

$$\frac{1}{\vec{a} \cdot \vec{b}}\left\{\left(\frac{d^2\vec{a}}{dt^2} \times \vec{b}\right) + 2\left(\frac{d\vec{a}}{dt} \times \frac{d\vec{b}}{dt}\right) + \left(\vec{a} \times \frac{d^2\vec{b}}{dt^2}\right)\right\}$$

2. If $\vec{r} = x\hat{i} + \sin x\hat{j} + \cos x\hat{k}$, then find the derivative of $\vec{r}$ with respect to x.

Ans: $\hat{i} + \cos x\hat{j} - \sin x\hat{k}$

3. Prove that the vector derivative of a vector function $\vec{r}$ is zero if and only if $\vec{r}$ is a constant vector.

4. Find $\dfrac{d}{dt}\left(\vec{A} \cdot \vec{B}\right)$, $\dfrac{d}{dt}\left(\vec{A} \times \vec{B}\right)$ and $\dfrac{d^2}{dt^2}\left(\vec{A} \times \vec{B}\right)$, if

(a) $\vec{A} = 3t^2\hat{i} + t\hat{j} - t^3\hat{k}$ and $\vec{B} = \sin t\hat{i} - 2\cos t\hat{j}$

Ans: $8t\sin t + (3t^2 - 2)\cos t$;

$$(2t^3\sin t - 6t\cos t)\hat{i} - (t^3\cos t + 3t^2\sin t)\hat{j} + \{11t\cos t - (6t^2 + 1)\sin t\}\hat{k};$$
$$\{(2t^3 - 6)\cos t + (6t^2 + 6t)\sin t\}\hat{i} - \{6t^2\cos t + (6t - t^3)\sin t\}\hat{j}$$
$$+10\{(10 - 6t^2)\cos t - 23t\sin t\}\hat{k}.$$

(b) $\vec{A} = t^2\hat{i} - t\hat{j} + (2t + 1)\hat{k}$ and $\vec{B} = (2t - 3)\hat{i} + \hat{j} - t\hat{k}$ at $t = 1$

Ans: -6; $7\hat{j} + 3\hat{k}$; $2\hat{i} + 14\hat{j} + 6\hat{k}$

(c) $\vec{A} = 3t\hat{i} + 4t\hat{j} - t^3\hat{k}$ and $\vec{B} = t^2\hat{i} - 8t^3\hat{j} + 3t\hat{k}$

Ans: $9t^2 - 140t^3$; $24t(1 - 2t^4)\hat{i} - t(18 + 5t^3)\hat{j} - 12t^2(8t + 1)\hat{k}$; $24(1 - 10t^4)\hat{i} - 2(9 + 10t^3)\hat{j} - 24t(12t - 1)\hat{k}$

(d) $\vec{A} = 3t^2\hat{i} - (4 + t)\hat{j} + (t^2 - 2t)\hat{k}$ and $\vec{B} = \sin t\hat{i} + 3e^{-t}\hat{j} - 3\cos t\hat{k}$ at $t = 0$

Ans: $3; 9\hat{i} + 4\hat{k}; -30\hat{i} + 14\hat{j} + 20\hat{k}$

5. If $\vec{R}$ is a unit vector in the direction of $\vec{r}$, then prove that

$$\vec{R} \times \frac{d\vec{R}}{dt} = \frac{1}{r^2}\vec{r} \times \frac{d\vec{r}}{dt}, \quad r = |\vec{r}|$$

6. If $\dfrac{d^2\vec{r}}{dt^2} = \vec{r}$, then show that $\vec{r} \times \dfrac{d\vec{r}}{dt}$ is a constant vector.

7. If $\dfrac{d\vec{a}}{dt} = \vec{r} \times \vec{a}$ and $\dfrac{d\vec{b}}{dt} = \vec{r} \times \vec{b}$, show that $\dfrac{d}{dt}(\vec{a} \times \vec{b}) = \vec{r} \times (\vec{a} \times \vec{b})$, where $\vec{r}$ is a constant vector and $\vec{a}$ and $\vec{b}$ are vector functions of a scalar variable t.

8. Find the unit vector in the direction of the tangent at any point on the curve

(a) $\vec{r} = (a\cos t, a\sin t, bt)$ *Ans*: $(-a\sin t, a\cos t, b)/\sqrt{a^2 + b^2}$

(b) $\vec{r} = (2\cos t, 2\sin t, 2t)$ *Ans*: $\sqrt{2}(-2\sin t, \cos t, 1)/2$

(c) $\vec{r} = (3\cos t, 3\sin t, 4t)$ *Ans*: $(-3\sin t, 3\cos t, 4)/5$

(d) $\vec{r} = (t^2 + 1, 4t - 3, 2t^2 - 6t)$ at $t = 1$ *Ans*: $\sqrt{6}(1, 2, -1)/6$

(e) $\vec{r} = (t^2 + 1, 4t, 2t^2 - 6t + 3)$ at $t = 1$ *Ans*: $\sqrt{6}(1, 2, -1)/6$

9. Find the unit vector in the direction of the normal at any point on the curve

(a) $\vec{r} = (2\cos t, 2\sin t, 2t)$ (b) $\vec{r} = (t\cos t, t\sin t, t)$

(c) $\vec{r} = (t^2 + 1, 4t - 3, 2t^2 - 6t)$ at $t = 1$ (d) $\vec{r} = (t^2 + 1, 4t, 2t^2 - 6t + 3)$ at $t = 1$

Ans: (a) $(-\cos t, -\sin t)$ (b) $\hat{j}$ (c) $\sqrt{6}(1, 0, 2)/6$ (d) $\sqrt{6}(1, 0, 2)/6$

10. Find the curvature (ρ) and torsion (τ) of the following curves:

(a) $\vec{r} = a(u)\hat{i} + b(u)\hat{j}$; *Ans*: $\rho = |\dot{a}\ddot{b} - \ddot{a}\dot{b}|/(\dot{a}^2 + \dot{b}^2)^{3/2}$, $\tau = 0$

(b) $\vec{r} = (a\cos 2t, a\sin 2t, 2a\sin t)$ at $t = 0$; *Ans*: $\rho = 1/2a$, $\tau = 0$

(c) $\vec{r} = a\cos t + a\sin t + \cosh at$; *Ans*: $\rho = \sqrt{a^2 + 1}\,\text{sech}^2 at/a$, $k = \sinh at\,\text{sech}^2 at$

(d) $\vec{r} = (e^t, e^{-t}, \sqrt{2}t)$ at $t = 0$; *Ans*: $\rho = 1/2\sqrt{2}$

(e) $\vec{r} = (\log\cos u, \log\sin u, \sqrt{2}u)$ *Ans*: $\rho = \sqrt{2}$ cosec u

(f) $\vec{r} = (\cos u, \sin u, -\log\cos u)$ at $t = \pi/4$; *Ans*: $\rho = \sqrt{3}/2$

(g) $\vec{r} = \tan^{-1} s, \sqrt{2}\log(s^2 + 1)/2, s - \tan^{-1} s$; *Ans*: $\rho = \frac{1}{\sqrt{2}}(s^2 + 1)$, $\tau = \frac{\sqrt{2}}{s^2+1}$

(h) $\vec{r} = a(3t - t^3, 3t^2, 3t + t^3)$; *Ans*: $\rho = \tau = 3a(1 + t^2)^2$

11. Prove that the rate of change of unit tangent vector with respect to the arc length measured from a fixed point on a space curve is perpendicular to the tangent vector.

12. For any space curve, prove that the normal plane is orthogonal to the osculating plane at any point.

13. Prove that the curvature κ at any point of the curve $\vec{r} = a(s)\hat{i} + b(s)\hat{j} + c(s)\hat{k}$ is given by $\kappa = \sqrt{\ddot{a}^2 + \ddot{b}^2 + \ddot{c}^2}$.

14. Prove that the moment of the velocity vector about the origin is twice the areal velocity.
15. Find the magnitude of the tangential and normal components of acceleration of a particle moving on a circle of radius ρ with constant angular velocity ω. *Ans*: 0, $\rho\omega^2$
16. Find the areal velocity at any time t of a particle moving along the space curve given by the vector equation
(a) $\vec{r} = 2\cos t\hat{i} + \sin t\hat{j}$ (b) $\vec{r} = \cosh t\hat{i} + \sinh t\hat{j} + \hat{k}$ (c) $\vec{r} = a\cos\omega t\hat{i} + b\sin\omega t\hat{j}$
Ans: (a) $\hat{k}$, (b) $(-\cosh t\hat{i} + \sinh t\hat{j} + \hat{k})/2$, (c) $ab\omega\hat{k}/2$
17. Prove that there is no normal component of velocity for the motion of a particle on a plane.
18. Prove that the component of tangential acceleration is the rate of increase of speed in the direction tangent to the path, but independent of the shape of the space curve.
19. Prove that for rectilinear motion, the acceleration will be purely tangential, whereas the acceleration will be purely normal if the speed is constant.
20. The position vector of a particle at time t is given by $\vec{r} = \cos(t-1)\hat{j} + \sin(t-1)\hat{j} + pt^3\hat{k}$. Find the condition imposed on p, such that at $t = 1$, the acceleration is normal to the positive vector. $p = \pm 1/\sqrt{6}$
21. A particle describes a circle $r = a\cos\theta$ with constant speed. Show that the acceleration is constant in magnitude and is directed towards the centre of the circle.

Long Answer Questions

1. Prove the following (where $r = |\vec{r}|$ and $\dfrac{dr}{dt} = r'$):
(a) $\dfrac{d}{dt}\left[r^2\,\vec{r} + (\vec{a}\cdot\vec{r})\,\vec{b}\right] = 2\,r\,r'\vec{r} + r^2\vec{r}' + (\vec{a}\cdot\vec{r}')\vec{b}$
(b) $\dfrac{d}{dt}\left[a\,\vec{r} + \vec{r}\,\vec{b}\right]^2 = 2\,(a\,\vec{r} + \vec{r}\,\vec{b})\cdot(a\,\vec{r}' + \vec{r}'\,\vec{b}')$
(c) $\dfrac{d}{dt}\left[\dfrac{1}{r}\,\vec{r}\right] = \dfrac{1}{r}\,\vec{r}' - \dfrac{r'}{r^2}\,\vec{r}$ (d) $\dfrac{d}{dt}\left(r^n\,\vec{r}\right) = nr^{n-1}\,r'\,\vec{r} + r^n\,\vec{r}$
(e) $\dfrac{d}{dt}\left[\dfrac{1}{r^2}\,\vec{r} + \dfrac{r}{\vec{a}\cdot\vec{r}}\,\vec{b}\right] = \dfrac{1}{r^2}\,\vec{r}' - \dfrac{2r'}{r^3}\,\vec{r} + \dfrac{r'}{\vec{a}\cdot\vec{r}}\,\vec{b} - \dfrac{r(\vec{a}\cdot\vec{r}')}{(\vec{a}\cdot\vec{r})^2}\,\vec{b}$
(f) $\dfrac{d}{dt}\left[r^3\,\vec{r} + \vec{a}\times\dfrac{\vec{r}}{dt}\right] = 3r^2r'\vec{r} + r^3\vec{r}' + \vec{a}\times\vec{r}''$
(g) $\dfrac{d}{dt}\left[\dfrac{1}{\vec{a}\cdot\vec{r}}\,\vec{r} - \dfrac{1}{r}\,(\vec{r}\times\vec{a})\right] = \dfrac{1}{\vec{a}\cdot\vec{r}}\,\vec{r}' - \dfrac{(\vec{a}\cdot\vec{r}')}{(\vec{a}\cdot\vec{r})^2}\vec{r} - \dfrac{1}{r}(\vec{r}'\times\vec{a}) - \dfrac{r'}{r^2}(\vec{r}\times\vec{a})$
2. Find the osculating plane, curvature and torsion of the curve $\vec{r} = a\cos 2u\hat{i} + a\sin 2u\hat{j} + 2a\sin u\hat{k}$ at the point $u = 0$. *Ans*: $y - z = 0$, $1/2a$, $3/8a$
3. Determine a unit normal to the surface $\vec{r} = a\cos u\sin v\hat{i} + a\sin u\sin v\hat{j} + a\cos v\hat{k}$, where $a > 0$. *Ans*: $\pm(\cos u\sin v\hat{i} + a\sin u\sin v\hat{j} + \cos v\hat{k})$.
4. Find an equation for the tangent plane to the surface $z = x^2 + y^2$ at the point $(1, -1, 2)$. *Ans*: $2x - 2y - z = 2$

5. Prove that the acceleration vector of a particle moving along a space curve always lies on the oscillating plane.
6. Describe and sketch the coordinate surfaces and coordinate curves for
   (a) cylindrical coordinates (b) spherical polar coordinates
7. Determine the transformation from
   (a) rectangular to cylindrical coordinates (b) spherical to rectangular coordinates
   (c) spherical to cylindrical coordinates (d) rectangular to spherical polar coordinates
8. Find the unit vectors $\hat{i}, \hat{j}$ and $\hat{k}$ in a rectangular coordinate system in terms of unit vectors $\hat{e}_\rho$, $\hat{e}_\theta$ and $\hat{e}_\phi$ of the spherical polar coordinate system.
9. Find the unit vectors $\hat{i}, \hat{j}$ and $\hat{k}$ in a rectangular coordinate system in terms of unit vectors $\hat{e}_\rho$, $\hat{e}_\phi$ and $\hat{e}_z$ of the cylindrical coordinate system.
10. If u, v and w are curvilinear coordinates, show that $\dfrac{\partial \vec{r}}{\partial u}, \dfrac{\partial \vec{r}}{\partial v}$ and $\dfrac{\partial \vec{r}}{\partial w}$ and $\vec{\nabla} u, \vec{\nabla} v$ and $\vec{\nabla} w$ are a reciprocal system of vectors.
11. If $\vec{r} = a\cos t\hat{i} + a\sin t\hat{j} + bt\hat{k}$, show that
    (a) $\left|\dfrac{d\vec{r}}{dt}\right|^2 = a^2 + b^2$ (b) $\left|\dfrac{d\vec{r}}{dt} \times \dfrac{d^2\vec{r}}{dt^2}\right|^2 = a^2(a^2+b^2)$
    (c) $\left[\dfrac{d\vec{r}}{dt}\ \dfrac{d^2\vec{r}}{dt^2}\ \dfrac{d^3\vec{r}}{dt^3}\right] = \dfrac{d\vec{r}}{dt} \cdot \dfrac{d^2\vec{r}}{dt^2} \times \dfrac{d^3\vec{r}}{dt^3} = a^2 b$
12. Prove the following:
    (a) $\dfrac{d}{dt}\left\{\vec{r} \cdot \dfrac{d\vec{r}}{dt}\right\} = \left(\dfrac{d\vec{r}}{dt}\right)^2 + \vec{r} \cdot \dfrac{d^2\vec{r}}{dt^2}$ (b) $\dfrac{d}{dt}\left\{\vec{r} \times \dfrac{d\vec{r}}{dt}\right\} = \vec{r} \times \dfrac{d^2\vec{r}}{dt^2}$
    (c) $\dfrac{d}{dt}\left\{\vec{r} \cdot \dfrac{d\vec{r}}{dt} \times \dfrac{d^2\vec{r}}{dt^2}\right\} = \left[\vec{r}, \dfrac{d\vec{r}}{dt}, \dfrac{d^3\vec{r}}{dt^3}\right]$
    (d) $\dfrac{d^2}{dt^2}\left\{\vec{r} \cdot \dfrac{d\vec{r}}{dt} \times \dfrac{d^2\vec{r}}{dt^2}\right\} = \left[\vec{r}, \dfrac{d^2\vec{r}}{dt^2}, \dfrac{d^3\vec{r}}{dt^3}\right] + \left[\vec{r}, \dfrac{d\vec{r}}{dt}, \dfrac{d^4\vec{r}}{dt^4}\right]$
13. If $\vec{a}$ and $\vec{b}$ are two vector functions of the scalar variables t, then prove that
    (a) $\dfrac{d}{dt}\left(\vec{a} \times \dfrac{d\vec{b}}{dt} - \dfrac{d\vec{a}}{dt} \times \vec{b}\right) = \vec{a} \times \dfrac{d^2\vec{b}}{dt^2} - \dfrac{d^2\vec{a}}{dt^2} \times \vec{b}$
    (b) $\dfrac{d}{dt}\left(\vec{a} \cdot \dfrac{d\vec{b}}{dt} - \dfrac{d\vec{a}}{dt} \cdot \vec{b}\right) = \vec{a} \cdot \dfrac{d^2\vec{b}}{dt^2} - \dfrac{d^2\vec{a}}{dt^2} \cdot \vec{b}$
    (c) $\dfrac{d}{dt}\left(\vec{a} \cdot \dfrac{d\vec{a}}{dt} \times \dfrac{d^2\vec{a}}{dt^2}\right) = \vec{a} \cdot \dfrac{d\vec{a}}{dt} \times \dfrac{d^3\vec{a}}{dt^3}$ (d) $\vec{a} \cdot \dfrac{d\vec{a}}{dt} = a\dfrac{da}{dt}$
14. Calculate $\dfrac{\partial \vec{F}}{\partial x}, \dfrac{\partial \vec{F}}{\partial y}, \dfrac{\partial \vec{F}}{\partial x} \times \dfrac{\partial \vec{F}}{\partial y}, \dfrac{\partial^2 \vec{F}}{\partial x^2}, \dfrac{\partial^2 \vec{F}}{\partial y^2}, \dfrac{\partial^2 \vec{F}}{\partial x^2} \times \dfrac{\partial^2 \vec{F}}{\partial y^2}\ \dfrac{\partial^2 \vec{F}}{\partial x\, \partial y}$ and $\dfrac{\partial^2 \vec{F}}{\partial y\, \partial x}$ when
    (a) $\vec{F} = x^2yz\hat{i} - 2xz^3\hat{j} + xz^2\hat{k}$ *Ans:* $2xyz\hat{i} - 2z^3\hat{j} + z^2\hat{k};\ x^2z\hat{i};$
    $x^2z^3\hat{j} + 2x^2z^4\hat{k};\ 2yz\hat{i};\ \vec{0};\ \vec{0};\ 2xz\hat{i};\ 2xz\hat{i}$
    (b) $\vec{F} = (3x^2 - xy^2)\hat{i} + (e^{xy} - x\sin y)\hat{j} + y^2\cos xk$
    Ans: $(6x - y^2)\hat{i} + (ye^{xy} - \sin y)\hat{j} - y^2\sin x\hat{k};$

$$-2xy\hat{i} + (xe^{xy} - x\cos y)\hat{j} + 2y\cos x\hat{k};$$
$$\{e^{xy}(2y^3\cos x + xy^2\sin x) - y^2(2\cos x\sin y + x\sin x\cos y)\}\hat{i}$$
$$+\{2y^2xy\sin x - (6x - y^2)\cos x\}\hat{j} + \{e^{xy}(6x^2 + xy^2) - (6x + y^2)x\cos y - 2xy\sin y\};$$
$$6\hat{i} + y^2e^{xy}\hat{j} - y^2\hat{k}; -2x\hat{i} + (x^2e^{xy} + x\sin y)\hat{j} + 2\cos x\hat{k};$$
$$\{(2 + x^2)y^2e^{xy}\cos x + xy^2\sin y\}\hat{i} + 2\cos x(xy^2 - 6)\hat{j} + \{xe^{xy}(6x + 2y^2) + x\sin y\}\hat{k};$$
$$-2y\hat{i} + \{e^{xy}(xy + 1) - \cos y\}\hat{j} - 2y\sin x\hat{k};$$ same

(c) $\vec{F} = (2x^2y - x^4)\hat{i} + (e^{xy} - y\sin x)\hat{j} + x^2\cos y\hat{k}$

Ans: $4x(y - x^2)\hat{i} + y(e^{xy} - \cos x)\hat{j} + 2x\cos y\hat{k};$
$2x^2\hat{i} + (xe^{xy} - \sin x)\hat{j} - x^2\sin y\hat{k}$
$\{x^2y\sin y(\cos x - e^{xy}) - 2x\cos y(xe^{xy} - \sin x)\}\hat{i} + \{2x^3\cos y + 4x^3(y - x^2)\sin y\}\hat{j} + \{2x^2y(e^{xy} + \cos x) + 4x(x^2 - y)\sin x - 4x^4e^{xy}\}\hat{k};$
$x^2e^{xy}\hat{j} - x^2\cos y\hat{k}; (4y - 12x^2)\hat{i} + (y^2e^{xy} + y\sin x)\hat{j} + 2\cos y\hat{k};$
$\{-x^2y(ye^{xy} + \sin x) - 2x^2e^{xy}\cos y\}\hat{i} + 4x^2(3x^2 - y)\cos y\hat{j} + 4x^2(y - 3x^2)e^{xy}\hat{k};$
$4x\hat{i} + \{(xy + 1)e^{xy} - \cos x\}\hat{j} - 2x\sin y\hat{k};$ same

(d) $\vec{F} = e^{xy}\hat{i} + (2x - y)\hat{j} + y\sin x\hat{k}$ *Ans*: $ye^{xy}\hat{i} + 2\hat{j} + y\cos x\hat{k}; xe^{xy}\hat{i} - \hat{j} + \sin x\hat{k};$
$(2\sin x + y\cos x)\hat{i} + ye^{xy}(x\cos x - \sin x)\hat{j} + e^{xy}(y + 2x)\hat{k};$
$x^2e^{xy}\hat{i}; ye^{xy}\hat{i} - y\sin x\hat{k}; -x^2ye^{xy}\sin x\hat{j}; (e^{xy} + xye^{xy})\hat{i} + \cos x\hat{k};$ same

15. Given an equation of space curve $x = t^2 + 1, y = 4t - 3, z = 2t^2 - 6t$, determine
    (a) the tangent vector to any point on the space curve. *Ans*: $2t\hat{i} + 4\hat{j} + (4t - 6)\hat{k}$
    (b) the unit tangent vector to the space curve at any point on the curve.
    (c) the unit tangent vector to the space curve at the point on the curve where $t = 2$.
    Ans: $(2\hat{i} + 2\hat{j} + \hat{k})/3$
    (d) the equation of the tangent at the point on the curve where $t = 2$.
    Ans: $\vec{R} = (5\hat{i} + 5\hat{j}4\hat{k}) + \lambda(2\hat{i} + 2\hat{j} + \hat{k})$
16. Find the unit tangent, unit normal, unit binormal, curvature, radius of curvature, torsion, radius of torsion at any point of the following curves:
    (a) $\vec{r} = \left(3t, 3t^2, 2t^3\right)$ (b) $\vec{r} = \left(a(3t - t^3), 3at^2, a(3t + t^3)\right)$
    (c) $\vec{r} = \left(a\cos 2t, a\sin 2t, 2a\sin t\right)$ (d) $\vec{r} = \left(a\cos t, a\sin t, a\log\sec t\right)$
    (e) $\vec{r} = \left(e^t, e^{-t}, \sqrt{2}\,t\right)$ (f) $\vec{r} = \left(2\log t, 4t, 2t^2 + 1\right)$
    (g) $\vec{r} = (e^t\cos t, e^t\sin t, e^t)$ at $t = 0$ *Ans*: $(1, 1, 1)/\sqrt{3}, (1, 1, 0)/\sqrt{2}, (-1, 1, \sqrt{2})/2$
17. Show that the radius of torsion = radius of curvature at any time t, for the space curve, given by the vector equation
    (a) $\vec{r} = t\hat{i} + t^2\hat{j} + \frac{2}{3}t^3\hat{k}$ *Ans*: $(1 + 2t^2)^2/2$
    (b) $\vec{r} = a(3t - t^3, 3t^2, 3t + t^3)$ *Ans*: $3a(1 + t^2)^2$
18. Show that for the curve
    (a) $\vec{r} = \tan^{-1} s\hat{i} + \frac{1}{\sqrt{2}}\log(1 + s^2)\hat{j} + (s - \tan^1)\hat{k}$, $\rho = (s^2 + 1)/\sqrt{2}$, and $\tau = \sqrt{2}/(s^2 + 1)$
    (b) $\vec{r} = (a\cos u, a\sin u, bu)$, $\rho = a(a^2 + b^2)^{-1}$ and $\tau = b(a^2 + b^2)^{-1}$

19. Show that the equations of the osculating plane, normal plane and rectifying plane to the
 (a) twisted cubic: $x = 2t, y = t^2, z = t^3/3$ at $t = 1$ are $x - 2y + 2z = 2/3$, $2x + 2y + z = 19/3$, $6x - 3y - 6z = 7$
 (b) curve: $\vec{r} = (a\cos 2t, a\sin 2t, 2a\sin t)$ at $t = 0$ are $y = z, x = a, x = a$
 (c) curve: $x = 3t - t^3, y = 3t^2, z = 3t + t^3$ at $t = 1$ are $y - z + 1 = 0, y + z = 7$ and $x = 2$
 (d) space curve: $\vec{r} = (t+1)\hat{i} + t^2\hat{j} + (1+t^2)\hat{k}$ at $t = 1$ are $y - z - 3 = 0$, $4x - y - z - 2 = 0, x + 2y + 2z - 8 = 0$
 (e) space curve: $\vec{r} = e^t\hat{i} + e^{-t}\hat{j} + 2t\hat{k}$ at $t = 1$ are $x + e^2y - ez + 4e = 0$, $(2e + \frac{1}{e})x + (e + \frac{2}{e})y + \frac{4}{e^2} + 2 = 0$
20. If the tangential and normal components of the acceleration of a moving particle on a plane are constant, show that the locus of the particle is a spiral.
21. If the radial and cross radial components of the velocity of a moving particle on a plane are identical, then find the locus of the particle in plane coordinates.
 Ans: $r = \mu e^{\theta}$; $a =$ constant.
22. Find the magnitudes of the tangential and normal components of acceleration, when a particle moves along the curve
 (a) $x = \sin t, y = \cos t, z = t$, at $t = \pi/2$ *Ans*: 0, 2
 (b) $x = t^3 - 4t, y = t^2 + 4t$ and $z = 8t^2 - 3t^3$, at $t = 2$ *Ans*: $16, 2\sqrt{73}$
 (c) $x = 1 - 4t + t^3, y = 2 + 4t + t^2$ and $z = 3 + 8t^2 - 3t^3$, at $t = 2$ *Ans*: $16, 2\sqrt{73}$
 (d) $x = t^3 - 4t, y = t^2 - 4t$ and $z = 3t^2 + 4t^3$, at $t = 2$
 Ans: $834\sqrt{229}/229, 30\sqrt{1601}/229$
 (e) $\vec{r} = \cosh t\hat{i} + \sinh t\hat{j} + t\hat{k}$, at any time t *Ans*: $\sqrt{2}\sinh t, 2\cosh^2 t$
23. Find the velocity and acceleration, when a particle moves along the curve
 (a) $x = 4\cos t, y = 4\sin t, z = 6t$ at $t = 0$ and $t = \pi/2$ *Ans*: $4\hat{j} + 6\hat{k}, -4\hat{i}; -4\hat{i} + 6\hat{k}, -4\hat{j}$
 (b) $x = e^{-t}, y = 2\cos 3t, z = 2\cos 3t$ at $t = 0$. *Ans*: $\sqrt{37}, \sqrt{325}$
 (c) $x = t^3, y = t^2, z = t$ at $t = 1$ along the direction of the vector $\hat{i} + \hat{j} + \hat{k}$. *Ans*: 6, 8
 (d) $x = t^3 + 1, y = t^2, z = 2t + 5$ at $t = 1$ along the direction of the vector $\hat{i} + \hat{j} + 3\hat{k}$
 Ans: $\sqrt{11}, 8\sqrt{11}$
 (e) $\vec{r} = (e^{-t} + t - 1)\hat{i} - \left(\frac{2}{3}t^3 + 2t^2\right)\hat{j} + 5(t - \sin t)\hat{k}$ at $t = 1$ in the direction of the vector $\hat{i} + \hat{j} + \hat{k}$ *Ans*: $5\cos 1 - \frac{1}{e}, \frac{1}{e} - 5\sin 1 - 8$
24. A particle moves such that its position vector is given by $\vec{r} = \cos\omega t\hat{i} + \sin\omega t\hat{j}$, where ω is constant. Show that the velocity $\vec{v}$ of the particle is perpendicular to $\vec{r}$ and that the acceleration is directed towards the origin and has magnitude proportional to the distance from the origin. Show further that $\vec{r} \times \vec{v}$ is a constant vector.
25. If $\vec{p}$ and $\vec{q}$ are constant vectors, and t is the variable time, then show that a particle, whose position vector at any instant is $\vec{r} = \vec{p}\cos at + \vec{q}\sin at$, moves in an ellipse whose centre is the origin. Show that the motion is due to a central force varying as distance.

5 Vector Differential Operator

In the previous chapter, the components of a vector function were taken as constants or the functions of scalar variable(s), for the discussion on vector functions. In this chapter, we shall introduce three important concepts: gradient, divergence and curl. These are differential operators which, as a result of operations on point functions, lead to new point functions that possess some important geometrical and physical properties. The operators are widely used in the study of hydrodynamics, electrostatics and other branches of physics and applied mathematics. Vector identities involving differential operators will be proved. The geometrical meaning of gradient and the physical significance of divergence and curl of vector point functions will be explained.

After the discussion on Cartesian components for various vector operations such as gradient, the divergence and Laplacian of a scalar field and the components of polar, cylindrical and spherical coordinates for these operations will be derived in terms of orthogonal curvilinear coordinates.

5.1 DIFFERENTIAL OPERATOR

The vector differential operator *del* or *nabla*, denoted by $\vec{\nabla}$, is defined by

$$\vec{\nabla} = \hat{i}\,\frac{\partial}{\partial x} + \hat{j}\,\frac{\partial}{\partial y} + \hat{k}\,\frac{\partial}{\partial z} = \sum \hat{i}\,\frac{\partial}{\partial x} \tag{5.1}$$

This possesses properties analogous to those of ordinary vectors, and it operates distributively. Clearly, it should be understood that the expression for $\vec{\nabla}$ holds only when we are dealing with Cartesian rectangular coordinates. The operators $\frac{\partial}{\partial x}$, $\frac{\partial}{\partial y}$ and $\frac{\partial}{\partial z}$ are the components of $\vec{\nabla}$, which has a dual character as an operator as well as a vector, and it operates upon a function (scalar and vector) distributively. The nabla operator was introduced by Hamilton to develop the theory of quaternions. The operator $\vec{\nabla}$ is also termed the *Hamiltonian operator* or *differential operator*.

Let $\vec{a}$ be any vector. The operator $\vec{a}\cdot\vec{\nabla}$ is defined by

$$\vec{a}\cdot\vec{\nabla} = \vec{a}\cdot\hat{i}\,\frac{\partial}{\partial x} + \vec{a}\cdot\hat{j}\,\frac{\partial}{\partial y} + \vec{a}\cdot\hat{k}\,\frac{\partial}{\partial z} \tag{5.2}$$

Note that the expression (5.2) of $\vec{a}\cdot\vec{\nabla}$ is an operator.

5.2 GRADIENT

Let a scalar point single-valued function $\phi(x, y, z)$ of the Cartesian coordinate (x, y, z) be defined and continuously differentiable in a certain region of space E^3. Then, in the Cartesian coordinate system, the *gradient* of the scalar point function ϕ, denoted by $\vec{\nabla}\phi$ or grad ϕ, is defined by the vector function

$$\vec{\nabla}\phi = \left(\hat{i}\frac{\partial}{\partial x} + \hat{j}\frac{\partial}{\partial y} + \hat{k}\frac{\partial}{\partial z}\right)\phi = \frac{\partial \phi}{\partial x}\,\hat{i} + \frac{\partial \phi}{\partial y}\,\hat{j} + \frac{\partial \phi}{\partial z}\,\hat{k} \tag{5.3}$$

where the partial derivatives of ϕ are assumed to be continuous. Now, $\vec{\nabla}\phi$ being a product of a vector $\vec{\nabla}$ by a scalar ϕ, defines a vector field.

(i) The gradient changes a scalar point function to a vector point function.

(ii) For a scalar function ϕ, using Eq. (5.2), we have

$$(\vec{a}\cdot\vec{\nabla})\phi = \vec{a}\cdot\hat{i}\,\frac{\partial \phi}{\partial x} + \vec{a}\cdot\hat{j}\,\frac{\partial \phi}{\partial y} + \vec{a}\cdot\hat{k}\,\frac{\partial \phi}{\partial z} \tag{5.4}$$

If $\hat{a}$ is a unit vector, then $\hat{a}\cdot\hat{i}$, $\hat{a}\cdot\hat{j}$ and $\hat{a}\cdot\hat{k}$ are the direction cosines of $\hat{a}$. Thus, $(\vec{a}\cdot\vec{\nabla})\phi$ represents the directional derivative of the function ϕ along the direction of the unit vector $\hat{a}$. Also

$$(\vec{a}\cdot\vec{\nabla})\phi = \vec{a}\cdot\vec{\nabla}\phi = \vec{a}\cdot \text{ grad }\phi \tag{5.5}$$

(iii) Using the total differential formula of ordinary real analysis, we have

$$\begin{aligned} d\phi &= \frac{\partial \phi}{\partial x}\,dx + \frac{\partial \phi}{\partial y}\,dy + \frac{\partial \phi}{\partial z}\,dz \\ &= \left(\frac{\partial \phi}{\partial x}\hat{i} + \frac{\partial \phi}{\partial y}\hat{j} + \frac{\partial \phi}{\partial z}\hat{k}\right)\cdot(dx\hat{i} + dy\hat{j} + dz\hat{k}) \\ &= \vec{\nabla}\phi\cdot d\vec{r} = \text{ grad }\phi\cdot d\vec{r} \end{aligned} \tag{5.6}$$

(iv) Let ϕ and ψ be two scalar point functions and ϕ be a scalar function of ψ, then

$$\begin{aligned} \vec{\nabla}\phi &= \frac{\partial \phi}{\partial x}\,\hat{i} + \frac{\partial \phi}{\partial y}\,\hat{j} + \frac{\partial \phi}{\partial z}\,\hat{k} \\ &= \frac{\partial \phi}{\partial \psi}\cdot\frac{\partial \psi}{\partial x}\,\hat{i} + \frac{\partial \phi}{\partial \psi}\cdot\frac{\partial \psi}{\partial y}\,\hat{j} + \frac{\partial \phi}{\partial \psi}\cdot\frac{\partial \psi}{\partial z}\,\hat{k} \\ &= \frac{d\phi}{d\psi}\left[\frac{\partial \phi}{\partial x}\,\hat{i} + \frac{\partial \phi}{\partial y}\,\hat{j} + \frac{\partial \phi}{\partial z}\,\hat{k}\right] = \frac{d\phi}{d\psi}\,\vec{\nabla}\phi \end{aligned}$$

(v) For the vector function $\vec{f}$, using Eq. (5.3), we have

$$(\vec{a}\cdot\vec{\nabla})\vec{f} = \vec{a}\cdot\hat{i}\frac{\partial \vec{f}}{\partial x} + \vec{a}\cdot\hat{j}\frac{\partial \vec{f}}{\partial y} + \vec{a}\cdot\hat{k}\frac{\partial \vec{f}}{\partial z}$$

If $\hat{a}$ is an unit vector, then $\hat{a}\cdot\hat{i}$, $\hat{a}\cdot\hat{j}$ and $\hat{a}\cdot\hat{k}$ are the direction cosines of $\hat{a}$. Thus, $(\vec{a}\cdot\vec{\nabla})\vec{f}$ stands for the directional derivatives of the respective functions along the direction of the unit vector. Corresponding to the relation (5.6), we have

$$d\vec{f} = \frac{\partial \vec{f}}{\partial x}\,dx + \frac{\partial \vec{f}}{\partial y}\,dy + \frac{\partial \vec{f}}{\partial z}\,dz$$

$$= d\vec{r} \cdot \hat{i}\frac{\partial \vec{f}}{\partial x} + d\vec{r} \cdot \hat{j}\frac{\partial \vec{f}}{\partial y} + d\vec{r} \cdot \hat{k}\frac{\partial \vec{f}}{\partial z} = d\vec{r} \cdot \vec{\nabla} \vec{f}$$

for any vector function $\vec{f}$.

Property 5.2.1

Linearity property: The gradient operator is linear, i.e., if $\phi(x, y, z)$ and $\psi(x, y, z)$ are the scalar point functions, and a and b are constant scalars, then

$$\vec{\nabla}(a\phi + b\psi) = a\vec{\nabla}\phi + b\vec{\nabla}\psi \tag{5.7}$$

Proof: Using the definition of gradient, we have

$$\begin{aligned}
\vec{\nabla}(a\phi + b\psi) &= \frac{\partial}{\partial x}(a\phi + b\psi)\hat{i} + \frac{\partial}{\partial y}(a\phi + b\psi)\hat{j} + \frac{\partial}{\partial z}(a\phi + b\psi)\hat{k} \\
&= \left[a\frac{\partial \phi}{\partial x} + b\frac{\partial \psi}{\partial x}\right]\hat{i} + \left[a\frac{\partial \phi}{\partial y} + b\frac{\partial \psi}{\partial y}\right]\hat{j} + \left[a\frac{\partial \phi}{\partial z} + b\frac{\partial \psi}{\partial z}\right]\hat{k} \\
&= a\left[\frac{\partial \phi}{\partial x}\hat{i} + \frac{\partial \phi}{\partial y}\hat{j} + \frac{\partial \phi}{\partial z}\hat{k}\right] + b\left[\frac{\partial \psi}{\partial x}\hat{i} + \frac{\partial \psi}{\partial y}\hat{j} + \frac{\partial \psi}{\partial z}\hat{k}\right] \\
&= a\vec{\nabla}\phi + b\vec{\nabla}\psi
\end{aligned}$$

This shows that the gradient operator is linear.

Example 5.2.1

Find $\vec{\nabla}\phi$ if $\phi(x, y, z) = 3x^2y - y^3z^2$ at the point $(1, -2, -1)$.

Solution

Using Eq. (5.3), we get

$$\begin{aligned}
\vec{\nabla}\phi &= \left(\hat{i}\frac{\partial}{\partial x} + \hat{j}\frac{\partial}{\partial y} + \hat{k}\frac{\partial}{\partial z}\right)(3x^2y - y^3z^2) \\
&= \hat{i}\frac{\partial}{\partial x}(3x^2y - y^3z^2) + \hat{j}\frac{\partial}{\partial y}(3x^2y - y^3z^2) + \hat{k}\frac{\partial}{\partial z}(3x^2y - y^3z^2) \\
&= 6xy\,\hat{i} + (3x^2 - 3y^2z^2)\,\hat{j} - 2y^3z\,\hat{k} \\
&= 6(1)(-2)\,\hat{i} + \{3(1)^2 - 3(-2)^2(-1)^2\}\,\hat{j} - 2(-2)^3(-1)\,\hat{k} \\
&= -12\hat{i} - 9\hat{j} - 16\hat{k} \text{ at } (1, -2, -1)
\end{aligned}$$

Example 5.2.2

If $\vec{r} = x\hat{i} + y\hat{j} + z\hat{k}$ and $r = |\vec{r}|$, evaluate $\vec{\nabla}\, r$.

Solution

Since $r = |\vec{r}| = \sqrt{x^2 + y^2 + z^2}$, so

$$\frac{\partial r}{\partial x} = \frac{\partial}{\partial x}(x^2 + y^2 + z^2)^{1/2} = \frac{2x}{2(x^2 + y^2 + z^2)^{1/2}} = \frac{x}{\sqrt{x^2 + y^2 + z^2}} = \frac{x}{r}$$

Similarly, $\dfrac{\partial r}{\partial y} = \dfrac{y}{r}$ and $\dfrac{\partial r}{\partial z} = \dfrac{z}{r}$. Using Eq. (5.3), we get

$$\vec{\nabla} r = \frac{\partial r}{\partial x}\hat{i} + \frac{\partial r}{\partial y}\hat{j} + \frac{\partial r}{\partial z}\hat{k} = \frac{x}{r}\,\hat{i} + \frac{y}{r}\,\hat{j} + \frac{z}{r}\,\hat{k} = \frac{1}{r}(x\hat{i} + y\hat{j} + z\hat{k}) = \frac{1}{r}\,\vec{r}$$

Example 5.2.3

Find $\vec{\nabla}\phi$ if $\phi = \ln|\vec{r}|$, where $\vec{r} = x\hat{i} + y\hat{j} + z\hat{k}$ and $r = |\vec{r}|$.

Solution

We have $r = |\vec{r}|$ and $\phi = \ln|\vec{r}| = \ln r = \frac{1}{2}\ln(x^2 + y^2 + z^2)$. Using Eq. (5.3), we get

$$\begin{aligned}\vec{\nabla}\phi &= \frac{1}{2}\vec{\nabla}\ln(x^2 + y^2 + z^2) = \frac{1}{2}\Big[\frac{\partial}{\partial x}\ln(x^2 + y^2 + z^2)\,\hat{i} \\ &\qquad + \frac{\partial}{\partial y}\ln(x^2 + y^2 + z^2)\,\hat{j} + \frac{\partial}{\partial z}\ln(x^2 + y^2 + z^2)\,\hat{k}\Big] \\ &= \frac{1}{2}\left[\frac{2x}{x^2 + y^2 + z^2}\,\hat{i} + \frac{2y}{x^2 + y^2 + z^2}\,\hat{j} + \frac{2z}{x^2 + y^2 + z^2}\,\hat{k}\right] \\ &= \frac{1}{x^2 + y^2 + z^2}(x\,\hat{i} + y\,\hat{j} + z\,\hat{k}) = \frac{1}{r^2}\,\vec{r}\end{aligned}$$

Example 5.2.4

If ϕ is a differentiable scalar function of x, y, z and t, where x, y and z are differentiable scalar functions of t, prove that

$$\frac{d\phi}{dt} = \frac{\partial\phi}{\partial t} + \vec{\nabla}\cdot\frac{d\vec{r}}{dt};\ \vec{r} = x\hat{i} + y\hat{j} + z\hat{k}$$

Solution

We have $\vec{r} = x(t)\hat{i} + y(t)\hat{j} + z(t)\hat{k} = \vec{r}(t)$, so

$$\frac{d\vec{r}}{dt} = \frac{dx}{dt}\hat{i} + \frac{dy}{dt}\hat{j} + \frac{dz}{dt}\hat{k}$$

We know from the calculus of scalar functions of more than one variable that $\phi = \phi(x, y, z, t)$ is a differentiable scalar function of t. Then, as ϕ is ultimately a differentiable function of t

$$\begin{aligned}\frac{d\phi}{dt} &= \frac{\partial\phi}{\partial t} + \frac{\partial\phi}{\partial x}\cdot\frac{dx}{dt} + \frac{\partial\phi}{\partial y}\cdot\frac{dy}{dt} + \frac{\partial\phi}{\partial z}\cdot\frac{dz}{dt} \\ &= \frac{\partial\phi}{\partial t} + \left(\frac{\partial\phi}{\partial x}\hat{i} + \frac{\partial\phi}{\partial y}\hat{j} + \frac{\partial\phi}{\partial z}\hat{k}\right)\cdot\left(\frac{dx}{dt}\hat{i} + \frac{dy}{dt}\hat{j} + \frac{dz}{dt}\hat{k}\right) \\ &= \frac{\partial\phi}{\partial t} + \vec{\nabla}\phi\cdot\frac{d\vec{r}}{dt}\end{aligned}$$

Example 5.2.5

If $\phi = \frac{1}{2}\ln(x^2 + y^2)$, show that $\vec{\nabla}\phi = \dfrac{\vec{r} - (\hat{k}\cdot\vec{r})\hat{k}}{\left[\vec{r} - (\hat{k}\cdot\vec{r})\hat{k}\right]\cdot\left[\vec{r} - (\hat{k}\cdot\vec{r})\hat{k}\right]}$.

Solution

Here, $\vec{r} = x\hat{i} + y\hat{j} + z\hat{k}$, so $\vec{r} \cdot \hat{k} = z = \hat{k} \cdot \vec{r}$. Now

$$\frac{\partial \phi}{\partial x} = \frac{1}{2}\frac{1}{x^2+y^2} \cdot 2x = \frac{x}{x^2+y^2} \;;\; \frac{\partial \phi}{\partial y} = \frac{y}{x^2+y^2} \text{ and } \frac{\partial \phi}{\partial z} = 0$$

Using Eq. (5.3), we get

$$\begin{aligned}\vec{\nabla}\phi &= \frac{\partial \phi}{\partial x}\hat{i} + \frac{\partial \phi}{\partial y}\hat{j} + \frac{\partial \phi}{\partial z}\hat{k} = \frac{1}{x^2+y^2}\left[x\hat{i} + y\hat{j}\right] \\ &= \frac{1}{[x\hat{i}+y\hat{j}]\cdot[x\hat{i}+y\hat{j}]}\left[\vec{r} - z\hat{k}\right] = \frac{1}{\left[\vec{r} - z\hat{k}\right]\cdot\left[\vec{r} - z\hat{k}\right]}(\vec{r} - z\hat{k}) \\ &= \frac{\vec{r} - (\hat{k}\cdot\vec{r})\hat{k}}{\left[\vec{r} - (\hat{k}\cdot\vec{r})\hat{k}\right]\cdot\left[\vec{r} - (\hat{k}\cdot\vec{r})\hat{k}\right]}\end{aligned}$$

Example 5.2.6

If $\vec{r} = x\hat{i} + y\hat{j} + z\hat{k}$ and $\vec{a}$ and $\vec{b}$ are constant vectors, prove that

(i) $\vec{\nabla}\phi(r) = \dfrac{\phi'(r)}{r} \cdot \vec{r}$ (ii) $\vec{\nabla}\phi(r) \times \vec{r} = \vec{0}$

(iii) $\vec{b} \cdot \vec{\nabla}\left[\vec{a} \cdot \vec{\nabla}\left(\dfrac{1}{r}\right)\right] = \dfrac{3(\vec{a}\cdot\vec{r})(\vec{b}\cdot\vec{r})}{r^5} - \dfrac{\vec{a}\cdot\vec{b}}{r^3}$

Solution

(i) Since $r = \sqrt{x^2+y^2+z^2}$, $\dfrac{\partial x}{\partial r} = \dfrac{x}{r}$, $\dfrac{\partial y}{\partial r} = \dfrac{y}{r}$ and $\dfrac{\partial z}{\partial r} = \dfrac{z}{r}$. Hence

$$\frac{\partial \phi}{\partial x} = \frac{\partial \phi}{\partial r} \cdot \frac{\partial r}{\partial x} = \phi'(r)\frac{x}{r} = \frac{\phi'(r)}{r}x$$

Similarly $\dfrac{\partial \phi}{\partial y} = \dfrac{\phi'(r)}{r}y$ and $\dfrac{\partial \phi}{\partial z} = \dfrac{\phi'(r)}{r}z$. Using the definition of gradient, we have

$$\begin{aligned}\vec{\nabla}\phi(r) &= \frac{\partial \phi}{\partial x}\hat{i} + \frac{\partial \phi}{\partial y}\hat{j} + \frac{\partial \phi}{\partial z}\hat{k} = \frac{x\phi'(r)}{r}\hat{i} + \frac{y\phi'(r)}{r}\hat{j} + \frac{z\phi'(r)}{r}\hat{k} \\ &= \frac{\phi'(r)}{r}\left(x\hat{i} + y\hat{j} + z\hat{k}\right) = \frac{\phi'(r)}{r}\vec{r}\end{aligned}$$

(ii) Using the first point, we obtain

$$\vec{\nabla}\phi(r) \times \vec{r} = \frac{\phi'(r)}{r}\vec{r} \times \vec{r} = \vec{0}$$

(iii) If $r = |\vec{r}|$, then $\vec{\nabla}r^{-1} = -(1/r^3)\vec{r}$. Hence, $\vec{a} \cdot \vec{\nabla}r^{-1} = -(\vec{a}\cdot\vec{r})/r^3$. Now

$$\frac{\partial}{\partial x}\left(\frac{\vec{a}\cdot\vec{r}}{r^3}\right) = \frac{1}{r^3}\left[\vec{a} \cdot \frac{\partial \vec{r}}{\partial x} + \frac{\partial \vec{a}}{\partial x} \cdot \vec{r}\right] - \frac{3\vec{a}\cdot\vec{r}}{r^4} \cdot \frac{\partial r}{\partial x} = \frac{\vec{a}\cdot\hat{i}}{r^3} - \frac{3x}{r^5}(\vec{a}\cdot\vec{r})$$

Similarly, we get

$$\frac{\partial}{\partial y}\left(\frac{\vec{a}\cdot\vec{r}}{r^3}\right) = \frac{\vec{a}\cdot\hat{j}}{r^3} - \frac{3x}{r^5}(\vec{a}\cdot\vec{r}) \text{ and } \frac{\partial}{\partial z}\left(\frac{\vec{a}\cdot\vec{r}}{r^3}\right) = \frac{\vec{a}\cdot\hat{k}}{r^3} - \frac{3x}{r^5}(\vec{a}\cdot\vec{r})$$

Using the definition of gradient, we get

$$\vec{\nabla}\left(\frac{\vec{a}\cdot\hat{k}}{r^3}\right) = \frac{\partial}{\partial x}\left(\frac{\vec{a}\cdot\vec{r}}{r^3}\right) + \frac{\partial}{\partial y}\left(\frac{\vec{a}\cdot\vec{r}}{r^3}\right) + \frac{\partial}{\partial z}\left(\frac{\vec{a}\cdot\vec{r}}{r^3}\right)$$

$$= \frac{1}{r^3}\left[(\vec{a}\cdot\hat{i})\hat{i} + (\vec{a}\cdot\hat{j})\hat{j} + (\vec{a}\cdot\hat{k})\hat{k}\right] - \frac{3}{r^5}(\vec{a}\cdot\vec{r})(x\hat{i} + y\hat{j} + z\hat{k})$$

Therefore

$$\vec{b}\cdot\vec{\nabla}\left[\vec{a}\cdot\vec{\nabla}r^{-1}\right] = -\vec{b}\cdot\vec{\nabla}\left[\frac{\vec{a}\cdot\vec{r}}{r^3}\right] = -\vec{b}\cdot\left[\frac{1}{r^3}\vec{a} - \frac{3(\vec{a}\cdot\vec{r})}{r^5}\vec{r}\right]$$

$$= -\frac{1}{r^3}(\vec{a}\cdot\vec{b}) + \frac{3}{r^5}(\vec{a}\cdot\vec{r})(\vec{b}\cdot\vec{r})$$

Example 5.2.7

Show that $\vec{\nabla}\left[\frac{(\vec{a}\cdot\vec{r})}{r^n}\right] = \frac{\vec{a}}{r^n} - \frac{n(\vec{a}\cdot\vec{r})\vec{r}}{r^{n+2}}$.

Solution

Let $\vec{a} = a_1\hat{i} + a_2\hat{j} + a_3\hat{k}$ and let $\phi = (\vec{a}\cdot\vec{r})/r^n$, then

$$\phi = \frac{(a_1\hat{i} + a_2\hat{j} + a_3\hat{k})\cdot(x\hat{i} + y\hat{j} + z\hat{k})}{r^n} = \frac{a_1x + a_2y + a_3z}{r^n}$$

We know $\frac{\partial r}{\partial x} = \frac{x}{r}$, $\frac{\partial r}{\partial y} = \frac{y}{r}$ and $\frac{\partial r}{\partial z} = \frac{z}{r}$. Using the partial differential formula

$$\frac{\partial}{\partial x}\left(\frac{a_1x + a_2y + a_3z}{r^n}\right) = \frac{r^n\cdot a_1 - (a_1x + a_2y + a_3z)nr^{n-1}\frac{\partial r}{\partial x}}{r^{2n}}$$

$$= \frac{r^n a_1 - n(a_1x + a_2y + a_3z)\cdot r^{n-1}(x/r)}{r^{2n}}$$

$$= \frac{a_1}{r^n} - \frac{n}{r^{n+2}}(a_1x + a_2y + a_3z)x$$

Therefore, using the definition of gradient, we get

$$\vec{\nabla}\phi = \left(\hat{i}\frac{\partial}{\partial x} + \hat{j}\frac{\partial}{\partial y} + \hat{k}\frac{\partial}{\partial z}\right)\left(\frac{a_1x + a_2y + a_3z}{r^n}\right)$$

$$= \sum\left[\frac{a_1}{r^n} - \frac{n(a_1x + a_2y + a_3z)x}{r^{n+2}}\right]\hat{i}$$

$$= \frac{1}{r^n}(a_1\hat{i} + a_2\hat{j} + a_3\hat{k}) - \frac{n(a_1x + a_2y + a_3z)}{r^{n+2}}(x\hat{i} + y\hat{j} + z\hat{k})$$

$$= \frac{1}{r^n}\vec{a} - \frac{n(\vec{a}\cdot)\vec{r}}{r^{n+2}}\vec{r}$$

DEDUCTION 5.2.1

Geometrical interpretation of gradient: Let $\vec{r} = x\,\hat{i} + y\,\hat{j} + z\,\hat{k}$ be the position vector of any point $P(x, y, z)$ (Fig. 5.1) on the level surface $\phi(x, y, z) = c$, where c is a constant with respect to some chosen origin. Let $Q(x + \Delta x, y + \Delta y, z + \Delta z)$ be any neighbouring point on the level surface, i.e., its position vector is $\vec{r} + \Delta\vec{r}$. Then

$\vec{PQ} = \vec{r} + \Delta\vec{r} - \vec{r} = \Delta\vec{r} = \Delta x\hat{i} + \Delta y\hat{j} + \Delta z\hat{k}$

As $Q \to P$, the straight line PQ tends to the tangent line to the level surface at P. Hence, the limit is the line vector along

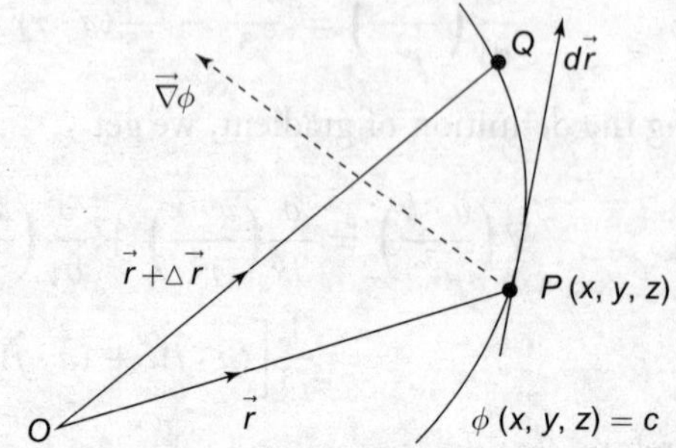

Figure 5.1 Geometrical interpretation of gradient

$$\lim_{Q \to P} \vec{PQ} = d\vec{r} = dx\,\hat{i} + dy\,\hat{j} + dz\,\hat{k}$$

This lies on the tangent plane at P to the level surface, and hence, the direction of $d\vec{r}$ is along the tangent to the surface at the point P. As $\phi(x, y, z) = c$, $d\phi = 0$, and hence, using Eq. (5.6), we get

$$\frac{\partial\phi}{\partial x}dx + \frac{\partial\phi}{\partial y}dy + \frac{\partial\phi}{\partial z}dz = 0$$

$$\text{or,} \quad \left(\frac{\partial\phi}{\partial x}\hat{i} + \frac{\partial\phi}{\partial y}\hat{j} + \frac{\partial\phi}{\partial z}\hat{k}\right) \cdot \left(dx\hat{i} + dy\hat{j} + dz\hat{k}\right) = 0$$

$$\text{or,} \quad \vec{\nabla}\phi \cdot d\vec{r} = 0$$

This shows that $\vec{\nabla}\phi$ is perpendicular to $d\vec{r}$ and is hence normal to the surface. i.e., $\frac{\partial\phi}{\partial x}, \frac{\partial\phi}{\partial y}, \frac{\partial\phi}{\partial z}$ represents the direction ratios of a line which is perpendicular to the tangent plane to the level surface at P. Thus, the vector $\vec{\nabla}\phi$ has the direction in the direction of the normal at the point (x, y, z) to the level surface.

(i) The angle between two space curves is defined as the angle between their normals. Let $\phi_1(x, y, z) = c_1$ and $\phi_2(x, y, z) = c_2$ be two space curves. If θ is the angle between them, then it is given by

$$\vec{\nabla}\phi_1 \cdot \vec{\nabla}\phi_2 = |\vec{\nabla}\phi_1|\,|\vec{\nabla}\phi_2| \cos\theta \qquad (5.8)$$

(ii) Let $S : \phi(x, y, z) = c$ be a given level surface and $\vec{r}_0 = x_0\hat{i} + y_0\hat{j} + z_0\hat{k}$ be the position vector of any point $P(x_0, y_0, z_0)$ on the level surface (Fig. 5.2). $\vec{\nabla}\phi$ has the direction in the direction of the normal PN to the surface. Let

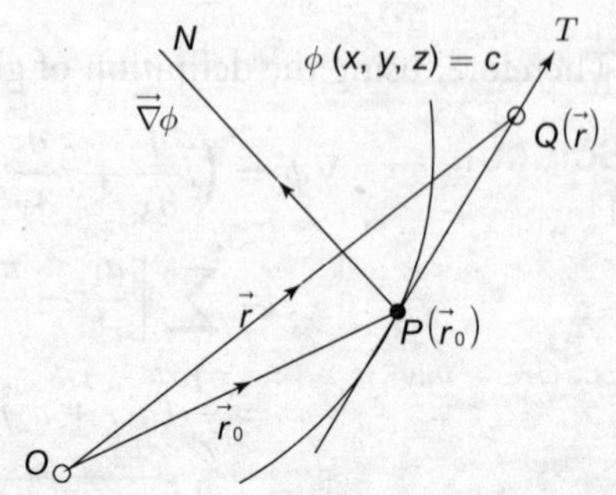

Figure 5.2 Tangent plane

$\vec{r} = x\hat{i} + y\hat{j} + z\hat{k}$ be the position vector of any point on the tangent plane PT. Then the equation of the tangent plane is given by

$$(\vec{r} - \vec{r}_0).(\vec{\nabla}\phi)_Q = 0 \qquad (5.9)$$

This relation is true for any point Q on the tangent plane, and hence, it represents the equation of the tangent plane to the surface $\phi(x, y, z) = c$ at $P(\vec{r}_0)$.

(iii) Let $\vec{r}$ be the position vector of any point on the normal line at the point (Fig. 5.3). Then, the equation of the normal is

$$(\vec{r} - \vec{r}_0) \times (\vec{\nabla}(\phi))_P = \vec{0} \qquad (5.10)$$

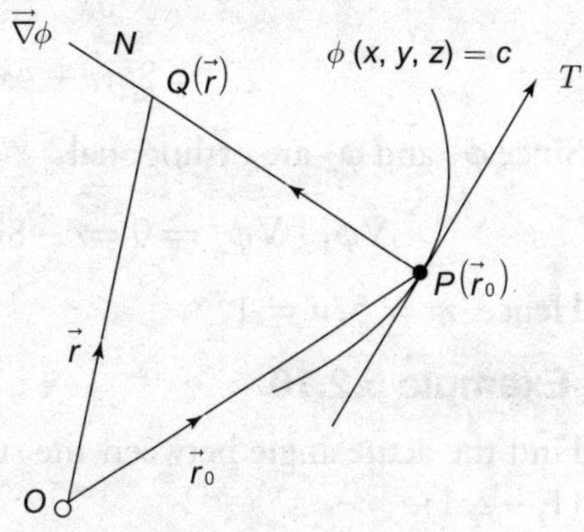

Figure 5.3 Normal line

This relation is true for any point Q on the normal line, and hence, it represents the equation of the normal line to the surface $\phi(x, y, z) = c$ at $P(\vec{r}_0)$.

Example 5.2.8

Find the unit outward drawn normal to the surface $xz^2 + x^2y = z - 1$ at $(1, -3, 2)$.

Solution

The equation of the given curve is of the form $\phi(x, y, z) = c$, where $\phi(x, y, z) = xz^2 + x^2y - z$ and $c = -1$. Therefore

$$\vec{\nabla}\phi = \vec{\nabla}(xz^2 + x^2y - z) = (z^2 + 2xy)\hat{i} + x^2\hat{j} + (2xz - 1)\hat{k}$$
$$= -2\hat{i} + \hat{j} + 3\hat{k} \quad \text{at } (1, -3, 2)$$

Then, a unit normal to the surface is given by

$$\hat{n} = \frac{1}{|\vec{\nabla}\phi|}\vec{\nabla}\phi = \frac{1}{\sqrt{14}}(-2\hat{i} + \hat{j} + 3\hat{k})$$

Another unit normal is $\frac{1}{\sqrt{14}}(2\hat{i} - \hat{j} - 3\hat{k})$ having the direction that is the opposite of that above.

Example 5.2.9

Find the constants m and n such that the surface $mx^2 - 2nyz = (m + 4)x$ will be orthogonal to the surface $4x^2y + z^3 = 4$ at the point $(1, -1, 2)$.

Solution

The point $P(1, -1, 2)$ lies on both surfaces. This point lies on $mx^2 - 2nyz = (m + 4)x$, so

$$m + 4n = m + 4 \Rightarrow 4n = 4 \Rightarrow n = 1$$

Let $\phi_1 = mx^2 - 2nxyz - (m+4)x = mx^2 - 2xyz - (m+4)x$; since $n = 1$ and $\phi_2 = 4x^2y + z^3 - 4$

$$\vec{\nabla}\phi_1 = \left(\hat{i}\frac{\partial}{\partial x} + \hat{j}\frac{\partial}{\partial y} + \hat{k}\frac{\partial}{\partial z}\right)(mx^2 - 2xyz - (m + 4)x)$$
$$= (2mx - m - 4)\hat{i} - (2z)\hat{j} - (2y)\hat{k} = (m - 4)\hat{i} - 4\hat{j} + 2\hat{k} \text{ at } (1, -1, 2)$$

$$\vec{\nabla}\phi_2 = \left(\hat{i}\frac{\partial}{\partial x} + \hat{j}\frac{\partial}{\partial y} + \hat{k}\frac{\partial}{\partial z}\right)(4x^2y + z^3 - 4)$$
$$= 8xy\hat{i} + 4x^2\hat{j} + 3z^2\hat{k} = -8\hat{i} + 4\hat{j} + 12\hat{k} \text{ at } (1, -1, 2)$$

Since ϕ_1 and ϕ_2 are orthogonal

$$\vec{\nabla}\phi_1 \cdot \vec{\nabla}\phi_2 = 0 \Rightarrow -8m + 32 - 16 + 24 = 0 \Rightarrow 8m = 40 \Rightarrow m = 5$$

Hence, $m = 5, n = 1$.

Example 5.2.10

Find the acute angle between the surfaces $xy^2z = 3x + z^2$ and $3x^2 - y^2 + 2z = 1$ at the point $(1, -2, 1)$.

Solution

The angle between two surfaces is defined as the angle between their normals at their point of intersection. A normal to $\phi_1(x, y, z) = xy^2z - 3x - z^2 = 0$ at $(1, -2, 1)$ is given by

$$\hat{n}_1 = \vec{\nabla}\phi_1 = \vec{\nabla}(xy^2z - 3x - z^2) = (y^2z - 3)\hat{i} + 2xyz\hat{j} + (xy^2 - 2z)\hat{k}$$
$$= \hat{i} - 4\hat{j} + 2\hat{k} \quad \text{at } (1, -2, 1)$$

Similarly, the normal to $\phi_2(x, y, z) = 3x^2 - y^2 + 2z = 1$ at $(1, -2, 1)$ is given by

$$\hat{n}_2 = \vec{\nabla}\phi_2 = \vec{\nabla}(3x^2 - y^2 + 2z) = 6x\hat{i} - 2y\hat{j} + 2\hat{k}$$
$$= 6\hat{i} + 4\hat{j} + 2\hat{k} \quad \text{at } (1, -2, 1)$$

Let θ be the angle between the normals. Then, by the definition of dot product of vectors, we have, $\hat{n}_1 \cdot \hat{n}_2 = |\hat{n}_1||\hat{n}_2|\cos\theta$, and so

$$\left|(\hat{i} - 4\hat{j} + 2\hat{k}) \cdot (6\hat{i} + 4\hat{j} + 2\hat{k})\right| = \left|\hat{i} - 4\hat{j} + 2\hat{k}\right|\left|6\hat{i} + 4\hat{j} + 2\hat{k}\right|\cos\theta$$
$$\Rightarrow |6 - 16 + 4| = \sqrt{(1)^2 + (-4)^2 + (2)^2}\sqrt{(6)^2 + (4)^2 + (2)^2}\cos\theta$$
$$\Rightarrow \cos\theta = \frac{6}{\sqrt{21}\sqrt{56}} = \frac{3}{7\sqrt{6}} \Rightarrow \theta = 79{\cdot}5'$$

Example 5.2.11

Find the equations for the tangent plane and normal line to the surface $z = x^2 + y^2$ at the point $(2, -1, 5)$.

Solution

For the given level surface $\phi(x, y, z) = x^2 + y^2 - z = 0$

$$\vec{\nabla}\phi = \frac{\partial\phi}{\partial x}\hat{i} + \frac{\partial\phi}{\partial y}\hat{j} + \frac{\partial\phi}{\partial z}\hat{k} = 2x\hat{i} + 2y\hat{j} - \hat{k} = 4\hat{i} - 2\hat{j} - \hat{k}$$

at the point $(2, -1, 5)$. Thus, the normal to the surface at $(2, -1, 5)$ is $4\hat{i} - 2\hat{j} - \hat{k}$. Let $\vec{r} = x\hat{i} + y\hat{j} + z\hat{k}$ be the position vector of any point on the tangent plane. Then, its equation according to Eq. (5.9) is given by

$$\left[(x\hat{\imath}+y\hat{\jmath}+z\hat{k})-(2\hat{\imath}-\hat{\jmath}+5\hat{k})\right]\cdot(4\hat{\imath}-2\hat{\jmath}-\hat{k})=0$$
$$\Rightarrow 4(x-2)-2(y+1)-(z-5)=0$$
$$\Rightarrow 4x-2y-z=5$$

If $\vec{r}=x\hat{\imath}+y\hat{\jmath}+z\hat{k}$ is the position vector of any point on the normal line, then its equation of the normal line, as in Eq. (5.10), is given by

$$\left[(x\hat{\imath}+y\hat{\jmath}+z\hat{k})-(2\hat{\imath}-\hat{\jmath}+5\hat{k})\right]\times(4\hat{\imath}-2\hat{\jmath}-\hat{k})=\vec{0}$$
$$\Rightarrow \frac{x-2}{4}=\frac{y+1}{-2}=\frac{z-5}{-1}$$

Theorem 5.1

The gradient of a constant scalar point function is a zero vector, and conversely.

Proof: Let us consider a constant scalar point function $\phi(x,y,z)$. Then, we have $\frac{\partial\phi}{\partial x}=0$, $\frac{\partial\phi}{\partial y}=0$ and $\frac{\partial\phi}{\partial z}=0$. Using the definition of gradient, we have

$$\vec{\nabla}\phi=\frac{\partial\phi}{\partial x}\hat{\imath}+\frac{\partial\phi}{\partial y}\hat{\jmath}+\frac{\partial\phi}{\partial z}\hat{k}=0\hat{\imath}+0\hat{\jmath}+0\hat{k}=\vec{0}$$

This shows that the gradient of a constant scalar point function is a zero vector.

Conversely, we suppose that $\vec{\nabla}\phi=\vec{0}$, which yields

$$\frac{\partial\phi}{\partial x}\hat{\imath}+\frac{\partial\phi}{\partial y}\hat{\jmath}+\frac{\partial\phi}{\partial z}\hat{k}=\vec{0}\Rightarrow\frac{\partial\phi}{\partial x}=\frac{\partial\phi}{\partial y}\frac{\partial\phi}{\partial z}=0$$

This implies that ϕ is independent of x,y and z, and hence, ϕ is a constant scalar point function. This completes the proof.

Deduction 5.2.2

Let $\mathcal{S}_1:\phi(x,y,z)=c$ (c = parameter) be a level surface passing through the point P and $\mathcal{S}_2:\phi(x,y,z)=c_1$ be the neighbouring level surface (Fig. 5.4).

Let $\hat{n}$ be the unit outward-drawn normal vector to the level surface $\mathcal{S}_1$ at P. We assume that ϕ is a continuously differentiable scalar point function. Then, the direction cosines of the outward normal to the level surface $\mathcal{S}_1$ at P and l, m and n are

$$l=\frac{1}{K}\frac{\partial\phi}{\partial x},\ m=\frac{1}{K}\frac{\partial\phi}{\partial y},\ n=\frac{1}{K}\frac{\partial\phi}{\partial z};$$

$$K=\sqrt{\left(\frac{\partial\phi}{\partial x}\right)^2+\left(\frac{\partial\phi}{\partial y}\right)^2+\left(\frac{\partial\phi}{\partial z}\right)^2}$$

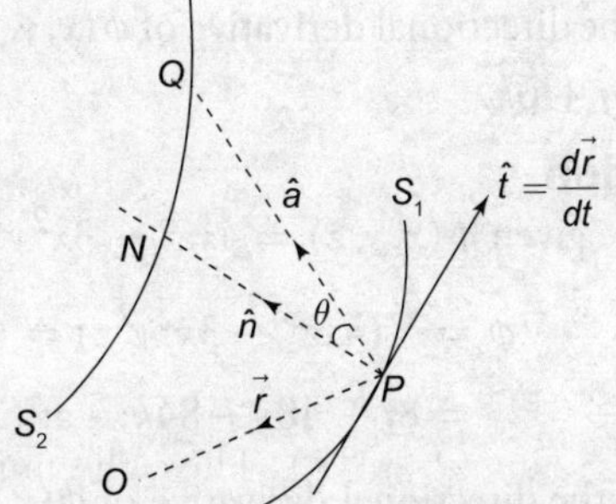

Figure 5.4 Geometrical interpretation of $\vec{\nabla}\phi$

Now, the directional derivative of ϕ along PN is

$$\frac{d\phi}{ds} = l\frac{\partial\phi}{\partial x} + m\frac{\partial\phi}{\partial y} + n\frac{\partial\phi}{\partial z} = \frac{1}{K}\frac{\partial\phi}{\partial x}\cdot\frac{\partial\phi}{\partial x} + \frac{1}{K}\frac{\partial\phi}{\partial y}\cdot\frac{\partial\phi}{\partial y} + \frac{1}{K}\frac{\partial\phi}{\partial z}\cdot\frac{\partial\phi}{\partial z}$$

$$= \frac{1}{K}\left[\left(\frac{\partial\phi}{\partial x}\right)^2 + \left(\frac{\partial\phi}{\partial y}\right)^2 + \left(\frac{\partial\phi}{\partial z}\right)^2\right] = K = |\vec{\nabla}\phi|$$

Therefore, the directional derivative of ϕ along PN is positive with magnitude $|\vec{\nabla}\phi|$, and hence, ϕ increases along PN. Further, the direction cosines of the vector $\vec{\nabla}\phi$ at P are the same as those of the outward normal PN. Now

(i) $|\vec{\nabla}\phi|$ is the *directional derivative* of ϕ in the direction of the normal to the level surface, i.e., $\vec{\nabla}\phi = \frac{d\phi}{dn}\,\hat{n}$, where $\hat{n}$ is a unit vector normal to the surface $\mathcal{S}_1$ in the direction of increasing ϕ and $\frac{d\phi}{dn}$ is the directional derivative of ϕ along $\hat{n}$, called the *normal derivative* of ϕ.

(ii) Let $\hat{a}$ be another unit vector through the point P that makes an angle θ with $\hat{n}$. If $|PN| = \Delta n$ and $|PQ| = \Delta s$, then $\Delta n = \Delta s\cos\theta$. Also, the directional derivative of ϕ at P along $\hat{a}$ is

$$\frac{d\phi}{ds} = \lim_{\Delta s\to 0}\frac{\Delta\phi}{\Delta s} = \lim_{\Delta n\to 0}\frac{\Delta\phi}{\Delta n}\lim_{\Delta s\to 0}\frac{\Delta n}{\Delta s}$$

$$= \frac{d\phi}{dn}\cos\theta = \frac{d\phi}{dn}(\hat{n}\cdot\hat{a}) = \left(\frac{d\phi}{dn}\hat{n}\right)\cdot\hat{a} = \vec{\nabla}\phi\cdot\hat{a} \tag{5.11}$$

Thus, $\frac{d\phi}{ds}$ is the projection of $\vec{\nabla}\phi$ in the direction of $\hat{a}$. Let $|\vec{\nabla}\phi| \neq 0$ and $\theta = 0$, i.e., PQ coincides with PN, i.e., $\vec{\nabla}\phi$ and $\hat{a}$ have the same direction. Since $\theta = 0$, $\cos\theta$ has the maximum value. So the directional derivative becomes maximum in the direction of $\vec{\nabla}\phi$.

Thus, the directional derivative of a scalar point function ϕ at a point in the level surface is maximum in the direction of the normal $\hat{n}$ to the surface at P, where $\hat{n} = \frac{1}{|\vec{\nabla}\phi|}\vec{\nabla}\phi$ and the maximum value of the directional derivative is $|\vec{\nabla}\phi|$.

Example 5.2.12

Find the directional derivative of $\phi(x, y, z) = 4xz^3 - 3x^2y^2z$ at the point $(2, -1, 2)$ in the direction $2\hat{i} - 3\hat{j} + 6\hat{k}$.

Solution

For the given $\phi(x, y, z) = 4xz^3 - 3x^2y^2z$

$$\vec{\nabla}\phi = \vec{\nabla}(4xz^3 - 3x^2y^2z) = (4z^3 - 6xy^2z)\hat{i} - (6x^2yz)\hat{j} + (12xz^2 - 3x^2y^2)\hat{k}$$

$$= 8\hat{i} + 48\hat{j} + 84\hat{k}; \quad \text{at } (2, -1, 2)$$

Then, the directional derivative of $\phi(x, y, z)$ in the direction of $\vec{a}$ is given by

$$\vec{\nabla}\phi\cdot\hat{a} = \vec{\nabla}\cdot\frac{1}{|\vec{a}|}\vec{a} = (8\hat{i} + 48\hat{j} + 84\hat{k})\cdot\frac{1}{\sqrt{2^2 + (-3)^2 + 6^2}}(2\hat{i} - 3\hat{j} + 6\hat{k})$$

$$= \frac{1}{7}\Big[8 \cdot 2 + 48 \cdot (-3) + 84 \cdot 6\Big] = 376/7$$

Example 5.2.13

Find the direction in which the directional derivative at $\phi(x, y) = (x^2 + y^2)/xy$ at $(1, 1)$ is zero, and hence, find the component of velocity at the vector $\vec{r} = (t^3 + 1)\hat{\imath} + t^2\hat{\jmath}$ in same direction at $t = 1$.

Solution

Given $\phi(x, y) = (x^2 + y^2)/xy$, so

$$\vec{\nabla}\phi = \Big(\hat{\imath}\frac{\partial}{\partial x} + \hat{\jmath}\frac{\partial}{\partial y} + \hat{k}\frac{\partial}{\partial z}\Big)\Big(\frac{x}{y} + \frac{y}{x}\Big) = \hat{\imath}\Big(\frac{1}{y} - \frac{y}{x^2}\Big) + \hat{\jmath}\Big(-\frac{x}{y^2} + \frac{1}{x}\Big)$$

$$= \hat{\imath}\Big(\frac{x^2 - y^2}{x^2 y}\Big) + \hat{\jmath}\Big(\frac{y^2 - x^2}{xy^2}\Big) = 0\hat{\imath} + 0\hat{\jmath} = \vec{0}, \text{ at } (1, 1)$$

Therefore, the directional derivative of ϕ at $(1, 1)$ is zero in any direction. Now, $\vec{r} = (t^3+1)\hat{\imath}+t^2\hat{\jmath}$. Thus, the velocity is given by

$$\vec{v} = \frac{d\vec{r}}{dt} = 3t^2\hat{\imath} + 2t\hat{\jmath} = 3\hat{\imath} + 2\hat{\jmath} \text{ at } t = 1$$

The component of velocity in same direction of velocity is

$$= (3\hat{\imath} + 2\hat{\jmath}) \cdot \frac{1}{\sqrt{9 + 4}}(3\hat{\imath} + 2\hat{\jmath}) = \frac{9 + 4}{\sqrt{13}} = \sqrt{13}$$

Example 5.2.14

Find the directional derivative of $\phi(x, y, z) = x^2yz + 4xz^2$ at $(1, -2, 1)$ in the direction at $2\hat{\imath} - \hat{\jmath} - 2\hat{k}$. Find the greatest rate of increases at ϕ.

Solution

Here, $\phi(x, y, z) = x^2yz + 4xz^2$

$$\vec{\nabla}\phi = \Big(\hat{\imath}\frac{\partial}{\partial x} + \hat{\jmath}\frac{\partial}{\partial y} + \hat{k}\frac{\partial}{\partial z}\Big)(x^2yz + 4xz^2)$$

$$= (2xyz + 4z^2)\hat{\imath} + (x^2z)\hat{\jmath} + (x^2y + 8xz)\hat{k} = \hat{\imath} + 6\hat{k} \text{ at } (1, -2, 1)$$

Let $\vec{a} = 2\hat{\imath} - \hat{\jmath} - 2\hat{k}$. Unit vector $\hat{a} = \dfrac{2\hat{\imath} - \hat{\jmath} - 2\hat{k}}{\sqrt{4 + 1 + 4}} = \dfrac{1}{3}(2\hat{\imath} - \hat{\jmath} - 2\hat{k})$. Therefore, directional derivative is given by

$$= \vec{\nabla} \cdot \hat{a} = (\hat{\imath} + 6\hat{k}) \cdot \frac{1}{3}(2\hat{\imath} - \hat{\jmath} - 2\hat{k}) = -\frac{1}{3} - 4 = -\frac{13}{3}$$

The greatest rate of increase at $\phi = |\hat{\imath} + 6\hat{k}| = \sqrt{1 + 36} = \sqrt{37}$.

Example 5.2.15

For the function $\phi(x, y) = x/(x^2 + y^2)$, find the magnitude at the directional derivative along a line making an angle of 30° with the positive x axis at $(0, 2)$.

Solution

Let $\phi(x, y) = x/(x^2 + y^2)$, then

$$\vec{\nabla}\phi = \hat{i}\frac{\partial\phi}{\partial x} + \hat{j}\frac{\partial\phi}{\partial y} = \left[\frac{1}{x^2 + y^2} + \frac{x(-2x)}{(x^2 + y^2)^2}\right]\hat{i} + \left[\frac{x(-2x)}{(x^2 + y^2)^2}\right]\hat{j}$$

$$= \frac{x^2 + y^2 - 2x^2}{(x^2 + y^2)^2}\hat{i} + \frac{(-2xy)}{(x^2 + y^2)^2}\hat{j}$$

$$= \frac{y^2 - x^2}{(x^2 + y^2)^2}\hat{i} - \frac{2xy}{(x^2 + y^2)^2}\hat{j} = \frac{1}{4}\hat{i} \text{ at } (0, 2)$$

Again, the unit vector $\hat{a}$ in the given direction is $\hat{a} = \cos 30^0\hat{i} + \sin 30^0\hat{j}$. Therefore the directional derivative in the direction of $\hat{ab}$ is given by

$$\vec{\nabla}\phi \cdot \hat{a} = \frac{1}{4}\hat{i} \cdot (\cos 30^0\hat{i} + \sin 30^0\hat{j}) = \frac{1}{4}\cos 30^0 = \frac{\sqrt{3}}{8}$$

Example 5.2.16

Find the directional derivative at $\vec{v}^2$, where $\vec{v} = xy^2\hat{i} + zy^2\hat{j} + xz^2\hat{k}$, at the point $(2, 0, 3)$ in the direction of the outward normal to the sphere $x^2 + y^2 + z^2 = 14$ at the point (3,2,1).

Solution

Since $\vec{v} = xy^2\hat{i} + zy^2\hat{j} + xz^2\hat{k}$

$$v^2 = \vec{v} \cdot \vec{v} = (xy^2\hat{i} + zy^2\hat{j} + xz^2\hat{k}) \cdot (xy^2\hat{i} + zy^2\hat{j} + xz^2\hat{k})$$

$$= x^2y^4 + z^2y^4 + x^2z^4$$

$$\Rightarrow \vec{\nabla}v^2 = \left(\hat{i}\frac{\partial}{\partial x} + \hat{j}\frac{\partial}{\partial y} + \hat{k}\frac{\partial}{\partial z}\right)(x^2y^4 + z^2y^4 + x^2z^4)$$

$$= (2xy^4 + 2xz^4)\hat{i} + (4x^2y^3 + 4z^2y^3)\hat{j} + (2zy^4 + 4x^2z^3)\hat{k}$$

$$= 162\hat{i} + 432\hat{k} = 54(3\hat{i} + 8\hat{k}) \text{ at } (2, 0, 3)$$

Let $\phi = x^2 + y^2 + z^2 - 14$.

$$\text{Normal vector } = \vec{\nabla}\phi = \left(\hat{i}\frac{\partial}{\partial x} + \hat{j}\frac{\partial}{\partial y} + \hat{k}\frac{\partial}{\partial z}\right)(x^2 + y^2 + z^2 - 14)$$

$$= 2x\hat{i} + 2y\hat{j} + 2z\hat{k} = 6\hat{i} + 4\hat{j} + 2\hat{k} \text{ at } (2, 0, 3)$$

$$\text{Unit normal vector } = \frac{1}{\sqrt{36 + 16 + 4}}(6\hat{i} + 4\hat{j} + 2\hat{k}) = \frac{1}{\sqrt{14}}(3\hat{i} + 2\hat{j} + \hat{k})$$

The directional derivative along the normal

$$= 54(3\hat{i} + 8\hat{k}) \cdot \frac{1}{\sqrt{14}}(3\hat{i} + 2\hat{j} + \hat{k}) = \frac{918}{\sqrt{14}}$$

Example 5.2.17

Find the directional derivative of $x^2y^2z^2$ at the point $(1, 1, -1)$ in the direction at the tangent to the curve $x = e^t, y = \sin 2t + 1$ and $z = 1 - \cos t$ at $t = 0$.

Solution

Let $\phi = x^2y^2z^2$. Then, using the definition of gradient

$$\vec{\nabla}\phi = (\hat{i}\frac{\partial}{\partial x} + \hat{j}\frac{\partial}{\partial y} + \hat{k}\frac{\partial}{\partial z})(x^2y^2z^2)$$
$$= 2xy^2z^2\hat{i} + 2x^2yz^2\hat{j} + 2x^2y^2z\hat{k} = 2\hat{i} + 2\hat{j} - 2\hat{k} \text{ at } (1, 1, -1)$$

Also, $\vec{r} = x\hat{i} + y\hat{j} + z\hat{k} = e^t\hat{i} + (\sin 2t + 1)\hat{j} + (1 - \cos t)\hat{k}$. Thus, the tangent vector is given by

$$\vec{t} = \frac{d\vec{r}}{dt} = e^t\hat{i} + 2\cos 2t\hat{j} + \sin t\hat{k} = \hat{i} + 2\hat{j} \text{ at } t = 0$$

Thus, the required direction derivative along the tangent

$$= \vec{\nabla}\phi \cdot \hat{t} = (2\hat{i} + 2\hat{j} - 2\hat{k}) \cdot \frac{1}{\sqrt{5}}(\hat{i} + 2\hat{j}) = \frac{2+4}{\sqrt{5}} = \frac{6}{\sqrt{5}} \text{ units}$$

Example 5.2.18

Calculate the normal derivative of the function $\phi = x^3 - y^2 + 2z^2$ at the point $(1, 2, 3)$.

Solution

Since $\phi = x^3 - y^2 + 2z^2$

$$\vec{\nabla}\phi = 3x^2\hat{i} - 2y\hat{j} + 4z\hat{k} = 3\hat{i} - 4\hat{j} + 12\hat{k} \text{ at } (1, 2, 3)$$

If $\hat{n}$ be the unit normal vector at a point to a given level surface, then

$$\vec{\nabla}\phi = \frac{d\phi}{dn}\hat{n} \Rightarrow \frac{d\phi}{dn} = |\vec{\nabla}\phi| = 13$$

This is the required normal derivative of the function ϕ at the point $(1, 2, 3)$.

Theorem 5.2

If $\phi(x, y, z)$ and $\psi(x, y, z)$ are the scalar point functions, then

(i) $\vec{\nabla}(\phi\psi) = \phi\vec{\nabla}\psi + \psi\vec{\nabla}\phi$ (ii) $\vec{\nabla}(\frac{\phi}{\psi}) = \frac{1}{\psi^2}\left[\psi\vec{\nabla}\phi - \phi\vec{\nabla}\psi\right]; \ \psi \neq 0$

Proof: Here, $\phi(x, y, z)$ and $\psi(x, y, z)$ are the two given scalar point functions.

(i) Using the definition of gradient, we get

$$\vec{\nabla}(\phi\psi) = \frac{\partial}{\partial x}(\phi\psi)\hat{i} + \frac{\partial}{\partial y}(\phi\psi)\hat{j} + \frac{\partial}{\partial z}(\phi\psi)\hat{k}$$
$$= \left(\phi\frac{\partial\psi}{\partial x} + \psi\frac{\partial\phi}{\partial x}\right)\hat{i} + \left(\phi\frac{\partial\psi}{\partial y} + \psi\frac{\partial\phi}{\partial y}\right)\hat{j} + \left(\phi\frac{\partial\psi}{\partial z} + \psi\frac{\partial\phi}{\partial z}\right)\hat{k}$$

$$= \phi\left(\frac{\partial\psi}{\partial x}\hat{i} + \frac{\partial\psi}{\partial y}\hat{j} + \frac{\partial\psi}{\partial z}\hat{k}\right) + \psi\left(\frac{\partial\phi}{\partial x}\hat{i} + \frac{\partial\phi}{\partial y}\hat{j} + \frac{\partial\phi}{\partial z}\hat{k}\right)$$
$$= \phi\vec{\nabla}\psi + \psi\vec{\nabla}\phi$$

(ii) Using the definition of gradient, we get

$$\vec{\nabla}(\frac{\phi}{\psi}) = \frac{\partial}{\partial x}(\frac{\phi}{\psi})\hat{i} + \frac{\partial}{\partial y}(\frac{\phi}{\psi})\hat{j} + \frac{\partial}{\partial z}(\frac{\phi}{\psi})\hat{k}$$
$$= \frac{1}{\psi^2}\left[\psi\frac{\partial\phi}{\partial x} - \phi\frac{\partial\psi}{\partial x}\right]\hat{i} + \frac{1}{\psi^2}\left[\psi\frac{\partial\phi}{\partial y} - \phi\frac{\partial\psi}{\partial y}\right]\hat{j} + \frac{1}{\psi^2}\left[\psi\frac{\partial\phi}{\partial z} - \phi\frac{\partial\psi}{\partial z}\right]\hat{k}$$
$$= \frac{1}{\psi^2}\left[\psi\left(\frac{\partial\phi}{\partial x}\hat{i} + \frac{\partial\phi}{\partial y}\hat{j} + \frac{\partial\phi}{\partial z}\hat{k}\right) - \phi\left(\frac{\partial\psi}{\partial x}\hat{i} + \frac{\partial\psi}{\partial y}\hat{j} + \frac{\partial\psi}{\partial z}\hat{k}\right)\right]$$
$$= \frac{1}{\psi^2}\left[\psi\vec{\nabla}\phi - \phi\vec{\nabla}\psi\right]$$

Example 5.2.19

Find the values of the constants a, b and c so that the directional derivative of $\phi(x, y, z) = axy^2 + byz + cx^3z^2$ at the point $(1, 2, -1)$ has the maximum magnitude of 64 in a direction parallel to the z axis.

Solution

The gradient of $\phi(x, y, z)$ is given by

$$\vec{\nabla}\phi = \vec{\nabla}\left(axy^2 + byz + cx^3z^2\right) = \left(ay^2 + 3cz^2x^2\right)\hat{i} + \left(2axy + bz\right)\hat{j} + \left(by + 2czx^3\right)\hat{k}$$
$$= (4a + 3c)\hat{i} + (4a - b)\hat{j} + 2(b - c)\hat{k} \quad \text{at} \quad (1, 2, -1)$$

Now, the maximum directional derivative has a direction parallel to the z axis; so, $4a + 3c = 0$ and $4a - b = 0$. Thus, at $(1, 2, -1)$, $\vec{\nabla}\phi = 2(b - c)\hat{k}$. Also, the maximum magnitude of the directional derivative is 64, so

$$\sqrt{0^2 + 0^2 + 4(b - c)^2} = 64 \Rightarrow |b - c| = 32, i.e.\ b - c = \pm 32$$

Taking $b - c = 32$, we get $b + 3c = 0$, $b = 4a$ and $b - c = 32$. Therefore, $c = -8$, $b = 24$ and $a = 6$.

Example 5.2.20

If the pressure p at a point(x, y, z) within a fluid is given by $p(x, y, z) = 3xz^2 + yz^2$:

(i) Find the surface of equipressure through the point $(1, 1, 1)$.

(ii) Find also the direction and magnitude of the greatest variation in pressure at that point.

Solution

(i) Let the surface of equipressure be given by $p(x, y, z) = c$, c being a constant. Since the surface of equipressure passes through the point $(1, 1, 1)$

$$p(1, 1, 1) = c \Rightarrow 3 + 1 = c \Rightarrow c = 4$$

Therefore, the surface of equipressure through the point $(1, 1, 1)$ is obtained as $3xz^2 + yz^2 = 2$.

(ii) Using the definition of gradient, we have

$$\vec{\nabla} p = \frac{\partial p}{\partial x}\hat{i} + \frac{\partial p}{\partial y}\hat{j} + \frac{\partial p}{\partial y}\hat{k} = 3z\hat{i} + z^2\hat{j} + (3x + 2yz)\hat{k}$$

$$= 3\hat{i} + \hat{j} + 5\hat{k}, \text{ at } (1, 1, 1)$$

$$\Rightarrow |\vec{\nabla} p| = \sqrt{3^2 + 1^2 + 5^2} = \sqrt{35}$$

Therefore, the greatest variation in pressure at the point $(1, 1, 1)$ occurs in the direction of the vector $\vec{\nabla} p$ with magnitude $\sqrt{35}$.

Example 5.2.21

Show that the directional derivative of a scalar function of any point along any tangent line to the level surface at any point is zero.

Solution

We consider a scalar point function $\phi(x, y, z)$ and the corresponding level surface as $\phi(x, y, z) = c$, where c is a parameter. We know that the directional derivative along any direction $\hat{a}$ is $\hat{a} \cdot \vec{\nabla}\phi$. The unit vector $\hat{a}$ is along the direction of the tangent line. Also, $\vec{\nabla}\phi$ has the direction along the normal.

Thus, the directional derivative of ϕ along $\hat{a}$ is

$$= \hat{a} \cdot \vec{\nabla}\phi = 0$$

DEDUCTION 5.2.3

Functional dependence: Let $u(x, y, z)$, $v(x, y, z)$ and $w(x, y, z)$ be differentiable functions of x, y and z. Using the definition (5.3) of gradient, we have

$$\vec{\nabla} u = \frac{\partial u}{\partial x}\hat{i} + \frac{\partial u}{\partial y}\hat{j} + \frac{\partial u}{\partial z}\hat{k}$$

We can obtain similar expressions for $\vec{\nabla} v$ and $\vec{\nabla} w$. Using the definition of scalar triple product, we get

$$\left[\vec{\nabla} u \; \vec{\nabla} v \; \vec{\nabla} w\right] = \vec{\nabla} u \cdot \vec{\nabla} v \times \vec{\nabla} w = \begin{vmatrix} \frac{\partial u}{\partial x} & \frac{\partial u}{\partial y} & \frac{\partial u}{\partial z} \\ \frac{\partial v}{\partial x} & \frac{\partial v}{\partial y} & \frac{\partial v}{\partial z} \\ \frac{\partial w}{\partial x} & \frac{\partial w}{\partial y} & \frac{\partial w}{\partial z} \end{vmatrix}$$

This is called the Jacobian of u, v and w with respect to x, y and z and is written as $J\left(\frac{u,v,w}{x,y,z}\right)$ or simply J. If the scalar functions u, v and w are functionally related, then $J = 0$, and so $\left[\vec{\nabla} u \; \vec{\nabla} v \; \vec{\nabla} w\right] = 0$, i.e., $\vec{\nabla} u$, $\vec{\nabla} v$ and $\vec{\nabla} w$ are coplanar.

Example 5.2.22

If $\phi_1(x, y, z)$ and $\phi_2(x, y, z)$ are differentiable scalar point functions, then they are functionally related by the function $f(\phi_1, \phi_2) = 0$ if and only if $\vec{\nabla}\phi_1 \times \vec{\nabla}\phi_2 = \vec{0}$.

Solution

Using the definition (5.3) of gradient, we have

$$\vec{\nabla}\phi_1 = \frac{\partial \phi_1}{\partial x}\hat{i} + \frac{\partial \phi_1}{\partial y}\hat{j} + \frac{\partial \phi_1}{\partial z}\hat{k};\ \vec{\nabla}\phi_2 = \frac{\partial \phi_2}{\partial x}\hat{i} + \frac{\partial \phi_2}{\partial y}\hat{j} + \frac{\partial \phi_2}{\partial z}\hat{k}$$

$$\Rightarrow \vec{\nabla}\phi_1 \times \vec{\nabla}\phi_2 = \begin{vmatrix} \hat{i} & \hat{j} & \hat{k} \\ \frac{\partial \phi_1}{\partial x} & \frac{\partial \phi_1}{\partial y} & \frac{\partial \phi_1}{\partial z} \\ \frac{\partial \phi_2}{\partial x} & \frac{\partial \phi_2}{\partial y} & \frac{\partial \phi_2}{\partial z} \end{vmatrix}$$

If the scalar point functions $\phi_1(x, y, z)$ and $\phi_2(x, y, z)$ are functionally related by the function $f(\phi_1, \phi_2) = 0$, then $\vec{\nabla}\phi_1$ and $\vec{\nabla}\phi_2$ are parallel, and so $\vec{\nabla}\phi_1 \times \vec{\nabla}\phi_2 = \vec{0}$.

Example 5.2.23

Prove that the scalar point functions $u = x + y + z$, $v = x^2 + y^2 + z^2$ and $w = yz + zx + xy$ are functionally related.

Solution

Using the definition of gradient, we have

$$\vec{\nabla}u = (\hat{i}\frac{\partial}{\partial x} + \hat{j}\frac{\partial}{\partial y} + \hat{k}\frac{\partial}{\partial z})(x + y + z) = \hat{i} + \hat{j} + \hat{k}$$

$$\vec{\nabla}v = (\hat{i}\frac{\partial}{\partial x} + \hat{j}\frac{\partial}{\partial y} + \hat{k}\frac{\partial}{\partial z})(x^2 + y^2 + z^2) = 2x\hat{i} + 2y\hat{j} + 2z\hat{k}$$

$$\vec{\nabla}w = (\hat{i}\frac{\partial}{\partial x} + \hat{j}\frac{\partial}{\partial y} + \hat{k}\frac{\partial}{\partial z})(yz + zx + xy) = (z + y)\hat{i} + (z + x)\hat{j} + (y + x)\hat{k}$$

Using the definition of box product, we have

$$\begin{aligned}
\left[\vec{\nabla}u\ \vec{\nabla}v\ \vec{\nabla}w\right] = \vec{\nabla}u \cdot (\vec{\nabla}v \times \vec{\nabla}w) &= \begin{vmatrix} 1 & 1 & 1 \\ 2x & 2y & 2z \\ y+z & z+x & y+x \end{vmatrix} \\
&= 2\begin{vmatrix} 1 & 1 & 1 \\ x & y & z \\ y+z & z+x & y+x \end{vmatrix} \\
&= 2\begin{vmatrix} 1 & 1 & 1 \\ x+y+z & x+y+z & x+y+z \\ y+z & z+x & y+x \end{vmatrix};\ R_2' = R_2 + R_3 \\
&= 2(x+y+z)\begin{vmatrix} 1 & 1 & 1 \\ 1 & 1 & 1 \\ y+z & z+x & y+x \end{vmatrix} = 0
\end{aligned}$$

Since the box product $\left[\vec{\nabla}u\ \vec{\nabla}v\ \vec{\nabla}w\right] = 0$, u, v and w are functionally related.

DEDUCTION 5.2.4

Maxima and minima: Let us find a point (x^*, y^*) that gives the extrema (maximum or minimum value) of a function $f(x, y)$ subject to the constraint $\phi(x, y, z) = c$, where c is a constant. This will occur only when $\vec{\nabla}f$ and $\vec{\nabla}\phi$ (directional derivatives) are orthogonal to the curve $\phi(x, y) = c$.

Therefore, $\vec{\nabla} f$ and $\vec{\nabla}\phi$ are parallel, i.e.

$$\vec{\nabla} f = \lambda \vec{\nabla}\phi, \quad \text{where } \lambda = \text{ constant} \tag{5.12}$$

The scalar λ, introduced in Eq. (5.12), is called the Lagrange multiplier. Eq. (5.12) gives

$$\frac{\partial f}{\partial x} = \lambda \frac{\partial \phi}{\partial x}; \quad \frac{\partial f}{\partial y} = \lambda \frac{\partial \phi}{\partial y} \tag{5.13}$$

together with the given constant $\phi(x, y) = c$. Solutions of the system Eq. (5.13) for x and y give (x^*, y^*), which gives the extreme $f(x^*, y^*)$.

Example 5.2.24

Minimise the function $f(x, y) = x^2 + 2y^2$, subject to the constraint $2x + y = 9$, by the vector method.

Solution

Let us find a point (x^*, y^*) that gives the minimum value of the given function $f(x, y) = x^2 + 2y^2$, subject to the constraint $\phi(x, y) = 2x + y = 9$. Using the condition (5.12) together with the constraint, we get

$$2x = 2\lambda, 4y = \lambda, 2x + y = 9$$
$$\Rightarrow x = 4y \text{ and } 2x + y = 9 \Rightarrow x = 4, y = 1$$

Therefore, $(x^*, y^*) = (4, 1)$ and $f(x^*, y^*) = 16 + 2 = 18$ is the minimum value of f subject to the constant $2x + y = 9$.

DEDUCTION 5.2.5

Let $\vec{\nabla}\phi = F_1(x, y, z)\hat{i} + F_1(x, y, z)\hat{i} + F_2(x, y, z)\hat{j} + F_3(x, y, z)\hat{k}$ be given. Then

$$\frac{\partial \phi}{\partial x} = F_1(x, y, z) \Longrightarrow \phi(x, y, z) = \int F_1(x, y, z)\partial x + \psi_1(y, z)$$
$$\frac{\partial \phi}{\partial y} = F_2(x, y, z) \Longrightarrow \phi(x, y, z) = \int F_2(x, y, z)\partial y + \psi_2(x, z)$$
$$\frac{\partial \phi}{\partial z} = F_3(x, y, z) \Longrightarrow \phi(x, y, z) = \int F_3(x, y, z)\partial z + \psi_3(x, y)$$

where $\psi_1(y, z)$, $\psi_2(x, z)$ and $\psi_3(x, y)$ are the functions of the corresponding variables and/or constants.

Example 5.2.25

If $\vec{\nabla}\phi = 2xyz^3\hat{i} + x^2z^3\hat{j} + 3x^2yz^2\hat{k}$ and $\phi(1, -2, 2) = 4$, find $\phi(x, y, z)$.

Solution

Given that $\vec{\nabla}\phi = 2xyz^3\hat{i} + x^2z^3\hat{j} + 3x^2yz^2\hat{k}$

$$\frac{\partial \phi}{\partial x} = 2xyz^3 \Longrightarrow \phi(x, y, z) = x^2yz^3 + \psi_1(y, z)$$
$$\frac{\partial \phi}{\partial y} = x^2z^3 \Longrightarrow \phi(x, y, z) = x^2yz^3 + \psi_2(x, z)$$

$$\frac{\partial \phi}{\partial z} = 3x^2yz^2 \Longrightarrow \phi(x,y,z) = x^2yz^3 + \psi_3(x,y)$$

Let us take $\psi_1(y,z) = \psi_2(x,z) = \psi_3(x,y) = c$, a scalar constant. Then, $\phi(x,y,z) = x^2yz^3 + c$. Given that $\phi(1,-2,2) = 4$, $c = 20$, and so we get $\phi(x,y,z) = x^2yz^3 + 20$.

5.2.1 Invariance of the Gradient

Let φ be a scalar point function of x, y and z which is invariant with respect to the rotation of the axes, given by the Eq. (2.15). Then, $\varphi(x,y,z) = \varphi'(x',y',z')$. From Eq. (2.15)

$$\frac{\partial \varphi}{\partial x} = \frac{\partial \varphi'}{\partial x'}\frac{\partial x'}{\partial x} + \frac{\partial \varphi'}{\partial y'}\frac{\partial y'}{\partial x} + \frac{\partial \varphi'}{\partial z'}\frac{\partial z'}{\partial x}$$

$$\Rightarrow \frac{\partial \varphi}{\partial x} = l_1\frac{\partial \varphi'}{\partial x'} + l_2\frac{\partial \varphi'}{\partial y'} + l_3\frac{\partial \varphi'}{\partial z'}$$

Similarly, $\dfrac{\partial \varphi}{\partial y} = m_1\dfrac{\partial \varphi'}{\partial x'} + m_2\dfrac{\partial \varphi'}{\partial y'} + m_3\dfrac{\partial \varphi'}{\partial z'}$ and $\dfrac{\partial \varphi}{\partial z} = n_1\dfrac{\partial \varphi'}{\partial x'} + n_2\dfrac{\partial \varphi'}{\partial y'} + n_3\dfrac{\partial \varphi'}{\partial z'}$. Let $\hat{i}', \hat{j}'$ and $\hat{k}'$ be any other set of mutually orthogonal unit vectors. Then, using the relation (2.17), we have

$$\begin{aligned}
\hat{i}\frac{\partial \varphi}{\partial x} + \hat{j}\frac{\partial \varphi}{\partial y} + \frac{\partial \varphi}{\partial z} &= \frac{\partial \varphi}{\partial x}(l_1\hat{i}' + l_2\hat{j}' + l_3\hat{k}') + \frac{\partial \varphi}{\partial y}(m_1\hat{i}' + m_2\hat{j}' + m_3\hat{k}') \\
&\quad + \frac{\partial \varphi}{\partial z}(n_1\hat{i}' + n_2\hat{j}' + n_3\hat{k}') \\
&= \frac{\partial \varphi}{\partial x}(l_1\,\hat{i} + m_1\,\hat{j} + n_1\,\hat{k}) + \frac{\partial \varphi}{\partial y}(l_2\,\hat{i} + m_2\,\hat{j} + n_2\,\hat{k}) \\
&\quad + \frac{\partial \varphi}{\partial z}(l_3\,\hat{i} + m_2\,\hat{j} + n_3\,\hat{k}) = \hat{i}'\frac{\partial \varphi'}{\partial x'} + \hat{j}'\frac{\partial \varphi'}{\partial y'} + \hat{k}'\frac{\partial \varphi'}{\partial z'}
\end{aligned}$$

i.e., gradφ = gradφ', which shows that gradφ is invariant with respect to the rotation of the axes, given by Eq. (2.15). Thus, the vector $\vec{\nabla}$ at any point is completely describable in terms of the level surface through the point and this description is independent of the choice of coordinate axes.

5.2.2 Gradient of a Vector Function

Let $\vec{A} = A_1\hat{i} + A_2\hat{j} + A_3\hat{k}$ be a vector point function. Then, grad $\vec{A}$ is defined as

$$\begin{aligned}
\vec{\nabla}\vec{A} &= \left(\hat{i}\frac{\partial}{\partial x} + \hat{j}\frac{\partial}{\partial y} + \hat{k}\frac{\partial}{\partial z}\right)(A_1\hat{i} + A_2\hat{j} + A_3\hat{k}) \\
&= \frac{\partial A_1}{\partial x}\hat{i}\,\hat{i} + \frac{\partial A_2}{\partial x}\hat{i}\,\hat{j} + \frac{\partial A_3}{\partial x}\hat{i}\,\hat{k} \\
&\quad + \frac{\partial A_1}{\partial x}\hat{j}\,\hat{i} + \frac{\partial A_2}{\partial x}\hat{j}\,\hat{j} + \frac{\partial A_3}{\partial x}\hat{j}\,\hat{k} \\
&\quad + \frac{\partial A_1}{\partial x}\hat{k}\,\hat{i} + \frac{\partial A_2}{\partial x}\hat{k}\,\hat{j} + \frac{\partial A_3}{\partial x}\hat{k}\,\hat{k}
\end{aligned} \tag{5.14}$$

The quantities $\hat{i}\,\hat{i}$, $\hat{i}\,\hat{j}$, etc. are called *unit dyads,* where $\hat{i}\,\hat{j} \neq \hat{j}\,\hat{i}$. The quantity (5.14) can be written in the form

$$\vec{\Phi} = A_{11}\hat{i}\,\hat{i} + A_{12}\hat{i}\,\hat{j} + A_{13}\hat{i}\,\hat{k} + A_{21}\hat{j}\,\hat{i} + A_{22}\hat{j}\,\hat{j} + A_{23}\hat{j}\,\hat{k} + A_{31}\hat{k}\,\hat{i} + A_{32}\hat{k}\,\hat{j} + A_{33}\hat{k}\,\hat{k} \tag{5.15}$$

This is called a *dyadic,* which is a generalisation of a vector. The coefficients $A_{11}, A_{12}, \cdots, A_{33}$ are called the *components of that dyadic* . An array of these nine components in the form

$$\begin{pmatrix} A_{11} & A_{12} & A_{13} \\ A_{21} & A_{22} & A_{23} \\ A_{31} & A_{32} & A_{33} \end{pmatrix} \tag{5.16}$$

is called a 3×3 matrix. Further generalisation leads to *triadics,* which are quantities consisting of 27 terms.

Result 15

The following results are important

(i) $\hat{i} \cdot \hat{i}\,\hat{i} = (\hat{i} \cdot \hat{i})\hat{i} = \hat{i} = \hat{i}(\hat{i} \cdot \hat{i}) = \hat{i}\,\hat{i} \cdot \hat{i}$ (ii) $\hat{i} \cdot \hat{j}\,\hat{j} = (\hat{i} \cdot \hat{j})\hat{j} = \vec{0} = \hat{j}(\hat{j} \cdot \hat{i}) = \hat{j}\,\hat{j} \cdot \hat{i}$

(iii) $\hat{i} \cdot \hat{i}\,\hat{j} = (\hat{i} \cdot \hat{i})\hat{j} = \hat{j}$; $\hat{i}\,\hat{j} \cdot \hat{i} = \hat{i}(\hat{j} \cdot \hat{i}) = \vec{0}$ (iv) $\hat{i} \cdot \hat{j}\,\hat{i} = (\hat{i} \cdot \hat{j})\hat{i} = \vec{0}$; $\hat{j}\,\hat{i} \cdot \hat{i} = \hat{j}(\hat{i} \cdot \hat{i}) = \hat{j}$

Result 16

Let $\vec{\Phi}$ be a given dyadic, as in Eq. (5.15). Let us take the dot product $\hat{i}$ with the elements of $\vec{\Phi}$, as

$$\hat{i} \cdot A_{11}\hat{i}\,\hat{i} = A_{11}(\hat{i} \cdot \hat{i})\hat{i} = A_{11}\hat{i}; \quad \text{as } \hat{i} \cdot \hat{i} = 1$$
$$\hat{i} \cdot A_{12}\hat{i}\,\hat{j} = A_{12}(\hat{i} \cdot \hat{i})\hat{j} = A_{12}\hat{j}; \quad \text{as } \hat{i} \cdot \hat{i} = 1$$
$$\hat{i} \cdot A_{13}\hat{i}\,\hat{k} = A_{13}(\hat{i} \cdot \hat{i})\hat{k} = A_{13}\hat{k}; \quad \text{as } \hat{i} \cdot \hat{i} = 1$$
$$\hat{i} \cdot A_{21}\hat{j}\,\hat{i} = A_{21}(\hat{i} \cdot \hat{j})\hat{i} = \vec{0}; \quad \text{as } \hat{i} \cdot \hat{j} = 0$$
$$\hat{i} \cdot A_{22}\hat{j}\,\hat{j} = \vec{0}; \quad \hat{i} \cdot A_{23}\hat{j}\,\hat{k} = \vec{0}; \quad \hat{i} \cdot A_{31}\hat{k}\,\hat{i} = \vec{0}$$
$$\hat{i} \cdot A_{32}\hat{k}\,\hat{j} = \vec{0}; \quad \hat{i} \cdot A_{33}\hat{k}\,\hat{k} = \vec{0}$$

Analogous interpretation to the terms of $\hat{j} \cdot \vec{\Phi}$ and $\hat{k} \cdot \vec{\Phi}$.

Result 17

Let $\vec{\alpha} = \alpha_1\hat{i} + \alpha_2\hat{j} + \alpha_3\hat{k}$ and $\vec{\Phi}$ be a given dyadic, as in Eq. (5.15). Assuming the distributive law to hold, we have

$$\begin{aligned} \vec{\alpha} \cdot \vec{\Phi} &= (\alpha_1\hat{i} + \alpha_2\hat{j} + \alpha_3\hat{k}) \cdot \Phi = \alpha_1\hat{i} \cdot \Phi + \alpha_2\hat{j} \cdot \Phi + \alpha_3\hat{k} \cdot \Phi \\ &= \alpha_1(A_{11}\hat{i} + A_{12}\hat{j} + A_{13}\hat{k}) + \alpha_2(A_{21}\hat{i} + A_{22}\hat{j} + A_{23}\hat{k}) \\ &\quad + \alpha_3(A_{31}\hat{i} + A_{32}\hat{j} + A_{33}\hat{k}) \\ &= (\alpha_1 A_{11} + \alpha_2 A_{21} + \alpha_3 A_{31})\hat{i} + (\alpha_1 A_{12} + \alpha_2 A_{22} + \alpha_3 A_{32})\hat{j} \\ &\quad + (\alpha_1 A_{13} + \alpha_2 A_{23} + \alpha_3 A_{33})\hat{k} \end{aligned} \tag{5.17}$$

which is a vector.

Example 5.2.26

If $\vec{A} = xz\hat{i} - y^2\hat{j} + yz^2\hat{k}$ and $\vec{B} = 2z^2\hat{i} - xy\hat{j} + y^3\hat{k}$, give a possible significance to $(\vec{A} \times \vec{\nabla})\vec{B}$ at the point $(1, -1, 1)$.

Solution

Since $\vec{A} = xz\hat{i} - y^2\hat{j} + yz^2\hat{k}$, using the definition of cross product

$$\vec{A} \times \vec{\nabla} = \begin{vmatrix} \hat{i} & \hat{j} & \hat{k} \\ xz & -y^2 & yz^2 \\ \dfrac{\partial}{\partial x} & \dfrac{\partial}{\partial y} & \dfrac{\partial}{\partial z} \end{vmatrix}$$

$$= \hat{i}\left(-y^2\frac{\partial}{\partial z} - yz^2\frac{\partial}{\partial y}\right) + \hat{j}\left(yz^2\frac{\partial}{\partial z} - xz\frac{\partial}{\partial z}\right) + \hat{k}\left(xz\frac{\partial}{\partial y} + y^2\frac{\partial}{\partial x}\right)$$

$$(\vec{A} \times \vec{\nabla})\vec{B} = \left[\hat{i}\left(-y^2\frac{\partial}{\partial z} - yz^2\frac{\partial}{\partial y}\right) + \hat{j}\left(yz^2\frac{\partial}{\partial z} - xz\frac{\partial}{\partial z}\right)\right.$$

$$\left. + \hat{k}\left(xz\frac{\partial}{\partial y} + y^2\frac{\partial}{\partial x}\right)\right](2z^2\hat{i} - xy\hat{j} + y^3\hat{k})$$

$$= -4y^2z\hat{i}\,\hat{i} + xyz^2\hat{i}\,\hat{j} - 3y^3z^2\hat{i}\,\hat{k} - 4xz^2\hat{j}\,\hat{i}$$

$$-y^2z^2\hat{j}\,\hat{j} - (x^2z + y^3)\hat{k}\,\hat{j} + 3xy^2z\hat{k}\,\hat{k}$$

$$= -4\hat{i}\,\hat{i} - \hat{i}\,\hat{j} + 3\hat{i}\,\hat{k} - 4\hat{i}\,\hat{j} - \hat{j}\,\hat{j} + 3\hat{k}\,\hat{k}, \text{ at } (1, -1, 1)$$

Example 5.2.27

Given a dyadic $\vec{\Phi} = \hat{i}\,\hat{i} + \hat{j}\,\hat{j} + \hat{k}\,\hat{k}$, evaluate $\vec{r} \cdot (\vec{\Phi} \cdot \vec{r})$ and $(\vec{r} \cdot \vec{\Phi}) \cdot \vec{r}$. What does $\vec{r} \cdot \vec{\Phi} \cdot \vec{r} = 4$ represent geometrically ?

Solution

Since $\vec{r} = x\hat{i} + y\hat{j} + z\hat{k}$, the vector quantity $\vec{r} \cdot \vec{\Phi}$ is given by

$$\vec{r} \cdot \vec{\Phi} = (x\hat{i} + y\hat{j} + z\hat{k}) \cdot (\hat{i}\,\hat{i} + \hat{j}\,\hat{j} + \hat{k}\,\hat{k})$$

$$= x\hat{i} \cdot \hat{i}\,\hat{i} + x\hat{i} \cdot \hat{j}\,\hat{j} + x\hat{i} \cdot \hat{k}\,\hat{k} + y\hat{j} \cdot \hat{i}\,\hat{i} + y\hat{j} \cdot \hat{j}\,\hat{j}$$

$$+ y\hat{j} \cdot \hat{k}\,\hat{k} + z\hat{k} \cdot \hat{i}\,\hat{i} + z\hat{k} \cdot \hat{j}\,\hat{j} + z\hat{k} \cdot \hat{k}\,\hat{k}$$

$$= x(\hat{i} \cdot \hat{i})\hat{i} + x(\hat{i} \cdot \hat{j})\hat{j} + x(\hat{i} \cdot \hat{k})\hat{k} + y(\hat{j} \cdot \hat{i})\hat{i} + y(\hat{j} \cdot \hat{j})\hat{j}$$

$$+ y(\hat{j} \cdot \hat{k})\hat{k} + z(\hat{k} \cdot \hat{i})\hat{i} + z(\hat{k} \cdot \hat{j})\hat{j} + z(\hat{k} \cdot \hat{k})\hat{k}$$

$$= x\hat{i} + y\hat{j} + z\hat{k}$$

Similarly, as $\vec{r} = x\hat{i} + y\hat{j} + z\hat{k}$, the vector quantity $\vec{\Phi} \cdot \vec{r}$ is given by

$$\vec{\Phi} \cdot \vec{r} = (\hat{i}\,\hat{i} + \hat{j}\,\hat{j} + \hat{k}\,\hat{k}) \cdot (x\hat{i} + y\hat{j} + z\hat{k})$$

$$= x\hat{i}\,\hat{i} \cdot \hat{i} + x\hat{i}\,\hat{j} \cdot \hat{j} + x\hat{i}\,\hat{k} \cdot \hat{k} + y\hat{j}\,\hat{i} \cdot \hat{i} + y\hat{j}\,\hat{j} \cdot \hat{j}$$

$$+ y\hat{j}\,\hat{k} \cdot \hat{k} + z\hat{k}\,\hat{i} \cdot \hat{i} + z\hat{k}\,\hat{j} \cdot \hat{j} + z\hat{k}\,\hat{k} \cdot \hat{k}$$

$$= x\hat{i}(\hat{i}\cdot\hat{i}) + x\hat{j}(\hat{i}\cdot\hat{j}) + x\hat{k}(\hat{i}\cdot\hat{k}) + y\hat{i}(\hat{j}\cdot\hat{i}) + y\hat{j}(\hat{j}\cdot\hat{j})$$
$$+y\hat{k}(\hat{j}\cdot\hat{k}) + z\hat{i}(\hat{k}\cdot\hat{i}) + z\hat{j}(\hat{k}\cdot\hat{j}) + z\hat{k}(\hat{k}\cdot\hat{k})$$
$$= x\hat{i} + y\hat{j} + z\hat{k}$$

Thus, the expressions for $\vec{r}\cdot(\vec{\Phi}\cdot\vec{r})$ and $(\vec{r}\cdot\vec{\Phi})\cdot\vec{r}$ become

$$\vec{r}\cdot(\vec{\Phi}\cdot\vec{r}) = (x\hat{i} + y\hat{j} + z\hat{k})\cdot(x\hat{i} + y\hat{j} + z\hat{k}) = x^2 + y^2 + z^2$$
$$(\vec{r}\cdot\vec{\Phi})\cdot\vec{r} = (x\hat{i} + y\hat{j} + z\hat{k})\cdot(x\hat{i} + y\hat{j} + z\hat{k}) = x^2 + y^2 + z^2$$

Since $\vec{r}\cdot(\vec{\Phi}\cdot\vec{r}) = (\vec{r}\cdot\vec{\Phi})\cdot\vec{r}$, there is no ambiguity in writing $\vec{r}\cdot\vec{\Phi}\cdot\vec{r}$. Now

$$\vec{r}\cdot\vec{\Phi}\cdot\vec{r} = 4 \Rightarrow x^2 + y^2 + z^2 = 4$$

This represents the equation of the sphere of radius 2 with the centre as origin.

Deduction 5.2.6

Gradient of a vector in polar coordinates: Consider polar coordinates (r, θ) such that $r = \sqrt{x^2 + y^2}$ and $\theta = \tan^{-1}(y/x)$ (Fig. 5.5). The unit base vectors $\hat{e}_r$ and $\hat{e}_\theta$ can be expressed in terms of the Cartesian base vectors $\hat{i}$ and $\hat{j}$ as

$$\hat{e}_r = \cos\theta\,\hat{i} + \sin\theta\,\hat{j};\quad \hat{e}_\theta = -\sin\theta\,\hat{i} + \cos\theta\,\hat{j}$$

Now, $d\hat{i} = \vec{0}$ and $d\hat{j} = \vec{0}$. The unit vectors $\hat{e}_r$ and $\hat{e}_\theta$ vary in direction as θ changes.

$$d\hat{e}_r = (-\sin\theta\,\hat{i} + \cos\theta\,\hat{j})d\theta = d\theta\,\hat{e}_\theta$$
$$d\hat{e}_\theta = (-\cos\theta\,\hat{i} - \sin\theta\,\hat{j})d\theta = -d\theta\,\hat{e}_r$$

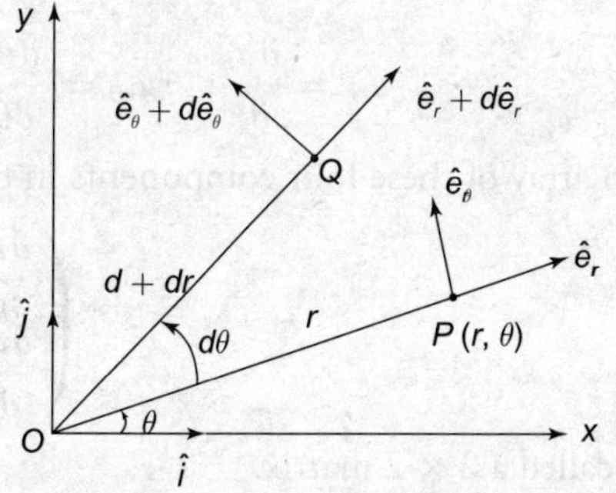

Figure 5.5 Gradient of a vector in polar coordinates

The position vector for the point P can be written as $\vec{r} = r\,\hat{e}_r$, where $r = |\vec{r}|$, the magnitude of $\vec{r}$. Thus

$$d\vec{r} = dr\,\hat{e}_r + r\,d\hat{e}_r = dr\,\hat{e}_r + r\,d\theta\,\hat{e}_\theta$$

In polar coordinates, a vector $\vec{u}$ can be written as

$$\vec{u}(r,\theta) = u_r(r,\theta)\,\hat{e}_r + u_\theta(r,\theta)\,\hat{e}_\theta \tag{5.18}$$

where $u_r(r,\theta)$ and $u_\theta(r,\theta)$ are the components of $\vec{u}(r,\theta)$ in the directions $\hat{e}_r$ and $\hat{e}_\theta$, respectively. Since $u_r = u_r(r,\theta)$ and $u_\theta = u_\theta(r,\theta)$, by the chain rule of differentiation, we get

$$du_r = \frac{\partial u_r}{\partial r}\,dr + \frac{\partial u_r}{\partial\theta}\,d\theta;\quad du_\theta = \frac{\partial u_\theta}{\partial r}\,dr + \frac{\partial u_\theta}{\partial\theta}\,d\theta \tag{5.19}$$

Therefore

$$d\vec{u} = du_r\,\hat{e}_r + u_r\,d\hat{e}_r + du_\theta\,\hat{e}_\theta + u_\theta\,d\hat{e}_\theta$$
$$= \left(\frac{\partial u_r}{\partial r}\,dr + \frac{\partial u_r}{\partial\theta}\,d\theta\right)\hat{e}_r + u_r\,d\theta\,\hat{e}_\theta + \left(\frac{\partial u_\theta}{\partial r}\,dr + \frac{\partial u_\theta}{\partial\theta}\,d\theta\right)\hat{e}_\theta - u_\theta\,d\theta\,\hat{e}_r$$
$$= \left[\frac{\partial u_r}{\partial r}\,dr + \left(\frac{\partial u_r}{\partial\theta} - u_\theta\right)d\theta\right]\hat{e}_r + \left[\frac{\partial u_\theta}{\partial r}\,dr + \left(\frac{\partial u_\theta}{\partial\theta} + u_r\right)d\theta\right]\hat{e}_\theta \tag{5.20}$$

By the definition (5.14) of gradient of the vector function $\vec{u}(r, \theta)$, we have

$$d\vec{u} = \vec{\nabla}\vec{u}\, d\vec{r} = \vec{\nabla}\vec{u}\Big[dr\, \hat{e}_r + r\, d\theta\, \hat{e}_\theta\Big] = dr\, \vec{\nabla}\vec{u}\, \hat{e}_r + r\, d\theta\, \vec{\nabla}\vec{u}\, \hat{e}_\theta$$
$$= dr\Big[u_{rr}\, \hat{e}_r + u_{\theta r}\hat{e}_\theta\Big] + r\, d\theta\Big[u_{r\theta}\, \hat{e}_r + u_{\theta\theta}\, \hat{e}_\theta\Big]$$
$$= \Big[u_{rr}\, dr + u_{r\theta}\, r\, d\theta\Big]\hat{e}_r + \Big[u_{\theta r}\, dr + u_{\theta\theta}\, r\, d\theta\Big]\hat{e}_\theta \tag{5.21}$$

where the coefficients u_{rr}, $u_{r\theta}$, $u_{\theta r}$ and $u_{\theta\theta}$ are the components of that dyadic $\vec{\nabla}\vec{u}$. Comparing Eqs (5.20) and (5.21), we get

$$u_{rr}\, dr + u_{r\theta}\, r\, d\theta = \frac{\partial u_r}{\partial r}\, dr + \left(\frac{\partial u_r}{\partial \theta} - u_\theta\right)d\theta$$
$$u_{\theta r}\, dr + u_{\theta\theta}\, r\, d\theta = \frac{\partial u_\theta}{\partial r}\, dr + \left(\frac{\partial u_\theta}{\partial \theta} + u_r\right)d\theta$$

These relations must hold for any values of dr and $d\theta$, so

$$u_{rr} = \frac{\partial u_r}{\partial r};\quad ru_{r\theta} = \frac{\partial u_r}{\partial \theta} - u_\theta;\quad u_{\theta r} = \frac{\partial u_\theta}{\partial r};\quad ru_{\theta\theta} = \frac{\partial u_\theta}{\partial \theta} + u_r \tag{5.22}$$

An array of these four components in the form

$$\begin{pmatrix} \dfrac{\partial u_r}{\partial r} & \dfrac{1}{r}\dfrac{\partial u_r}{\partial \theta} - \dfrac{u_\theta}{r} \\ \dfrac{\partial u_\theta}{\partial r} & \dfrac{1}{r}\dfrac{\partial u_\theta}{\partial \theta} + \dfrac{u_r}{r} \end{pmatrix} \tag{5.23}$$

is called a 2 × 2 matrix.

DEDUCTION 5.2.7

Gradient of a vector in cylindrical coordinates: Consider cylindrical coordinates (r, θ, z) (Fig. 5.6). The unit base vectors $\hat{e}_r, \hat{e}_\theta$ and $\hat{e}_z$ can be expressed in terms of the Cartesian base vectors $\hat{i}, \hat{j}$ and $\hat{k}$ as

$$\hat{e}_r = \cos\theta\, \hat{i} + \sin\theta\, \hat{j};$$
$$\hat{e}_\theta = -\sin\theta\, \hat{i} + \cos\theta\, \hat{j};$$
$$\hat{e}_z = \hat{k}$$

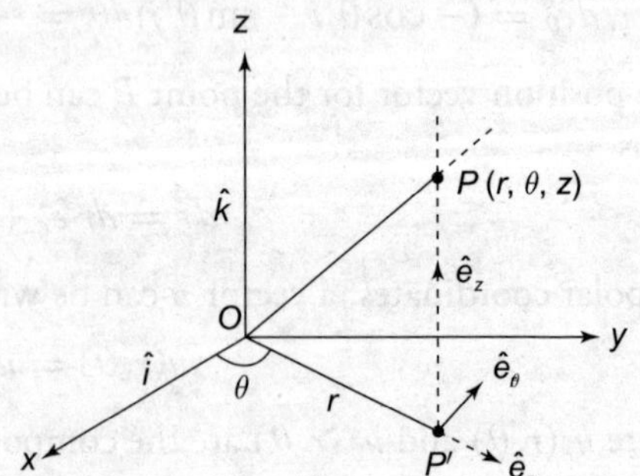

Figure 5.6 Gradient of a vector in cylindrical coordinates

Now, $d\hat{i} = \vec{0}, d\hat{j} = \vec{0}$ and $d\hat{k} = d\hat{e}_z = \vec{0}$. The unit vectors $\hat{e}_r$ and $\hat{e}_\theta$ vary in direction as θ changes.

$$d\hat{e}_r = (-\sin\theta\, \hat{i} + \cos\theta\, \hat{j})d\theta = d\theta\, \hat{e}_\theta$$
$$d\hat{e}_\theta = (-\cos\theta\, \hat{i} - \sin\theta\, \hat{j})d\theta = -d\theta\, \hat{e}_r$$

The position vector for the point P can be written as $\vec{r} = r\, \hat{e}_r + z\hat{k}$, where $r = |\vec{r}|$, the magnitude of $\vec{r}$. Thus

$$d\vec{r} = dr\, \hat{e}_r + r\, d\hat{e}_r + dz\, \hat{e}_z = dr\, \hat{e}_r + r\, d\theta\, \hat{e}_\theta + dz\, \hat{e}_z$$

In cylindrical coordinates, a vector $\vec{u}$ can be written as

$$\vec{u}(r,\theta,z) = u_r(r,\theta,z)\,\hat{e}_r + u_\theta(r,\theta,z)\,\hat{e}_\theta + u_z(r,\theta,z)\,\hat{e}_z \tag{5.24}$$

where $u_r(r,\theta,z), u_\theta(r,\theta,z)$ and $u_z(r,\theta,z)$ are the components of $\vec{u}(r,\theta,z)$ in the directions $\hat{e}_r$, $\hat{e}_\theta$ and $\hat{e}_z$, respectively. Since $u_r = u_r(r,\theta,z)$, $u_\theta = u_\theta(r,\theta,z)$ and $u_z = u_z(r,\theta,z)$, by the chain rule of differentiation, we get

$$\left.\begin{aligned} du_r &= \frac{\partial u_r}{\partial r}\,dr + \frac{\partial u_r}{\partial \theta}\,d\theta + \frac{\partial u_r}{\partial z}\,dz \\ du_\theta &= \frac{\partial u_\theta}{\partial r}\,dr + \frac{\partial u_\theta}{\partial \theta}\,d\theta + \frac{\partial u_\theta}{\partial z}\,dz \\ du_z &= \frac{\partial u_z}{\partial r}\,dr + \frac{\partial u_z}{\partial \theta}\,d\theta + \frac{\partial u_z}{\partial z}\,dz \end{aligned}\right\} \tag{5.25}$$

Therefore

$$\begin{aligned} d\vec{u} &= du_r\,\hat{e}_r + u_r\,d\hat{e}_r + du_\theta\,\hat{e}_\theta + u_\theta\,d\hat{e}_\theta + du_z\,\hat{e}_z \\ &= \left(\frac{\partial u_r}{\partial r}\,dr + \frac{\partial u_r}{\partial \theta}\,d\theta + \frac{\partial u_r}{\partial z}\,dz\right)\hat{e}_r + u_r\,d\theta\,\hat{e}_\theta \\ &\quad + \left(\frac{\partial u_\theta}{\partial r}\,dr + \frac{\partial u_\theta}{\partial \theta}\,d\theta + \frac{\partial u_\theta}{\partial z}\,dz\right)\hat{e}_\theta + \left(\frac{\partial u_z}{\partial r}\,dr + \frac{\partial u_z}{\partial \theta}\,d\theta + \frac{\partial u_z}{\partial z}\,dz\right)\hat{e}_z \\ &= \left[\frac{\partial u_r}{\partial r}\,dr + \left(\frac{\partial u_r}{\partial \theta} - u_\theta\right)d\theta + \frac{\partial u_r}{\partial z}\,dz\right]\hat{e}_r + \left[\frac{\partial u_\theta}{\partial r}\,dr + \left(\frac{\partial u_\theta}{\partial \theta} + u_r\right)d\theta \right. \\ &\quad \left. + \frac{\partial u_\theta}{\partial z}\,dz\right]\hat{e}_\theta + \left[\frac{\partial u_r}{\partial r}\,dr + \frac{\partial u_z}{\partial \theta}\,d\theta + \frac{\partial u_z}{\partial z}\,dz\right]\hat{e}_z \end{aligned} \tag{5.26}$$

By the definition (5.14) of gradient of the vector function $\vec{u}(r,\theta)$, we have

$$\begin{aligned} d\vec{u} &= \vec{\nabla}\vec{u}\,d\vec{r} = \vec{\nabla}\vec{u}\Big[dr\,\hat{e}_r + r\,d\theta\,\hat{e}_\theta + dz\hat{e}_z\Big] \\ &= dr\,\vec{\nabla}\vec{u}\,\hat{e}_r + r\,d\theta\,\vec{\nabla}\vec{u}\,\hat{e}_\theta + dz\,\vec{\nabla}\vec{u}\,\hat{e}_z \\ &= dr\Big[u_{rr}\,\hat{e}_r + u_{\theta r}\hat{e}_\theta + u_{zr}\hat{e}_z\Big] + r\,d\theta\Big[u_{r\theta}\,\hat{e}_r + u_{\theta\theta}\,\hat{e}_\theta + u_{z\theta}\hat{e}_z\Big] \\ &\quad + dz\Big[u_{rz}\hat{e}_r + u_{\theta z}\hat{e}_\theta + u_{zz}\hat{e}_z\Big] \\ &= \Big[u_{rr}\,dr + u_{r\theta}\,r\,d\theta + u_{rz}\,dz\Big]\hat{e}_r + \Big[u_{\theta r}\,dr \\ &\quad + u_{\theta\theta}\,r\,d\theta + u_{\theta z}\,dz\Big]\hat{e}_\theta + dz\Big[u_{zr}\,dr + ru_{z\theta}\,d\theta + u_{zz}\,dz\Big]\hat{e}_z \end{aligned} \tag{5.27}$$

where the coefficients $u_{rr}, u_{r\theta}, u_{\theta r}, u_{\theta\theta}, \cdots$ are the components of that dyadic $\vec{\nabla}\vec{u}$. Comparing Eqs (5.26) and (5.27), we get

$$u_{rr}\,dr + u_{r\theta}\,r\,d\theta + u_{rz}\,dz = \frac{\partial u_r}{\partial r}\,dr + \left(\frac{\partial u_r}{\partial \theta} - u_\theta\right)d\theta + \frac{\partial u_r}{\partial z}\,dz$$

$$u_{\theta r}\,dr + u_{\theta\theta}\,r\,d\theta + u_{\theta z}\,dz = \frac{\partial u_\theta}{\partial r}\,dr + \left(\frac{\partial u_\theta}{\partial \theta} + u_r\right)d\theta + \frac{\partial u_\theta}{\partial z}\,dz$$

$$u_{zr}\,dr + ru_{z\theta}\,d\theta + u_{zz}\,dz = \frac{\partial u_z}{\partial r}\,dr + \frac{\partial u_z}{\partial \theta}d\theta + \frac{\partial u_z}{\partial z}\,dz$$

These equations must be valid for arbitrary values of $dr, d\theta$ and $d\phi$. Therefore

$$u_{rr} = \frac{\partial u_r}{\partial r}; \quad ru_{r\theta} = \frac{\partial u_r}{\partial \theta} - u_\theta; \quad u_{rz} = \frac{\partial u_r}{\partial z}$$
$$u_{\theta r} = \frac{\partial u_\theta}{\partial r}; \quad ru_{\theta\theta} = \frac{\partial u_\theta}{\partial \theta} + u_r; \quad u_{\theta z} = \frac{\partial u_\theta}{\partial z}$$
$$u_{zr} = \frac{\partial u_z}{\partial r}; \quad ru_{z\theta} = \frac{\partial u_z}{\partial \theta}; \quad u_{zz} = \frac{\partial u_z}{\partial z}$$

In cylindrical coordinates (r, θ, z), the matrix of the components of ∇u can be written as

$$\begin{pmatrix} \frac{\partial u_r}{\partial r} & \frac{1}{r}\frac{\partial u_r}{\partial \theta} - \frac{u_\theta}{r} & \frac{\partial u_r}{\partial z} \\ \frac{\partial u_\theta}{\partial r} & \frac{1}{r}\frac{\partial u_\theta}{\partial \theta} + \frac{u_r}{r} & \frac{\partial u_\theta}{\partial z} \\ \frac{\partial u_z}{\partial r} & \frac{1}{r}\frac{\partial u_z}{\partial \theta} & \frac{\partial u_z}{\partial z} \end{pmatrix}$$

DEDUCTION 5.2.8

Gradient of a vector in spherical coordinates: Let the coordinates of the point P be (r, θ, ϕ)

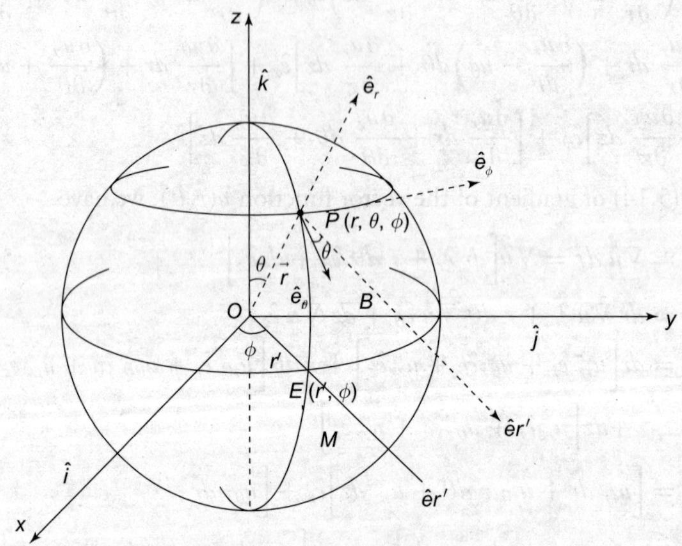

Figure 5.7 Gradient of a vector in spherical polar coordinates

(Fig. 5.7) and $\hat{e}_r, \hat{e}_\theta$ and $\hat{e}_\phi$ be the unit vectors in the direction of increasing r, θ and ϕ, respectively. Let $\hat{i}, \hat{j}$ and $\hat{k}$ be the unit vectors in the direction of increasing x, y and z, respectively. Thus

$$\hat{e}_{r'} = \cos\phi\ \hat{i} + \sin\phi\ \hat{j}; \quad \hat{e}_\phi = -\sin\phi\ \hat{i} + \cos\phi\ \hat{j}$$

where $\hat{e}_{r'}$ is the unit vector in the direction r' in the xy plane. Thus

$$d\hat{e}_{r'} = (-\sin\phi\ \hat{i} + \cos\phi\ \hat{j})d\phi = d\phi\ \hat{e}_\phi$$
$$d\hat{e}_\phi = (-\cos\phi\ \hat{i} - \sin\phi\ \hat{j})d\phi = -d\phi\ \hat{e}_r$$

Also, the unit vectors $\hat{e}_r$ and $\hat{e}_\theta$ can be written as

$$\hat{e}_r = \cos\theta\hat{k} + \sin\theta\hat{e}_{r'}; \;\; \hat{e}_\theta = \cos\theta\,\hat{e}_{r'} - \sin\theta\,\hat{k}; \;\; \hat{e}_{r'} = \sin\theta\hat{e}_r + \cos\theta\hat{e}_\theta$$

$$\Rightarrow \; d\hat{e}_r = (-\sin\theta\hat{k} + \cos\theta\hat{e}_{r'})d\theta + \sin\theta d\hat{e}_{r'} = d\theta\,\hat{e}_\theta + \sin\theta\,d\phi\,\hat{e}_\phi$$

$$d\hat{e}_\theta = (-\sin\theta\,\hat{e}_{r'} - \cos\theta\,\hat{k})d\theta + \cos\theta\,d\hat{e}_{r'} = -d\theta\,\hat{e}_r + \cos\theta\,d\phi\,\hat{e}_\phi$$

$$d\hat{e}_\phi = -d\phi\,\hat{e}_r = -(\sin\theta\,\hat{e}_r + \cos\theta\,\hat{e}_\theta)d\phi$$

The position vector for the point P can be written as $\vec{r} = r\,\hat{e}_r$, where $r = |\vec{r}|$, the magnitude of $\vec{r}$. Thus

$$\begin{aligned} d\vec{r} &= dr\,\hat{e}_r + r\,d\hat{e}_r = dr\,\hat{e}_r + r\left[d\theta\,\hat{e}_\theta + \sin\theta\,d\phi\,\hat{e}_\phi\right] \\ &= dr\,\hat{e}_r + r\,d\theta\,\hat{e}_\theta + r\,\sin\theta\,d\phi\,\hat{e}_\phi \end{aligned}$$

In spherical polar coordinates, a vector $\vec{u}$ can be written as

$$\vec{u}(r,\theta,\phi) = u_r(r,\theta,\phi)\,\hat{e}_r + u_\theta(r,\theta,\phi)\,\hat{e}_\theta + u_\phi(r,\theta,\phi)\,\hat{e}_\phi \tag{5.28}$$

where $u_r(r,\theta,\phi), u_\theta(r,\theta,\phi)$ and $u_\phi(r,\theta,\phi)$ are the components of $\vec{u}(r,\theta,\phi)$ in the directions $\hat{e}_r, \hat{e}_\theta$ and $\hat{e}_\phi$, respectively. Since $u_r = u_r(r,\theta,\phi), u_\theta = u_\theta(r,\theta,\phi)$ and $u_\phi = u_\phi(r,\theta,\phi)$, by the chain rule of differentiation, we get

$$\left.\begin{aligned} du_r &= \frac{\partial u_r}{\partial r}\,dr + \frac{\partial u_r}{\partial \theta}\,d\theta + \frac{\partial u_r}{\partial \phi}\,d\phi \\ du_\theta &= \frac{\partial u_\theta}{\partial r}\,dr + \frac{\partial u_\theta}{\partial \theta}\,d\theta + \frac{\partial u_\theta}{\partial \phi}\,d\phi \\ du_\phi &= \frac{\partial u_\phi}{\partial r}\,dr + \frac{\partial u_\phi}{\partial \theta}\,d\theta + \frac{\partial u_\phi}{\partial \phi}\,d\phi \end{aligned}\right\} \tag{5.29}$$

Therefore

$$\begin{aligned} d\vec{u} &= du_r\,\hat{e}_r + u_r\,d\hat{e}_r + du_\theta\hat{e}_\theta + u_\theta\,d\hat{e}_\theta + du_\phi\,\hat{e}_\phi + u_\phi\,d\hat{e}_\phi \\ &= \Big(\frac{\partial u_r}{\partial r}\,dr + \frac{\partial u_r}{\partial \theta}\,d\theta + \frac{\partial u_r}{\partial \phi}\,d\phi\Big)\hat{e}_r + u_r\Big(d\theta\,\hat{e}_\theta + \sin\theta\,d\phi\,\hat{e}_\phi\Big) \\ &\quad + \Big(\frac{\partial u_\theta}{\partial r}\,dr + \frac{\partial u_\theta}{\partial \theta}\,d\theta + \frac{\partial u_\theta}{\partial \phi}\,d\phi\Big)\hat{e}_\theta + u_\theta\Big(-d\theta\,\hat{e}_r + \cos\theta\,d\phi\,\hat{e}_\phi\Big) \\ &\quad + \Big(\frac{\partial u_\phi}{\partial r}\,dr + \frac{\partial u_\phi}{\partial \theta}\,d\theta + \frac{\partial u_\phi}{\partial \phi}\,d\phi\Big)\hat{e}_\phi + u_\phi\Big(-\sin\theta\,\hat{e}_r d\phi - \cos\theta\,\hat{e}_\theta d\phi\Big) \\ &= \left[\frac{\partial u_r}{\partial r}\,dr + \Big(\frac{\partial u_\theta}{\partial \theta} - u_\theta\Big)d\theta + \Big(\frac{\partial u_r}{\partial \phi} - u_\phi\,\sin\theta\Big)d\phi\right]\hat{e}_r \\ &\quad + \left[\frac{\partial u_\theta}{\partial r}\,dr + \Big(\frac{\partial u_\theta}{\partial \theta} + u_r\Big)d\theta + \Big(\frac{\partial u_\theta}{\partial \phi} - u_\phi\,\cos\theta\Big)d\phi\right]\hat{e}_\theta \\ &\quad + \left[\frac{\partial u_\phi}{\partial r}\,dr + \frac{\partial u_\phi}{\partial \theta}d\theta + \Big(\frac{\partial u_\phi}{\partial \phi} + u_r\sin\theta + u_\theta\,\cos\theta\Big)d\phi\right]\hat{e}_\phi. \end{aligned} \tag{5.30}$$

By the definition (5.14) of gradient of the vector function $\vec{u}(r,\theta,\phi)$, we have

$$d\vec{u} = \vec{\nabla}\vec{u}\,d\vec{r} = \vec{\nabla}\vec{u}\Big[dr\,\hat{e}_r + r\,d\theta\,\hat{e}_\theta + r\,\sin\theta\,d\phi\,\hat{e}_\phi\Big]$$

$$
\begin{aligned}
&= dr\ \vec{\nabla}\vec{u}\hat{e}_r + r\ d\theta\ \vec{\nabla}\vec{u}\hat{e}_\theta + r\ \sin\theta\ d\phi\ \vec{\nabla}\vec{u}\hat{e}_\phi \\
&= dr(u_{rr}\ \hat{e}_r + u_{\theta r}\ \hat{e}_\theta + u_{\phi r}\hat{e}_\phi) + r\ d\theta(u_{r\theta}\ \hat{e}_r + u_{\theta\theta}\hat{e}_\theta + u_{\phi\theta}\hat{e}_\phi \\
&\quad +r\ \sin\theta\ d\phi\ (u_{r\phi}\hat{e}_r + u_{\theta\phi}\hat{e}_\theta + u_{\phi\phi}\hat{e}_\phi) \\
&= (u_{rr}\ dr + ru_{r\theta}\ d\theta + r\sin\theta\, u_{r\phi}\ d\phi)\hat{e}_r \\
&\quad +(u_{\theta r}\ dr + ru_{\theta\theta}\ d\theta + r\sin\theta\ u_{\theta\phi}\ d\phi)\hat{e}_\theta \\
&\quad +(u_{\phi r}\ dr + ru_{\phi\theta}\ d\theta + r\sin\theta\ u_{\phi\phi}\ d\phi)\hat{e}_\phi
\end{aligned}
\tag{5.31}
$$

where the coefficients $u_{rr}, u_{r\theta}, u_{\theta r}, u_{\theta\theta}, \cdots$ are the components of that dyadic $\vec{\nabla}\vec{u}$. Comparing Eqs (5.30) and (5.31), we get

$$
\begin{aligned}
&u_{rr}\ dr + ru_{r\theta}\ d\theta + r\sin\theta\, u_{r\phi}\ d\phi \\
&\quad = \frac{\partial u_r}{\partial r}\ dr + \left(\frac{\partial u_\theta}{\partial \theta} - u_\theta\right)d\theta + \left(\frac{\partial u_r}{\partial \phi} - u_\phi\ \sin\theta\right)d\phi \\
&u_{\theta r}\ dr + ru_{\theta\theta}\ d\theta + r\sin\theta\ u_{\theta\phi}\ d\phi \\
&\quad = \frac{\partial u_\theta}{\partial r}\ dr + \left(\frac{\partial u_\theta}{\partial \theta} + u_r\right)d\theta + \left(\frac{\partial u_\theta}{\partial \phi} - u_\phi\ \cos\theta\right)d\phi \\
&u_{\phi r}\ dr + ru_{\phi\theta}\ d\theta + r\sin\theta\ u_{\phi\phi}\ d\phi \\
&\quad = \frac{\partial u_\phi}{\partial r}\ dr + \frac{\partial u_\phi}{\partial \theta}d\theta + \left(\frac{\partial u_\phi}{\partial \phi} + u_r\sin\theta + u_\theta\ \cos\theta\right)d\phi
\end{aligned}
$$

These relations must hold for any value of $dr, d\theta$ and $d\phi$, so

$$
\begin{aligned}
&u_{rr} = \frac{\partial u_r}{\partial r};\quad ru_{r\theta} = \frac{\partial u_r}{\partial \theta} - u_\theta;\quad r\sin\theta\, u_{r\phi} = \frac{\partial u_r}{\partial \phi} - u_\phi\sin\theta \\
&u_{\theta r} = \frac{\partial u_\theta}{\partial r};\quad ru_{\theta\theta} = \frac{\partial u_\theta}{\partial \theta} + u_r;\quad r\sin\theta\, u_{\theta\phi} = \frac{\partial u_\theta}{\partial \phi} - u_\phi\cos\theta \\
&u_{\phi r} = \frac{\partial u_\phi}{\partial r};\quad ru_{\phi\theta} = \frac{\partial u_\phi}{\partial \theta};\quad r\sin\theta\, u_{\phi\phi} = \frac{\partial u_\phi}{\partial \phi} + u_r\sin\theta + u_\theta\cos\theta
\end{aligned}
$$

An array of these four components in the form

$$
\begin{pmatrix}
\dfrac{\partial u_r}{\partial r} & \dfrac{1}{r}\left[\dfrac{\partial u_\theta}{\partial \theta} - u_\theta\right] & \dfrac{1}{r\sin\theta}\left[\dfrac{\partial u_r}{\partial \phi} - u_\phi\sin\theta\right] \\
\dfrac{\partial u_\theta}{\partial r} & \dfrac{1}{r}\left[\dfrac{\partial u_\theta}{\partial \theta} + u_r\right] & \dfrac{1}{r\sin\theta}\left[\dfrac{\partial u_\theta}{\partial \phi} - u_\phi\sin\theta\right] \\
\dfrac{\partial u_\phi}{\partial r} & \dfrac{1}{r}\dfrac{\partial u_\phi}{\partial \theta} & \dfrac{1}{r\sin\theta}\left[\dfrac{\partial u_\phi}{\partial \phi} + u_r\sin\theta + u_\theta\cos\theta\right]
\end{pmatrix}
\tag{5.32}
$$

is called a 3×3 matrix.

5.2.3 Gradient in Curvilinear Coordinates

Let ϕ be a continuously differentiable scalar point function of orthogonal curvilinear coordinates u, v and w. Then

$$
d\phi = \frac{\partial \phi}{\partial u}\ du + \frac{\partial \phi}{\partial v}\ dv + \frac{\partial \phi}{\partial w}\ dw \tag{5.33}
$$

Since $\vec{\nabla}\phi$ is a vector, it can be expressed as a linear combination of orthogonal unit vectors $\hat{e}_1, \hat{e}_2$ and $\hat{e}_3$. Let

$$\vec{\nabla}\phi = \lambda_1\,\hat{e}_1 + \lambda_2\,\hat{e}_2 + \lambda_3\,\hat{e}_3 \tag{5.34}$$

where λ_1, λ_2 and λ_3 are unknown scalars to be determined. Now, using Eq. (4.51), we get

$$\begin{aligned} d\phi = \vec{\nabla}\phi \cdot d\vec{r} &= \left[\lambda_1\,\hat{e}_1 + \lambda_2\,\hat{e}_2 + \lambda_3\,\hat{e}_3\right]\cdot\left[h_1\,du\,\hat{e}_1 + h_2\,dv\,\hat{e}_2 + h_3\,dw\,\hat{e}_3\right] \\ &= h_1\,\lambda_1\,du + h_2\,\lambda_2\,dv + h_3\,\lambda_3\,dw \end{aligned} \tag{5.35}$$

Comparing Eqs (5.33) and (5.35), we get

$$\frac{\partial\phi}{\partial u} = h_1\,\lambda_1, \quad \frac{\partial\phi}{\partial v} = h_2\,\lambda_2, \quad \frac{\partial\phi}{\partial w} = h_3\,\lambda_3$$

$$\Rightarrow \lambda_1 = \frac{1}{h_1}\frac{\partial\phi}{\partial u}, \quad \lambda_2 = \frac{1}{h_2}\frac{\partial\phi}{\partial v}, \quad \lambda_3 = \frac{1}{h_3}\frac{\partial\phi}{\partial w}$$

$$\Rightarrow \vec{\nabla}\phi = \frac{1}{h_1}\frac{\partial\phi}{\partial u}\,\hat{e}_1 + \frac{1}{h_2}\frac{\partial\phi}{\partial v}\,\hat{e}_2 + \frac{1}{h_3}\frac{\partial\phi}{\partial w}\,\hat{e}_3 \text{ using Eq.(5.34)} \tag{5.36}$$

This is the expression of $\vec{\nabla}\phi$ in terms of orthogonal curviliear coordinates. From Eq. (5.36), we get

$$\vec{\nabla} = \frac{1}{h_1}\,\hat{e}_1\,\frac{\partial}{\partial u} + \frac{1}{h_2}\,\hat{e}_2\,\frac{\partial}{\partial v} + \frac{1}{h_3}\,\hat{e}_3\,\frac{\partial}{\partial w} \tag{5.37}$$

RESULT 18

Taking $\phi = u$, since u, v and w are independent, we have $\vec{\nabla}u = \frac{1}{h_1}\,\hat{e}_1$. Similarly, $\vec{\nabla}v = \frac{1}{h_2}\,\hat{e}_2$ and $\vec{\nabla}w = \frac{1}{h_3}\,\hat{e}_3$. Using these notations

(i) $\hat{e}_1 = \hat{e}_2 \times \hat{e}_3 = h_2h_3\left(\vec{\nabla}v \times \vec{\nabla}w\right)$

(ii) $\hat{e}_2 = \hat{e}_3 \times \hat{e}_1 = h_3h_1\left(\vec{\nabla}w \times \vec{\nabla}u\right)$

(iii) $\hat{e}_3 = \hat{e}_1 \times \hat{e}_2 = h_1h_2\left(\vec{\nabla}u \times \vec{\nabla}v\right)$

Using these notations, $\vec{\nabla}\phi$ can be written in the form

$$\vec{\nabla}\phi = \frac{\partial\phi}{\partial u}\,\vec{\nabla}u + \frac{\partial\phi}{\partial v}\,\vec{\nabla}v + \frac{\partial\phi}{\partial w}\,\vec{\nabla}w \tag{5.38}$$

Now, $\vec{\nabla}u, \vec{\nabla}v$ and $\vec{\nabla}w$ are the normals to the coordinate surfaces $u = c_1, v = c_2$ and $w = c_3$, respectively.

DEDUCTION 5.2.9

Gradient in cylindrical coordinates: Here, $u = \rho,\ v = \phi$ and $w = z$; $\hat{e}_1 = \hat{e}_\rho, \hat{e}_2 = \hat{e}_\phi$ and $\hat{e}_3 = \hat{e}_z$; $h_1 = h_\rho = 1,\ h_2 = h_\phi = \rho$, and $h_3 = h_z = 1$. Let f be a scalar valued function. Therefore, using Eq. (5.36), we get

$$\vec{\nabla}f = \frac{1}{h_1}\frac{\partial f}{\partial u}\,\hat{e}_1 + \frac{1}{h_2}\frac{\partial f}{\partial v}\,\hat{e}_2 + \frac{1}{h_3}\frac{\partial f}{\partial w}\,\hat{e}_3$$

$$= \frac{1}{1}\frac{\partial f}{\partial \rho}\hat{e}_\rho + \frac{1}{\rho}\frac{\partial f}{\partial \phi}\hat{e}_\phi + \frac{1}{1}\frac{\partial f}{\partial z}\hat{e}_z$$

$$= \frac{\partial f}{\partial \rho}\hat{e}_\rho + \frac{1}{\rho}\frac{\partial f}{\partial \phi}\hat{e}_\phi + \frac{\partial f}{\partial z}\hat{e}_z \tag{5.39}$$

DEDUCTION 5.2.10

Gradient in spherical polar coordinates: Here, $u = \rho, v = \theta$ and $w = \phi$; $\hat{e}_1 = \hat{e}_\rho, \hat{e}_2 = \hat{e}_\theta$, and $\hat{e}_3 = \hat{e}_\phi$; $h_1 = h_\rho = 1, h_2 = h_\theta = \rho$ and $h_3 = h_\phi = \rho \sin\theta$. Let f be a scalar valued function. Therefore, using Eq. (5.36), we get

$$\vec{\nabla} f = \frac{1}{h_1}\frac{\partial f}{\partial u}\hat{e}_1 + \frac{1}{h_2}\frac{\partial f}{\partial v}\hat{e}_2 + \frac{1}{h_3}\frac{\partial f}{\partial w}\hat{e}_3$$

$$= \frac{1}{1}\frac{\partial f}{\partial \rho}\hat{e}_\rho + \frac{1}{\rho}\frac{\partial f}{\partial \theta}\hat{e}_\theta + \frac{1}{\rho\sin\theta}\frac{\partial f}{\partial \phi}\hat{e}_\phi$$

$$= \frac{\partial f}{\partial \rho}\hat{e}_\rho + \frac{1}{\rho}\frac{\partial f}{\partial \theta}\hat{e}_\theta + \frac{1}{\rho\sin\theta}\frac{\partial f}{\partial \phi}\hat{e}_\phi \tag{5.40}$$

5.3 DIVERGENCE

In the Cartesian coordinate system, let $\vec{F}(x,y,z) = F_1(x,y,z)\hat{i} + F_2(x,y,z)\hat{j} + F_3(x,y,z)\hat{k}$ be any given vector point function that is continuously differentiable at each point (x,y,z) in a certain region of space E^3. In the Cartesian coordinate system, the scalar point function

$$\vec{\nabla}\cdot\vec{F} = (\frac{\partial}{\partial x}\hat{i} + \frac{\partial}{\partial y}\hat{j} + \frac{\partial}{\partial z}\hat{k})\cdot(F_1\hat{i} + F_2\hat{j} + F_3\hat{k})$$

$$= \frac{\partial F_1}{\partial x} + \frac{\partial F_2}{\partial y} + \frac{\partial F_3}{\partial z} \tag{5.41}$$

where the partial derivatives of F_1, F_2 and F_3 are assumed to be continuous. This is called the divergence of $\vec{F}$ and is written as $\vec{\nabla}\cdot\vec{F}$ or $div\vec{F}$.

(i) The operator symbol can be written as $\vec{\nabla}\cdot \approx \sum \hat{i}\cdot\frac{\partial}{\partial x}$.

(ii) Divergence changes a vector point function into a scalar point function.

(iii) From the physical point of view, the divergence of a vector at a point is a measure of how much the vector $\vec{F}$ diverges, i.e., spreads out, at that point (Fig. 5.8). A point of large positive divergence is known as a *source*, whereas a point of large negative divergence is known as a *sink*.

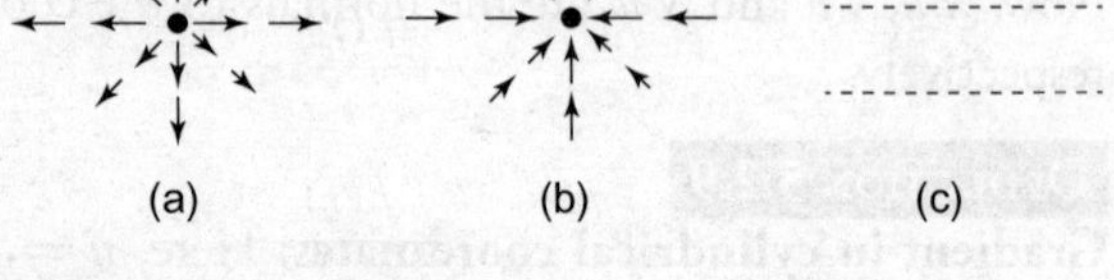

Figure 5.8 Physical concept of divergence

(iv) A vector point function $\vec{F}(x,y,z)$ is *solenoidal* if divergence is zero, i.e.,

$$\text{div}\,\vec{F} = \vec{\nabla}\cdot\vec{F} = 0 \tag{5.42}$$

For example, the magnetic field is solenoidal, as div$\vec{H} = 0$, where $\vec{H}$ is the magnetic field of force. In the case of fluid motion, if $\vec{v}$ is the velocity of the fluid and div $\vec{v} = 0$ throughout the region of the fluid, fluid motion is said to be *incompressible.*

(v) Using the definition of gradient and divergence, $\vec{\nabla} \cdot \vec{F}$ can be written as

$$\vec{\nabla} \cdot \vec{F} = (\vec{\nabla} F_1) \cdot \hat{i} + (\vec{\nabla} F_2) \cdot \hat{j} + (\vec{\nabla} F_3) \cdot \hat{k}$$

The notion of divergence is important because of its widespread use in physics and engineering problems. Note that $\vec{\nabla} \cdot \vec{F} \neq \vec{F} \cdot \vec{\nabla}$.

Example 5.3.1

Find $\vec{\nabla} \cdot \vec{F}$ if $\vec{F} = x^2 z\, \hat{i} - 2y^3 z^2\, \hat{j} + xy^2 z\, \hat{k}$, at the point $(1, -1, 1)$.

Solution

Using Eq. (5.41), we get

$$\begin{aligned}
\vec{\nabla} \cdot \vec{F} &= (\frac{\partial}{\partial x}\hat{i} + \frac{\partial}{\partial y}\hat{j} + \frac{\partial}{\partial z}\hat{k}) \cdot (x^2 z\, \hat{i} - 2y^3 z^2\, \hat{j} + xy^2 z\, \hat{k}) \\
&= \frac{\partial}{\partial x}(x^2 z) + \frac{\partial}{\partial y}(-2y^3 z^2) + \frac{\partial}{\partial z}(xy^2 z) \\
&= 2xz - 6y^2 z^2 + xy^2 = -3 \text{ at } (1, -1, 1)
\end{aligned}$$

Example 5.3.2

Prove that $\vec{\nabla} \cdot \left(\frac{f(r)}{r}\,\vec{r}\right) = \frac{1}{r^2}\frac{d}{dr}\left[r^2 f(r)\right]$.

Solution

Using Eq. (5.41), we get

$$\begin{aligned}
\vec{\nabla} \cdot \left(\frac{f(r)}{r}\,\vec{r}\right) &= \left(\hat{i}\,\frac{\partial}{\partial x} + \hat{j}\,\frac{\partial}{\partial y} + \hat{k}\,\frac{\partial}{\partial z}\right) \cdot \frac{f(r)}{r}\,\vec{r} \\
&= \left(\hat{i}\,\frac{\partial}{\partial x} + \hat{j}\,\frac{\partial}{\partial y} + \hat{k}\,\frac{\partial}{\partial z}\right) \cdot \frac{f(r)}{r}\left[x\hat{i} + y\hat{j} + z\hat{k}\right] \\
&= \frac{\partial}{\partial x}\left[x\frac{f(r)}{r}\right] + \frac{\partial}{\partial y}\left[y\frac{f(r)}{r}\right] + \frac{\partial}{\partial z}\left[z\frac{f(r)}{r}\right] \\
&= \frac{f(r)}{r} + \frac{xf'(r)}{r}\frac{x}{\sqrt{x^2 + y^2 + z^2}} - \frac{xf(r)}{r^2}\frac{x}{\sqrt{x^2 + y^2 + z^2}} \\
&\quad + \frac{f(r)}{r} + \frac{yf'(r)}{r}\frac{y}{\sqrt{x^2 + y^2 + z^2}} - \frac{yf(r)}{r^2}\frac{y}{\sqrt{x^2 + y^2 + z^2}} \\
&\quad + \frac{f(r)}{r} + \frac{zf'(r)}{r}\frac{z}{\sqrt{x^2 + y^2 + z^2}} - \frac{zf(r)}{r^2}\frac{z}{\sqrt{x^2 + y^2 + z^2}} \\
&= 3\frac{f(r)}{r} + \left[\frac{f'(r)}{r^2} - \frac{f(r)}{r^3}\right](x^2 + y^2 + z^2)
\end{aligned}$$

$$= f'(r) + \frac{2f(r)}{r} = \frac{2rf(r) + r^2 f'(r)}{r^2} = \frac{1}{r^2}\frac{d}{dr}\left[r^2 f(r)\right]$$

Example 5.3.3

Prove that $\vec{b} \cdot \vec{\nabla}\left[\vec{a} \cdot \vec{\nabla}\left(\frac{1}{r}\right)\right] = \frac{3(\vec{a} \cdot \vec{r})(\vec{b} \cdot \vec{r})}{r^5} - \frac{\vec{a} \cdot \vec{b}}{r^3}$, where $\vec{a}$ and $\vec{b}$ are constant vectors and $\vec{r} = x\hat{i} + y\hat{j} + z\hat{k}, r = |\vec{r}|$.

Solution

Since $r = |\vec{r}|$, using Eq. (5.3) as the definition of gradient, we get

$$\begin{aligned}\vec{\nabla}\left(\frac{1}{r}\right) &= \left(\hat{i}\frac{\partial}{\partial x} + \hat{j}\frac{\partial}{\partial y} + \hat{k}\frac{\partial}{\partial z}\right)\left(\frac{1}{r}\right) = -\frac{1}{r^2}\left(\hat{i}\frac{\partial r}{\partial x} + \hat{j}\frac{\partial r}{\partial y} + \hat{k}\frac{\partial r}{\partial z}\right)\\ &= -\frac{1}{r^2}\left[\hat{i}\frac{2x}{2\sqrt{x^2+y^2+z^2}} + \hat{j}\frac{2y}{2\sqrt{x^2+y^2+z^2}} + \hat{k}\frac{2z}{2\sqrt{x^2+y^2+z^2}}\right]\\ &= -\frac{1}{r^3}\left[x\hat{i} + y\hat{j} + z\hat{k}\right] = -\frac{1}{r^3}\vec{r}\end{aligned}$$

Let $\vec{a} = a_1\hat{i} + a_2\hat{j} + a_3\hat{k}$, then

$$\begin{aligned}\vec{a} \cdot \vec{\nabla}\left(\frac{1}{r}\right) &= \vec{a} \cdot \left[-\frac{1}{r^3}\vec{r}\right] = -\frac{1}{r^3}\left[a_1\hat{i} + a_2\hat{j} + a_3\hat{k}\right] \cdot \left[x\hat{i} + y\hat{j} + z\hat{k}\right]\\ &= -\frac{1}{r^3}(a_1 x + a_2 y + a_3 z) = -\frac{\vec{a} \cdot \vec{r}}{r^3}\end{aligned}$$

Using Eq. (5.3) as the definition of gradient, we get

$$\begin{aligned}\vec{\nabla}\left[\vec{a} \cdot \vec{\nabla}\left(\frac{1}{r}\right)\right] &= \left(\hat{i}\frac{\partial}{\partial x} + \hat{j}\frac{\partial}{\partial y} + \hat{k}\frac{\partial}{\partial z}\right)\left(-\frac{\vec{a} \cdot \vec{r}}{r^3}\right)\\ &= -\hat{i}\frac{\partial}{\partial x}\left(\frac{\vec{a} \cdot \vec{r}}{r^3}\right) - \hat{j}\frac{\partial}{\partial y}\left(\frac{\vec{a} \cdot \vec{r}}{r^3}\right) - \hat{k}\frac{\partial}{\partial z}\left(\frac{\vec{a} \cdot \vec{r}}{r^3}\right)\\ &= -\left[\frac{a_1}{r^3} - \frac{3(\vec{a} \cdot \vec{r})}{r^4}\frac{\partial r}{\partial x}\right]\hat{i} - \left[\frac{a_2}{r^3} - \frac{3(\vec{a} \cdot \vec{r})}{r^4}\frac{\partial r}{\partial y}\right]\hat{j} - \left[\frac{a_3}{r^3} - \frac{3(\vec{a} \cdot \vec{r})}{r^4}\frac{\partial r}{\partial z}\right]\hat{k}\\ &= -\left[\frac{a_1}{r^3} - \frac{3(\vec{a} \cdot \vec{r})}{r^4}\frac{x}{r}\right]\hat{i} - \left[\frac{a_2}{r^3} - \frac{3(\vec{a} \cdot \vec{r})}{r^4}\frac{y}{r}\right]\hat{j} - \left[\frac{a_3}{r^3} - \frac{3(\vec{a} \cdot \vec{r})}{r^4}\frac{z}{r}\right]\hat{k}\\ &= -\frac{1}{r^3}(a_1\hat{i} + a_2\hat{j} + a_3\hat{k}) + \frac{3(\vec{a} \cdot \vec{r})}{r^5}(x\hat{i} + y\hat{j} + z\hat{k})\\ &= -\frac{1}{r^3}\vec{a} + \frac{3(\vec{a} \cdot \vec{r})}{r^5}\vec{r}\end{aligned}$$

Therefore, the definition of dot product gives

$$\begin{aligned}\vec{b} \cdot \vec{\nabla}\left[\vec{a} \cdot \vec{\nabla}\left(\frac{1}{r}\right)\right] &= \vec{b} \cdot \left[-\frac{1}{r^3}\vec{a} + \frac{3(\vec{a} \cdot \vec{r})}{r^5}\vec{r}\right]\\ &= -\frac{1}{r^3}(\vec{b} \cdot \vec{a}) + \frac{3(\vec{a} \cdot \vec{r})}{r^5}(\vec{b} \cdot \vec{r}) = \frac{3(\vec{a} \cdot \vec{r})(\vec{b} \cdot \vec{r})}{r^5} - \frac{\vec{a} \cdot \vec{b}}{r^3}\end{aligned}$$

Example 5.3.4

Find the directional derivative of div($\vec{u}$) at the point (1, 2, 2) in the direction of the outer normal of the sphere $x^2 + y^2 + z^2 = 9$ for $\vec{u} = x^4\hat{i} + y^4\hat{j} + z^4\hat{k}$.

Solution

Sin ce $\vec{u} = x^4\hat{i} + y^4\hat{j} + z^4\hat{k}$, using the definition of divergence,we get

$$\vec{\nabla} \cdot \vec{u} = 4(x^3 + y^3 + z^3)$$

Let $\phi = x^2 + y^2 + z^2 - 9$. Then, at (1, 2, 2)

$$\text{Outer normal of the sphere} = \vec{\nabla}\phi = \left(\hat{i}\frac{\partial}{\partial x} + \hat{j}\frac{\partial}{\partial y} + \hat{k}\frac{\partial}{\partial z}\right)(x^2 + y^2 + z^2 - 9)$$

$$= 2x\hat{i} + 2y\hat{j} + 2z\hat{k} = 2\hat{i} + 4\hat{j} + 4\hat{k}$$

$$\text{The directional derivative } = \vec{\nabla}(\vec{\nabla} \cdot \vec{u}) = \left(\hat{i}\frac{\partial}{\partial x} + \hat{j}\frac{\partial}{\partial y} + \hat{k}\frac{\partial}{\partial z}\right)4(x^3 + y^3 + z^3)$$

$$= 12(x^2\hat{i} + y^2\hat{j} + z^2\hat{k}) = 12(\hat{i} + 4\hat{j} + 4\hat{k})$$

Directional derivative along the outer normal

$$= 12(\hat{i} + 4\hat{j} + 4\hat{k}) \cdot \frac{1}{\sqrt{4 + 16 + 16}}(2\hat{i} + 4\hat{j} + 4\hat{k})$$

$$= \frac{12(2 + 16 + 16)}{6} = 68$$

Thus, the vector field determined by $\vec{F}$ is solenoidal.

Example 5.3.5

Determine the constant a so that the vector $\vec{F}(x, y, z) = (x + 2y)\hat{i} + (y - 2z)\hat{j} + (x + az)\hat{k}$ is solenoidal.

Solution

Using Eq. (5.41) as the definition of divergence, we get

$$\vec{\nabla} \cdot \vec{F} = \left(\frac{\partial}{\partial x}\hat{i} + \frac{\partial}{\partial y}\hat{j} + \frac{\partial}{\partial z}\hat{k}\right) \cdot \left[(x + 2y)\hat{i} + (y - 2z)\hat{j} + (x + az)\hat{k}\right]$$

$$= \frac{\partial}{\partial x}(x + 3y) + \frac{\partial}{\partial y}(y - 2x) + \frac{\partial}{\partial z}(x + az)$$

$$= 1 + 1 + a = 2 + a$$

Thus, the vector field determined by $\vec{F}$ is solenoidal, if $\vec{\nabla} \cdot \vec{F} = 2 + a = 0$ i.e., $a = -2$.

Example 5.3.6

Show that the vector field determined by $r^n\vec{r}$, where $\vec{r} = x\hat{i} + y\hat{j} + z\hat{k}$ and $r = |\vec{r}|$, is solenoidal for $n = -3$.

Solution

Using Eq. (5.41) as the definition of divergence, we get

$$\begin{aligned}
\vec{\nabla}\cdot(r^n\vec{r}) &= \left(\frac{\partial}{\partial x}\hat{i}+\frac{\partial}{\partial y}\hat{j}+\frac{\partial}{\partial z}\hat{k}\right)\cdot(r^n x\,\hat{i}+r^n y\,\hat{j}+r^n z\,\hat{k})\\
&= \frac{\partial}{\partial x}(r^n x)+\frac{\partial}{\partial y}(r^n y)+\frac{\partial}{\partial z}(r^n z)\\
&= r^n + x\frac{\partial r^n}{\partial x}+r^n+y\frac{\partial r^n}{\partial y}+r^n+z\frac{\partial r^n}{\partial z}\\
&= 3r^n+\left[x\frac{n}{2}\cdot 2x+y\frac{n}{2}\cdot 2y+z\frac{n}{2}\cdot 2z\right](x^2+y^2+z^2)^{n/2-1}\\
&= 3r^n+n\left[x^2+y^2+z^2\right](x^2+y^2+z^2)^{n/2-1}\\
&= 3r^n+n(x^2+y^2+z^2)^{n/2}=3r^n+nr^n=(3+n)r^n
\end{aligned}$$

Thus, the vector field determined by $r^n\vec{r}$ is solenoidal, if $\vec{\nabla}\cdot(r^n\vec{r})=(3+n)r^n=0$ i.e., $n=-3$.

Example 5.3.7

If $\vec{F}(x,y,z)=r^n(\vec{\alpha}\times\vec{r})$, then prove that $\vec{F}$ is solenoidal, where $\vec{\alpha}$ is a constant vector, $\vec{r}=x\hat{i}+y\hat{j}+z\hat{k}$ and $r=|\vec{r}|$.

Solution

Let $\vec{\alpha}=\alpha_1\hat{i}+\alpha_2\hat{j}+\alpha_3\hat{k}$ be a constant vector, where α_1,α_2 and α_3 are constants. Now, we find that

$$\begin{aligned}
\vec{\alpha}\times\vec{r} &= \begin{vmatrix}\hat{i} & \hat{j} & \hat{k}\\ \alpha_1 & \alpha_2 & \alpha_3\\ x & y & z\end{vmatrix}\\
&= (\alpha_2 z-\alpha_3 y)\hat{i}+(\alpha_3 x-\alpha_1 z)\hat{j}+(\alpha_1 y-\alpha_2 x)\hat{k}
\end{aligned}$$

Hence, by the definition of divergence of vector point function $\vec{F}$

$$\begin{aligned}
\vec{\nabla}\cdot\vec{F} &= \frac{\partial}{\partial x}\left\{r^n(\alpha_2 z-\alpha_3 y)\right\}+\frac{\partial}{\partial y}\left\{r^n(\alpha_3 x-\alpha_1 z)\right\}+\frac{\partial}{\partial z}\left\{r^n(\alpha_1 y-\alpha_2 x)\right\}\\
&= (\alpha_2 z-\alpha_3 y)nr^{n-1}\frac{\partial r}{\partial x}+(\alpha_3 x-\alpha_1 z)nr^{n-1}\frac{\partial r}{\partial y}+(\alpha_1 y-\alpha_2 x)nr^{n-1}\frac{\partial r}{\partial z}\\
&= nr^{n-1}\left[(\alpha_2 z-\alpha_3 y)\frac{x}{r}+(\alpha_3 x-\alpha_1 z)\frac{y}{r}+(\alpha_1 y-\alpha_2 x)\frac{z}{r}\right]\\
&= nr^{n-1}\left[(\alpha_2 xz-\alpha_3 xy)\frac{x}{r}+(\alpha_3 xy-\alpha_1 zy)\frac{y}{r}+(\alpha_1 yz-\alpha_2 xz)\frac{z}{r}\right]=0
\end{aligned}$$

Therefore, by the definition of a solenoidal vector, it follows that the vector $\vec{F}$ is solenoidal.

DEDUCTION 5.3.1

Physical interpretation of divergence: The physical interpretation of divergence can be found in fluid mechanics as well as in electricity and magnetism. We consider a region in the three-dimensional Euclidean space E^3 filled with a moving fluid of density $\rho(x,y,z,t)$ at any time t (Fig. 5.9).

The fluid flows in such a manner that the velocity vector $\vec{v}(x, y, z, t)$ associated with the fluid at P is given by

$$\vec{v}(x, y, z, t) = v_1(x, y, z, t)\hat{i} + v_2(x, y, z, t)\hat{j} + v_3(x, y, z, t)\hat{k}$$

Further, we consider a vector field $\vec{F}(x, y, z, t) = F_1(x, y, z, t)\hat{i}+F_2(x, y, z, t)\hat{j}+ F_3(x, y, z, t)\hat{k}$ such that

$$\vec{F}(x, y, z, t) = \rho(x, y, z, t)\ \vec{v}(x, y, z, t)$$

This is the amount of fluid flowing at a unit time across a unit area at the point P in the direction of $\vec{v}(x, y, z, t)$. Thus, we have

$$F_1 = \rho v_1, F_2 = \rho v_2, F_3 = \rho v_3$$

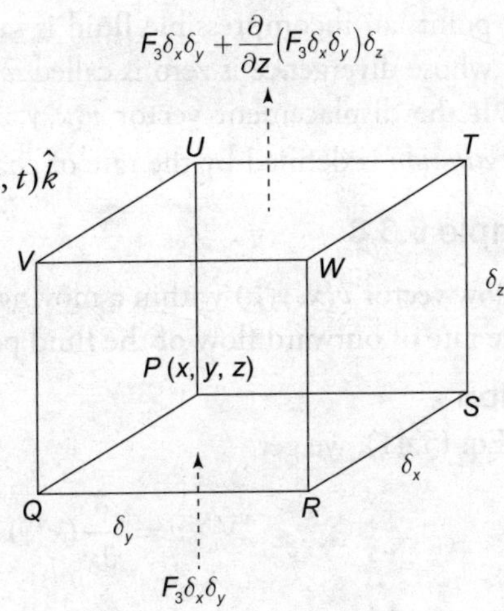

Figure 5.9 Physical interpretation of divergence

Now, at P, we consider a small rectangular parallelepiped $PQRSTUVW$ with sides δx, δy and δz which are parallel to the axes of x, y and z, respectively. The area of the face $PQRS$ is $\delta x \delta y$. So the amount of fluid entering the rectangular parallelepiped across the face $PQRS$ perpendicular to the z axis at (x, y, z) is $F_3 \delta x \delta y$ per unit time. Therefore, the amount of fluid leaving the volume per unit time across the opposite face $TUVW$ at a distance δz is

$$F_3\ \delta x\ \delta y + \frac{\partial}{\partial z}(F_3\ \delta x\ \delta y)\ \delta z$$

where the last term represents the change of amount $F_3\ \delta x\ \delta y$ due to the change δz in z. Thus, the amount of fluid flowing out of the parallelepiped in the direction of the z axis is $\frac{\partial}{\partial z}(F_3\ \delta x\ \delta y)\ \delta z$.

Similarly, if we consider the flow across the faces $QRWV$ and $RSTW$, we find that the amounts of fluid flowing out of the parallelepiped parallel to the axes of x and y, are respectively,

$$\frac{\partial}{\partial x}(F_1\ \delta y\ \delta z)\ \delta x \text{ and } \frac{\partial}{\partial y}(F_2\ \delta z\ \delta x)\ \delta y$$

Therefore, the total amount of flow out of the parallelepiped of volume $\delta x\ \delta y\ \delta z$ is

$$\begin{aligned}
&\frac{\partial}{\partial x}(F_1\ \delta y\ \delta z)\ \delta x + \frac{\partial}{\partial y}(F_2\ \delta z\ \delta x)\ \delta y + \frac{\partial}{\partial z}(F_3\ \delta x\ \delta y)\ \delta z \\
&= \left[\frac{\partial F_1}{\partial x} + \frac{\partial F_2}{\partial y} + \frac{\partial F_3}{\partial z}\right]\delta x\ \delta y\ \delta z = \vec{\nabla} \cdot \vec{F}\ \delta x\ \delta y\ \delta z
\end{aligned}$$

This shows that $\vec{\nabla} \cdot \vec{F}$ represents the rate of outward flow of the fluid per unit time per unit volume.

(i) If $\vec{\nabla} \cdot \vec{F} > 0$ in the neighbourhood of a point P, then the outflow is positive and we call P a *source*. If $\vec{\nabla} \cdot \vec{F} < 0$, the point P is a *sink* (inflow, not outflow).

(ii) If there is no gain of fluid anywhere, then $\vec{\nabla} \cdot \vec{F} = 0$. This is called the continuity equation for an incompressible fluid. Since fluid is neither created nor destroyed at any point, an incompressible fluid is said to have no sources or sinks Fig. 5.8(c). A vector $\vec{F}$ whose divergence is zero is called *solenoidal*.

(iii) If the displacement vector $\vec{v}(x, y, z)$ takes place within a deformable elastic solid, then *dilation* is defined by the rate of change of volume per unit volume $= \vec{\nabla} \cdot \vec{v}$.

Example 5.3.8

If the flow vector $\vec{v}(x, y, z)$ within a moving fluid is given by $\vec{v}(x, y, z) = x^4 y\hat{i} + y^4 z\hat{j} + z^4 x\hat{k}$, then find the rate of outward flow of the fluid per unit volume per unit time at the point $(1, 1, 1)$.

Solution

Using Eq. (5.41), we get

$$\vec{\nabla} \cdot \vec{v} = \frac{\partial}{\partial x}(x^4 y) + \frac{\partial}{\partial y}(y^4 z) + \frac{\partial}{\partial z}(z^4 x)$$
$$= 4(x^3 y + y^3 z + z^3 x) = 12 \text{ at } (1, 1, 1)$$

Hence, the rate of outward flow of the fluid per unit volume per unit time at the point $(1, 1, 1)$ is 12. Since $\vec{\nabla} \cdot \vec{v} > 0$, the point $(1, 1, 1)$ is a source.

Example 5.3.9

If the displacement vector $\vec{v}(x, y, z)$ within a deformable elastic solid is given by $\vec{v}(x, y, z) = xy\hat{i} + yz\hat{j} + zx\hat{k}$, then find the dilation at the point $(2, 1, 1)$.

Solution

Dilation is defined by the rate of change of volume per unit volume $= \vec{\nabla} \cdot \vec{v}$. Using Eq. (5.41), we get

$$\vec{\nabla} \cdot \vec{v} = \frac{\partial}{\partial x}(xy) + \frac{\partial}{\partial y}(yz) + \frac{\partial}{\partial z}(zx) = y + z + x = 1 + 1 + 2 = 4$$

Example 5.3.10

Prove that the divergence of a constant vector point function is zero.

Solution

Let us consider a point vector function $\vec{F}(x, y, z) = F_1(x, y, z)\hat{i} + F_2(x, y, z)\hat{j} + F_3(x, y, z)\hat{k}$. Since the vector point function $\vec{F}(x, y, z)$ is constant, F_1, F_2 and F_3 are constants. Hence, $\dfrac{\partial F_1}{\partial x} = 0$, $\dfrac{\partial F_2}{\partial y} = 0$, $\dfrac{\partial F_3}{\partial z} = 0$. So, by Eq. (5.41)

$$\vec{\nabla} \cdot \vec{F} = \frac{\partial F_1}{\partial x} + \frac{\partial F_2}{\partial y} + \frac{\partial F_3}{\partial z} = 0$$

This shows that the divergence of a constant vector point function is zero.

Theorem 5.3

The divergence of differential operators is distributive with respect to sum and difference, i.e.,

$$\text{div } (\vec{F} + \vec{G}) = \text{div } \vec{F} + \text{div } \vec{G} \tag{5.43}$$

Proof: Let $\vec{F} = F_1\hat{i}+F_2\hat{j}+F_3\hat{k}$ and $\vec{G} = G_1\hat{i}+G_2\hat{j}+G_3\hat{k}$ be two given continuously differentiable vector point functions. Then

$$\begin{aligned}
\vec{\nabla}\cdot(\vec{F}+\vec{G}) &= (\frac{\partial}{\partial x}\hat{i}+\frac{\partial}{\partial y}\hat{j}+\frac{\partial}{\partial z}\hat{k})\cdot\left[(F_1+G_1)\hat{i}+(F_2+G_2)\hat{j}+(F_3+G_3)\hat{k}\right]\\
&= \frac{\partial}{\partial x}(F_1+G_1)+\frac{\partial}{\partial y}(F_2+G_2)+\frac{\partial}{\partial z}(F_3+G_3)\\
&= \left(\frac{\partial F_1}{\partial x}+\frac{\partial F_2}{\partial y}+\frac{\partial F_3}{\partial z}\right)+\left(\frac{\partial G_1}{\partial x}+\frac{\partial G_2}{\partial y}+\frac{\partial G_3}{\partial z}\right)\\
&= \left(\frac{\partial}{\partial x}\hat{i}+\frac{\partial}{\partial y}\hat{j}+\frac{\partial}{\partial z}\hat{k}\right)\cdot(F_1\hat{i}+F_2\hat{j}+F_3\hat{k})\\
&\quad+\left(\frac{\partial}{\partial x}\hat{i}+\frac{\partial}{\partial y}\hat{j}+\frac{\partial}{\partial z}\hat{k}\right)\cdot(G_1\hat{i}+G_2\hat{j}+G_3\hat{k}) = \vec{\nabla}\cdot\vec{F}+\vec{\nabla}\cdot\vec{G}
\end{aligned}$$

Hence, the theorem. In general, if $\vec{F}$ and $\vec{G}$ are the vector point functions, then

$$\vec{\nabla}\cdot(a\vec{F}+b\vec{G}) = a\vec{\nabla}\cdot\vec{F}+b\vec{\nabla}\cdot\vec{G} \tag{5.44}$$

where a and b are constant scalars. This property is called the *linearity property of divergence.*

Theorem 5.4

Let $\vec{F}$ be a vector point function and $\phi(x, y, z)$ be a scalar point. Then

$$\vec{\nabla}\cdot(\phi\vec{F}) = \vec{\nabla}\phi\cdot\vec{F}+\phi(\vec{\nabla}\cdot\vec{F}) \tag{5.45}$$

Proof: Let $\vec{F} = F_1\hat{i}+F_2\hat{j}+F_3\hat{k}$ be a vector point function and $\phi(x, y, z)$ be a scalar point. Then, using the definition of divergence

$$\begin{aligned}
\vec{\nabla}\cdot(\phi\vec{F}) &= \vec{\nabla}\cdot(\phi F_1\hat{i}+\phi F_2\hat{j}+\phi F_3\hat{k})\\
&= \frac{\partial}{\partial x}(\phi F_1)+\frac{\partial}{\partial y}(\phi F_2)+\frac{\partial}{\partial z}(\phi F_3)\\
&= \left(\frac{\partial\phi}{\partial x}F_1+\frac{\partial\phi}{\partial y}F_2+\frac{\partial\phi}{\partial z}F_3\right)+\phi\left(\frac{\partial F_1}{\partial x}+\frac{\partial F_2}{\partial y}+\frac{\partial F_3}{\partial z}\right)\\
&= \left(\frac{\partial\phi}{\partial x}\hat{i}+\frac{\partial\phi}{\partial y}\hat{j}+\frac{\partial\phi}{\partial z}\hat{k}\right)\cdot\left(F_1\hat{i}+F_2\hat{j}+F_3\hat{k}\right)\\
&\quad+\phi\left(\frac{\partial}{\partial x}\hat{i}+\frac{\partial}{\partial y}\hat{j}+\frac{\partial}{\partial z}\hat{k}\right)\cdot\left(F_1\hat{i}+F_2\hat{j}+F_3\hat{k}\right)\\
&= (\vec{\nabla}\phi)\cdot\vec{F}+\phi(\vec{\nabla}\cdot\vec{F})
\end{aligned}$$

Hence, the theorem is proved.

Example 5.3.11

Prove that $\nabla\left(\frac{1}{r}\vec{r}\right) = \frac{2}{r}$, where $\vec{r} = x\hat{i}+y\hat{j}+z\hat{k}$ and $r = |\vec{r}|$.

Solution

Using the relation, we find that

$$\vec{\nabla}\left(\frac{1}{r}\right) = \frac{\partial}{\partial x}\left(\frac{1}{r}\right)\hat{i} + \frac{\partial}{\partial y}\left(\frac{1}{r}\right)\hat{j} + \frac{\partial}{\partial z}\left(\frac{1}{r}\right)\hat{k}$$

$$= -\frac{1}{r^2}\left[\frac{\partial r}{\partial x}\hat{i} - \frac{1}{r^2}\frac{\partial r}{\partial y}\hat{j} - \frac{1}{r^2}\frac{\partial r}{\partial z}\hat{k}\right] = -\frac{1}{r^2}\left(\frac{x}{r}\hat{i} + \frac{y}{r}\hat{j} + \frac{z}{r}\hat{k}\right) = -\frac{1}{r^3}\vec{r}$$

Also $\vec{\nabla} \cdot \vec{r} = 3$. We now recall the formula (5.45) and find that

$$\vec{\nabla} \cdot \left(\frac{1}{r}\vec{r}\right) = \frac{1}{r}\vec{\nabla} \cdot \vec{r} + \vec{r} \cdot \vec{\nabla}\left(\frac{1}{r}\right)$$

$$= \frac{1}{r} \cdot 3 + \vec{r}\left(-\frac{1}{r^3}\vec{r}\right) = \frac{3}{r} - \frac{1}{r} = \frac{2}{r}; \quad \text{as } \vec{r} \cdot \vec{r} = r^2$$

This proves the result.

Example 5.3.12

Prove that $\vec{\nabla} \cdot (r^{-3}\vec{r}) = 0$, i.e., $r^{-3}\vec{r}$ is a solenoidal vector.

Solution

Let $\phi = r^{-3}$ and $\vec{F} = \vec{r}$ in the result (5.45). Then

$$\vec{\nabla} \cdot (\phi\vec{F}) = \vec{\nabla}\,(r^{-3}) \cdot \vec{r} + r^{-3}(\vec{\nabla} \cdot \vec{r})$$

$$= \left[\frac{\partial \phi}{\partial x}\hat{i} + \frac{\partial \phi}{\partial y}\hat{j} + \frac{\partial \phi}{\partial z}\hat{k}\right] \cdot (x\hat{i} + y\hat{j} + z\hat{k}) + r^{-3}\left[\vec{\nabla} \cdot (x\hat{i} + y\hat{j} + z\hat{k})\right]$$

$$= \left[x\frac{\partial}{\partial x}(r^{-3}) + y\frac{\partial}{\partial y}(r^{-3}) + \frac{\partial}{\partial z}(r^{-3})\right] + r^{-3}\left[\frac{\partial x}{\partial x} + \frac{\partial y}{\partial y} + \frac{\partial z}{\partial z}\right]$$

$$= -3r^{-4}\left[x\frac{\partial r}{\partial x} + y\frac{\partial r}{\partial y} + z\frac{\partial r}{\partial z}\right] + 3r^{-3}$$

$$= -3r^{-4}\left[x \cdot xr^{-1} + y \cdot yr^{-1} + z \cdot zr^{-1}\right] + 3r^{-3} = 0$$

Theorem 5.5

If $\vec{F}$ and $\vec{G}$ are continuously differentiable vector point functions of position (x, y, z), then

$$\vec{\nabla} \cdot (\vec{F} \times \vec{G}) = (\vec{\nabla} \times \vec{F}) \cdot \vec{G} - (\vec{\nabla} \times \vec{G}) \cdot \vec{F} \tag{5.46}$$

Proof: Using summation conversion, we have

$$\vec{\nabla} \cdot (\vec{F} \times \vec{G}) = \sum \hat{i} \cdot \frac{\partial}{\partial x}(\vec{F} \times \vec{G}) = \sum \hat{i} \cdot \left[\frac{\partial \vec{F}}{\partial x} \times \vec{G} + \vec{F} \times \frac{\partial \vec{G}}{\partial x}\right]$$

$$= \sum \hat{i} \cdot \frac{\partial \vec{F}}{\partial x} \times \vec{G} + \sum \hat{i} \cdot \vec{F} \times \frac{\partial \vec{G}}{\partial x}$$

$$= \sum \hat{i} \times \frac{\partial \vec{F}}{\partial x} \cdot \vec{G} - \sum \hat{i} \times \frac{\partial \vec{G}}{\partial x} \cdot \vec{F}$$

$$= \left(\sum \hat{i} \times \frac{\partial \vec{F}}{\partial x}\right) \cdot \vec{G} - \left(\sum \hat{i} \times \frac{\partial \vec{G}}{\partial x}\right) \cdot \vec{F}$$

$$= (\vec{\nabla} \times \vec{F}) \cdot \vec{G} - (\vec{\nabla} \times \vec{G}) \cdot \vec{F}$$

5.3.1 Invariance of Divergence

Let $\hat{i}, \hat{j}, \hat{k}$ and $\hat{i}', \hat{j}', \hat{k}'$ be two sets of mutually orthogonal unit vectors with common origin 0. Let $\vec{F}$ be a vector point function of x, y, z which is invariant with respect to the rotation of axes, given by Eq. (2.15). Then, $\vec{F}(x,y,z) = \vec{F}'(x',y',z')$. From Eq. (2.15)

$$\frac{\partial \vec{F}}{\partial x} = \frac{\partial \vec{F}'}{\partial x'}\frac{\partial x'}{\partial x} + \frac{\partial \vec{F}'}{\partial y'}\frac{\partial y'}{\partial x} + \frac{\partial \vec{F}'}{\partial z'}\frac{\partial z'}{\partial x}$$

$$\Rightarrow \frac{\partial \vec{F}}{\partial x} = l_1\frac{\partial \vec{F}'}{\partial x'} + l_2\frac{\partial \vec{F}'}{\partial y'} + l_3\frac{\partial \vec{F}'}{\partial z'}$$

Similarly, $\frac{\partial \vec{F}}{\partial y} = m_1\frac{\partial \vec{F}'}{\partial x'} + m_2\frac{\partial \vec{F}'}{\partial y'} + m_3\frac{\partial \vec{F}'}{\partial z'}$ and $\frac{\partial \vec{F}}{\partial z} = n_1\frac{\partial \vec{F}'}{\partial x'} + n_2\frac{\partial \vec{F}'}{\partial y'} + n_3\frac{\partial \vec{F}'}{\partial z'}$. Then, using the relations (2.16) and (2.17), we have

$$\hat{i}\frac{\partial \vec{F}}{\partial x} + \hat{j}\frac{\partial \vec{F}}{\partial y} + \hat{k}\frac{\partial \vec{F}}{\partial z} = \frac{\partial \vec{F}}{\partial x}(l_1\hat{i}' + l_2\hat{j}' + l_3\hat{k}') + \frac{\partial \vec{F}}{\partial y}(m_1\hat{i}' + m_2\hat{j}' + m_3\hat{k}')$$

$$+\frac{\partial \vec{F}}{\partial z}(n_1\hat{i}' + n_2\hat{j}' + n_3\hat{k}')$$

$$= \frac{\partial \vec{F}}{\partial x}(l_1\,\hat{i} + m_1\,\hat{j} + n_1\,\hat{k}) + \frac{\partial \vec{F}}{\partial y}(l_2\,\hat{i} + m_2\,\hat{j} + n_2\,\hat{k})$$

$$+\frac{\partial \vec{F}}{\partial z}(l_3\,\hat{i} + m_2\,\hat{j} + n_3\,\hat{k}) = \hat{i}'\frac{\partial \vec{F}'}{\partial x'} + \hat{j}'\frac{\partial \vec{F}'}{\partial y'} + \hat{k}'\frac{\partial \vec{F}'}{\partial z'}$$

i.e., div$\vec{F}$ = div$\vec{F}'$, which shows that div$\vec{F}$ is invariant with respect to the rotation of axes, given by Eq. (2.15). Therefore, $\vec{\nabla} \cdot \vec{F}$ is independent of the choice of three mutually orthogonal unit vectors.

5.3.2 Divergence in Curvilinear Coordinates

Let $\vec{F} = \vec{F}(u, v, w)$ be a continuously differentiable vector point function of orthogonal curvilinear coordinates u, v and w. Then, we can express $\vec{F}$ as $\vec{F} = F_1\hat{e}_1 + F_2\hat{e}_2 + F_3\hat{e}_3$, where F_1, F_2 and F_3 are the components of $\vec{F}$ along the unit vectors $\hat{e}_1, \hat{e}_2$ and $\hat{e}_3$, respectively. Using $\hat{e}_1 = \hat{e}_2 \times \hat{e}_3 = h_2h_3\left(\vec{\nabla}v \times \vec{\nabla}w\right)$, we get

$$\vec{\nabla} \cdot (F_1\hat{e}_1) = \vec{\nabla} \cdot (F_1\hat{e}_2 \times \hat{e}_3) = \vec{\nabla} \cdot (F_1h_2h_3\vec{\nabla}v \times \vec{\nabla}w)$$

$$= \vec{\nabla} \cdot (F_1h_2h_3) \cdot (\vec{\nabla}v \times \vec{\nabla}w) + (F_1h_2h_3)\vec{\nabla} \cdot (\vec{\nabla}v \times \vec{\nabla}w)$$

$$= \vec{\nabla} \cdot (F_1h_2h_3) \cdot \left(\frac{1}{h_2}\hat{e}_2 \times \frac{1}{h_3}\,\hat{e}_3\right) + F_1h_2h_3\left[\vec{\nabla} \times \vec{\nabla}v \cdot \vec{\nabla}w - \vec{\nabla} \times \vec{\nabla}w \cdot \vec{\nabla}v\right]$$

$$= \vec{\nabla} \cdot (F_1 h_2 h_3) \cdot \left(\frac{1}{h_2\, h_3}\, \hat{e}_1 \right) + \vec{0}, \text{ as } \vec{\nabla} \times \vec{\nabla}\phi = \vec{0}, \hat{e}_1 = \hat{e}_2 \times \hat{e}_3$$

$$= \left[\frac{\partial}{\partial u}(F_1 h_2 h_3)\, \vec{\nabla} u + \frac{\partial}{\partial v}(F_1 h_2 h_3)\, \vec{\nabla} v + \frac{\partial}{\partial w}(F_1 h_2 h_3)\, \vec{\nabla} w \right] \cdot \left(\frac{1}{h_2\, h_3}\, \hat{e}_1 \right)$$

$$= \left[\frac{\partial}{\partial u}(F_1 h_2 h_3) \frac{1}{h_1} \hat{e}_1 + \frac{\partial}{\partial v}(F_1 h_2 h_3) \frac{1}{h_2} \hat{e}_2 + \frac{\partial}{\partial w}(F_1 h_2 h_3) \frac{1}{h_3} \hat{e}_3 \right] \cdot \left(\frac{1}{h_2 h_3} \hat{e}_1 \right)$$

$$= \frac{1}{h_1\, h_2\, h_3} \frac{\partial}{\partial u}(F_1 h_2 h_3)$$

Similarly, $\vec{\nabla} \cdot (F_2 \hat{e}_2) = \dfrac{1}{h_1 h_2 h_3} \dfrac{\partial}{\partial v}(F_2 h_3 h_1)$ and $\vec{\nabla} \cdot (F_3 \hat{e}_3) = \dfrac{1}{h_1 h_2 h_3} \dfrac{\partial}{\partial w}(F_3 h_1 h_2)$. Thus

$$\vec{\nabla} \cdot \vec{F} = \vec{\nabla} \cdot (F_1 \hat{e}_1) + \vec{\nabla} \cdot (F_2 \hat{e}_2) + \vec{\nabla} \cdot (F_3 \hat{e}_3)$$

$$= \frac{1}{h_1 h_2 h_3} \left[\frac{\partial}{\partial u}(F_1 h_2 h_3) + \frac{\partial}{\partial v}(F_2 h_3 h_1) + \frac{\partial}{\partial w}(F_3 h_1 h_2) \right] \tag{5.47}$$

This is the required expression for divergence in an orthogonal curvilinear coordinate system.

Deduction 5.3.2

Divergence in cylindrical coordinates: Here, $u = \rho$, $v = \phi$ and $w = z$; $\hat{e}_1 = \hat{e}_\rho$, $\hat{e}_2 = \hat{e}_\phi$ and $\hat{e}_3 = \hat{e}_z$; $h_1 = h_\rho = 1$, $h_2 = h_\phi = \rho$ and $h_3 = h_z = 1$. Let $\vec{F}$ be a scalar valued function. Therefore, using Eq. (5.47), we get

$$\vec{\nabla} \cdot \vec{F} = \frac{1}{h_1 h_2 h_3} \left[\frac{\partial}{\partial u}(F_1 h_2 h_3) + \frac{\partial}{\partial v}(F_2 h_3 h_1) + \frac{\partial}{\partial w}(F_3 h_1 h_2) \right]$$

$$= \frac{1}{1\ \rho\ 1} \left[\frac{\partial}{\partial \rho}(F_\rho\ \rho\ 1) + \frac{\partial}{\partial \phi}(F_\phi\ 1\ 1) + \frac{\partial}{\partial z}(F_z\ 1\ \rho) \right]$$

$$= \frac{1}{\rho} \left[\frac{\partial}{\partial \rho}(F_\rho\ \rho) + \frac{\partial}{\partial \phi}(F_\phi) + \frac{\partial}{\partial z}(F_z\ \rho) \right] \tag{5.48}$$

Deduction 5.3.3

Divergence in spherical polar coordinates: Here, $u = \rho$, $v = \theta$ and $w = \phi$; $\hat{e}_1 = \hat{e}_\rho$, $\hat{e}_2 = \hat{e}_\theta$ and $\hat{e}_3 = \hat{e}_\phi$; $h_1 = h_\rho = 1$, $h_2 = h_\theta = \rho$ and $h_3 = h_\phi = \rho \sin\theta$. Let $\vec{F}$ be a scalar valued function. Therefore, using Eq. (5.47), we get

$$\vec{\nabla} \cdot \vec{F} = \frac{1}{h_1 h_2 h_3} \left[\frac{\partial}{\partial u}(F_1 h_2 h_3) + \frac{\partial}{\partial v}(F_2 h_3 h_1) + \frac{\partial}{\partial w}(F_3 h_1 h_2) \right]$$

$$= \frac{1}{1\ \rho\ \rho \sin\theta} \left[\frac{\partial}{\partial \rho}(F_\rho\ \rho\ \rho \sin\theta) + \frac{\partial}{\partial \theta}(F_\theta\ \rho \sin\theta\ 1) + \frac{\partial}{\partial \phi}(F_\phi\ 1\ \rho) \right]$$

$$= \frac{1}{\rho^2 \sin\theta} \left[\frac{\partial}{\partial \rho}(F_\rho\ \rho^2 \sin\theta) + \frac{\partial}{\partial \theta}(F_\theta \rho \sin\theta) + \frac{\partial}{\partial \phi}(F_\phi\ \rho) \right] \tag{5.49}$$

5.4 CURL

Let $\vec{A}(x,y,z) = A_1(x,y,z)\hat{i} + A_2(x,y,z)\hat{j} + A_3(x,y,z)\hat{k}$ be a differentiable vector field. Then the curl or rotation of $\vec{A}$, denoted by $\vec{\nabla} \times \vec{A}$ or *curl* $\vec{A}$ or *rot* $\vec{A}$, is defined by the vector quantity

$$\vec{\nabla} \times \vec{A} = \left(\hat{i}\frac{\partial}{\partial x} + \hat{j}\frac{\partial}{\partial y} + \hat{k}\frac{\partial}{\partial z}\right) \times (A_1\hat{i} + A_2\hat{j} + A_3\hat{k})$$

$$= \begin{vmatrix} \hat{i} & \hat{j} & \hat{k} \\ \frac{\partial}{\partial x} & \frac{\partial}{\partial y} & \frac{\partial}{\partial z} \\ A_1 & A_2 & A_3 \end{vmatrix} = \begin{vmatrix} \frac{\partial}{\partial y} & \frac{\partial}{\partial z} \\ A_2 & A_3 \end{vmatrix}\hat{i} - \begin{vmatrix} \frac{\partial}{\partial x} & \frac{\partial}{\partial z} \\ A_1 & A_3 \end{vmatrix}\hat{j} + \begin{vmatrix} \frac{\partial}{\partial x} & \frac{\partial}{\partial y} \\ A_1 & A_2 \end{vmatrix}\hat{k}$$

$$= \left(\frac{\partial A_3}{\partial y} - \frac{\partial A_2}{\partial z}\right)\hat{i} + \left(\frac{\partial A_1}{\partial z} - \frac{\partial A_3}{\partial x}\right)\hat{j} + \left(\frac{\partial A_2}{\partial x} - \frac{\partial A_1}{\partial y}\right)\hat{k}. \tag{5.50}$$

Example 5.4.1

If $\vec{A}(x,y,z) = xz^3\hat{i} - 2x^2yz\hat{j} + 2yz^4\hat{k}$, find curl $\vec{A}$ (or $\vec{\nabla} \times \vec{A}$) at the point $(1, -1, 1)$.

Solution

Using Eq. (5.50) as the definition of curl, we get

$$\vec{\nabla} \times \vec{A} = (\hat{i}\frac{\partial}{\partial x} + \hat{j}\frac{\partial}{\partial y} + \hat{k}\frac{\partial}{\partial z}) \times (xz^3\hat{i} - 2x^2yz\hat{j} + 2yz^4\hat{k})$$

$$= \begin{vmatrix} \hat{i} & \hat{j} & \hat{k} \\ \frac{\partial}{\partial x} & \frac{\partial}{\partial y} & \frac{\partial}{\partial z} \\ xz^3 & -2x^2yz & 2yz^4 \end{vmatrix}$$

$$= \left[\frac{\partial}{\partial y}(2xz^4) - \frac{\partial}{\partial z}(-2x^2yz)\right]\hat{i} + \left[\frac{\partial}{\partial z}(xz^3) - \frac{\partial}{\partial x}(2yz^4)\right]\hat{j}$$

$$+ \left[\frac{\partial}{\partial x}(-2x^2yz) - \frac{\partial}{\partial y}(xz^3)\right]\hat{k} = (2z^4 + 2x^2y)\hat{i} + 3xz^2\hat{j} - 4xyz\hat{k}$$

$$= 3\hat{j} + 4\hat{k} \text{ at } (1, -1, 1)$$

Example 5.4.2

If $\vec{A} = x^2y\hat{i} - 2xz\hat{j} + 2yz\hat{k}$, find curl $\vec{A}$ or $\vec{\nabla} \times (\vec{\nabla} \times \vec{A})$ at $(1, -1, 1)$.

Solution

Using Eq. (5.50) as the definition of curl, we get

$$\vec{\nabla} \times \vec{A} = \begin{vmatrix} \hat{i} & \hat{j} & \hat{k} \\ \frac{\partial}{\partial x} & \frac{\partial}{\partial y} & \frac{\partial}{\partial z} \\ x^2y & -2xz & 2yz \end{vmatrix}$$

$$= (2z + 2x)\hat{i} - (2z + x^2)\hat{k}$$

Again, using Eq. (5.50) as the definition of curl, we get

$$\vec{\nabla} \times (\vec{\nabla} \times \vec{A}) = (\hat{i}\frac{\partial}{\partial x} + \hat{j}\frac{\partial}{\partial y} + \hat{k}\frac{\partial}{\partial z}) \times \left[(2z + 2x)\hat{i} - (2z + x^2)\hat{k}\right]$$

$$= \left[\frac{\partial}{\partial y}(-x^2 - 2z) - \frac{\partial}{\partial z}(0)\right]\hat{i} + \left[\frac{\partial}{\partial z}(2x + 2z) - \frac{\partial}{\partial x}(-x^2 - 2z)\right]\hat{j}$$

$$+\left[\frac{\partial}{\partial x}(0) - \frac{\partial}{\partial y}(2x + 2z)\right]\hat{k} = (2 + 2x)\hat{j} = 4\hat{j} \text{ at } (1, -1, 1)$$

Example 5.4.3

Prove that $\vec{\nabla} \times (\vec{\nabla}\phi) = \vec{0}$, i.e., curl grad $\phi = \vec{0}$.

Solution

The gradient of ϕ is given by Eq. (5.1). Therefore, using Eq. (5.50), the curl of the gradient of ϕ is

$$\vec{\nabla} \times (\vec{\nabla}\phi) = (\hat{i}\frac{\partial}{\partial x} + \hat{j}\frac{\partial}{\partial y} + \hat{k}\frac{\partial}{\partial z}) \times \left[\hat{i}\frac{\partial \phi}{\partial x} + \hat{j}\frac{\partial \phi}{\partial y} + \hat{k}\frac{\partial \phi}{\partial z}\right]$$

$$= \begin{vmatrix} \hat{i} & \hat{j} & \hat{k} \\ \dfrac{\partial}{\partial x} & \dfrac{\partial}{\partial y} & \dfrac{\partial}{\partial z} \\ \dfrac{\partial \phi}{\partial x} & \dfrac{\partial \phi}{\partial y} & \dfrac{\partial \phi}{\partial z} \end{vmatrix}$$

$$= \left[\frac{\partial}{\partial y}\left(\frac{\partial \phi}{\partial z}\right) - \frac{\partial}{\partial z}\left(\frac{\partial \phi}{\partial y}\right)\right]\hat{i}$$

$$+\left[\frac{\partial}{\partial z}\left(\frac{\partial \phi}{\partial x}\right) - \frac{\partial}{\partial x}\left(\frac{\partial \phi}{\partial z}\right)\right]\hat{j} + \left[\frac{\partial}{\partial x}\left(\frac{\partial \phi}{\partial y}\right) - \frac{\partial}{\partial y}\left(\frac{\partial \phi}{\partial x}\right)\right]\hat{k}$$

$$= \left[\frac{\partial^2 \phi}{\partial y \partial z} - \frac{\partial^2 \phi}{\partial z \partial y}\right]\hat{i} + \left[\frac{\partial^2 \phi}{\partial z \partial x} - \frac{\partial^2 \phi}{\partial x \partial z}\right]\hat{j} + \left[\frac{\partial^2 \phi}{\partial x \partial y} - \frac{\partial^2 \phi}{\partial y \partial x}\right]\hat{k} = \vec{0}$$

provided we assume that ϕ has continuous second partial derivatives so that the order of differentiation is immaterial.

Example 5.4.4

Verify the formula $\vec{\nabla} \times (\vec{\nabla}\phi) = \vec{0}$ for the function $\phi(x, y, z) = x^2yz + z^2$.

Solution

We have $\phi(x, y, z) = 2x^2yz + z^2$, $\dfrac{\partial \phi}{\partial x} = 2xyz$, $\dfrac{\partial \phi}{\partial y} = x^2z$ and $\dfrac{\partial \phi}{\partial x} = x^2y + 2z$. Then, using the above relations, we find that

$$\vec{\nabla}\phi = \frac{\partial \phi}{\partial x}\hat{i} + \frac{\partial \phi}{\partial y}\hat{j} + \frac{\partial \phi}{\partial z}\hat{k}$$

$$= 2xy\hat{i} + x^2z\hat{j} + (x^2y + 2z)\hat{k}$$

Hence, we obtain

$$\vec{\nabla} \times (\vec{\nabla}\phi) = \begin{vmatrix} \hat{i} & \hat{j} & \hat{k} \\ \dfrac{\partial}{\partial x} & \dfrac{\partial}{\partial y} & \dfrac{\partial}{\partial z} \\ 2xyz & x^2z & (x^2y + 2z) \end{vmatrix}$$

$$= (x^2 - x^2)\hat{i} + (2xy - 2xy)\hat{j} + (2xy - 2xy)\hat{k} = \vec{0}$$

Example 5.4.5

Prove that $\vec{\nabla} \cdot (\vec{\nabla} \times \vec{A}) = 0$, i.e., div curl $\vec{A} = 0$.

Solution

The curl of $\vec{A}$ is given by Eq. (5.50). Therefore, using Eq. (5.41), the divergence of curl of $\vec{A}$ is

$$\vec{\nabla} \cdot (\vec{\nabla} \times \vec{A}) = \vec{\nabla} \cdot \left[\left(\frac{\partial A_3}{\partial y} - \frac{\partial A_2}{\partial z}\right)\hat{i} + \left(\frac{\partial A_1}{\partial z} - \frac{\partial A_3}{\partial x}\right)\hat{j} + \left(\frac{\partial A_2}{\partial x} - \frac{\partial A_1}{\partial y}\right)\hat{k} \right]$$

$$= \frac{\partial}{\partial x}\left(\frac{\partial A_3}{\partial y} - \frac{\partial A_2}{\partial z}\right) + \frac{\partial}{\partial y}\left(\frac{\partial A_1}{\partial z} - \frac{\partial A_3}{\partial x}\right) + \frac{\partial}{\partial z}\left(\frac{\partial A_2}{\partial x} - \frac{\partial A_1}{\partial y}\right)$$

$$= \frac{\partial^2 A_3}{\partial x \partial y} - \frac{\partial^2 A_2}{\partial x \partial z} + \frac{\partial^2 A_1}{\partial y \partial z} - \frac{\partial A_3}{\partial y \partial x} + \frac{\partial^2 A_2}{\partial z \partial x} - \frac{\partial^2 A_1}{\partial z \partial y} = 0$$

assuming that $\vec{A}$ has continuous second order partial derivatives.

Example 5.4.6

Verify that $\vec{\nabla} \times \vec{\nabla} r^n = \vec{0}$.

Solution

Using definition (5.1), the gradient of ϕ is given by

$$\vec{\nabla} r^n = \frac{\partial r^n}{\partial x}\hat{i} + \frac{\partial r^n}{\partial y}\hat{j} + \frac{\partial r^n}{\partial z}\hat{k} = \sum \frac{n}{2}\left(x^2 + y^2 + z^2\right)^{\frac{n}{2}-1} 2x\hat{i}$$

$$= nr^{n-2}\left(x\hat{i} + y\hat{j} + z\hat{k}\right) = nr^{n-2}\vec{r}$$

The curl of a vector is given by Eq. (5.50). Therefore

$$\vec{\nabla} \times \vec{\nabla} r^n = \begin{vmatrix} \hat{i} & \hat{j} & \hat{k} \\ \dfrac{\partial}{\partial x} & \dfrac{\partial}{\partial y} & \dfrac{\partial}{\partial z} \\ nr^{n-2}x & nr^{n-2}y & nr^{n-2}z \end{vmatrix}$$

$$= \sum \left[\frac{\partial}{\partial y}(nr^{n-2}z) - \frac{\partial}{\partial z}(nr^{n-2}y) \right]\hat{i}$$

$$= \sum \left[n(n-2)r^{n-3}\frac{\partial r}{\partial y}z - n(n-2)r^{n-3}\frac{\partial r}{\partial z}y \right]\hat{i}$$

$$= \sum \left[n(n-2)r^{n-3}\frac{yz}{r} - n(n-2)r^{n-3}\frac{zy}{r} \right]\hat{i} = \vec{0}$$

DEDUCTION 5.4.1

Physical concept: Let $\vec{\omega}=\omega_1\hat{i}+\omega_2\hat{j}+\omega_3\hat{k}$ be the angular velocity of a rigid body moving about a fixed point. Then, the linear velocity $\vec{v}$ of any particle of the body with position vector $\vec{r}$ relative to a fixed point is $\vec{\omega}\times\vec{r}$. Now

$$\vec{\nabla}\times\vec{v}=\vec{\nabla}\times(\vec{\omega}\times\vec{r})=\vec{\nabla}\times\begin{vmatrix}\hat{i} & \hat{j} & \hat{k}\\ \omega_1 & \omega_2 & \omega_3\\ x & y & z\end{vmatrix}$$

$$=\vec{\nabla}\times\left[(\omega_2 z-\omega_3 y)\hat{i}+(\omega_3 x-\omega_1 z)\hat{j}+(\omega_1 y-\omega_2 x)\hat{k}\right]$$

$$=\begin{vmatrix}\hat{i} & \hat{j} & \hat{k}\\ \dfrac{\partial}{\partial x} & \dfrac{\partial}{\partial y} & \dfrac{\partial}{\partial z}\\ \omega_2 z-\omega_3 y & \omega_3 x-\omega_1 z & \omega_1 y-\omega_2 x\end{vmatrix}$$

$$=2(\omega_1\hat{i}+\omega_2\hat{j}+\omega_3\hat{k})=2\vec{\omega}$$

i.e., the curl of the velocity of any particle of a rigid body is equal to twice the angular velocity of the body. This result indicates that the *curl* of a vector field is related to the rotational properties of the field. If the field $\vec{F}$ is one caused by a moving fluid, then a paddle wheel placed at different points in the field would tend to rotate in those regions where $\vec{\nabla}\times\vec{F}\neq\vec{0}$. But at those points where $\vec{\nabla}\times\vec{F}=\vec{0}$, there would be no rotation. The field $\vec{F}$ is then called *irrotational.* A field which is not irrotational is often called a *vortex field.*

Example 5.4.7

Find the constants a, b and c so that the vector $\vec{F}=(x+2y+az)\hat{i}+(bx-3y-z)\hat{j}+(4x+cy+2z)\hat{k}$ is irrotational. Hence, find the potential function V.

Solution

A vector $\vec{F}$ is called irrotational if $\vec{\nabla}\times\vec{F}=\vec{0}$. Now

$$\vec{\nabla}\times\vec{F}=\begin{vmatrix}\hat{i} & \hat{j} & \hat{k}\\ \dfrac{\partial}{\partial x} & \dfrac{\partial}{\partial y} & \dfrac{\partial}{\partial z}\\ x+2y+az & bx-3y-z & 4x+cy+2z\end{vmatrix}=\vec{0}$$

$$\Rightarrow (c+1)\hat{i}+(a-4)\hat{j}+(b-2)\hat{k}=\vec{0}$$

$$\Rightarrow a=4, b=2, c=-1$$

For this value of a, b and c, $\vec{F}=(x+2y+4z)\hat{i}+(2x-3y-z)\hat{j}+(4x-y+2z)\hat{k}$. Thus, there exists a scalar potential function V, such that $\vec{F}=-\vec{\nabla}V$, i.e.

$$(x+2y+4z)\hat{i}+(2x-3y-z)\hat{j}+(4x-y+2z)\hat{k}=-\left[\frac{\partial V}{\partial x}\hat{i}+\frac{\partial V}{\partial y}\hat{j}+\frac{\partial V}{\partial z}\hat{k}\right]$$

$$\Rightarrow \frac{\partial V}{\partial x}=x+2y+4z; \qquad V=\frac{x^2}{2}+2xy+4zx+\psi_1(y,z)$$

$$\frac{\partial V}{\partial y} = 2x - 3y - z; \qquad V = 2xy - \frac{3}{2}y^2 - zy + \psi_2(x, z)$$

$$\frac{\partial V}{\partial z} = 4x - y + 2z; \qquad V = 4zx - yz + z^2 + \psi_3(x, y)$$

$\psi_1(y, z), \psi_2(x, z)$ and $\psi_3(x, y)$ are arbitrary functions with respect to the variables. Thus, the scalar potential function is $V(x, y, z) = \frac{x^2}{2} - \frac{3}{2}y^2 + z^2 + 2xy + 4xz - yz + d$, where d, a constant, is independent of x, y and z.

Example 5.4.8

Prove that the vector $f(r)\vec{r}$, where $f(r)$ is differentiable, is irrotational.

Solution

We know that $\frac{\partial r}{\partial x} = \frac{x}{r}, \frac{\partial r}{\partial y} = \frac{y}{r}, \frac{\partial r}{\partial z} = \frac{z}{r}$. Using Eq. (5.50), we get

$$\begin{aligned}
\vec{\nabla} \times \left[f(r)\vec{r}\right] &= \vec{\nabla} \times \left[f(r)\,(x\hat{i} + y\hat{j} + z\hat{k})\right] \\
&= \begin{vmatrix} \hat{i} & \hat{j} & \hat{k} \\ \frac{\partial}{\partial x} & \frac{\partial}{\partial y} & \frac{\partial}{\partial z} \\ xf(r) & yf(r) & zf(r) \end{vmatrix} \\
&= \left[\frac{\partial}{\partial y}\{zf(r)\} - \frac{\partial}{\partial z}\{yf(r)\}\right]\hat{i} + \left[\frac{\partial}{\partial z}\{xf(r)\} - \frac{\partial}{\partial x}\{zf(r)\}\right]\hat{j} \\
&\quad + \left[\frac{\partial}{\partial x}\{yf(r)\} - \frac{\partial}{\partial y}\{xf(r)\}\right]\hat{k} \\
&= \left[zf'(r)\frac{\partial r}{\partial y} - yf'(r)\frac{\partial r}{\partial z}\right]\hat{i} + \left[xf'(r)\frac{\partial r}{\partial z} - zf'(r)\frac{\partial r}{\partial x}\right]\hat{j} \\
&\quad + \left[yf'(r)\frac{\partial r}{\partial x} - zf'(r)\frac{\partial r}{\partial y}\right]\hat{k} \\
&= f'(r)\left[\frac{zy}{r} - \frac{yz}{r}\right]\hat{i} + f'(r)\left[\frac{xz}{r} - \frac{zx}{r}\right]\hat{j} + f'(r)\left[\frac{yx}{r} - \frac{xy}{r}\right]\hat{k} = \vec{0}
\end{aligned}$$

Hence, the vector $f(r)\vec{r}$ is irrotational.

Theorem 5.6

If $\vec{A}$ is a differentiable vector function and ϕ is a differentiable scalar function of position, then

$$\vec{\nabla} \times (\phi\vec{A}) = (\vec{\nabla}\phi) \times \vec{A} + \phi(\vec{\nabla} \times \vec{A}) \tag{5.51}$$

Proof: Let $\vec{A}(x, y, z) = A_1\hat{i} + A_2\hat{j} + A_3\hat{k}$ be a differentiable vector function. Then

$$\vec{\nabla} \times (\phi\vec{A}) = \vec{\nabla} \times (\phi A_1\hat{i} + \phi A_2\hat{j} + \phi A_3\hat{k}) = \begin{vmatrix} \hat{i} & \hat{j} & \hat{k} \\ \frac{\partial}{\partial x} & \frac{\partial}{\partial y} & \frac{\partial}{\partial z} \\ \phi A_1 & \phi A_2 & \phi A_3 \end{vmatrix}$$

$$\therefore \vec{\nabla} \times (\phi\vec{A}) = \left[\frac{\partial}{\partial y}(\phi A_3) - \frac{\partial}{\partial z}(\phi A_2)\right]\hat{i} + \left[\frac{\partial}{\partial z}(\phi A_1) - \frac{\partial}{\partial x}(\phi A_3)\right]\hat{j}$$
$$+ \left[\frac{\partial}{\partial x}(\phi A_2) - \frac{\partial}{\partial y}(\phi A_1)\right]\hat{k}$$
$$= \phi\left[\left(\frac{\partial A_3}{\partial y} - \frac{\partial A_2}{\partial z}\right)\hat{i} + \left(\frac{\partial A_1}{\partial z} - \frac{\partial A_3}{\partial x}\right)\hat{j} + \left(\frac{\partial A_2}{\partial x} - \frac{\partial A_1}{\partial y}\right)\hat{k}\right]$$
$$+ \left[\left(\frac{\partial \phi}{\partial y}A_3 - \frac{\partial \phi}{\partial z}A_2\right)\hat{i} + \left(\frac{\partial \phi}{\partial z}A_1 - \frac{\partial \phi}{\partial x}A_3\right)\hat{j} + \left(\frac{\partial \phi}{\partial x}A_2 - \frac{\partial \phi}{\partial y}A_1\right)\hat{k}\right]$$
$$= \phi(\vec{\nabla} \times \vec{A}) + \begin{vmatrix} \hat{i} & \hat{j} & \hat{k} \\ \frac{\partial}{\partial x} & \frac{\partial}{\partial y} & \frac{\partial}{\partial z} \\ A_1 & A_2 & A_3 \end{vmatrix} = \phi(\vec{\nabla} \times \vec{A}) + (\vec{\nabla}\phi) \times \vec{A}$$

Example 5.4.9

Prove that $\vec{\nabla}(f(r)\,\vec{r}) = \vec{0}$.

Solution

Let $\phi = f(r)$ and $\vec{A} = \vec{r}$. Then, using Eq. (5.51), we get

$$\vec{\nabla} \times (f(r))\vec{r} = \vec{\nabla} f(r) \times \vec{r} + f(r)\vec{\nabla} \times \vec{r} = \vec{\nabla} f(r) \times \vec{r}; \qquad \text{as } \vec{\nabla} \times \vec{r} = \vec{0}$$
$$= \left\{\frac{\partial}{\partial x} f(r)\hat{i} + \frac{\partial}{\partial y} f(r)\hat{j} + \frac{\partial}{\partial z} f(r)\hat{k}\right\} \times \vec{r}$$
$$= \{f'(r)\frac{x}{r}\hat{i} + f'(r)\frac{y}{r}\hat{j} + f'(r)\frac{z}{r}\hat{k}\} \times \vec{r}$$
$$= \frac{f'(r)}{r}(x\hat{i} + y\hat{j} + z\hat{k}) \times \vec{r} = \frac{f'(r)}{r}\vec{r} \times \vec{r} = \vec{0}$$

Example 5.4.10

Show that $\vec{\nabla} \times \left[(\vec{a}\cdot\vec{r})\vec{a}\right] = \vec{0}$, where $\vec{a}$ is a constant vector.

Solution

From the identity (5.51), for a continuous differentiable scalar function ϕ and vector function $\vec{F}$, we have

$$\vec{\nabla} \times \left[(\vec{a}\cdot\vec{r})\vec{a}\right] = \vec{\nabla}(\vec{a}\cdot\vec{r}) \times \vec{a} + (\vec{a}\cdot\vec{r})(\vec{\nabla} \times \vec{a})$$
$$= \left[\sum \hat{i}\frac{\partial}{\partial x}\right](\vec{a}\cdot\vec{r}) \times \vec{a} + \vec{0}, \text{ as } \vec{a} \text{ is constant}$$
$$= \left[\sum \hat{i}\left(\frac{\partial \vec{a}}{\partial x}\cdot\vec{r} + \vec{a}\cdot\frac{\partial \vec{r}}{\partial x}\right)\right] \times \vec{a}$$
$$= \left[\sum \hat{i}(\vec{a}\cdot\hat{i})\right] \times \vec{a}\ ;\ \vec{a} = \text{constant}, \frac{\partial \vec{a}}{\partial x} = \vec{0}$$
$$= \left[(\vec{a}\cdot\hat{i})\hat{i} + (\vec{a}\cdot\hat{j})\hat{j} + (\vec{a}\cdot\hat{k})\hat{k}\right] \times \vec{a} = \vec{a} \times \vec{a} = \vec{0}$$

Theorem 5.7

If $\vec{A}$ and $\vec{B}$ be differentiable vector functions, then

$$\vec{\nabla} \times (\vec{A} \times \vec{B}) = (\vec{\nabla} \cdot \vec{B})\vec{A} - (\vec{\nabla} \cdot \vec{A})\vec{B} + (\vec{B} \cdot \vec{\nabla})\vec{A} - (\vec{A} \cdot \vec{\nabla})\vec{B} \tag{5.52}$$

Proof: Here, $\vec{A}$ and $\vec{B}$ are two differentiable vector functions. Then

$$\begin{aligned}
\vec{\nabla} \times (\vec{A} \times \vec{B}) &= \sum \hat{i} \times \frac{\partial}{\partial x}(\vec{A} \times \vec{B}) = \sum \hat{i} \times \left[\frac{\partial \vec{A}}{\partial x} \times \vec{B} + \vec{A} \times \frac{\partial \vec{B}}{\partial x}\right] \\
&= \sum \hat{i} \times \left(\vec{A} \times \frac{\partial \vec{B}}{\partial x}\right) + \sum \hat{i} \times \left(\frac{\partial \vec{A}}{\partial x} \times \vec{B}\right) \\
&= \sum \left[\left(\hat{i} \cdot \frac{\partial \vec{B}}{\partial x}\right)\vec{A} - \left(\hat{i} \cdot \vec{A}\right)\frac{\partial \vec{B}}{\partial x}\right] + \sum \left[\left(\hat{i} \cdot \vec{B}\right)\frac{\partial \vec{A}}{\partial x} - \left(\hat{i} \cdot \frac{\partial \vec{A}}{\partial x}\right)\vec{B}\right] \\
&= \left(\sum \hat{i} \cdot \frac{\partial \vec{B}}{\partial x}\right)\vec{A} - \left(\sum \hat{i} \cdot \frac{\partial \vec{A}}{\partial x}\right)\vec{B} + \sum \left(\hat{i} \cdot \vec{B}\right)\frac{\partial \vec{A}}{\partial x} - \left(\hat{i} \cdot \vec{A}\right)\frac{\partial \vec{B}}{\partial x} \\
&= \left(\sum \hat{i} \cdot \frac{\partial \vec{B}}{\partial x}\right)\vec{A} - \left(\sum \hat{i} \cdot \frac{\partial \vec{A}}{\partial x}\right)\vec{B} + \sum \left(\vec{B} \cdot \hat{i}\right)\frac{\partial \vec{A}}{\partial x} - \left(\vec{A} \cdot \hat{i}\right)\frac{\partial \vec{B}}{\partial x} \\
&= (\vec{\nabla} \cdot \vec{B})\vec{A} - (\vec{\nabla} \cdot \vec{A})\vec{B} + (\vec{B} \cdot \vec{\nabla})\vec{A} - (\vec{A} \cdot \vec{\nabla})\vec{B}
\end{aligned}$$

DEDUCTION 5.4.2

If $\vec{\nabla} \times \vec{F} = \vec{0}$, then we can find ϕ so that $\vec{F} = \vec{\nabla}\phi$. A vector field $\vec{F}$ which can be derived from a scalar field ϕ so that $\vec{F} = \vec{\nabla}\phi$, is called a *conservative vector field* and ϕ is called the *scalar potential.*

Example 5.4.11

Prove that $\vec{F} = (2xy + z^3)\hat{i} + x^2\hat{j} + 3xz^2\hat{k}$ is a conservative force field. Find the scalar potential function V. Find also the work done in moving an object in this field from $(1, -2, 1)$ to $(3, 1, 4)$.

Solution

For the given force field $\vec{F}$, we have

$$\vec{\nabla} \times \vec{F} = \begin{vmatrix} \hat{i} & \hat{j} & \hat{k} \\ \dfrac{\partial}{\partial x} & \dfrac{\partial}{\partial y} & \dfrac{\partial}{\partial z} \\ 2xy + z^3 & x^2 & 3xz^2 \end{vmatrix}$$

$$= (0 - 0)\hat{i} - (3z^2 - 3z^2)\hat{j} + (2x - 2x)\hat{k} = \vec{0}$$

Thus, the given force field is conservative. There exists a scalar potential function V, such that $\vec{F} = -\vec{\nabla}V$, i.e.

$$(2xy + z^3)\hat{i} + x^2\hat{j} + 3xz^2\hat{k} = -\left[\frac{\partial V}{\partial x}\hat{i} + \frac{\partial V}{\partial y}\hat{j} + \frac{\partial V}{\partial z}\hat{k}\right]$$

$$\Rightarrow \frac{\partial V}{\partial x} = -2xy - z^3; \quad V = -x^2y - xz^3 + \psi_1(y, z)$$

$$\frac{\partial V}{\partial y} = -x^2; \qquad V = -x^2y + \psi_2(x, z)$$

$$\frac{\partial V}{\partial z} = -3xz^2; \qquad V = -xz^3 + \psi_3(x,y)$$

Thus, the scalar potential function is $V(x,y,z) = -x^2y - xz^3 + c$, where c is independent of x, y and z. Now, the work done in moving an object in this field from $A(1,-2,1)$ to $B(3,1,4)$ is $W_{AB} = V_A - V_B =$202 units.

Theorem 5.8

Exact differential form: The necessary and sufficient condition that $F_1\,dx + F_2\,dy + F_3\,dz$ is an exact differential is that $\vec{\nabla} \times \vec{F} = \vec{0}$, where $\vec{F} = F_1\hat{i} + F_2\hat{j} + F_3\hat{k}$.

Proof: First, let $F_1\,dx + F_2\,dy + F_3\,dz$ be an exact differential $d\phi$, then

$$F_1\,dx + F_2\,dy + F_3\,dz = \frac{\partial\phi}{\partial x}\,dx + \frac{\partial\phi}{\partial y}\,dy + \frac{\partial\phi}{\partial z}\,dz$$

$$\Rightarrow F_1 = \frac{\partial\phi}{\partial x},\quad F_2 = \frac{\partial\phi}{\partial y},\quad F_3 = \frac{\partial\phi}{\partial z}$$

$$\Rightarrow \vec{F} = F_1\hat{i} + F_2\hat{j} + F_3\hat{k} = \frac{\partial\phi}{\partial x}\hat{i} + \frac{\partial\phi}{\partial y}\hat{j} + \frac{\partial\phi}{\partial z}\hat{k} = \vec{\nabla}\phi$$

Therefore, $\vec{\nabla} \times \vec{F} = \vec{0}$. Conversely, let $\vec{\nabla} \times \vec{F} = \vec{0}$, then there exists a scalar function $\phi(x,y,z)$ such that $\vec{F} = \vec{\nabla}\phi$. The given differential $F_1\,dx + F_2\,dy + F_3\,dz$ can be written as

$$\begin{aligned} F_1\,dx + F_2\,dy + F_3\,dz &= (F_1\hat{i} + F_2\hat{j} + F_3\hat{k})\cdot(dx\hat{i} + dy\hat{j} + dz\hat{k}) \\ &= \vec{F}\cdot d\vec{r};\quad \vec{r} = x\hat{i} + y\hat{j} + z\hat{k} \\ &= \vec{\nabla}\phi\cdot d\vec{r} = \frac{\partial\phi}{\partial x}\,dx + \frac{\partial\phi}{\partial y}\,dy + \frac{\partial\phi}{\partial z}\,dz = d\phi \end{aligned}$$

Hence, the theorem. The solution of $F_1\,dx + F_2\,dy + F_3\,dz = 0$ is $\phi(x,y,z) = c$.

Example 5.4.12

Show that $(y^2 + z^3)\,dx + (2xy - 5z)\,dy + (3xz^2 - 5y)\,dz$ is an exact differential. Hence, solve the differential equation $(y^2 + z^3)\,dx + (2xy - 5z)\,dy + (3xz^2 - 5y)\,dz = 0$.

Solution

The given differential can be written as $\vec{F}\cdot d\vec{r}$, where $\vec{F} = (y^2+z^3)\,\hat{i} + (2xy-5z)\,\hat{j} + (3xz^2-5y)\,\hat{k}$ and $\vec{r} = x\hat{i} + y\hat{j} + z\hat{k}$. Now

$$\vec{\nabla}\times\vec{F} = \begin{vmatrix} \hat{i} & \hat{j} & \hat{k} \\ \dfrac{\partial}{\partial x} & \dfrac{\partial}{\partial y} & \dfrac{\partial}{\partial z} \\ y^2 + z^3 & 2xy - 5z & 3xz^2 - 5y \end{vmatrix}$$

$$= (-5+5)\hat{i} + (3z^2 - 3z^2)\hat{j} + (2y - 2y)\hat{k} = \vec{0}$$

Thus, the given differential is an exact one. So there exists a scalar function $\phi(x,y,z)$ such that

$$\vec{F} = \vec{\nabla}\phi = \frac{\partial\phi}{\partial x}\hat{i} + \frac{\partial\phi}{\partial y}\hat{j} + \frac{\partial\phi}{\partial z}\hat{k}$$

$$\frac{\partial \phi}{\partial x} = y^2 + z^3 \Longrightarrow \phi(x, y, z) = xy^2 + xz^3 + \psi_1(y, z)$$
$$\frac{\partial \phi}{\partial y} = 2xy - 5z \Longrightarrow \phi(x, y, z) = xy^2 - 5yz + \psi_2(x, z)$$
$$\frac{\partial \phi}{\partial z} = 3xz^2 - 5y \Longrightarrow \phi(x, y, z) = xz^3 - 5yz + \psi_3(x, y)$$

Let us take $\psi_1(y, z) = -5yz + c, \psi_2(x, z) = xz^3 + c$ and $\psi_3(x, y) = xy^2 + c$. Then, we get $\phi(x, y, z) = xy^2 - 5yz + xz^3 + c$, where c is an arbitrary constant. The solution is $xy^2 - 5yz + xz^3 = c_1$.

5.4.1 Invariance of Curl

Let $\vec{F}$ be a vector point function of x, y and z which is invariant with respect to the rotation of the axes, given by Eq. (2.15). Then, $\vec{F}(x, y, z) = \vec{F}'(x', y', z')$. From Eq. (2.15)

$$\frac{\partial \vec{F}}{\partial x} = \frac{\partial \vec{F}'}{\partial x'}\frac{\partial x'}{\partial x} + \frac{\partial \vec{F}'}{\partial y'}\frac{\partial y'}{\partial x} + \frac{\partial \vec{F}'}{\partial z'}\frac{\partial z'}{\partial x}$$
$$\Rightarrow \frac{\partial \vec{F}}{\partial x} = l_1\frac{\partial \vec{F}'}{\partial x'} + l_2\frac{\partial \vec{F}'}{\partial y'} + l_3\frac{\partial \vec{F}'}{\partial z'}$$

Similarly, $\frac{\partial \vec{F}}{\partial y} = m_1\frac{\partial \vec{F}'}{\partial x'} + m_2\frac{\partial \vec{F}'}{\partial y'} + m_3\frac{\partial \vec{F}'}{\partial z'}$ and $\frac{\partial \vec{F}}{\partial z} = n_1\frac{\partial \vec{F}'}{\partial x'} + n_2\frac{\partial \vec{F}'}{\partial y'} + n_3\frac{\partial \vec{F}'}{\partial z'}$. Then, using the relations (2.16) and (2.17), we have

$$\hat{i} \times \frac{\partial \vec{F}}{\partial x} + \hat{j} \times \frac{\partial \vec{F}}{\partial y} + \hat{k} \times \frac{\partial \vec{F}}{\partial z}$$
$$= (\hat{i}l_1 + \hat{j}m_1 + \hat{k}n_1) \times \frac{\partial \vec{F}'}{\partial x'} + (\hat{i}l_2 + \hat{j}m_2 + \hat{k}n_2) \times \frac{\partial \vec{F}'}{\partial y'}$$
$$+(\hat{i}l_3 + \hat{j}m_3 + \hat{k}n_3) \times \frac{\partial \vec{F}'}{\partial z'}$$
$$= \hat{i}' \times \frac{\partial \vec{F}'}{\partial x'} + \hat{j}' \times \frac{\partial \vec{F}'}{\partial y'} + \hat{k}' \times \frac{\partial \vec{F}'}{\partial z'}$$

i.e., curl $\vec{F}$ = curl $\vec{F}'$, which shows that curl $\vec{F}$ is invariant with respect to the rotation of axes, given by Eq. (2.15). Thus, $\vec{\nabla} \times \vec{F}$ is independent of choice of three mutually orthogonal unit vectors, and therefore, the curl of a vector point function is a point function.

5.4.2 Curl in Curvilinear Coordinates

For any continuously differentiable vector point function $\vec{F}$ of orthogonal cuvilinear coordinates u, v and w, $\vec{F} = F_1\hat{e}_1 + F_2\hat{e}_2 + F_3\hat{e}_3$, where F_1, F_2 and F_3 are the components of $\vec{F}$ along the unit vectors $\hat{e}_1, \hat{e}_2$ and $\hat{e}_3$, respectively. Now

$$\vec{\nabla} \times (F_1\hat{e}_1) = \vec{\nabla} \times (h_1\, F_1\vec{\nabla}u) = \vec{\nabla}(h_1\, F_1) \times \vec{\nabla}u + h_1\, F_1\vec{\nabla} \times \vec{\nabla}u$$
$$= \vec{\nabla}(h_1\, F_1) \times \vec{\nabla}u + \vec{0}, \text{ as } \vec{\nabla} \times \vec{\nabla}\phi = \vec{0}$$

$$= \left[\frac{\partial}{\partial u}(F_1h_1)\,\vec{\nabla}u + \frac{\partial}{\partial v}(F_1h_1)\,\vec{\nabla}v + \frac{\partial}{\partial w}(F_1h_1)\,\vec{\nabla}w\right] \times \vec{\nabla}u$$

$$= \vec{0} + \frac{\partial}{\partial v}(F_1h_1)\,\vec{\nabla}v \times \vec{\nabla}u + \frac{\partial}{\partial w}(F_1h_1)\,\vec{\nabla}w \times \vec{\nabla}u$$

$$= \frac{1}{h_1h_2}\frac{\partial}{\partial v}(F_1h_1)\,\hat{e}_2 \times \hat{e}_1 + \frac{1}{h_3h_1}\frac{\partial}{\partial w}(F_1h_1)\,\hat{e}_3 \times \hat{e}_1$$

$$= -\frac{1}{h_1h_2}\frac{\partial}{\partial v}(F_1h_1)\,\hat{e}_3 + \frac{1}{h_3h_1}\frac{\partial}{\partial w}(F_1h_1)\,\hat{e}_2$$

Similarly

$$\vec{\nabla} \times (F_2\hat{e}_2) = \vec{\nabla} \times (h_2\,F_2\vec{\nabla}v) = -\frac{1}{h_2h_3}\frac{\partial}{\partial w}(F_2h_2)\,\hat{e}_1 + \frac{1}{h_2h_1}\frac{\partial}{\partial u}(F_2h_2)\,\hat{e}_3$$

$$\vec{\nabla} \times (F_3\hat{e}_3) = \vec{\nabla} \times (h_3\,F_3\vec{\nabla}w) = -\frac{1}{h_3h_1}\frac{\partial}{\partial u}(F_3h_3)\,\hat{e}_2 + \frac{1}{h_2h_3}\frac{\partial}{\partial v}(F_3h_3)\,\hat{e}_1$$

Therefore, $\vec{\nabla} \times \vec{F}$ is given by

$$\vec{\nabla} \times \vec{F} = \vec{\nabla} \times (F_1\hat{e}_1) + \vec{\nabla} \times (F_2\hat{e}_2) + \vec{\nabla} \times (F_3\hat{e}_3)$$

$$= \frac{1}{h_2h_3}\left[\frac{\partial}{\partial v}(F_3h_3) - \frac{\partial}{\partial w}(F_2h_2)\right]\hat{e}_1 - \frac{1}{h_3h_1}\left[\frac{\partial}{\partial u}(F_3h_3) - \frac{\partial}{\partial w}(F_1h_1)\right]\hat{e}_2$$

$$+ \frac{1}{h_1h_2}\left[\frac{\partial}{\partial u}(F_2h_2) - \frac{\partial}{\partial v}(F_1h_1)\right]\hat{e}_3$$

$$= \frac{1}{h_1\,h_2\,h_3}\begin{vmatrix} h_1\hat{e}_1 & h_2\hat{e}_2 & h_3\hat{e}_3 \\ \dfrac{\partial}{\partial u} & \dfrac{\partial}{\partial v} & \dfrac{\partial}{\partial w} \\ h_1\,F_1 & h_2\,F_2 & h_3\,F_3 \end{vmatrix} \tag{5.53}$$

This is the required expression (5.53) of $\vec{\nabla} \times \vec{F}$ in terms of the orthogonal cuvilinear coordinates u, v and w.

DEDUCTION 5.4.3

Curl in cylindrical coordinates: Here, $u = \rho$, $v = \phi$ and $w = z$; $\hat{e}_1 = \hat{e}_\rho$, $\hat{e}_2 = \hat{e}_\phi$ and $\hat{e}_3 = \hat{e}_z$; $h_1 = h_\rho = 1$, $h_2 = h_\phi = \rho$ and $h_3 = h_z = 1$. Let $\vec{F}$ be a scalar valued function. Therefore, using Eq. (5.53), we get

$$\vec{\nabla} \times \vec{F} = \frac{1}{h_1\,h_2\,h_3}\begin{vmatrix} h_1\hat{e}_1 & h_2\hat{e}_2 & h_3\hat{e}_3 \\ \dfrac{\partial}{\partial u} & \dfrac{\partial}{\partial v} & \dfrac{\partial}{\partial w} \\ h_1\,F_1 & h_2\,F_2 & h_3\,F_3 \end{vmatrix}$$

$$= \frac{1}{1\,\rho\,1}\begin{vmatrix} 1\hat{e}_\rho & \rho\hat{e}_\phi & 1\hat{e}_z \\ \dfrac{\partial}{\partial \rho} & \dfrac{\partial}{\partial \phi} & \dfrac{\partial}{\partial z} \\ 1\,F_\rho & \rho\,F_\phi & 1\,F_z \end{vmatrix} = \frac{1}{\rho}\begin{vmatrix} \hat{e}_\rho & \rho\hat{e}_\phi & \hat{e}_z \\ \dfrac{\partial}{\partial \rho} & \dfrac{\partial}{\partial \phi} & \dfrac{\partial}{\partial z} \\ F_\rho & \rho\,F_\phi & F_z \end{vmatrix}$$

$$= \frac{1}{\rho}\left[\left(\frac{\partial F_z}{\partial \phi} - \frac{\partial}{\partial z}(\rho F_\phi)\right)\hat{e}_\rho + \left(\rho\frac{\partial F_\rho}{\partial z} - \rho\frac{\partial F_z}{\partial \rho}\right)\hat{e}_\phi + \left(\frac{\partial}{\partial \rho}(\rho F_\phi) - \frac{\partial F_\rho}{\partial \phi}\right)\hat{e}_z\right] \tag{5.54}$$

DEDUCTION 5.4.4

Curl in spherical polar coordinates: Here, $u = \rho$, $v = \theta$ and $w = \phi$; $\hat{e}_1 = \hat{e}_\rho$, $\hat{e}_2 = \hat{e}_\theta$ and $\hat{e}_3 = \hat{e}_\phi$; $h_1 = h_\rho = 1$, $h_2 = h_\theta = \rho$ and $h_3 = h_\phi = \rho\sin\theta$. Let $\vec{F}$ be a scalar valued function. Therefore, using Eq. (5.47), we get

$$\vec{\nabla}\times\vec{F} = \frac{1}{h_1\,h_2\,h_3}\begin{vmatrix} h_1\hat{e}_1 & h_2\hat{e}_2 & h_3\hat{e}_3 \\ \dfrac{\partial}{\partial u} & \dfrac{\partial}{\partial v} & \dfrac{\partial}{\partial w} \\ h_1\,F_1 & h_2\,F_2 & h_3\,F_3 \end{vmatrix}$$

$$= \frac{1}{\rho^2\sin\theta}\begin{vmatrix} \hat{e}_\rho & \rho\hat{e}_\theta & \rho\sin\theta\,\hat{e}_\phi \\ \dfrac{\partial}{\partial\rho} & \dfrac{\partial}{\partial\theta} & \dfrac{\partial}{\partial\phi} \\ F_\rho & \rho\,F_\phi & \rho\sin\theta\,F_z \end{vmatrix} \tag{5.55}$$

5.5 LAPLACIAN OPERATOR

We have already discussed three very effective and useful operators, namely, gradient, divergence and curl. Another important and widely used operator is the Laplacian, which plays a significant role in the study of sound waves, electricity and magnetism.

In the three-dimensional Cartesian coordinate system of E^3, the operator *Laplacian* is denoted by ∇^2 or $\vec{\nabla}\cdot\vec{\nabla}$, which is the dot product of the vector operators $\vec{\nabla}$ and $\vec{\nabla}$ and is defined by

$$\vec{\nabla}\cdot\vec{\nabla} = \nabla^2 = \frac{\partial^2}{\partial x^2} + \frac{\partial^2}{\partial y^2} + \frac{\partial^2}{\partial z^2} \tag{5.56}$$

This operator can be used in scalar point functions as well as in vector point functions. Let $\phi(x, y, z)$ be an arbitrary scalar point function. Using the definition, the gradient of ϕ is given by Eq. (5.1). Therefore, using Eq. (5.41), the divergence of the gradient of ϕ is

$$\vec{\nabla}\cdot\vec{\nabla}\phi = \nabla^2\phi = \left(\frac{\partial}{\partial x}\hat{i} + \frac{\partial}{\partial y}\hat{j} + \frac{\partial}{\partial z}\hat{k}\right)\cdot\left(\frac{\partial\phi}{\partial x}\hat{i} + \frac{\partial\phi}{\partial y}\hat{j} + \frac{\partial\phi}{\partial z}\hat{k}\right)$$

$$= \frac{\partial}{\partial x}\left(\frac{\partial\phi}{\partial x}\right) + \frac{\partial}{\partial y}\left(\frac{\partial\phi}{\partial y}\right) + \frac{\partial}{\partial z}\left(\frac{\partial\phi}{\partial z}\right)$$

$$= \left(\frac{\partial^2}{\partial x^2} + \frac{\partial^2}{\partial y^2} + \frac{\partial^2}{\partial z^2}\right)\phi = \frac{\partial^2\phi}{\partial x^2} + \frac{\partial^2\phi}{\partial y^2} + \frac{\partial^2\phi}{\partial z^2}$$

Thus, if we consider $\phi(x, y, z)$ as a scalar point function and $\vec{F}(x, y, z)$ as a vector point function, we have

$$\nabla^2\phi = \frac{\partial^2\phi}{\partial x^2} + \frac{\partial^2\phi}{\partial y^2} + \frac{\partial^2\phi}{\partial z^2},\quad \nabla^2\vec{F} = \frac{\partial^2\vec{F}}{\partial x^2} + \frac{\partial^2\vec{F}}{\partial y^2} + \frac{\partial^2\vec{F}}{\partial z^2}$$

Thus, the Laplacian operator transforms a scalar into another scalar and a vector into another vector.

(i) The divergence of gradient of the scalar function ϕ is $\nabla^2\phi$.
(ii) The equation $\nabla^2\phi = 0$ is called *Laplace's equation.* In other words, ϕ is the solution of the Laplace equation and is known as the *harmonic scalar point function.*

Example 5.5.1

Prove that $\nabla^2 r^n = n(n+1)r^{n-2}$, where n is a constant and $r = |\vec{r}|$.

Solution

Using definition (5.1), the gradient of $\phi = r^n$ is given by

$$\vec{\nabla}\phi = \vec{\nabla}r^n = nr^{n-1}\left[\frac{\partial r}{\partial x}\hat{i} + \frac{\partial r}{\partial y}\hat{j} + \frac{\partial r}{\partial z}\hat{k}\right] = nr^{n-2}\vec{r}$$

Therefore, using Eq. (5.41), the divergence of gradient of ϕ is

$$\begin{aligned}
\nabla^2 r^n = \vec{\nabla}\cdot\left(\vec{\nabla}r^n\right) &= \left[\frac{\partial}{\partial x}\hat{i} + \frac{\partial}{\partial y}\hat{j} + \frac{\partial}{\partial z}\hat{k}\right]\cdot\left[nr^{n-2}\left(x\hat{i} + y\hat{j} + z\hat{k}\right)\right] \\
&= n\left[\frac{\partial}{\partial x}\left(xr^{n-2}\right) + \frac{\partial}{\partial y}\left(yr^{n-2}\right) + \frac{\partial}{\partial z}\left(zr^{n-2}\right)\right] \\
&= n\Big[r^{n-2} + x(n-2)r^{n-3}\frac{\partial r}{\partial x} + r^{n-2} + y(n-2)r^{n-3}\frac{\partial r}{\partial y} \\
&\qquad + r^{n-2} + z(n-2)r^{n-3}\frac{\partial r}{\partial z}\Big] \\
&= n\left[3r^{n-2} + (n-2)r^{n-3}\left\{x\cdot\frac{x}{r} + y\cdot\frac{y}{r} + z\cdot\frac{z}{r}\right\}\right] \\
&= 3nr^{n-2} + n(n-2)r^{n-4}\left(x^2 + y^2 + z^2\right) \\
&= \{3n + n(n-2)\}r^{n-2} = n(n+1)r^{n-2}
\end{aligned}$$

Example 5.5.2

Show that $\phi = r^{-1}$ is a solution of the Laplace equation $\nabla^2\phi = 0$.

Solution

As $\vec{r} = x\hat{i} + y\hat{j} + z\hat{k}$, so $r = |\vec{r}| = \sqrt{x^2 + y^2 + z^2}$. Using definition (5.1), the gradient of $\phi = r^{-1}$ is given by

$$\begin{aligned}
\vec{\nabla}r^{-1} &= \frac{\partial}{\partial x}\left(\frac{1}{r}\right)\hat{i} + \frac{\partial}{\partial y}\left(\frac{1}{r}\right)\hat{j} + \frac{\partial}{\partial z}\left(\frac{1}{r}\right)\hat{k} \\
&= \left(x^2 + y^2 + z^2\right)^{-\frac{3}{2}}\left[-\frac{1}{2}2x\hat{i} - \frac{1}{2}2y\hat{j} - \frac{1}{2}2z\hat{k}\right] \\
&= -\left(x^2 + y^2 + z^2\right)^{-\frac{3}{2}}\left(x\hat{i} + y\hat{j} + z\hat{k}\right) = -\frac{1}{r^3}\vec{r}
\end{aligned}$$

Therefore, using Eq. (5.41), the divergence of gradient of $\phi = r^{-1}$ is

$$\begin{aligned}
\nabla^2 r^{-1} = \vec{\nabla}\cdot\left(\vec{\nabla}r^{-1}\right) &= -\left[\frac{\partial}{\partial x}\left(xr^{-3}\right) + \frac{\partial}{\partial y}\left(yr^{-3}\right) + \frac{\partial}{\partial z}\left(zr^{-3}\right)\right] \\
&= -\Big[r^{-3} + x\left(\frac{-3}{2}\right)\left(x^2 + y^2 + z^2\right)^{-\frac{5}{2}}2x + r^{-3} \\
&\qquad + y\left(\frac{-3}{2}\right)\left(x^2 + y^2 + z^2\right)^{-\frac{5}{2}}2y + r^{-3} + z\left(\frac{-3}{2}\right)\left(x^2 + y^2 + z^2\right)^{-\frac{5}{2}}2z\Big]
\end{aligned}$$

$$= -3r^{-3} + 3x^2r^{-5} + 3y^2r^{-5} + 3z^2r^{-5}$$
$$= -3r^{-3} + 3r^{-5}(x^2 + y^2 + z^2) = -3r^{-3} + 3r^{-3} = 0$$

Thus, it follows that $\phi = r^{-1}$ is the solution of the Laplace equation and is known as the harmonic scalar point function.

Example 5.5.3

Prove that $\nabla^2 f(r) = \dfrac{d^2f}{dr^2} + \dfrac{2}{r}\dfrac{df}{dr} = f''(r) + \dfrac{2}{r}f'(r)$.

Solution

Using definition (5.1), we get

$$\vec{\nabla} f(r) = \left(\hat{i}\,\frac{\partial}{\partial x} + \hat{j}\,\frac{\partial}{\partial y} + \hat{k}\,\frac{\partial}{\partial z}\right) f(r)$$
$$= \hat{i}\,\frac{f'(r)x}{\sqrt{x^2+y^2+z^2}} + \hat{j}\,\frac{f'(r)y}{\sqrt{x^2+y^2+z^2}} + \hat{k}\,\frac{f'(r)z}{\sqrt{x^2+y^2+z^2}}$$
$$= \frac{f'(r)}{r}\left[x\hat{i} + y\hat{j} + z\hat{k}\right] = \frac{f'(r)}{r}\,\vec{r}$$

We know $\dfrac{\partial r}{\partial x} = \dfrac{x}{r}$, $\dfrac{\partial r}{\partial y} = \dfrac{y}{r}$ and $\dfrac{\partial r}{\partial z} = \dfrac{z}{r}$. Therefore

$$\nabla^2 f(r) = \vec{\nabla}\cdot\left[\vec{\nabla} f(r)\right] = \left(\hat{i}\,\frac{\partial}{\partial x} + \hat{j}\,\frac{\partial}{\partial y} + \hat{k}\,\frac{\partial}{\partial z}\right)\cdot \frac{f'(r)}{r}\,\vec{r}$$
$$= \left(\hat{i}\,\frac{\partial}{\partial x} + \hat{j}\,\frac{\partial}{\partial y} + \hat{k}\,\frac{\partial}{\partial z}\right)\cdot \frac{f'(r)}{r}\left[x\hat{i} + y\hat{j} + z\hat{k}\right]$$
$$= \frac{\partial}{\partial x}\left[x\frac{f'(r)}{r}\right] + \frac{\partial}{\partial y}\left[y\frac{f'(r)}{r}\right] + \frac{\partial}{\partial z}\left[z\frac{f'(r)}{r}\right]$$
$$= \frac{f'(r)}{r} + \frac{xf''(r)}{r}\frac{\partial r}{\partial x} - \frac{xf'(r)}{r^2}\frac{\partial r}{\partial x}$$
$$+ \frac{f'(r)}{r} + \frac{yf''(r)}{r}\frac{\partial r}{\partial y} - \frac{yf'(r)}{r^2}\frac{\partial r}{\partial y}$$
$$+ \frac{f'(r)}{r} + \frac{zf''(r)}{r}\frac{\partial r}{\partial z} - \frac{zf'(r)}{r^2}\frac{\partial r}{\partial z}$$
$$= 3\frac{f'(r)}{r} + \left[\frac{f''(r)}{r^2} - \frac{f'(r)}{r^3}\right](x^2 + y^2 + z^2)$$
$$= f''(r) + \frac{2f'(r)}{r} = \frac{d^2f}{dr^2} + \frac{2}{r}\frac{df}{dr}$$

In particular, if $f(r) = \ln r$, then

$$\nabla^2(\ln r) = \frac{d^2}{dr^2}(\ln r) + \frac{2}{r}\frac{d}{dr}(\ln r) = -\frac{1}{r^2} + \frac{2}{r}\cdot\frac{1}{r} = \frac{1}{r^2}$$

Example 5.5.4

Prove that $\nabla^2\left(\frac{x}{r^2}\right) = -\frac{2x}{r^4}$.

Solution

Using definition (5.1), we get

$$\begin{aligned}
\vec{\nabla} f(r) &= \left(\hat{i}\,\frac{\partial}{\partial x} + \hat{j}\,\frac{\partial}{\partial y} + \hat{k}\,\frac{\partial}{\partial z}\right)\left(\frac{x}{r^2}\right) \\
&= \hat{i}\left[1 \cdot \frac{1}{r^2} - \frac{2x}{r^3} \cdot \frac{\partial x}{\partial r}\right] + \hat{j}\left[-\frac{2x}{r^3} \cdot \frac{\partial y}{\partial r}\right] + \hat{k}\left[-\frac{2x}{r^3} \cdot \frac{\partial z}{\partial r}\right] \\
&= \hat{i}\left[1 \cdot \frac{1}{r^2} - \frac{2x}{r^3} \cdot \frac{x}{r}\right] + \hat{j}\left[-\frac{2x}{r^3} \cdot \frac{y}{r}\right] + \hat{k}\left[-\frac{2x}{r^3} \cdot \frac{z}{r}\right] \\
&= \left(\frac{1}{r^2} - \frac{2x^2}{r^4}\right)\hat{i} - \frac{2xy}{r^4}\,\hat{j} - \frac{2xz}{r^4}\,\hat{k}
\end{aligned}$$

As $\nabla^2\left(\frac{x}{r^2}\right) = \vec{\nabla} \cdot \left[\vec{\nabla}\left(\frac{x}{r^2}\right)\right]$, therefore

$$\begin{aligned}
\nabla^2\left(\frac{x}{r^2}\right) &= \left(\hat{i}\,\frac{\partial}{\partial x} + \hat{j}\,\frac{\partial}{\partial y} + \hat{k}\,\frac{\partial}{\partial z}\right) \cdot \left[\left(\frac{1}{r^2} - \frac{2x^2}{r^4}\right)\hat{i} - \frac{2xy}{r^4}\,\hat{j} - \frac{2xz}{r^4}\,\hat{k}\right] \\
&= \frac{\partial}{\partial x}\left(\frac{1}{r^2} - \frac{2x^2}{r^4}\right) + \frac{\partial}{\partial y}\left(-\frac{2xy}{r^4}\right) + \frac{\partial}{\partial z}\left(-\frac{2xz}{r^4}\right) \\
&= \left[-\frac{2}{r^3} \cdot \frac{x}{r} - \frac{4x}{r^4} + \frac{8x^2}{r^5} \cdot \frac{x}{r}\right] - 2x\left[\frac{1}{r^4} - \frac{4y}{r^5} \cdot \frac{y}{r}\right] - 2x\left[\frac{1}{r^4} - \frac{4z}{r^5} \cdot \frac{z}{r}\right] \\
&= -\frac{6x}{r^4} + \frac{8x^3}{r^6} - 2x\left[\frac{1}{r^4} - \frac{4y^2}{r^6}\right] - 2x\left[\frac{1}{r^4} - \frac{4z^2}{r^6}\right] \\
&= -2x\left[\frac{3}{r^4} + \frac{1}{r^4} + \frac{1}{r^4} - \frac{4}{r^6}\,r^2\right] \\
&= -\frac{2x}{r^4}
\end{aligned}$$

Example 5.5.5

Show that $\frac{x}{r^3}$ is a harmonic function.

Solution

We know $\frac{\partial r}{\partial x} = \frac{x}{r}$, $\frac{\partial r}{\partial y} = \frac{y}{r}$ and $\frac{\partial r}{\partial z} = \frac{z}{r}$. Then

$$\frac{\partial}{\partial x}\left(\frac{x}{r^3}\right) = \frac{1}{r^3} - \frac{3x^2}{r^5},\ \frac{\partial}{\partial y}\left(\frac{x}{r^3}\right) = -\frac{3xy}{r^5},\ \frac{\partial}{\partial z}\left(\frac{x}{r^3}\right) = -\frac{3xz}{r^5}$$

$$\frac{\partial^2}{\partial x^2}\left(\frac{x}{r^3}\right) = \frac{\partial}{\partial x}\left(\frac{1}{r^3} - \frac{3x^2}{r^5}\right) = -\frac{9x}{r^5} + \frac{15x^3}{r^7}$$

$$\frac{\partial^2}{\partial y^2}\left(\frac{x}{r^3}\right) = \frac{\partial}{\partial y}\left(-\frac{3xy}{r^5}\right) = -\frac{3x}{r^5} + \frac{15xy^2}{r^7}$$

$$\frac{\partial^2}{\partial z^2}\left(\frac{x}{r^3}\right) = \frac{\partial}{\partial x}\left(-\frac{3xz}{r^5}\right) = -\frac{3x}{r^5} + \frac{15xz^2}{r^7}$$

Adding the last three relations, we obtain the result.

Example 5.5.6

Let $\vec{r} = x\hat{i} + y\hat{j} + z\hat{k}$ and $r = |\vec{r}|$. If a scalar field ϕ and a vector field $\vec{u}$ satisfy $\vec{\nabla}\phi = \vec{\nabla} \times \vec{u} + f(r)\,\vec{r}$, where f is an arbitrary differentiable function, then show that $\nabla^2\phi = r\,f'(r) + 3\,f(r)$.

Solution

Given that $\vec{\nabla}\phi = \vec{\nabla} \times \vec{u} + f(r)\,\vec{r}$. Now

$$\begin{aligned}
\nabla^2\phi &= \vec{\nabla}\cdot(\vec{\nabla}\phi) = \vec{\nabla}\cdot\left[\vec{\nabla}\times\vec{u} + f(r)\,\vec{r}\right] \\
&= \vec{\nabla}\cdot(\vec{\nabla}\times\vec{u}) + \vec{\nabla}\cdot(f(r)\,\vec{r}) = 0 + \vec{\nabla}\cdot(f(r)\,\vec{r}) \\
&= f(r)\left[\vec{\nabla}\cdot\vec{r}\right] + \left[\vec{\nabla}f(r)\right]\cdot\vec{r} \\
&= f(r)\,\left(\frac{\partial}{\partial x}\hat{i} + \frac{\partial}{\partial y}\hat{j} + \frac{\partial}{\partial z}\hat{k}\right)\cdot(x\hat{i} + y\hat{j} + z\hat{k}) \\
&\quad + \left[\left(\hat{i}\,\frac{\partial}{\partial x} + \hat{j}\,\frac{\partial}{\partial y} + \hat{k}\,\frac{\partial}{\partial z}\right)f(r)\right]\cdot\vec{r} \\
&= f(r)\left(\frac{\partial x}{\partial x} + \frac{\partial y}{\partial y} + \frac{\partial z}{\partial z}\right) + \left[\frac{f'(r)}{r}\,(x\hat{i} + y\hat{j} + z\hat{k})\right]\cdot\vec{r} \\
&= 3\,f(r) + \frac{f'(r)}{r}\,\vec{r}\cdot\vec{r} = 3\,f(r) + r\,f'(r)
\end{aligned}$$

DEDUCTION 5.5.1

In this result (5.45), set $\phi = U$ and $\vec{F} = \vec{\nabla}V$, then

$$\vec{\nabla}\cdot(U\,\vec{\nabla}V) = \vec{\nabla}U\cdot\vec{\nabla}V + U(\vec{\nabla}\cdot\vec{\nabla}V) = \vec{\nabla}U\cdot\vec{\nabla}V + U\nabla^2 V$$

Interchanging U and V yields

$$\vec{\nabla}\cdot(V\,\vec{\nabla}U) = \vec{\nabla}V\cdot\vec{\nabla}U + V(\vec{\nabla}\cdot\vec{\nabla}U) = \vec{\nabla}V\cdot\vec{\nabla}U + V\nabla^2 U$$

Subtracting, we get

$$\begin{aligned}
\vec{\nabla}\cdot(U\,\vec{\nabla}V) - \vec{\nabla}\cdot(V\,\vec{\nabla}U) &= \vec{\nabla}\cdot(U\vec{\nabla}V - V\vec{\nabla}U) \\
&= \vec{\nabla}U\cdot\vec{\nabla}V + U\nabla^2V - \left[\vec{\nabla}V\cdot\vec{\nabla}U + V\nabla^2U\right] \\
&= U\nabla^2V - V\nabla^2U
\end{aligned}$$

Example 5.5.7

Show that if $\phi(x, y, z)$ is any solution of Laplace's equation, then $\vec{\nabla}\phi$ is a vector which is both solenoidal and irrotational.

Solution

Given that ϕ satisfies Laplace's equation, i.e., $\vec{\nabla}^2\phi = \vec{\nabla}\cdot(\vec{\nabla}\phi) = 0$. Since divergence of the vector $\vec{\nabla}\phi$ is zero, $\vec{\nabla}\phi$ is a solenoidal vector. We assume that ϕ has continuous second partial derivatives so that the order of differentiation is commutative. Also

$$\vec{\nabla} \times (\vec{\nabla}\phi) = \begin{vmatrix} \hat{i} & \hat{j} & \hat{k} \\ \dfrac{\partial}{\partial x} & \dfrac{\partial}{\partial y} & \dfrac{\partial}{\partial z} \\ \dfrac{\partial \phi}{\partial x} & \dfrac{\partial \phi}{\partial y} & \dfrac{\partial \phi}{\partial z} \end{vmatrix}$$

$$= \left[\frac{\partial}{\partial y}\left(\frac{\partial \phi}{\partial z}\right) - \frac{\partial}{\partial z}\left(\frac{\partial \phi}{\partial y}\right)\right]\hat{i} + \left[\frac{\partial}{\partial z}\left(\frac{\partial \phi}{\partial x}\right) - \frac{\partial}{\partial x}\left(\frac{\partial \phi}{\partial z}\right)\right]\hat{j}$$

$$+ \left[\frac{\partial}{\partial x}\left(\frac{\partial \phi}{\partial y}\right) - \frac{\partial}{\partial y}\left(\frac{\partial \phi}{\partial x}\right)\right]\hat{k}$$

$$= \left[\frac{\partial^2 \phi}{\partial y \partial z} - \frac{\partial^2 \phi}{\partial z \partial y}\right]\hat{i} + \left[\frac{\partial^2 \phi}{\partial z \partial x} - \frac{\partial^2 \phi}{\partial x \partial z}\right]\hat{j} + \left[\frac{\partial^2 \phi}{\partial x \partial y} - \frac{\partial^2 \phi}{\partial y \partial x}\right]\hat{k} = \vec{0}$$

Thus, the curl of the gradient of ϕ is zero. Since $\vec{\nabla} \times (\vec{\nabla}\phi) = \vec{0}$, $\vec{\nabla}\phi$ is also irrotational.

Theorem 5.9

If $\vec{A}$ be a differentiable vector function, then

$$\vec{\nabla} \times (\vec{\nabla} \times \vec{A}) = \vec{\nabla}(\vec{\nabla} \cdot \vec{A}) - \nabla^2 \vec{A} \tag{5.57}$$

Proof: Let $\vec{A}(x, y, z) = A_1\hat{i} + A_2\hat{j} + A_3\hat{k}$, where A_1, A_2 and A_3 are real-valued differentiable functions of real variables x, y and z. Then

$$\vec{\nabla} \times (\vec{\nabla} \times \vec{A}) = \begin{vmatrix} \hat{i} & \hat{j} & \hat{k} \\ \dfrac{\partial}{\partial x} & \dfrac{\partial}{\partial y} & \dfrac{\partial}{\partial z} \\ \dfrac{\partial A_3}{\partial y} - \dfrac{\partial A_2}{\partial z} & \dfrac{\partial A_1}{\partial z} - \dfrac{\partial A_3}{\partial x} & \dfrac{\partial A_2}{\partial x} - \dfrac{\partial A_1}{\partial y} \end{vmatrix}$$

$$= \sum\left[\hat{i}\left\{\frac{\partial}{\partial y}\left(\frac{\partial A_2}{\partial x} - \frac{\partial A_1}{\partial y}\right) - \frac{\partial}{\partial z}\left(\frac{\partial A_1}{\partial z} - \frac{\partial A_3}{\partial x}\right)\right\}\right]$$

$$= \sum\left[\hat{i}\left\{\left(\frac{\partial^2 A_2}{\partial y \partial x} - \frac{\partial^2 A_1}{\partial y^2}\right) - \left(\frac{\partial^2 A_1}{\partial z^2} - \frac{\partial^2 A_3}{\partial z \partial x}\right)\right\}\right]$$

$$= \sum\left[\hat{i}\left\{\left(\frac{\partial^2 A_2}{\partial y \partial x} + \frac{\partial^2 A_3}{\partial z \partial x}\right) - \left(\frac{\partial^2 A_1}{\partial z^2} + \frac{\partial^2 A_1}{\partial y^2}\right)\right\}\right]$$

$$= \sum\left[\hat{i}\left\{\frac{\partial}{\partial x}\left(\frac{\partial A_2}{\partial y} + \frac{\partial A_3}{\partial z}\right) - \left(\frac{\partial^2 A_1}{\partial y^2} + \frac{\partial^2 A_1}{\partial z^2}\right)\right\}\right]$$

$$= \sum\left[\hat{i}\left\{\frac{\partial}{\partial x}\left(\frac{\partial A_1}{\partial x} + \frac{\partial A_2}{\partial y} + \frac{\partial A_3}{\partial z}\right) - \left(\frac{\partial^2 A_1}{\partial x^2} + \frac{\partial^2 A_1}{\partial y^2} + \frac{\partial^2 A_1}{\partial z^2}\right)\right\}\right]$$

$$= \sum\left[\hat{i}\left\{\frac{\partial}{\partial x}(\vec{\nabla} \cdot \vec{A}) - \vec{\nabla}^2 A_1\right\}\right]$$

$$= \sum\left[\hat{i}\left\{\frac{\partial}{\partial x}(\vec{\nabla} \cdot \vec{A})\right\}\right] - \vec{\nabla}^2 \sum A_1\hat{i} = \vec{\nabla}(\vec{\nabla} \cdot \vec{A}) - \vec{\nabla}^2 \vec{A}$$

Example 5.5.8

Prove that $\vec{\nabla} \times \left\{\frac{1}{r^3}(\vec{a} \times \vec{r})\right\} = -\frac{1}{r^3}\vec{a} + \frac{3}{r^5}(\vec{a}\cdot\vec{r})\,\vec{r}$, where $\vec{a}$ is a constant vector.

Solution

Using the formulae (5.51) and (5.52), we get

$$\begin{aligned}
\vec{\nabla} \times \left\{\frac{1}{r^3}(\vec{a} \times \vec{r})\right\} &= \vec{\nabla}\left(\frac{1}{r^3}\right)(\vec{a} \times \vec{r}) + \frac{1}{r^3}\,\vec{\nabla} \times (\vec{a} \times \vec{r}) \\
&= -\frac{3}{r^5}\,\vec{r} \times (\vec{a} \times \vec{r}) + \frac{1}{r^3}\left[\vec{a}\,\vec{\nabla}\cdot\vec{r} - (\vec{a}\cdot\vec{\nabla})\vec{r}\right] \\
&= -\frac{3}{r^5}\left[(\vec{r}\cdot\vec{r})\vec{a} - (\vec{a}\cdot\vec{r})\vec{r}\right] + \frac{1}{r^3}(3\vec{a} - \vec{a}) \\
&= -\frac{3}{r^5}\left[r^2\vec{a} - (\vec{a}\cdot\vec{r})\vec{r}\right] + \frac{2}{r^3}\vec{a} = -\frac{1}{r^3}\,\vec{a} + \frac{3}{r^5}(\vec{a}\cdot\vec{r})\,\vec{r}
\end{aligned}$$

Example 5.5.9

If $\vec{\nabla}\cdot\vec{E} = 0$, $\vec{\nabla}\cdot\vec{H} = 0$, $\vec{\nabla} \times \vec{E} = -\frac{\partial \vec{H}}{\partial t}$ and $\vec{\nabla} \times \vec{H} = \frac{\partial \vec{E}}{\partial t}$, then show that $\nabla^2\vec{H} = \frac{\partial^2 \vec{H}}{\partial t^2}$ and $\nabla^2\vec{E} = \frac{\partial^2 \vec{E}}{\partial t^2}$.

Solution

Given that $\vec{E}$ and $\vec{H}$ are solenoidal vector fields. Replacing the continuously differentiable function $\vec{F}$ with $\vec{E}$ in the identity (5.57), we get

$$\vec{\nabla} \times (\vec{\nabla} \times \vec{E}) = \vec{\nabla}(\vec{\nabla}\cdot\vec{E}) - \nabla^2\vec{E} = -\nabla^2\vec{E} \quad \text{as } \vec{\nabla}\cdot\vec{E} = 0$$

$$\Rightarrow\ -\nabla^2\vec{E} = \vec{\nabla} \times \left(-\frac{\partial \vec{H}}{\partial t}\right) = -\frac{\partial}{\partial t}(\vec{\nabla} \times \vec{H}); \text{ as } \vec{\nabla} \times \vec{E} = -\frac{\partial \vec{H}}{\partial t}$$

$$\Rightarrow\ \nabla^2\vec{E} = \frac{\partial}{\partial t}\left(\frac{\partial \vec{E}}{\partial t}\right) = \frac{\partial^2 \vec{E}}{\partial t^2}; \text{ as } \vec{\nabla} \times \vec{H} = \frac{\partial \vec{E}}{\partial t}$$

Further replacing $\vec{F}$ with $\vec{H}$ in the identity (5.57), we get

$$\vec{\nabla} \times (\vec{\nabla} \times \vec{H}) = \vec{\nabla}(\vec{\nabla}\cdot\vec{H}) - \nabla^2\vec{H} = -\nabla^2\vec{H} \quad \text{as } \vec{\nabla}\cdot\vec{H} = 0$$

$$\Rightarrow\ -\nabla^2\vec{H} = \vec{\nabla} \times \left(-\frac{\partial \vec{E}}{\partial t}\right) = -\frac{\partial}{\partial t}(\vec{\nabla} \times \vec{E})$$

$$\Rightarrow\ \nabla^2\vec{H} = \frac{\partial}{\partial t}\left(\frac{\partial \vec{H}}{\partial t}\right) = \frac{\partial^2 \vec{H}}{\partial t^2}; \text{ as } \vec{\nabla} \times \vec{E} = -\frac{\partial \vec{H}}{\partial t}$$

Example 5.5.10

If $\vec{H}, \vec{E}$ and $\vec{v}$ are continuously differentiable vector functions of x, y and z and t and ρ are continuously differentiable scalar functions of x, y, z and t, $\vec{\nabla}\cdot\vec{E} = \rho$, $\vec{\nabla}\cdot\vec{H} = 0$, $\vec{\nabla} \times \vec{H} = \frac{1}{c}\left(\frac{\partial \vec{E}}{\partial t} + \rho\vec{v}\right)$, $\vec{\nabla} \times \vec{E} = -\frac{1}{c}\frac{\partial \vec{H}}{\partial t}$, where c is a constant, prove that

(i) $\nabla^2\vec{E} - \dfrac{1}{c^2}\dfrac{\partial^2\vec{E}}{\partial t^2} = \vec{\nabla}\rho + \dfrac{1}{c}\dfrac{\partial}{\partial t}(\rho\vec{v})$ (ii) $\nabla^2\vec{H} = -\dfrac{1}{c^2}\dfrac{\partial^2\vec{H}}{\partial t^2} = -\dfrac{1}{c}\vec{\nabla}\times(\rho\vec{v})$

Solution

We use the vector identity (5.57) for any continuously differentiable vector function $\vec{F}$ of x, y, z and t.

(i) Then, from Eq. (5.57), we get

$$\begin{aligned}
\nabla^2\vec{E} &= \vec{\nabla}(\vec{\nabla}\cdot\vec{E}) - \vec{\nabla}\times(\vec{\nabla}\times\vec{E}) \\
&= \vec{\nabla}\rho - \vec{\nabla}\times\left(-\frac{1}{c}\frac{\partial\vec{H}}{\partial t}\right); \text{ as } \vec{\nabla}\cdot\vec{E} = \rho, \vec{\nabla}\times\vec{E} = -\frac{1}{c}\frac{\partial\vec{H}}{\partial t} \\
&= \vec{\nabla}\rho + \frac{1}{c}\frac{\partial}{\partial t}(\vec{\nabla}\times\vec{H}) \\
&= \vec{\nabla}\rho + \frac{1}{c^2}\frac{\partial}{\partial t}\left(\frac{\partial\vec{E}}{\partial t} + \rho\vec{v}\right); \text{ as } \vec{\nabla}\times\vec{H} = \frac{1}{c}\left(\frac{\partial\vec{E}}{\partial t} + \rho\vec{v}\right) \\
&= \nabla^2\vec{E} - \frac{1}{c^2}\frac{\partial^2\vec{E}}{\partial t^2} = \vec{\nabla}\rho + \frac{1}{c}\frac{\partial}{\partial t}(\rho\vec{v})
\end{aligned}$$

(ii) Also, from Eq. (5.57), we get

$$\begin{aligned}
\nabla^2\vec{H} &= \vec{\nabla}(\vec{\nabla}\cdot\vec{H}) - \vec{\nabla}\times(\vec{\nabla}\times\vec{H}) \\
&= -\frac{1}{c}\vec{\nabla}\times\left(\frac{\partial\vec{E}}{\partial t} + \rho\vec{v}\right); \text{ as } \vec{\nabla}\cdot\vec{H} = 0, \vec{\nabla}\times\vec{H} = \frac{1}{c}\left(\frac{\partial\vec{E}}{\partial t} + \rho\vec{v}\right) \\
&= -\frac{1}{c}\left[\vec{\nabla}\times\frac{\partial\vec{E}}{\partial t} + \vec{\nabla}\times(\rho\vec{v})\right] \\
&= -\frac{1}{c}\left[\frac{\partial}{\partial t}\vec{\nabla}\times\vec{E} + \vec{\nabla}\times(\rho\vec{v})\right] \\
&= -\frac{1}{c}\left[\frac{\partial}{\partial t}\left(-\frac{1}{c}\frac{\partial\vec{H}}{\partial t}\right) + \vec{\nabla}\times(\rho\vec{v})\right] \\
&= \nabla^2\vec{H} - \frac{1}{c^2}\frac{\partial^2\vec{H}}{\partial t^2} = -\frac{1}{c}\vec{\nabla}\times(\rho\vec{v})
\end{aligned}$$

5.5.1 Laplacian in Curvilinear Coordinates

For any continuously differentiable scalar point function f of orthogonal curvilinear coordinates u, v and w, the gradient is given by Eq. (5.36). Now

$$\begin{aligned}
\nabla^2 f = \vec{\nabla}\cdot\vec{\nabla}f &= \vec{\nabla}\cdot\left[\frac{1}{h_1}\frac{\partial f}{\partial u}\hat{e}_1 + \frac{1}{h_2}\frac{\partial f}{\partial v}\hat{e}_2 + \frac{1}{h_3}\frac{\partial f}{\partial w}\hat{e}_3\right] \\
&= \frac{1}{h_1h_2h_3}\left[\frac{\partial}{\partial u}\left(\frac{h_2h_3}{h_1}\frac{\partial f}{\partial u}\right) + \frac{\partial}{\partial v}\left(\frac{h_3h_1}{h_2}\frac{\partial f}{\partial v}\right) + \frac{\partial}{\partial w}\left(\frac{h_1h_2}{h_3}\frac{\partial f}{\partial w}\right)\right] \qquad (5.58)
\end{aligned}$$

[using Eq.(5.47)]

Hence, the Laplacian operator is given by

$$\nabla^2 \equiv \frac{1}{h_1h_2h_3}\left[\frac{\partial}{\partial u}\left(\frac{h_2h_3}{h_1}\right)\frac{\partial}{\partial u}+\frac{h_2h_3}{h_1}\frac{\partial^2}{\partial u^2}+\frac{\partial}{\partial v}\left(\frac{h_3h_1}{h_2}\right)\frac{\partial}{\partial v}\right.$$
$$\left.+\frac{h_3h_1}{h_2}\frac{\partial^2}{\partial v^2}+\frac{\partial}{\partial w}\left(\frac{h_1h_2}{h_3}\right)\frac{\partial}{\partial w}+\frac{h_1h_2}{h_3}\frac{\partial^2}{\partial w^2}\right] \tag{5.59}$$

DEDUCTION 5.5.2

Laplacian in cylindrical coordinates: Here, $u=\rho$, $v=\phi$ and $w=z$; $\hat{e}_1=\hat{e}_\rho$, $\hat{e}_2=\hat{e}_\phi$ and $\hat{e}_3=\hat{e}_z$; $h_1=h_\rho=1$, $h_2=h_\phi=\rho$ and $h_3=h_z=1$. Let f be any continuously differentiable scalar point function of orthogonal cylindrical coordinates ρ,ϕ and z. Therefore, using Eq. (5.58), we

$$\nabla^2 f=\frac{1}{h_1h_2h_3}\left[\frac{\partial}{\partial u}\left(\frac{h_2h_3}{h_1}\frac{\partial f}{\partial u}\right)+\frac{\partial}{\partial v}\left(\frac{h_3h_1}{h_2}\frac{\partial f}{\partial v}\right)+\frac{\partial}{\partial w}\left(\frac{h_1h_2}{h_3}\frac{\partial f}{\partial w}\right)\right]$$
$$=\frac{1}{\rho}\left[\frac{\partial}{\partial \rho}\left(\rho\frac{\partial f}{\partial \rho}\right)+\frac{\partial}{\partial \phi}\left(\frac{1}{\rho}\frac{\partial f}{\partial \phi}\right)+\frac{\partial}{\partial z}\left(\rho\frac{\partial f}{\partial z}\right)\right] \tag{5.60}$$

DEDUCTION 5.5.3

Laplacian in spherical polar coordinates: Here, $u=\rho$, $v=\theta$ and $w=\phi$; $\hat{e}_1=\hat{e}_\rho$, $\hat{e}_2=\hat{e}_\theta$ and $\hat{e}_3=\hat{e}_\phi$; $h_1=h_\rho=1$, $h_2=h_\theta=\rho$ and $h_3=h_\phi=\rho\sin\theta$. Let f be any continuously differentiable scalar point function of orthogonal cylindrical coordinates ρ,θ and ϕ. Therefore, using Eq. (5.58), we get

$$\nabla^2 f=\frac{1}{h_1h_2h_3}\left[\frac{\partial}{\partial u}\left(\frac{h_2h_3}{h_1}\frac{\partial f}{\partial u}\right)+\frac{\partial}{\partial v}\left(\frac{h_3h_1}{h_2}\frac{\partial f}{\partial v}\right)+\frac{\partial}{\partial w}\left(\frac{h_1h_2}{h_3}\frac{\partial f}{\partial w}\right)\right]$$
$$=\frac{1}{\rho^2\sin\theta}\left[\frac{\partial}{\partial \rho}\left(\rho^2\sin\theta\frac{\partial f}{\partial \rho}\right)+\frac{\partial}{\partial \theta}\left(\sin\theta\frac{\partial f}{\partial \theta}\right)+\frac{\partial}{\partial \phi}\left(\frac{1}{\sin\theta}\frac{\partial f}{\partial \phi}\right)\right] \tag{5.61}$$

EXERCISES

Multiple Choice Questions

1. Let r be the distance of a point $P(x,y,z)$ from the origin O. Then, $\vec{\nabla} r$ is a vector
   (a) orthogonal to OP (b) normal to the level surface of r at P
   (c) normal to the surface of revolution generated by OP about the x axis
   (d) normal to the surface of revolution generated by OP about the y axis
2. $\vec{\nabla}\log|\vec{r}|$ will be equal to
   (a) $\vec{r}$ (b) $\frac{1}{r^3}\vec{r}$ (c) $\frac{1}{r^2}\vec{r}$ (d) $\frac{1}{r^4}\vec{r}$
3. The unit normal vector(s) to the label surface $x^2y+2xz=4$ at the point $(2,-2,3)$ is/are
   (a) $-\frac{1}{3}\hat{i}+\frac{2}{3}\hat{j}+\frac{2}{3}\hat{k}$ and $\frac{1}{3}\hat{i}-\frac{2}{3}\hat{j}-\frac{2}{3}\hat{k}$ (b) $\frac{1}{3}\hat{i}+\frac{2}{3}\hat{j}-\frac{2}{3}\hat{k}$ and $\frac{1}{3}\hat{i}-\frac{2}{3}\hat{j}-\frac{2}{3}\hat{k}$
   (c) $-\frac{1}{3}\hat{i}+\frac{2}{3}\hat{j}+\frac{2}{3}\hat{k}$ only (d) $\frac{1}{3}\hat{i}-\frac{2}{3}\hat{j}-\frac{2}{3}\hat{k}$ only

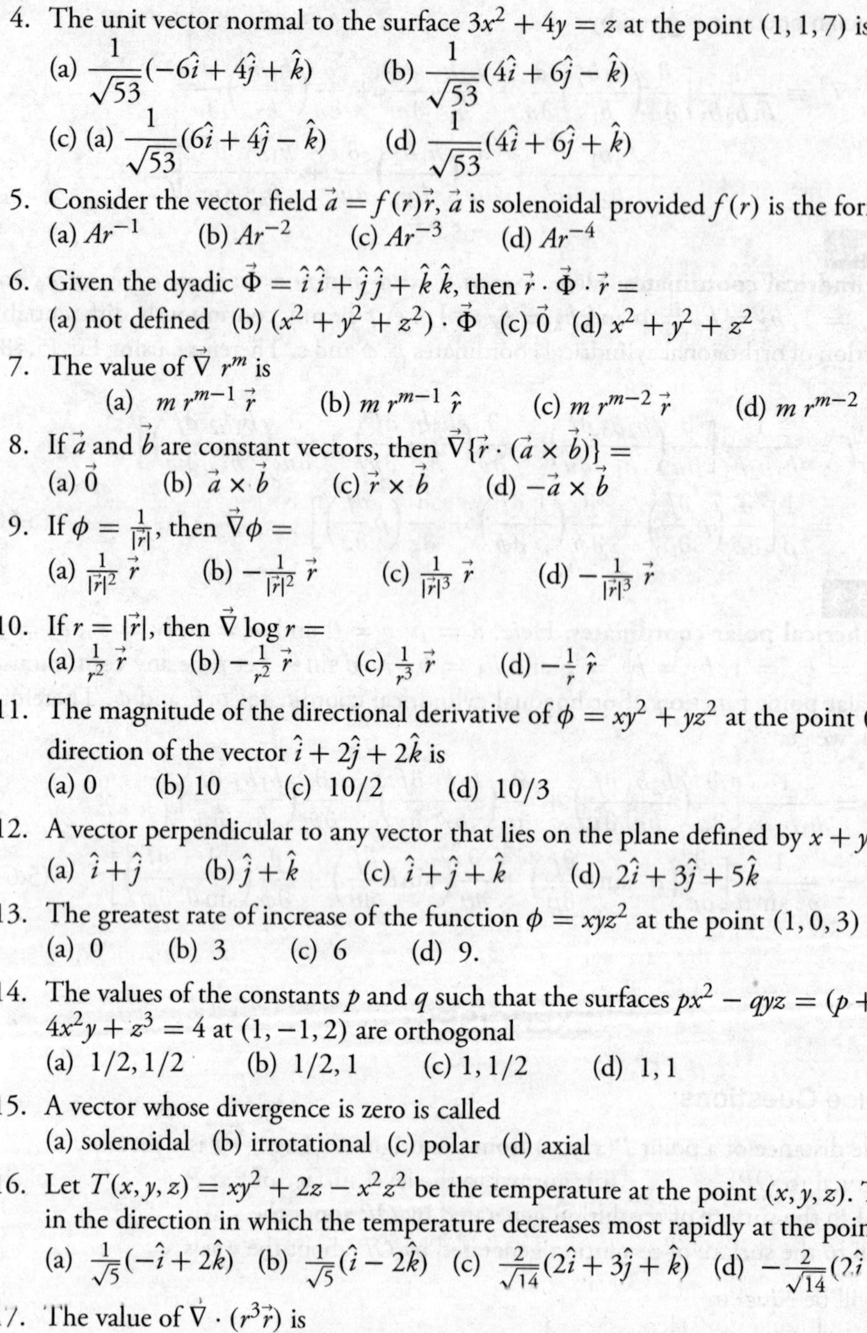

4. The unit vector normal to the surface $3x^2 + 4y = z$ at the point $(1, 1, 7)$ is
(a) $\frac{1}{\sqrt{53}}(-6\hat{i} + 4\hat{j} + \hat{k})$ (b) $\frac{1}{\sqrt{53}}(4\hat{i} + 6\hat{j} - \hat{k})$
(c) (a) $\frac{1}{\sqrt{53}}(6\hat{i} + 4\hat{j} - \hat{k})$ (d) $\frac{1}{\sqrt{53}}(4\hat{i} + 6\hat{j} + \hat{k})$

5. Consider the vector field $\vec{a} = f(r)\vec{r}$, $\vec{a}$ is solenoidal provided $f(r)$ is the form of
(a) Ar^{-1} (b) Ar^{-2} (c) Ar^{-3} (d) Ar^{-4}

6. Given the dyadic $\vec{\Phi} = \hat{i}\,\hat{i} + \hat{j}\,\hat{j} + \hat{k}\,\hat{k}$, then $\vec{r} \cdot \vec{\Phi} \cdot \vec{r} =$
(a) not defined (b) $(x^2 + y^2 + z^2) \cdot \vec{\Phi}$ (c) $\vec{0}$ (d) $x^2 + y^2 + z^2$

7. The value of $\vec{\nabla}\, r^m$ is
(a) $m\, r^{m-1}\, \vec{r}$ (b) $m\, r^{m-1}\, \hat{r}$ (c) $m\, r^{m-2}\, \vec{r}$ (d) $m\, r^{m-2}\, \hat{r}$

8. If $\vec{a}$ and $\vec{b}$ are constant vectors, then $\vec{\nabla}\{\vec{r} \cdot (\vec{a} \times \vec{b})\} =$
(a) $\vec{0}$ (b) $\vec{a} \times \vec{b}$ (c) $\vec{r} \times \vec{b}$ (d) $-\vec{a} \times \vec{b}$

9. If $\phi = \frac{1}{|\vec{r}|}$, then $\vec{\nabla}\phi =$
(a) $\frac{1}{|\vec{r}|^2}\, \vec{r}$ (b) $-\frac{1}{|\vec{r}|^2}\, \vec{r}$ (c) $\frac{1}{|\vec{r}|^3}\, \vec{r}$ (d) $-\frac{1}{|\vec{r}|^3}\, \vec{r}$

10. If $r = |\vec{r}|$, then $\vec{\nabla} \log r =$
(a) $\frac{1}{r^2}\, \vec{r}$ (b) $-\frac{1}{r^2}\, \vec{r}$ (c) $\frac{1}{r^3}\, \vec{r}$ (d) $-\frac{1}{r}\, \hat{r}$

11. The magnitude of the directional derivative of $\phi = xy^2 + yz^2$ at the point $(2, -1, 1)$ in the direction of the vector $\hat{i} + 2\hat{j} + 2\hat{k}$ is
(a) 0 (b) 10 (c) 10/2 (d) 10/3

12. A vector perpendicular to any vector that lies on the plane defined by $x + y + z = 5$ is
(a) $\hat{i} + \hat{j}$ (b) $\hat{j} + \hat{k}$ (c) $\hat{i} + \hat{j} + \hat{k}$ (d) $2\hat{i} + 3\hat{j} + 5\hat{k}$

13. The greatest rate of increase of the function $\phi = xyz^2$ at the point $(1, 0, 3)$ is
(a) 0 (b) 3 (c) 6 (d) 9.

14. The values of the constants p and q such that the surfaces $px^2 - qyz = (p + 2)x$ and $4x^2y + z^3 = 4$ at $(1, -1, 2)$ are orthogonal
(a) 1/2, 1/2 (b) 1/2, 1 (c) 1, 1/2 (d) 1, 1

15. A vector whose divergence is zero is called
(a) solenoidal (b) irrotational (c) polar (d) axial

16. Let $T(x, y, z) = xy^2 + 2z - x^2z^2$ be the temperature at the point (x, y, z). The unit vector in the direction in which the temperature decreases most rapidly at the point $A(1, 0, -1)$ is
(a) $\frac{1}{\sqrt{5}}(-\hat{i} + 2\hat{k})$ (b) $\frac{1}{\sqrt{5}}(\hat{i} - 2\hat{k})$ (c) $\frac{2}{\sqrt{14}}(2\hat{i} + 3\hat{j} + \hat{k})$ (d) $-\frac{2}{\sqrt{14}}(2\hat{i} + 3\hat{j} + \hat{k})$

17. The value of $\vec{\nabla} \cdot (r^3\vec{r})$ is
(a) 0 (b) $2r^2$ (c) $3r^3$ (d) $6r^3$

18. The value of $\vec{\nabla} \cdot (\frac{1}{r}\vec{r})$ is
(a) $1/r$ (b) $-1/r$ (c) $2/r$ (d) $-2/r$

19. The value of $\vec{\nabla} \cdot (\frac{1}{r^3}\vec{r})$ is
(a) $3/r^4$ (b) $-3/r^4$ (c) 0 (d) $4/r^4$

20. The value of $\vec{\nabla} \times (\frac{1}{r}\vec{r})$ is
(a) $\vec{0}$ (b) $\vec{r}$ (c) $\hat{r}$ (d) $1/r\,\hat{r}$

21. The value of $\vec{\nabla} \times (\frac{1}{r^3}\vec{r})$ is
(a) $3/r^4\hat{r}$ (b) $-3/r^4\hat{r}$ (c) $\vec{0}$ (d) $4/r^4\vec{r}$

22. $\vec{\nabla} \cdot (r^n r^{-1})$ is solenoidal if
(a) $n = 1$ (b) $n = 2$ (c) $n = 3$ (d) $n = -3$

23. Find the value of p, q and r such that $\vec{F} = (3x - 4y + pz)\hat{i} + (rx + 5y - 2z)\hat{j} + (x - qy + 7z)\hat{k}$ is irrotational:
(a) $1, 2, -4$ (b) $-4, 2, 1$ (c) $2, 1, -4$ (d) $-4, 1, 2$

24. The unit vector normal to the surface $x^2 + y^2 - z = 1$ at the point $P(1, 1, 1)$ is
(a) $\frac{1}{\sqrt{3}}(\hat{i} + \hat{j} - \hat{k})$ (b) $\frac{1}{\sqrt{6}}(2\hat{i} + \hat{j} - \hat{k})$ (c) $\frac{1}{\sqrt{6}}(\hat{i} + 2\hat{j} - \hat{k})$ (d) $\frac{1}{3}(2\hat{i} + 2\hat{j} - \hat{k})$

25. If $\vec{r} = x\hat{i} + y\hat{j}$, then
(a) $\vec{\nabla} \cdot \vec{r} = 0$ and $\vec{\nabla}|\vec{r}| = \vec{r}$ (b) $\vec{\nabla} \cdot \vec{r} = 2$ and $\vec{\nabla}|\vec{r}| = \vec{r}$
(c) $\vec{\nabla} \cdot \vec{r} = 2$ and $\vec{\nabla}|\vec{r}| = 1/r\,\hat{r}$ (d) $\vec{\nabla} \cdot \vec{r} = 3$ and $\vec{\nabla}|\vec{r}| = 1/r\,\hat{r}$

26. If $\vec{F} = 3x^2z\hat{i} - 3y^3z^2\hat{j} + xyz^3\hat{k}$, then $\vec{\nabla} \cdot \vec{F}$ at $(-1, 1, 1)$ is
(a) 13 (b) -13 (c) 15 (d) -15

27. If $\vec{F} = (3x^2y - z)\hat{i} + (xz^3 + y)\hat{j} + 2x^2z\hat{k}$, then $\vec{\nabla}\left(\vec{\nabla} \cdot \vec{F}\right)$ at the point $(2, 1, -1)$ is
(a) $\vec{0}$ (b) $12\hat{i} + 14\hat{j}$ (c) $14\hat{i} + 12\hat{j}$ (d) $14\hat{i} + 12\hat{k}$

28. If the surface $5x^2 - 2byz = 9x$ be orthogonal to the surface $4x^2y + z^3 = 4$ at the point $(1, -1, 2)$ then b is equal to
(a) 0 (b) 1 (c) 2 (d) 3

29. The position vector of a particle at time t is $\vec{r} = \cos(t-1)\hat{i} + \sinh(t-1)\hat{j} + at^2\hat{k}$. If at $t = 1$, the acceleration of the particle is perpendicular to its position vector, and a is equal to
(a) 0 (b) 1 (c) $1/2$ (d) $1/\sqrt{2}$

30. If $u = x^2 - y^2 + z^2$ and $\vec{r} = x\hat{i} + y\hat{j} + z\hat{k}$, then $\vec{\nabla} \cdot (u\vec{r})$ is equal to
(a) $5u$ (b) $5|\vec{r}|$ (c) $5(u - |\vec{r}|)$ (d) $5(u + |\vec{r}|)$

31. If $\vec{r} = x\hat{i} + y\hat{j} + z\hat{k}$ and $r = |\vec{r}|$, then $\vec{\nabla}\phi(r)$ is
(a) $\phi'(r)\vec{r}$ (b) $\frac{\phi(r)}{r}\vec{r}$ (c) $\frac{\phi'(r)}{r}\vec{r}$ (d) None of these

32. The value of $\vec{\nabla} \cdot (\vec{r} \times \vec{a})$, where $\vec{a}$ is a constant vector is
(a) 0 (b) $-2a$ (c) $2a$ (d) a

33. If $\vec{a}$ and $\vec{b}$ are irrotational, then $\vec{a} \times \vec{b}$ is
(a) neither irrotational nor solenoidal (b) solenoidal but not irrotational
(c) both solenoidal and irrotational (d) irrotational but not solenoidal

34. If $\vec{F} = 3x^2\hat{i} + 5xy^2\hat{j} + xyz^3\hat{k}$, then $\vec{\nabla} \times \vec{F}$ at $(1, 2, 3)$ is
(a) $27\hat{i} - 54\hat{j} + 20\hat{k}$ (b) $27\hat{i} + 54\hat{j} + 20\hat{k}$
(c) $-27\hat{i} - 54\hat{j} + 20\hat{k}$ (d) $27\hat{i} - 54\hat{j} - 20\hat{k}$

35. Consider the vector field $\vec{F} = (ax + y + a)\hat{i} + \hat{j} - (x + y)\hat{k}$, where a is a constant. If $\vec{F} \cdot (\vec{\nabla} \times \vec{F}) = 0$, then the value of a is
(a) 1 (b) 0 (c) -1 (d) $3/2$

36. If $\vec{v}$ is a vector having fixed direction, then the value of $\vec{\nabla} \times \vec{v}$ is
(a) 0 (b) along $\vec{v}$ (c) perpendicular to $\vec{v}$ (d) none of these

37. For the circular motion of a particle, we have $\vec{v} = \vec{\omega} \times \vec{r}$, assuming that $\vec{\omega}$ is a constant, we have $\vec{\nabla} \cdot \vec{v} =$
(a) 2ω (b) ω (c) -2ω (d) 0

38. Two surfaces $\phi(x, y, z) = c_1$ and $\phi(x, y, z) = c_2$ will cut orthogonally, if
(a) $\nabla\phi \cdot \nabla\psi = 0$ (b) $\nabla^2\phi = 0, \nabla^2\psi = 0$
(c) $\nabla^2\phi = 0, \nabla^2\psi \neq 0$ (d) $\nabla^2\phi \neq 0, \nabla^2\psi = 0$

39. The curl of the velocity of any particle of a rigid body is equal to
(a) the linear velocity vector of the body
(b) twice the linear velocity vector of the body
(c) the angular velocity vector of the body
(d) twice of the angular velocity vector of the body

40. The value of $\vec{a} \cdot \vec{\nabla}(1/r)$ is
(a) $\frac{\vec{a} \cdot \vec{r}}{r^2}$ (b) $\frac{\vec{a} \cdot \vec{r}}{r^2}$ (c) $-\frac{\vec{a} \cdot \vec{r}}{r^3}$ (d) $\frac{\vec{a} \cdot \vec{r}}{r^3}$

41. If $\vec{A} = x\hat{i} + y\hat{j} + z\hat{k}$, then $\nabla^2\vec{A}$ equals
(a) 1 (b) 3 (c) -3 (d) 0

42. The condition for which $\phi = ax^2 + by^2 + cz^2$ is harmonic is
(a) $a + b + c = 0$ (b) $a - b + c = 0$ (c) $a + b - c = 0$ (d) $a - b - c = 0$

43. If $\vec{\nabla} \times \vec{a} = \vec{0}$ then $\vec{a}$ is
(a) solenoidal (b) irrotational (c) polar (d) axial

44. The function r^{-1} is
(a) solenoidal (b) axial (c) polar (d) harmonic

45. Let $\vec{r} = x\hat{i} + y\hat{j} + z\hat{k}$. If $\phi(x, y, z)$ is a solution of the Laplace equation, then the vector field $\vec{\nabla}\phi + \vec{r}$ is
(a) neither solenoidal nor irrotational (b) solenoidal but not irrotational
(c) both solenoidal and irrotational (d) irrotational but not solenoidal

46. The value of c for which there exists a twice differentiable vector field $\vec{F}$ with $\vec{\nabla} \times \vec{F} = 2x\hat{i} - 7y\hat{j} + cz\hat{k}$ is
(a) 0 (b) 2 (c) 5 (d) 7

47. If $\phi(x, y, z)$ is any solution of Laplace's equation, then $\vec{\nabla}\phi$ is a vector which is
(a) solenoidal (b) irrotational (c) both solenoidal and irrotational (d) none

48. If $\vec{\omega}$ is a constant vector and $\vec{v} = \vec{\omega} \times \vec{r}$, then
(a) $\vec{\nabla} \cdot \vec{v} = 0$ (b) $\vec{\nabla} \cdot \vec{v} \neq 0$ (c) $\vec{\nabla} \times \vec{v} = \vec{0}$ (d) $\vec{\nabla} \times \vec{v} = 2\vec{\omega}$

49. If $\nabla^2 f(r) = 0$, then $f(r)$ equals to
(a) $A + Br$ (b) $A + B/r$ (c) $A/r + B/r^2$ (d) $A + Br^2$

50. $\nabla^2(\ln r)$ equals to
(a) $\frac{1}{r}$ (b) $\frac{1}{r^2}$ (c) $2r^{-1}$ (d) $3r^{-2}$

51. The vector $\sin y\hat{i} + \sin x\hat{j} + e^z\hat{k}$ is
(a) solenoidal (b) irrotational
(c) both solenoidal and irrotational (d) neither solenoidal nor irrotational

Answers

1. b	2. c	3. a	4. c	5. c	6. d	7. c	8. b	9. d	10. a	11. d	12. c	13. d
14. b	15. a	16. b	17. d	18. c	19. c	20. a	21. c	22. d	23. a	24. d	25. c	
26. b	27. c	28. b	29. d	30. a	31. c	32. a	33. b	34. a	35. a	36. c	37. d	
38. a	39. d	40. c	41. d	42. a	43. b	44. d	45. d	46. c	47. c	48. a	49. b	
50. b	51. c											

Short Answer Questions

1. If ϕ and ψ are scalar point functions in three-dimensional space, both having first order continuous partial derivatives, then show that
(a) $\vec{\nabla}(c\,\phi) = c\,\vec{\nabla}\phi$, for any scalar c (b) $\vec{\nabla}(\phi \pm \psi) = \vec{\nabla}\phi \pm \vec{\nabla}\psi$

2. If $\phi(x, y) = \dfrac{y}{x^2 + y^2}$, find the magnitude of the directional derivative making a 30° angle with the positive x axis at the point $(0, 1)$. *Ans*: $-1/2$

3. Find $\vec{\nabla}\phi$, where $\phi(x, y) = x^2 + y^2$ and determine its magnitude and direction at $(3, 4)$. *Ans*: $10, \tan^{-1}(\frac{4}{3})$

4. Let R be the distance from a fixed point $A(a, b, c)$ to any point $P(x, y, z)$. Prove that $\vec{\nabla}R$ is a unit vector in the direction $\vec{AP} = \vec{R}$.

5. Show that a necessary and sufficient condition for a scalar point function ϕ to be constant is that $\vec{\nabla}\phi = \vec{0}$.

6. Prove that the gradient of a constant scalar point function is a zero vector, and conversely.

7. Prove that $\vec{\nabla}\phi$ is invariant of the choice of the basis.

8. Let ρ and p be two scalar point functions such that ρ is a function of p; show that $\vec{\nabla}\rho = \dfrac{d\rho}{dp}\vec{\nabla}p$.

9. Prove that gradient is a linear vector differential operator.

10. Prove that the sum of any two scalar point functions is equal to the vector sum of their gradients.

11. Prove that the differential operation gradient/ divergence/ curl is distributive with respect to the sum and difference of function.

12. Find the directional derivative of
(a) $x^2 + y^2 + z^2$ along the vector $\hat{j} - \hat{k}$ at the point $(1, -1, 2)$ *Ans*: $-3\sqrt{2}$

(b) $xy^2 + yz^2$ along the vector $\vec{i} + 2\hat{j} + 2\hat{k}$ at the point $(2, -1, 1)$ *Ans*: -3

(c) $xy^2 + yz^2 + zx^2$ along the direction of the coordinate axes at the point $(1, -2, 5)$.
Ans: $14, 21, -19$

13. Determine whether $\vec{F}$ is a gradient of some scalar function ϕ. If it is so, find ϕ:

(a) $\vec{F} = yz\hat{i} + xz\hat{j} + xy\hat{k}$ *Ans*: $xyz + c$

(b) $\vec{F} = 2xyz\hat{i} + x^2z\hat{j} + (x^2y + 1)\hat{k}$ *Ans*: $(x^2y + 1)z + c$

14. If the vector $\vec{F}$ has a fixed direction, then prove that $\vec{\nabla} \times \vec{F}$ is orthogonal to $\vec{F}$.

15. Show that the directional derivative of a scalar function of any point along any tangent line to the level surface at the point is zero.

16. Find the value of the directional derivative of

(a) $\phi(x, y, z) = xyz$ at the point $(2, 2, 2)$ along the directions $\hat{i}$ and $\hat{j}$ *Ans*: $4, 4$

(b) $\phi(x, y, z) = x^2y^3z^4$ at the point $A(2, 3, -1)$ along the direction with makes equal angles with the positive x, y and z axis *Ans*: 0

17. If $\vec{r} = x\hat{i} + y\hat{j} + z\hat{k}$, $r = |\vec{r}|$, prove the following:

(a) $\vec{\nabla}(x^3 + y^3 + z^3 - 3xyz) = 6(x + y + z)$ (b) $\vec{\nabla}(\log r) = \dfrac{1}{r^2}\vec{r}$

(c) $\vec{\nabla} f(r) = f'(r)\,\vec{\nabla} r = \dfrac{f'(r)}{r}\vec{r}$ (d) $\vec{A}\cdot\vec{\nabla}\left(\dfrac{1}{r}\right) = -\dfrac{\vec{A}\cdot\vec{r}}{r^3}$ (e) $\vec{\nabla}|\vec{r}|^3 = 3r\,\vec{r}$

(f) $\vec{\nabla}(\vec{r}\cdot\vec{r}) = 2\vec{r}$ (g) $\vec{\nabla}(\vec{c}\cdot\vec{r}) = \vec{c}$; $\vec{c}$ is a constant vector

(h) $\vec{\nabla}\left|\vec{c}\times\vec{r}\right|^n = n\left|\vec{c}\times\vec{r}\right|^{n-2}\vec{c}\times(\vec{r}\times\vec{c})$ (i) $\vec{c}\cdot\vec{\nabla}(\vec{c}\cdot\vec{\nabla}r^{-1}) = \dfrac{3(\vec{c}\cdot\vec{r})^2 - c^2r^2}{r^5}$

(j) $\vec{\nabla}\left(3r^2 - 4\sqrt{r} + \dfrac{6}{\sqrt[3]{r}}\right) = (6 - 2r^{-3/2} - 2r^{-7/3})\vec{r}$

(k) If $\vec{A} = \dfrac{1}{r}\vec{r}$, then $\vec{\nabla}(\vec{\nabla}\cdot\vec{A}) = -2r^{-3}\,\vec{r}$ (l) $\vec{\nabla} r^m = mr^{m-2}\,\vec{r}$

(m) $\vec{\nabla} r = \dfrac{1}{r}\vec{r}$ and $\vec{\nabla} f(r) = \dfrac{f'(r)}{r}\vec{r}$

(n) $\vec{b}\cdot\vec{\nabla}\left[\vec{a}\cdot\vec{\nabla}\dfrac{1}{r}\right] = \dfrac{3(a\cdot\vec{r})(\vec{b}\cdot\vec{r})}{r^5} - \dfrac{\vec{a}\cdot\vec{b}}{r^3}$, where $\vec{a}, \vec{b}$ are constant vectors.

18. Prove that $\vec{\nabla}\displaystyle\int \phi(u)\, du = \phi(u)\,\vec{\nabla} u$.

19. (a) Define divergence of a vector point function. Prove that the function $\vec{\nabla}\cdot\vec{F}$ is a point function.

(b) Prove that $\vec{\nabla}\cdot\vec{F}$ is independent of the choice of three mutually orthogonal unit vectors.

(c) Prove that the divergence of a constant vector point function is zero.

20. Prove that a solenoidal can be expressed as a curl of another vector.

21. Prove the following:

(a) $\vec{\nabla}\cdot(x^2z\hat{i} - 2y^3z^2\hat{j} + xy^2z\hat{k}) = -3$ at $(1, -1, 1)$

(b) $\vec{\nabla}\cdot\vec{r} = 3$ and $\vec{\nabla}\cdot(r^m\vec{r}) = (m + 3)r^m$

(c) $\vec{\nabla}\cdot(2x^2z\hat{i} - xy^2z\hat{j} + 3yz^2\hat{k}) = 4xz - 2xyz + 6yz$

(d) $\vec{\nabla} \cdot \left[(\vec{r} \times \vec{a}) \times \vec{b}\right] = -2\vec{b} \cdot \vec{a}$

(e) $\vec{\nabla} \cdot \left\{\dfrac{f(r)}{r}\vec{r}\right\} = \dfrac{1}{r^2}\dfrac{d}{dr}(r^2 f)$ and $\vec{\nabla} f(r) \times \vec{r} = \vec{0}$

(f) $\vec{\nabla} \cdot (\vec{\nabla} r^m) = m(m+1)r^{n-2}$ and $\vec{\nabla} \times (\vec{\nabla} r^m) = 0$

22. Prove the following:

(a) $\vec{\nabla} \times \left[e^{xyz}\,(\hat{i} + \hat{j} + \hat{k})\right] = 2e(-\hat{j} + \hat{k})$ at $(1, -1, -1)$

(b) $\vec{\nabla} \times \left\{(\vec{a} \cdot \vec{r})\,\vec{a}\right\} = \vec{0}$, where $\vec{a}$ is a constant vector (c) $\vec{\nabla} \times \left(\dfrac{1}{r^2}\vec{r}\right) = \vec{0}$

(d) $\vec{\nabla} \times \left(\dfrac{1}{r^3}\vec{r}\right) = \vec{0}$ (e) $\vec{\nabla} \times \left(\phi \vec{\nabla} \phi\right) = \vec{0}$ (f) $\vec{\nabla} \times \left(\vec{r}\phi(r)\right) = \vec{0}$

(g) $\vec{\nabla} \cdot \vec{r} = 3$, $\vec{\nabla} \cdot \vec{c} = 0$, $\vec{\nabla} \times \vec{r} = \vec{0}$, $\vec{\nabla} \times \vec{c} = \vec{0}$, where $\vec{c}$ is a constant vector

23. Prove the following:

(a) $\vec{\nabla} \cdot \vec{\nabla}\phi = \dfrac{\partial^2 \phi}{\partial x^2} + \dfrac{\partial^2 \phi}{\partial y^2} + \dfrac{\partial^2 \phi}{\partial z^2}$ (b) $\vec{\nabla} \times (\vec{\nabla}\phi) = \vec{0}$

(c) $\vec{\nabla} \cdot (\vec{\nabla} \times \vec{f}) = 0$ (d) $\vec{\nabla} \times (\vec{\nabla} \times \vec{f}) = \vec{\nabla}(\vec{\nabla} \cdot \vec{f}) - \vec{\nabla}^2 \vec{f}$

(e) $\vec{\nabla} \cdot \vec{\nabla} \times \vec{\nabla}(\vec{a}\phi) + \nabla^2\, \vec{\nabla} \cdot (\vec{a}\phi) = \vec{a} \cdot \vec{\nabla}(\nabla^2 \phi)$, where ϕ is a scalar point function

24. Prove the following:

(a) $\nabla^2 f(r) = f''(r) + \dfrac{2}{r} f'(r)$ (b) $\nabla^2\left(\dfrac{x}{r^2}\right) = -\dfrac{2x}{r^4}$ (c) $\vec{\nabla}^2\left(\dfrac{x}{r^3}\right) = 0$

(d) $\vec{\nabla}^2\left(\ln r\right) = r^{-2}$ (e) $\vec{\nabla}^2\left(\dfrac{1}{r^2}\vec{r}\right) = -2r^{-3}\vec{r}$

(f) $\vec{\nabla}^2\left[\vec{\nabla} \cdot \left(\dfrac{1}{r^2}\vec{r}\right)\right] = 2r^{-4}$ (g) $\vec{\nabla}^2 r^n = n(n+1)r^{n-2}$

(h) $\vec{\nabla}^2 \phi = 6z + 24xy - 2z^3 - 6y^2 z$, where $\phi = 3x^2 z - y^2 z^3 + 4x^3 y + 2x - 3y - 5$

25. (a) Define curl of a vector point function. Prove that $\vec{\nabla} \times \vec{f}$ is a point function.

(b) A vector function $\vec{f}$ is the product of a scalar function and the gradient of a scalar function, prove that $\vec{f} \cdot \vec{\nabla} \times \vec{f} = 0$.

(c) If $\vec{f} = \vec{\nabla}\left(\vec{a} \cdot \vec{\nabla} r^{-1}\right)$, show that $\vec{\nabla} \cdot \vec{f} = 0$ and $\vec{f} = \vec{\nabla} \times \vec{g}$, where $\vec{g} = -\vec{a} \times \vec{\nabla} r^{-1}$.

26. Prove the following:

(a) $\vec{\nabla}\left[\vec{r}\;\vec{a}\;\vec{b}\right] = \vec{a} \times \vec{b}$, where $\vec{a}$, $\vec{b}$ are constant vectors

(b) $\vec{\nabla} \cdot \left\{(\vec{r} \times \vec{a}) \times \vec{b}\right\} = 0$ and $\vec{\nabla} \times \left\{(\vec{r} \times \vec{a}) \times \vec{b}\right\} = 2\vec{b} \times \vec{a}$

(c) $\vec{\nabla} \cdot \left\{\vec{r} \times \vec{c}\right\} = 0$ but $\vec{\nabla} \times (\vec{r} \times \vec{c}) = -2\vec{c}$

(d) $\vec{\nabla}\left\{(\vec{r} \times \vec{a}) \cdot (\vec{r} \times \vec{b})\right\} = (\vec{b} \times \vec{r}) \times \vec{a} + (\vec{a} \times \vec{r}) \times \vec{b}$

(e) $\vec{\nabla}^2 u = 0$ if $u = x^2 - y^2 + 4z$

(f) $\vec{\nabla} \cdot (\vec{\nabla}\phi) = 12xy^2 z^4 + 4x^3 z^4 + 24x^3 y^2 z^2$, if $\phi = 2x^3 y^2 z^4$

(g) $\vec{\nabla} \cdot \left[(\vec{r} \times \vec{a}) \times (\vec{r} \times \vec{b})\right] = 2\left[\vec{a} \cdot (\vec{b} \times \vec{r}) + \vec{b} \cdot (\vec{r} \times a)\right]$

27. If $\vec{a}$ be a constant vector and $\vec{r} = x\hat{\imath} + y\hat{\jmath} + z\hat{k}$, prove that the following vector point functions are solenoidal
(a) $\vec{a} \times \vec{r}$ (b) $r^n(\vec{a} \times \vec{r})$ (c) $(x+3y)\hat{\imath} + (y-3z)\hat{\jmath} + (x-2z)\hat{k}$
(d) $yz\hat{\imath} + zx\hat{\jmath} + xy\hat{k}$ (e) $3y^4z^2\hat{\imath} + 4x^3z^2\hat{\jmath} - 3x^2y^2\hat{k}$
(f) $x(y-z)\hat{\imath} + y(z-x)\hat{\jmath} + z(x-y)\hat{k}$

28. Find the value of p such that the following vectors are solenoidal:
(a) $\vec{F}(x,y,z) = (px+y)\hat{\imath} + (y-3z)\hat{\jmath} + (x+pz)\hat{k}$ *Ans*: $:p = -1/2$
(b) $\vec{F}(x,y,z) = \dfrac{1}{x^2+y^2}(x\hat{\imath} + py\hat{\jmath}) + \hat{k}$ *Ans*: $:p = 1$

29. Prove that an irrotational vector can be expressed as a gradient of a scalar.

30. Prove that the following vector point functions are irrotational:
(a) $\vec{F}(x,y,z) = (x^2+xy^2)\hat{\imath} + (y^2+x^2y)\hat{\jmath}$ *Ans*: $:\dfrac{x^3}{3} + \dfrac{y^3}{3} + \dfrac{x^2y^2}{2} + c$
(b) $\vec{F}(x,y,z) = (6xy+z^3)\hat{\imath} + (3x^2-z)\hat{\jmath} + (3xz^2-y)\hat{k}$ *Ans*: $:3x^2y - yz + xz^3 + c.$
(c) $\vec{F} = \dfrac{1}{r^3}\vec{r}$ *Ans*: $:-r^{-1}$
Also find the scalar potential function.

31. Prove that the following vector point functions are solenoidal and irrotational:
(a) $\vec{F}(x,y,z) = y\hat{\imath} + x\hat{\jmath} + 2\hat{k}$
(b) $\vec{F}(x,y,z) = (y^2 - z^2 + 3yz - 2x)\hat{\imath} + (3xz + 2xy)\hat{\jmath} + (3xy - 2xz + 2z)\hat{k}$

32. Find ϕ, if
(a) $\vec{\nabla}\phi = 2r^4\,\vec{r}$ *Ans*: $r^3/3 + c$
(b) $\vec{\nabla}\phi = \dfrac{1}{r^5}\,\vec{r}$ and $\phi(1) = 0$ *Ans*: $\dfrac{1}{3}(1 - \dfrac{1}{r^3})$
(c) $\vec{\nabla}\phi = (2-r)e^{-r}\,\vec{r}$ *Ans*: re^{-r}
(d) $\vec{\nabla}\phi = (y^2 - 2xyz^3)\hat{\imath} + (3 + 2xy - x^2z^3)\hat{\jmath} + (6z^3 - 3x^2yz^2)\hat{k}$
Ans: $xy^2 - x^2yz^3 + 3y + 3z^4/2 + c$
(e) $\vec{\nabla}\phi = 2xyz^3\hat{\imath} + x^2z^3\hat{\jmath} + 3x^2yz^2\hat{k}$ and $\phi(1,-2,2) = 4$ *Ans*: $x^2yz^3 + 20$

33. (a) Find the directional derivative of $\phi = x^2 + y^2 + z^2$ at (x,y,z) along the x axis. *Ans*: $2x$
(b) Find the directional derivative of $\phi = x^3 + y^3 + z^3$ at $(1,-1,2)$ in the direction of the vector $\hat{\imath} + 2\hat{\jmath} + \hat{k}$. *Ans*: $21/\sqrt{6}$
(c) Find the directional derivative of $\phi = xy + yz + zx$ at $(1,2,0)$ in the direction of the vector $\hat{\imath} + 2\hat{\jmath} + 2\hat{k}$. *Ans*: $10/3$
(d) Find the directional derivative of $\phi = x^2 - y^2 + 2z^2$ at the point $P(1,2,3)$ in the direction of the straight line PQ, where Q is the point $(5,0,4)$. *Ans*: $28/\sqrt{21}$

34. (a) Find the values of the constants a, b and c so that the directional derivative of the function $\phi = ax^2 - by^2 + cz^2$ at $(1,1,2)$ has a maximum magnitude of 4 in the direction parallel to the y axis. *Ans*: $a = 0, b = 2, c = 0$

(b) Find the values of the constants a, b and c so that the directional derivative of $\phi = axy^2 + byz + cz^2x^3$ at $(1, 2, -1)$ has a maximum magnitude of 64 in the direction parallel to the z axis. *Ans*: $a = \pm 6, b = \pm 24, c = \mp 8$

35. Find the gradient of f at the point $(1, -2, -1)$, where $f = 3x^2y - y^3z^2$.

Ans: $-(12\hat{i} + 9\hat{j} + 16\hat{k})$

36. What is the greatest rate of increase of $f = xyz^2$ at the point $(1, 0, 3)$? *Ans*: 9

37. (a) Find a unit normal to the surface $x^2 - y^2 + z = 2$ at $(1, -1, 2)$. *Ans*: $(2\hat{i} + 2\hat{j} + \hat{k})/3$

(b) Find the unit vector normal to the surface $z = x^2 + y^2$ at $(-1, -2, 5)$.

Ans: $-(2\hat{i} + 4\hat{j} + \hat{k})/\sqrt{21}$

(c) Find the unit vector normal to the surface $z = x^2 + y^2$ at $(1, 2, 5)$.

Ans: $(2\hat{i} + 4\hat{j} - \hat{k})/\sqrt{21}$

38. If $\vec{F} = x^2\hat{i} + y^2\hat{j} + z^2\hat{k}$, then show that $\vec{\nabla} \cdot \vec{F} = 2(x + y + z)$ and $\vec{\nabla} \times \vec{F} = \vec{0}$.

39. Are the following forces conservative? Give reasons.

(a) $\vec{F} = m\omega^2\,\vec{r}$ (b) $\vec{F} = \vec{A} \times \vec{r}$ (c) $\vec{F} = r^2\,\vec{r}$

(d) $\vec{F} = r^5\,\vec{r}$ (e) $\vec{F} = -kr^3\,\vec{r}$ (f) $\vec{F} = -kr^{-n}\,\vec{r}$, k and n are constants

If conservative, find the potential function.

40. If the vector $\vec{F}$ has a fixed direction, then prove that $\vec{\nabla} \times \vec{F}$ is orthogonal to $\vec{F}$.

41. If the Laplacian of a scalar point function vanishes, then prove that its gradient is solenoidal as well as irrotational.

42. For the function $\phi(x, y) = x/(x^2 + y^2)$, find the magnitude of the directional derivative making an angle of 45° with the positive direction of the x axis at the point $(1, 2)$.

Ans: $-1/25\sqrt{2}$

43. If $\phi = (x^2 + y^2 + z^2)e^{-\sqrt{x^2+y^2+z^2}}$; prove that $|\vec{\nabla}\phi| = (2 - r)re^{-r}$.

44. Verify the following results:

(a) If $\vec{F} = x^2y\hat{i} + y^2z\hat{j} + z^2x\hat{k}$, then $\vec{\nabla} \cdot (\vec{\nabla} \times \vec{F}) = 0$

(b) If $\vec{F} = \vec{\nabla}(x^3 + y^3 + z^3 - 3xyz)$, then $\vec{\nabla} \cdot \vec{F} = 6(x + y + z)$ and $\vec{\nabla} \times \vec{F} = 0$

(c) If $\vec{F} = (x^3 - y^3)\hat{i} + (y^3 - z^3)\hat{j} + (z^3 - x^3)k$, then $\vec{\nabla} \cdot (\vec{\nabla} \times \vec{F}) = 0$

(d) If $\vec{F} = x^3y^3\hat{i} + (z^3 - 3xy)\hat{j} + yz\hat{k}$, then $\vec{\nabla} \cdot (\vec{\nabla} \times \vec{F}) = 0$

(e) If $\vec{F} = x^2y\hat{i} + y^2z\hat{j} + z^2x\hat{k}$, then $\vec{\nabla} \cdot (\vec{\nabla} \times \vec{F}) = 0$

(f) If $\vec{F} = (x + y + 1)\hat{i} + \hat{j} + (-x - y)\hat{k}$, then $\vec{F} \cdot (\vec{\nabla} \times \vec{F}) = 0$

45. (a) If ϕ and ψ are differentiable scalar fields, prove that $\vec{\nabla}\phi \times \vec{\nabla}\psi$ is solenoidal.

(b) Find the most general differential function $f(r)$ so that $f(r)\,\vec{r}$ is solenoidal.

Ans: $f(r) = cr^{-3}$.

(c) Prove that the vector $f(r)\,\vec{r}$ is irrotational, where $f(r)$ is differentiable.

(d) If $\vec{A}$ and $\vec{B}$ are irrotational, then prove that $\vec{A} \times \vec{B}$ is a solinoidal vector.

(e) Prove that $r^n\vec{r}$ is irrotational for all values of n but it is solinoidal only when $n = -3$.

(f) If $\vec{\nabla}^2\phi = 0$, show that $\vec{\nabla}\phi$ is both solenoidal and irrotational.

(g) Prove that the vector $\vec{F} = r^{-2}\,\vec{r}$ is irrotational. Find ϕ such that $\vec{F} = -\vec{\nabla}\phi$ and such that $\phi(a) = 0, a > 0$. *Ans*: $\phi = \ln(a/r)$.

(h) Prove that the vector $\vec{F} = r^{-3}\vec{r}$ is harmonic.

46. Give an example of a vector function $\vec{F}$ such that
(a) $\vec{F} \cdot (\vec{\nabla} \times \vec{F}) = 0$. (b) $\vec{\nabla} \cdot (\vec{\nabla} \times \vec{F}) = 0$. (c) $\vec{\nabla} \times (\vec{\nabla} \times \vec{F}) = \vec{\nabla}(\vec{\nabla} \cdot \vec{F}) - \nabla^2 \vec{F}$.
Ans: (a) $xyz(\hat{i} + \hat{j} + \hat{k})$ (b) $x^2\hat{i} + y^2\hat{j} + z^2\hat{k}$ (c) $x^2\hat{i} + y^2\hat{j} + z^2\hat{k}$

47. Is there a differentiable vector function $\vec{F}$ such that
(a) $\vec{\nabla} \times \vec{F} = \vec{r}$ (b) $\vec{\nabla} \times \vec{F} = 2\hat{i} + \hat{j} + 3\hat{k}$ *Ans*: (a) No (b) $3x\hat{i} + (2y - x)\hat{k} + \vec{\nabla}\phi$

48. Prove the following:
(a) $\nabla^2\left(\frac{1}{r}\right) = 0$ (b) $\nabla^2\left(\frac{y}{r^3}\right) = 0$ (c) $\nabla^2\left(\frac{x}{r^3}\right) = 0$
(d) $\nabla^2\phi = 0$, where $\phi = x^2 - y^2 + z$ (e) $\nabla^2\phi = \lambda \log r + \mu$ (f) $\nabla^2(r^{-3}\vec{r}) = 0$

49. Prove that in any orthogonal curvilinear coordinate system
(a) div curl $\vec{F} = 0$ (b) curl grad $\vec{F} = \vec{0}$

50. Prove that the scalar function $\phi(r) = \lambda \log r + \mu$ satisfies Laplace's equation $\nabla^2\phi(r) = 0$, where λ and μ are constants.

51. Find the condition for which $\phi = \lambda x^2 + \mu y^2 + \nu z^2$ is harmonic. *Ans*: $\lambda + \mu + \nu = 0$

Long Answer Questions

1. Prove the following:
(a) $\vec{a} \cdot \vec{\nabla}(\vec{b} \cdot \vec{\nabla} r^{-1}) = \dfrac{3(\vec{a} \cdot \vec{r})(\vec{b} \cdot \vec{r})}{r^5} - \dfrac{\vec{a} \cdot \vec{b}}{r^3}$
(b) $\vec{\nabla} \cdot \left\{r^n \vec{c} \times \vec{r}\right\} = 0$; and $\vec{\nabla} \cdot \left(\frac{1}{r}\vec{r}\right) = \frac{2}{r}$
(c) $\vec{\nabla} \times \left\{r^n \vec{c} \times \vec{r}\right\} = (n+2)r^n\vec{c} - nr^{n-2}(\vec{r} \cdot \vec{c})$
(d) $\vec{\nabla} \times \left\{\frac{1}{r^3}(\vec{a} \times \vec{r})\right\} = -\frac{1}{r^3}\vec{a} + \frac{3}{r^5}(\vec{a} \cdot \vec{r})\,\vec{r}$, where $\vec{a}$ is a constant vector
(e) $\vec{\nabla}^2 r^n = n(n+1)r^{n-2}$ and $\vec{\nabla}^2(r^n\vec{r}) = n(n+3)r^{n-2}\vec{r}$ (f) $\vec{\nabla}^2\left\{\vec{\nabla} \cdot \left(\frac{1}{r^2}\vec{r}\right)\right\} = \frac{2}{r^4}$

2. If the flow vector $\vec{u}(x, y, z)$ within a moving fluid is given by $\vec{u}(x, y, z) = x^4y\hat{i} + y^4z\hat{j} + z^4x\hat{k}$, find the rate of outward flow of the fluid per unit volume per unit time at the point $(1, 1, 1)$. *Ans*: :12
Hint: Calculate div$\vec{u}$ at $(1, 1, 1)$

3. Find the directional derivative of
(a) $f = x^3 + y^3 + z^3$ at $(1, -1, 2)$ in the direction $\hat{j} - \hat{k}$ *Ans*: $-9\sqrt{2}/2$
(b) $f = xy + yz + zx$ at $(1, 2, 0)$ in the direction $\hat{i} + 2\hat{j} + 2\hat{k}$ *Ans*: 10/3

4. Find the maximum value of the directional derivative and the direction (in which it occurs) of
(a) $f = x^2 - y^2 + z^2$ at $(1, 3, 2)$ *Ans*: $\sqrt{56}, (2, 6, 4)/\sqrt{56}$
(b) $f = x^2yz$ at $(1, 4, 1)$ *Ans*: 9

5. Calculate the normal derivative of
(a) $f = yz + zx + xy$ at $(-1, 1, 1)$ *Ans*: 2

(b) $f = xyz$ at $(-1, 1, 1)$ *Ans:* $\sqrt{3}$

(c) $f = x^3 - y^2 + z$ at $(-1, 1, 1)$ *Ans:* $\sqrt{14}$

6. Find the unit normal to the level surface

(a) $x^2 + y - z = 1$ at the point $(1, 0, 0)$ *Ans:* $(2\hat{i} + \hat{j} - \hat{k})/\sqrt{6}$

(b) $x^2yz + 4xz^2 = 6$ at $(1, -2, -1)$ *Ans:* $\frac{1}{\sqrt{165}}(8\hat{i} - \hat{j} - 10\hat{k})$

7. Find the equations of the tangent plane and normal line to the surface

(a) $xyz = 4$ at the point $\hat{i} + 2\hat{j} + 2\hat{k}$ *Ans:* $2x + y + z = 6, \frac{x-1}{2} = \frac{y-2}{1} = \frac{z-2}{1}$

(b) $x^2 + y^2 - z = 0$ at $(2, -1, 5)$ *Ans:* $4x - 2y - z = 5, \frac{x-2}{4} = \frac{y+1}{-2} = \frac{z-5}{-1}$

(c) $yz - zx + xy = -5$ at $(1, -1, 2)$ *Ans:* $3x - 3y + 2z = 10$

8. Find the angle between the surfaces

(a) $x^2 + y^2 + z^2 = 9$ and $z = x^2 + y^2 - 3$ at $(2, -1, 2)$ *Ans:* $\cos^{-1}(8/3\sqrt{21})$

(b) $z = x^2 + y^2$ and $z = \left(x - \frac{\sqrt{6}}{6}\right)^2 + \left(y - \frac{\sqrt{6}}{6}\right)^2$ at $\left(\frac{\sqrt{6}}{12}, \frac{\sqrt{6}}{12}, \frac{1}{12}\right)$ *Ans:* $60°$

9. Find the divergence and curl of the following vector point functions:

(a) $\vec{F} = xy^2\hat{i} + 2x^2yz\hat{j} - 3yz^2\hat{k}$ *Ans:* $y^2 + 2x^2z - 6yz, -(3z^2 + 2x^2y)\hat{i} + (4xyz - 2xy)\hat{k}$.

(b) $\vec{F} = \vec{\nabla}(x^3 + y^3 + z^3 - 3xyz)$ *Ans:* $6(x + y + z), 0$

(c) $\vec{F} = (y^2 + z^2 - x^2)\hat{i} + (z^2 + x^2 - y^2)\hat{j} + (x^2 + y^2 - z^2)\hat{k}$

Ans: $-2(x + y + z), 2(y - z)\hat{i} + 2(z - x)\hat{j} + (x - y)\hat{k}$

(d) $\vec{F} = xz^3\hat{i} - 2x^2yz\hat{j} + 2yz^4\hat{k}$, at the point$(1, -1, 1)$ *Ans:* $-3, 3\hat{j} + 4\hat{k}$

10. Find the direction in which the directional derivative at $f(x, y) = \dfrac{x^2 - y^2}{xy}$ at $(1, 1)$ is zero.

Ans: $\frac{1}{\sqrt{2}}(\hat{i} + \hat{j})$

11. Find the directional derivative at the scalar function $\phi(x, y, z) = xyz$ in the direction of the outer normal to the surface $z = xy$ at the point $(3, 1, 3)$. *Ans:* $\dfrac{27}{\sqrt{11}}$

12. Find the value of the directional derivative of $f = xyz$, at the point $(1, -1, 1)$, along the direction which makes equal angles with the positive direction of the x, y and z axes.

Ans: $\sqrt{3}$

13. The temperature at points in space is given by $T(x, y, z) = x^2 + y^2 - z^2$. A mosquito located at $(1, 1, 2)$ desires to fly in such a direction that it will get warm as soon as possible. In what direction should it move ? *Ans:* $\frac{1}{3}(2\hat{i} + 2\hat{j} - \hat{k})$

14. (a) The pressure p at a point (x, y, z) within a fluid is given by $p(x, y, z) = yz + z^2x$. Find the surface of equipressure through the point $(1, 1, 1)$. Find also the direction and magnitude of the greatest variation of pressure at that point.

Ans: $yz + xz^2 = 2, \hat{i} + \hat{j} + 3\hat{k}, \sqrt{11}$

(b) $p(x, y, z) = xy + yz + z^2$ point $(1, 2, 1)$. *Ans:* $xy + yz + z^2 = 5, \frac{1}{\sqrt{6}}(\hat{i} + \hat{j} + 2\hat{k}), 2\sqrt{6}$

15. Determine $\vec{\nabla} \cdot \left(\frac{\vec{r}}{r^n}\right), n > 0$ and $\vec{r} = x\hat{i} + y\hat{j} + z\hat{k}$. *Ans:* $(3 - n)/r^n$

16. If $u = x + y + z$, $v = x^2 + y^2 + z^2$ and $w = yz + zx + xy$, prove that $\vec{\nabla} u$, $\vec{\nabla} v$ and $\vec{\nabla} w$ are coplanar vectors.

17. Find the nature the force, conservative or non-conservative, given by
(a) $\vec{F} = (y^2 - x^2)\hat{i} + 2xy\hat{j}$ (b) $\vec{F} = (2xy + yz^2)\,\hat{i} + (x^2 + xz^2)\,\hat{j} + 2xyz\,\hat{k}$
If conservative, find the potential function.

18. Find constants a, b and c so that the force field defined by
(a) $\vec{F} = (x + 2y + az)\hat{i} + (bx - 3y - z)\hat{j} + (4x + cy + 2z)\hat{k}$ is conservative.
Ans: $a = 4, b = 2, c = -1$
(b) $\vec{F} = (x + 2y + az)\hat{i} + (bx + 3y + 2z)\hat{j} + (3x + cy + z)\hat{k}$ is conservative.
Ans: $a = 3, b = 2, c = 2$

19. Find the work required to move a particle from the point $(1, -1, 2)$ to $(2, 3, -1)$ in a force field with potential $V = x^3 - y^3 + 2xy - y^2 + 4x$. *Ans*: 15 units

20. Let $\vec{r} = x\hat{i} + y\hat{j} + z\hat{k}$, $r = |\vec{r}|$ and $f(r)$ be scalar functions possessing first and second order derivatives. Prove that $\nabla^2 f(r) = \dfrac{d^2 f}{dr^2} + \dfrac{2}{r}\dfrac{df}{dr}$ and hence show that if $\nabla^2 f(r) = 0$, then $f(r) = c_1 + c_2/r$, where c_1 and c_2 are arbitrary constants.

21. Let $\vec{f} = \vec{\nabla} \times (\vec{r}\phi_1 + \vec{r} \times \vec{\nabla}\phi_2)$, where ϕ_1 and ϕ_2 are two scalar functions such that $(\nabla^2 + k^2)\phi_1 = 0$ and $(\nabla^2 + k^2)\phi_2 = 0$, and show that $\vec{\nabla} \cdot \vec{f} = 0$ and $(\nabla^2 + k^2)\vec{f} = 0$.

22. (a) Interpret the symbol $\vec{F} \cdot \vec{\nabla}$.
(b) Give a possible meaning to $(\vec{F} \cdot \vec{\nabla})\vec{G}$.
(c) Is it possible to write $\vec{F} \cdot \vec{\nabla}\vec{G}$ without ambiguity?

23. If $\vec{F} = 2yz\hat{i} - x^2y\hat{j} + xz^2\hat{k}$, $\vec{G} = x^2\hat{i} + yz\hat{j} - xy\hat{k}$ and $\phi = 2x^2yz^3$, show that
(a) $(\vec{F} \cdot \vec{\nabla})\phi = 8xy^2z^4 - 2x^4yz^3 + 6x^3yz^4 = \vec{F} \cdot \vec{\nabla}\phi$
(b) $(\vec{G} \cdot \vec{\nabla})\vec{F} = (2yz^2 - 2xy^2)\hat{i} - (2x^3y + x^2yz)\hat{j} + (x^2z^2 - 2x^2yz)\hat{k} = \vec{G} \cdot \vec{\nabla}\vec{F}$
(c) $(\vec{F} \times \vec{\nabla})\phi = -(6x^4y^2z^2 + 2x^3z^5)\hat{i} + (4x^2yz^5 - 12x^2y^2z^3)\hat{j} + (4x^2yz^4 + 4x^3y^2z^3)\hat{k} = \vec{F} \times \vec{\nabla}\phi$

24. If $\vec{F} = x^2z\hat{i} + yz^3\hat{j} - 3xy\hat{k}$, $\vec{G} = y^2\hat{i} - yz\hat{j} + 2x\hat{k}$ and $\phi = 2x^2 + yz$, show that
(a) $(\vec{F} \cdot \vec{\nabla})\phi = 4x^3z + yz^4 - 3xy^2 = \vec{F} \cdot \vec{\nabla}\phi$
(b) $(\vec{G} \cdot \vec{\nabla})\vec{F} = 2y^2z^3\hat{i} - (3xy^2 - yz^4)\hat{j} + 2x^2z\hat{k} = \vec{G} \cdot \vec{\nabla}\vec{F}$

25. If $\vec{F} = yz^2\hat{i} - 3xz^2\hat{j} + 2xyz\hat{k}$, $\vec{G} = 3x\hat{i} + 4z\hat{j} - xy\hat{k}$ and $\phi = xyz$, show that
(a) $(\vec{F} \times \vec{\nabla})\phi = -5x^2yz^2\hat{i} + xy^2z^2\hat{j} + 4xyz^3\hat{k} = \vec{F} \times \vec{\nabla}\phi$
(b) $(\vec{\nabla} \times \vec{F}) \times \vec{G} = 16z^3\hat{i} + (8x^2yz - 12xz^2)\hat{j} + 32xz^2\hat{k}$
(c) $\vec{G} \cdot (\vec{\nabla} \times \vec{F}) = 24x^2z + 4xyz^2$

26. If $\vec{F} = xz^2\hat{i} + 2y\hat{j} - 3xz\hat{k}$ and $\vec{G} = 3xz\hat{i} + 2yz\hat{j} - z^2\hat{k}$, show that at the point $(1, -1, 2)$
(a) $(\vec{F} \times \vec{\nabla}) \times \vec{G} = 4\hat{j} + 76\hat{k}$ (b) $\vec{F} \times (\vec{\nabla} \times \vec{G}) = 18\hat{i} - 12\hat{j} + 16\hat{k}$

27. Evaluate:
(a) $\vec{\nabla} \cdot (\vec{a}\, \vec{b} \cdot \vec{r})$ (b) $\vec{\nabla} \times \left[\vec{a} \times (\vec{b} \times \vec{r})\right]$ (c) $\vec{\nabla}\left(\vec{a} \cdot \frac{1}{r^n}\vec{r}\right)$
(d) $\vec{\nabla} \cdot \left(\vec{a} \times \frac{1}{r^n}\vec{r}\right)$ (e) $\vec{\nabla} \times \left(\vec{a} \times \frac{1}{r^n}\vec{r}\right)$ (f) $\vec{\nabla} \times (\vec{r}\phi(r))$ and $(\vec{a} \cdot \vec{r})\vec{a}$

28. Are the following forces are irrotational? If so, find the scalar potential function.
(a) $\vec{F} = (6xy + z^3)\hat{i} + (3x^2 - z)\hat{j} + (3xz^2 - y)\hat{k}$ *Ans*: $3x^2y + xz^3 - yz + c$
(b) $\vec{F} = (x + 2y + 3z)\hat{i} + (2x + 3y + 2z)\hat{j} + (3x + 2y + z)\hat{k}$
Ans: $\frac{1}{2}x^2 + 2xy + 3xz + \frac{3}{2}y^2 + 2yz + \frac{1}{2}z^2$
(c) $\vec{F} = (2x - yz)\hat{i} + (2y - zx)\hat{j} + (2z - xy)\hat{k}$ *Ans*: $x^2 + y^2 + z^2 - xyz + c$

29. (a) If $\phi = 3x^2y$, $\psi = xz^2 - 2y$, evaluate $\vec{\nabla}\left[(\vec{\nabla}\phi) \cdot (\vec{\nabla}\psi)\right]$
Ans: $(6yz^2 - 12x)\hat{i} + 6xz^2\hat{j} + 12xyz\hat{k}$
(b) If $\vec{F} = (3x^2y - z)\hat{i} + (xz^3 + y^4)\hat{j} - 2x^3z^2\hat{k}$, find $\vec{\nabla}(\vec{\nabla} \cdot \vec{F})$ at the point $(2, -1, 0)$
Ans: $-6\hat{i} + 24\hat{j} - 32\hat{k}$

30. If $\vec{a}$ and $\vec{b}$ are constant vectors and $\vec{r} = x\hat{i} + y\hat{j} + z\hat{k}$, with $r = |\vec{r}|$, prove that
(a) $\vec{\nabla} \times \left\{\frac{1}{r^3}(\vec{a} \times \vec{r})\right\} = -\frac{1}{r^3}\vec{a} + \frac{3}{r^5}(\vec{a} \cdot \vec{r})\,\vec{r} = \vec{\nabla}\left\{\vec{a} \cdot \vec{\nabla}\left(\frac{1}{r}\right)\right\}$
(b) $\vec{a} \cdot \vec{\nabla}\left\{\vec{b} \cdot \vec{\nabla}\left(\frac{1}{r}\right)\right\} = \frac{3(\vec{a} \cdot \vec{r})(\vec{b} \cdot \vec{r})}{r^5} - \frac{\vec{a} \cdot \vec{b}}{r^3}$ (c) $\vec{\nabla} \cdot (\vec{a} \times \vec{r}) = \vec{r} \cdot (\vec{\nabla} \times \vec{a})$
(d) $\vec{\nabla} \cdot \{\vec{r} \times (\vec{a} \times \vec{r})\} = -2\vec{r} \cdot \vec{a}$ and $\vec{\nabla} \times \{\vec{r} \times (\vec{a} \times \vec{r})\} = 3\vec{r} \times \vec{a}$

31. If $\vec{F} = \left(y\frac{\partial f}{\partial z} - z\frac{\partial f}{\partial y}\right)\hat{i} + \left(z\frac{\partial f}{\partial x} - x\frac{\partial f}{\partial z}\right)\hat{j} + \left(x\frac{\partial f}{\partial y} - y\frac{\partial f}{\partial x}\right)\hat{k}$, prove that
(a) $\vec{F} = \vec{r} \times \vec{\nabla}f$ (b) $\vec{F} \cdot \vec{r} = 0$ (c) $\vec{F} \cdot \vec{\nabla}f = 0$

32. If f is a scalar point function, then express $\nabla^2 f$ in
(a) cylindrical coordinates (b) spherical polar coordinates

33. If the Laplacian of a scalar point function vanishes identically, then prove that its gradient is solenoidal as well as irrotational.

34. Prove the following identities:
(a) $\vec{\nabla}(\phi\psi) = \phi\,\vec{\nabla}\psi + \psi\,\vec{\nabla}\phi$
(b) $\vec{\nabla}(\vec{f} \cdot \vec{g}) = \vec{f} \times (\vec{\nabla} \times \vec{g}) + \vec{g} \times (\vec{\nabla} \times \vec{f}) + (\vec{f} \cdot \vec{\nabla})\vec{g} + (\vec{g} \cdot \vec{\nabla})\vec{f}$
(c) $\vec{\nabla} \cdot (\phi\vec{f}) = \phi\,\vec{\nabla} \cdot \vec{f} + \vec{f} \cdot \vec{\nabla}\phi$ (d) $\vec{\nabla} \cdot (\vec{f} \times \vec{g}) = \vec{g} \cdot (\vec{\nabla} \times \vec{f}) - \vec{f} \cdot (\vec{\nabla} \times \vec{g})$
(e) $\vec{\nabla} \times (\phi\vec{f}) = \vec{\nabla}\phi \times \vec{f} + \phi\,\vec{\nabla} \times \vec{f}$
(f) $\vec{\nabla} \times (\vec{f} \times \vec{g}) = \vec{f}\,\vec{\nabla} \cdot \vec{g} - \vec{g}\,\vec{\nabla} \cdot \vec{f} + (\vec{g} \cdot \vec{\nabla})\vec{f} - (\vec{f} \cdot \vec{\nabla})\vec{g}$
(g) $\vec{\nabla} \times (\psi\,\vec{\nabla}\phi) = \vec{\nabla}\phi \times \vec{\nabla}\psi = -\vec{\nabla} \times (\phi\,\vec{\nabla}\psi)$, where ϕ and ψ are two scalar function

35. Express the following vectors in terms of spherical polar and cylindrical polar coordinates:
(a) $\vec{A} = 3y\hat{i} + x^2\hat{j} - z^2\hat{k}$ (b) $\vec{A} = xyz^2\hat{i} + yz\hat{j} + xy\hat{k}$
(c) $\vec{A} = xy\hat{i} + 2yz\hat{j} + (z^2 - yz)\hat{k}$ (d) $\vec{A} = z\hat{i} - 2x\hat{j} + y\hat{k}$

36. Transform the following functions into the Cartesian system:
(a) $\vec{A} = \rho\hat{e}_\rho + \rho\hat{e}_\phi$ (b) $\vec{A} = 3a\rho^2 \sin\theta \cos\phi\,\hat{e}_\rho + 2a^2 r \cos\theta \sin\phi\,\hat{e}_\theta + r^3\hat{e}_\phi$

6 Vector Integration

The procedures of integral calculus can be directly applied to vector integration. In general, a vector integral can be converted into a scalar integral, and this, in turn, can be evaluated by the usual methods. In this chapter, we shall study three special types of integrals involving vector functions: line integrals, surface integrals and volume integrals. In the process of integration, we actually look for a vector function whose derivative is given. The process of integration helps us to study several important problems of geometry and the physical sciences. However, before we examine these problems, we need to understand some basic concepts such as smooth curve, smooth closed curve, simply connected region, multiply connected region and smooth surface.

6.1 DEFINITIONS OF BASIC CONCEPTS

Smooth curve: Let $\vec{r} = \vec{r}(u) = x(u)\hat{i} + y(u)\hat{j} + z(u)\hat{k}$ denote the position vector of a point $P(x, y, z)$ in a three-dimensional Euclidean space. If we vary the scalar u continuously, the point P will describe a space curve $\mathcal{C}$. If $\vec{r}$ possesses a continuous first order derivative $\frac{d\vec{r}}{du}(\neq \vec{0})$ for all values of u in a specified interval, then the curve $\mathcal{C}$ is called a *smooth curve* or a *regular curve*.

For example, $\vec{r} = 4u^2\hat{i} + 5(u-1)\hat{j} + 8u^3\hat{k}$, $1 \leq u \leq 2$, represents a smooth curve in a three-dimensional space. An arc of a circle is a smooth curve.

Piecewise smooth curve: If a curve is not smooth but can be broken into a finite number of smooth curves, then such a type of curve is known as a *piecewise smooth curve*.

For example, the curve $\vec{r} = u\,\hat{i} + \frac{1}{u}\,\hat{j}$ is not smooth in $(-\infty, \infty)$ as the derivative does not exist at $u = 0$. However, the curve can be divided into two pieces such that each piece is smooth in the intervals $(-\infty, 0)$ and $(0, \infty)$. Therefore, this is a piecewise smooth curve.

Simple closed curve: A smooth curve is said to be a *simple closed curve* if it does not intersect itself anywhere. A simple closed curve does not have multiple points. For example, $\vec{r} = \cos u\hat{i} + \sin u\hat{j} + 0\,\hat{k}$, $0 \leq u \leq 2\pi$, is a simple closed curve.

Jordan curve: A plane curve $\vec{r} = \vec{r}(u) = x(u)\hat{i} + y(u)\hat{j}$, $a \leq u \leq b$, without multiple points, is said to be a *Jordan curve* or *arc* if $x(u)$ and $y(u)$ are continuous. For example, $\vec{r} = \cos u\,\hat{i} + \sin u\,\hat{j}$, $0 \leq u \leq \pi$ is a Jordan curve.

Connected region: A region is said to be *connected* if any two points of it can be joined by an arc, and the arc lies wholly within the region, that is, every point of the arc belongs to the region; otherwise, it is *disconnected*.

(i) **Simply connected region**: A region R is said to be a *simply connected region* if any closed curve lying entirely in R can be contracted continuously to a point without any portion of the curve extending out of R in that process of contraction. Circular, cubical and spherical regions are examples of simply connected regions.

(ii) **Multiply connected region**: A region R which is not simply connected is called a *multiply connected region*. For example, the regions bounded by two concentric circles or spheres are multiply connected regions.

Smooth surface: A surface represented by $\vec{r} = \vec{r}(u, v)$ is said to be a *smooth surface* if $\vec{r}$ possesses continuous partial derivatives of any order. For example, the surface of a sphere is a smooth surface.

Closed surface: A surface which encloses a finite volume is known as a *closed surface*. For example, the surface of a sphere is a closed surface, but a hemisphere without its diametral plane is not a closed surface.

6.2 ORDINARY INTEGRAL

Let us consider a vector function $\vec{f}(t)$ of a scalar variable t. Like the calculus of scalar variables, we can find another vector function $\vec{g}(t)$, such that $\vec{f}(t) = \vec{g}'(t) = \frac{d}{dt}\vec{g}(t)$; then

$$\int \vec{f}(t)\, dt = \int \frac{d}{dt}\vec{g}(t)\, dt = \vec{g}(t) + \vec{c} \tag{6.1}$$

Here, $\vec{c}$ is an arbitrary constant vector independent of t. $\vec{g}(t) + \vec{c}$ is called the *indefinite integral of* $\vec{f}(t)$ with respect to t. Thus, the process of integration involves a search for the function $\vec{g}(t)$, called the *integral*, whose derivative with respect to t is the function $\vec{f}(t)$, called the *integrand*.

The definite integral between the limits $t = a$ and $t = b$ is given by

$$\int_a^b \vec{f}(t)\, dt = \vec{g}(t) + \vec{c}\Big]_a^b = \vec{g}(b) - \vec{g}(a)$$

If $f_1(t), f_2(t)$ and $f_3(t)$ are considered as continuous functions of a single variable t defined on a specified closed interval a, b and $\vec{f}(t) = f_1(t)\hat{i} + f_2(t)\hat{j} + f_3(t)\hat{k}$, then

$$\int \vec{f}(t)dt = \hat{i}\int f_1(t)dt + \hat{j}\int f_2(t)dt + \hat{k}\int f_3(t)dt + \vec{c}$$

This is called an *indefinite integral* of $\vec{f}(t)$. In the case of a definite integral between the limits $t = a$ and $t = b$, we have

$$\int_a^b \vec{f}(t)dt = \hat{i}\int_a^b f_1(t)dt + \hat{j}\int_a^b f_2(t)dt + \hat{k}\int_a^b f_3(t)dt$$

General Rules of Vector Integration

Using the various formulae of vector differentiation, we can write several formulae for vector integration.

1. If a is an arbitrary constant scalar and $\vec{f}(t)$ is an integrable vector function, then
$$\int a\vec{f}(t)\,dt = a\int \vec{f}(t)\,dt.$$
2. If $\vec{f}(t)$ and $\vec{g}(t)$ are two integrable vector functions, then
$$\int \left[\vec{f}(t) \pm \vec{g}(t)\right] dt = \int \vec{f}(t)\,dt \pm \int \vec{g}(t)\,dt.$$
3. If $\vec{f}(t)$ and $\vec{g}(t)$ are two integrable vector functions, then
$$\int \left[\vec{f}(t)\cdot\frac{d\vec{g}(t)}{dt} + \frac{d\vec{f}(t)}{dt}\cdot\vec{g}(t)\right] dt = \vec{f}(t)\cdot\vec{g}(t) + c,\ c = \text{constant}.$$
4. If $\vec{f}(t)$ is an integrable vector function, then
$$\int \left[\vec{f}(t)\times\frac{d^2\vec{f}(t)}{dt^2}\right] dt = \vec{f}(t)\times\frac{d\vec{f}(t)}{dt} + \vec{c},\ \vec{c} = \text{constant vector}.$$
5. If $\vec{f}(t)$ is an integrable vector function and $\vec{a}$ is a constant vector, then
$$\int \left[\vec{a}\times\frac{d\vec{f}(t)}{dt}\right] dt = \vec{a}\times\vec{f}(t) + \vec{c},\ \vec{c} = \text{constant vector}.$$

Some Standard Results

From the definition, we can easily write

1. $\frac{d}{dt}(\vec{r}\cdot\vec{a}) = \vec{r}\cdot\frac{d\vec{a}}{dt} + \frac{d\vec{r}}{dt}\cdot\vec{a} \Rightarrow \int\left[\vec{r}\cdot\frac{d\vec{a}}{dt} + \frac{d\vec{r}}{dt}\cdot\vec{a}\right]dt = \vec{r}\cdot\vec{a} + c$
2. $\frac{d\vec{r}^2}{dt} = \frac{d}{dt}(\vec{r}\cdot\vec{r}) = 2\vec{r}\cdot\frac{d\vec{r}}{dt} \Rightarrow \int\left(2\vec{r}\cdot\frac{d\vec{r}}{dt}\right)dt = \vec{r}^2 + c$
3. $\frac{d}{dt}\left|\frac{d\vec{r}}{dt}\right|^2 = \frac{d}{dt}\left\{\frac{d\vec{r}}{dt}\cdot\frac{d\vec{r}}{dt}\right\} = 2\frac{d\vec{r}}{dt}\cdot\frac{d^2\vec{r}}{dt^2} \Rightarrow \int 2\frac{d\vec{r}}{dt}\cdot\frac{d^2\vec{r}}{dt^2}\,dt = \left|\frac{d\vec{r}}{dt}\right|^2 + c$
4. $\frac{d}{dt}\left(\frac{1}{r}\vec{r}\right) = \frac{1}{r}\frac{d\vec{r}}{dt} - \frac{1}{r^2}\frac{dr}{dt}\vec{r} \Rightarrow \int\left(\frac{1}{r}\frac{d\vec{r}}{dt} - \frac{1}{r^2}\frac{dr}{dt}\vec{r}\right)dt = \frac{1}{r}\vec{r} + \vec{c}$
5. $\frac{d}{dt}(\vec{r}\times\vec{a}) = \vec{r}\times\frac{d\vec{a}}{dt} + \frac{d\vec{r}}{dt}\times\vec{a} \Rightarrow \int\left[\vec{r}\times\frac{d\vec{a}}{dt} + \frac{d\vec{r}}{dt}\times\vec{a}\right]dt = \vec{r}\times\vec{a} + \vec{c}$
6. $\frac{d}{dt}\left(\vec{a}\times\vec{r}\right) = \vec{a}\times\frac{d\vec{r}}{dt} \Rightarrow \int\left(\vec{a}\times\frac{d\vec{r}}{dt}\right)dt = \vec{a}\times\vec{r} + \vec{c},\ \vec{a} = \text{a constant vector}$
7. $\frac{d}{dt}\left(\vec{r}\times\frac{d\vec{r}}{dt}\right) = \vec{r}\times\frac{d^2\vec{r}}{dt^2} \Rightarrow \int\left(\vec{r}\times\frac{d^2\vec{r}}{dt^2}\right)dt = \vec{r}\times\frac{d\vec{r}}{dt} + \vec{c}$

The constant of integration has the same nature (scalar or vector) as the integrand. The constants in the first three integrals are scalars and those in the last four are vectors.

Example 6.2.1

If $\vec{r}(t) = 5t^2\hat{i} + t\hat{j} - t^3\hat{k}$, find $\int_1^2 (\vec{r}\times\frac{d^2\vec{r}}{dt^2})dt$.

Solution

Here, $\vec{r}(t) = 5t^2\hat{i} + t\hat{j} - t^3\hat{k}$. Now

$$\frac{d}{dt}\left(\vec{r}\times\frac{d\vec{r}}{dt}\right)=\frac{d\vec{r}}{dt}\times\frac{d\vec{r}}{dt}+\vec{r}\times\frac{d^2\vec{r}}{dt^2}=\vec{r}\times\frac{d^2\vec{r}}{dt^2}$$

$$=\begin{vmatrix}\hat{i} & \hat{j} & \hat{k}\\ 5t^2 & t & -t^3\\ 10t & 1 & -3t^2\end{vmatrix}=-2t^3\hat{i}+5t^4\hat{j}-5t^2\hat{k}$$

So, the value of the required integral becomes

$$\int_1^2(\vec{r}\times\frac{d^2\vec{r}}{dt^2})dt=\int_1^2\frac{d}{dt}(\vec{r}\times\frac{d\vec{r}}{dt})dt=\left[\vec{r}\times\frac{d\vec{r}}{dt}\right]_1^2$$

$$=\left[-2t^3\hat{i}+5t^4\hat{j}-5t^2\hat{k}\right]_1^2=-14\hat{i}+75\hat{j}-15\hat{k}$$

Example 6.2.2

The acceleration of a particle at time $t\geq 0$ is given by $\vec{a}=e^{-t}\hat{i}-6(t+1)\hat{j}+3\sin t\hat{k}$. If the velocity $\vec{v}$ and displacement $\vec{r}$ are zero at $t=0$, find $\vec{v}$ and $\vec{r}$ at any time.

Solution

Since $\vec{a}=\dfrac{d\vec{v}}{dt}$, the velocity $\vec{v}$ at time $t\geq 0$ is given by

$$\vec{v}=\int\vec{a}\,dt=\int_0^t\left[e^{-t}\hat{i}-6(t+1)\hat{j}+3\sin t\hat{k}\right]dt$$

$$=\left[-e^{-t}\hat{i}-3(t^2+2t)\hat{j}-3\cos t\hat{k}\right]_0^t$$

$$=(1-e^{-t})\hat{i}-3(t^2+2t)\hat{j}+3(1-\cos t)\hat{k}$$

Since $\vec{v}=\dfrac{d\vec{r}}{dt}$, the position $\vec{r}$ at time $t\geq 0$ is given by

$$\vec{r}=\int\vec{v}\,dt=\int_0^t\left[(1-e^{-t})\hat{i}-3(t^2+2t)\hat{j}+3(1-\cos t)\hat{k}\right]dt$$

$$=\left[(t+e^{-t})\hat{i}-(t^3+3t^2)\hat{j}+3(t-\sin t)\hat{k}\right]_0^t$$

$$=(t-1+e^{-t})\hat{i}-(t^3+3t^2)\hat{j}+3(t-\sin t)\hat{k}$$

Example 6.2.3

The position vector $\vec{r}$ of a moving particle at time t satisfies the equation $\dfrac{d^2\vec{r}}{dt^2}=6t\hat{i}-24t^2\hat{j}+4\sin t\hat{k}$. If $\vec{r}=2\hat{i}+\hat{j}$ and $\dfrac{d\vec{r}}{dt}=-\hat{i}-3\hat{k}$ when $t=0$, find $\vec{v}$ and $\vec{r}$ at any time.

Solution

By integrating the given differential equation, we get

$$\left.\frac{d\vec{r}}{dt}\right|_{-\hat{i}-3\hat{k}}^{\vec{v}}=\left.\vec{v}\right|_{-\hat{i}-3\hat{k}}^{\vec{v}}=\int_0^t\left[6t\hat{i}-24t^2\hat{j}+4\sin t\hat{k}\right]dt$$

$$\Rightarrow \vec{v} - (-\hat{i} - 3\hat{k}) = \left[3t^2\hat{i} - 8t^3\hat{j} - 4\cos t\hat{k}\right]_0^t$$

$$\Rightarrow \vec{v} = (3t^2 - 1)\hat{i} - (8t^3 + 3)\hat{j} + 4(1 - \cos t)\hat{k}$$

This gives the expression for velocity at any time t. Further integration gives

$$\vec{r}\Big|_{2\hat{i}+\hat{j}}^{\vec{r}} = \int_0^t \left[(3t^2 - 1)\hat{i} - (8t^3 + 3)\hat{j} + 4(1 - \cos t)\hat{k}\right]dt$$

$$\Rightarrow \vec{r} = 2\hat{i} + \hat{j} + \left[(3t^2 - 1)\hat{i} - (8t^3 + 3)\hat{j} + 4(1 - \cos t)\hat{k}\right]_0^t$$

$$= (t^3 - t + 2)\hat{i} + (1 - 2t^4)\hat{j} + (t - 4\sin t)\hat{k}$$

This gives the position vector of the particle at any time t.

Example 6.2.4

Let $\vec{r}$ be the position vector of a point P measured from an origin O and $\hat{r}$ be a unit vector in the direction of $\vec{r}$. The equation of motion of a particle P of mass m is given by

$$m\frac{d^2\vec{r}}{dt^2} = f(r)\,\hat{r};\, r = |\vec{r}| \text{ and } \hat{r} = \frac{1}{|\vec{r}|}\vec{r}$$

where $f(r)$ is a function of the distance $r = OP$.

(i) Show that $\vec{r} \times \dfrac{d\vec{r}}{dt} = \vec{h}$, where $\vec{h}$ is a constant vector. Interpret this result geometrically.

(ii) Interpret physically the cases $f(r) < 0$ and $f(r) > 0$.

Solution

(i) Since $\vec{a}$ and $\hat{a}$ are collinear, $\vec{a} \times \hat{a} = \vec{0}$. $\vec{r}$ is a position vector of the scalar variable t. Now

$$\frac{d}{dt}\left[m\vec{r} \times \frac{d\vec{r}}{dt}\right] = m\frac{d\vec{r}}{dt} \times \frac{d\vec{r}}{dt} + m\vec{r} \times \frac{d\vec{r}}{dt^2}$$

$$= m\vec{r} \times f(r)\hat{r} = mf(r)\vec{r} \times \hat{r} = \vec{0}$$

$$\Rightarrow \vec{r} \times \frac{d\vec{r}}{dt} = \vec{h}, \text{ where } \vec{h} = \text{a constant vector}$$

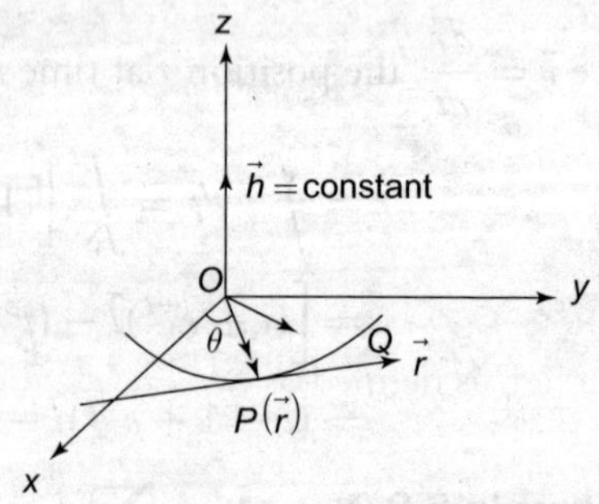

Figure 6.1 Areal velocity

Let P and Q be the positions of the moving particle at time t and $t + \Delta t$, $\vec{r}$ and $\vec{r} + \Delta\vec{r}$ be their respective position vectors relative to the origin O (Fig. 6.1). Let $\Delta\vec{A}$ denote the vector area of ΔOPQ. Then

$$\vec{A} = \frac{1}{2}\vec{r} \times (\vec{r} + \Delta\vec{r}) = \frac{1}{2}\vec{r} \times \Delta\vec{r}, \text{ as } \vec{r} \times \vec{r} = \vec{0}$$

$$\Rightarrow \lim_{\Delta t \to 0}\frac{\Delta\vec{A}}{\Delta t} = \lim_{\Delta t \to 0}\frac{1}{2}\vec{r} \times \frac{\Delta\vec{r}}{\Delta t}$$

$$\Rightarrow \frac{d\vec{A}}{dt} = \frac{1}{2}\vec{r} \times \frac{d\vec{r}}{dt} = \frac{1}{2}\vec{r} \times \vec{v} = \frac{1}{2}\vec{h}$$

where $\vec{v}$ is the instantaneous velocity of the particle. Now, $\frac{d\vec{a}}{dt}$ represents the areal velocity, so

$$\vec{h} = 2\frac{d\vec{A}}{dt} = 2 \times \text{ Areal velocity}$$

Since $\vec{r} \cdot \vec{h} = \vec{r} \cdot (\vec{r} \times \vec{r}) = 0$, the motion takes place on a plane, which may be taken as the xy plane.

(ii) If $f(r) < 0$, the acceleration $\frac{d^2\vec{r}}{dt^2}$ is in the direction opposite of that of $\hat{r}$. Hence, the force is directed towards O and the particle always moves towards O. If $f(r) > 0$, the force is directed away from O and the particle is under the influence of a repulsive force at O. A force directed towards or away from the fixed point O and having a magnitude that depends only on the distance r from O is called a *central force.*

6.3 LINE INTEGRALS

Let $\vec{r}(u) = x(u)\,\hat{i} + y(u)\,\hat{j} + z(u)\,\hat{k}$ represent a simple smooth space curve Γ with end points A and B corresponding to the values $u = a$ and $u = b$ of the parameter u.

Let us consider a real-valued function $\phi(x, y, z)$ which is single-valued and continuous on the curve Γ (Fig. 6.2). The curve Γ is divided into n parts by the $(n + 1)$ points

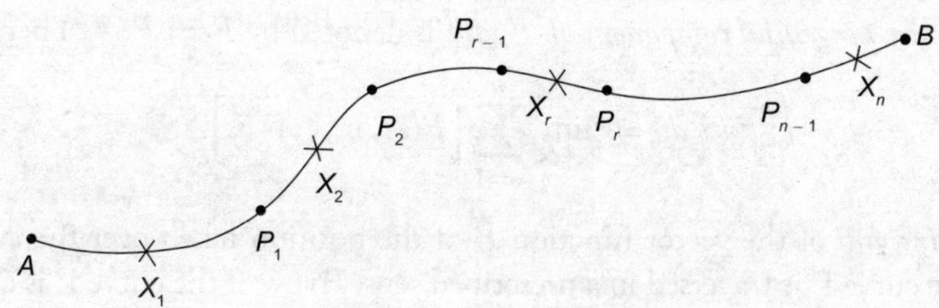

Figure 6.2 Regular arc

$P_0, P_1, \cdots, P_{r-1}, P_r, \cdots P_n$. Let us denote the line segment $P_{r-1}P_r$ by $\triangle S_r$ for $r = 1, 2, \cdots, n$. Let $X_r(\xi_r, \eta_r, \zeta_r)$ be a point on the line segment $P_{r-1}P_r$. Then, the *line integral* of $\phi(x, y, z)$ over the curve Γ is defined as

$$\int_\Gamma \phi(x, y, z)\, dS = \lim_{\triangle S_r \to 0} \sum_{r=1}^{n} \phi(\xi_r, \eta_r, \zeta_r) \triangle S_r$$

$$= \lim_{n \to \infty} \sum_{r=1}^{n} \phi(\xi_r, \eta_r, \zeta_r) \triangle S_r \quad (6.2)$$

The limit is independent of the mode of subdivision of the curve. Here, we consider the integral of the type

Figure 6.3 Regular arc

$$\int_\Gamma \phi\, d\vec{r} = \hat{i}\int_\gamma \phi(x, y, z)dx + \hat{j}\int_\gamma \phi(x, y, z)dy + \hat{k}\int_\gamma \phi(x, y, z)dz \qquad (6.3)$$

where $\phi = \phi(x, y, z)$ is a scalar point function representing a scalar field and Γ is some contour along which integration is to be carried out. In particular, $\phi(x, y, z) = 1$ everywhere on Γ, then the integral (6.3) represents the arc length of Γ or the displacement from A to B of the curve (Fig. 6.3).

6.3.1 Tangential Line Integral

In general, any integral to be evaluated along a curve is called the *tangential line integral* or simply the *line integral*. Let $P(\xi_r, \eta_r, \zeta_r)$ be a general point on the curve Γ, as in Fig. 6.4.

We assume that the curve Γ is composed of n (finite number) continuous arcs $\triangle S_1, \triangle S_2, \cdots, \triangle S_n$ by introducing the points (ξ_r, η_r, ζ_r); $r = 0, 1, 2, \cdots, n$ on it such that A and B are end points. Let $\hat{t}_r$ be the unit tangent vector to Γ at P. Let $\vec{F}(x, y, z) = F_1\hat{i} + F_2\hat{j} + F_3\hat{k}$ be a single-valued, continuous vector point function defined on the rectifiable curve Γ such that there is a definite tangent at every point of it as Γ is traversed from one end to the other. The orthogonal projection of $\vec{F}$ on the unit tangent vector $\hat{t}$ is called the *tangential component of* $\vec{F}$ and is denoted by $F_t = \vec{F} \cdot \hat{t}$. Then

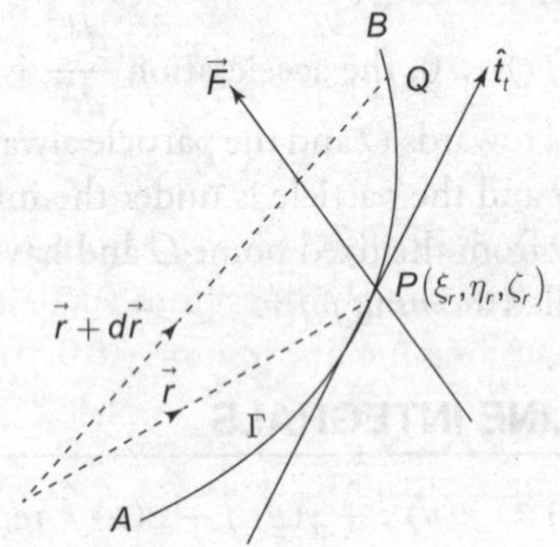

Figure 6.4 Tangential line integral

$$\int_{\Gamma} \vec{F} \cdot \hat{t}\, du = \lim_{n \to \infty} \sum_{r=1}^{n} \left[\vec{F}(\xi_r, \eta_r, \zeta_r) \cdot \hat{t}_r \right] \triangle S_r \tag{6.4}$$

This is the *line integral* of the vector function $\vec{F}$ of the position taken over the curve Γ from A and B, where the curve Γ is traversed in a prescribed sense. Now, if the curve Γ is closed, then the above integral is denoted by the symbol $\oint \vec{F} \cdot \hat{t}\, du$.

DEDUCTION **6.3.1**

Cartesian form of the line integral: Consider a vector point function $\vec{F}(x, y, z) = F_1(x, y, z)\hat{i} + F_2(x, y, z)\hat{j} + F_3(x, y, z)\hat{k}$, where $F_1(x, y, z), F_2(x, y, z)$ and $F_3(x, y, z)$ are integrable scalar point functions of x, y and z. If $\hat{t}$ is the unit tangent vector for the space curve Γ, then

$$\hat{t} = \frac{dx}{du}\hat{i} + \frac{dy}{du}\hat{j} + \frac{dz}{du}\hat{k}$$

$$\vec{F} \cdot \hat{t} = F_1(x, y, z)\frac{dx}{du} + F_2(x, y, z)\frac{dy}{du} + F_3(x, y, z)\frac{dz}{du}$$

This is the Cartesian form of the line integral. Alternatively, if $\vec{r} = x\hat{i} + y\hat{j} + z\hat{k}$ is the position vector of any point on the curve, then Eq. (6.4) can be written as

$$\int_{\Gamma} (\vec{F} \cdot \hat{t})\, du = \int_{\Gamma} \vec{F} \cdot d\vec{r}$$

Thus, the line integral can be written as

$$\int_\gamma \vec{F}\cdot\hat{t}\,du = \int_\Gamma \left[F_1(x,y,z)\frac{dx}{du} + F_2(x,y,z)\frac{dy}{du} + F_3(x,y,z)\frac{dz}{du}\right] du$$

$$= \int_\Gamma F_1(x,y,z)\,dx + F_2(x,y,z)\,dy + F_3(x,y,z)\,dz \qquad (6.5)$$

Example 6.3.1

If $\vec{F}(x,y,z) = (2y+3)\hat{i} + xz\hat{j} + (yz-x)\hat{k}$, evaluate the line integral $\int_C \vec{F}\cdot d\vec{r}$ along the following path C:

(i) $x = 2t^2, y = t, z = t^3$ from $t = 0$ to $t = 1$
(ii) The straight lines from $(0,0,0)$ to $(0,0,1)$, then to $(0,1,1)$, and then to $(2,1,1)$
(iii) Along the line joining $(0,0,0)$ to $(2,1,1)$

Solution

(i) Along the path $C : x = 2t^2, y = t$ and $z = t^3$, we have

$$\vec{F} = (2y+3)\hat{i} + xz\hat{j} + (yz-x)\hat{k} = (2t+3)\hat{i} + 2t^5\hat{j} + (t^4 - 2t^2)\hat{k}$$

Also along this path $C : \vec{r}(t) = x\hat{i} + y\hat{j} + z\hat{k} = 2t^2\hat{i} + t\hat{j} + t^3\hat{k}$ so that $d\vec{r} = (4t\hat{i} + \hat{j} + 3t^2\hat{k})dt$, and therefore

$$\vec{F}\cdot d\vec{r} = \left[(2t+3)\hat{i} + 2t^5\hat{j} + (t^4-2t^2)\hat{k}\right]\cdot\left[4t\hat{i} + \hat{j} + 3t^2\hat{k}\right]dt$$

$$= \left[4t(2t+3) + 2t^5 + (t^4-2t^2)\cdot 3t^2\right]dt$$

$$= (8t^2 + 12t + 2t^5 + 3t^6 - 6t^4)dt$$

Therefore, along this path C

$$\int_C \vec{F}\cdot d\vec{r} = \int_0^1 (8t^2 + 12t + 2t^5 + 3t^6 - 6t^4)dt$$

$$= \left[\frac{8}{3}t^3 + 6t^2 + \frac{1}{3}t^6 + \frac{3}{7}t^7 - \frac{6}{5}t^5\right]_0^1 = \frac{288}{35}$$

(ii) The path C consists of straight lines from $A(0,0,0)$ to $B(0,0,1)$, then to $C(0,1,1)$ and then to $D(2,1,1)$ (Fig. 6.5). Along the line AB, $x = 0, y = 0$ but $0 \le z \le 1$. Now

$$\int_{AB} \vec{F}\cdot d\vec{r} = \int_{AB} \left[(2y+3)dx + xzdy + (yz-x)dz\right] = 0$$

Along the line BC, $x = 0$ but $0 \le y \le 1$ and $z = 1$, so $dz = 0$. Now

$$\int_{BC} \vec{F}\cdot d\vec{r} = \int_{BC} (2y+3)dx + xzdy + (yz-x)dz = 0$$

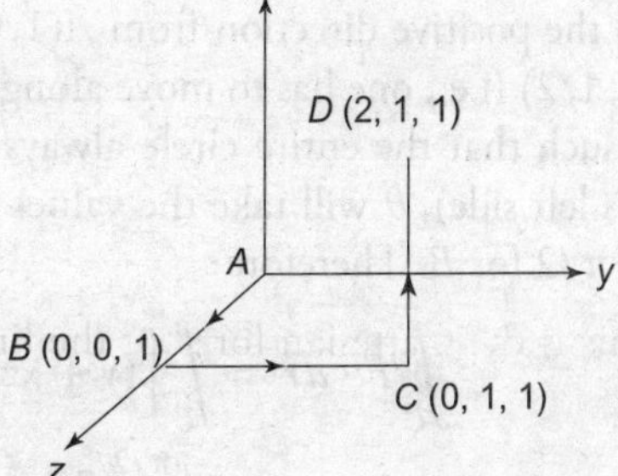

Figure 6.5 The line joining A, B, C and D

Along the line CD, $0 \leq x \leq 2, y = 1, z = 1$. Now

$$\int_{CD} \vec{F} \cdot d\vec{r} = \int_{CD} (2y+3)dx + xzdy + (yz - x)dz = \int_0^2 5dx = 10$$

Thus, the required line integral is

$$\int_C \vec{F} \cdot d\vec{r} = \int_{AB} \vec{F} \cdot d\vec{r} + \int_{BC} \vec{F} \cdot d\vec{r} + \int_{CD} \vec{F} \cdot d\vec{r} = 0 + 0 + 10 = 10$$

(iii) The parametric equation of the line joining $(0, 0, 0)$ and $(2, 1, 1)$ is

$$\frac{x-0}{0-2} = \frac{y-0}{0-1} = \frac{z-0}{0-1} = t \Rightarrow x = t, y = \frac{t}{2}, z = \frac{t}{2}$$

Thus, along the line $x = t, y = t/2$ and $z = t/2$, where $0 \leq t \leq 2$, we get

$$\begin{aligned} \vec{F} \cdot d\vec{r} &= (2y+3)dx + xzdy + (yz - x)dz \\ &= (t+3)dt + \frac{t^2}{4}dt + \frac{1}{2}(\frac{t^2}{4} - t)dt = \left(\frac{3}{8}t^2 + \frac{1}{2}t + 3\right)dt \end{aligned}$$

Thus, along the line joining $(0, 0, 0)$ to $(2, 1, 1)$

$$\int_C \vec{F} \cdot d\vec{r} = \int_0^2 \left(\frac{3}{8}t^2 + \frac{1}{2}t + 3\right)dt = \left[\frac{t^3}{8} + \frac{t^2}{4} + 3t\right]_0^2 = 8$$

Example 6.3.2

Evaluate $\int_C \vec{F} \cdot d\vec{r}$ along the curve $C : x^2 + y^2 = 1, z = 2$ in the positive direction from $A(1, 0, 2)$ to $B(0, 1, 2)$, where $\vec{F} = (y + xz^2)\hat{i} + (z - y)\hat{j} + (xy - z)\hat{k}$.

Solution

We perform integration on the plane $z = 2$, parallel to the xy plane, so we take $\vec{r} = x\hat{i} + y\hat{j}$ so that $d\vec{r} = dx\hat{i} + dy\hat{j}$ (Fig. 6.6). Let $\mathcal{C}$ be the arc AB of the circle $x^2 + y^2 = 1, z = 2$. The parametric equations of the arc are $x = \cos\theta, y = \sin\theta$ and $z = 2$. Since the integration must be performed along the curve $\mathcal{C}$ in the positive direction from $A(1, 0, 2)$ to $B(0, 1, 2)$ (i.e., one has to move along the arc AB such that the entire circle always lies on one's left side), θ will take the values 0 for A and $\pi/2$ for B. Therefore

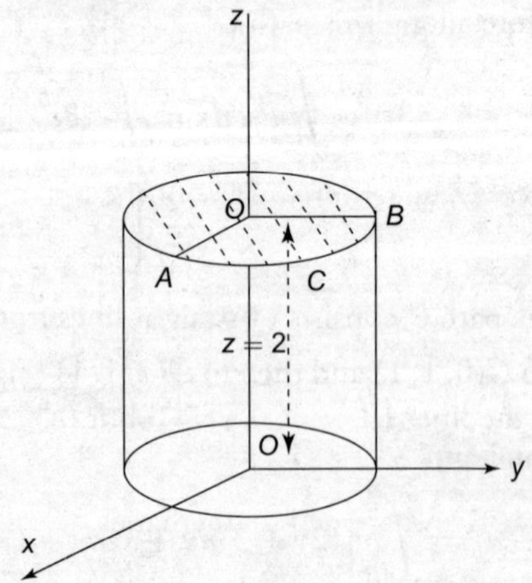

Figure 6.6 Line of integration

$$\begin{aligned} \oint_C \vec{F} \cdot d\vec{r} &= \int_C \left[(y + xz^2)dx + (z - y)dy\right] \\ &= \int_0^{\pi/2} \left[(\sin\theta + 4\cos\theta)(-\sin\theta) + (2 - \sin\theta)\cos\theta\right]d\theta \end{aligned}$$

$$= \int_0^{\pi/2} \left[2\cos\theta - \sin^2\theta - 5\sin\theta\cos\theta\right] d\theta$$

$$= \frac{1}{2}\int_0^{\pi/2} \left[4\cos\theta - (1-\cos 2\theta) - 5\sin 2\theta\right] d\theta = -\frac{1}{4}(\pi+2)$$

Example 6.3.3

Let $\vec{F} = 2z\hat{i} + 4x\hat{j} + 5y\hat{k}$, and let C be the curve of the intersection of the plane $z = x + 4$ and the cylinder $x^2 + y^2 = 4$, oriented counter-clockwise. Find the value of the line integral $\oint_C \vec{F}\cdot d\vec{r}$.

Solution

The curve of intersection of the plane $z = x + 4$ and the cylinder $x^2 + y^2 = 4$ is shown in Fig. 6.7. Here, we need to evaluate the line integral over the curve of intersection of the plane $z = x+4$ and the cylinder $x^2 + y^2 = 4$, oriented counter-clockwise. Let $x = 2\cos\theta$, $y = 2\sin\theta$ and $z = 2\cos\theta + 4$, $0 \le \theta \le 2\pi$ be the parametric equation of the curve. Since $\vec{F} = 2z\hat{i} + 4x\hat{j} + 5y\hat{k}$

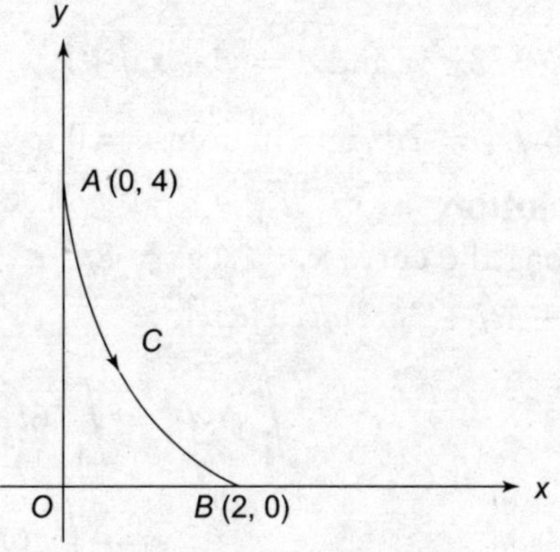

Figure 6.7 Line integral

$$\begin{aligned}\vec{F}\cdot d\vec{r} &= 2z\,dx + 4x\,dy + 5y\,dz \\ &= \Big[2(2\cos\theta+4)(-2\sin\theta) + 8\cos\theta(2\cos\theta) \\ &\quad +10\sin\theta(-2\sin\theta)\Big]d\theta \\ &= \Big[-8\cos\theta\sin\theta - 16\sin\theta \\ &\quad +16\cos^2\theta - 20\sin^2\theta\Big]d\theta \\ &= \Big[-4\sin 2\theta - 16\sin\theta + 18\cos 2\theta - 2\Big]d\theta\end{aligned}$$

Therefore, the required value of the line integral becomes

$$\oint_C \vec{F}\cdot d\vec{r} = \int_0^{2\pi}\Big[-4\sin 2\theta - 16\sin\theta + 18\cos 2\theta - 2\Big]d\theta$$

$$= \Big[2\cos 2\theta + 16\cos\theta + 9\sin 2\theta - 2\theta\Big]_0^{2\pi} = -4\pi$$

RESULT 19

In physics and applied mathematics, we often come across integrals of the following types:

1. If $\vec{A}$ is a continuous vector point function, then the *vector line integral* is defined as

$$\int_\Gamma \vec{A}\times d\vec{r} = \int_\Gamma \vec{A}\times\hat{t}\,ds$$

If $\vec{A} = A_1\hat{i} + A_2\hat{j} + A_3\hat{k}$ and $\vec{r} = x\hat{i} + y\hat{j} + z\hat{k}$, then

$$\int_\Gamma \vec{A} \times d\vec{r} = \hat{i} \int_\Gamma (A_2dz - A_3dy) + \hat{j} \int_\Gamma (A_3dx - A_1dz) + \hat{k} \int_\Gamma (A_1dy - A_2dx)$$

2. If ϕ is a continuous scalar point function, representing a scalar field, then

$$\int_\Gamma \phi \, d\vec{r} = \hat{i} \int_\Gamma \phi \, dx + \hat{j} \int_\Gamma \phi \, dy + \hat{k} \int_\Gamma \phi \, dz$$

Γ is the contour along which integration is carried out.

Note that $\hat{i}, \hat{j}$ and $\hat{k}$ are constant unit vectors, i.e., their magnitudes and directions do not depend on the position vector $\vec{r}$, and hence, these are taken out of the integral sign.

Example 6.3.4

If $\phi = 3x^2yz$ and $\vec{A} = y\hat{i} - xz\hat{j} + x^2\hat{k}$, then evaluate $\int_\Gamma \phi \, d\vec{r}$ and $\int_\Gamma \vec{A} \times d\vec{r}$, where Γ is the curve $x = t, y = 2t^2, z = t^3$ from $t = 0$ to $t = 1$.

Solution

Along the curve $x = t, y = 2t^2, z = t^3$, $\phi = 3x^2yz = 3 \cdot t^2 \cdot 2t^2 \cdot t^3 = 6t^7$ and $d\vec{r} = (\hat{i} + 4t\hat{j} + 3t^2\hat{k})dt$. Therefore

$$\int_\Gamma \phi \, d\vec{r} = \int_0^1 6t^7 \, (\hat{i} + 4t\hat{j} + 3t^2\hat{k})dt$$

$$= \hat{i} \int_0^1 6t^7 \, dt + 24\hat{j} \int_0^1 t^8 \, dt + 18\hat{k} \int_0^1 t^9 \, dt$$

$$= \left[6\frac{t^8}{8}\hat{i} + 24\frac{t^9}{9}\hat{j} + 18 \, \frac{t^{10}}{10}\hat{k} \right]_0^1 = \frac{3}{4}\hat{i} + \frac{8}{3}\hat{j} + \frac{9}{5}\hat{k}$$

Along the curve $x = t, y = 2t^2, z = t^3$, $\vec{A} = y\hat{i} - xz\hat{j} + x^2\hat{k} = 2t^2\hat{i} - t^4\hat{j} + t^2\hat{k}$ and $d\vec{r} = (\hat{i} + 4t\hat{j} + 3t^2\hat{k})dt$. Therefore

$$\vec{A} \times d\vec{r} = \begin{vmatrix} \hat{i} & \hat{j} & \hat{k} \\ 2t^2 & -t^4 & t^2 \\ dt & 4t \, dt & 3t^2 \, dt \end{vmatrix}$$

$$= \left[-(3t^6 + 4t^3)\hat{i} + (t^2 - 6t^4)\hat{j} + (t^4 + 8t^3)\hat{k} \right] dt$$

$$\int_\Gamma \vec{A} \times d\vec{r} = \int_0^1 \left[-(3t^6 + 4t^3)\hat{i} + (t^2 - 6t^4)\hat{j} + (t^4 + 8t^3)\hat{k} \right] dt$$

$$= \left[-\left(3\frac{t^7}{7} + 4\frac{t^4}{4}\right)\hat{i} + \left(\frac{t^3}{3} - 6\frac{t^5}{5}\right)\hat{j} + \left(\frac{t^5}{5} + 8\frac{t^4}{4}\right)\hat{k} \right]_0^1 = -\frac{10}{7}\hat{i} - \frac{13}{15}\hat{j} + \frac{11}{5}\hat{k}$$

Example 6.3.5

Evaluate the vector line integral $\int_C \vec{F} \times d\vec{r}$, where $\vec{F} = z\,\hat{i}$ and C is the part of the circular helix $\vec{r} = b\cos t\hat{i} + b\sin t\hat{j} + ct\,\hat{k}$ between the points $(-b, 0, \pi c)$ and $(b, 0, 0)$.

Solution

For the circular helix $\vec{r} = b\cos t\hat{i} + b\sin t\hat{j} + ct\,\hat{k}$, we have

$$\frac{d\vec{r}}{dt} = -b\sin t\hat{i} + b\cos t\hat{j} + c\,\hat{k}$$

$$\vec{F} \times \frac{d\vec{r}}{dt} = \begin{vmatrix} \hat{i} & \hat{j} & \hat{k} \\ z & 0 & 0 \\ -b\sin t & b\cos t & c \end{vmatrix} = -c^2 t\,\hat{j} + bct\cos t\,\hat{k}$$

Thus, the required vector line integral becomes

$$\int_C \vec{F} \times d\vec{r} = \int_C \vec{F} \times \frac{d\vec{r}}{dt}\,dt = \int_\pi^0 \left[-c^2 t\,\hat{i} + bct\cos t\,\hat{k} \right] dt$$

$$= \left[-c^2 \frac{t^2}{2}\hat{j} + bc(t\sin t + \cos t)\hat{k} \right]_\pi^0 = \frac{1}{2}\pi^2 c^2 \hat{j} + 2bc\hat{k}$$

6.3.2 Properties of Line Integrals

Property 6.3.1

Let $\vec{A}$, a continuous vector point function, represent the velocity of a fluid. The line integral along a simple closed curve Γ (not intersecting anywhere) is given by

$$\oint_\Gamma \vec{A} \cdot d\vec{r} = \oint_\Gamma (A_1 dx + A_2 dy + A_3 dz) \tag{6.6}$$

The integral (6.6) is called the *circulation* of $\vec{A}$ about Γ, which can be discussed in aerodynamics and fluid mechanics. The field $\vec{A}$ defined over a region $\mathcal{R}$ is said to be *irrotational* if its circulation around all the closed curves in $\mathcal{R}$ is zero.

Example 6.3.6

If $\vec{A} = (y - 2x)\hat{i} + (3x + 2y)\hat{j}$, compute the circulation of $\vec{A}$ around the circle Γ in the xy plane with the centre at the origin and radius 2 and where Γ is traversed in the positive direction.

Solution

Here, we need to evaluate a line integral around the circular path $x^2 + y^2 = 4$ described in the anti-clockwise sense. Also, let the parametric equation of the circle be $x = 2\cos t, y = 2\sin t, 0 \le t \le 2\pi$. Now, since $\vec{r}$ is the position vector of any point on the circle, we have $\vec{r} = 2\cos t\,\hat{i} + 2\sin t\,\hat{j}$. Therefore, the circulation of $\vec{A}$ around the circle is given by Eq.(6.6), as

$$\oint_\Gamma \vec{A} \cdot d\vec{r} = \oint_\Gamma (y - 2x)dx + (3x + 2y)dy$$

$$= \int_0^{2\pi} \left[-(2\sin t - 4\cos t)2\sin t + (6\cos t + 4\sin t)2\cos t \right] dt$$

$$= \left[10t - 4\cos 2t + 7\sin 2t \right]_0^{2\pi} = 20\pi$$

The result shows that the value of the line integral depends only on the initial and terminal values of t, and hence, it is independent of the path of integration.

Example 6.3.7

Find the circulation of $\vec{A} = (2x + y^2)\hat{i} + (3y - 4x)\hat{j}$ round the curve Γ, where Γ is the curve $y = x^2$ from $(0, 0)$ to $(1, 1)$ and the curve $y^2 = x$ from $(1, 1)$ to $(0, 0)$.

Solution

We need to evaluate the circulation of $\vec{A}$ along two curves (Fig. 6.8), $\Gamma_1 : y = x^2$ from $(0, 0)$ to $(1, 1)$ and $\Gamma_2 : y^2 = x$ from $(1, 1)$ to $(0, 0)$. Since $\vec{A} = (2x + y^2)\hat{i} + (3y - 4x)\hat{j}$, along curve Γ_1

$$\int_{\Gamma_1} \vec{A} \cdot d\vec{r} = \int_0^1 \left[(2x + x^4)dx + (3x^2 - 4x) \cdot 2xdx\right]$$

$$= \int_0^1 \left[2x - 8x^2 + 6x^3 + x^4\right]dx$$

$$= \left[2\frac{x^2}{2} - 8\frac{x^3}{3} + 6\frac{x^4}{4} + \frac{x^5}{5}\right]_0^1 = \frac{1}{30}$$

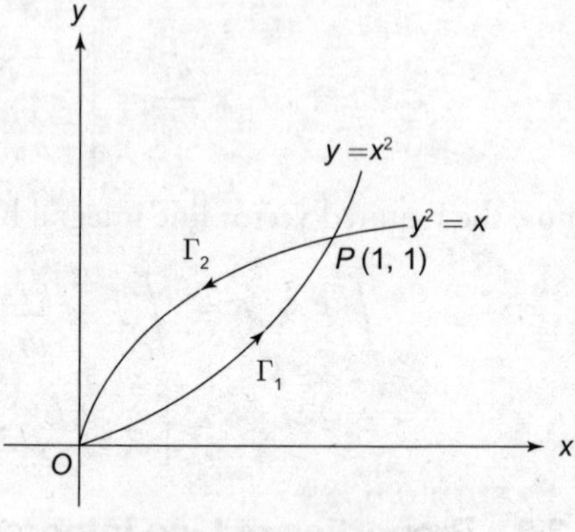

Figure 6.8 Circulation

Along curve Γ_2

$$\int_{\Gamma_2} \vec{A} \cdot d\vec{r} = \int_1^0 \left[(2y^2 + y^2) \cdot 2ydy + (3y - 4y^2)dy\right]$$

$$= \int_1^0 \left[3y - 4y^2 + 6y^3\right]dy$$

$$= \left[3\frac{y^2}{2} - 4\frac{y^3}{3} + 6\frac{y^4}{4}\right]_1^0 = -\frac{5}{3}$$

Therefore, the circulation of $\vec{A}$ around the circle is given by Eq.(6.6), as

$$\oint_{\Gamma} \vec{A} \cdot d\vec{r} = \int_{\Gamma_1} \vec{A} \cdot d\vec{r} + \int_{\Gamma_2} \vec{A} \cdot d\vec{r} = \frac{1}{30} - \frac{5}{3} = -\frac{49}{30}$$

Property 6.3.2

If the vector $\vec{F} = F_1\hat{i} + F_2\hat{j} + F_3\hat{k}$ is the force acting on a particle moving along the curve Γ, then the line integral

$$\int_{\Gamma} \vec{F} \cdot d\vec{r} = \int_{\Gamma} (F_1dx + F_2dy + F_3dz) \tag{6.7}$$

represents the work done by the force $\vec{F}$ in describing the curve Γ.

Example 6.3.8

Find the total work done in moving a particle in the force field

$$\vec{F} = (2x - y + z)\hat{i} + (x + y - z)\hat{j} + (3x - 2y - 5z)\hat{k}$$

along the circle Γ in the xy plane $x^2 + y^2 = 9, z = 0$.

Solution

The parametric equation of the circle Γ in the xy plane $x^2 + y^2 = 9, z = 0$ is $x = 3\cos t, y = 3\sin t$, where $0 \le t \le 2\pi$. So

$$\begin{aligned}\vec{F}\cdot d\vec{r} &= (2x - y + z)dx + (x + y - z)dy + (3x - 2y - 5z)dz\\ &= -(6\cos t - 3\sin t)3\sin t dt + 3(3\cos t + 3\sin t)\cos t dt\\ &= (-18\sin t\cos t + 9\sin^2 t + 9\cos^2 t + 9\sin t\cos t)dt\\ &= (9 - 9\sin t\cos t)dt = \left(9 - \frac{9}{2}\sin 2t\right)dt\end{aligned}$$

So the work done (W) is given by

$$W = \int_\Gamma \vec{F}\cdot d\vec{r} = \int_0^{2\pi}\left(9 - \frac{9}{2}\sin 2t\right)dt = \left[9t + \frac{9}{4}\cos 2t\right]_0^{2\pi} = 18\pi$$

Example 6.3.9

Let $\vec{F} = xy\hat{i} + y\hat{j} - yz\hat{k}$ denote the force field on a particle traversing the path L from $(0, 0, 0)$ to $(1, 1, 1)$ along the curve of intersection of the cylinder $y = x^2$ and the plane $z = x$. Find the work done by $\vec{F}$.

Solution

The work done by the force $\vec{F}$ is given by

$$\begin{aligned}W &= \oint_C \vec{F}\cdot d\vec{r} = \oint_C xydx + ydy - yzdz\\ &= \int_0^1 x\cdot x^2\, dx + x^2\cdot 2x\, dx - x^2\cdot x\, dx = \int_0^1 2x^3\, dx = \frac{1}{2}\end{aligned}$$

Property 6.3.3

A necessary and sufficient condition that the line integral of a vector point function $\vec{F}$ with continuous derivatives be independent of a path is that $\vec{F} = \vec{\nabla}\phi$, where ϕ is a single-valued scalar function having continuous partial derivatives.

Proof: First, let the function $\vec{F}$ be the gradient of a differentiable scalar point function ϕ, i.e., $\vec{F} = \vec{\nabla}\phi$. Then, the line integral becomes

$$\begin{aligned}\int_\Gamma \vec{F}\cdot d\vec{r} &= \int_\Gamma \vec{\nabla}\phi\cdot d\vec{r}\\ &= \int_A^B\left(\frac{\partial\phi}{\partial x}\hat{i} + \frac{\partial\phi}{\partial y}\hat{j} + \frac{\partial\phi}{\partial z}\hat{k}\right)\cdot\left(\hat{i}dx + \hat{j}dy + \hat{k}dz\right)\\ &= \int_A^B\left(\frac{\partial\phi}{\partial x}dx + \frac{\partial\phi}{\partial y}dy + \frac{\partial\phi}{\partial z}dz\right)\\ &= \int_A^B d\phi = \phi(B) - \phi(A)\end{aligned}$$

where Γ is the curve joining the points A and B (Fig. 6.9). Thus, we see that the line integral $\int_\Gamma \vec{F}\cdot d\vec{r}$ depends only the end points A and B of Γ and not on the choice of the curve joining those points, provided ϕ is single valued at points A and B.

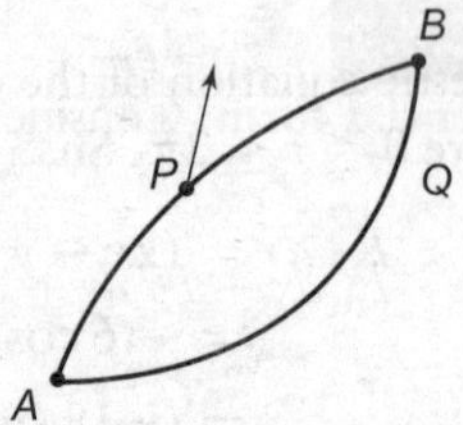

Figure 6.9 Independent of the path

Condition necessary: With reference to the fundamental basis $\hat{i}, \hat{j}$ and $\hat{k}$, let $\vec{F} = F_1\hat{i} + F_2\hat{j} + F_3\hat{k}$. Assume that $\int \vec{F}\cdot d\vec{r}$ is independent of the choice of the path Γ joining any two points $A(x_1, y_1, z_1)$, which is fixed and $P(x, y, z)$, respectively. Then

$$\phi(x,y,z) = \int_{A(x_1,y_1,z_1)}^{P(x,y,z)} \vec{F}\cdot d\vec{r} = \int_{(x_1,y_1,z_1)}^{(x,y,z)} (F_1dx + F_2dy + F_3dz)$$

This is some function of the position $P(x, y, z)$ and is independent of the path joining the points (x_1, y_1, z_1) and (x, y, z). If the line integral is independent of the path, then

$$\phi(x,y,z) = \int_A^P \vec{F}\cdot d\vec{r} = \int_A^P F\cdot\frac{dr}{ds}ds$$

$$\Rightarrow \frac{d\phi}{ds} = \vec{F}\cdot\frac{d\vec{r}}{ds} \text{ by differentiation}$$

$$\Rightarrow \vec{\nabla}\phi\frac{d\vec{r}}{ds} = \vec{F}\cdot\frac{d\vec{r}}{ds} \Rightarrow \left(\vec{\nabla}\phi - \vec{F}\right)\cdot\frac{d\vec{r}}{ds} = 0$$

This relation must hold irrespective of $\frac{d\vec{r}}{ds}$, therefore, $\vec{F} = \vec{\nabla}\phi$.

Example 6.3.10

If $\vec{F} = (4xy - 3x^2z^2)\hat{i} + 2x^2\hat{j} - 2x^3z\hat{k}$, prove that the line integral $\int_\Gamma \vec{F}\cdot d\vec{r}$ is independent of the curve γ joining the two given points.

Solution

Let $\vec{F} = \vec{\nabla}\phi = (4xy - 3x^2z^2)\hat{i} + 2x^2\hat{j} - 2x^3z\hat{k}$, so

$$\frac{\partial\phi}{\partial x} = 4xy - 3x^2z^2 \Longrightarrow \phi(x,y,z) = 2x^2y - x^3z^2 + \psi_1(y,z)$$

$$\frac{\partial\phi}{\partial y} = 2x^2 \Longrightarrow \phi(x,y,z) = 2x^2y + \psi_2(x,z)$$

$$\frac{\partial\phi}{\partial z} = -2x^3z \Longrightarrow \phi(x,y,z) = -x^3z^2 + \psi_3(x,y)$$

Let us take $\psi_1(y,z) = c$, $\psi_2(x,z) = -x^3z^2 + c$, $\psi_3(x,y) = 2x^2y + c$, where c is a scalar constant; then $\phi(x,y,z) = 2x^2y - x^3z^2 + c$. Thus there exists a function ϕ such that $\vec{F} = \vec{\nabla}\phi$ is satisfied. Hence, the line integral $\int_\Gamma \vec{F}\cdot d\vec{r}$ is independent of the curve γ joining the two given points.

Property 6.3.4

Exact differential form: Consider a differential $F_1\,dx + F_2\,dy + F_3\,dz$. This differential can be written as

$$\begin{aligned} F_1\,dx + F_2\,dy + F_3\,dz &= (F_1\hat{i} + F_2\hat{j} + F_3\hat{k}) \cdot (dx\hat{i} + dy\hat{j} + dz\hat{k}) \\ &= \vec{F} \cdot d\vec{r}; \quad \vec{r} = x\hat{i} + y\hat{j} + z\hat{k} \end{aligned}$$

Hence, the given differential is exact if the line integral $\int \vec{F} \cdot d\vec{r}$ is independent of the path, where $\vec{F} = F_1\hat{i} + F_2\hat{j} + F_3\hat{k}$. Thus, there exists a function $\phi(x, y, z)$ such that $\vec{F} = \vec{\nabla}\phi$. The solution of $F_1\,dx + F_2\,dy + F_3\,dz = 0$ is $\phi(x, y, z) = c$.

Example 6.3.11

Show that $(2x\cos y + z\sin y)dx + (xz\cos y - x^2\sin y)\;dy + x\sin y\;dz$ is an exact differential. Hence, solve $(2x\cos y + z\sin y)dx + (xz\cos y - x^2\sin y)\;dy + x\sin y\;dz = 0$.

Solution

The given differential can be written as $\vec{F} \cdot d\vec{r}$, where $\vec{F} = (2x\cos y + z\sin y)\;\hat{i} + (xz\cos y - x^2\sin y)\;\hat{j} + x\sin y\;\hat{k}$ and $\vec{r} = x\hat{i} + y\hat{j} + z\hat{k}$. Now

$$\begin{aligned} \vec{\nabla} \times \vec{F} &= \begin{vmatrix} \hat{i} & \hat{j} & \hat{k} \\ \dfrac{\partial}{\partial x} & \dfrac{\partial}{\partial y} & \dfrac{\partial}{\partial z} \\ 2x\cos y + z\sin y & xz\cos y - x^2\sin y & x\sin y \end{vmatrix} \\ &= (x\cos y - x\cos y)\hat{i} + (\sin y - \sin y)\hat{j} \\ &\quad + (z\cos y - 2x\sin y + 2x\sin y - z\cos y)\hat{k} = \vec{0} \end{aligned}$$

Thus, the given differential is an exact one. So, there exists a scalar function $\phi(x, y, z)$ such that

$$\vec{F} = \vec{\nabla}\phi = \frac{\partial \phi}{\partial x}\hat{i} + \frac{\partial \phi}{\partial y}\hat{j} + \frac{\partial \phi}{\partial z}\hat{k}$$

$$\frac{\partial \phi}{\partial x} = 2x\cos y + z\sin y \Longrightarrow \phi(x, y, z) = x^2\cos y + xz\sin y + \psi_1(y, z)$$

$$\frac{\partial \phi}{\partial y} = xz\cos y - x^2\sin y \Longrightarrow \phi(x, y, z) = xz\sin y + x^2\cos y + \psi_2(x, z)$$

$$\frac{\partial \phi}{\partial z} = x\sin y \Longrightarrow \phi(x, y, z) = xz\sin y + \psi_3(x, y)$$

Therefore, $\phi(x, y, z) = x^2\cos y + xz\sin y + c$, where c is an arbitrary constant. The solution is $x^2\cos y + xz\sin y = c_1$.

Definition 6.3.1

If the line integral $\int_\Gamma \vec{F} \cdot d\vec{r}$ of a vector field $\vec{F}$ depends on $\vec{F}$, and on the end points A and B of Γ, but is independent of the nature of the path Γ joining A and B, then the vector field $\vec{F}$ is called a *conservative force field*. It follows that if $\vec{F} = \vec{\nabla}\phi$, then $\vec{F}$ is conservative. Gravitational and electrostatic fields are common examples of conservative fields.

Theorem 6.1

A necessary and sufficient condition that the vector field defined by the vector point function $\vec{F}$ with continuous derivatives in an open region $\mathcal{R}$ be conservative is that

$$\text{curl}\,\vec{F} = \vec{\nabla} \times \vec{F} = \vec{0}$$

Proof: If $\vec{F}$ be a conservative force field then there exists a scalar potential function ϕ with continuous partial derivatives such that $\vec{F} = \vec{\nabla}\phi$. Since the scalar function ϕ has continuous partial derivatives, mixed derivatives like $\dfrac{\partial^2\phi}{\partial x\partial y} = \dfrac{\partial^2\phi}{\partial y\partial x}$ hold. Now

$$\vec{\nabla} \times \vec{F} = \vec{\nabla} \times \vec{\nabla}\phi = \vec{\nabla} \times \left(\frac{\partial\phi}{\partial x}\hat{i} + \frac{\partial\phi}{\partial y}\hat{j} + \frac{\partial\phi}{\partial z}\hat{k}\right)$$

$$= \begin{vmatrix} \hat{i} & \hat{j} & \hat{k} \\ \dfrac{\partial}{\partial x} & \dfrac{\partial}{\partial y} & \dfrac{\partial}{\partial z} \\ \dfrac{\partial\phi}{\partial x} & \dfrac{\partial\phi}{\partial y} & \dfrac{\partial\phi}{\partial z} \end{vmatrix}$$

$$= \hat{i}\left(\frac{\partial^2\phi}{\partial y\partial z} - \frac{\partial^2\phi}{\partial y\partial z}\right) - \hat{j}\left(\frac{\partial^2\phi}{\partial x\partial z} - \frac{\partial^2\phi}{\partial x\partial z}\right) + \hat{k}\left(\frac{\partial^2\phi}{\partial x\partial y} - \frac{\partial^2\phi}{\partial x\partial y}\right) = \vec{0}$$

Thus, the condition is necessary. Conversely, let $\vec{\nabla} \times \vec{F} = \vec{0}$. Referring to the fundamental basis, let $\vec{F} = F_1\hat{i} + F_2\hat{j} + F_3\hat{k}$, then

$$\vec{\nabla} \times \vec{F} = \vec{0} \Rightarrow \begin{vmatrix} \hat{i} & \hat{j} & \hat{k} \\ \dfrac{\partial}{\partial x} & \dfrac{\partial}{\partial y} & \dfrac{\partial}{\partial z} \\ F_1 & F_2 & F_3 \end{vmatrix} = \vec{0} \Rightarrow \frac{\partial F_3}{\partial y} = \frac{\partial F_2}{\partial z}, \quad \frac{\partial F_3}{\partial x} = \frac{\partial F_1}{\partial z}, \quad \frac{\partial F_2}{\partial x} = \frac{\partial F_1}{\partial y}$$

We need to show that $\vec{F} = \vec{\nabla}\phi$. Let Γ be a curve joining two points $A(x_1, y_1, z_1)$ and $P(x, y, z)$. Then, the work done in moving a particle from A to P in the force field $\vec{F}$ is

$$\int_\Gamma \vec{F}\cdot d\vec{r} = \int_\Gamma (F_1dx + F_2dy + F_3dz)$$

Let us choose a particular path with the following straight line segments: (x_1, y_1, z_1) to (x, y_1, z_1); then from (x, y_1, z_1) to (x, y, z_1) and finally from (x, y, z_1) to (x, y, z). Let the work done be $\phi(x, y, z)$ the value of $\int_\Gamma \vec{F}\cdot d\vec{r}$ along this particular path. Then

$$\phi(x, y, z) = \int_{x_1}^{x} F_1dx + \int_{y_1}^{y} F_2dy + \int_{z_1}^{z} F_3dz$$

By differentiation, it follows that

$$\begin{aligned}
\frac{\partial \phi}{\partial z} &= F_3(x,y,z) \\
\frac{\partial \phi}{\partial y} &= F_2(x,y,z_1) + \int_{z_1}^{z} \frac{\partial F_3(x,y,z)}{\partial y} dz \\
&= F_2(x,y,z_1) + \int_{z_1}^{z} \frac{\partial F_2}{\partial z} dz \\
&= F_2(x,y,z_1) + F_2(x,y,z) - F_2(x,y,z_1) = F_2(x,y,z) \\
\frac{\partial \phi}{\partial x} &= F_1(x,y_1,z_1) + \int_{y_1}^{y} \frac{\partial F_2(x,y,z)}{\partial x} dy + \int_{z_1}^{z} \frac{\partial F_3(x,y,z)}{\partial x} dz \\
&= F_1(x,y_1,z_1) + \int_{y_1}^{y} \frac{\partial F_1(x,y,z)}{\partial y} dy + \int_{z_1}^{z} \frac{\partial F_1(x,y,z)}{\partial z} dz \\
&= F_1(x,y_1,z_1) + F_1(x,y,z_1) - F_1(x,y_1,z_1) + F_1(x,y,z) - F_1(x,y,z_1) \\
&= F_1(x,y,z)
\end{aligned}$$

Therefore, $\vec{F} = F_1\hat{i} + F_2\hat{j} + F_3\hat{k}$ becomes

$$\vec{F} = \frac{\partial \phi}{\partial x}\hat{i} + \frac{\partial \phi}{\partial y}\hat{j} + \frac{\partial \phi}{\partial z}\hat{k} = \vec{\nabla}\phi$$

This shows that the force field $\vec{F}$ is conservative.

RESULT 20

The conservative forces may be distinguished by any one of the following equivalent features:

(i) From the definition (5.4.2) of *irrotational vector*, we conclude that every conservative force field vector is irrotational. Thus

Irrotational ≡ Every circulation is zero
Conservative ≡ Line integral independent of the path

(ii) For a conservative force field $\vec{F}$, as $\vec{\nabla} \times \vec{F} = \vec{0}$, i.e., one can write $\vec{F}$ as the gradient of some scalar point function. Since the gradient points towards the direction of increasing potential and forces cause the system to move towards the lower potential, $\vec{F} = \vec{\nabla}\,\phi(\vec{r})$. The scalar point function ϕ introduced here is called the *potential energy* of the particle at that point. Thus, a conservative force is also central in nature.

The work done in moving a particle in this field from A to B is $W_{AB} = \phi(B) - \phi(A)$.

(iii) The work done by the force is independent of the path. Since the work done is zero around any closed path in the conservative force field and does not depend on the length of the path, we may carry out integration over the perimeter of area $d\vec{S}$.

(iv) For a conservative force $\vec{F}$, from (Fig. 6.9), we get

$$P\int_A^B \vec{F}\cdot d\vec{r} = Q\int_A^B \vec{F}\cdot d\vec{r}$$

$$\Rightarrow P\int_A^B \vec{F}\cdot d\vec{r} + Q\int_B^A \vec{F}\cdot d\vec{r} = 0$$

$$\Rightarrow \oint \vec{F} \cdot d\vec{r} = 0 \tag{6.8}$$

Thus, if the force is conservative, the work done on the particle around a closed path in the force field is zero.

(v) Referring to the fundamental basis, let $\vec{F} = F_1\hat{i} + F_2\hat{j} + F_3\hat{k}$ be the Cartesian component of the force $\vec{F}$. Then, $F_1\ dx + F_2\ dy + F_3\ dz$ is an *exact differential.*

(vi) At every point $T + \phi =$ constant for a particle, where T is the kinetic energy.

Example 6.3.12

(i) Prove that $\vec{F} = (y^2 \cos x + z^3)\hat{i} + (2y \sin x - 4)\hat{j} + (3xz^2 + 2)\hat{k}$ is a conservative force field.

(ii) Find the scalar potential for $\vec{F}$.

(iii) Find the work done in moving an object in this field from $(0, 1, -1)$ to $(\pi/2, -1, 2)$.

Solution

(i) To show that the given vector field $\vec{F}$ is conservative, we have

$$\vec{\nabla} \times \vec{F} = \begin{vmatrix} \hat{i} & \hat{j} & \hat{k} \\ \dfrac{\partial}{\partial x} & \dfrac{\partial}{\partial y} & \dfrac{\partial}{\partial z} \\ y^2 \cos x + z^3 & 2y \sin x - 4 & 3xz^2 + 2 \end{vmatrix}$$

$$= (0 - 0)\hat{i} + (3z^2 - 3z^2)\hat{j} + (2y \cos x - 2y \cos x)\hat{k} = \vec{0}$$

Hence, by the necessary and sufficient condition, the given force field $\vec{F}$ is conservative.

(ii) Since $\vec{\nabla} \times \vec{F} = \vec{0}$, there exists a scalar function $\phi(x, y, z)$ such that

$$\vec{F} = \vec{\nabla}\phi = \frac{\partial \phi}{\partial x}\hat{i} + \frac{\partial \phi}{\partial y}\hat{j} + \frac{\partial \phi}{\partial z}\hat{k}$$

$$\frac{\partial \phi}{\partial x} = y^2 \cos x + z^3 \Longrightarrow \phi(x, y, z) = y^2 \sin x + xz^3 + \psi_1(y, z)$$

$$\frac{\partial \phi}{\partial y} = 2y \sin x - 4 \Longrightarrow \phi(x, y, z) = y^2 \sin x - 4y + \psi_2(x, z)$$

$$\frac{\partial \phi}{\partial z} = 3xz^2 + 2 \Longrightarrow \phi(x, y, z) = xz^3 + 2z + \psi_3(x, y)$$

Therefore, $\phi(x, y, z) = y^2 \sin x + xz^3 - 4y + 2z + c$, where c is an arbitrary constant.

(iii) The work done in moving an object in this field from $(0, 1, -1)$ to $(\pi/2, -1, 2)$is given by

$$\begin{aligned} \phi(\pi/2, -1, 2) - \phi(0, 1, -1) &= (-1)^2 \sin(\pi/2) + (\pi/2)2^3 - 4 \cdot (-1) + 2 \cdot 2 \\ &\quad - \left[1^2 \sin 0 + 0 \cdot (-1)^3 - 4 \cdot 1 + 2(-1)\right] \\ &= (1 + 4\pi + 4 + 4) - (0 + 0 - 4 - 2) \\ &= 15 + 4\pi \end{aligned}$$

Example 6.3.13

Prove the principle of energy using the vector method.

Solution

If a particle is moved from a point A to a point B under the action of a force $\vec{F}$, then the work done by the particle is

$$\int_A^B \vec{F}\cdot d\vec{r} = \int_A^B m\frac{d\vec{v}}{dt}\cdot\frac{d\vec{r}}{dt}\,dt = \int_A^B m\vec{v}\cdot\frac{d\vec{v}}{dt}\,dt$$

$$= \frac{m}{2}\int_A^B \frac{d}{dt}(\vec{v}\cdot\vec{v})dt = \frac{m}{2}\int_A^B dv^2 = \frac{1}{2}\,m\,v_B^2 - \frac{1}{2}\,m\,v_A^2$$

Thus, the work done by the particle in moving from A to B under the force $\vec{F}$ is equal to the change in kinetic energy, which is the *principle of energy*.

Example 6.3.14

Determine whether the force represented by $\vec{F} = p\left[(x^2+y^2)\hat{i} + 2xy\hat{j}\right]$ is conservative or not. Here $p = 1Nm^{-2}$. Calculate the work done by this force in moving a particle from the origin $O(0,0,0)$ to the point $D(1,1,0)$ on the $z=0$ plane along the paths $OABD$ and OD, as shown in Fig. 6.10, where the coordinates are measured in metres.

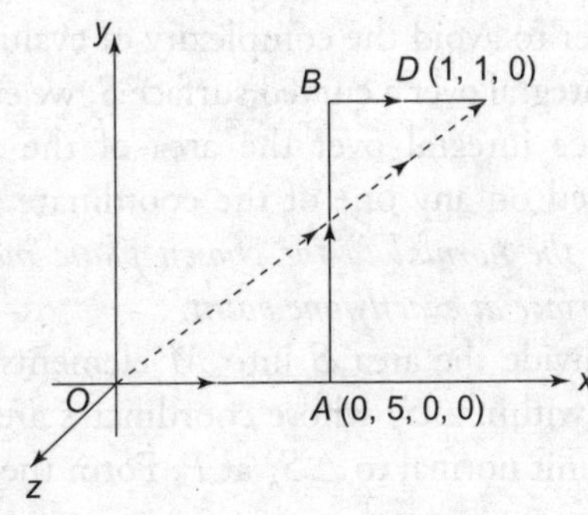

Figure 6.10 Line integral

Solution

Here, it is given that $\vec{F} = p\left[(x^2+y^2)\hat{i} + 2xy\hat{j}\right]$. Now

$$\vec{\nabla}\times\vec{F} = p\begin{vmatrix} \hat{i} & \hat{j} & \hat{k} \\ \dfrac{\partial}{\partial x} & \dfrac{\partial}{\partial y} & \dfrac{\partial}{\partial z} \\ x^2+y^2 & 2xy & 0 \end{vmatrix} = \vec{0}$$

Therefore, by definition, $\vec{F}$ is a conservative force field. So there exists a scalar function $\phi(x,y,z)$ such that

$$\vec{F} = \vec{\nabla}\phi = \frac{\partial\phi}{\partial x}\hat{i} + \frac{\partial\phi}{\partial y}\hat{j} + \frac{\partial\phi}{\partial z}\hat{k}$$

$$\frac{\partial\phi}{\partial x} = (x^2+y^2) \Longrightarrow \phi(x,y,z) = \frac{x^3}{3} + xy^2 + \psi_1(y,z)$$

$$\frac{\partial\phi}{\partial y} = 2xy \Longrightarrow \phi(x,y,z) = xy^2 + \psi_2(x,z)$$

Therefore $\phi(x,y,z) = \frac{x^3}{3} + xy^2 + c$, where c is an arbitrary constant. Also, as $\vec{F}$ is an exact function, the work done depends only on the initial and final points (path independent). Thus, the required work done by $\vec{F}$ is

$$W = \phi(1,1,0) - \phi(0,0,0) = \left(\frac{1}{3}+1\right) - (0+0) = \frac{4}{3}J$$

6.4 SURFACE INTEGRALS

Let S be a two-sided surface, as shown in Fig. 6.11. Associated with the differential of surface area dS is a vector $d\vec{S}$ whose magnitude is dS and whose direction is that of $\hat{n}$. Then, $d\vec{S} = \vec{n}\, dS$. Let $\vec{F}(\vec{r})$ be a vector point function defined on the two-sided surface $\mathcal{S}$. Then, the integral

$$\int\int_S \vec{F}\cdot d\vec{S} = \int\int_S \vec{F}\cdot \hat{n}\, dS \qquad (6.9)$$

is defined as the *surface integral* or *flux* of $\vec{F}$ over S. In order to avoid the complexity of evaluating a surface integral over a curved surface S, we evaluate the surface integral over the area of the surface S projected on any one of the coordinate planes, provided *the normal to the chosen plane meets the curved surface at exactly one point.*

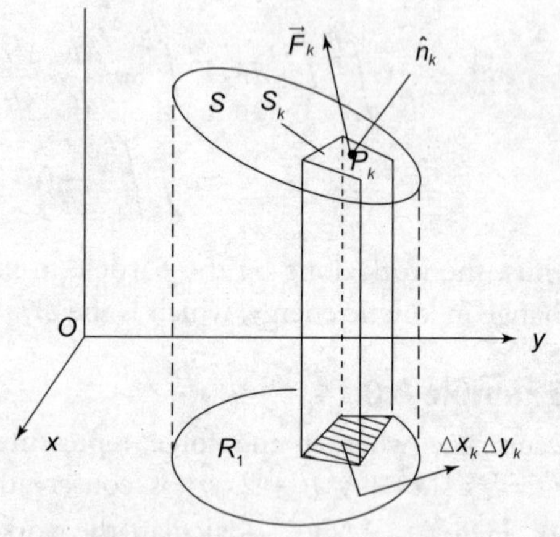

Figure 6.11 Surface integrals

Subdivide the area $\mathcal{S}$ into M elements of area $\triangle\mathcal{S}_k$, where $k = 1, 2, \cdots, M$. Choose any point P_k within $\triangle\mathcal{S}_k$ whose coordinates are (x_k, y_k, z_k). Define $\vec{F}(x_k, y_k, z_k) = \vec{F}_k$. Let $\hat{n}_k$ be the positive unit normal to $\triangle\mathcal{S}_k$ at P. Form the sum

$$\sum_{k=1}^{M} \vec{F}(x_k, y_k, z_k)\cdot \hat{n}_k \,\triangle\mathcal{S}_k = \sum_{k=1}^{M} \vec{F}_k\cdot \hat{n}_k \,\triangle\mathcal{S}_k \qquad (6.10)$$

where $\vec{F}_k \cdot \hat{n}_k$ is the normal component of $\vec{F}_k$ at P_k. Now, take the limit of this sum as $M \to \infty$ in such a way that the largest dimension of each $\triangle\mathcal{S}_k$ approaches zero. This limit, if it exists, is called the *surface integral* of the normal component of $\vec{A}$ over $\mathcal{S}$ and is written as

$$\int\int_S \vec{F}\cdot d\vec{S} = \lim_{M\to\infty} \sum_{k=1}^{M} \vec{F}_k\cdot \hat{n}_k \,\triangle\mathcal{S}_k \qquad (6.11)$$

Suppose the surface $\mathcal{S}$ has a projection R_1 on the *xy*-plane. The projection of $\triangle\mathcal{S}_k$ on the *xy* plane is $|(\hat{n}_k\triangle\mathcal{S}_k)\cdot\hat{k}|$ or $|\hat{n}_k\cdot\hat{k}|\triangle\mathcal{S}_k$, which is equal to $\triangle x_k\triangle y_k$, so that $\triangle\mathcal{S}_k = \dfrac{\triangle x_k \triangle y_k}{|\hat{n}_k\cdot\hat{k}|}$. Thus, Eq.(6.11) becomes

$$\int\int_S \vec{F}\cdot d\vec{S} = \lim_{M\to\infty} \sum_{k=1}^{M} \vec{F}_k\cdot \hat{n}_k \,\frac{\triangle x_k\triangle y_k}{|\hat{n}_k\cdot\hat{k}|}$$

By the fundamental theorem of integral calculus, the limit of this sum is $M \to \infty$ in such a manner that the largest $\triangle x_k \triangle y_k$ approaches zero as

$$\int\int_S \vec{F}\cdot d\vec{S} = \int\int_S \vec{F}\cdot \hat{n}\, dS = \int\int_{R_1} \vec{F}\cdot \hat{n}\frac{dxdy}{|\hat{n}\cdot\hat{k}|} \qquad (6.12)$$

where $\hat{n}$ is the outward drawn normal to the surface. Similarly, let the surface $\mathcal{S}$ have a projection R_2 on the yz plane and R_3 on the zx plane; then

$$\int\int_S \vec{F}\cdot\hat{n}\,dS = \int\int_{R_2} \vec{F}\cdot\hat{n}\frac{dydz}{|\hat{n}\cdot\hat{i}|} = \int\int_{R_3} \vec{F}\cdot\hat{n}\frac{dxdz}{|\hat{n}\cdot\hat{j}|} \tag{6.13}$$

Similarly, if the surface $\mathcal{S}$ has a projection R_1 on the xy plane, then

$$\int\int_S \phi\,\hat{n}\,dS = \int\int_{R_1} \phi\,\hat{n}\,\frac{dx\,dy}{|\hat{n}\cdot\hat{k}|} \tag{6.14}$$

Other forms of vector surface integrals are

$$\int\int_S \phi\,dS, \quad \int\int_S \phi\,\hat{n}\,dS, \quad \int\int_S \vec{F}\times d\vec{S} \tag{6.15}$$

where ϕ is a scalar function. These integrals are defined in terms of limits of sums.

Example 6.4.1

Evaluate $\int\int_S \vec{F}\cdot\hat{n}dS$, where $\vec{F} = 18z\hat{i} - 12\hat{j} + 3y\hat{k}$ and $\mathcal{S}$ is a part of the plane $2x+3y+6z = 12$, which is located in the first octant.

Solution

The region of integration is bounded by four planes, as shown in Fig. 6.12. We have to calculate the transformed surface integral over the projected area of the given area, either in the xy plane, or yz plane or xz plane. Suppose that surface $\mathcal{S}$ has a projection $R_1 = \{(x,y) : 2x + 3y \leq 12\}$ on the xy plane, as shown in Fig. 6.12. For the given surface $\phi(x,y,z) = 2x + 3y + 6z = 12$

$$\vec{\nabla}\phi = \vec{\nabla}(2x + 3y + 6z) = 2\hat{i} + 3\hat{j} + 6\hat{k}$$

at any point of $\mathcal{S}$. Thus, the outward unit normal to any point of the surface $\mathcal{S}$ is

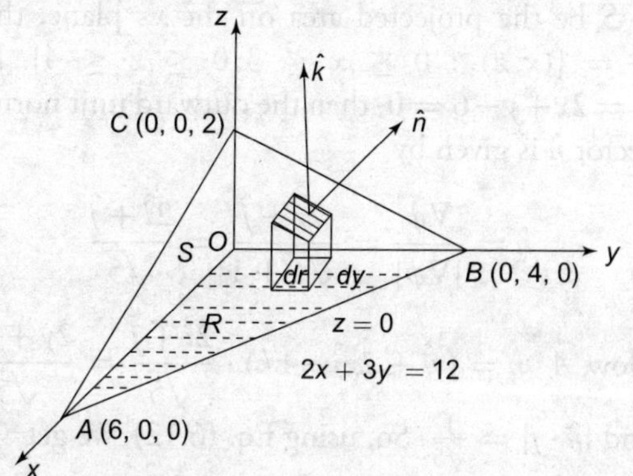

Figure 6.12 Surface integrals

$$\hat{n} = \frac{1}{\sqrt{2^2+3^2+6^2}}(2\hat{i} + 3\hat{j} + 6\hat{k}) = \frac{2}{7}\hat{i} + \frac{3}{7}\hat{j} + \frac{6}{7}\hat{k}$$

$$\Rightarrow \hat{n}\cdot\hat{k} = \left(\frac{2}{7}\hat{i} + \frac{3}{7}\hat{j} + \frac{6}{7}\hat{k}\right)\cdot\hat{k} = \frac{6}{7}$$

Using the fact that $z = (12 - 2x - 3y)/6$ from the equation of the surface $\mathcal{S}$, we get

$$\vec{F}\cdot\hat{n} = (18z\hat{i} - 12\hat{j} + 3y\hat{k})\cdot\left(\frac{2}{7}\hat{i} + \frac{3}{7}\hat{j} + \frac{6}{7}\hat{k}\right)$$

$$= \frac{36z - 36 + 18y}{7} = \frac{36 - 12x}{7}$$

Using Eq. (6.12), we get

$$\int\int_S \vec{F}\cdot\hat{n}\,dS = \int\int_{R_1} \vec{F}\cdot\hat{n}\frac{dxdy}{|\hat{n}\cdot\hat{k}|} = \int\int_{R_1}\left(\frac{36-12x}{7}\right)\frac{7}{6}\,dx\,dy$$

$$= \int\int_{R_1}(6-2x)dx\,dy == \int_{x=0}^{6}\left[\int_{y=0}^{(12-2x)/3}(6-2x)dy\right]dx$$

$$= \int_0^6\left(24-12x+\frac{4x^2}{3}\right)dx = 24$$

Example 6.4.2

Evaluate $\int\int_S \vec{A}\cdot\hat{n}\,dS$, where $\vec{A} = y\hat{i}+2x\hat{j}-z\hat{k}$ and $\mathcal{S}$ is the surface of the plane $2x+y=6$ included in the first octant cut off by the plane $z=4$.

Solution

Here, the plane is perpendicular to the xy plane, so we have to calculate the transformed surface integral over the projected area of the given area either on the yz plane or on the xz plane (Fig. 6.13). If $\mathcal{S}$ be the projected area on the xz plane, then $\mathcal{S} = \{(x,z) : 0 \le x \le 3; 0 \le z \le 4\}$. Let $\phi = 2x+y-6 = 0$, then the outward unit normal vector $\hat{n}$ is given by

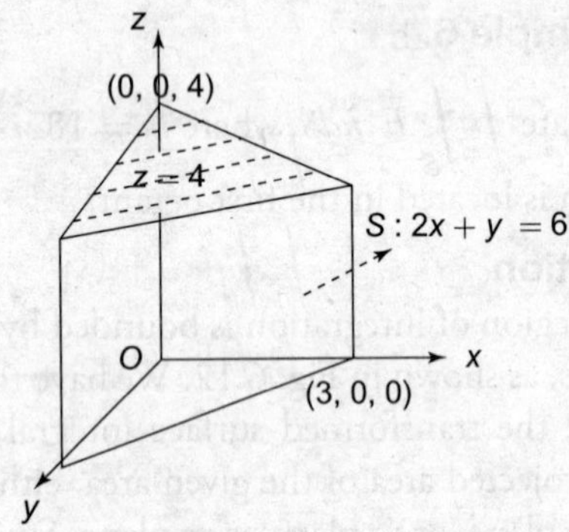

Figure 6.13 Region of integration

$$\hat{n} = \frac{\vec{\nabla}\phi}{|\vec{\nabla}\phi|} = \frac{2\hat{i}+\hat{j}}{\sqrt{4+1}} = \frac{2\hat{i}+\hat{j}}{\sqrt{5}}$$

Now, $\vec{A}\cdot\hat{n} = (y\hat{i}+2x\hat{j}-z\hat{k})\cdot\dfrac{2\hat{i}+\hat{j}}{\sqrt{5}} = \dfrac{2y+2x}{\sqrt{5}}$

and $|\hat{n}\cdot\hat{j}| = \frac{1}{\sqrt{5}}$. So, using Eq. (6.12), we get

$$\int\int_S \vec{A}\cdot\hat{n}\,dS = \int\int_S \frac{2(x+y)}{\sqrt{5}}\frac{\sqrt{5}}{1}\,dxdz$$

$$= 2\int_{x=0}^{3}\int_{z=0}^{4}(x+6-2x)\,dx\,dz$$

$$= 8\int_0^3 (6-x)\,dx$$

$$= 8\left[6x-\frac{x^2}{2}\right]_0^3 = 108$$

Example 6.4.3

Evaluate $\int\int_S \vec{F}\cdot\hat{n}dS$ over the entire surface of the region above the xy plane bounded by the cone $z^2 = x^2+y^2$ and the plane $z=4$, where $\vec{F} = 4xz\hat{i}+xyz^2\hat{j}+3z\hat{k}$.

Solution

The surface S consists two parts (Fig. 6.14): S_1 is the curved surface $\varphi(x, y, z) = 0$, where $\varphi(x, y, z) = x^2 + y^2 - z^2$, and S_2 is the plane surface $z = 4$.

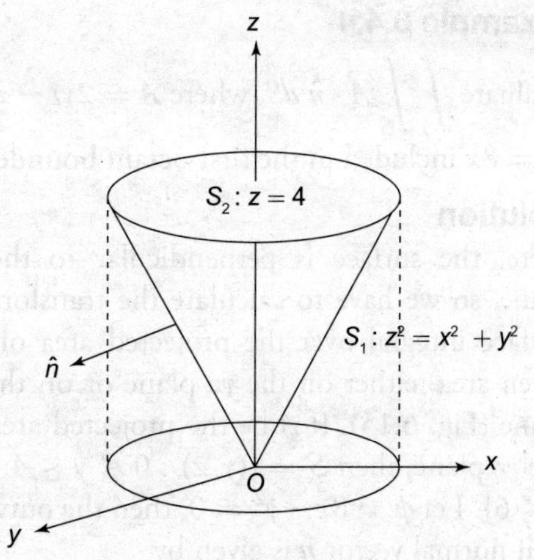

Figure 6.14 Surface integrals

$$\begin{aligned}\hat{n}_1 &= \text{ Unit normal to the surface } S_1\\ &= \frac{1}{|\vec{\nabla}\varphi|}\,\vec{\nabla}\varphi = \frac{1}{\sqrt{x^2+y^2+z^2}}\,(x\hat{i} + y\hat{j} - z\hat{k})\end{aligned}$$

Also, $\hat{n}_2 =$ unit normal to the surface $S_2 = \hat{k}$. Let R be the projection of the surface area S_1 on the xy plane, which is the circular region $x^2 + y^2 = 4^2$. Thus

$$\begin{aligned}\int\int_{S_1} \vec{F}\cdot\hat{n}_1 dS_1 &= \int\int_R \frac{1}{|\vec{n}_1\cdot\hat{k}|}\vec{F}\cdot\hat{n}_1 dxdy\\ &= \int\int_R \frac{4x^2z + xy^2z^2 - 3z^2}{z}dxdy\\ &= \int\int_R 4x^2 + (xy^2 - 3)zdxdy\\ &= \int\int_R \left[4x^2 + (xy^2 - 3)\sqrt{x^2+y^2}\right]dxdy;\end{aligned}$$

$$x = r\cos\theta, y = r\sin\theta; J = r$$

$$\begin{aligned}&= \int_{r=0}^{4}\int_{\theta=0}^{2\pi} 4r^2\cos^2\theta + (r^3\cos\theta\sin^2\theta - 3)rr\, d\theta\, dr\\ &= \int_{r=0}^{4}\int_{\theta=0}^{2\pi} (4r^3\cos^2\theta - 3r^2)d\theta\, dr\\ &= \int_0^4 (4r^3\pi - 3r^2\times 2\pi)dr = 256\pi - 128\pi = 128\pi\end{aligned}$$

On $S_/2, z = 4$, so

$$\begin{aligned}\int\int_{S_2} \vec{F}\cdot\hat{n}_2 dS_2 = \int\int_{S_2} 3z\, dx\, dy &= 12\int\int_{S_2} dxdy\\ &= 12\times \text{ area of the surface } S_2\\ &= 12\times\pi\cdot 4^2 = 192\pi\end{aligned}$$

Therefore, the required surface integral is

$$\begin{aligned}\int\int_S \vec{F}\cdot\hat{n}dS &= \int\int_{S_1}\vec{F}\cdot\hat{n}_1 dS_1 + \int\int_{S_2}\vec{F}\cdot\hat{n}_2 dS_2\\ &= 128\pi + 192\pi = 320\pi\end{aligned}$$

Example 6.4.4

Evaluate $\int\int_S \vec{A} \cdot \hat{n}\, dS$, where $\vec{A} = 2y\hat{\imath} - z\hat{\jmath} + x^2\hat{k}$ and $\mathcal{S}$ is the surface of the parabolic cylinder $y^2 = 8x$ included in the first octant bounded by the planes $y = 4$ and $z = 6$.

Solution

Here, the surface is perpendicular to the xy plane, so we have to calculate the transformed surface integral over the projected area of the given area, either on the yz plane or on the xz plane (Fig. 6.15). If A be the projected area on the yz plane, then $\mathcal{S} = \{(y, z) : 0 \leq y \leq 4; 0 \leq z \leq 6\}$. Let $\phi = 8x - y^2 = 0$, then the outward unit normal vector $\hat{n}$ is given by

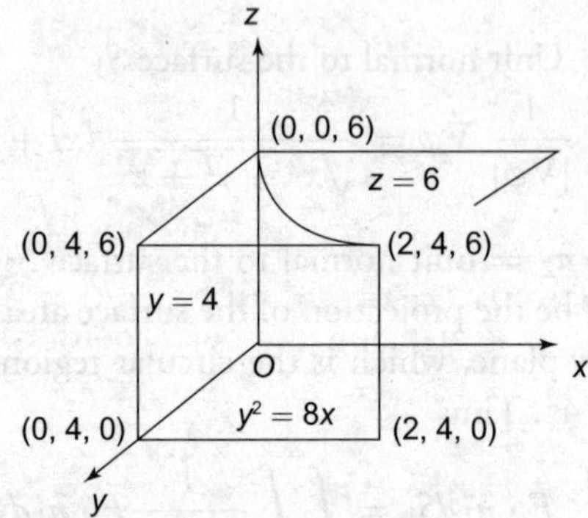

Figure 6.15 Region of integration

$$\hat{n} = \frac{1}{|\vec{\nabla}\phi|}\vec{\nabla}\phi = \frac{8\hat{\imath} - 2y\hat{\jmath}}{\sqrt{64 + 4y^2}} = \frac{4\hat{\imath} - y\hat{\jmath}}{\sqrt{16 + y^2}}$$

Now, $\vec{A} \cdot \hat{n} = (2y\hat{\imath} - z\hat{\jmath} - x^2\hat{k} \cdot \dfrac{4\hat{\imath} - y\hat{\jmath}}{\sqrt{16 + y^2}} = \dfrac{8y + yz}{\sqrt{16 + y^2}}$ and $|\hat{n} \cdot \hat{\imath}| = \dfrac{4}{\sqrt{16 + y^2}}$. So

$$\int_S \vec{A} \cdot \hat{n}\, dS = \int\int_S \frac{y(8 + z)}{\sqrt{16 + y^2}} \frac{\sqrt{16 + y^2}}{4}\, dydz$$

$$= \frac{1}{4}\int_{y=0}^{4}\int_{z=0}^{6} y(8 + z)\, dy\, dz$$

$$= \frac{1}{4}\int_0^4 \left[8yz + y\frac{z^2}{2}\right]_0^6 dy = \frac{1}{4}\int_0^4 66y\, dy = 132$$

Example 6.4.5

Let $\mathcal{S}$ be the surface $\{(x, y, z) \in \Re^3 : x^2 + y^2 + 2z = 2, z \geq 0\}$, and let $\hat{n}$ be the outward unit normal to $\mathcal{S}$. If $\vec{F} = y\hat{\imath} + xz\hat{\jmath} + (x^2 + y^2)\hat{k}$, then evaluate the integral $\int\int_S \vec{F} \cdot \hat{n}\, dS$.

Solution

Here, $\phi(x, y, z) = x^2 + y^2 + 2z$. The unit normal to the surface is given by

$$\hat{n} = \frac{1}{|\vec{\nabla}\phi|}\vec{\nabla}\phi = \frac{1}{\sqrt{x^2 + y^2 + 1}}(x\hat{\imath} + y\hat{\jmath} + \hat{k})$$

Now, suppose that surface $\mathcal{S}$ has a projection R_1 on the xy plane. Using Eq. (6.12), we get

$$\int\int_S \vec{F} \cdot \hat{n}\, dS = \int\int_S \left[xy + xyz + (x^2 + y^2)\right] dS$$

$$= \int\int_S \left[xy + xyz + (x^2 + y^2)\right] \frac{dx\, dy}{|\hat{n} \cdot \hat{k}|}$$

$$= \int\int_S \left[xy + xy\left(2 - \frac{x^2+y^2}{2}\right) + (x^2+y^2)\right]\sqrt{x^2+y^2+1}\,dx\,dy$$

$$\left(\text{as } x^2+y^2+2z = 2 \Rightarrow 2z = 2-(x^2+y^2)\right)$$

$$= \int\int_{x^2+y^2\le 2} \left[xy\left(2 - \frac{x^2+y^2}{2}\right) + (x^2+y^2)\right]\sqrt{x^2+y^2+1}\,dx\,dy$$

$$= \int_0^{\sqrt{2}}\left[\int_0^{2\pi}\left\{r^2\sin\theta\cos\theta\left(2-\frac{r^2}{2}\right)+r^2\right\}d\theta\right]\sqrt{r^2+1}\,r\,dr$$

$$\left(\text{changing into polar coordinates } x = r\cos\theta, y = r\sin\theta\right)$$

$$= \frac{1}{2}\int_0^{\sqrt{2}} r^2\left(2-\frac{r^2}{2}\right)\sqrt{r^2+1}\,dr\int_0^{2\pi}\sin 2\theta\,d\theta + \int_0^{2\pi}d\theta\int_0^{\sqrt{2}} r^3\sqrt{r^2+1}\,dr$$

$$= 0 + 2\pi\int_0^{\sqrt{2}}(r^2+1-1)\sqrt{r^2+1}\,r\,dr;\quad r^2+1 = t^2$$

$$= 2\pi\int_1^{\sqrt{3}}(t^2-1)\cdot t\cdot t\,dt = 2\pi\left[\frac{t^5}{5}-\frac{t^3}{3}\right]_1^{\sqrt{3}} = \frac{4\pi}{15}(6\sqrt{3}+1)$$

Example 6.4.6

Evaluate $\int\int_S \vec{F}\cdot\hat{n}dS$, where $\vec{F} = z\hat{i}+x\hat{j}-3y^2z\hat{k}$ and S is the surface of the cylinder $x^2+y^2 = 16$ included in the first octant between $z = 0$ and $z = 5$.

Solution

Here, the surface is perpendicular to the xy plane, so we have to calculate the transformed surface integral over the projected area of the given area, either on the yz plane or on the xz plane (Fig. 6.16). Suppose the surface S has a projection $R_3 = \{(x,z) : 0 \le x \le 4, 0 \le z \le 5\}$ on the xz plane, as shown in Fig. 6.16. For the given surface $\phi(x,y,z) = x^2+y^2 = 16$

$$\vec{\nabla}\phi = \vec{\nabla}(x^2+y^2) = 2x\hat{i}+2y\hat{j}$$

at any point of S. Thus, the outward unit normal to any point of the surface S is

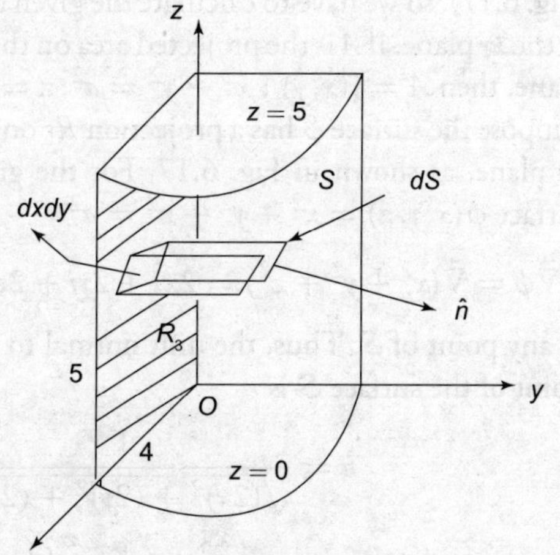

Figure 6.16 Surface integrals

$$\hat{n} = \frac{1}{\sqrt{(2x)^2+(2y)^2}}(2x\hat{i}+2y\hat{j}) = \frac{x}{4}\hat{i}+\frac{y}{4}\hat{j}$$

$$\Rightarrow \hat{n}\cdot\hat{j} = \left(\frac{x}{4}\hat{i}+\frac{y}{4}\hat{j}\right)\cdot\hat{j} = \frac{y}{4}$$

$$\text{and } \vec{F}\cdot\hat{n} = (z\hat{i}+x\hat{j}-3y^2z\hat{k})\cdot\left(\frac{x}{4}\hat{i}+\frac{y}{4}\hat{j}\right) = \frac{x}{4}(z+y)$$

Using the fact that $y = \sqrt{16 - x^2}$ from the equation of surface $\mathcal{S}$, and using Eq. (6.12), we get

$$\int\int_S \vec{F}\cdot\hat{n}\,dS = \int\int_{R_3} \vec{F}\cdot\hat{n}\frac{dxdz}{|\hat{n}\cdot\hat{j}|} = \int\int_{R_3} \frac{xz + xy}{y}\,dx\,dz$$

$$= \int_{z=0}^{5}\left[\int_{x=0}^{4}\left(\frac{xz}{\sqrt{16 - x^2}} + x\right)dx\right]dz$$

$$= \int_0^5 \left[-z\sqrt{16 - x^2} + \frac{x^2}{2}\right]_0^4 dz = \int_0^5 (4z + 8)dz = 90$$

Example 6.4.7

Evaluate $\int\int_S (\vec{\nabla}\times\vec{F})\cdot\hat{n}\,dS$, where $\vec{F} = y\hat{i} + (x - 2xz)\hat{j} - xy\hat{k}$ and $\mathcal{S}$ is the surface of the sphere $x^2 + y^2 + z^2 = a^2$ above the xy plane.

Solution

Here, $\vec{F} = y\hat{i} + (x - 2xz)\hat{j} - xy\hat{k}$, so using the definition Eq. (5.55) of curl, we get

$$\vec{\nabla}\times\vec{F} = \begin{vmatrix} \hat{i} & \hat{j} & \hat{k} \\ \frac{\partial}{\partial x} & \frac{\partial}{\partial y} & \frac{\partial}{\partial z} \\ y & x - 2xz & -xy \end{vmatrix} = \left[\frac{\partial}{\partial y}(-xy) - \frac{\partial}{\partial z}(x - 2xz)\right]\hat{i}$$

$$+\left[\frac{\partial y}{\partial z} - \frac{\partial}{\partial x}(-xy)\right]\hat{j} + \left[\frac{\partial}{\partial x}(x - 2xz) - \frac{\partial y}{\partial y}\right]\hat{k} = x\hat{i} + y\hat{j} - 2z\hat{k}$$

Here, the surface is above the xy plane (Fig. 6.17), so we have to calculate the given area in the xy plane. If A is the projected area on the xy plane, then $A = \{(x, y) : x^2 + y^2 = a^2, z = 0\}$. Suppose the surface $\mathcal{S}$ has a projection R_1 on the xy plane, as shown in Fig. 6.17. For the given surface $\phi(x, y, z) = x^2 + y^2 + z^2 = a^2$

$$\vec{\nabla}\phi = \vec{\nabla}(x^2 + y^2 + z^2) = 2x\hat{i} + 2y\hat{j} + 2z\hat{k}$$

at any point of $\mathcal{S}$. Thus, the unit normal to any point of the surface $\mathcal{S}$ is

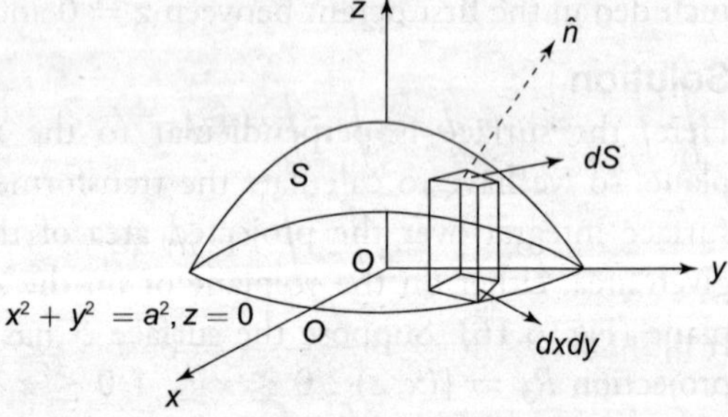

Figure 6.17 Region of integration

$$\hat{n} = \frac{1}{\sqrt{(2x)^2 + (2y)^2 + (2z)^2}}(2x\hat{i} + 2y\hat{j} + 2z\hat{k}) = \frac{x}{a}\hat{i} + \frac{y}{a}\hat{j} + \frac{z}{a}\hat{k}$$

$$\Rightarrow \hat{n}\cdot\hat{k} = \left(\frac{x}{a}\hat{i} + \frac{y}{a}\hat{j} + \frac{z}{a}\hat{k}\right)\cdot\hat{k} = \frac{z}{a}$$

Using the fact that $z = \sqrt{a^2 - x^2 - y^2}$ from the equation of the surface $\mathcal{S}$, and using Eq. (6.12), we get

$$\int\int_S (\vec{\nabla}\times\vec{F})\cdot\hat{n}\,dS = \int\int_{\mathcal{R}_1} (\vec{\nabla}\times\vec{F})\cdot\hat{n}\,\frac{dx\,dy}{|\hat{n}\cdot\hat{k}|}$$

$$= \int\int_{\mathcal{R}_1} (x\hat{i} + y\hat{j} - 2z) \cdot \left(\frac{x}{a}\hat{i} + \frac{y}{a}\hat{j} + \frac{z}{a}\hat{k}\right) \frac{dx\,dy}{z/a}$$

$$= \int_{-a}^{a} \left[\int_{-\sqrt{a^2-x^2}}^{\sqrt{a^2-x^2}} \frac{3(x^2+y^2) - 2a^2}{\sqrt{a^2-x^2-y^2}}\,dy\right] dx$$

$$= \int_0^{2\pi} \left[\int_0^a \frac{3r^2 - 2a^2}{\sqrt{a^2 - r^2}}\,r\,dr\right] d\theta; x = r\cos\theta, y = r\sin\theta$$

$$= \int_0^{2\pi} \left[\int_0^a \frac{3(r^2 - a^2) + a^2}{\sqrt{a^2 - r^2}}\,r\,dr\right] d\theta$$

$$= \int_0^{2\pi} \left[\int_0^a \left(-3r\sqrt{a^2 - r^2} + \frac{a^2 r}{\sqrt{a^2 - r^2}}\right) dr\right] d\theta$$

$$= \int_0^{2\pi} \left[(a^2 - r^2)^{3/2} - a^2\sqrt{a^2 - r^2}\right]_0^a d\theta = \int_0^{2\pi} (a^3 - a^3)d\theta = 0$$

Example 6.4.8

Evaluate $\int\int_{\mathcal{S}} \vec{F} \cdot \hat{n} dS$, where $\vec{F} = 4xz\hat{i} - y^2\hat{j} + yz\hat{k}$ and $\mathcal{S}$ is the surface of the cube bounded by $x = 0, x = 1, y = 0, y = 1, z = 0$ and $z = 1$.

Solution

The surface $\mathcal{S}$ is a cube, as shown in Fig. 6.18. It consists of six faces: *DEFG*, *ABCO*, *ABEF*, *OGDC*, *BCDE* and *AFGO*. On the face *DEFG*, $\hat{n} = \hat{i}, x = 1$, so

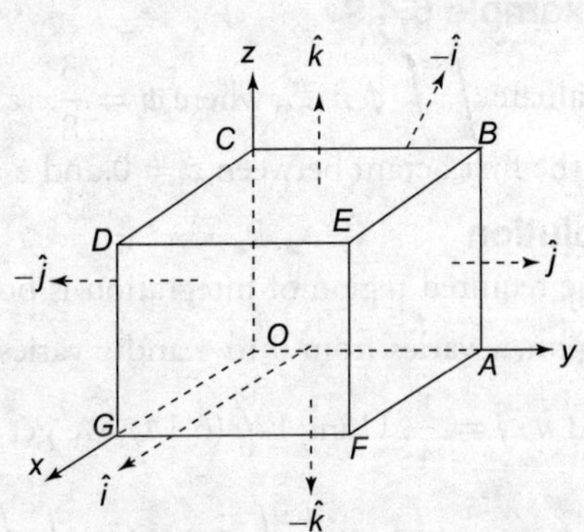

Figure 6.18 Surface integrals

$$\int\int_{DEFG} \vec{A} \cdot \hat{n}\,ds = \int_0^1\int_0^1 (4z\hat{i} - y^2\hat{j} + yz\hat{k}) \cdot \hat{i}\,dy\,dz$$

$$= \int_0^1\int_0^1 4z\,dy\,dz = 2$$

On the face *ABCO*, $\hat{n} = -\hat{i}, x = 0$, so

$$\int\int_{ABCO} \vec{A} \cdot \hat{n}\,ds = \int_0^1\int_0^1 (-y^2\hat{j} + yz\hat{k}) \cdot (-\hat{i})\,dy\,dz = 0$$

On the face *ABEF*, $\hat{n} = \hat{j}, y = 1$, so

$$\int\int_{ABEF} \vec{A} \cdot \hat{n}\,ds = \int_0^1\int_0^1 (4xz\hat{i} - \hat{j} + z\hat{k}) \cdot \hat{j}\,dx\,dz$$

$$= -\int_0^1\int_0^1 dx\,dz = -1$$

On the face *OGDC*, $\hat{n} = -\hat{j}, y = 0$, so

$$\int\int_{OGDC} \vec{A} \cdot \hat{n}\,ds = \int_0^1\int_0^1 (4xz\hat{i}) \cdot (-\hat{j})\,dx\,dz = 0$$

On the face $BCDE$, $\hat{n} = \hat{k}, z = 1$, so

$$\int\int_{BCDE} \vec{A} \cdot \hat{n}\, ds = \int_0^1 \int_0^1 (4x\hat{i} - y^2\hat{j} + y\hat{k}) \cdot \hat{k}\, dx\, dy$$

$$= \int_0^1 \int_0^1 y\, dx\, dy = 1/2$$

On the face $AFGO$, $\hat{n} = -\hat{k}, z = 0$, so

$$\int\int_{AFGO} \vec{A} \cdot \hat{n}\, ds = \int_0^1 \int_0^1 (-y^2\hat{j}) \cdot (-\hat{k})\, dx\, dy = 0$$

Adding all the results, we get

$$\int\int_S \vec{F} \cdot \hat{n} dS = \Big[\int\int_{DEFG} + \int\int_{ABCO} + \int\int_{ABEF} + \int\int_{OGDC} + \int\int_{BCDE} + \int\int_{AFGO}\Big] \vec{A} \cdot \hat{n}\, ds$$

$$= 2 + 0 + (-1) + 0 + \frac{1}{2} + 0 = \frac{3}{2}$$

Example 6.4.9

Evaluate $\int\int_S \phi\, \hat{n}\, dS$, where $\phi = \frac{3}{8}xyz$ and S is the surface of the cylinder $x^2 + y^2 = 16$ included in the first octant between $z = 0$ and $z = 5$.

Solution

The required region of integration is bounded, as shown in Fig. 6.16, where, to cover the whole region, x varies from 0 to 4 and z varies from 0 to 5. From Example 6.4.6, we get $\hat{n} = \frac{x}{4}\hat{i} + \frac{y}{4}\hat{j}$ and $\hat{n} \cdot \hat{j} = \frac{y}{4}$. Using Eq. (6.14), we get

$$\int\int_S \phi\, \hat{n}\, dS = \int\int_{R_3} \phi\, \hat{n} \frac{dx\, dz}{|\hat{n} \cdot \hat{j}|}$$

$$= \int\int_{R_3} \frac{3}{8} xz(x\hat{i} + y\hat{j})\, dx\, dz$$

$$= \frac{3}{8} \int_{z=0}^{5} \Big[\int_{x=0}^{4} (x^2 z\hat{i} + xz\sqrt{16 - x^2}\,\hat{j}) dx\Big] dz$$

$$= \frac{3}{8} \int_0^5 z\Big(\frac{64}{3}\hat{i} + \frac{64}{3}\hat{j}\Big) dz = 100(\hat{i} + \hat{j})$$

Example 6.4.10

Find the heat flux across the surface $x^2 + y^2 = 1 - z$ above the xy plane, when the temperature distribution is given by $T(x, y, z) = y + z$.

Solution

The heat flux across the surface $\mathcal{S}$ is $\displaystyle\iint_S \vec{F}\cdot\hat{n}dS$,

where $\vec{F} = \vec{\nabla}T$, i.e.

$$\vec{F} = \frac{\partial}{\partial x}(y+z)\hat{i} + \frac{\partial}{\partial y}(y+z)\hat{j} + \frac{\partial}{\partial z}(y+z)\hat{k} = \hat{j} + \hat{k}$$

Here, the surface is above the xy plane (Fig. 6.19), so we have to calculate the transformed surface integral over the projected area of the given area, either on the yz plane or on the xz plane. Suppose the surface $\mathcal{S}$ has a projection R_1 on the xy plane, as shown in Fig. 6.19. For the given surface $\phi(x, y, z) = x^2 + y^2 + z - 1$

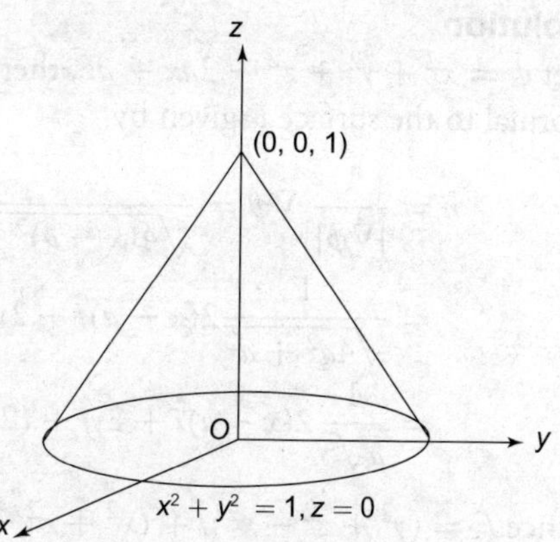

Figure 6.19 Region of integration

$$\vec{\nabla}\phi = \vec{\nabla}(x^2 + y^2 + z) = 2x\hat{i} + 2y\hat{j} + \hat{k}$$

at any point of $\mathcal{S}$. Thus, the unit normal to any point of the surface $\mathcal{S}$ is

$$\hat{n} = \frac{1}{\sqrt{(2x)^2 + (2y)^2 + 1^2}}(2x\hat{i} + 2y\hat{j} + \hat{k}) = 2x\hat{i} + 2y\hat{j} + \hat{k}$$

$$\Rightarrow \hat{n}\cdot\hat{i} = \frac{2x}{\sqrt{4x^2 + 4y^2 + 1}}; \quad \vec{F}\cdot\hat{n} = \frac{2y+1}{\sqrt{4x^2 + 4y^2 + 1}}$$

Using Eq. (6.12), we get

$$\begin{aligned}
\iint_S \vec{F}\cdot\hat{n}\,dS &= \iint_{\mathcal{R}_1} (\vec{\nabla}\times\vec{F})\cdot\hat{n}\,\frac{dy\,dz}{|\hat{n}\cdot\hat{i}|} \\
&= \int_{y=0}^{1}\int_{z=0}^{1-y^2} \frac{2y+1}{\sqrt{4x^2+4y^2+1}}\,\frac{dy\,dz}{2x/\sqrt{4x^2+4y^2+1}} \\
&= \int_{y=0}^{1}\int_{z=0}^{1-y^2} \frac{2y+1}{\sqrt{1-z^2-y^2}}\,dz\,dy \\
&= \frac{1}{2}\int_0^1 (2y+1)\Big[(-2)\sqrt{1-y^2-z}\Big]_0^{1-y^2} dy \\
&= \int_0^1 (2y+1)\sqrt{1-y^2}\,dy = 2\int_0^1 y\sqrt{1-y^2}\,dy + \int_0^1 \sqrt{1-y^2}dy \\
&= \Big[-\frac{2}{3}(1-y^2)^{3/2} + y\sqrt{1-y^2} + \sin^{-1}y\Big]_0^1 = \frac{2}{3} + \frac{\pi}{2}
\end{aligned}$$

Example 6.4.11

Evaluate the surface integral $\displaystyle\iint_S (\vec{\nabla}\times\vec{F})\cdot\hat{n}\,dS$, where $\vec{F} = (y^2+z^2-x^2)\hat{i} + (x^2+z^2-y^2)\hat{j} + (y^2+x^2-z^2)\hat{k}$ and $\mathcal{S}$ is the surface of the sphere $x^2+y^2+z^2-2ax+az=0$ above the xy plane.

Solution

Let $\phi = x^2 + y^2 + z^2 - 2ax + az$, then $\vec{\nabla}\phi = 2(x-a)\hat{i} + 2y\hat{j} + (2z+a)\hat{k}$. The outward unit normal to the surface is given by

$$\hat{n} = \frac{1}{|\vec{\nabla}\phi|}\vec{\nabla}\phi = \frac{1}{\sqrt{4(x-a)^2 + 4y^2 + (2z+a)^2}} 2(x-a)\hat{i} + 2y\hat{j} + (2z+a)\hat{k}$$

$$= \frac{1}{\sqrt{4a^2 + a^2}} 2(x-a)\hat{i} + 2y\hat{j} + (2z+a)\hat{k}$$

$$= \frac{1}{a\sqrt{5}} 2(x-a)\hat{i} + 2y\hat{j} + (2z+a)\hat{k}$$

Since $\vec{F} = (y^2 + z^2 - x^2)\hat{i} + (x^2 + z^2 - y^2)\hat{j} + (y^2 + x^2 - z^2)\hat{k}$, by the definition of curl, we get

$$\vec{\nabla} \times \vec{F} = \begin{vmatrix} \hat{i} & \hat{j} & \hat{k} \\ \dfrac{\partial}{\partial x} & \dfrac{\partial}{\partial y} & \dfrac{\partial}{\partial z} \\ y^2 + z^2 - x^2 & x^2 + z^2 - y^2 & y^2 + x^2 - z^2 \end{vmatrix}$$

$$= 2(y-z)\hat{i} + 2(z-x)\hat{j} + 2(x-y)\hat{k}$$

On the plane $z = 0$, we have

$$(\vec{\nabla} \times \vec{F}) \cdot \hat{n} = \frac{2}{a\sqrt{5}}\Big[2(x-a)(y-z) + 2y(z-x) + (2z+a)(x-y)\Big]_{z=0} = \frac{2}{\sqrt{5}}(x - 3y)$$

Considering the surface integral over the projected area of the given area on the plane $z = 0$, we get

$$\iint_S (\vec{\nabla} \times \vec{F}) \cdot \hat{n}\, dS = \iint_S (\vec{\nabla} \times \vec{F}) \cdot \hat{n} \frac{dx\, dy}{|\hat{n} \cdot \hat{k}|}$$

$$= 2\int_0^{2a} \left[\int_{-\sqrt{a^2-(x-a)^2}}^{\sqrt{a^2-(x-a)^2}} (x - 3y)\, dy\right] dx$$

$$= 4\int_0^{2a} x\sqrt{a^2 - (x-a)^2}\, dx; \quad x = a + a\sin\theta$$

$$= 4a^3 \int_{-\pi/2}^{\pi/2} (1 + \sin\theta)\cos^2\theta\, d\theta = 2\pi a^3$$

6.4.1 Normal Surface Integrals

Let the equation of the surface be given by $G = z - g(x, y) = 0$. Now

$$\vec{\nabla} G = \hat{i}\frac{\partial G}{\partial x} + \hat{j}\frac{\partial G}{\partial y} + \hat{k}\frac{\partial G}{\partial z} = -\hat{i}\frac{\partial z}{\partial x} - \hat{j}\frac{\partial z}{\partial y} + \hat{k}$$

From differential geometry, we obtain the result that $-\dfrac{\partial z}{\partial x}, -\dfrac{\partial z}{\partial y}, 1$ are the direction ratios of the normal to the surface at any point (x, y, z) on the surface. We know that the vector field $\vec{\nabla} G$ is normal to the surface, so

$$\hat{n} = \left(-\hat{i}\frac{\partial z}{\partial x} - \hat{j}\frac{\partial z}{\partial y} + \hat{k}\right)\left\{1 + \left(\frac{\partial z}{\partial x}\right)^2 + \left(\frac{\partial z}{\partial y}\right)^2\right\}^{-1/2}$$

$$\Rightarrow \hat{n}\cdot\hat{k} = \left\{1 + \left(\frac{\partial z}{\partial x}\right)^2 + \left(\frac{\partial z}{\partial y}\right)^2\right\}^{-1/2}$$

Hence, the surface area is given by the double integral

$$\int\int_S dS = \int\int_A \frac{dxdy}{|\hat{n}\cdot\hat{k}|} = \int\int_R \sqrt{1 + \left(\frac{\partial z}{\partial x}\right)^2 + \left(\frac{\partial z}{\partial y}\right)^2}\, dx\, dy \qquad (6.16)$$

For the function $\phi(x, y, z)$, we have

$$\int\int_S \phi\, dS = \int\int_{S'} \phi\left(x, y, g(x, y)\right)\left\{1 + \left(\frac{\partial z}{\partial x}\right)^2 + \left(\frac{\partial z}{\partial y}\right)^2\right\}^{-1/2} dS'$$

RESULT 21

Let the equation of the curve be $F(x, y, z) = 0$; then

$$\vec{\nabla}F = \frac{\partial F}{\partial x}\hat{i} + \frac{\partial F}{\partial y}\hat{j} + \frac{\partial F}{\partial z}\hat{k}$$

$$\hat{n} = \frac{1}{|\vec{\nabla}F|}\,\vec{\nabla}F = \frac{1}{\sqrt{(\frac{\partial F}{\partial x})^2 + (\frac{\partial F}{\partial y})^2 + (\frac{\partial F}{\partial z})^2}}$$

$$\Rightarrow \hat{n}\cdot\hat{k} = \left\{(\frac{\partial F}{\partial x})^2 + (\frac{\partial F}{\partial y})^2 + (\frac{\partial F}{\partial z})^2\right\}^{-\frac{1}{2}}\frac{\partial F}{\partial z}$$

Hence, the surface area is given by

$$\int\int_S dS = \int\int_A \frac{dx\,dy}{|\hat{n}\cdot\hat{k}|} = \int\int_R \frac{\sqrt{(\frac{\partial F}{\partial x})^2 + (\frac{\partial F}{\partial y})^2 + (\frac{\partial F}{\partial z})^2}}{\left|\frac{\partial F}{\partial z}\right|} dx\,dy$$

Example 6.4.12

Find the area of the part of the surface of the paraboloid $4ax = y^2 + z^2$ enclosed by the cylinder $y^2 = ax$ and the plane $x = 3a$.

Solution

The region of integration is shown in Fig. 6.20. From the equation of the paraboloid $4ax = y^2 + z^2$, we get $\frac{\partial z}{\partial x} = \frac{2a}{z}, \frac{\partial z}{\partial y} = -\frac{y}{z}$, therefore

$$1 + \left(\frac{\partial z}{\partial x}\right)^2 + \left(\frac{\partial z}{\partial y}\right)^2 = 1 + \frac{4a^2}{z^2} + \frac{y^2}{z^2} = \frac{y^2 + z^2 + 4a^2}{z^2}$$

$$= \frac{4ax + 4a^2}{4ax - y^2} = \frac{4a(x + a)}{4ax - y^2}$$

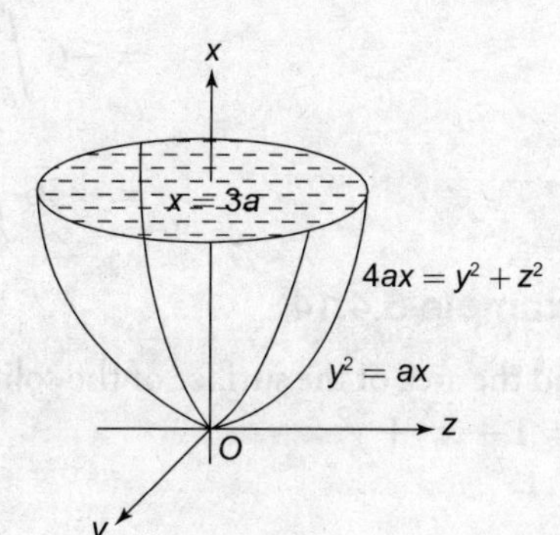

Figure 6.20 Region of integration

Thus, the required surface area of the paraboloid bounded by $y^2 = ax$ and the plane $x = 3a$ is

$$= \int_0^{3a}\int_{-\sqrt{ax}}^{\sqrt{ax}} \sqrt{1+\left(\frac{\partial z}{\partial x}\right)^2+\left(\frac{\partial z}{\partial y}\right)^2}\, dxdy$$

$$= 2\sqrt{a}\int_0^{3a}\int_{-\sqrt{ax}}^{\sqrt{ax}} \sqrt{\frac{x+a}{4ax-y^2}}\, dxdy$$

$$= 2\sqrt{a}\int_0^{3a} \sqrt{x+a}\left[\sin^{-1}\frac{y}{2\sqrt{ax}}\right]_{-\sqrt{ax}}^{\sqrt{ax}} dx$$

$$= 2\sqrt{a}\frac{\pi}{3}\int_0^{3a} \sqrt{x+a}dx = \frac{2\pi}{3}\sqrt{a}\left[\frac{2}{3}(x+a)^{3/2}\right]_0^{3a}$$

$$= \frac{4\pi}{9}\sqrt{a}\left[(4a)^{3/2}-a^{3/2}\right] = \frac{4\pi}{9}\sqrt{a}\cdot 7a^{3/2} = \frac{28}{9}\pi a^2$$

Example 6.4.13

Find the surface area of the sphere $x^2 + y^2 + z^2 = 9$ lying inside the cylinder $x^2 + y^2 = 3y$.

Solution

From the equation of the sphere $x^2 + y^2 + z^2 = 9$, we have $\frac{\partial z}{\partial x} = -\frac{x}{z}, \frac{\partial z}{\partial y} = -\frac{y}{z}$ and so

$$1+\left(\frac{\partial z}{\partial x}\right)^2+\left(\frac{\partial z}{\partial y}\right)^2 = \frac{z^2+x^2+y^2}{z^2} = \frac{9}{z^2} = \frac{9}{9-(x^2+y^2)}$$

The bounding surface in $x^2 + y^2 = 3y$, i.e., $r = 3\sin\theta$. Thus, the required area is given by

$$A = \int\int \sqrt{1+\left(\frac{\partial z}{\partial x}\right)^2+\left(\frac{\partial z}{\partial y}\right)^2}\, dxdy$$

$$= 4\int_{\theta=0}^{\pi/2}\int_{r=0}^{3\sin\theta} \frac{3}{\sqrt{9-r^2}}\, rdrd\theta$$

$$= 6\int_{\theta=0}^{\pi/2}\left[\int_{r=0}^{3\sin\theta} \frac{2r}{\sqrt{9-r^2}}\right] dr\, d\theta$$

$$= -6\int_{\theta=0}^{\pi/2}\left[\frac{(9-r^2)^{-1/2+1}}{-\frac{1}{2}+1}\right]_0^{3\sin\theta} d\theta$$

$$= -36\int_0^{\pi/2} (\cos\theta - 1)\, d\theta = 18(\pi - 2)$$

Example 6.4.14

Find the area of the surface of the solid bounded by the cone $z = 3 - \sqrt{x^2+y^2}$ and the paraboloid $z = 1 + x^2 + y^2$.

Solution

The region of integration is shown in Fig. 6.21. Let us find the point of intersection of the cone $z = 3 - \sqrt{x^2 + y^2}$ and the paraboloid $z = 1 + x^2 + y^2$. Now

$z = 1 + (z - 3)^2 \Rightarrow z^2 - 7z + 10 = 0 \Rightarrow z = 2, 5$

So, $z = 2$ and the curve of intersection is given by $x^2 + y^2 = 1$. For $z = 1 + x^2 + y^2$, we have $\frac{\partial z}{\partial x} = 2x, \frac{\partial z}{\partial y} = 2y$. The surface area of the parabolic part is

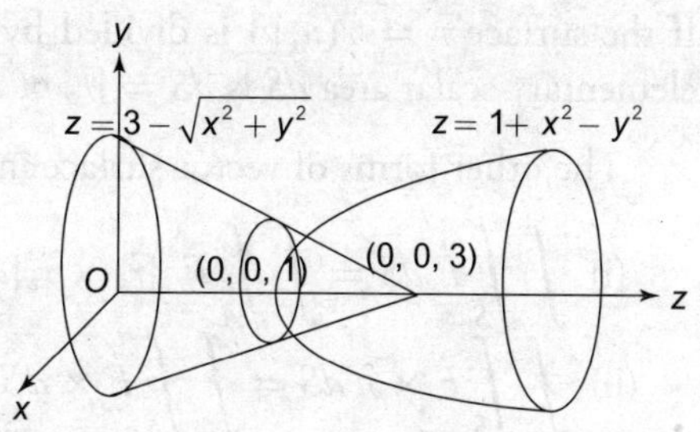

Figure 6.21 Region of integration

$$A_1 = \int\int \sqrt{1 + \left(\frac{\partial z}{\partial x}\right)^2 + \left(\frac{\partial z}{\partial y}\right)^2}\, dxdy$$

$$= \int\int_{x^2+y^2=1} \sqrt{1 + 4x^2 + 4y^2}\, dx\, dy; \quad x = r\cos\theta, y = r\sin\theta$$

$$= \int_0^{2\pi} \int_0^1 \sqrt{1 + 4r^2}\, r\, dr\, d\theta = 2\pi \int_0^1 \sqrt{1 + 4r^2} r dr = \frac{5\sqrt{5}}{6}\pi$$

The surface area of the conical part is

$$A_2 = \int\int \sqrt{1 + \left(\frac{\partial z}{\partial x}\right)^2 + \left(\frac{\partial z}{\partial y}\right)^2}\, dxdy$$

$$= \int\int_{x^2+y^2=1} \sqrt{1 + \frac{x^2}{x^2 + y^2} + \frac{y^2}{x^2 + y^2}}\, dx\, dy; \quad x = r\cos\theta, y = r\sin\theta$$

$$= \sqrt{2} \int_0^{2\pi} \int_0^1 r\, dr\, d\theta = \sqrt{2}\pi$$

Therefore, the required surface area of the solid is

$$A = A_1 + A_2 = \frac{5\sqrt{5}}{6}\pi + \sqrt{2}\pi = \left(\frac{5\sqrt{5}}{6} + \sqrt{2}\right)\pi$$

DEDUCTION 6.4.1

If we take the equation of the surface as $\vec{r} = \vec{r}(u, v)$, then we have already seen that the outward drawn normal vector is given by

$$\hat{n} = \frac{1}{|\vec{r}_u \times \vec{r}_v|}\left[\vec{r}_u \times \vec{r}_v\right]; \quad \vec{r}_u = \frac{\partial \vec{r}}{\partial u}, \vec{r}_v = \frac{\partial \vec{r}}{\partial v}$$

So, when the equation of the surface is given in parametric form, the above surface integral (6.9) takes the form

$$\int\int_S \vec{F} \cdot d\vec{S} = \int\int_S \vec{F} \cdot \frac{1}{|\vec{r}_u \times \vec{r}_v|}\left[\vec{r}_u \times \vec{r}_v\right] dS \qquad (6.17)$$

If the surface $\vec{r} = \vec{r}(u, v)$ is divided by several u-curves and v-curves into subregions, then the elementary scalar area dS is $dS = |\vec{r}_u \times \vec{r}_v| du\, dv$.

The other forms of vector surface integrals are

(i) $\displaystyle\int\int_S \vec{F}\, dS = \int\int_A \vec{F}\ |\vec{r}_u \times \vec{r}_v| du\, dv$

(ii) $\displaystyle\int\int_S \vec{F} \times \hat{n}\, dS = \int\int_S \vec{F} \times d\vec{S} = \int\int_A \vec{F} \times \hat{n}\ |\vec{r}_u \times \vec{r}_v| du\, dv$

(iii) $\displaystyle\int\int_S f\, d\vec{S} = \int\int_A f\, \hat{n}\ |\vec{r}_u \times \vec{r}_v| du\, dv$

where f is a scalar point function.

Example 6.4.15

Evaluate the surface integral $\displaystyle\int\int_S \sqrt{b^4x^2 + a^4y^2}\, dS$, where S is the curved surface of the cylinder $\dfrac{x^2}{a^2} + \dfrac{y^2}{b^2} = 1, 0 \le z \le c$.

Solution

The parametric equation of this surface is $\vec{r} = (a\cos u, b\sin u, v), 0 \le u \le 2\pi, 0 \le v \le c$. In this case,

$$|r_u \times r_v| = \Big|(-a\sin u, b\cos u, 0) \times (0, 0, 1)\Big| = \sqrt{b^2\cos^2 u + a^2\sin^2 u}$$

Hence, using Eq. (6.17), we get

$$\begin{aligned}
\int\int_S \sqrt{b^4x^2 + a^4y^2}\, dS &= \int\int_S \sqrt{b^4x^2 + a^4y^2} |r_u \times r_v| du\, dv \\
&= \int_0^c \left[\int_0^{2\pi} \sqrt{b^4a^2\cos^2 u + a^4b^2\sin^2 u}\sqrt{b^2\cos^2 u + a^2\sin^2 u} du\right] dv \\
&= ab\int_{v=0}^{c}\int_{u=0}^{2\pi} (b^2\cos^2 u + a^2\sin^2 u) du\, dv \\
&= \frac{abc}{2}\int_0^{2\pi} \{b^2(1 + \cos 2u) + a^2(1 - \cos 2u)\} du \\
&= \frac{abc}{2}\left[(a^2 + b^2)u + \frac{b^2 - a^2}{2}\sin 2u\right]_0^{2\pi} = \pi abc(a^2 + b^2)
\end{aligned}$$

Example 6.4.16

Evaluate $\displaystyle\int\int_S (x^2 + y^2) d\vec{S}$, where S is the sphere $x^2 + y^2 + z^2 = 1$.

Solution

The outward drawn unit normal to the sphere $\phi(x, y, z) = x^2 + y^2 + z^2 = 1$ is given by

$$\hat{n} = \frac{1}{|\vec{\nabla}\phi|}\vec{\nabla}\phi = \frac{1}{2\sqrt{x^2 + y^2 + z^2}} 2(x\hat{i} + y\hat{j} + z\hat{k}) = x\hat{i} + y\hat{j} + z\hat{k}$$

The parametric equation of the surface is $\vec{r} = (\sin u \cos v, \sin u \sin v, \cos u)$, where $o \leq u \leq \pi$ and $0 \leq v \leq 2\pi$. Now

$$\vec{r}_u \times \vec{r}_v = \begin{vmatrix} \hat{i} & \hat{j} & \hat{k} \\ \cos u \cos v & \cos u \sin v & -sinv \\ -\sin u \sin v & \sin u \cos v & 0 \end{vmatrix}$$

$$= \sin^2 u(\cos v\hat{i} - \sin v\hat{j}) + \sin u \cos u(\cos^2 v + \sin^2 v)\hat{k}$$

$$= \sin^2 u(\cos v\hat{i} - \sin v\hat{j}) + \sin u \cos u\hat{k}$$

$$\Rightarrow |\vec{r}_u \times \vec{r}_v| = \left|\sin^2 u(\cos v\hat{i} - \sin v\hat{j}) + \sin u \cos u\hat{k}\right|$$

$$= \sqrt{\sin^4 u(\cos^2 v + \sin^2 v) + \sin^2 u \cos^2 u} = \sin u$$

Hence, using Eq. (6.17), we get

$$\int\int_S (x^2 + y^2)d\vec{S} = \int\int_S (x^2 + y^2)\hat{n}\,|\vec{r}_u \times \vec{r}_v|\,du\,dv$$

$$= \int\int_S (\sin^2 u \cos^2 v + \sin^2 u \sin^2 v)\vec{r}\,\sin u\,du\,dv$$

$$= \int_{u=0}^{\pi}\int_{v=0}^{2\pi} \sin^3 u\Big[\sin u \cos v\hat{i} + \sin u \sin v\hat{j} + \cos u\hat{k}\Big]dv\,du$$

$$= \int_{u=0}^{\pi}\int_{v=0}^{2\pi} \Big[\sin^4 u \cos v\hat{i} + \sin^4 u \sin v\hat{j} + \sin^3 u \cos u\hat{k}\Big]dv\,du$$

$$= \int_0^{\pi} \Big[\sin^4 u \sin v\hat{i} - \sin^4 u \cos v\hat{j} + \sin^3 u \cos u\, v\hat{k}\Big]_0^{2\pi} du$$

$$= \int_0^{\pi} \sin^3 u \cos u \cdot 2\pi\,\hat{k}\,du = 2\pi\left[\frac{\sin^4 u}{4}\right]_0^{\pi}\hat{k} = \vec{0}$$

6.5 VOLUME INTEGRALS

Consider a closed surface in space enclosing a volume V. Let there be a real-valued function $f(x, y, z)$ of three independent variables x, y and z defined over a three-dimensional region V in space enclosed by a surface S. We divide the region V into n subregions of elementary volumes $\Delta V_1, \Delta V_2, \Delta V_3, \cdots, \Delta V_n$. Let $P(\xi_n, \eta_r, \zeta_r)$ be any point inside or on the boundary of the region ΔV_r. Then, we form the sum

$$S_n = \sum_{r=1}^{n} f(\xi_n, \eta_r, \zeta_r)\,\Delta V_r$$

If $\lim_{n\to\infty} S_n$ exists ($n \to \infty$ in such a way that each ΔV_r shrinks to a point) and the limit is independent of the mode of sub-division of V and the choice of the points (ξ_n, η_r, ζ_r), then this limit is called the *triple* or *volume integral* of $f(x, y, z)$ over the volume V and is denoted by

$$\iiint_V f(x,y,z)\,dV = \iiint_V f(x,y,z)\,dx\,dy\,dz$$

If the sub-division is made by planes parallel to the coordinate planes (each ΔV_r is a cube, say), then the volume integral can be evaluated as an iterated integral.

(i) Let $\vec{F}(x,y,z) = F_1(x,y,z)\hat{i} + F_2(x,y,z)\hat{j} + F_3(x,y,z)\hat{k}$ be a single-valued and continuous vector point function defined on V. Then, the *vector volume integral* of the vector $\vec{F}$ is defined as

$$\iiint_V \vec{F}(x,y,z)\,dV = \lim_{n\to\infty}\sum_{r=1}^{n}\vec{F}(\xi_n,\eta_r,\zeta_r)\,\Delta V_r$$
$$= \hat{i}\iiint_V F_1\,dV + \hat{j}\iiint_V F_2\,dV + \hat{k}\iiint_V F_3\,dV$$

This is an example of a space or volume integral. (6.18)

(ii) Let $\phi(x,y,z)$ be a scalar point function, then

$$\iiint_V \phi(x,y,z)\,dV = \iiint_V \phi(x,y,z)\,dx\,dy\,dz \tag{6.19}$$

This is an example of a space or volume integral.

Example 6.5.1

Evaluate the vector volume integral $\iiint_V (x\hat{i} + y\hat{k} + z\hat{k})dV$, where V is the closed region bounded by the cylinder $z = x^2$ and the planes $x = 0, y = 0, y = 6$ and $z = 4$.

Solution

The required region of integration is shown in Fig. 6.22. It is evident that the limits of integration of x, y and z will be from $x = 0$ to $x = \sqrt{z}, y = 0$ to $y = 6$ and $z = 0$ to $z = 4$. Hence, the required volume integral is

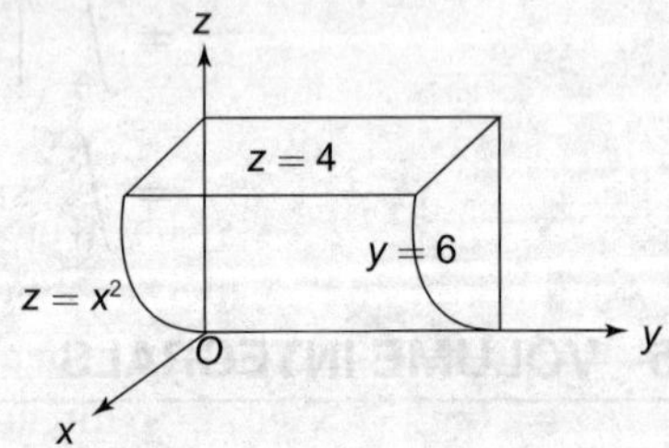

Figure 6.22 Region of integration

$$\iiint_V (x\hat{i} + y\hat{k} + z\hat{k})dV$$
$$= \int_{z=0}^{4}\int_{y=0}^{6}\int_{x=0}^{\sqrt{z}} (x\hat{i} + y\hat{j} + z\hat{k})dxdydz$$
$$= \int_{y=0}^{6}\int_{z=0}^{4}\left[\frac{z}{2}\hat{i} + y\sqrt{z}\hat{j} + z\sqrt{z}\hat{k}\right]dzdy$$
$$= \int_{y=0}^{6}\left[\frac{z^2}{4}\hat{i} + \frac{2}{3}yz\sqrt{z}\hat{j} + \frac{2}{5}z^2\sqrt{z}\hat{k}\right]_0^4 dy$$
$$= \int_{y=0}^{6}\left[4\hat{i} + \frac{16}{3}y\hat{j} + \frac{64}{5}\hat{k}\right]dy$$

$$= \left[4y\hat{\imath} + \frac{8}{3}y^2\hat{\jmath} + \frac{64}{5}y\hat{k}\right]_0^6 = 24\hat{\imath} + 96\hat{\jmath} + \frac{384}{5}\hat{k}$$

Example 6.5.2

Let $\vec{F} = 2xz\hat{\imath} - x\hat{\jmath} + y^2\hat{k}$ and let V be the region bounded by the surfaces $x = 0, y = 0, y = 6, z = x^2$ and $z = 4$. Evaluate the vector volume integral $\iiint_V \vec{F}\,dV$.

Solution

The required region of integration is shown in Fig. 6.22. To cover the region V, we first keep x and y fixed and integrate from $z = x^2$ to $z = 4$. Then, we keep x fixed and integrate from $y = 0$ to $y = 6$, and finally integrate from $x = 0$ to $x = 2$. We have $dV = dxdydz$. The vector integral

$$\iiint_V \vec{F}\,dV = \int_V (2xz\hat{\imath} - x\hat{\jmath} + y^2\hat{k})dxdydz$$

$$= \int_{x=0}^{2}\left[\int_{y=0}^{6}\left\{\int_{z=x^2}^{4}(2xz\hat{\imath} - x\hat{\jmath} + y^2\hat{k})dz\right\}dy\right]dx$$

$$= \int_{x=0}^{2}\left[\int_{y=0}^{6}\{(16x - x^5)\hat{\imath} + (x^3 - 4x)\hat{\jmath} + (4 - x^2)y^2\hat{k}\}dy\right]dx$$

$$= \int_{x=0}^{2}(96x - 6x^5)\hat{\imath} + (6x^3 - 24x)\hat{\jmath} + 72(4 - x^2)\hat{k}dx = 128\hat{\imath} - 24\hat{\jmath} + 384\hat{k}$$

Example 6.5.3

Show that $\iiint_V \vec{\nabla}\cdot\vec{F}dV = \frac{8}{3}$, over the region bounded by the coordinate planes and the plane $2x + 2y + z = 4$, where $\vec{F} = (2x^2 - z)\hat{\imath} - 2xy\hat{\jmath} - 4x\hat{k}$.

Solution

Here, $\vec{\nabla}\cdot\vec{F} = 4x - 2x = 2x$. The region of integration is bounded by the four planes, as shown in Fig. 6.23. From the equation on the plane, when $z = 0, x + y = 2$; therefore $y = 2 - x$. We are to integrate with respect to y from 0 to $(2 - x)$ and with respect to x from 0 to 2. Thus, the given integral is equal to

$$\iiint_V \vec{\nabla}\cdot\vec{F}dV = \int_{x=0}^{2}\int_{y=0}^{2-x}\int_{z=0}^{4-2x-2y} 2x\,dxdydz$$

$$= \int_{x=0}^{2}\int_{y=0}^{2-x} 2x(4 - 2x - 2y)dxdy$$

$$= \int_0^2\left[(8 - 4x^2)(2 - x) - 2x(2 - x)^2\right]dx$$

$$= \int_0^2 2x(4 - 4x + x^2)dx = 16 - \frac{1}{3}\cdot 64 + 8 = \frac{8}{3}$$

Figure 6.23 Volume integrals

Example 6.5.4

Evaluate $\iiint_V \vec{\nabla} \times \vec{F}\, dV$, where V is the closed region bounded by the surfaces $x = 0, x = 2,$ $y = 0, y = 6, z = x^2, z = 4$ and $\vec{F} = y\hat{i} + 2x\hat{j} - z\hat{k}$.

Solution

Given $\vec{F} = y\hat{i} + 2x\hat{j} - z\hat{k}$, so

$$\vec{\nabla} \times \vec{F} = \begin{vmatrix} \hat{i} & \hat{j} & \hat{k} \\ \dfrac{\partial}{\partial x} & \dfrac{\partial}{\partial y} & \dfrac{\partial}{\partial z} \\ y & 2x & -z \end{vmatrix} = \hat{k}$$

The required region of integration is shown in Fig. 6.22. We are to integrate with respect to x from 0 to 2, y from 0 to 6 and with respect to z from x^2 to 4. Thus, the required integral is equal to

$$\iiint_V \vec{\nabla} \times \vec{F}\, dV = \hat{k} \int_0^2 \left[\int_0^6 \left\{ \int_{x^2}^4 dz \right\} dy \right] dx$$

$$= \hat{k} \int_0^2 \left[\int_0^6 (4 - x^2)\, dy \right] dx$$

$$= 6\hat{k} \int_0^2 (4 - x^2) dx = 32\hat{k}$$

Example 6.5.5

Show that $\iiint_V \phi\, dV = \dfrac{1}{4}$, where $\phi = 15xy$ and V denotes the closed region bounded by the planes $x + 4y + z = 2, x = 0, y = 0$ and $z = 0$.

Solution

The region of integration is bounded by four planes, as shown in Fig. 6.24. To cover the whole region, x varies from 0 to 2, y varies from 0 to $(2 - x)/4$ and z varies from 0 to $2 - x - 4y$. Thus, the required integral is equal to

Figure 6.24 Volume integrals

$$\iiint_V \phi\, dV = \int_{x=0}^{2} \int_{y=0}^{(2-x)/4} \int_{z=0}^{2-x-4y} 15xy\, dx\, dy\, dz$$

$$= \int_{x=0}^{2} \int_{y=0}^{(2-x)/4} 15xy(2 - x - 4y) dx\, dy$$

$$= \int_0^2 \left[15x(2 - x)\frac{y^2}{2} - 20xy^3 \right]_0^{(2-x)/4} dx$$

$$= \int_0^2 \left[15x(2 - x)\frac{(x - 2)^2}{32} + 10x\frac{(x - 2)^3}{32} \right] dx$$

$$= -\frac{5}{32}\int_0^2 x(x-2)^3\,dx = -\frac{5}{32}\left[\frac{x^5}{5} - \frac{3x^4}{2} + 4x^3 - 4x^2\right]_0^2 = \frac{1}{4}$$

Example 6.5.6

Find the surface area and volume of the region common to the intersecting cylinders $x^2 + y^2 = a^2$ and $x^2 + z^2 = a^2$.

Solution

Here, we need to evaluate the surface area and the volume of the region common to the intersecting cylinders $x^2 + y^2 = a^2$ and $x^2 + z^2 = a^2$ lying in the first octant (Fig. 6.25). The surface S is shown, which is the intersection of the cylinders $x^2 + y^2 = a^2$ and $x^2 + z^2 = a^2$ which is included in the first octant. For the surface $z^2 = a^2 - x^2$, we have $\frac{\partial z}{\partial x} = \frac{x}{z}$, $\frac{\partial z}{\partial y} = 0$, and so

$$1 + \left(\frac{\partial z}{\partial x}\right)^2 + \left(\frac{\partial z}{\partial y}\right)^2 = 1 + \frac{x^2}{z^2} = \frac{z^2 + x^2}{z^2}$$

$$= \frac{a^2}{a^2 - x^2}$$

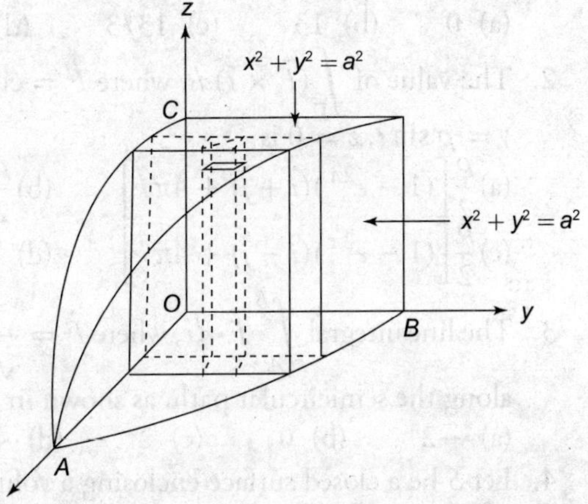

Figure 6.25 Intersection of the cylinders $x^2 + y^2 = a^2$, $x^2 + z^2 = a^2$ in the first octant

The projection of the given surface on $z = 0$ is $x^2 = a^2$, so the limits of x for the first quadrant are 0 to a. Also, for the cylinder $x^2 + y^2 = a^2$, we have $y^2 = a^2 - x^2$. Using Eq. (6.16), the required area is

$$A = \iint \sqrt{1 + \left(\frac{\partial z}{\partial x}\right)^2 + \left(\frac{\partial z}{\partial y}\right)^2}\,dxdy$$

$$= 8\int_0^a \left[\int_0^{\sqrt{a^2-x^2}} \frac{a}{\sqrt{a^2 - x^2}}\,dy\right] dx = 8a^2$$

We consider a thin column parallel to the z axis as an elementary volume $dx\,dy\,dz$. We integrate this elementary volume with respect to z from the xy plane, i.e., $z = 0$ to the surface of the cylinder $x^2 + z^2 = a^2$, i.e., to $z = \sqrt{a^2 - x^2}$ along this column. Then, we integrate with respect to y from the x axis, i.e., $y = 0$ to the circle $x^2 + y^2 = a^2$ in the xy plane, i.e., to $y = \sqrt{a^2 - x^2}$ and finally integrate with respect to x from $x = 0$ to $x = a$. Then, from symmetry, the required volume is

$$V = 8\int_0^a \left[\int_0^{\sqrt{a^2-x^2}} \left\{\int_0^{\sqrt{a^2-x^2}} dz\right\} dy\right] dx$$

$$= 8\int_0^a \left[\int_0^{\sqrt{a^2-x^2}} \sqrt{a^2 - x^2}\,dy\right] dx$$

$$= 8\int_0^a (a^2 - x^2)\,dx = 8\left[a^2x - \frac{x^3}{3}\right]_0^a = \frac{16}{3}a^3$$

EXERCISES

Multiple Choice Questions

1. If $\vec{F} = (3x^2 + 6y)\hat{i} - 14yz\hat{j} + 20xz^2\hat{k}$ and curve C is the straight line joining $(0, 0, 0)$ and $(1, 1, 1)$, then the value of $\int_C \vec{F} \cdot d\vec{r}$ is
(a) 0 (b) 13 (c) 13/3 (d) 13/4

2. The value of $\int_\Gamma (\vec{F} \times \hat{t})ds$, where $\vec{F} = \cos\hat{i} + \sin t\hat{j} + e^t\hat{k}$ and Γ is the circle $x = \rho\cos t$, $y = \rho\sin t, z = 0$ is
(a) $\frac{\rho}{2}\left[(1 - e^{2\pi})(\hat{i} + \hat{j}) + 4\pi\hat{k}\right]$ (b) $\frac{\rho}{2}\left[(1 - e^{-2\pi})(\hat{i} + \hat{j}) + 4\pi\hat{k}\right]$
(c) $\frac{\rho}{2}\left[(1 - e^{2\pi})(\hat{i} - \hat{j}) + 4\pi\hat{k}\right]$ (d) $\frac{\rho}{2}\left[(1 - e^{2\pi})(\hat{i} + \hat{j}) + 4\pi\hat{k}\right]$

3. The line integral $\int_A^B \vec{F} \cdot d\vec{r}$, where $\vec{F} = \frac{x}{\sqrt{x^2 + y^2}}\hat{i} + \frac{y}{\sqrt{x^2 + y^2}}\hat{j}$ along the semicircular path, as shown in the figure is
(a) −2 (b) 0 (c) 2 (d) 4

4. Let S be a closed surface enclosing a volume V and $\vec{F} = x\hat{i} + 2y\hat{j} + 3z\hat{k}$. Also, $\int\int_S \vec{F} \cdot \hat{n}dS = 64\pi$, then $V =$
(a) 64π (b) 384π (c) $(32/3)\pi$ (d) none

5. The heat flux across the surface $x^2 + y^2 = 1 - z$ above the xy plane, where the temperature distribution $T(x, y, z) = (y + z)$, is
(a) 0 (b) π (c) 2π (d) 3π

6. The value of $\int\int_S \vec{r} \cdot \hat{n}dS$ over the sphere of the radius a and the centre $(-2, 0, -3)$ is
(a) 0 (b) $4\pi a^3$ (c) $6\pi a^3$ (d) $24\pi a^3$

7. If $\int\int_S \vec{r} \cdot \hat{n}dS = V$, where the surface S is a unit cube bounded by a coordinate plane and the planes $x = 1, y = 1, z = 1$, then $V =$
(a) 0 (b) 1 (c) 2 (d) 3

8. The value of $\int\int_S (x^2 + y^2)dxdy$, where S is the region bounded by $y = x^2, x = 2, y = 1$, is
(a) 0 (b) 1008 (c) 105 (d) 1008/105

9. If $\vec{r}(t) = 2t^2\hat{i} + t\hat{j} - 3t^3\hat{k}$; then $\int_0^1 (\vec{r} \times \frac{d^2\vec{r}}{dt^2})dt$ is
(a) $6\hat{i} - 6\hat{j} + 1\hat{k}$ (b) $-6\hat{i} + 6\hat{j} - 2\hat{k}$ (c) $6\hat{i} - 6\hat{j} + 3\hat{k}$ (d) $6\hat{i} - 6\hat{j} + 24\hat{k}$

10. The value of $\int_1^2 \left(\frac{d\vec{r}}{dt} \cdot \frac{d\vec{r}}{dt}\right) dt$, where

$$\vec{r}(t) = \begin{cases} 2\hat{i} + \hat{j} - 3\hat{k}; & \text{when } t = 1 \\ \hat{i} + \hat{j} + 2\hat{k}; & \text{when } t = 2 \end{cases} \textit{ is}$$

(a) 4 (b) −4 (c) 0 (d) 2

11. Considering the vector $\vec{A} = \frac{1}{r^3}\vec{r}$, the surface integral of this vector over the surface of a cube of size a and centred at the origin is
(a) 0 (b) 2π (c) $2\pi a^3$ (d) 4π

12. The value of the line integral $\oint_C \frac{-ydx + xdy}{x^2 + y^2}$, where C is the unit centre at O is
(a) 0 (b) 1 (c) 2 (d) ∞

13. If $\vec{A} = xy\hat{i} + x^2\hat{j}$, then $\oint_C \vec{A} \cdot d\vec{l}$ over the path shown in the figure below is
(a) 0 (b) $2/\sqrt{3}$ (c) 1 $2\sqrt{3}$

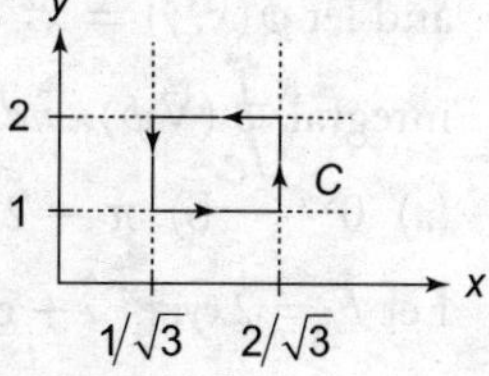

14. Consider the vector $\vec{V} = \frac{1}{r^3}\vec{r}$. Which one of the following is incorrect?
(a) the value of the line integral of this vector around any closed curve is zero.
(b) this vector can be written as the gradiant of some scalar function.
(c) the line integral of this vector from point P to the point Q is independent of the path taken.
(d) this vector can represent the magnetic field of some current distribution.

15. If $\vec{R}(t) = 2t^2\hat{i} + t\hat{j} - 3t^3\hat{k}$ and $\vec{S}(t) = t\hat{i} + t^2\hat{j} + t^3\hat{k}$, then $\int_0^1 \vec{R} \cdot \vec{S}\, dt$ and $\int_0^1 \vec{R} \times \vec{S}\, dt$ are
(a) $\frac{9}{28}, \frac{24}{5}\hat{i} - \frac{40}{3}\hat{j} + \frac{64}{5}\hat{k}$ (b) $\frac{9}{28}, \frac{1}{3}\hat{i} + \frac{1}{2}\hat{j} - \frac{1}{2}\hat{k}$
(c) $\frac{9}{28}, -24\hat{i} - \frac{40}{3}\hat{j} + \frac{64}{5}\hat{k}$ (d) $\frac{9}{28}, \frac{7}{10}\hat{i} - \frac{14}{15}\hat{j} + \frac{1}{15}\hat{k}$

16. The value of the integral $\int_\Gamma \vec{A} \cdot d\vec{r}$, where
$\vec{A} = c{-}3a\sin^2\theta\cos\theta\hat{i} + a(2\sin\theta - 3\sin^3\theta)\hat{j} + b\sin 2\theta\hat{k}$ and
$\Gamma : \vec{r} = a\cos\theta\hat{i} + a\sin\theta\hat{j} + b\theta\hat{k}$ from $\pi/4$ to $\pi/2$ is
(a) $c(3a^2 + 2b^2)/2$ (b) $c(3a^2 - 2b^2)/2$ (c) $c(a^2 + b^2)/2$ (d) $c(a^2 - b^2)/2$

17. The value of the integral $\int_\Gamma \vec{A} \cdot d\vec{r}$, where $\vec{A} = (x^2 + y^2)\hat{i} - 2xy\hat{j}$ and Γ, the rectangle bounded by $y = 0, x = a, y = b, x = 0$, is
(a) $2ab^2$ b) $-2ab^2$ (c) $2a^2b$ (d) $-2a^2b$

18. The value of the integral $\int_\Gamma \vec{A} \cdot d\vec{r}$, where $\vec{A} = 3xy\hat{i} - y^2\hat{j}$ and $\Gamma : y = 2x^2$ from $(0, 0)$ to $(1, 2)$ is
(a) $7/6$ (b) $-7/6$ (c) $6/7$ (d) $-6/7$

19. The value of the integral $\vec{A} = y\hat{i} + z\hat{j} + x\hat{k}$ over the circle $x^2 + y^2 = 4, z = 0$ is
(a) -2π (b) 2π (c) 4π d) -4π

20. Let $\vec{F} = ay\hat{i} + z\hat{j} + x\hat{k}$ and C be the positively oriented closed curve given by $x^2 + y^2 = 1, z = 0$. If $\oint_C \vec{F} \cdot d\vec{r} = \pi$, then the value of a is
(a) -1 (b) 0 (c) $1/2$ (d) 1

21. Let $\vec{F} = (ae^x \sin y - 4x)\hat{\imath} + (2y + e^x \cos y)\hat{\jmath} + az\hat{k}$, where a is a constant. If the line integral $\oint_C \vec{F} \cdot d\vec{r}$ over every closed curve C is zero, then a is equal to
(a) -1 (b) -2 (c) 0 (d) 1

22. Let C denote the boundary of the semi-circular disk $D = \{(x, y) \in \Re^2 : x^2 + y^2 \leq 1, y \geq 0\}$ and let $\phi(x, y) = x^2 + y$ for $(x, y) \in D$. If $\hat{n}$ is the outward unit normal to C, then the integral $\oint_C (\vec{\nabla}\phi) \cdot \hat{n} dS$, evaluated counter-clockwise over C is equal to
(a) 0 b) $\pi - 2$ (c) π (d) $\pi + 2$

23. Let $\vec{F} = 2xyze^{x^2}\hat{\imath} + ze^{x^2}\hat{\jmath} + ye^{x^2}\hat{k}$ be the gradient of a scalar function. The value of $\oint_C \vec{F} \cdot d\vec{r}$ along the oriented path L from $(0, 0, 0)$ to $(1, 0, 2)$ and then to $(1, 1, 2)$ is
(a) 0 b) $2e$ (c) e (d) e^2

24. The work done by the force $\vec{F} = 4y\hat{i} - 3xy\hat{j} + z^2\hat{k}$ in moving a particle over the circular path $x^2 + y^2 = 1, z = 0$ from $(1, 0, 0)$ to $(0, 1, 0)$ is
(a) $\pi + 1$ b) $-\pi - 1$ (c) $-\pi + 1$ (d) $\pi - 1$

25. Let $\vec{F} = xy\hat{i} + y\hat{j} - yz\hat{k}$ denote the force field on a particle traversing the path L from $(0, 0, 0)$ to $(1, 1, 1)$ along the curve of intersection of the cylinder $y = x^2$ and the plane $z = x$. The work done by $\vec{F}$ is
(a) 0 b) 1/2 (c) 1/4 (d) 1

26. The area of the surface bounded by $z + x^2 + y^2 - 1 = 0, 0 \leq x \leq 1, 0 \leq y \leq 1, z \geq 0$ is
(a) $\pi(\sqrt{5} - 1)/12$ b) $\pi(\sqrt[2]{5^3} - 1)/24$ (c) $\pi(\sqrt{5} + 1)/12$ (d) $\pi(\sqrt[2]{5^3} + 1)/24$

27. Let V be the volume of a region bounded by a smooth closed surface S. Let $\vec{r}$ denote the position vector and $\hat{n}$ denote the outward unit normal to S. Then, the integral $\iint_S \vec{r} \cdot \hat{n}\, dS$ equals
(a) V (b) $V/3$ c) $3V$ (d) 0

28. Let $B = \{(x, y, z) : x, y, z \in \Re \text{ and } x^2 + y^2 + z^2 \leq 4\}$. Let $v(x, y, z) = x\hat{i} + y\hat{j} + z\hat{k}$ be a vector-valued function defined in B. If $r^2 = x^2 + y^2 + z^2$, the value of the integral $\iiint_B \vec{\nabla} \cdot (r^2 v(x, y, z))\, dV$ is
(a) 16π (b) 32π (c) 64π d) 128π

29. Let S be the surface bounding the region $x^2 + y^2 \leq 1, x \geq 0, y \geq 0, |z| \leq 1$, and $\hat{n}$ be the outward unit normal to S. Then $\iint_S \left[(\sin^2 x)\hat{i} + 2y\hat{j} - z(1 + \sin 2x)\hat{k}\right] \cdot \hat{n}\, dS$ equals
(a) 1 (b) $\pi/2$ (c) π (d) 2π

30. Let $V = \left\{(x, y, z) \in \Re^3 : \frac{1}{4} \leq x^2 + y^2 + z^2 \leq 1\right\}$ and $\vec{F} = \frac{1}{(x^2+y^2+z^2)^2}(x\hat{i} + y\hat{j} + z\hat{k})$ for $(x, y, z) \in V$. Let $\hat{n}$ denote the outward unit normal vector to the boundary of V and S denote the part $\left\{(x, y, z) \in \Re^3 : x^2 + y^2 + z^2 = \frac{1}{4}\right\}$ of the boundary V. Then $\iint_S \vec{F} \cdot \hat{n}\, dS$ is equal to
(a) -8π (b) -4π (c) 4π (d) 8π

Answers

1. c 2. a 3. c 4. c 5. b 6. b 7. b 8. d 9. b 10. b 11. a 12. a 13. c
14. d 15. d 16. c 17. b 18. b 19. d 20. a 21. a 22. b 23. b 24. b 25. b
26. b 27. c 28. d 29. c 30. a

Short Answer Questions

1. The acceleration of a particle at any time $t \geq 0$ is given by $\vec{a} = 12\cos 2t\hat{i} - 8\sin 2t\hat{j} + 16t\hat{k}$. If the velocity $\vec{v}$ and displacement $\vec{r}$ are zero at $t = 0$, find $\vec{v}$ and $\vec{r}$ at any time.
2. Find the values of $\vec{r}$ satisfying the equation
   (a) $\dfrac{d^2\vec{r}}{dt^2} = \vec{a}$, where $\vec{a}$ is a constant vector.
   (b) $\dfrac{d^2\vec{r}}{dt^2} = \vec{a}t + \vec{b}$, where $\vec{a}$ and $\vec{b}$ are constant vectors.
   (c) $\dfrac{d^2\vec{r}}{dt^2} + \omega^2\vec{r} = \vec{0}$, where ω is a constant different from zero.
3. Show that $\displaystyle\int_2^3 \left(\vec{r}\cdot\frac{d\vec{r}}{dt}\right) dt = 10$, where
$$\vec{r}(t) = \begin{cases} 2\hat{i} - \hat{j} + 2\hat{k}; & \text{when } t = 2 \\ 4\hat{i} - 2\hat{j} + 3\hat{k}; & \text{when } t = 3 \end{cases}$$
4. If $\vec{a}(t) = 2\hat{i} + t\hat{j} - \hat{k}$, $\vec{b}(t) = t\hat{i} + 2\hat{j} + 3\hat{k}$ and $\vec{c}(t) = 2\hat{i} - 3\hat{j} + 4\hat{k}$, show that
   (a) $\displaystyle\int_1^2 \vec{a}\cdot(\vec{b}\times\vec{c})\, dt = \frac{253}{6}$ (b) $\displaystyle\int_1^2 \vec{a}\times(\vec{b}\times\vec{c})\, dt = -12\hat{i} - 6\hat{j} - 26\hat{k}$
5. Define the normal surface integral of a vector function $\vec{F}(\vec{r})$ over a surface S.
6. What is meant by flux of $\vec{F}$ across S?
7. If $\vec{A} = xy\hat{i} + (x^2 + y^2)\hat{j}$, then evaluate $\displaystyle\int_\Gamma \vec{A}\cdot d\vec{r}$, where
   (a) Γ is the arc of the parabola $y = x^2 - 4$ from $(2, 0)$ to $(4, 12)$ *Ans: 732*
   (b) Γ is the part of the x axis from $x = 2$ to $x = 4$, then a part of the vertical line $x = 4$ from $y = 0$ to $y = 12$ *Ans: 768*
8. Prove that $\displaystyle\iiint_V \vec{r}\cdot\hat{n}\, dS = 3V$, where V is the volume enclosed by the surface S. What is the value of V if the surface of integration is that of the sphere of radius R.
9. If $\vec{A} = \vec{\nabla}\phi$ and $\nabla^2\phi = \rho$, a specified scalar point function, show that
$$\iint_S \frac{\partial\phi}{\partial n}\, dS = \iiint_V \rho\, dV$$
   Hint: $\dfrac{\partial\phi}{\partial n} = \vec{\nabla}\phi\cdot\hat{n}$
10. Evaluate $\int\int\int_V \vec{F}\, dV$, where V is the region bounded by the faces $x = 0, x = 2$; $y = 0, y = 2$; and $z = 0, z = 2$, where $\vec{F} = 2zx\hat{i} - x\hat{j} + y\hat{k}$. *Ans:* $8(2\hat{i} - \hat{j} + \hat{k})$

Long Answer Questions

1. Evaluate $\int_C \vec{A} \cdot d\vec{r}$, where
   (a) C is the straight line from $(0, 0, 0)$ to $(0, 0, 1)$, then to $(0, 1, 1)$ and then $(2, 1, 1)$ and $\vec{A} = (2y + 3)\hat{\imath} + xz\hat{\jmath} + (yz - x)\hat{k}$. *Ans:* 10
   (b) $\vec{A} = (xy - z)\hat{\imath} + (yz - x)\hat{\jmath} + (zx - y)\hat{k}$ and C is the path (see figure given below)described by the triangle OPQ whose vertices are $O(0, 0, 0)$, $P(1, 1, 1)$ and $Q(2, 2, 2)$ traversed in counter-clockwise direction. *Ans: 0*

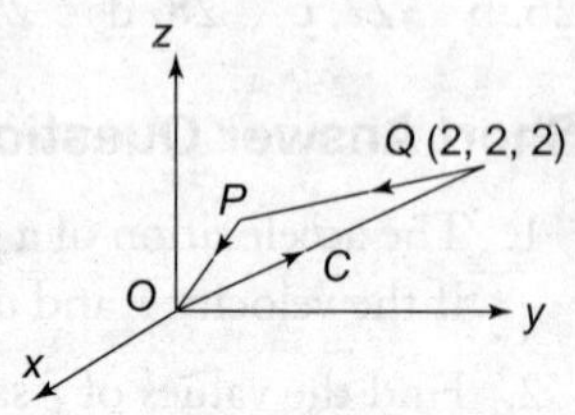

2. Evaluate the value of the line integral
   (a) $\int_C \left[(x + 2y)\hat{\imath} + (x^3 - 3xy)\hat{\jmath}\right] \cdot d\vec{r}$, where C is the closed curve in the first quadrant, formed by the straight line $x = 1$, the x axis and the arc of the parabola $y^2 = x$ described in the anti-clockwise sense. *Ans:* 103/84
   (b) $\int_C \left[(x + y)\hat{\imath} + (y^2 + z^2)\hat{\jmath} + (x^3 + z^3)\hat{k}\right] \cdot d\vec{r}$ along the curve $C : \vec{r} = \hat{\imath} + t\,\hat{\jmath} + t^2\,\hat{k}$, $0 \leq t \leq 2$. *Ans:* 1156/15
   (c) $\int_C \left[(2x - y + z)\hat{\imath} + (x + y - z^2)\hat{\jmath} + (3x - 2y + 4z)\hat{k}\right] \cdot d\vec{r}$ along the curve $C : x^2 + y^2 = 9, z = 0$ described in the anti-clockwise sense. *Ans:* 18π
   (d) $\int_C \left[\frac{1}{x^2 + 2y^2}(y\hat{\imath} + x\hat{\jmath})\right] \cdot d\vec{r}$ along the elliptic path $C : x^2 + 2y^2 = 1$ described in the anti-clockwise sense. *Ans:* 0
3. Evaluate the line integral of $yz\hat{\imath} + zx\hat{\jmath} + xy\hat{k}$ over
   (a) the straight line joining the points $(b, 0, 0)$ and $(b, 0, 2\pi c)$. *Ans:*0
   (b) the circular helix $b\cos t\hat{\imath} + b\sin t\hat{\jmath} + ct\hat{k}$ which joins the two points $(b, 0, 0)$ and $(b, 0, 2\pi c)$. *Ans:*2
4. Evaluate $\int_C \vec{A} \cdot d\vec{r}$, where $\vec{A} = (x^2 + y^2)\hat{\imath} - 2xy\hat{\jmath}$ and C is the boundary of the rectangle in the xy plane contained by $y = 0, x = a, y = b, x = 0$. *Ans:*$-2ab^2$
5. Find the work done by a particle under the force $\vec{F}$ moving round the
   (a) circle $x^2 + y^2 = 4, z = 0$, where $\vec{F} = (2x - y + 4z)\hat{\imath} + (x + y - z^2)\hat{\jmath} + (3x - 2y + 4z^3)\hat{k}$ *Ans:*-8π
   (b) curve $x = t^2 + 1, y = 2t^2, z = t^3, 0 \geq t \geq 2$, where $\vec{F} = 3xy\hat{\imath} - 5z\hat{\jmath} + 10z\hat{k}$ *Ans:*303 *units*
6. Find the work done in moving a particle in the force field $\vec{F}$ along
   (a) the space curve $x = 2t^2, y = t, z = 4t^2 - t$ from $t = 0$ to $t = 1$ *Ans:* 14.2
   (b) the curve defined by $x^2 = 4y, 3x^3 = 8z$ from $x = 0$ to $x = 2$ and $\vec{F} = 3x^2\hat{\imath} + (2xz - y)\hat{\jmath} + z\hat{k}$ *Ans: 16*
7. Evaluate $\int_\Gamma yzdx + (xz + 1)dy + xydz$ along any path Γ joining the points $(1, 0, 0)$ to $(2, 1, 4)$. *Ans.:*9

8. Evaluate $\int\int_S \vec{F}\cdot\hat{n}dS$ for each of the following cases:

   (a) $\vec{F}=(x+y^2)\hat{i}-2x\hat{j}+2yz\hat{k}$ and S is the surface of the plane $2x+y+2z=6$ in the first octant *Ans:*81

   (b) $\vec{F}=6z\hat{i}-4\hat{j}+y\hat{k}$ and S is the surface of the plane $2x+6y+3z=10$ which is located in the first octant *Ans:*25/2

9. Find $\int\int_S \vec{F}\cdot\hat{n}dS$, where $\vec{F}=r^2\vec{r}$ and S is the surface of the sphere $x^2+y^2+z^2=a^2$. Compute the integral directly and also by using the divergence theorem.

10. Evaluate $\int\int_S (\vec{\nabla}\times\vec{A})\cdot\hat{n}dS$ for each of the following cases:

    (a) $\vec{A}=(x+2y)\hat{i}-3z\hat{j}+x\hat{k}$ and S is the surface of the plane $2x+6y+3z=10$ which is located in the first octant *Ans:*1

    (b) $\vec{A}=(x+2y)\hat{i}-3z\hat{j}+x\hat{k}$ and S is the surface of the plane $2x+y=6$ in the first octant cut off by the plane $z=4$ *Ans:*9/2

11. Evaluate $\int\int_S \phi\,\hat{n}\,dS$ for each of the following cases:

    (a) $\phi=\frac{2}{3}xyz$ and S is the surface of the cylinder $x^2+y^2=9$ included in the first octant between $z=0$ and $z=3$ *Ans:*$27(\hat{i}+\hat{j})$

    (b) $\phi=\frac{9}{64}xyz^2$ and S is the surface of the cylinder $x^2+y^2=25$ included in the first octant between $z=0$ and $z=4$ *Ans:*$125(\hat{i}+\hat{j})$

12. Find the surface area of

    (a) plane $x+2y+2z=12$ cut off by the planes $x=0, x=1, y=0, y=1$ Ans:3/2

    (b) plane $x+2y+2z=12$ cut off by $x=0, y=0$ and $x^2+y^2=16$ *Ans:*6π

    (c) region common to the intersecting cylinders $x^2+y^2=a^2$ and $x^2+z^2=a^2$ *Ans:*$8a^2$

13. Evaluate $\int\int\int_V \phi\hat{n}dV$ for each of the following cases:

    (a) $\phi=45x^2y$ and V is the closed region bounded by the planes $4x+2y+z=8, x=0, y=0, z=0.$ *Ans:*128

    (b) $\phi=15xy$ and V is the closed region bounded by the planes $x+4y+z=2, x=0, y=0, z=0.$ *Ans:*1/4

14. Evaluate $\int\int\int_V \vec{\nabla}\times\vec{F}dV$ for each of the following cases:

    (a) $\vec{F}=(3x^2-8z)\hat{i}-2xy\hat{j}-8x\hat{k}$, where V is the closed region bounded by the planes $x=0, y=0, z=0$ and $2x+2y+z=4$ *Ans:*$-8/3\,\hat{k}$

    (b) $\vec{F}=(2x^2-3z)\hat{i}-2xy\hat{j}-4x\hat{k}$, where V is the closed region bounded by the planes $x=0, y=0, z=0$ and $2x+2y+z=4$ *Ans:*$8(\hat{j}-\hat{k})/3$

15. Evaluate $\int\int\int_V \vec{\nabla}\cdot\vec{F}dV$ for each of the following cases:

    (a) $\vec{F}=y\hat{i}+2x\hat{j}-z\hat{k}$, where V is the closed region bounded by the surfaces $x=0, x=2, y=0, y=2, z=x^2$ and $z=4$ *Ans:*$-32/3$

(b) $\vec{F} = (3x^2 - 8z)\hat{\imath} - 2xy\hat{\jmath} - 8x\hat{k}$, where V is the closed region bounded by the planes $x = 0, y = 0, z = 0$ and $2x + 2y + z = 4$ *Ans:*16/3

(c) $\vec{F} = (2x^2 - 3z)\hat{\imath} - 2xy\hat{\jmath} - 4x\hat{k}$, where V is the closed region bounded by the planes $x = 0, y = 0, z = 0$ and $2x + 2y + z = 4$ *Ans:*8/3

16. Evaluate $\int\int\int_V (x\hat{\imath} + 2y\hat{\jmath})dV$, where V is the closed region bounded by the cylinder $z = 9 - x^2$ and the planes $x = 0, y = 0, y = 1$ and $z = 0$. *Ans:* $\frac{81}{4}\hat{\imath} + 18\hat{\jmath}$

7 Integral Theorems

We have already discussed line, surface and volume integrals in the previous chapter. In this chapter, we will find out how those integrals are related, as sometimes, it is easier to solve a surface integral rather than a volume integral or a line integral related to it, and conversely. For this purpose, two very useful theorems of vector integral calculus, namely, the Gauss divergence theorem and Stokes' theorem with related results are studied. Green's theorem in the plane is, of course, a special case of Stokes' theorem. Verification of each theorem is shown using illustrative examples.

7.1 GREEN'S THEOREM IN THE PLANE

Let $\mathcal{R}$ be a closed bounded region in the xy plane whose boundary C consists of a finite number of smooth curves. If the functions P and Q are continuous functions of x and y, having continuous partial derivatives $\frac{\partial P}{\partial y}$ and $\frac{\partial Q}{\partial x}$ in $\mathcal{R}$, then

$$\oint_C \left(Pdx + Qdy\right) = \int\int_{\mathcal{R}} \left(\frac{\partial Q}{\partial x} - \frac{\partial P}{\partial y}\right) dx\, dy \tag{7.1}$$

where C is traversed in the anti-clockwise (+ve) direction.

Proof: Let us consider a simple closed curve C enclosing the simply connected region $\mathcal{R}$ of the xy plane such that any line parallel to any one coordinate axis meets the curve C at a maximum of two points. Let the region $\mathcal{R}$ be bounded by the closed curve C : $ABEDA$ (Fig. 7.1). Let the equations of the curves ABE and EDA be $y = y_1(x)$ and $y = y_2(x)$, respectively, such that the functions $y_1(x)$ and $y_2(x)$ are single valued and continuous in $\left[a, b\right]$. Then

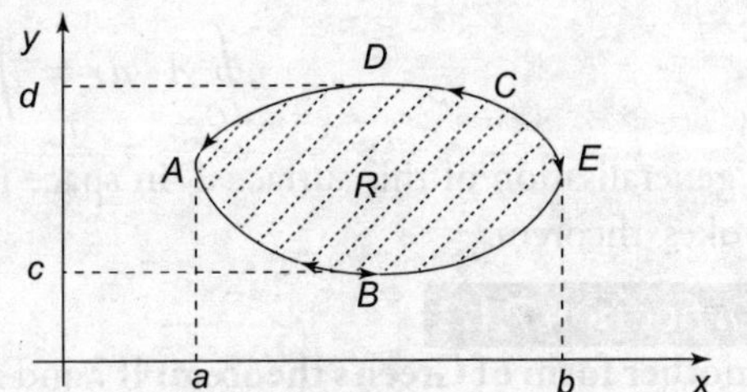

Figure 7.1 Green's theorem

$$\int\int_{\mathcal{R}} \frac{\partial P}{\partial y}\, dx\, dy = \int_a^b \left[\int_{y_1(x)}^{y_2(x)} \frac{\partial P}{\partial y}\, dy\right] dx = \int_a^b \left[P(x, y_2(x)) - P(x, y_1(x))\right] dx$$

$$= -\int_a^b P(x, y_1(x))dx - \int_b^a P(x, y_2(x))dx = -\oint_C P(x, y)dx \tag{7.2}$$

Again, if we assume the equations of the curves DAB and BED as $x = x_1(y)$ and $x = x_2(y)$, respectively, such that $x_1(y)$ and $x_2(y)$ are single valued and continuous in $\left[c, d\right]$, then

$$\int\int_{\mathcal{R}} \frac{\partial Q}{\partial x}\, dx\, dy = \int_c^d \left[\int_{x_1(y)}^{x_2(y)} \frac{\partial Q}{\partial x}\, dx\right] dy = \int_c^d \left[Q(x_2(y), y) - P(x_1(y), y)\right] dy$$

$$= \int_d^c Q(x_1(y), y)dy + \int_c^d Q(x_2(y), y)dy = \oint_{\mathcal{C}} Q(x, y)dy \qquad (7.3)$$

From Eqs (7.2) and (7.3), we get

$$\oint_{\mathcal{C}} \left(Pdx + Qdy\right) = \int\int_{\mathcal{R}} \left(\frac{\partial Q}{\partial x} - \frac{\partial P}{\partial y}\right) dx\, dy$$

DEDUCTION 7.1.1

Vector form of Green's theorem: If $\vec{A} = P(x, y)\hat{\imath} + Q(x, y)\hat{\jmath}$ is a continuously differentiable vector function of points in $\mathcal{R}$ (Fig. 7.2), then

$$\vec{\nabla} \times \vec{A} = \begin{vmatrix} \hat{\imath} & \hat{\jmath} & \hat{k} \\ \dfrac{\partial}{\partial x} & \dfrac{\partial}{\partial y} & \dfrac{\partial}{\partial z} \\ P & Q & 0 \end{vmatrix}$$

$$= \left(\frac{\partial Q}{\partial x} - \frac{\partial P}{\partial y}\right)\hat{k}$$

$$\Rightarrow (\vec{\nabla} \times \vec{A}) \cdot \hat{k} = \frac{\partial Q}{\partial x} - \frac{\partial P}{\partial y} \qquad (7.4)$$

Figure 7.2 Vector form of Green's theorem

where $\hat{k}$ is the unit positive normal to the xy plane. Let $\vec{r} = x\hat{\imath} + y\hat{\jmath}$, then $Pdx + Qdy = (P\hat{\imath} + Q\hat{\jmath}) \cdot (dx\hat{\imath} + dy\hat{\jmath}) = \vec{A} \cdot d\vec{r}$. Hence, Green's theorem in the plane takes the form

$$\oint_{\mathcal{C}} \vec{A} \cdot d\vec{r} = \int\int_{\mathcal{R}} (\vec{\nabla} \times \vec{A}) \cdot \hat{k}\, dx\, dy \qquad (7.5)$$

A generalisation of this surface $\mathcal{S}$ in space having a curve $\mathcal{C}$ as boundary leads quite naturally to Stokes' theorem.

DEDUCTION 7.1.2

Another form of Green's theorem: If $\hat{t}$ and $\hat{n}$ be the unit tangent and unit outward normal vectors to curve $\mathcal{C}$ in the xy plane, then we have $\hat{t} = \hat{k} \times \hat{n}$. Now

$$P\, dx + Q\, dy = \vec{A} \cdot d\vec{r} = \vec{A} \cdot \hat{t}\, ds = \left[\vec{A}\,\hat{k}\,\hat{n}\right] ds$$

$$= \left[\hat{n}\,\vec{A}\,\hat{k}\right] ds = (\vec{A} \times \hat{k})\, ds$$

$$\text{and } \vec{A} \times \hat{k} = (P\hat{\imath} + Q\hat{\jmath}) \times \hat{k} = Q\hat{\imath} - P\hat{\jmath}$$

$$\Rightarrow \vec{\nabla} \cdot (\vec{A} \times \hat{k}) = \frac{\partial Q}{\partial x} - \frac{\partial P}{\partial y}$$

Hence, Green's theorem in the plane takes the form

$$\oint (\vec{A} \times \hat{k}) \cdot \hat{n}\, ds = \int\int_{\mathcal{R}} \vec{\nabla} \cdot (\vec{A} \times \hat{k})\, dx\, dy. \tag{7.6}$$

DEDUCTION 7.1.3

Physical interpretation: The work done by the force $\vec{F}$ in moving a particle along a closed curve Γ is determined by the curl of $\vec{F}$. Now

$$\vec{\nabla} \times \vec{F} = \vec{0} \Rightarrow \frac{\partial Q}{\partial x} = \frac{\partial P}{\partial y}$$

$\Rightarrow$ the integral along the closed path is zero

$\Rightarrow$ the work done is independent of the curve joining the points

$\Rightarrow$ the force field is conservative

Example 7.1.1

Verify Green's theorem for the function $\vec{A} = (3x^2 - 8y^2)\hat{\imath} + (4y - 6xy)\hat{\jmath}$, where $\mathcal{R}$ is the region bounded by $y = \sqrt{x}$ and $y = x^2$.

Solution

Here, $\vec{A} = (3x^2 - 8y^2)\hat{\imath} + (4y - 6xy)\hat{\jmath}$. The region is enclosed by the curves $y = \sqrt{x}$ and $y = x^2$ (Fig. 7.3). Solving the equation of the parabolas, we obtain the points of intersection as $(0, 0)$ and $(1, 1)$. $\mathcal{C}$ is the boundary of the region $\mathcal{R}$ described in the counter-clockwise manner, which consists of two points $\mathcal{C}_1$ and $\mathcal{C}_2$.

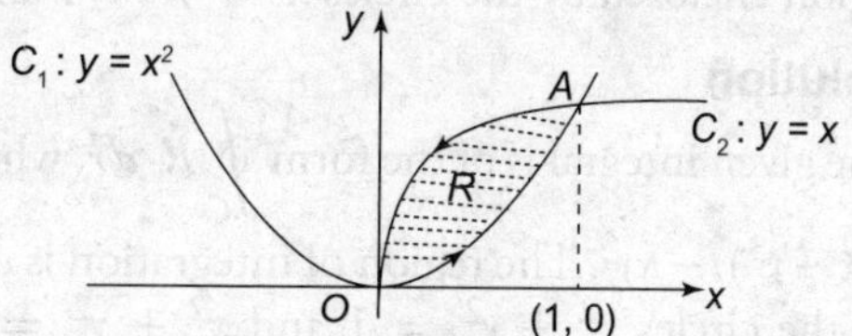

Figure 7.3 Profiles of $y = \sqrt{x}$, $y = x^2$

Evaluation of the line integral: Along the curve $\mathcal{C}_1 : y = x^2$, $dy = 2x\, dx$ and so

$$\int_{\mathcal{C}_1} \vec{A} \cdot d\vec{r} = \int_{\mathcal{C}_1} (3x^2 - 8y^2)dx + (4y - 6xy)dy$$

$$= \int_0^1 (3x^2 - 8x^4)dx + (4x^2 - 6x^3)2xdx$$

$$= \int_0^1 \Big(3x^2 - 20x^4 + 8x^3)dx = 1 - 4 + 2 = -1$$

Along the curve $\mathcal{C}_2 : y = \sqrt{x}$, $dy = dx/2\sqrt{x}$ and so

$$\int_{\mathcal{C}_2} \vec{A} \cdot d\vec{r} = \int_{\mathcal{C}_2} (3x^2 - 8y^2)dx + (4y - 6xy)dy$$

$$= \int_1^0 (3x^2 - 8x)dx + (4\sqrt{x} - 6x\sqrt{x})\frac{1}{2\sqrt{x}}dx$$

$$= \int_1^0 \left(3x^2 - 11x + 2\right)dx = 5/2$$

Therefore, the line integral becomes

$$\oint_C \vec{A} \cdot d\vec{r} = \int_{C_1} \vec{A} \cdot d\vec{r} + \int_{C_2} \vec{A} \cdot d\vec{r} = -1 + \frac{5}{2} = 3/2$$

Evaluation of the surface integral: Since $P = 3x^2 - 8y^2$ and $Q = 4y - 6xy$ are continuous functions having continuous partial derivatives in $\mathcal{R}$

$$\int\int_{\mathcal{R}} \left(\frac{\partial Q}{\partial x} - \frac{\partial P}{\partial y}\right) dx\, dy = 10 \int_0^1 dx \int_{x^2}^{\sqrt{x}} ydy = 5 \int_0^1 \left(x - x^4\right)dx = 3/2$$

This is equal to the given line integral. Hence, Green's theorem (7.1) is verified.

Example 7.1.2

Verify Green's theorem in the plane for $\int_C (2x - y^3)dx - xy\, dy$, where C is the boundary of the region enclosed by the circles $x^2 + y^2 = 1$ and $x^2 + y^2 = 9$.

Solution

The given integral is of the form $\oint_C \vec{F} \cdot d\vec{r}$, where $\vec{F} = (2x - y^3)\hat{\imath} - xy\hat{\jmath}$. The region of integration is enclosed by the circles $x^2 + y^2 = 1$ and $x^2 + y^2 = 9$ (Fig. 7.4), which is not simply connected. The boundary C has two simple closed curves $C_1 : x^2 + y^2 = 9$ (described in the counter-clockwise manner) and $C_2 : x^2 + y^2 = 1$ (described in the clockwise manner) so that the region $\mathcal{R}$ is always on the left side while moving along the curves C_1 and C_2.

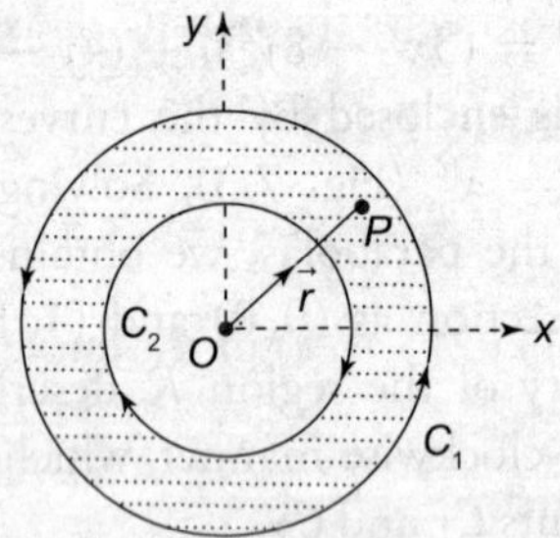

Figure 7.4 Region enclosed by the circles C_1 and C_3

Evaluation of the line integral: On the surface C_1, we have $x = 3\cos t, y = 3\sin t, 0 \le t \le 2\pi$, so the line integral becomes

$$\int_{C_1} \vec{F} \cdot d\vec{r} = \int_0^{2\pi} \left[(6\cos t - 27\sin^3 t)(-3\sin t) - 3\cos t \cdot 3\sin t(3\cos t)\right]dt$$

$$= \int_0^{2\pi} \left[-9\sin 2t + 81\sin^4 t - 27\cos^2 t\ \sin t\right] dt = \frac{243}{4}\pi$$

On the surface $C_2 : x = \cos t, y = \sin t$, where t varies from 2π to 0, so the integral becomes

$$\int_{C_2} \vec{F} \cdot d\vec{r} = \int_{2\pi}^0 \left[(2\cos t - \sin^3 t)(-\sin t) - \cos t \cdot \sin t(\cos t)\right]dt$$

$$= -\int_0^{2\pi} \left[-\sin 2t + \sin^4 t - \cos^2 t\ \sin t\right] dt = -\frac{3}{4}\pi$$

Since $\mathcal{C}$ is the boundary bounded by the circles $\mathcal{C}_1$ and $\mathcal{C}_2$

$$\oint_{\mathcal{C}} \vec{F} \cdot d\vec{r} = \int_{\mathcal{C}_1} \vec{F} \cdot d\vec{r} + \int_{\mathcal{C}_2} \vec{F} \cdot d\vec{r} = \frac{243}{4}\,\pi - \frac{3\pi}{4} = 60\,\pi$$

Evaluation of the surface integral: Since $P = 2x - y^3$ and $Q = -xy$ are continuous functions having continuous partial derivatives in $\mathcal{R}$

$$\begin{aligned}
\int\!\!\int_{\mathcal{R}} \left(\frac{\partial Q}{\partial x} - \frac{\partial P}{\partial y}\right) dx\,dy &= \int\!\!\int_{\mathcal{R}} \left[\frac{\partial}{\partial x}(-xy) - \frac{\partial}{\partial y}(2x - y^3)\right] dx\,dy \\
&= \int\!\!\int_{\mathcal{R}} (-y + 3y^2)\,dx\,dy; \quad x = r\cos\theta, y = r\sin\theta \\
&= \int_1^3 \left[\int_0^{2\pi} (-r\sin\theta + 3r^2\sin^2\theta\,d\theta)\right] r\,dr \\
&= \int_0^{2\pi} \left[-\frac{26}{3}\sin\theta + 60\sin^2\theta\right] d\theta \\
&= \left[\frac{26}{3}\cos\theta + 30\left(\theta - \frac{1}{2}\sin 2\theta\right)\right]_0^{2\pi} = 60\pi
\end{aligned}$$

Hence, Green's theorem (7.1) is verified.

Example 7.1.3

Let $\vec{F} = (3y^2 + 2z^2)\hat{\imath} + (6x - 10z)\hat{\jmath} + (4xz - 5y^2)\hat{k}$. Evaluate $\oint_{\mathcal{C}} \vec{F} \cdot d\vec{r}$ along the portion from $(1, 0, 1)$ to $(3, 4, 5)$ of the curve $\mathcal{C}$, which is the intersection of the two surfaces $z^2 = x^2 + y^2$ and $z = y + 1$.

Solution

Here, the given integral is $\oint_{\mathcal{C}} (3y^2 + 2z^2)dx + (6x - 10z)dy + (4xz - 5y^2)dz$ and the curves are $z^2 = x^2 + y^2$ and $z = y + 1$, i.e., $(y + 1)^2 = x^2 + y^2$, i.e., $x^2 = 2y + 1$, i.e., $y = \frac{1}{2}(x^2 - 1) \Rightarrow z = \frac{1}{2}(x^2 + 1)$. Therefore

$$\begin{aligned}
\oint_{\mathcal{C}} \vec{F} \cdot d\vec{r} &= \oint_{\mathcal{C}} (3y^2 + 2z^2)dx + (6x - 10z)dy + (4xz - 5y^2)dz \\
&= \oint_{\mathcal{C}} (3y^2 + 2z^2)dx + \oint_{\mathcal{C}} (6x - 10z)dy + \oint_{\mathcal{C}} (4xz - 5y^2)dz \\
&= \int_1^3 \left[\frac{3}{4}(x^2 - 1)^2 + \frac{2}{4}(x^2 + 1)^2\right] dx + \int_0^4 \left[6y\sqrt{2y + 1} - 10y(y + 1)\right] dy \\
&\quad + \int_1^5 \left[4z\sqrt{2z - 1} - 5(z - 1)^2\right] dz \\
&= \int_1^3 \left[\frac{5}{4}x^4 - \frac{x^2}{2} + \frac{5}{4}\right] dx + \int_0^4 \left[6y\sqrt{2y + 1} - 10y(y + 1)\right] dy \\
&\quad + \int_1^5 \left[4z\sqrt{2z - 1} - 5(z - 1)^2\right] dz
\end{aligned}$$

$$= \frac{x^5}{4} - \frac{x^2}{6} + \frac{5x}{4}\Big|_1^3 + 6\Big[\frac{y(2y+1)^{3/2}}{3} - \frac{(2y+1)^{5/2}}{15}\Big]_0^4 - 10\Big[\frac{y^3}{3} + \frac{y^2}{2}\Big]_0^4$$

$$+4\Big[\frac{z(2z-1)^{3/2}}{3} - \frac{(2z-1)^{5/2}}{15}\Big]_1^5 - 5\Big[\frac{(z-1)^3}{3}\Big]_1^5 = -108$$

Example 7.1.4

Use Green's theorem to evaluate the integral $\oint_C x^2\,dx + (x+y^2)\,dy$, where C is the closed curve given by $y=0, y=x$ and $y^2 = 2-x$ in the first quadrant, oriented counter-clockwise.

Solution

Here, $P(x,y) = x^2$ and $Q(x,y) = x + y^2$. The closed curve given by $y = 0, y = x$ and $y^2 = 2 - x$ in the first quadrant, as shown in Fig. 7.5. Now, by Green's theorem

Figure 7.5 Closed curve $y=0, y=x$ and $y^2 = 2-x$

$$\int x^2\,dx + (x+y^2)\,dy = \oint P\,dx + Q\,dy$$

$$= \int\int_S \left(\frac{\partial Q}{\partial x} - \frac{\partial P}{\partial y}\right) dx\,dy$$

$$= \int\int_S (1-0)\,dx\,dy$$

$$= \int_0^1 \Big[\int_0^x dy\Big] dx + \int_1^2 \Big[\int_0^{\sqrt{2-x}} dy\Big] dx = \int_0^1 x\,dx + \int_1^2 \sqrt{2-x}\,dx$$

$$= \frac{x^2}{2}\Big|_0^1 + \Big[-\frac{2}{3}(2-x)^{3/2}\Big]_1^2 = \frac{1}{2} + \frac{2}{3} = \frac{7}{6}$$

Example 7.1.5

Use Green's theorem to evaluate the line integral $\oint_C (x^2 - 2xy)dx + (x^2y+3)dy$, around the boundary C of the region defined by $y^2 = 8x, x = 2$.

Solution

Let $P = x^2 - 2xy$ and $Q = x^2y + 3$. Then, both P and Q together with their first order partial derivatives are continuous in the region R enclosed by C in the xy plane (Fig. 7.6). The line $x = 2$ meets the parabola $y^2 = 8x$ at points $A(2,-4)$ and $B(2,4)$. The boundary C of the region consists of $\widehat{OA}$, the line segment AB and $\widehat{BO}$. So, using the Cartesian form of Green's theorem (7.1), we get

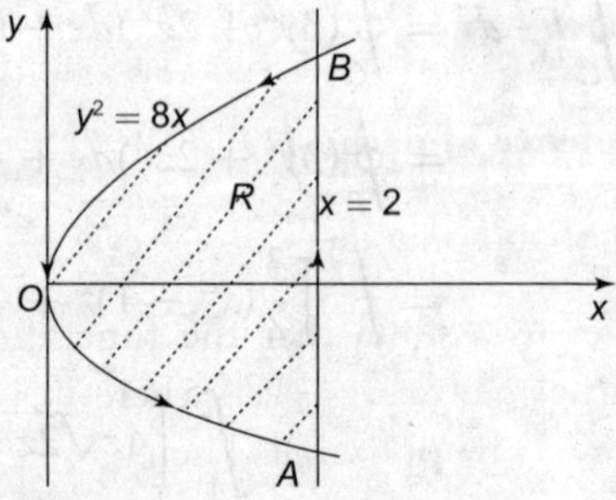

Figure 7.6 The curve $y^2 = 8x, x = 2$

$$\oint_C \Big(Pdx + Qdy\Big) = \oint_C (x^2 - 2xy)dx + (x^2y+3)dy$$

$$= \iint_{\mathcal{R}} \left(\frac{\partial}{\partial x}(x^2 y + 3) - \frac{\partial}{\partial y}(x^2 - 2xy) \right) dx\, dy$$

$$= \int_0^2 \left[\int_{-2\sqrt{2x}}^{2\sqrt{2x}} 2x(y+1)\, dy \right] dx$$

$$= 2\int_0^2 x\left[\frac{y^2}{2} + y\right]_{-2\sqrt{2x}}^{2\sqrt{2x}} dx = 8\sqrt{2}\int_0^2 x^{3/2}\, dx = 128/5$$

Example 7.1.6

Evaluate by Green's theorem $\oint_{\mathcal{C}} \frac{1}{x^2+y^2}(x\, dy - y\, dx)$, where $\mathcal{C} = \mathcal{C}_1 \cup \mathcal{C}_2$; $\mathcal{C}_1 : x^2 + y^2 = 1$ and $\mathcal{C}_2 : x = -2, y = -2, x = 2, y = 2$.

Solution

Let $P = x/(x^2 + y^2)$ and $Q = -y/(x^2 + y^2)$. Then, both P and Q together with their first order partial derivatives are continuous in the region R enclosed by $\mathcal{C}$ in the xy plane (Fig. 7.7). Also, $\frac{\partial P}{\partial y} = -2y$ and $\frac{\partial Q}{\partial x} = 2x$. So, using the Cartesian form of Green's theorem (7.1), we get

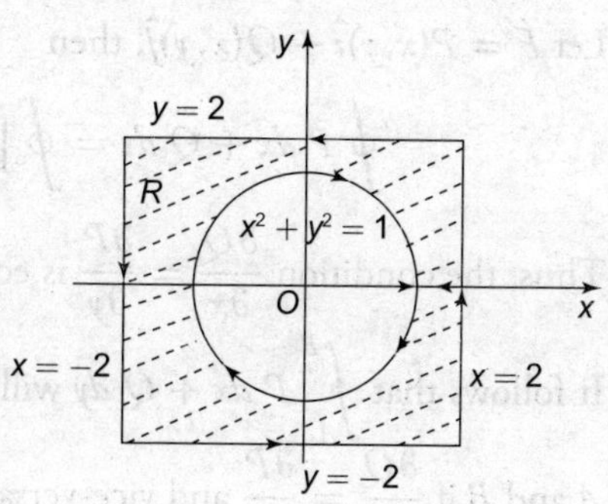

Figure 7.7 The region

$$\oint_{\mathcal{C}} \left(Pdx + Qdy\right) = \iint_{\mathcal{R}} \left[\frac{\partial}{\partial x}\left(\frac{x}{x^2+y^2}\right) - \frac{\partial}{\partial y}\left(-\frac{y}{x^2+y^2}\right)\right] dx\, dy$$

$$= \iint_{S} \left[\frac{1}{x^2+y^2} - \frac{x \cdot 2x}{(x^2+y^2)^2} + \frac{1}{x^2+y^2} - \frac{y \cdot 2y}{(x^2+y^2)^2}\right] dx\, dy$$

$$= \iint_{S} \left[\frac{x^2+y^2-2x^2+x^2+y^2-2y^2}{(x^2+y^2)^2}\right] dx\, dy = 0$$

Theorem 7.1

Independence of path: $\oint P\, dx + Q\, dy = 0$ around every closed curve $\mathcal{C}$ in a simply connected region $\mathcal{R}$ if and only if $\frac{\partial P}{\partial y} = \frac{\partial Q}{\partial x}$ everywhere in the region $\mathcal{R}$.

Proof: Let us assume that the functions P and Q are continuous and have continuous first order partial derivatives everywhere in the region $\mathcal{R}$ bounded by the curve $\mathcal{C}$, so that Green's theorem (7.1) is applicable. Then

$$\oint_{\mathcal{C}} \left(P\, dx + Q\, dy\right) = \iint_{\mathcal{R}} \left(\frac{\partial Q}{\partial x} - \frac{\partial P}{\partial y}\right) dx\, dy$$

$$= 0; \qquad \text{if, } \frac{\partial P}{\partial y} = \frac{\partial Q}{\partial x} \text{ in } \mathcal{R}$$

Conversely, let $\oint P\,dx + Q\,dy = 0$ for all curves of $\mathcal{C}$. If $\dfrac{\partial Q}{\partial x} - \dfrac{\partial P}{\partial y} > 0$ at a point P, then by the neighbourhood property and the continuity of the derivatives, it follows that $\dfrac{\partial Q}{\partial x} - \dfrac{\partial P}{\partial y} > 0$ in some region A surrounding P. If Γ is the boundary of A, then

$$\oint_\Gamma \left(P\,dx + Q\,dy\right) = \int\!\!\int_A \left(\frac{\partial Q}{\partial x} - \frac{\partial P}{\partial y}\right) dx\,dy > 0$$

This contradicts the assumption that the line integral is zero around every closed curve. Similarly, the assumption $\dfrac{\partial Q}{\partial x} - \dfrac{\partial P}{\partial y} < 0$ leads to a contradiction. Thus, $\dfrac{\partial Q}{\partial x} - \dfrac{\partial P}{\partial y} = 0$ at all points.

(i) Let $\vec{F} = P(x,y)\hat{\imath} + Q(x,y)\hat{\jmath}$, then

$$\oint P\,dx + Q\,dy = \oint \left[P\,\hat{\imath} + Q\,\hat{\jmath}\right]\cdot\left[dx\,\hat{\imath} + dy\,\hat{\jmath}\right] = \oint \vec{F}\cdot d\vec{r}$$

Thus, the condition $\dfrac{\partial Q}{\partial x} = \dfrac{\partial P}{\partial y}$ is equivalent to $\vec{\nabla}\times\vec{F} = \vec{0}$.

(ii) It follows that $\displaystyle\int_A^B P\,dx + Q\,dy$ will be independent of the path in $\mathcal{R}$ joining two points A and B if $\dfrac{\partial Q}{\partial x} = \dfrac{\partial P}{\partial y}$ and vice-versa.

Example 7.1.7

Evaluate $\displaystyle\int_{(0,0)}^{(2,1)} (10x^4 - 2xy^3)dx - 3x^2y^2\,dy$ along the path $x^4 - 6xy^3 = 4y^2$.

Solution

Here, $P = 10x^4 - 2xy^3$, so $\dfrac{\partial P}{\partial y} = -6xy$ and $Q = -3x^2y^2$, and so $\dfrac{\partial Q}{\partial x} = -6xy$. Since $\dfrac{\partial P}{\partial y} = \dfrac{\partial Q}{\partial x}$, the integral is independent of the path and the differential $(10x^4 - 2xy^3)dx - 3x^2y^2\,dy$ is an exact differential. Thus

$$\int_{(0,0)}^{(2,1)} (P\,dx + Q\,dy) = \int_{(0,0)}^{(2,1)} d(2x^5 - x^2y^3) = \left[2x^5 - x^2y^3\right]_{(0,0)}^{(2,1)} = 60$$

Example 7.1.8

Let $\vec{F} = \dfrac{1}{x^2+y^2}(-y\hat{\imath} + x\hat{\jmath})$. Calculate $\vec{\nabla}\times\vec{F}$. Evaluate $\oint \vec{F}\cdot d\vec{r}$ around any closed path and explain the results.

Solution

Here, $\vec{F} = \dfrac{1}{x^2+y^2}(-y\hat{\imath} + x\hat{\jmath}) = -\dfrac{y}{r^2}\hat{\imath} + \dfrac{x}{r^2}\hat{\jmath}$. Using the definition of curl

$$\vec{\nabla} \times \vec{F} = \begin{vmatrix} \hat{i} & \hat{j} & \hat{k} \\ \dfrac{\partial}{\partial x} & \dfrac{\partial}{\partial y} & \dfrac{\partial}{\partial z} \\ -y/r^2 & x/r^2 & 0 \end{vmatrix} = \vec{0}$$

in any region excluding $(0, 0)$. Let $x = \rho \cos \theta$ and $y = \rho \sin \theta$, where (ρ, θ) are polar coordinates. Then

$$dx = -\rho \sin \theta + d\rho \ \cos \theta \text{ and } dy = \rho \cos \theta d\theta + d\rho \sin \theta$$

$$\Rightarrow \ \vec{F} \cdot d\vec{r} = \frac{-y\, dx + x\, dy}{x^2 + y^2} = d\theta = d\left(\tan^{-1} \frac{y}{x}\right)$$

For a closed curve $ABCDA$ (Fig. 7.8(a)) surrounding the origin, $\theta = 0$ at A and $\theta = 2\pi$ after a complete circuit back to A. In this case

$$\oint \vec{F} \cdot d\vec{r} = \int_0^{2\pi} d\theta = \Big[\theta\Big]_0^{2\pi} = 2\pi$$

For a closed curve $PQRSP$ (Fig. 7.8(b)) not surrounding the origin, $\theta = \theta_0$ at P and $\theta = \theta_0$ after

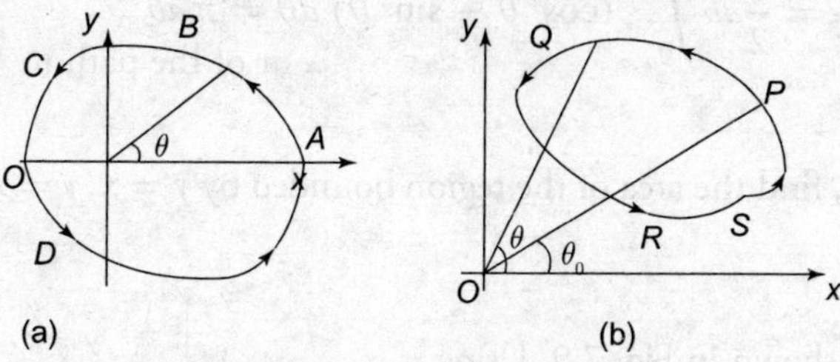

Figure 7.8 Closed curve

a complete circuit back to P. In this case

$$\oint \vec{F} \cdot d\vec{r} = \int_{\theta_0}^{\theta_0} d\theta = \Big[\theta\Big]_{\theta_0}^{\theta_0} = 0$$

Since $\vec{F} = P\hat{i} + Q\hat{j}$, $\vec{\nabla} \times \vec{F} = \vec{0}$ is equivalent to $\dfrac{\partial P}{\partial y} = \dfrac{\partial Q}{\partial x}$ and Theorem 7.1. However, no contradiction exists as $P = -\dfrac{y}{r}$ and $Q = \dfrac{x}{r}$ do not have continuous derivatives throughout any region including $(0, 0)$.

RESULT 22

In Green's theorem, let $P = -y$ and $Q = x$. Also, let A be the area enclosed by the closed curve C. Then, using Green's theorem (7.1), we get

$$\oint_C -y\, dx + x\, dy = \int\int_A \Big[1 - (-1)\Big]\, dx\, dy = 2\, A$$

$$\Rightarrow \ A = \frac{1}{2} \oint_C x\, dy - y\, dx \tag{7.7}$$

In polar coordinates, $x = r\cos\theta$ and $y = r\sin\theta$, then $dx = dr\cos\theta - r\sin\theta\, d\theta$ and $dy = dr\sin\theta + r\cos\theta\, d\theta$. Thus, Eq. (7.7) becomes

$$A = \frac{1}{2}\oint_C \left[r\cos\theta(dr\sin\theta + r\cos\theta\, d\theta) - r\sin\theta(dr\cos\theta - r\sin\theta\, d\theta)\right]$$

$$= \frac{1}{2}\oint_C \left[r\cos\theta \ \sin\theta\, dr + r^2\cos^2\theta\, d\theta - r\sin\theta \ \cos\theta\, dr + r^2\sin^2\theta\, d\theta\right]$$

$$= \frac{1}{2}\oint_C \left[r^2(\cos^2\theta + \sin^2\theta)\right] d\theta = \frac{1}{2}\oint_C r^2\, d\theta \tag{7.8}$$

Example 7.1.9

Show that the area of the ellipse $x = a\cos\theta$ and $y = b\sin\theta$ is πab.

Solution

Here, $x = a\cos\theta$ and $y = b\sin\theta$. Using Eq. (7.7), we get the required area as

$$A = \frac{1}{2}\int_0^{2\pi} \left[a\cos\theta\, b\cos\theta - b\sin\theta(-a\sin\theta)\right] d\theta$$

$$= \frac{1}{2}ab\int_0^{2\pi} (\cos^2\theta + \sin^2\theta)\, d\theta = \pi ab$$

Example 7.1.10

Using Green's theorem, find the area of the region bounded by $y = x, y = x^{-1}, x = 4y$ in the first quadrant.

Solution

The required region is shown in Fig. 7.9. Using Green's theorem, the area enclosed by a curve C is given by Eq. (7.7). Now, along the line OA, $C_1 : x = 4y, dx = 4dy$ and $0 \le x \le 2$. So

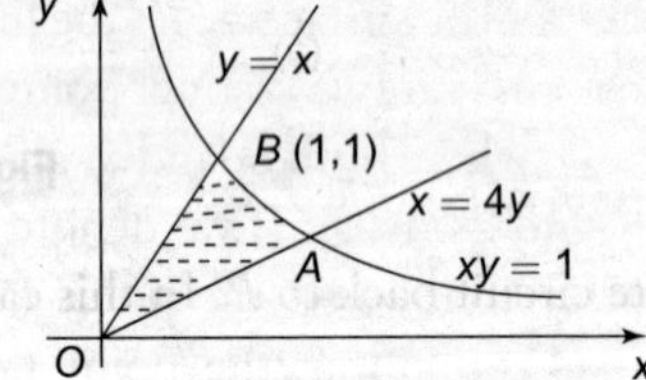

Figure 7.9 S is the shaded region

$$\frac{1}{2}\oint_{C_1} x\,dy - y\,dx = \frac{1}{2}\int_0^2 \left[\frac{x}{4} - \frac{x}{4}\right] dx = 0$$

Along the line AB, $C_2 : xy = 1, y\,dx = -x\,dy$ and x varies from 2 to 1. So

$$\frac{1}{2}\oint_{C_2} x\,dy - y\,dx = -\frac{1}{2}\int_2^1 \frac{2}{x}\, dx = \log 2$$

Finally, along the line BO, $C_3 : y = x, dx = dy$ and x varies from 1 to 0. So

$$\frac{1}{2}\oint_{C_3} x\,dy - y\,dx = \frac{1}{2}\int_1^0 0\, dx = 0$$

Adding, we get the required area as

$$A = \frac{1}{2}\oint_C x\,dy - y\,dx = 0 + \log 2 + 0 = \log 2$$

RESULT 23

If the arc length dr is replaced by dS (differential of surface area), curve $\mathcal{C}$ by $\mathcal{S}$ and the region $\mathcal{R}$ by volume V enclosed by $\mathcal{S}$, Green's theorem in space is of the form

$$\int\int_{\mathcal{C}} \vec{A} \cdot \hat{n}\, dS = \int\int\int_{V} \vec{\nabla} \cdot \vec{A}\, dV$$

This is the expression for Gauss's theorem of divergence. Thus, a generalised Green's theorem (7.5) leads to Gauss's divergence theorem.

7.2 GAUSS'S DIVERGENCE THEOREM

Gauss's divergence theorem is one of the most important theorems of vector analysis and it relates volume integrals with surface integrals.

Statement: Let V be the volume bounded by a closed piecewise, smooth surface $\mathcal{S}$. Let $\vec{A}$ be a vector function of position which has continuous first partial derivatives in V. Then

$$\int\int\int_{V} \vec{\nabla} \cdot \vec{A}\, dV = \int\int_{S} \vec{A} \cdot \hat{n}\, dS = \oint\oint_{S} \vec{A} \cdot d\vec{S} \tag{7.9}$$

where $\hat{n}$ is the outward drawn unit normal vector to the surface $\mathcal{S}$.

Proof: Let us consider a closed surface $\mathcal{S}$ enclosing the volume V in such a way that any straight line parallel to the coordinate axes cuts the surface $\mathcal{S}$ at a maximum of two points (Fig. 7.10). For the surface $\mathcal{S}$, let the line parallel to the z-axis cut it at two distinct points (x, y, z_1) and (x, y, z_2), where both z_1 and z_2 are functions of x and y and $z_1 < z_2$. If we denote the lower portion of $\mathcal{S}$ containing the point (x, y, z_1) as $\mathcal{S}_1$ and that containing the point (x, y, z_2) as $\mathcal{S}_2$, then the closed surface for which $z_1 = z_2$ separates $\mathcal{S}_1$ from $\mathcal{S}_2$. From Fig. 7.10, it is clear that the normal $\hat{n}_2$ to the surface $\mathcal{S}_2$ makes an acute angle with $\hat{k}$ but the normal $\hat{n}_1$ to the surface $\mathcal{S}_1$ makes an obtuse angle with $\hat{k}$. Hence, we can write

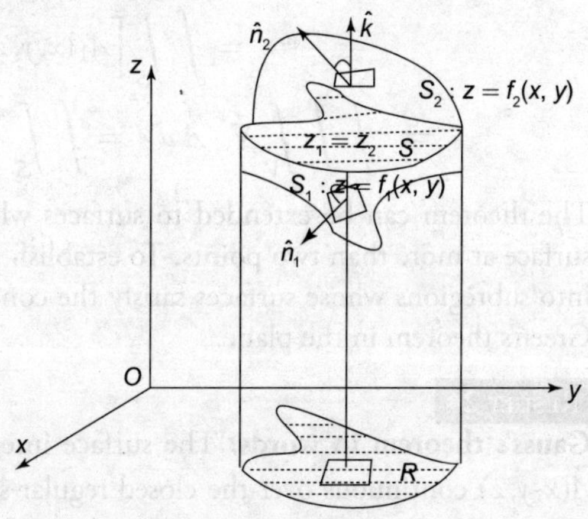

Figure 7.10 Gauss's divergence theorem

$$dx\, dy = \hat{n}_2 \cdot \hat{k}\, dS_2 \text{ and } dx\, dy = -\hat{n}_1 \cdot \hat{k}\, dS_1$$

Let $\vec{A} = A_1\hat{i} + A_2\hat{j} + A_3\hat{k}$ and consider the integral $\displaystyle\int\int\int_V \frac{\partial A_3}{\partial z}\, dxdydz$. If $\mathcal{R}$ be the projection of the surface $\mathcal{S}$ on the xy plane, then

$$\int\int\int_{V} \frac{\partial A_3}{\partial z}\, dxdydz = \int\int_{\mathcal{R}} \left[\int_{z=f_1(x,y)}^{z=f_2(x,y)} \frac{\partial A_3}{\partial z}\, dz\right] dx\, dy$$

$$= \int\int_{\mathcal{R}} \Big[A_3(x,y,z)\Big]_{f_1(x,y)}^{f_2(x,y)} dx\, dy$$
$$= \int\int_{\mathcal{R}} \Big[A_3(x,y,f_2) - A_3(x,y,f_1)\Big] dx\, dy$$
$$= \int\int_{S_2} \Big[A_3(x,y,z)\hat{k}\cdot\hat{n}_2\Big] dS_2 + \int\int_{S_1} \Big[A_3(x,y,z)\hat{k}\cdot\hat{n}_1\Big] dS_1$$
$$= \int\int_{S} \Big[A_3(x,y,z)\hat{k}\cdot\hat{n}\Big] dS$$

Similarly, considering the projections of $\mathcal{S}$ on the *xz* plane and *yz* plane, we obtain

$$\int\int\int_V \frac{\partial A_2}{\partial y}\, dxdydz = \int\int_S \Big[A_2(x,y,z)\hat{j}\cdot\hat{n}\Big] dS$$
$$\text{and} \quad \int\int\int_V \frac{\partial A_1}{\partial x}\, dxdydz = \int\int_S \Big[A_1(x,y,z)\hat{i}\cdot\hat{n}\Big] dS$$

Adding the above results, we get

$$\int\int\int_V \left[\frac{\partial A_1}{\partial x} + \frac{\partial A_2}{\partial y} + \frac{\partial A_3}{\partial z}\right] dxdydz$$
$$= \int\int_S \Big[A_1(x,y,z)\hat{i} + A_2(x,y,z)\hat{j} + A_3(x,y,z)\hat{k}\Big]\cdot\hat{n} dS$$
$$\Rightarrow \int\int\int_V \vec{\nabla}\cdot\vec{A}\, dV = \int\int_S \vec{A}\cdot\hat{n}\, dS$$

The theorem can be extended to surfaces where lines parallel to the coordinate axes meet the surface at more than two points. To establish this extension, subdivide the region bounded by $\mathcal{S}$ into subregions whose surfaces satisfy the condition. The procedure is analogous to that used in Green's theorem in the plane.

RESULT 24

Gauss's theorem in words: The surface integral of the normal component of vector function $\vec{A}(x,y,z)$ continuous over the closed regular surface $\mathcal{S}$ and its interior is equal to the integral of divergence of $\vec{A}$ taken over the volume V enclosed by the surface; it is assumed that $\vec{\nabla}\cdot\vec{A}$ is continuous in V.

Example 7.2.1

Verify the divergence theorem for $\vec{A} = 2x^2y\hat{i} - y^2\hat{j} + 4xz^2\hat{k}$ taken over the region in the first octant bounded by $y^2 + z^2 = 9$ and $x = 2$.

Solution

Evaluation of the surface integral: Suppose that the surface $\mathcal{S}$ has a projection R_3 on the *xz* plane, as shown in Fig. 7.11. The surface $\mathcal{S}$ of the cylinder consists of a base $S_1(x = 0)$, a top $S_2(x = 2)$ and convex position $S_3(y^2 + z^2 = 9)$. On $S_1(x = 0)$: $\hat{n} = -\hat{i}$ so, $\vec{A}\cdot\hat{n} = -2x^2y$. Therefore

$$\int\int_{S_1} \vec{A}\cdot\hat{n}dS_1 = -2\int_0^3 dy\int_0^{\sqrt{9-y^2}} x^2ydz = 0, \text{ as } x=0$$

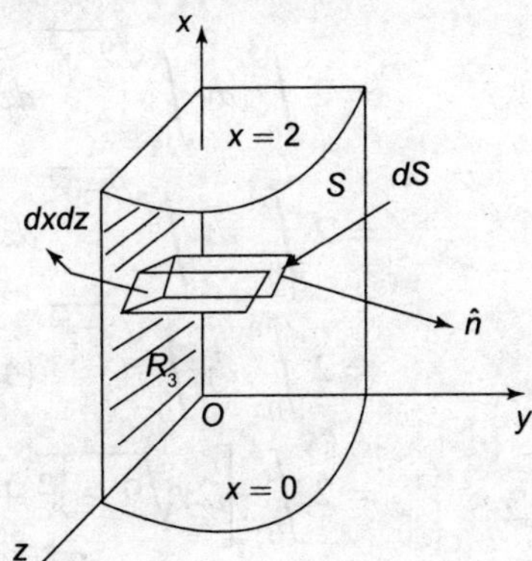

Figure 7.11 Surface integrals

On $S_2(x=2)$, $\hat{n}=\hat{i}$, so, $\vec{A}\cdot\hat{n} = 2x^2y$. Therefore

$$\int\int_{S_2} \vec{A}\cdot\hat{n}dS_2 = 8\int_0^3 dy\int_0^{\sqrt{9-y^2}} ydz$$

$$= 8\int_0^3 y\sqrt{9-y^2}dy; \quad \text{put } 9-y^2=t^2$$

$$= 8\int_0^3 t^2dt = \frac{8}{3}\left[t^3\right]_0^3 = 72$$

On $S_3(y^2+z^2=9)$: a normal to $y^2+z^2=9$ is $\vec{\nabla}(y^2+z^2) = 2(y\hat{j}+z\hat{k})$. Thus, the unit vector is given by

$$\hat{n} = \frac{1}{|\vec{\nabla}\phi|}\vec{\nabla}\phi = \frac{1}{\sqrt{y^2+z^2}}(\widehat{yj}+\widehat{zk}) = \frac{1}{3}(\widehat{yj}+\widehat{zk}) \Rightarrow \vec{A}\cdot\hat{n} = \frac{1}{3}(-y^3+4xz^3)$$

Projecting S_3 on the xy plane, the surface integral becomes

$$\int\int_{S_3} \vec{A}\cdot\hat{n}dS_3 = \int\int_{R_1} \vec{A}\cdot\hat{n}\frac{dx\,dy}{|\hat{n}\cdot\hat{k}|} = \int_0^2 dx\int_0^3\left[4x(9-y^2) - \frac{y^3}{\sqrt{9-y^2}}\right]dy$$

$$= \int_0^3\left[2x^2(9-y^2) - x\frac{y^3}{\sqrt{9-y^2}}\right]_0^2 dy$$

$$= \int_0^3\left[8(9-y^2) - \frac{2y^3}{\sqrt{9-y^2}}\right]dy$$

$$= \left[8\left(9y-\frac{y^3}{3}\right) + 18\sqrt{9-y^2} - \frac{2}{3}(9-y^2)^{3/2}\right]_0^3 = 108$$

Thus, the surface integral becomes

$$\int\int_S \vec{A}\cdot\hat{n}dS = \int\int_{S_1} \vec{A}\cdot\hat{n}\,dS_1 + \int\int_{S_2} \vec{A}\cdot\hat{n}dS_2 + \int\int_{S_3} \vec{F}\cdot\hat{n}dS_3$$

$$= 0 + 72 + 108 = 180$$

Evaluation of the volume integral: The volume integral is

$$\int\int\int_V \vec{\nabla}\cdot\vec{A}\,dV = \int\int\int_V\left[\frac{\partial}{\partial x}(2x^2y) + \frac{\partial}{\partial y}(-y^2) + \frac{\partial}{\partial z}(4xz^2)\right]dV$$

$$= \iiint_V [4xy - 2y + 8xz]\, dV$$

$$= 2\int_0^3 dy \int_0^{\sqrt{9-y^2}} dz \int_0^2 (2xy - y + 4xz) dx$$

$$= 2\int_0^3 dy \int_0^{\sqrt{9-y^2}} dz \left[x^2y - xy + 2x^2 z\right]_0^2$$

$$= 2\int_0^3 dy \int_0^{\sqrt{9-y^2}} (4y - 2y + 8z)\, dz = 2\int_0^3 dy \left[2yz + 4z^2\right]_0^{\sqrt{9-y^2}}$$

$$= 2\int_0^3 \left[2y\sqrt{9-y^2} + 4(9-y^2)\right] dy \qquad \text{put } \; 9 - y^2 = t^2$$

$$= 4\int_0^3 t^2 dt + 2\left[36y - \frac{4}{3}y^3\right]_0^3 = \left[\frac{4t^3}{3}\right]_0^3 + 2\,(108 - 36) = 180$$

Hence, Gauss's divergence theorem (7.9) is verified.

Example 7.2.2

Verify the divergence theorem for $\vec{A} = 4x\hat{i} - 2y\hat{j} + z\hat{k}$ taken over the region bounded by $x^2 + y^2 = 9$, $z = 0$ and $z = 4$.

Solution

Evaluation of the surface integral: The surface $\mathcal{S}$ is shown in Fig. 7.12. The surface $\mathcal{S}$ of the cylinder consists of a base $\mathcal{S}_1(z = 0)$, a top $\mathcal{S}_2(z = 4)$ and convex position $\mathcal{S}_3(x^2 + y^2 = 9)$. On $\mathcal{S}_1(z = 0)$: $\hat{n} = -\hat{k}$ so, $\vec{A} \cdot \hat{n} = 0$. Therefore

$$\iint_{\mathcal{S}_1} \vec{A} \cdot \hat{n}\; dS_1 = 0, \text{ as on } \mathcal{S}_1, z = 0$$

i.e., the surface integral vanishes. On $\mathcal{S}_2(z = 4)$, $\hat{n} = \hat{k}$, so $\vec{A} \cdot \hat{n} = 4$. Therefore

$$\iint_{\mathcal{S}_2} \vec{A} \cdot \hat{n}\, dS_2 = 4 \times 4 \int_0^3 dx \int_0^{\sqrt{9-x^2}} dy$$

$$= 16 \int_0^3 \sqrt{9 - x^2} dx$$

$$= 16 \left[\frac{x}{2}\sqrt{9 - x^2} + \frac{9}{2}\sin^{-1}\frac{x}{3}\right]_0^3 = 36\pi$$

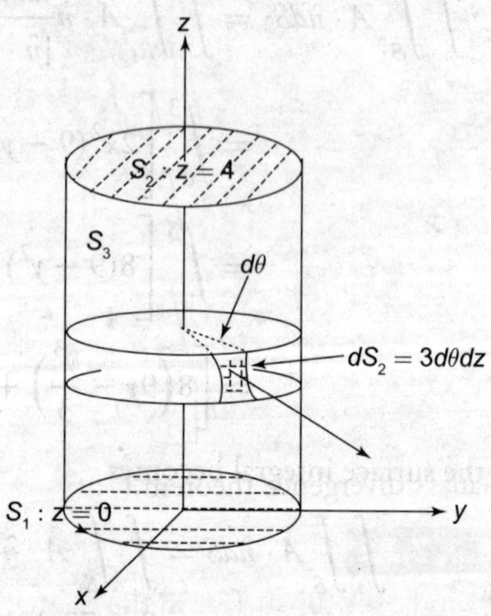

Figure 7.12 Surface integrals

On $\mathcal{S}_3(x^2+y^2=9)$: a normal to $x^2+y^2=9$ is $\vec{\nabla}(x^2+y^2)=2(x\hat{i}+y\hat{j})$. Thus, the unit normal vector is given by

$$\hat{n}=\frac{1}{|\vec{\nabla}\phi|}\vec{\nabla}\phi=\frac{1}{\sqrt{x^2+y^2}}(x\hat{i}+y\hat{k})=\frac{1}{3}(x\hat{i}+y\hat{j})\Rightarrow\vec{A}\cdot\hat{n}=\frac{2}{3}(2x^2-y^2)$$

To calculate the surface integral over the curved surface, we consider $x=3\cos\theta, y=3\sin\theta$ and consider a small element $dS_3=3d\theta dz$ on the surface at a point $(3\cos\theta, 3\sin\theta)$. Hence, the surface integral over the curved surface $\mathcal{S}_3$ is equal to

$$\int\int_{S_3}\vec{A}\cdot\hat{n}\,dS_3=\int_{z=0}^{4}\left[\int_{\theta=0}^{2\pi}(36\cos^2\theta-18\sin^2\theta)d\theta\right]dz=72\pi$$

Hence, the surface integral over the whole surface of the given region is

$$\begin{aligned}\int\int_{S}\vec{A}\cdot\hat{n}dS&=\int\int_{S_1}\vec{A}\cdot\hat{n}\,dS_1+\int\int_{S_2}\vec{A}\cdot\hat{n}dS_2+\int\int_{S_3}\vec{F}\cdot\hat{n}dS_3\\&=0+36\pi+72\pi=108\pi\end{aligned}$$

Evaluation of the volume integral: If V is the volume of the region bounded by the surface $\mathcal{S}:x^2+y^2=9, z=0$ and $z=4$, then the volume integral is

$$\begin{aligned}\int\int\int_V\vec{\nabla}\cdot\vec{A}\,dV&=\int\int\int_V\left[\frac{\partial}{\partial x}(4x)+\frac{\partial}{\partial y}(-2y)+\frac{\partial}{\partial z}(z)\right]dV\\&=3\int\int\int_V dx\,dy\,dz=3\int_{x=-3}^{3}\left[\int_{y=-\sqrt{9-x^2}}^{\sqrt{9-x^2}}\left\{\int_{z=0}^{4}dz\right\}dy\right]dx\\&=12\int_{x=-3}^{3}\left[\int_{y=-\sqrt{9-x^2}}^{\sqrt{9-x^2}}dy\right]dx\\&=24\int_{x=-3}^{3}\sqrt{9-x^2}\,dx=48\int_{x=0}^{3}\sqrt{9-x^2}\,dx\\&=48\left[\frac{x}{2}\sqrt{9-x^2}+\frac{9}{2}\sin^{-1}\frac{x}{3}\right]_0^3=108\pi\end{aligned}$$

Hence, Gauss's divergence theorem (7.9) is verified.

DEDUCTION 7.2.1

Let $\vec{A}=x\hat{i}$ and $\mathcal{S}$ be the surface in the yz plane. By using Gauss's divergence theorem (7.9), we get

$$\int\int_{S}\vec{A}\cdot\hat{n}dS=\int\int_{S}x\,dy\,dz=\int\int\int_V dV=V$$

Similarly, we get $\int\int_S y\,dz\,dx=V=\int\int_S z\,dx\,dy$. Adding, we get

$$V+V+V=\int\int_S(x\,dy\,dz+y\,dz\,dx+z\,dx\,dy)$$

$$\Rightarrow \; V = \frac{1}{3}\int\int_S (x\,dy\,dz + y\,dz\,dx + z\,dx\,dy) \tag{7.10}$$

Example 7.2.3

Using Gauss's divergence theorem, evaluate the integral $\int\int_S \vec{A}\cdot\hat{n}\,dS$, where $\vec{A} = 4xz\hat{i} - y^2\hat{j} + 4yz\hat{k}$, S is the surface of the solid bounded by the sphere $x^2 + y^2 + z^2 = 10$ and the paraboloid $x^2 + y^2 = z - 2$, and $\hat{n}$ is the outward unit normal vector to S.

Solution

Here, V (Fig. 7.13) is the volume of the solid $ABCD$ bounded by the sphere $x^2 + y^2 + z^2 = 10$ and the paraboloid $x^2 + y^2 = z - 2$ above the xy plane. The region is bounded by $x^2 + y^2 + z^2 = 10$ and $x^2 + y^2 = z - 2$. Given $\vec{A} = 4xz\vec{i} - y^2\vec{j} + 4yz\vec{k}$, using the definition of divergence, $\vec{\nabla}\cdot\vec{A} = 4z + 2y$.

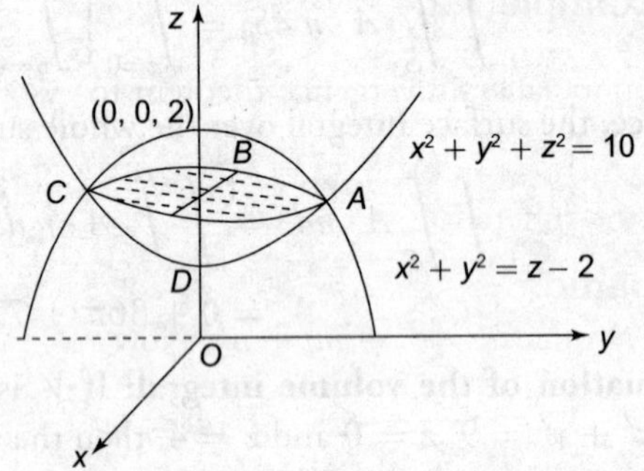

Figure 7.13 Region of integration

To evaluate the volume integral, we first integrate with respect to z parallel to the z axis from $z = x^2 + y^2 + 2(\geq 2)$ to $z = \sqrt{10 - x^2 - y^2}$. The region bounded by the sphere $x^2 + y^2 + z^2 = 10$ and the paraboloid $x^2 + y^2 = z - 2$ is given by

$$\begin{aligned}
& x^2 + y^2 + (x^2 + y^2 + 2)^2 = 10 \\
\Rightarrow\; & (x^2 + y^2) + (x^2 + y^2)^2 + 4(x^2 + y^2) + 4 = 10 \\
\Rightarrow\; & (x^2 + y^2 + 6)(x^2 + y^2 - 1) = 0; \text{ as } x^2 + y^2 + 6 \neq 0 \\
\Rightarrow\; & x^2 + y^2 = 1
\end{aligned}$$

Thus, we integrate with respect to y from $y = -\sqrt{1 - x^2}$ to $y = \sqrt{1 - x^2}$ parallel to the y axis. Finally, we integrate with respect to x from $x = -1$ to $x = 1$ along the x axis. From Gauss's divergence theorem (7.9), we get

$$\begin{aligned}
\int\int_S \vec{A}\cdot\hat{n}\,dS &= \int\int\int_V (\vec{\nabla}\cdot\vec{A})dV = \int\int\int_V (4x + 2y)dx\,dy\,dz \\
&= \int_{-1}^{1}\left[\int_{-\sqrt{1-x^2}}^{\sqrt{1-x^2}} \left\{\int_{x^2+y^2+2}^{\sqrt{10-x^2-y^2}} (4z + 2y)dz\right\}dy\right]dx \\
&= \int_{-1}^{1}\left[\int_{-\sqrt{1-x^2}}^{\sqrt{1-x^2}} 2\left\{(10 - x^2 - y^2) - (x^2 + y^2 + 2)^2\right\}dy\right]dx
\end{aligned}$$

$$= 4\int_0^1 \left[\int_0^{\sqrt{1-x^2}} 2\left\{6 - 5(x^2+y^2) - (x^2+y^2)^2\right\} dy\right] dx$$

$$\left[\text{as } y \text{ is an odd function}\right]$$

$$= 8\int_0^1 \left[\int_0^{\pi/2} (6 - 5r^2 - r^4) r\, d\theta\right] dr; \; x = r\cos\theta, y = r\sin\theta$$

$$= 8\cdot\frac{\pi}{2}\left[3 - \frac{5}{4} - \frac{1}{6}\right] = \frac{19}{3}\pi$$

Example 7.2.4

Apply Gauss's divergence theorem to evaluate $\int\int_S (lx^2 + my^2 + nz^2) dS$, taken over the sphere $(x-a)^2 + (y-b)^2 + (z-c)^2 = \rho^2$; l, m and n being the direction cosines of the external normal to the sphere.

Solution

Given that $\hat{n} = l\hat{i} + m\hat{j} + n\hat{k}$. Now

$$\vec{F}\cdot\hat{n} = (x^2\hat{i} + y^2\hat{j} + z^2\hat{k})\cdot(l\hat{i} + m\hat{j} + n\hat{k})$$

so that $\vec{F} = x^2\hat{i} + y^2\hat{j} + z^2\hat{k}$. The parametric equation of the sphere is $x = a + r\sin\theta\cos\phi$, $y = b + r\sin\theta\sin\phi$, $z = c + r\cos\theta$ and to cover the whole sphere, then $0 \le r \le \rho$, $0 \le \theta \le \pi$ and $0 \le \phi \le 2\pi$. Therefore, using Gauss's divergence theorem

$$\int\int_S \vec{F}\cdot\hat{n} dS = \int\int\int_V \vec{\nabla}\cdot(x^2\hat{i} + y^2\hat{j} + z^2\hat{k}) dV$$

$$= 2\int\int\int_V (x+y+z) dx\, dy\, dz, \text{ using Eq. (7.9)}$$

$$= 2\int_{\phi=0}^{2\pi}\int_{\theta=0}^{\pi}\int_{r=0}^{\rho} (a+b+c) r^2 \sin\theta\, dr\, d\theta\, d\phi$$

$$+2\int_0^{2\pi}\int_0^{\pi}\int_0^{\rho} r(\sin\theta\cos\phi + \sin\theta\sin\phi + \cos\theta) r^2 \sin\theta\, dr d\theta\, d\phi$$

$$= 2(a+b+c)\cdot 2\pi\cdot\Big[-\cos\theta\Big]_0^{\pi}\cdot\frac{\rho^3}{3} + 0 = \frac{8\pi}{3}(a+b+c)\rho^3$$

RESULT 25

Cartesian representation of Gauss's divergence theorem: Let $\vec{A}(x,y,z) = A_1(x,y,z)\hat{i} + A_2(x,y,z)\hat{j} + A_3(x,y,z)\hat{k}$, then

$$\text{div}\,\vec{A} = \vec{\nabla}\cdot\vec{A} = \frac{\partial A_1}{\partial x} + \frac{\partial A_2}{\partial y} + \frac{\partial A_3}{\partial z}$$

The unit normal to $\mathcal{S}$ is $\hat{n} = n_1\hat{i} + n_2\hat{j} + n_3\hat{k}$. Let α, β and γ be the angles which $\hat{n}$ makes with the positive x, y and z axes or the $\hat{i}, \hat{j}$ and $\hat{k}$ directions, respectively. Now

$$n_1 = \hat{n}\cdot\hat{i} = \cos\alpha, \quad n_2 = \hat{n}\cdot\hat{j} = \cos\beta, \quad n_3 = \hat{n}\cdot\hat{k} = \cos\gamma$$

The quantities $\cos\alpha, \cos\beta$ and $\cos\gamma$ are called the direction cosines of $\hat{n}$. Let

$$d\vec{S} = \hat{n}dS = (\cos\alpha\hat{i} + \cos\beta\hat{j} + \cos\gamma\hat{k})\, dS$$

Thus, $dS\cos\alpha, dS\cos\beta$ and $dS\cos\gamma$ are the orthogonal projections of the elementary area dS on the yz plane, the zx plane and the xy plane, respectively. If we make a subdivision of $\mathcal{S}$ by planes parallel to the yz plane, the zx plane and the xy plane, then the projections on the coordinate planes will be rectangular with sides dy and dz on the yz plane, dz and dx on the zx plane, and dx and dy on the xy plane. Hence, the projected surface elements are $dy\, dz$, $dz\, dx$ and $dx\, dy$. Therefore

$$\begin{aligned}\vec{A}\cdot\hat{n} &= \left[A_1\hat{i} + A_2\hat{j} + A_3\hat{k}\right]\cdot\left[n_1\hat{i} + n_2\hat{j} + n_3\hat{k}\right]\\ &= \left[A_1\hat{i} + A_2\hat{j} + A_3\hat{k}\right]\cdot\left[\cos\alpha\hat{i} + \cos\beta\hat{j} + \cos\gamma\hat{k}\right]\\ &= A_1\cos\alpha + A_2\cos\beta + A_3\cos\gamma\end{aligned}$$

Thus, Gauss's divergence theorem (7.9) becomes

$$\begin{aligned}\iiint_V \left[\frac{\partial A_1}{\partial x} + \frac{\partial A_2}{\partial y} + \frac{\partial A_3}{\partial z}\right] dxdydz &= \iint_S \left[A_1\cos\alpha + A_2\cos\beta + A_3\cos\gamma\right] dS\\ &= \iint_S A_1\, dy\, dz + A_2\, dz\, dx + A_3\, dx\, dy \qquad (7.11)\end{aligned}$$

Example 7.2.5

By using Gauss's divergence theorem, evaluate $I = \iint_S xz^2 dydz + (x^2y - z^3)dzdx + (2xy + y^2z)dxdy$, where $\mathcal{S}$ is the surface of the hemispherical region bounded by $z = \sqrt{a^2 - x^2 - y^2}$ and $z = 0$.

Solution

Here, $\mathcal{S}$ is a closed hemispherical surface bounding the volume V (Fig. 7.14). The given integral is of the form $\iint_S A_1\, dy\, dz + A_2\, dz\, dx + A_3\, dx\, dy$. Using Eq. (7.11), we get

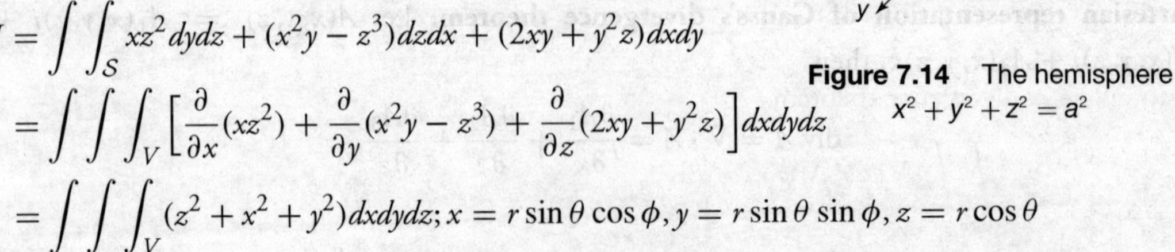

$$\begin{aligned}I &= \iint_S xz^2 dydz + (x^2y - z^3)dzdx + (2xy + y^2z)dxdy\\ &= \iiint_V \left[\frac{\partial}{\partial x}(xz^2) + \frac{\partial}{\partial y}(x^2y - z^3) + \frac{\partial}{\partial z}(2xy + y^2z)\right] dxdydz\\ &= \iiint_V (z^2 + x^2 + y^2)dxdydz; x = r\sin\theta\cos\phi, y = r\sin\theta\sin\phi, z = r\cos\theta\end{aligned}$$

Figure 7.14 The hemisphere $x^2 + y^2 + z^2 = a^2$

$$= \int_{\phi=0}^{2\pi} \int_{\theta=0}^{\pi/2} \int_{0}^{a} r^2 \cdot r^2 \sin\theta \; dr d\theta d\phi = \frac{2\pi a^5}{5}$$

Result 26

Let V be the region enclosed by the surface $\mathcal{S}$ and $\hat{n}$ be the unit outward drawn normal to it. Since $\vec{r} = x\hat{i} + y\hat{j} + z\hat{k}$

$$\vec{\nabla} \cdot \vec{r} = \frac{\partial x}{\partial x} + \frac{\partial y}{\partial y} + \frac{\partial z}{\partial z} = 3$$

Now, using Gauss's divergence theorem (7.9), we get

$$\int\int_{\mathcal{S}} \vec{r} \cdot \hat{n} \, dS = \int\int\int_{V} \vec{\nabla} \cdot \vec{r} \, dV = 3 \int\int\int dV = 3V \tag{7.12}$$

Example 7.2.6

If $\mathcal{S}$ is any closed surface enclosing a volume V and $\vec{A} = ax\hat{i} + by\hat{j} + cz\hat{k}$, then prove that $\int\int_{\mathcal{S}} \vec{A}.\widehat{n}ds = (a+b+c)V$.

Solution

Using Gauss's divergence theorem (7.9), we have

$$\int\int_{\mathcal{S}} \vec{A} \cdot \widehat{n} \, dS = \int\int\int_{V} (\vec{\nabla} \cdot \vec{A}) \, dV = \int\int\int_{V} \left[\frac{\partial}{\partial x}(ax) + \frac{\partial}{\partial y}(by) + \frac{\partial}{\partial z}(cz) \right] dV$$

$$= (a+b+c) \int\int\int_{V} dV = (a+b+c)V$$

since V is the volume bounded by the closed surface $\mathcal{S}$.

Result 27

Let V be the region enclosed by the surface $\mathcal{S}$ and $\hat{n}$ be the unit outward drawn normal to it. Using Gauss's divergence theorem (7.9), we get

$$\int\int\int_{V} \vec{\nabla} \cdot \hat{n} \, dV = \int\int_{\mathcal{S}} \hat{n} \cdot \hat{n} \, dS = \int\int_{\mathcal{S}} dS = S \tag{7.13}$$

Result 28

Let V be the region enclosed by the surface $\mathcal{S}$ and $\hat{n}$ be the unit outward drawn normal to it. Let us take the result

$$\vec{\nabla} \cdot (\vec{\nabla}\phi \times \vec{\nabla}\psi) = \vec{\nabla}\psi \cdot (\vec{\nabla} \times \vec{\nabla}\phi) - \vec{\nabla}\phi \cdot (\vec{\nabla} \times \vec{\nabla}\psi)$$

$$= 0, \text{ as } \vec{\nabla} \times \vec{\nabla}\psi = \vec{0}$$

Using Gauss's divergence theorem (7.9), we get

$$\int\int_{\mathcal{S}} (\vec{\nabla}\phi \times \vec{\nabla}\psi) \cdot \hat{n} \, dS = \int\int\int_{V} \vec{\nabla} \cdot (\vec{\nabla}\phi \times \vec{\nabla}\psi) \, dV = 0 \tag{7.14}$$

RESULT 29

Let V be the region enclosed by the surface $\mathcal{S}$ and $\hat{n}$ be the unit outward drawn normal to it. Using the definition of divergence, we get

$$\vec{\nabla}\left(\frac{1}{r^2}\vec{r}\right) = \vec{\nabla}\left(\frac{1}{r^2}\right)\cdot\vec{r} + \frac{1}{r^2}\vec{\nabla}\cdot\vec{r}$$

$$= -2\frac{1}{r^3}\,\frac{1}{r}\vec{r}\cdot\vec{r} + \frac{1}{r^2}\vec{\nabla}\cdot\vec{r} = -\frac{2r^2}{r^4} + \frac{3}{r^2} = \frac{1}{r^2}$$

Using Gauss's divergence theorem (7.9), we get

$$\iint_S \frac{\vec{r}\cdot\hat{n}}{r^2}\,dS = \iiint_V \vec{\nabla}\cdot\left(\frac{1}{r^2}\vec{r}\right)dV = \iiint_V \frac{dV}{r^2} \tag{7.15}$$

Theorem 7.2

If ϕ be a scalar point function defined in the region V enclosed by a surface $\mathcal{S}$ such that its first order partial derivatives be finite, single-valued and continuous throughout V and $\mathcal{S}$, then

$$\iiint_V \vec{\nabla}\phi\ dV = \iint_S \phi\ \hat{n}\ dS \tag{7.16}$$

where $\hat{n}$ is the unit outward vector to $\mathcal{S}$.

Proof: In Gauss's divergence theorem (7.9), let us take $\vec{A} = \phi\,\vec{C}$, where $\vec{C}$ is a non-null constant vector and ϕ is a scalar function. Therefore

$$\iiint_V \vec{\nabla}\cdot\vec{A}\,dV = \iint_S \vec{A}\cdot\hat{n}\,dS = \iint_S \phi\vec{C}\cdot\hat{n}\,dS$$

$$\Rightarrow \iiint_V \vec{\nabla}\cdot(\phi\vec{C})\,dV = \iint_S \vec{C}\cdot(\phi\hat{n})\,dS$$

$$\Rightarrow \iiint_V \vec{C}\cdot\vec{\nabla}\phi\,dV = \iint_S \vec{C}\cdot(\phi\hat{n})\,dS$$

$$\text{as } \vec{\nabla}\cdot(\phi\vec{C}) = \vec{C}\cdot\vec{\nabla}\phi$$

$$\Rightarrow \vec{C}\cdot\iiint_V \vec{\nabla}\phi\,dV = \vec{C}\cdot\iint_S \phi\hat{n}\,dS$$

As $\vec{C}$ is an arbitrary constant vector, we have

$$\iiint_V \vec{\nabla}\phi\,dV = \iint_S \phi\ \hat{n}\,dS$$

In particular, if ϕ be a constant, then $\vec{\nabla}\phi = 0$, and hence, $\iint_S \hat{n}\,dS = \iint_S d\vec{S} = \vec{0}$.

RESULT 30

Let V be the region enclosed by the surface $\mathcal{S}$ and $\hat{n}$ be the unit outward drawn normal to it. Let $\vec{A}$ be an arbitrary constant vector, then

$$\vec{A}\cdot\oint\int_S \hat{n}\,dS = \oint\int_S \vec{A}\cdot\hat{n}\,dS = \iiint \vec{\nabla}\cdot\vec{A}\,dV \text{ using Eq. (7.9)}$$

$$\Rightarrow \vec{A} \cdot \oint\int_S \hat{n}\, dS = 0, \text{ as } \vec{\nabla} \cdot \vec{A} = 0$$

Since $\vec{A}$ is arbitrary, it follows that

$$\oint\int_S \hat{n}\, dS \int\int_S d\vec{S} = \vec{0} \tag{7.17}$$

Theorem 7.3

If $\vec{A}$ be a vector point function defined in the region V enclosed by a surface $\mathcal{S}$ such that its finite order partial derivatives be finite, single-valued and continuous throughout V and $\mathcal{S}$, then

$$\int\int\int_V (\nabla \times \vec{A})\, dV = \int\int_S (\widehat{n} \times \vec{A})\, dS \tag{7.18}$$

where $\hat{n}$ is the unit outward vector to $\mathcal{S}$.

Proof: If $\vec{B}$ is an arbitrary constant vector

$$\vec{\nabla} \cdot (\vec{A} \times \vec{B}) = \vec{B} \cdot (\nabla \times \vec{A}) - \vec{A} \cdot (\nabla \times \vec{B}) = \vec{B} \cdot (\nabla \times \vec{A})$$

In Gauss's divergence theorem (7.9), let us take $\vec{F} = \vec{A} \times \vec{B}$, where $\vec{B}$ is an arbitrary constant vector. Then, using the above result, we get

$$\int\int\int_V \vec{\nabla} \cdot (\vec{A} \times \vec{B})\, dV = \int\int_S (\vec{A} \times \vec{B}) \cdot \hat{n}\, dS$$

$$\Rightarrow \int\int\int_V \vec{B} \cdot (\nabla \times \vec{A})\, dV = \int\int_S \vec{B} \cdot (\widehat{n} \times \vec{A})\, dS$$

$$\text{as } (\vec{A} \times \vec{B}) \cdot \hat{n} = \left[\hat{A}\, \hat{B}\, \hat{n}\right] = \left[\hat{B}\, \hat{n}\, \hat{A}\right] = \vec{B} \cdot (\widehat{n} \times \vec{A})$$

$$\Rightarrow \vec{B} \cdot \int\int\int_V (\vec{\nabla} \times \vec{A})\; dV = \vec{B} \cdot \int\int_S (\widehat{n} \times \vec{A})\; dS$$

Since $\vec{B}$ is an arbitrary constant vector, we have

$$\int\int\int_V (\vec{\nabla} \times \vec{A})\; dV = \int\int_S (\widehat{n} \times \vec{A})\; dS$$

Let $\vec{A}$ be in the direction of $\hat{n}$, then $\widehat{n} \times \vec{A} = \vec{0}$, and the result (7.18) becomes

$$\int\int\int_V (\vec{\nabla} \times \vec{A})\; dV = \int\int_S (\widehat{n} \times \vec{A})\; dS = 0$$

Result 31

Let V be the region enclosed by the surface $\mathcal{S}$ and $\hat{n}$ be the unit outward drawn normal to it. Let $\vec{A}$ be an arbitrary constant vector. Using Gauss's divergence theorem (7.9), we get

$$\int\int_S (\vec{\nabla} \times \vec{A})\hat{n}\, dS = \int\int\int_V \vec{\nabla} \cdot (\vec{\nabla} \times \vec{A})\, dV = 0 \tag{7.19}$$

since $\vec{\nabla} \cdot (\vec{\nabla} \times \vec{A}) = 0$ for all vectors $\vec{A}$.

RESULT 32

If $\vec{C}$ be a given vector, using the definition of vector triple product, we get

$$\vec{\nabla} \times (\vec{C} \times \vec{r}) = \vec{C}\vec{\nabla} \cdot \vec{r} - (\vec{C} \cdot \vec{\nabla})\vec{r} = 3\vec{C} - \vec{C}$$

Using Gauss's divergence theorem (7.9) and the result (7.18), we get

$$\begin{aligned} \int\int_S \hat{n} \times (\vec{C} \times \vec{r}) \; dS &= \int\int\int_V \vec{\nabla} \times (\vec{C} \times \vec{r}) \; dV \\ &= 2\vec{C}\int\int\int_V dV = 2V\vec{C} \end{aligned} \quad (7.20)$$

DEDUCTION 7.2.2

If ϕ be a scalar field with continuous second order derivatives, then we can write $\vec{F} = \vec{\nabla}\phi$. Substituting it in Eq. (7.9), we get

$$\int\int_S \vec{F} \cdot \hat{n} \, dS = \int\int\int_V \vec{\nabla} \cdot \vec{F} \, dV$$

$$\text{or,} \quad \int\int_S \vec{\nabla}\phi \cdot \hat{n} \, dS = \int\int\int_V \vec{\nabla} \cdot \vec{\nabla}\phi \, dV$$

$$\text{or,} \quad \int\int_S \frac{\partial \phi}{\partial n} \, dS = \int\int\int_V \nabla^2 \phi \, dV \quad (7.21)$$

where $\vec{\nabla} \cdot \vec{\nabla}$ is the Laplacian operator ∇^2 and $\dfrac{\partial \phi}{\partial n}$ is the directional derivative of ϕ in the direction of the outer normal to the surface S.

DEDUCTION 7.2.3

Let $\triangle V$ be the volume enclosed by the surface $\triangle S$. From the result (7.16), we see that

$$\int\int_S \phi \;\; \hat{n} \, dS = \int\int\int_V \vec{\nabla} \;\; \phi \, dV$$

$$\text{or,} \quad \int\int_{\triangle S} \phi \;\; \hat{n} \, dS = \int\int\int_{\triangle V} \vec{\nabla} \;\; \phi \, dV$$

By the mean value theorem of integral calculus, we have

$$\int\int\int_{\triangle V} \vec{\nabla}\phi \, dV = \left[(\vec{\nabla}\,\phi)_P + \varepsilon\right]\triangle V$$

where $(\vec{\nabla}\phi)_P$ is the value of $(\vec{\nabla}\phi)$ (independent of $\triangle V$) at the point P and ε is a function of $\triangle V$ such that $\varepsilon \to 0$ as $\triangle V \to 0$. Thus

$$\left[(\vec{\nabla}\phi)_P + \varepsilon\right]\triangle V = \int\int_{\triangle S} \phi \hat{n} ds$$

$$\text{or,} \quad (\vec{\nabla}\phi)_P + \varepsilon = \frac{1}{\triangle V}\int\int_{\triangle S} \phi \;\; \hat{n} \, dS$$

$$\text{or,} \quad \left(\vec{\nabla}\phi\right)_P = \lim_{\triangle V \to 0} \frac{1}{\triangle V}\int\int_{\triangle S} \phi \;\; \hat{n} \, dS \quad (7.22)$$

DEDUCTION 7.2.4

From Gauss's divergence theorem (7.9), it follows that

$$\iiint_{\triangle V} \vec{\nabla}\cdot\vec{F}\,dV = \iint_{\triangle S} \vec{F}\cdot\widehat{n}\,dS$$

By the mean value theorem of integral calculus, we have

$$\iiint_{\triangle V} \vec{\nabla}\cdot\vec{F}\,dV = \left[(\vec{\nabla}\cdot\vec{F})_P + \varepsilon\right]\triangle V$$

where $(\vec{\nabla}\cdot\vec{F})_P$ is the value of $(\vec{\nabla}\cdot\vec{F})$ (independent of $\triangle V$) at the point P and ε is a function of $\triangle V$ such that $\varepsilon \to 0$ as $\triangle V \to 0$. Thus

$$\left[(\vec{\nabla}\cdot\vec{F})_P + \varepsilon\right]\triangle V = \iint_{\triangle S} \vec{F}\cdot\widehat{n}\ dS$$

$$\text{or,}\quad (\vec{\nabla}\cdot\vec{F})_P + \varepsilon = \frac{1}{\triangle V}\iint_{\triangle S} \vec{F}\cdot\widehat{n}\ dS$$

$$\text{or,}\quad (\vec{\nabla}\cdot\vec{F})_P = \lim_{\triangle V\to 0}\frac{1}{\triangle V}\iint_{\triangle S} \vec{F}\cdot\widehat{n}\ dS \tag{7.23}$$

Using the result (7.18), $\displaystyle\iiint_{\triangle V} (\vec{\nabla}\times\vec{F})\,dV = -\iint_{\triangle S} (\vec{F}\times\widehat{n})\,dS$, we get

$$\left(\vec{\nabla}\times\vec{F}\right)_P = \lim_{\triangle V\to 0}\frac{1}{\triangle V}\iint_{\triangle S} (\widehat{n}\times\vec{F})\,dS \tag{7.24}$$

Theorem 7.4

If ϕ and ψ and their first order partial derivatives are single-valued and continuous functions of points in a region of volume V bounded by a closed surface S and $\hat{n}$ is the unit outward drawn normal to S, then

$$\iiint_V \left(\phi\,\vec{\nabla}^2\psi - \psi\,\vec{\nabla}^2\phi\right)dV = \iint_S \left(\phi\vec{\nabla}\psi - \psi\vec{\nabla}\phi\right).\widehat{n}\,dS \tag{7.25}$$

Proof: Let ϕ and ψ be scalar point functions which together with their derivatives in any direction are continuous within the region V, bounded by a closed surface S. In Gauss's divergence theorem, let us take $\vec{F} = \phi\vec{\nabla}\psi$, then

$$\iint_S \vec{F}\cdot\widehat{n}\,dS = \iiint_V \vec{\nabla}\cdot\vec{F}\,dV$$

$$\Rightarrow \iint_S (\phi\vec{\nabla}\psi)\cdot\widehat{n}\,dS = \iiint_V \vec{\nabla}\cdot(\phi\vec{\nabla}\psi)\,dV$$

$$= \iiint_V \left[\phi(\vec{\nabla}\cdot\vec{\nabla}\psi) + (\vec{\nabla}\phi)\cdot(\vec{\nabla}\psi)\right]dV$$

$$\Rightarrow \iint_S (\phi\vec{\nabla}\psi)\cdot\widehat{n}\,dS = \iiint_V \left[(\phi\vec{\nabla}^2\psi) + (\vec{\nabla}\phi)\cdot(\vec{\nabla}\psi)\right]dV \tag{7.26}$$

This relation (7.26) is known as *Green's first identity* or theorem. Interchanging ϕ and ψ in (7.26), we get

$$\iint_S (\psi\vec{\nabla}\phi)\cdot\widehat{n}\, dS = \iiint_V \left[(\psi\vec{\nabla}^2\phi) + (\vec{\nabla}\psi)\cdot(\vec{\nabla}\phi)\right] dV \tag{7.27}$$

Subtracting (7.27) from (7.26) we get

$$\iiint_V \left(\phi\vec{\nabla}^2\psi - \psi\vec{\nabla}^2\phi\right) dV = \iint_S \left(\phi\vec{\nabla}\psi - \psi\vec{\nabla}\phi\right)\cdot\widehat{n}\, dS \tag{7.28}$$

This relation (7.28) is known as *Green's second identity* or *Green's theorem in symmetrical form.*

Deduction 7.2.5

Another form of Eq. (7.28): Using the definition of gradient of a scalar point function, we have

$$\vec{\nabla}\psi = \frac{\partial\psi}{\partial n}\,\widehat{n}; \qquad \vec{\nabla}\phi = \frac{\partial\phi}{\partial n}\,\widehat{n}$$

where $\dfrac{\partial\psi}{\partial n}$ and $\dfrac{\partial\phi}{\partial n}$ are the directional derivatives of ψ and ϕ, respectively, in the direction of $\widehat{n}$. Hence, from (7.25), Green's second identity can be written in the form

$$\iiint_V \left(\phi\vec{\nabla}^2\psi - \psi\vec{\nabla}^2\phi\right) dV = \iint_S \left(\phi\frac{\partial\psi}{\partial n} - \psi\frac{\partial\phi}{\partial n}\right)\widehat{n}\cdot\widehat{n}\, dS$$

$$= \iint_S \left(\phi\frac{\partial\psi}{\partial n} - \psi\frac{\partial\phi}{\partial n}\right) dS \tag{7.29}$$

(i) In particular, if ϕ and ψ are both harmonic functions so that $\vec{\nabla}^2\psi = 0$ and $\vec{\nabla}^2\phi = 0$, then the identity (7.29) becomes

$$\iint_S \left(\phi\frac{\partial\psi}{\partial n} - \psi\frac{\partial\phi}{\partial n}\right) dS = 0 \tag{7.30}$$

(ii) Taking $\phi = k$ (constant) in the result (7.30), we obtain

$$\iint_S \frac{\partial\psi}{\partial n}\, dS = 0 \tag{7.31}$$

(iii) Further, if ϕ and ψ are such that $\dfrac{\partial\psi}{\partial n} = \dfrac{\partial\phi}{\partial n}$ on the surface S, from the same result (7.30), we obtain

$$\iint_S (\phi - \psi)\,\frac{\partial\psi}{\partial n}\, dS = 0 \tag{7.32}$$

Since ϕ and ψ are arbitrary harmonic functions, we must have $\phi - \psi = \text{constant}$.

(iv) Again, using $\phi = \psi$ in Green's first identity (7.26), we obtain

$$\iint_S (\phi\vec{\nabla}\phi)\cdot\widehat{n}\, dS = \iiint_V \left[(\phi\vec{\nabla}^2\phi) + (\vec{\nabla}\phi)\cdot(\vec{\nabla}\phi)\right] dV$$

$$\Rightarrow \iint_S (\phi\vec{\nabla}\phi)\cdot\widehat{n}\, dS = \iiint_V \left[(\phi\vec{\nabla}^2\phi) + \left|\vec{\nabla}\phi\right|^2\right] dV \tag{7.33}$$

If ϕ be a harmonic function such that its normal derivative vanishes on the boundary, then from the result (7.33), we get

$$\iiint_V \left|\vec{\nabla}\phi\right|^2 dV = \iint_S (\phi\vec{\nabla}\phi)\cdot\widehat{n}\, dS = \iint_S \phi\,\frac{\partial\phi}{\partial n}\, dS = 0 \qquad (7.34)$$

This gives $\vec{\nabla}\phi = \vec{0}$ at every point, implying that ϕ is constant through the region V.

Example 7.2.7

If ϕ is harmonic in V, then prove that $\iint_S \frac{d\phi}{dn}\, dS = 0$, where S is the surface enclosing V.

Solution

Using Gauss's divergence theorem

$$\iint_S \frac{d\phi}{dn}\, dS = \iint_S \left(\frac{d\phi}{dn}\hat{n}\right)\cdot\hat{n}\, dS = \iint_S (\vec{\nabla}\phi)\cdot\hat{n}\, dS$$
$$= \iiint_V \vec{\nabla}\cdot\vec{\nabla}\phi dV = \iiint \nabla^2\phi dV = 0$$

DEDUCTION 7.2.6

Let $\vec{C}$ be any arbitrary constant vector so that $\vec{\nabla}\cdot\vec{C} = 0$. Then

$$\vec{C}\cdot\oint\oint_S d\vec{S} = \oint\oint_S \vec{C}\cdot d\vec{S}$$
$$= \iiint_V \vec{\nabla}\cdot\vec{C}\; dV \quad \text{by Eq.(7.9)}$$
$$= 0 \quad \text{as } \vec{\nabla}\cdot\vec{C} = 0$$

Thus, the relation holds for any arbitrary vector $\vec{C}$. This is possible only when

$$\oint\oint_S d\overrightarrow{S} = 0$$

As *div* (*curl* $\vec{F}$) $= 0$, for any closed surface S, we have

$$\oint\oint_S (\vec{\nabla}\times\vec{F})\cdot\hat{n}\; dS = 0$$

Theorem 7.5

Gauss's theorem: If S is a closed surface enclosing a volume V and $\hat{n}$ is a unit outward normal vector to S, then

$$\iint_S \frac{\hat{n}\cdot\vec{r}}{r^3}\, dS = 0, \text{ if the origin } O \text{ lies outside } S$$
$$= 4\pi, \text{ if the origin } O \text{ lies inside } S$$

where $\vec{r}$ is the position vector of any point (x, y, z) measured from the origin O.

Proof: Using the definition of divergence, we get

$$\vec{\nabla}\cdot(\frac{1}{r^3}\vec{r}) = \nabla(\frac{1}{r^3})\cdot\vec{r} + \frac{1}{r^3}\vec{\nabla}\cdot\vec{r} = -\frac{3r^2}{r^5} + \frac{3}{r^3} = 0$$

By Gauss's divergence theorem, we have

$$\int\int_S \frac{\hat{n}\cdot\vec{r}}{r^3}\,dS = \int\int\int_V \vec{\nabla}\cdot(\frac{1}{r^3}\vec{r})dV = 0$$

within V and $r = |\vec{r}| \neq 0$, i.e., O lies outside the surface S. On the other hand, when the origin O lies inside the surface S (Fig. 7.15), we enclose it by a small sphere of radius ε with its centre at O. Let us take the surface of the sphere as S_1. Then, within the region bounded by S and S_1, the divergence theorem can be applied. Let V_1 be the region bounded by S and S_1.

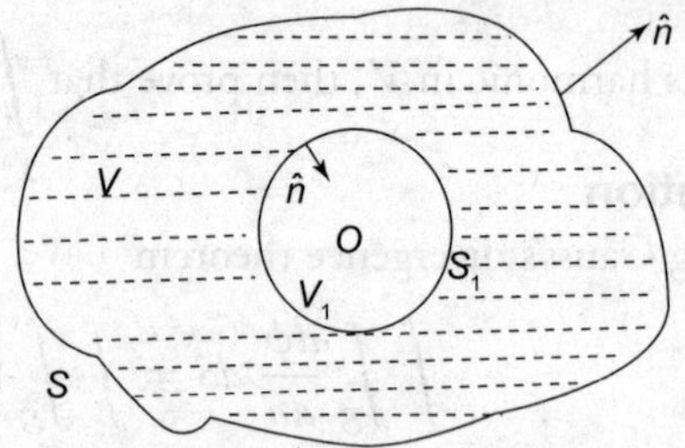

Figure 7.15 Gauss's theorem

$$\int\int_{S+S_1} \frac{\hat{n}\cdot\vec{r}}{r^3}\,dS = \int\int_S \frac{\hat{n}\cdot\vec{r}}{r^3}\,dS + \int\int_{S_1} \frac{\hat{n}\cdot\vec{r}}{r^3}\,dS$$

$$\Rightarrow \int\int\int_{V_1} \vec{\nabla}\cdot(\frac{1}{r^3}\vec{r})dV = \int\int_S \frac{\hat{n}\cdot\vec{r}}{r^3}\,dS + \int\int_{S_1} \frac{\hat{n}\cdot\vec{r}}{r^3}\,dS$$

$$\Rightarrow 0 = \int\int_S \frac{\hat{n}\cdot\vec{r}}{r^3}\,dS + \int\int_{S_1} \frac{\hat{n}\cdot\vec{r}}{r^3}\,dS$$

since, in V_1, $r = |\vec{r}| \neq 0$ and the function $\frac{1}{r^3}\vec{r}$ is continuously differentiable everywhere in V_1. Hence

$$\int\int_S \frac{\hat{n}\cdot\vec{r}}{r^3}\,dS = -\int\int_{S_1} \frac{\hat{n}\cdot\vec{r}}{r^3}\,dS$$

Now, the unit normal to the surface of the sphere is given by $\hat{n} = -\frac{1}{r}\vec{r}$ as it is the opposite direction of $\vec{r}$. So, on the surface S_1, $r = \varepsilon$ and

$$\int\int_{S_1} \frac{\hat{n}\cdot\vec{r}}{r^3}\,dS = -\int\int_{S_1} \frac{\vec{r}\cdot\vec{r}}{r^4}\,dS = -\int\int_{S_1} \frac{dS}{r^3}$$

$$= -\frac{1}{\varepsilon^2}\int\int_{S_1} dS = -\frac{1}{\varepsilon^2}\cdot 4\pi\varepsilon^2 = -4\pi$$

$$\Rightarrow \int\int_S \frac{\hat{n}\cdot\vec{r}}{r^3}\,dS = -\int\int_{S_1} \frac{\hat{n}\cdot\vec{r}}{r^3}\,dS = 4\pi$$

This result is known as *Gauss's theorem.*

Example 7.2.8

Let $W = \{(x, y, z) \in \Re^3 : 1 \leq x^2 + y^2 + z^2 \leq 4\}$ and $F : W \rightarrow \Re^3$ be defined by $\vec{F}(x, y, z) =$

$\dfrac{(x, y, z)}{(x^2+y^2+z^2)^{3/2}}$ for $(x, y, z) \in W$. If ∂W denotes the boundary of W oriented by the outward normal $\hat{n}$ to W, then evaluate $\displaystyle\iint_{\partial S} \vec{F} \cdot \hat{n}\, dS$.

Solution

Here, $\vec{F}(x, y, z) = \dfrac{(x, y, z)}{(x^2+y^2+z^2)^{3/2}} = \dfrac{1}{r^3}\vec{r}$, where $\vec{r} = (x, y, z), r^2 = x^2+y^2+z^2$. Using the definition of divergence, we get

$$\vec{\nabla} \cdot (\frac{1}{r^3}\vec{r}) = \nabla(\frac{1}{r^3}) \cdot \vec{r} + \frac{1}{r^3}\vec{\nabla} \cdot \vec{r} = -\frac{3r^2}{r^5} + \frac{3}{r^3} = 0$$

By Gauss's divergence theorem, we have

$$\iint_{\partial S} \vec{F} \cdot \hat{n}\, dS = \iiint_W \vec{\nabla} \cdot \vec{F}\, dV = 0$$

Example 7.2.9

Evaluate $\displaystyle\iint_S x^2(x\, dy\, dz + y\, dx\, dz + z\, dx\, dy)$, where $\mathcal{S}$ is the surface of the cylinder $x^2+y^2 = a^2$ bounded by the planes $z = 0$ and $z = b$.

Solution

The surface $\mathcal{S}$ of the cylinder $x^2+y^2 = a^2$ bounded by the planes $z = 0$ and $z = b$ is shown in Fig. 7.16.

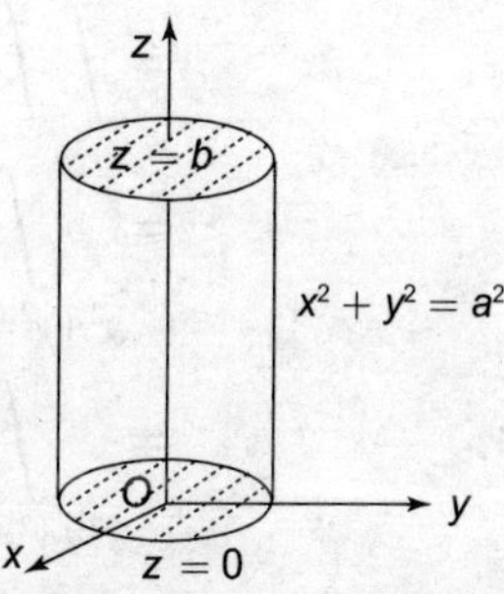

Figure 7.16 The cylinder

The given surface integral is of the form $\displaystyle\iint_S \vec{F} \cdot d\vec{S}$ or $\displaystyle\iint_S \vec{F} \cdot \hat{n}\, dS$, where $\vec{F} = x^3\,\hat{i} + x^2y\,\hat{j} + x^2z\,\hat{k}$.

Therefore, using the definition of divergence

$$\vec{\nabla} \cdot \vec{F} = \frac{\partial}{\partial x}(x^3) + \frac{\partial}{\partial y}(x^2y) + \frac{\partial}{\partial z}(x^2z) = 5x^2$$

Here, V is the volume of the region enclosed by the surface $\mathcal{S}$. Using Gauss's divergence theorem, we get

$$\begin{aligned}
\iint_S \vec{F} \cdot dS &= \iiint_V \vec{\nabla} \cdot \vec{F}\, dV = 5\int_0^b \left[\int_{-a}^{a} \left\{\int_{-\sqrt{a^2-x^2}}^{\sqrt{a^2-x^2}} x^2\, dy\right\} dx\right] dz \\
&= 20\, b \int_0^a \left[\int_0^{\sqrt{a^2-x^2}} x^2\, dy\right] dx \\
&= 20\, b \int_0^a x^2\sqrt{a^2-x^2}\, dx;\quad x = a\sin\theta \\
&= 5a^4b \int_0^{\pi/2} \sin^2 2\theta\, d\theta = \frac{5a^4b}{2}\int_0^{\pi/2} (1-\cos 4\theta)d\theta
\end{aligned}$$

$$= \frac{5a^4 b}{2}\left[\theta - \frac{\sin 4\theta}{4}\right]_0^{\pi/2} = \frac{5}{4}a^4 b\pi$$

Example 7.2.10

Evaluate $\iint_S \vec{F} \cdot \hat{n}\, dS$ over the entire surface of the region above the xy plane bounded by the cone $z^2 = x^2 + y^2$ and the plane $z = 4$ if $\vec{F} = 4xz\hat{i} + 2xyz^2\hat{j} + 3z\hat{k}$.

Solution

The region of integration is given in Fig. 7.17.

Since $\vec{F} = 4xz\hat{i} + 2xyz^2\hat{j} + 3z\hat{k}$, using the definition of divergence

$$\vec{\nabla} \cdot \vec{F} = \frac{\partial}{\partial x}(4xz) + \frac{\partial}{\partial y}(xyz^2) + \frac{\partial}{\partial z}(3z)$$

$$= 4z + xz^2 + 3$$

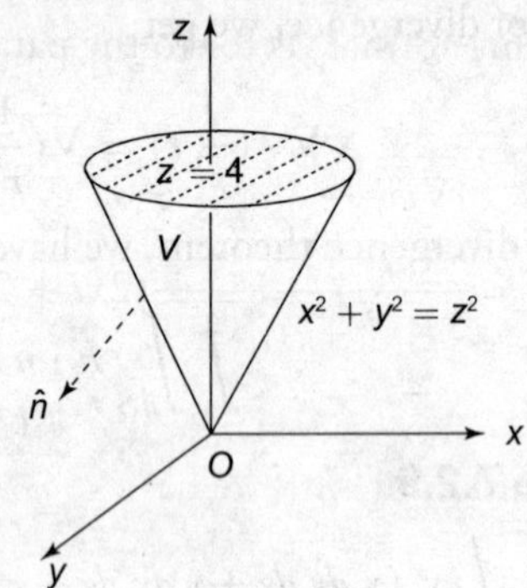

Figure 7.17 The cone $z^2 = x^2 + y^2$ and the plane $z = 4$

To evaluate the integral, we use Gauss's divergence theorem, which states that

$$\iint_S \vec{F} \cdot \hat{n}\, dS = \iiint_V (\vec{\nabla} \cdot \vec{F})\, dV$$

$$= \iiint_V (4z + xz^2 + 3)dx\, dy\, dz$$

$$= \iint_{x^2+y^2=16} \int_{\sqrt{x^2+y^2}}^{4} (4z + xz^2 + 3)dx\, dy\, dz$$

$$= \iint_{x^2+y^2=16} \left[2z^2 + \frac{2}{3}xz^3 + 3z\right]_{\sqrt{x^2+y^2}}^{4} dydx$$

$$= \iint_{x^2+y^2=16} \left[44 + \frac{128}{3}x - \left\{2(x^2 + y^2) + \frac{2}{3}x(x^2 + y^2)^{3/2} + 3(x^2 + y^2)^{1/2}\right\}\right] dydx; \quad x = r\cos\theta, y = r\sin\theta$$

$$= \int_0^{2\pi}\int_0^4 \left[44 + \frac{128}{3}r\cos\theta - 2r^2 - \frac{2}{3}r\cos\theta r^3 - 3r\right] rdrd\theta$$

$$= \int_0^4 \left[44\theta + \frac{128}{3}r\sin\theta - (2r^2 + 3r)\theta - \frac{2}{3}r^4\sin\theta\right]_0^{2\pi} rdr$$

$$= \int_0^4 \left[88\pi - (2r^2 + 3r)2\pi\right] r\, dr = 320\pi$$

Example 7.2.11

Using Gauss's divergence theorem, evaluate the surface integral $\int\int_S (\vec{\nabla} \times \vec{F}) \cdot \hat{n}\, dS$ where, $\vec{F} = (x^2 + y - 4)\hat{i} + 3xy\hat{j} + (2xz + z^2)\hat{k}$ and S is the surface of the paraboloid $x^2 + y^2 = 4 - z$ above the xy plane.

Solution

The region of integration is given in Fig. 7.18. If $\hat{n}$ be the unit normal vector to the paraboloid, then

$$\hat{n} = \frac{1}{|\vec{\nabla}\phi|}\vec{\nabla}\phi$$

$$= \frac{1}{\sqrt{4x^2 + 4y^2 + 1}}(2x\hat{i} + 2y\hat{j} + \hat{k})$$

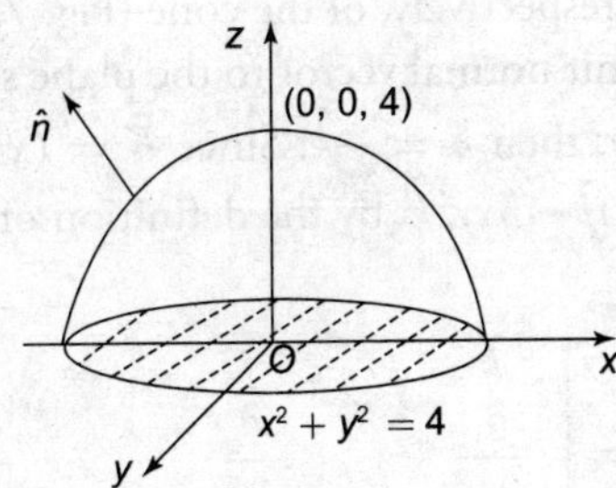

Figure 7.18 The cone $z^2 = x^2 + y^2$ and the plane $z = 4$

Since $\vec{F} = (x^2 + y - 4)\hat{i} + 3xy\hat{j} + (2xz + z^2)\hat{k}$, by the definition of curl, we have

$$\vec{\nabla} \times \vec{F} = \begin{vmatrix} \hat{i} & \hat{j} & \hat{k} \\ \dfrac{\partial}{\partial x} & \dfrac{\partial}{\partial y} & \dfrac{\partial}{\partial z} \\ x^2 + y - 4 & 3xy & 2xz + z^2 \end{vmatrix}$$

$$= -2z\hat{j} + (3y - 1)\hat{k}$$

$$\Rightarrow (\vec{\nabla} \times \vec{F}) \cdot \hat{n} = \left[-2z\hat{j} + (3y - 1)\hat{k}\right] \cdot \frac{2x\hat{i} + 2y\hat{j} + \hat{k}}{\sqrt{4x^2 + 4y^2 + 1}}$$

$$= \frac{1}{\sqrt{4x^2 + 4y^2 + 1}}(-4yz + 3y - 1)$$

Now, considering the surface integral over the projected area of the paraboloid on the xy plane, we get

$$\int\int_S (\vec{\nabla} \times \vec{F}) \cdot \hat{n}\, dS = \int\int_R (\vec{\nabla} \times \vec{F}) \cdot \hat{n}\, \frac{dx\, dy}{|\hat{n} \cdot \hat{k}|}$$

$$= \int_{-2}^{2}\left[\int_{-\sqrt{4-x^2}}^{\sqrt{4-x^2}} \frac{-4yz + 3y - 1}{\sqrt{4x^2 + 4y^2 + 1}}(\sqrt{4x^2 + 4y^2 + 1})\, dx\right] dy$$

$$= 2\int_{-2}^{2}(-1)\sqrt{4 - x^2}\, dx = -4\int_0^2 \sqrt{4 - x^2}\, dx$$

$$= -4\left[\frac{x}{2}\sqrt{4 - x^2} + 2\sin^{-1}\frac{x}{2}\right]_0^2 = -4\pi$$

Example 7.2.12

Evaluate the surface integral $\iint_S (\vec{\nabla} \times \vec{F}) \cdot \hat{n}\, dS$, where $\vec{F} = (x - z)\hat{i} + (x^3 + yz)\hat{j} - 3xy^2\hat{k}$, and S is the surface of the cone $z = 2 - \sqrt{x^2 + y^2}$ above the xy plane.

Solution

Let SC and SP be the curved surface and plane surface, respectively, of the cone (Fig. 7.19). If $\hat{n}$ be the unit normal vector to the plane surface of the cone, then $\hat{n} = -\hat{k}$. Since $\vec{F} = (x - z)\hat{i} + (x^3 + yz)\hat{j} - 3xy^2\hat{k}$, by the definition of curl, we have

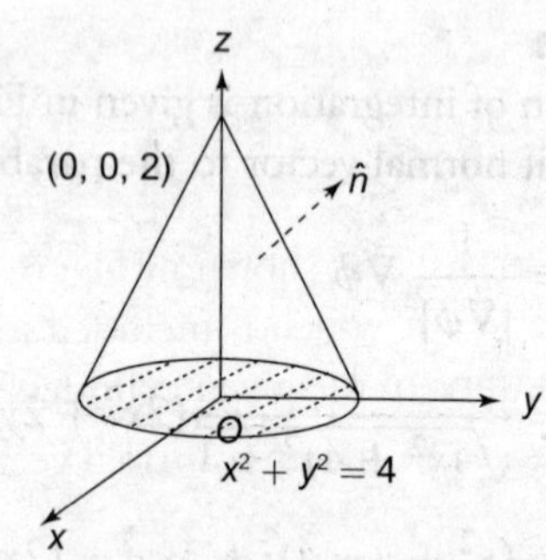

Figure 7.19 The cone $z = 2 - \sqrt{x^2 + y^2}$ above the xy plane

$$\vec{\nabla} \times \vec{F} = \begin{vmatrix} \hat{i} & \hat{j} & \hat{k} \\ \dfrac{\partial}{\partial x} & \dfrac{\partial}{\partial y} & \dfrac{\partial}{\partial z} \\ x - z & x^3 + yz & -3xy^2 \end{vmatrix}$$

$$= (-y - 6xy)\hat{i} + (3y^2 + 1)\hat{j} + 3x^2\hat{k}$$

$$\Rightarrow \quad (\vec{\nabla} \times \vec{F}) \cdot \hat{n} = \left[(-y - 6xy)\hat{i} + (3y^2 + 1)\hat{j} + 3x^2\hat{k}\right] \cdot (-\hat{k}) = -3x^2$$

Since $\vec{\nabla} \cdot (\vec{\nabla} \times \vec{F}) = 0$, using Gauss's divergence theorem, we get

$$\iint_S (\vec{\nabla} \times \vec{F}) \cdot \hat{n}\, dS = \iiint_V \vec{\nabla} \cdot (\vec{\nabla} \times \vec{F})\, dV = 0$$

$$\Rightarrow \iint_S (\vec{\nabla} \times \vec{F}) \cdot \hat{n}\, dS = \iint_{SC} (\vec{\nabla} \times \vec{F}) \cdot \hat{n}\, dS + \iint_{SP} (\vec{\nabla} \times \vec{F}) \cdot \hat{n}\, dS = 0$$

$$\Rightarrow \iint_{SC} (\vec{\nabla} \times \vec{F}) \cdot \hat{n}\, dS = -\iint_{SP} (\vec{\nabla} \times \vec{F}) \cdot \hat{n}\, dS$$

Thus, the surface integral becomes

$$\iint_{SC} (\vec{\nabla} \times \vec{F}) \cdot \hat{n}\, dS = -\int_{-2}^{2} \left[\int_{-\sqrt{4-x^2}}^{\sqrt{4-x^2}} (-3\,x^2)\, dy\right] dx$$

$$= 12 \int_0^2 \left[\int_0^{\sqrt{4-x^2}} x^2\, dy\right] dx$$

$$= 12 \int_0^2 x^2 \sqrt{4 - x^2}\, dx = 12\pi$$

Example 7.2.13

Evaluate the surface integral $\iint_S (\vec{\nabla} \times \vec{F}) \cdot \hat{n}\, dS$, where $\vec{F} = 2zy\hat{i} - (x + 3y)\hat{j} + (x^2 + z)\hat{k}$, and

$\mathcal{S}$ is the surface of intersection of the cylinders $x^2 + y^2 = a^2$ and $x^2 + z^2 = a^2$ which is included in the first octant.

Solution

Since $\vec{F} = 2zy\hat{i} - (x + 3y)\hat{j} + (x^2 + z)\hat{k}$, by the definition of curl, we have

$$\vec{\nabla} \times \vec{F} = \begin{vmatrix} \hat{i} & \hat{j} & \hat{k} \\ \dfrac{\partial}{\partial x} & \dfrac{\partial}{\partial y} & \dfrac{\partial}{\partial z} \\ 2zy & -x - 3y & x^2 + z \end{vmatrix} = -2(x - y)\hat{j} - (2z + 1)\hat{k}$$

Here, the surface $\mathcal{S}$ is shown in Fig. 6.27, which is the intersection of the cylinders $x^2 + y^2 = a^2$ and $x^2 + z^2 = a^2$ which is included in the first octant. Now, let the projections of the surface of the intersections of the given cylinders on the xy, xz and yz planes be $\mathcal{S}_1, \mathcal{S}_2$ and $\mathcal{S}_3$, respectively. If $\hat{n}_1, \hat{n}_2$ and $\hat{n}_3$ be the unit normal vectors to the surfaces $\mathcal{S}_1, \mathcal{S}_2$ and $\mathcal{S}_3$, then $\hat{n}_1 = -\hat{k}, \hat{n}_2 = -\hat{j}$ and $\hat{n}_3 = -\hat{i}$. Since $\vec{\nabla} \cdot (\vec{\nabla} \times \vec{F}) = 0$, using Gauss's divergence theorem, we get

$$\int\int_{\mathcal{S}} (\vec{\nabla} \times \vec{F}) \cdot \hat{n}\, dS = \int\int\int_V \vec{\nabla} \cdot (\vec{\nabla} \times \vec{F})\, dV = 0$$

Thus, the surface integral becomes

$$= -\int\int_{\mathcal{S}_1} (\vec{\nabla} \times \vec{F}) \cdot (-\hat{k})\, dS - \int\int_{\mathcal{S}_2} (\vec{\nabla} \times \vec{F}) \cdot (-\hat{j})\, dS - \int\int_{\mathcal{S}_3} (\vec{\nabla} \times \vec{F}) \cdot (-\hat{i})\, dS$$

$$= -\int_0^a \left[\int_0^{\sqrt{a^2-x^2}} dy\right] dx - 2\int_0^a \left[\int_0^{\sqrt{a^2-z^2}} x\, dx\right] dz - 0$$

$$= -\frac{\pi\, a^2}{4} - \frac{2a^3}{3} = -\frac{a^2}{12}(3\pi + 8a)$$

Example 7.2.14

Show that $\displaystyle\int\int_{\mathcal{S}} \frac{dS}{\sqrt{a^2x^2 + b^2y^2 + c^2z^2}}$ over the surface of the ellipsoid $ax^2 + by^2 + cz^2 = 1$ is equal to three times the volume of the ellipsoid.

Solution

Let $\phi = ax^2 + by^2 + cz^2 - 1$, then $\vec{\nabla}\phi = ax\hat{i} + by\hat{j} + cz\hat{k}$. The outward unit normal to the surface is given by

$$\hat{n} = \frac{1}{|\vec{\nabla}\phi|}\, \vec{\nabla}\phi = \frac{1}{\sqrt{a^2x^2 + b^2y^2 + c^2z^2}}(ax\hat{i} + by\hat{j} + cz\hat{k})$$

On the surface of the ellipsoid $ax^2 + by^2 + cz^2 = 1$, the integrand can be written as

$$\frac{1}{\sqrt{a^2x^2 + b^2y^2 + c^2z^2}} = \frac{ax^2 + by^2 + cz^2}{\sqrt{a^2x^2 + b^2y^2 + c^2z^2}}$$

$$= (x\hat{i} + y\hat{j} + z\hat{k}) \cdot \frac{ax\hat{i} + by\hat{j} + cz\hat{k}}{\sqrt{a^2x^2 + b^2y^2 + c^2z^2}} = \vec{F} \cdot n$$

Thus, the given integral is of the form $\int\int_S \vec{\nabla}\vec{F}\cdot\hat{n}\,dS$. Using Gauss's divergence theorem, we get

$$\int\int_S \frac{dS}{\sqrt{a^2x^2+b^2y^2+c^2z^2}} = \int\int_S \vec{\nabla}\vec{F}\cdot\hat{n}\,dS = \int\int\int_V \vec{\nabla}\cdot\vec{F}\,dV$$

$$= \int\int\int_V \vec{\nabla}\cdot(x\hat{i}+y\hat{j}+z\hat{k})\,dV = 3\int\int\int_V dV = 3V$$

Example 7.2.15

For the vector field $\vec{F} = xz^2\hat{i} - yz^2\hat{j} + z(x^2-y^2)\hat{k}$,

(i) Calculate the volume integral of the divergence of $\vec{F}$ over the region defined by $-a \le x \le a$, $-b \le y \le b$ and $0 \le z \le c$.

(ii) Calculate the flux of $\vec{F}$ out of the region through the surface $z = c$. Hence, deduce the net flux through the rest of the boundary of the region.

Solution

(i) The region of integration $-a \le x \le a$, $-b \le y \le b$ and $0 \le z \le c$ is given in Fig. 7.20. Since $\vec{F} = xz^2\hat{i} - yz^2\hat{j} + z(x^2-y^2)\hat{k}$, using the definition of divergence

$$\vec{\nabla}\cdot\vec{F} = \frac{\partial}{\partial x}(xz^2) - \frac{\partial}{\partial y}(yz^2) + \frac{\partial}{\partial z}\{z(x^2-y^2)\}$$

$$= x^2 - y^2$$

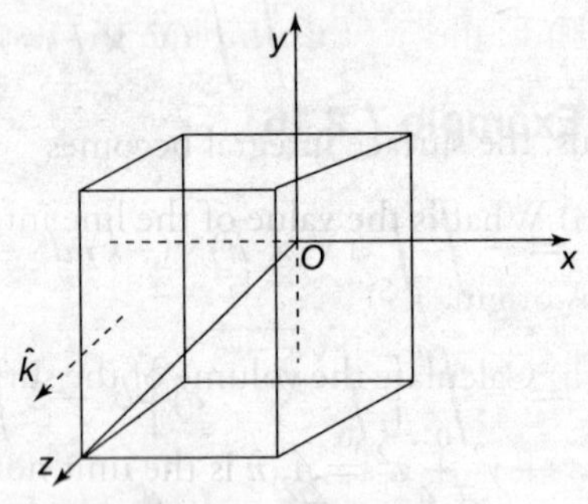

Figure 7.20 The region

Therefore, the volume integral of the divergence of $\vec{F}$ over the region defined by $-a \le x \le a$, $-b \le y \le b$ and $0 \le z \le c$ is given by

$$\int\int\int_V \vec{\nabla}\cdot\vec{F} = \int_{-a}^{a}\left[\int_{-b}^{b}\left\{\int_0^c \vec{\nabla}\cdot\vec{F}\,dz\right\}dy\right]dx$$

$$= \int_{-a}^{a}\left[\int_{-b}^{b}\left\{\int_0^c (x^2-y^2)\,dz\right\}dy\right]dx$$

$$= c\int_{-a}^{a}\left[x^2y - \frac{y^3}{3}\right]_{-b}^{b} dx = c\int_{-a}^{a}\left(2bx^2 - \frac{2b^3}{3}\right)dx$$

$$= c\left[\frac{2bx^3}{3} - \frac{2b^3x}{3}\right]_{-a}^{a} = \frac{4}{3}abc(a^2-b^2)$$

(ii) The flux of $\vec{F}$ out of the region through the surface $z = c$ is

$$\int\int_S \vec{F}\cdot d\vec{S} = \int\int_S \vec{F}\cdot\hat{n}\,dx\,dy = \int_{-a}^{a}\left[\int_{-b}^{b}\vec{F}\cdot\hat{k}\,dy\right]dx$$

$$= \int_{-a}^{a}\left[\int_{-b}^{b} z(x^2-y^2)\,dy\right]dx$$

$$= c\int_{-a}^{a}\left[x^2y - \frac{y^3}{3}\right]_{-b}^{b} dx = c\int_{-a}^{a}\left(2bx^2 - \frac{2b^3}{3}\right) dx$$

$$= c\left[\frac{2bx^3}{3} - \frac{2b^3x}{3}\right]_{-a}^{a} = \frac{4}{3}abc(a^2 - b^2)$$

From Gauss's divergence theorem, we get (RS = rest surface)

$$\iiint_V \vec{\nabla}\cdot\vec{F}\,dV = \iint_S \vec{F}\cdot\hat{n}\,dS$$

$$\Rightarrow \iint_{z=c} \vec{F}\cdot\hat{n}\,dS + \iint_{RS}\vec{F}\cdot\hat{n}\,dS = \iiint_V \vec{\nabla}\cdot\vec{F}\,dV$$

$$\Rightarrow \iint_{RS}\vec{F}\cdot\hat{n}\,dS = \iiint_V \vec{\nabla}\cdot\vec{F}\,dV - \iint_{z=c}\vec{F}\cdot\hat{n}\,dS$$

$$\Rightarrow \iint_{RS}\vec{F}\cdot\hat{n}\,dS = \frac{4}{3}abc(a^2 - b^2) - \frac{4}{3}abc(a^2 - b^2) = 0$$

Thus, the net flux through the rest of the boundary of the region is zero.

Example 7.2.16

(i) What is the value of the line integral $\oint_C \frac{x\,dy - y\,dx}{x^2 + y^2}$, where C is the unit sphere with its centre as origin.

(ii) Calculate the volume of the surface integral $\iint_S \vec{F}\cdot\hat{n}\,dS$, where S is the surface of the sphere $x^2 + y^2 + z^2 = 4$, $\hat{n}$ is the unit normal to the sphere and $\vec{F} = x\hat{i} + y\hat{j} + z\hat{k}$.

Solution

(i) The given integral is of the form $\oint \vec{G}\cdot d\vec{r}$, where

$$\vec{G} = -\frac{y}{x^2+y^2}\hat{i} + \frac{x}{x^2+y^2}\hat{j} = -\frac{y}{r}\hat{i} + \frac{y}{r}\hat{j};\;\; \vec{r} = x\hat{i} + y\hat{j}$$

Using the definition of curl, we get

$$\vec{\nabla}\times\vec{G} = \left[\frac{\partial}{\partial x}\hat{i} + \frac{\partial}{\partial y}\hat{j} + \frac{\partial}{\partial z}\hat{k}\right]\times\frac{1}{x^2+y^2}(-y\hat{i} + x\hat{j})$$

$$= \frac{y^2 - x^2 + x^2 - y^2}{(x^2+y^2)^2}\hat{k} = \vec{0}, \text{ provided } x^2 + y^2 \neq 0$$

Since $\vec{\nabla}\times\vec{G} = \vec{0}$, there exists a scalar function ϕ such that $\vec{G} = \vec{\nabla}\phi$. Now, from the total differential formula, we get

$$d\phi = \frac{\partial\phi}{\partial x}dx + \frac{\partial\phi}{\partial y}dy = \left(\frac{\partial\phi}{\partial x}\hat{i} + \frac{\partial\phi}{\partial y}\hat{j}\right)\cdot(dx\hat{i} + dy\hat{j})$$

$$= \vec{\nabla}\phi\cdot d\vec{r} = \vec{G}\cdot d\vec{r} = \frac{x\,dy - y\,dx}{x^2+y^2} = d\left(\tan^{-1}\frac{y}{x}\right)$$

$$\left[\phi\right]_{\text{at centre}} = \left[\tan^{-1}\frac{y}{x}\right]_{(0,0)} = 0$$

(ii) Using Gauss's divergence theorem, we get

$$\int\int_S \vec{F}\cdot\hat{n}\,dS = \int\int\int_V (\vec{\nabla}\cdot\vec{F})\,dV$$

$$= \int\int\int_V \left(\hat{i}\frac{\partial}{\partial x} + \hat{j}\frac{\partial}{\partial y} + \hat{k}\frac{\partial}{\partial z}\right)\cdot(x\hat{i} + y\hat{j} + z\hat{k})\,dV$$

$$= \int\int\int_V 3\,dV = 3\int_0^2\left[\int_0^{\pi/2}\left\{\int_0^{2\pi} r^2\sin\theta\,d\phi\right\}d\theta\right]dr$$

$$= 3\times\left[\frac{r^3}{3}\right]_0^2\times\left[-\cos\theta\right]_0^{\pi/2}\times 2\pi = 16\pi$$

Example 7.2.17

Let $\mathcal{S}_1$ be the hemisphere $x^2 + y^2 + z^2 = 1, z > 0$ and $\mathcal{S}_2$ be the closed disc $x^2 + y^2 \leq 1$ in the xy plane. Using Gauss's divergence theorem, evaluate $\int\int_{\mathcal{S}} \vec{F}\cdot\hat{n}\,dS$, where $\vec{F} = z^2x\hat{i} + \left(\frac{y^3}{3} + \tan z\right)\hat{j} + (x^2z + y^2)\hat{k}$ and $\mathcal{S} = \mathcal{S}_1 \cup \mathcal{S}_2$. Also, evaluate $\oint_{\mathcal{S}_2} \vec{F}\cdot\hat{n}\,dS$.

Solution

In Fig. 7.21, $\mathcal{S}_1$ is the hemisphere $x^2+y^2+z^2 = 1, z > 0$ and $\mathcal{S}_2$ is the closed disc $x^2 + y^2 \leq 1$ in the xy plane. Given $\vec{F} = z^2x\hat{i} + \left(\frac{y^3}{3} + \tan z\right)\hat{j} + (x^2z + y^2)\hat{k}$, using the definition of divergence, we get

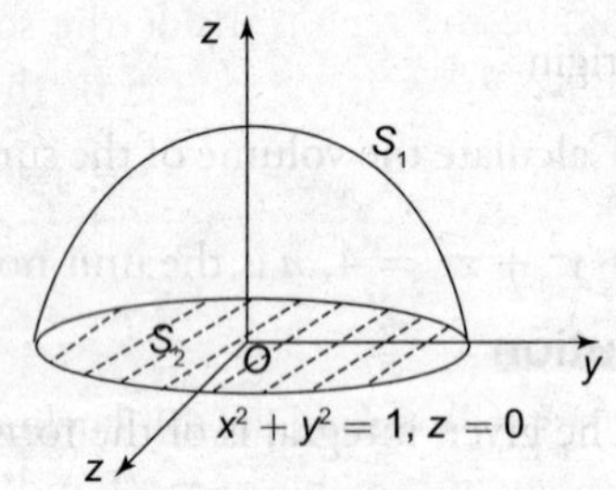

Figure 7.21 Region of integration

$$\vec{\nabla}\cdot\vec{F} = \left(\hat{i}\frac{\partial}{\partial x} + \hat{j}\frac{\partial}{\partial y} + \hat{k}\frac{\partial}{\partial z}\right)$$

$$\cdot\left[z^2x\hat{i} + \left(\frac{y^3}{3} + \tan z\right)\hat{j} + (x^2z + y^2)\hat{k}\right]$$

$$= \frac{\partial}{\partial x}(xz^2) + \frac{\partial}{\partial y}\left(\frac{y^3}{3} + \tan z\right) + \frac{\partial}{\partial z}(x^2z + y^2)$$

$$= z^2 + y^2 + x^2$$

On the surface $\mathcal{S}_1$, let us take the spherical polar transformation $x = r\sin\theta\ \cos\phi$, $y = r\sin\theta\ \sin\phi$, $z = r\cos\theta$. Then, using Gauss's divergence theorem, we get

$$\int\int_S \vec{F}\cdot\hat{n}\,dS = \int\int\int_V (\vec{\nabla}\cdot\vec{F})\,dV$$

$$= \int\int\int_V (x^2 + y^2 + z^2)dx\,dy\,dz$$

$$= \int_0^1\left[\int_0^{\pi/2}\left\{\int_0^{2\pi} r^2\cdot r^2\ \sin\theta\,d\phi\right\}d\theta\right]dr = \frac{2\pi}{5}$$

On the surface $\mathcal{S}_2$, $\hat{n} = -\hat{k}$. Let us take the spherical polar transformation $x = r\cos\theta, y = r\sin\theta$. Thus, the surface integral $z = 0$ on $\mathcal{S}_2$ is given by

$$\int\int_{\mathcal{S}_2} \vec{F}\cdot\hat{n}\,dS = -\int\int_{\mathcal{S}_2}\left[z^2x\hat{\imath} + \left(\frac{y^3}{3} + \tan z\right)\hat{\jmath} + (x^2z + y^2)\hat{k}\right]\cdot\hat{k}\,dS$$

$$= -\int\int_{\mathcal{S}_2}(x^2z + y^2)\,dS = -\int\int_{\mathcal{S}_2} y^2\,dx\,dy$$

$$= -\int_0^{2\pi}\left[\int_0^1 r^2\sin^2\theta\, r\,dr\right]d\theta = -\frac{\pi}{4}$$

Since $\mathcal{S} = \mathcal{S}_1 \cup \mathcal{S}_2$

$$\int\int_{\mathcal{S}} \vec{F}\cdot\hat{n}\,dS = \int\int_{\mathcal{S}_1} \vec{F}\cdot\hat{n}\,dS + \int\int_{\mathcal{S}_2} \vec{F}\cdot\hat{n}\,dS$$

$$\Rightarrow \int\int_{\mathcal{S}_1} \vec{F}\cdot\hat{n}\,dS = \int\int_{\mathcal{S}} \vec{F}\cdot\hat{n}\,dS - \int\int_{\mathcal{S}_2} \vec{F}\cdot\hat{n}\,dS$$

$$= \frac{2\pi}{5} + \frac{\pi}{4} = \frac{13\pi}{20}$$

Example 7.2.18

Let W be the region inside the solid cylinder $x^2 + y^2 \leq 4$ between the plane $z = 0$ and the parabola $z = x^2 + y^2$. Let $\mathcal{S}$ be the boundary of W. Using Gauss's divergence theorem, evaluate $\int\int_{\mathcal{S}} \vec{F}\cdot\hat{n}\,dS$, where $\vec{F} = (x^2 + y^2 - 4)\hat{\imath} + 3xy\hat{\jmath} + (2xz + z^2)\hat{k}$ and $\hat{n}$ is the outward unit normal to $\mathcal{S}$.

Solution

The region W inside the solid cylinder $x^2+y^2 \leq 4$ between the plane $z = 0$ and the parabola $z = x^2 + y^2$ is shown in Fig. 7.22. Applying Gauss's divergence theorem, we get

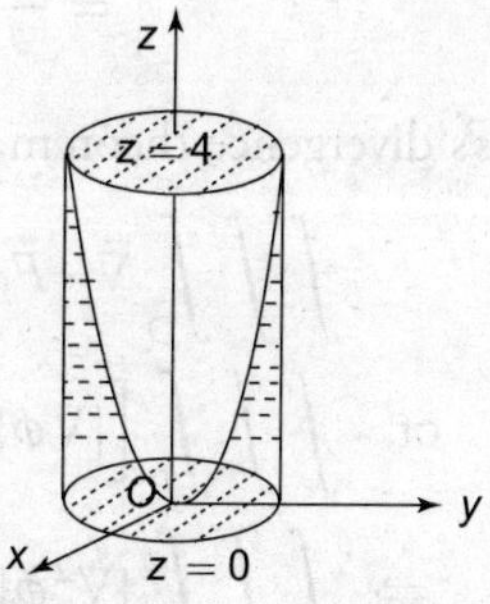

Figure 7.22 The region W

$$\int\int_{\mathcal{S}} \vec{F}\cdot\hat{n}\,dS = \int\int\int_V \vec{\nabla}\cdot\vec{F}\,dV$$

$$= \int_0^4\int_{-2}^2\int_{-\sqrt{4-y^2}}^{\sqrt{4-y^2}}(7x + 2z)dx\,dy\,dz$$

$$= 2\int_0^4\int_{-2}^2\int_0^{\sqrt{4-y^2}} 2z dx\,dy\,dz$$

$$= 4\int_0^4\int_{-2}^2 z\sqrt{4-y^2}\,dy\,dz = 8\int_0^4\int_0^2 z\sqrt{4-y^2}\,dy\,dz$$

$$= 8\int_0^4 z\left[\frac{y}{2}\sqrt{4-y^2} + 2\sin^{-1}\frac{y}{2}\right]_0^2 dz = 64\pi$$

Example 7.2.19

Let $\mathcal{D}$ be the region bounded by the concentric spheres $\mathcal{S}_1 : x^2+y^2+z^2 = a^2$ and $\mathcal{S}_2 : x^2+y^2+z^2 = b^2$, where $a < b$. Let $\hat{n}_1$ be the unit normal to $\mathcal{S}_1$ directed away from the origin. If $\nabla^2\phi = 0$ in $\mathcal{D}$ and $\phi = 0$ on $\mathcal{S}_2$, then show that

$$\iiint_{\mathcal{D}} \left|\vec{\nabla}\phi\right|^2 dV + \iint_{\mathcal{S}_1} \phi(\vec{\nabla}\phi)\cdot\hat{n}_1\, dS_1 = 0$$

Solution

Let $\vec{F} = \phi\,\vec{\nabla}\,\phi$, where ϕ is a continuously differentiable scalar point function of points on the surfaces $\mathcal{S}_1$ and $\mathcal{S}_2$ enclosing the region $\mathcal{D}$ (Fig. 7.23). Since $\nabla^2\phi = 0$ on $\mathcal{D}$, using the definition of divergence

$$\vec{\nabla}\cdot\vec{F} = \vec{\nabla}\cdot(\phi\vec{\nabla}\phi) = \vec{\nabla}\phi\cdot\vec{\nabla}\phi + \phi\vec{\nabla}^2\phi = |\vec{\nabla}^2\phi|$$

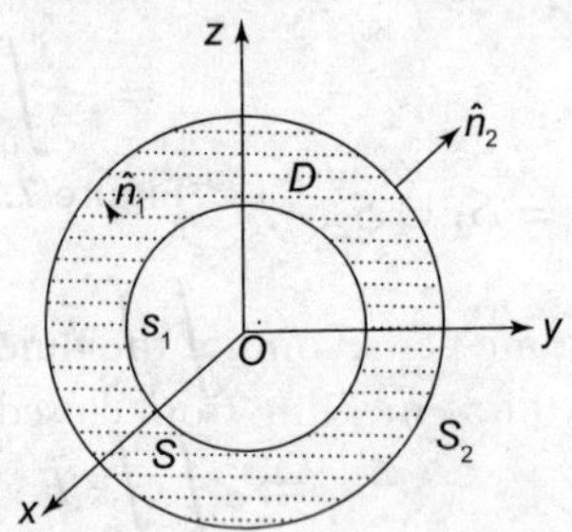

Figure 7.23 $\mathcal{S}$ is the shaded region

Let $\hat{n}$ be the unit normal(in the positive direction, i.e., outward) to the surface $\mathcal{S}$ enclosing the region $\mathcal{D}$. Now, $\hat{n}_1$ is the unit normal to the surface $\mathcal{S}_1$ directed away from the origin, i.e., it is inward normal, and hence, $-\hat{n}_1$ is the unit outward (i.e. positive) normal to $\mathcal{S}_1$ and $\hat{n}_2$ is the unit outward normal to $\mathcal{S}_2$. Now

$$\iint_{\mathcal{S}} \vec{F}\cdot\hat{n}\,dS = \iint_{\mathcal{S}_1} (\phi\vec{\nabla}\phi)\cdot(-\hat{n}_1)dS_1 + \iint_{\mathcal{S}_2} (\phi\vec{\nabla}\phi)\cdot\hat{n}_2\,dS_1$$

$$= -\iint_{\mathcal{S}_1} (\phi\vec{\nabla}\phi)\cdot\hat{n}_1 dS_1 \text{ as } \phi = 0 \text{ on } \mathcal{S}_2$$

Using Gauss's divergence theorem, we have

$$\iiint_{\mathcal{D}} \vec{\nabla}\cdot\vec{F}dV = \iiint_{\mathcal{D}} \vec{\nabla}\cdot(\phi\,\vec{\nabla}\phi)\,dV = \iint_{\mathcal{S}} \vec{F}\cdot\hat{n}\,dS$$

$$\text{or,}\quad \iiint_{\mathcal{D}} \left[|\vec{\nabla}\phi|^2\,dV = -\iint_{\mathcal{S}_1} (\phi\vec{\nabla}\phi)\cdot\hat{n}_1 dS_1\right.$$

$$\Rightarrow \iiint_{\mathcal{D}} |\nabla^2\phi|\,dV + \iint_{\mathcal{S}_1} (\phi\,\vec{\nabla}\phi)\cdot\hat{n}\,dS = 0$$

RESULT 33

Physical interpretation of the divergence theorem: A fluid moves so that its velocity at any point is $\vec{v}(x,y,z)$. Let $\vec{A}$ = velocity $\vec{v}$ at any point of a moving fluid. From Fig. 7.24(a)

Volume of fluid crossing dS in Δt seconds

= Volume contained in cylinder of base dS and slant height $\vec{v}\,\Delta t$

$= (\vec{v}\,\Delta t)\cdot\hat{n}dS = \vec{v}\cdot\hat{n}dS\,\Delta t$

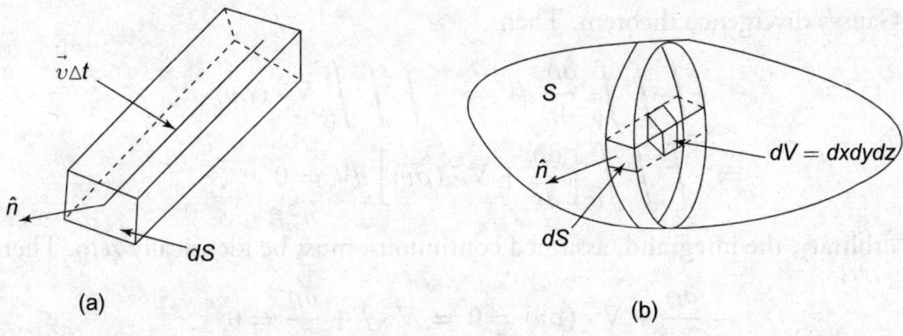

Figure 7.24 Gauss's divergence theorem

Then, volume per second of the fluid crossing $dS = \vec{v} \cdot \hat{n} dS$. From Fig. 7.24(b), total volume per second of fluid emerging from closed surface S

$$= \int\int_S \vec{v} \cdot \hat{n}\ dS$$

Now, $\vec{\nabla} \cdot \vec{v}\ dV$ is the volume per second of fluid emerging from a volume element dV. Total volume per second of fluid emerging from all volume elements in S

$$= \int\int_V \vec{\nabla} \cdot \vec{v}\ dV$$

Therefore, $\displaystyle\int\int_S \vec{v} \cdot \hat{n}\ dS = \int\int_V \vec{\nabla} \cdot \vec{v}\ dV.$

Example 7.2.20

A fluid of density $\rho(x, y, z, t)$ moves with velocity $\vec{v}(x, y, z, t)$. If there are no sources or sinks, prove that $\vec{\nabla} \cdot \vec{J} + \dfrac{\partial \rho}{\partial t} = 0$, where $\vec{J} = \rho\,\vec{v}$.

Solution

Consider an arbitrary surface enclosing a volume V of the fluid. At any time, the mass of fluid within V is

$$M = \int\int\int_V \rho\, dV$$

The time rate of increase of this mass is

$$\frac{\partial M}{\partial t} = \frac{\partial}{\partial t}\int\int\int_V \rho\, dV = \int\int\int_V \frac{\partial \rho}{\partial t}\, dV$$

The mass of fluid per unit time leaving V is $\displaystyle\int\int_S \rho\,\vec{v} \cdot \hat{n}\ dS$ and the time rate of increase in mass is therefore

$$= -\int\int_S \rho\,\vec{v} \cdot \hat{n}\ dS = -\int\int\int_V \vec{\nabla} \cdot (\rho\vec{v})\, dV$$

This is by Gauss's divergence theorem. Then

$$\iiint_V \frac{\partial \rho}{\partial t}\, dV = -\iiint_V \vec{\nabla} \cdot (\rho \vec{v})\, dV$$
$$\Rightarrow \iiint_V \left[\frac{\partial \rho}{\partial t} + \vec{\nabla} \cdot (\rho \vec{v})\right] dV = 0$$

Since V is arbitrary, the integrand, assumed continuous, must be identically zero. Then

$$\frac{\partial \rho}{\partial t} + \vec{\nabla} \cdot (\rho \vec{v}) = 0 \Rightarrow \vec{\nabla} \cdot \vec{J} + \frac{\partial \rho}{\partial t} = 0$$

where $\vec{J} = \rho\, \vec{v}$. The equation is called the *equation of continuity.* This equation also arises in electromagnetic theory, where ρ is the *charge density* and $\vec{J} = \rho\, \vec{v}$ is the *current density.*

If ρ is a constant, the fluid is incompressible and $\vec{\nabla} \cdot \vec{v} = 0$, i.e., $\vec{v}$ is solenoidal.

Example 7.2.21

If the temperature at any point (x, y, z) of a solid at time t is $T(x, y, z, t)$ and if κ, ρ and c are, respectively, the thermal conductivity, density and specific heat of the solid, assumed constant, show that $\frac{\partial T}{\partial t} = k\, \nabla^2 T$, where $k = \kappa / \rho\, c$.

Solution

Let V be an arbitrary volume lying within the solid, and let S denote its surface. The heat contained in a volume V is given by $H = \iiint_V c\, \rho\, T\, dV$. Then, the time rate of increase of heat is

$$\frac{\partial H}{\partial t} = \frac{\partial}{\partial t} \iiint_V c\, \rho\, T\, dV = \iiint_V c\, \rho \frac{\partial T}{\partial t}\, dV$$

The total heat across S, or the quantity of heat leaving S per unit time, is $\iint_S (-\kappa\, \vec{\nabla} T) \cdot \hat{n}\, dS$ and so the quantity of heat entering S per unit time is

$$= -\iint_S (-\kappa\, \vec{\nabla} T) \cdot \hat{n}\, dS = \iiint_V \vec{\nabla} \cdot (\kappa\, \vec{\nabla} T)\, dV$$

This is by Gauss's divergence theorem. Then

$$\iiint_V c\, \rho \frac{\partial T}{\partial t}\, dV = \iiint_V \vec{\nabla} \cdot (\kappa\, \vec{\nabla} T)\, dV$$
$$\Rightarrow \iiint_V \left[c\, \rho \frac{\partial T}{\partial t} - \vec{\nabla} \cdot (\kappa\, \vec{\nabla} T)\right] dV = 0$$

Since V is arbitrary, the integrand, assumed continuous, must be identically zero. Then

$$c\, \rho \frac{\partial T}{\partial t} - \vec{\nabla} \cdot (\kappa\, \vec{\nabla} T) = 0 \qquad \Rightarrow \qquad \frac{\partial T}{\partial t} = \frac{\kappa}{c\, \rho} \vec{\nabla} \cdot (\vec{\nabla} T) = k \nabla^2 T$$

if κ, ρ and c are constants. The quantity $k = \kappa / \rho\, c$ is called the *diffusivity* . For steady state heat flow, (i.e., T is independent of time) the equation reduces to Laplace's equation, $\nabla^2 T = 0$.

7.3 STOKES' THEOREM

Statement: Let $\mathcal{S}$ be any open, two-sided surface bounded by a closed, non-intersecting curve $\mathcal{C}$. Then, the line integral of the tangential component of a vector $\vec{F}$ taken around the closed curve $\mathcal{C}$ is equal to the surface integral of the normal component of curl $\vec{F}$ taken over any surface $\mathcal{S}$, with $\mathcal{C}$ as boundary. Thus, if $\vec{F}$ has continuous first order derivatives, then

$$\oint_{\mathcal{C}} \vec{F} \cdot d\vec{r} = \int\int_{\mathcal{S}} (\vec{\nabla} \times \vec{F}) \cdot \hat{n}\, dS = \int\int_{\mathcal{S}} (\vec{\nabla} \times \vec{F}) \cdot d\vec{S}$$
$$= \int\int_{\mathcal{S}} (\hat{n} \cdot \vec{\nabla} \times \vec{F})\, dS = \int\int_{\mathcal{S}} (\hat{n} \times \vec{\nabla}) \cdot \vec{F}\, dS \quad (7.35)$$

where $\hat{n}$ is the unique normal vector at any point of $\mathcal{S}$ drawn in the way in which a right-handed screw would move when rotated based on a description of $\mathcal{C}$.

Proof: Let $\mathcal{S}$ be a regular surface whose projected area on the coordinate planes is a region bounded by simple closed curves (Fig. 7.25). Let us take the equation of $\mathcal{S}$ as $z = f(x, y)$, where f is a single-valued, continuous function having continuous partial derivatives. Consider first the integral

$$\int\int_{\mathcal{S}} \left[\vec{\nabla} \times (F_1\hat{\imath})\right] \cdot \hat{n}\, dS$$

where F_1 is the component of $\vec{F}$ along $\hat{\imath}$. If $\hat{r} = x\hat{\imath} + y\hat{\jmath} + z\hat{k}$ is the position vector of any point on $\mathcal{S}$, then

$$\frac{\partial \vec{r}}{\partial y} = \hat{\jmath} + \frac{\partial \vec{z}}{\partial y}\hat{k} = \hat{\jmath} + \frac{\partial f}{\partial y}\hat{k}$$

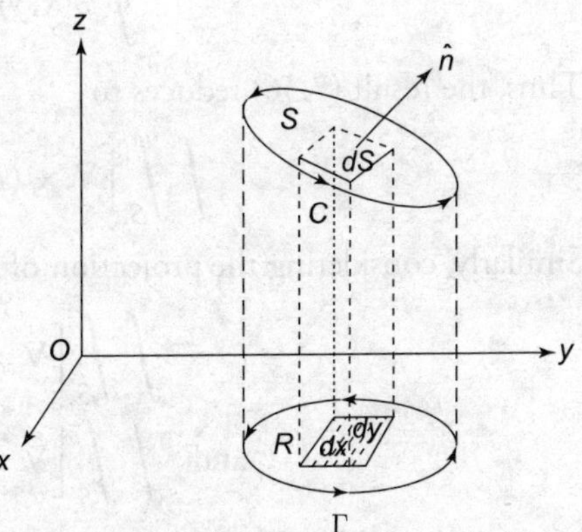

Figure 7.25 Stokes' theorem

as $\mathcal{S} : z = f(x, y)$ is a vector in the direction of the tangent to $\mathcal{S}$, and thus, perpendicular to $\hat{n}$, so that

$$\hat{n} \cdot \frac{\partial \vec{r}}{\partial y} = 0 \Rightarrow \hat{n} \cdot \hat{\jmath} + \frac{\partial z}{\partial y}\hat{n} \cdot \hat{k} = 0 \Rightarrow \hat{n} \cdot \hat{\jmath} = -\frac{\partial z}{\partial y}\hat{n} \cdot \hat{k}$$

Now, for the points on $\mathcal{S}, F_1(x, y, z) = F_1(x, y, f(x, y)) = h(x, y)$, using elementary calculus, we have

$$\frac{\partial h}{\partial y} = \frac{\partial F_1}{\partial y} + \frac{\partial F_1}{\partial z}\frac{\partial z}{\partial y}$$

Let $\mathcal{R}$ be the projected area of $\mathcal{S}$ on the xy plane bounded by the closed curve Γ. Again, using the result

$$\vec{\nabla} \times (F_1\hat{\imath}) = \begin{vmatrix} \hat{\imath} & \hat{\jmath} & \hat{k} \\ \dfrac{\partial}{\partial x} & \dfrac{\partial}{\partial y} & \dfrac{\partial}{\partial z} \\ F_1 & 0 & 0 \end{vmatrix} = \frac{\partial F_1}{\partial z}\hat{\jmath} - \frac{\partial F_1}{\partial y}\hat{k}$$

From Eq. (7.35), we get

$$\int\int_S \left[\vec{\nabla}\times(F_1\hat{i})\right]\cdot\hat{n}\,dS = \int\int_S \left(\vec{\nabla}\cdot\hat{j}\frac{\partial F_1}{\partial z} - \hat{n}\cdot\hat{k}\frac{\partial F_1}{\partial y}\right)dS$$

$$= -\hat{n}\cdot\hat{k}\int\int_S \left(\frac{\partial F_1}{\partial z}\frac{\partial z}{\partial y} + \frac{\partial F_1}{\partial y}\right)dS$$

$$= -\hat{n}\cdot\hat{k}\int\int_S \frac{\partial h}{\partial y}\,dS = -\int\int_{\mathcal{R}} \frac{\partial h}{\partial y}\,dx\,dy$$

$$= \oint_\Gamma h(x,y)\,dx\,dy \quad \text{by Green's theorem} \tag{7.36}$$

Now, the value of $h(x,y)$ at the point on Γ and that of $F_1(x,y,z)$ at any point on $\mathcal{C}$ are the same and the variation dx of x on both the curves $\mathcal{C}$ and Γ are also equal. Thus

$$\oint h(x,y)\,dx\,dy = \oint_{\mathcal{C}} F_1(x,y,z)\,dx$$

Thus, the result (7.36) reduces to

$$\int\int_S \left[\vec{\nabla}\times(F_1\hat{i})\right]\cdot\hat{n}\,dS = \oint_{\mathcal{C}} F_1(x,y,z)\,dx$$

Similarly, considering the projection of $\mathcal{S}$ on the yz and zx planes, we obtain

$$\int\int_S \left[\vec{\nabla}\times(F_2\hat{j})\right]\cdot\hat{n}\,dS = \oint_{\mathcal{C}} F_2(x,y,z)\,dy$$

$$\text{and}\quad \int\int_S \left[\vec{\nabla}\times(F_3\hat{k})\right]\cdot\hat{n}\,dS = \oint_{\mathcal{C}} F_3(x,y,z)\,dz$$

respectively. Adding, we get

$$\int\int_S \left[\vec{\nabla}\times(F_1\hat{i}+F_2\hat{j}+F_3\hat{k})\right]\cdot\hat{n}\,dS = \oint_{\mathcal{C}} (F_1\,dx + F_2\,dy + F_3\,dz)$$

$$\Rightarrow \int\int_S \left[\vec{\nabla}\times F\right]\cdot\hat{n}\,dS = \oint_{\mathcal{C}} \vec{F}\cdot d\vec{r}$$

where $\mathcal{C}$ is the boundary of $\mathcal{S}$.

RESULT 34

Stokes' theorem in words: The line integral of the tangential component of a vector $\vec{F}$ taken around a simple closed curve $\mathcal{C}$ is equal to the surface integral of the normal component of the curl of $\vec{F}$ taken over any surface $\mathcal{S}$, having $\mathcal{C}$ as its boundary.

RESULT 35

Since $d\vec{r} = \dfrac{d\vec{r}}{ds}\,ds = \hat{t}\,ds$, where $\hat{t}$ is the unit tangent vector and ds is the differential of arc lengths of $\mathcal{C}$, Eq. (7.35) reduces to

$$\int\int_S \vec{\nabla}\times\vec{F}\cdot\hat{n}\,dS = \oint_{\mathcal{C}} \vec{F}\cdot\hat{t}\,dS$$

Thus, the line integral of the tangential component of $\vec{F}$ taken around a simple closed curve C is equal to the surface integral of the normal component of the curl of $\vec{F}$ taken over the surface S, having C as its boundary.

RESULT 36

Cartesian representation of Stokes' theorem: As in the derivation of Eq. (7.11), let $\vec{A}(x,y,z) = A_1\hat{i} + A_2\hat{j} + A_3\hat{k}$, so that

$$\vec{A} \cdot d\vec{r} = A_1 dx + A_2 dy + A_3 dz$$

Also, let the unit normal to S be

$$\hat{n} = n_1\hat{i} + n_2\hat{j} + n_3\hat{k} = \cos\alpha\,\hat{i} + \cos\beta\,\hat{j} + \cos\gamma\,\hat{k}$$

where α, β and γ are the angles which $\hat{n}$ makes with the positive x, y and z axes or the $\hat{i}, \hat{j}$ and $\hat{k}$ directions, respectively. Then, using definition (5.55)

$$\vec{\nabla} \times \vec{A} = \left(\frac{\partial A_3}{\partial y} - \frac{\partial A_2}{\partial z}\right)\hat{i} + \left(\frac{\partial A_1}{\partial z} - \frac{\partial A_3}{\partial x}\right)\hat{j} + \left(\frac{\partial A_2}{\partial x} - \frac{\partial A_1}{\partial y}\right)\hat{k}$$

$$\Rightarrow (\vec{\nabla} \times \vec{A}) \cdot \hat{n} = \left(\frac{\partial A_3}{\partial y} - \frac{\partial A_2}{\partial z}\right)\cos\alpha + \left(\frac{\partial A_1}{\partial z} - \frac{\partial A_3}{\partial x}\right)\cos\beta + \left(\frac{\partial A_2}{\partial x} - \frac{\partial A_1}{\partial y}\right)\cos\gamma$$

Thus, Stokes' theorem (7.35) takes the form

$$\oint_C A_1 dx + A_2 dy + A_3 dz$$

$$= \int\!\!\int_S \left[\left(\frac{\partial A_3}{\partial y} - \frac{\partial A_2}{\partial z}\right)\cos\alpha + \left(\frac{\partial A_1}{\partial z} - \frac{\partial A_3}{\partial x}\right)\cos\beta + \left(\frac{\partial A_2}{\partial x} - \frac{\partial A_1}{\partial y}\right)\cos\gamma\right] dS$$

$$= \int\!\!\int_S \left[\left(\frac{\partial A_3}{\partial y} - \frac{\partial A_2}{\partial z}\right)dy\,dz + \left(\frac{\partial A_1}{\partial z} - \frac{\partial A_3}{\partial x}\right)dz\,dx + \left(\frac{\partial A_2}{\partial x} - \frac{\partial A_1}{\partial y}\right)dy\,dx\right] \quad (7.37)$$

Example 7.3.1

Verify Stokes' theorem (Eq. 7.35) for the function $\vec{F} = (2y+z)\hat{i} + (x-z)\hat{j} + (y-x)\hat{k}$ over the triangle ABC cut from the plane $x+y+z=1$ by the coordinate planes.

Solution

The plane cuts the x axis at $A(1,0,0)$, y axis at $B(0,1,0)$ and z axis at $C(0,0,1)$ (Fig. 7.26). Thus, the plane having vertices $A(1,0,0), B(0,1,0)$ and $C(0,0,1)$ is given in Fig. 7.26. The equation of the lines AB, BC and CA are, respectively, $y = 1 - x$, $y = 1 - z$ and $x + z = 1$. Here, S is the surface of the plane $x + y + z = 1$, $x \geq 0, y \geq 0, z \geq 0$ and C is the boundary of S traversed in the counter-clockwise direction.

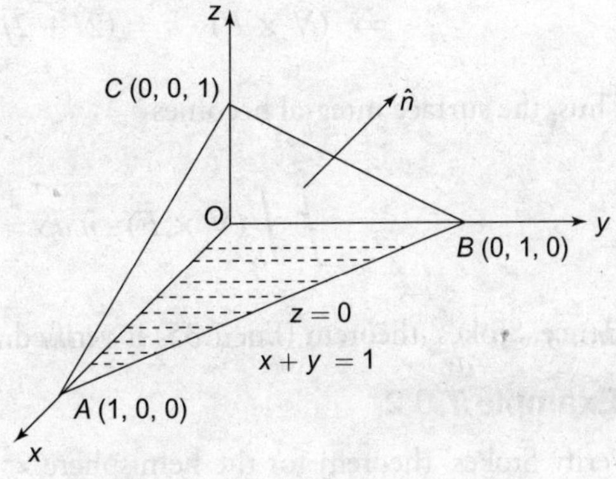

Figure 7.26 Verification of Stokes' theorem

Evaluation of line integral: Along the line AB

(Fig. 7.26), $x+y=1$, $dy=-dx$ and $z=0$, x varies from 1 to 0. So, along the line AB, the line integral is

$$\int_{AB} \vec{F}\cdot d\vec{r} = \int_1^0 \Big[2(1-x)-x\Big]\,dx = -2+\frac{3}{2} = -\frac{1}{2}$$

Along the line BC (Fig. 7.26), $y+z=1$, $dy=-dz$ and $x=0$, y varies from 1 to 0. So, along the line BC, the line integral is

$$\int_{BC} \vec{F}\cdot d\vec{r} = \int_1^0 \Big[y-1-y\Big]\,dy = 1$$

Along the line CA (Fig. 7.26), $x+z=1$, $dx=-dz$ and $y=0$, x varies from 0 to 1. So, along the line CA, the line integral is

$$\int_{CA} \vec{F}\cdot d\vec{r} = \int_0^1 \Big[1-x+x\Big]\,dx = 1$$

Adding all these results, we get

$$\oint_C \vec{F}\cdot d\vec{r} = \oint_C (2y+z)dx + (x-z)dy + (y-x)dz$$

$$= -\frac{1}{2}+1+1 = 3/2$$

Evaluation of the surface integral: Let $\phi = x+y+z-1$. If $\hat{n}$ be the unit normal vector to the surface, then

$$\hat{n} = \frac{1}{|\vec{\nabla}\phi|}\,\vec{\nabla}\phi = \frac{1}{\sqrt{3}}(\hat{\imath}+\hat{\jmath}+\hat{k})$$

Since $\vec{F} = (2y+z)\hat{\imath} + (x-z)\hat{\jmath} + (y-x)\hat{k}$, by the definition of curl, we have

$$\vec{\nabla}\times\vec{F} = \begin{vmatrix} \hat{\imath} & \hat{\jmath} & \hat{k} \\ \dfrac{\partial}{\partial x} & \dfrac{\partial}{\partial y} & \dfrac{\partial}{\partial z} \\ 2y+z & x-z & y-x \end{vmatrix} = 2\hat{\imath}+2\hat{\jmath}-\hat{k}$$

$$\Rightarrow (\vec{\nabla}\times\vec{F})\cdot\hat{n} = (2\hat{\imath}+2\hat{\jmath}-\hat{k})\cdot\frac{1}{\sqrt{3}}(\hat{\imath}+\hat{\jmath}+\hat{k}) = \frac{3}{\sqrt{3}}$$

Thus, the surface integral becomes

$$\int\!\!\int_S (\vec{\nabla}\times\vec{F})\cdot\hat{n}\,dS = \int_0^1\Big[\int_0^{1-x}\frac{3}{\sqrt{3}}\sqrt{3}\,dy\Big]dx = \frac{3}{2}$$

Hence, Stokes' theorem (Eq. 7.35) is verified.

Example 7.3.2

Verify Stokes' theorem for the hemisphere $x^2+y^2+z^2=9$, $z\geq 0$ and the vector field $\vec{F} = (z^2-y)\hat{\imath} + (x-2yz)\hat{\jmath} + (2xz-y^2)\hat{k}$.

Solution

Here, $\mathcal{S}$ is the upper half of the sphere $x^2+y^2+z^2=1, z \geq 0$. The boundary $\mathcal{C}: x^2+y^2+z^2=9, z=0$ of $\mathcal{S}$ is a circle in the xy plane of radius 3 and with centre at origin O. Let $x=3\cos\theta$, $y=3\sin\theta(0 \leq \theta \leq 2\pi)$ be the parametric equation of $\mathcal{C}$. Then

$$\begin{aligned}\oint_C \vec{F}\cdot d\vec{r} &= \oint_C (0^2-y)dx+(x-2y\cdot 0)dy+(2xz-y^2)0\\ &= \oint_C (-y\,dx+x\,dy);\quad x=3\cos\theta, y=3\sin\theta\\ &= \int_0^{2\pi} (9\sin^2\theta+9\cos^2\theta)\,d\theta = 18\pi\end{aligned}$$

The unit vector $\hat{n}$ to the surface $z=0$ is $\hat{n}=\hat{k}$. Using Eq. (5.55) as the definition of curl, we get

$$\vec{\nabla}\times\vec{F} = \begin{vmatrix}\widehat{i} & \widehat{j} & \widehat{k}\\ \dfrac{\partial}{\partial x} & \dfrac{\partial}{\partial y} & \dfrac{\partial}{\partial z}\\ z^2-y & x-2yz & 2xz-y^2\end{vmatrix} = 2k$$

$$\Rightarrow\quad (\vec{\nabla}\times\vec{F})\cdot\hat{n} = 2\hat{k}\cdot\hat{k} = 2$$

Here, $\mathcal{S}$ is the great sphere of radius 3. Let $\mathcal{R}$ be the projection of $\mathcal{S}$ on the xy plane. Therefore

$$\begin{aligned}\int\int_S (\vec{\nabla}\times\vec{F})\cdot\hat{n}\,dS &= \int_{-3}^{3}\left[\int_{-\sqrt{9-x^2}}^{\sqrt{9-x^2}} 2dy\right]dx\\ &= 8\int_0^3\left[\int_0^{\sqrt{9-x^2}} dy\right]dx = 8\int_0^3 \sqrt{9-x^2}dx\\ &= 8\left[\frac{x\sqrt{9-x^2}}{2}+\frac{9}{2}\sin^{-1}\frac{x}{3}\right]_0^3 = 8\times\frac{9}{2}\times\frac{\pi}{2} = 18\pi\end{aligned}$$

Hence, Stokes' theorem (Eq. 7.35) is verified.

RESULT 37

Let $\mathcal{S}$ be the surface bounded by the simple closed curve $\mathcal{C}$ and $\hat{n}$ be the unit outward drawn normal to it. Noting the fact $\vec{\nabla}\times\vec{r}=\vec{0}$ and using Stokes' theorem (Eq. 7.35), we get

$$\oint_C \vec{r}\cdot d\vec{r} = \int\int_S (\vec{\nabla}\times\vec{r})\cdot\hat{n}\,dS = 0 \qquad (7.38)$$

RESULT 38

Let us remember the following results:

(i) $\vec{\nabla}(\phi\,\psi) = \phi\,\vec{\nabla}\psi+\psi\vec{\nabla}\phi$

(ii) $\vec{\nabla}\times(\phi\,\vec{\nabla}\psi) = \vec{\nabla}\phi\times\vec{\nabla}\psi+\phi\vec{\nabla}\times(\vec{\nabla}\psi) = \vec{\nabla}\phi\times(\vec{\nabla}\psi)$

Let $\mathcal{S}$ be the surface bounded by the simple closed curve $\mathcal{C}$ and $\hat{n}$ be the unit outward drawn normal to it. We have $\vec{\nabla}\times(\vec{\nabla}f)=\vec{0}$, for the scalar point function f. Now, using the first result

(i) and Stokes' theorem (Eq. 7.35), we get

$$\oint_C \vec{\nabla}(\phi\,\psi)\cdot d\vec{r} = \int\!\!\int_S \vec{\nabla}\times\vec{\nabla}(\phi\,\psi)\cdot\hat{n}\,dS = 0$$

$$\Rightarrow \oint_C \vec{\nabla}\left[\phi\,\vec{\nabla}\psi + \psi\vec{\nabla}\phi\right]\cdot d\vec{r} = 0$$

$$\Rightarrow \oint_C \phi\,\vec{\nabla}\psi\cdot d\vec{r} = -\oint_C \psi\,\vec{\nabla}\phi\cdot d\vec{r} \tag{7.39}$$

Putting $\phi = \psi$ in the result (7.39), we get

$$\oint_C \phi\,\vec{\nabla}\phi\cdot d\vec{r} = -\oint_C \phi\,\vec{\nabla}\phi\cdot d\vec{r}$$

$$\Rightarrow 2\oint_C \phi\,\vec{\nabla}\phi\cdot d\vec{r} = 0 \Rightarrow \oint_C \phi\,\vec{\nabla}\phi\cdot d\vec{r} = 0 \tag{7.40}$$

Now, using the second result (ii) and Stokes' theorem (Eq. 7.35), we get

$$\int\!\!\int_S (\vec{\nabla}\phi\times\vec{\nabla}\psi)\cdot\hat{n}\,dS = \int\!\!\int_S \vec{\nabla}\times(\phi\,\vec{\nabla}\psi)\cdot\hat{n}\,dS = \oint_C \phi\vec{\nabla}\psi\cdot d\vec{r} \tag{7.41}$$

Result 39

Let $\vec{A}$ be an arbitrary constant vector. Then, we have

$$\vec{\nabla}\times(\vec{A}\times\vec{r}) = \vec{A}(\vec{\nabla}\cdot\vec{r}) - (\vec{A}\cdot\vec{\nabla})\vec{r} = 3\vec{A} - \vec{A} = 2\vec{A}$$

Let S be the surface bounded by the simple closed curve C and $\hat{n}$ be the unit outward drawn normal to it. Now, using Stokes' theorem (Eq. 7.35) on $\vec{A}\times\vec{r}$, we get

$$\oint_C (\vec{A}\times\vec{r})\cdot d\vec{r} = \int\!\!\int_S \vec{\nabla}\times(\vec{A}\times\vec{r})\cdot\hat{n}\,dS$$

$$\Rightarrow \oint_C \vec{A}\cdot(\vec{r}\times d\vec{r}) = \int\!\!\int_S 2\vec{A}\cdot\hat{n}\,dS$$

$$\Rightarrow \vec{A}\cdot\left[\oint_C \vec{r}\times d\vec{r} - \int\!\!\int_S 2\hat{n}\,dS\right] = 0$$

Since $\vec{A}$ is an arbitrary constant vector

$$\oint_C \vec{r}\times d\vec{r} = 2\int\!\!\int_S \hat{n}\,dS = 2\int\!\!\int_S d\vec{S} \tag{7.42}$$

Result 40

Let S be the surface bounded by the simple closed curve C and $\hat{n}$ be the unit outward drawn normal to it. Let $\vec{C}$ be an arbitrary constant vector so that $\vec{\nabla}\times\vec{C} = \vec{0}$. Let f be a continuously differentiable scalar point function. Then

$$\vec{\nabla}\times(f\vec{C}) = f\vec{\nabla}\times\vec{C} + \vec{\nabla}f\times\vec{C} = \vec{\nabla}f\times\vec{C}$$

Now, using Stokes' theorem (Eq. 7.35) on $\vec{F} = f\vec{C}$, we get

$$\oint_C f\,\vec{C}\cdot d\vec{r} = \int\int_S \vec{\nabla}\times(f\vec{C})\cdot\hat{n}\,dS$$

$$\Rightarrow \vec{C}\cdot\oint_C f\,d\vec{r} = \int\int_S (\vec{\nabla}f\times\vec{C})\cdot\hat{n}\,dS$$

$$\Rightarrow \vec{C}\cdot\oint_C f\,d\vec{r} = \int\int_S (\vec{C}\cdot\hat{n}\times\vec{\nabla}f)dS = \vec{C}\cdot\int\int_S (\hat{n}\times\vec{\nabla})f\,dS$$

Since $\vec{C}$ is an arbitrary constant vector

$$\int\int_S (\hat{n}\times\vec{\nabla})f\,dS = \oint_C f\,d\vec{r} \tag{7.43}$$

RESULT 41

Let S be the surface bounded by the simple closed curve C and $\hat{n}$ be the unit outward drawn normal to it. Let $\vec{C}$ be an arbitrary constant vector. Now, using Stokes' theorem (Eq. 7.35) on $\vec{F} = \vec{A}\times\vec{C}$, we get

$$\int\int_S \left[\vec{\nabla}\times(\vec{A}\times\vec{C})\right]\cdot\hat{n}\,dS = \oint_C (\vec{A}\times\vec{C})\cdot d\vec{r}$$

$$\Rightarrow \int\int_S \left[\vec{A}(\vec{\nabla}\cdot\vec{C}) - (\vec{C}\cdot\vec{\nabla})\cdot\vec{A} + (\vec{C}\cdot\vec{\nabla})\vec{A} - (\vec{A}\cdot\vec{\nabla})\vec{C}\right]\cdot\hat{n}\,dS = \oint_C (\vec{A}\times\vec{C})\cdot d\vec{r}$$

$$\Rightarrow \int\int_S \left[-\vec{C}(\vec{\nabla}\cdot\vec{A}) + (\vec{C}\cdot\vec{\nabla})\vec{A}\right]\cdot\hat{n}\,dS = -\oint_C \vec{C}\cdot(\vec{A}\times d\vec{r})$$

$$\Rightarrow \int\int_S \left[\vec{C}\cdot(\vec{\nabla}\cdot\vec{A})\hat{n} - \vec{C}\cdot\vec{\nabla}(\vec{A}\cdot\hat{n})\right]dS = -\oint_C \vec{C}\cdot(\vec{A}\times d\vec{r})$$

$$\left[\text{here}\,\vec{\nabla}\text{ does not operate on }\hat{n}\right]$$

$$\Rightarrow \vec{C}\cdot\int\int_S \left[\vec{\nabla}(\vec{A}\cdot\hat{n}) - (\vec{\nabla}\cdot\vec{A})\hat{n}\right]dS + \vec{C}\cdot\oint_C (\vec{A}\times d\vec{r}) = 0$$

$$\Rightarrow \vec{C}\cdot\left[\int\int_S \left\{(\hat{n}\times\vec{\nabla})\times\vec{A}\right\}dS + \oint_C \vec{A}\times d\vec{r}\right] = 0$$

Since $\vec{C}$ is an arbitrary vector, we have

$$\int\int_S \left\{(\hat{n}\times\vec{\nabla})\times\vec{A}\right\}dS = -\oint_C \vec{A}\times d\vec{r} = \oint_C d\vec{r}\times\vec{A} \tag{7.44}$$

Example 7.3.3

Verify Stokes' theorem for $\vec{F} = \sin z\hat{i} - \cos x\hat{j} + \sin y\hat{k}$, where S is the surface of the rectangle $0 \le x \le \pi, 0 \le y \le 1, z = 3$.

Solution

Here, S is the surface of the rectangle $0 \le x \le \pi$, $0 \le y \le 1$, $z = 3$ (Fig. 7.27).

Let $\mathcal{C}$ be the boundary of the surface $\mathcal{S}$, the rectangle $ABCD$ traversed in the couter-clockwise sense. Since $\vec{F} = \sin z\hat{\imath} - \cos x\hat{\jmath} + \sin y\hat{k}$, $\vec{F}\cdot d\vec{r} = \sin z\, dx - \cos x\, dy + \sin y\, dz$. On the line segment AB, $x = \pi, z = 3$ and $0 \le y \le 1$

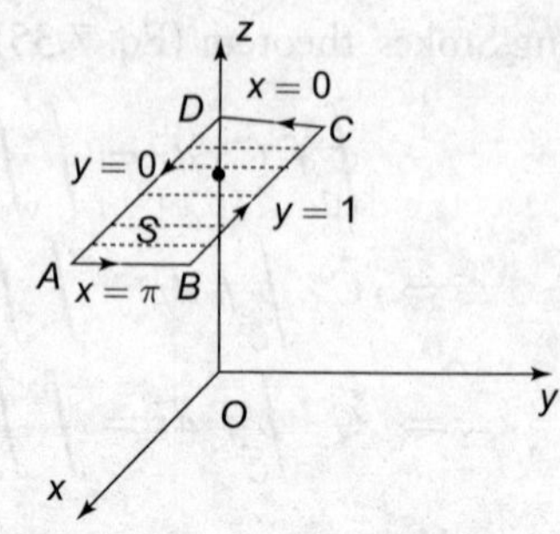

Figure 7.27 Rectangle $0 \le x \le \pi, 0 \le y \le 1, z = 3$

$$\int_{AB} \vec{F}\cdot d\vec{r} = \int_0^1 -\cos\pi\; dy = 1$$

On the line segment BC, $y = 1, z = 3$ and x varies from π to 0, so

$$\int_{BC} \vec{F}\cdot d\vec{r} = \int_\pi^0 \sin 3\; dx = -\pi \sin 3$$

On the line segment CD, $x = 0, z = 3$ and y varies from 1 to 0, so

$$\int_{CD} \vec{F}\cdot d\vec{r} = \int_1^0 -dy = 1$$

On the line segment DA, $y = 0, z = 3$ and $0 \le x \le \pi$, so

$$\int_{DA} \vec{F}\cdot d\vec{r} = \int_0^\pi \sin 3\; dx = \pi \sin 3$$

Adding the results, we get

$$\begin{aligned}\oint \vec{F}\cdot d\vec{r} &= \int_{AB} \vec{F}\cdot d\vec{r} + \int_{BC} \vec{F}\cdot d\vec{r} + \int_{CD} \vec{F}\cdot d\vec{r} + \int_{DA} \vec{F}\cdot d\vec{r}\\ &= 1 - \pi\sin 3 + 1 + \pi \sin 3 = 2\end{aligned}$$

Since $\vec{F} = \sin z\hat{\imath} - \cos x\hat{\jmath} + \sin y\hat{k}$, using the definition of curl, we get

$$\vec{\nabla}\times\vec{F} = \begin{vmatrix} \hat{\imath} & \hat{\jmath} & \hat{k} \\ \dfrac{\partial}{\partial x} & \dfrac{\partial}{\partial y} & \dfrac{\partial}{\partial z} \\ \sin z & -\cos x & \sin y \end{vmatrix} = \cos y\hat{\imath} + \cos z\hat{\jmath} + \sin x\hat{k}$$

Now, $\hat{n}$ is the outward unit normal to the surface $S(z = 3)$, i.e., $\hat{n} = \hat{k}$, so that $(\vec{\nabla}\times\vec{F})\cdot\hat{n} = (\vec{\nabla}\times\vec{F})\cdot\hat{k} = \sin x$. Thus

$$\iint_S (\vec{\nabla}\times\vec{F})\cdot\hat{n}\; dS = \int_0^\pi \left[\int_0^1 \sin x\; dy\right] dx = 2$$

Hence, Stokes' theorem is verified.

Example 7.3.4

Let $\mathcal{C}$ be the curve in $\Re^3$ given by $x^2 + y^2 = a^2, z = 0$ traced counter-clockwise, and let $\vec{F} = x^2y^3\,\hat{\imath} + \hat{\jmath} + z\hat{k}$. Using Stokes' theorem, evaluate $\oint \vec{F}\cdot d\vec{r}$.

Solution

Here, C : enclosing circle is $x^2 + y^2 + z^2 = a^2, z = 0$, i.e., $x^2 + y^2 = 9$ (Fig. 7.28). Using Eq. (5.55) as the definition of curl, we get

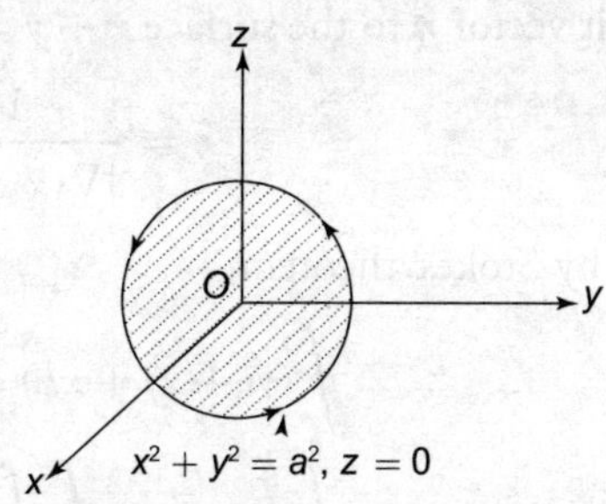

Figure 7.28 Verification of Stokes' theorem

$$\vec{\nabla} \times \vec{F} = \begin{vmatrix} \hat{i} & \hat{j} & \hat{k} \\ \dfrac{\partial}{\partial x} & \dfrac{\partial}{\partial y} & \dfrac{\partial}{\partial z} \\ x^2y^3 & 1 & z \end{vmatrix} = -3x^2y^2\,\hat{k}$$

Here, S is the great sphere of radius a. The unit vector $\hat{n}$ to the surface $z = 0$ is $\hat{n} = \hat{k}$. Let $x = \cos\theta, y = \sin\theta$; then by Stoke's theorem

$$\oint_C \vec{F} \cdot d\vec{r} = \int\int_S (\vec{\nabla} \times \vec{F}) \cdot \hat{n}\, dS = \int\int_S (-3x^2y^2)\,\hat{k} \cdot \hat{k}\, dS$$

$$= -3 \int_{-a}^{a} \left[\int_{-\sqrt{a^2-x^2}}^{\sqrt{a^2-x^2}} y^2\, dy \right] x^2\, dx$$

$$= -3 \cdot 4 \int_0^a \left[\int_0^{\sqrt{a^2-x^2}} x^2y^2\, dy \right] dx$$

$$= -4 \int_0^a \left(a^2 - x^2\right)^{3/2} x^2\, dx$$

$$= -4a^6\, \frac{3 \cdot 1}{6 \cdot 4 \cdot 2} \times \frac{\pi}{2} = -\frac{\pi}{8} a^6$$

Example 7.3.5

Using Stokes' theorem, evaluate the line integral $\oint_{\mathcal{L}} (y\hat{i} + z\hat{j} + x\hat{k}) \cdot d\vec{r}$, where $\mathcal{L}$ is the intersection of $x^2 + y^2 + z^2 = 1$ and $x + y = 0$ traversed in the clockwise direction when viewed from the point $(1, 1, 0)$.

Solution

The sphere has the centre at $(0, 0, 0)$ and radius unity (Fig. 7.29). Since the plane passes through the point $(0, 0, 0)$, the centre of the sphere, the circle C is a great circle of the sphere. Let $\vec{F} = y\hat{i} + z\hat{j} + x\hat{k}$. Using Eq. (5.55) as the definition of curl, we get

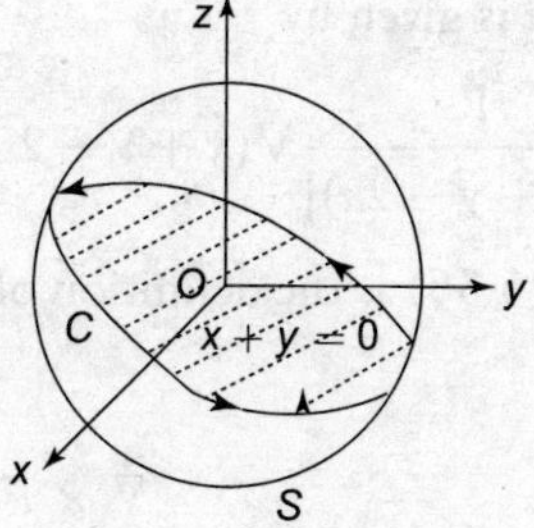

Figure 7.29 Verification of Stokes' theorem

$$\vec{\nabla} \times \vec{F} = \begin{vmatrix} \hat{i} & \hat{j} & \hat{k} \\ \dfrac{\partial}{\partial x} & \dfrac{\partial}{\partial y} & \dfrac{\partial}{\partial z} \\ y & z & x \end{vmatrix} = -(\hat{i} + \hat{j} + \hat{k})$$

The unit vector $\hat{n}$ to the surface $x + y = 0$ is given by

$$\hat{n} = \frac{1}{|\vec{\nabla}(x+y)|}\vec{\nabla}(x+y) = \frac{1}{\sqrt{2}}(\hat{i}+\hat{j})$$

Hence, by Stokes' theorem

$$\oint_C (y\hat{i} + z\hat{j} + x\hat{k}) \cdot d\vec{r} = \oint_C \vec{F} \cdot d\vec{r} = \int\int_S (\vec{\nabla} \times \vec{F}) \cdot \hat{n}\, dS$$

$$\Rightarrow \oint_C \vec{F} \cdot d\vec{r} = \int\int_S -(\hat{i}+\hat{j}+\hat{k}) \cdot \left[\frac{1}{\sqrt{2}}(\hat{i}+\hat{j})\right] dS$$

$$\Rightarrow \oint_C \vec{F} \cdot d\vec{r} = \int\int_S \frac{1+1}{\sqrt{2}}\, dS = \sqrt{2}\int\int_S dS$$

where S is the great circle whose radius is 1. Therefore

$$\oint_C \vec{F} \cdot d\vec{r} = \sqrt{2}\int\int_S dS = \sqrt{2} \cdot \pi \cdot 1^2 = \sqrt{2}\,\pi$$

Example 7.3.6

Apply Stokes' theorem to prove that $\oint_C y\,dx + z\,dy + x\,dz = -2\sqrt{2}\,\pi\, a^2$, where C is the curve given by $x^2+y^2+z^2-2ax-2ay=0$, $x+y=2a$, and begins at the point $(2a, 0, 0)$ and traverses at first, below the z plane.

Solution

Since the plane $x + y = 2a$ (Fig. 7.30) passes through the point $(a, a, 0)$, the centre of the sphere, the circle C is the great circle of the sphere cut by the plane of radius $\sqrt{2}\,a$. The integral is of the form $\oint_C \vec{F} \cdot d\vec{r}$, where $\vec{F} = y\hat{i} + z\hat{j} + x\hat{k}$. The unit vector $\hat{n}$ to the surface $x + y = 2a$ is given by

$$\hat{n} = \frac{1}{|\vec{\nabla}(x+y-2a)|}\vec{\nabla}(x+y-2a) = \frac{1}{\sqrt{2}}(\hat{i}+\hat{j})$$

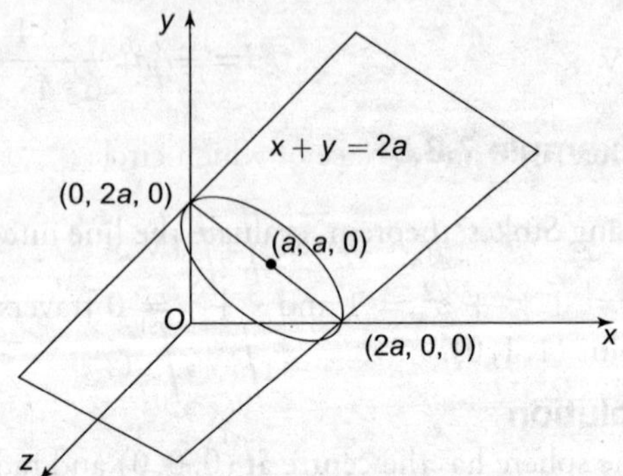

Figure 7.30 Stokes' theorem

Using Eq. (5.55) as the definition of curl, we get

$$\vec{\nabla} \times \vec{F} = \begin{vmatrix} \hat{i} & \hat{j} & \hat{k} \\ \frac{\partial}{\partial x} & \frac{\partial}{\partial y} & \frac{\partial}{\partial z} \\ y & z & x \end{vmatrix} = -(\hat{i}+\hat{j}+\hat{k})$$

$$(\vec{\nabla} \times \vec{F}) \cdot \hat{n} = -(\hat{i}+\hat{j}+\hat{k}) \cdot \frac{1}{\sqrt{2}}(\hat{i}+\hat{j}) = -\sqrt{2}$$

Here, S is any surface of which circle C is the boundary. Using Stokes' theorem, we get

$$\oint_C \vec{F}\cdot d\vec{r} = \int\int_S (\vec{\nabla}\times\vec{F})\cdot\hat{n}\,dS = \int\int_S -(\hat{\imath}+\hat{\jmath}+\hat{k})\cdot\frac{\hat{\imath}+\hat{\jmath}}{\sqrt{2}}\,dS$$
$$= -\sqrt{2}\,\pi(\sqrt{2}\,a)^2 = -2\sqrt{2}\,\pi\,a^2$$

Example 7.3.7

Evaluate $\oint_C \vec{F}\cdot d\vec{r}$, where $\vec{F} = -y^2\hat{\imath} + x\hat{\jmath} + z^2\hat{k}$ and C is the curve of intersection of the plane $y+z=2$ and the cylinder $x^2+y^2=1$.

Solution

The region of integration is shown in Fig. 7.31. The unit vector $\hat{n}$ to the surface $y+z=2$ is given by

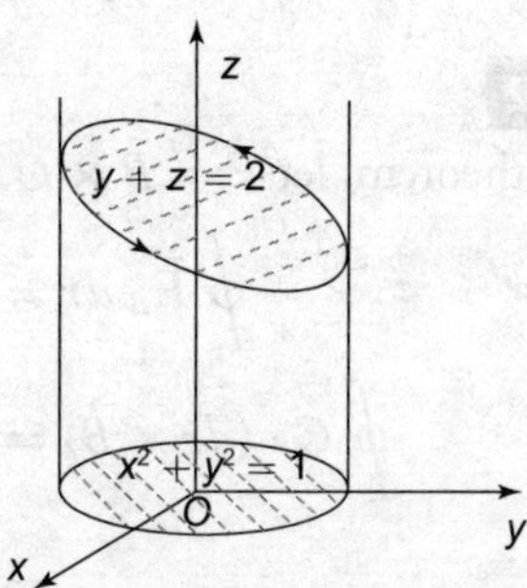

Figure 7.31 Stokes' theorem

$$\hat{n} = \frac{1}{|\vec{\nabla}(y+z-2)|}\vec{\nabla}(y+z-2) = \frac{1}{\sqrt{2}}(\hat{\jmath}+\hat{k})$$

Using Eq. (5.55) as the definition of curl, we get

$$\vec{\nabla}\times\vec{F} = \begin{vmatrix} \hat{\imath} & \hat{\jmath} & \hat{k} \\ \dfrac{\partial}{\partial x} & \dfrac{\partial}{\partial y} & \dfrac{\partial}{\partial z} \\ -y^2 & x & z^2 \end{vmatrix} = (1+2y)\hat{k}$$

$$(\vec{\nabla}\times\vec{F})\cdot\hat{n} = (1+2y)\hat{k}\cdot\frac{1}{\sqrt{2}}(\hat{\jmath}+\hat{k}) = \frac{(1+2y)}{\sqrt{2}}$$

Here, S is any surface of which circle C is the boundary. Using Stokes' theorem, we get

$$\oint_C \vec{F}\cdot d\vec{r} = \int\int_S (\vec{\nabla}\times\vec{F})\cdot\hat{n}\,dS = \int\int_S \frac{(1+2y)}{\sqrt{2}}\,\frac{dx\,dy}{1/\sqrt{2}}$$
$$= \int_0^{2\pi}\int_0^1 (1+2r\sin\theta)\,r\,dr\,d\theta = \int_0^1 r\Big[\theta - 2r\cos\theta\Big]_0^{2\pi} dr$$
$$= \int_0^1 r(2\pi - 2r + 2r)dr = \pi$$

Example 7.3.8

Using Stokes' theorem or otherwise, evaluate $\oint_C \left[(2x-y)dx - yz^2dy - y^2zdz\right]$ where C is the circle $x^2+y^2=1$, corresponding to the surface of the sphere at unit radius.

Solution

The integral is of the form $\oint_C \vec{F}\cdot d\vec{r}$, where $\vec{F} = (2x-y)\hat{\imath} - yz^2\hat{\jmath} - y^2z\hat{k}$. The unit vector $\hat{n}$ to the circle $x^2+y^2=1$ is $\hat{n}=\hat{k}$. Using Eq. (5.55) as the definition of curl, we get

$$\vec{\nabla} \times \vec{F} = \begin{vmatrix} \hat{i} & \hat{j} & \hat{k} \\ \dfrac{\partial}{\partial x} & \dfrac{\partial}{\partial y} & \dfrac{\partial}{\partial z} \\ 2x - y & -yz^2 & -y^2 z \end{vmatrix} = \hat{k}$$

$$(\vec{\nabla} \times \vec{F}) \cdot \hat{n} = \hat{k} \cdot \hat{k} = 1$$

Here, $\mathcal{S}$ is any surface of which circle $\mathcal{C}$ is the boundary. Using Stokes' theorem, we get

$$\oint_C \vec{F} \cdot d\vec{r} = \int\int_S (\vec{\nabla} \times \vec{F}) \cdot \hat{n}\, dS = \int\int_S dx\, dy = \pi$$

RESULT 42

In Stokes' theorem, let $\vec{F} = \vec{B} \times \vec{C}$, where $\vec{C}$ is a constant vector. Thus

$$\oint \vec{F} \cdot d\vec{r} = \oint d\vec{r} \cdot (\vec{B} \times \vec{C}) = \int\int_S \left[\vec{\nabla} \times (\vec{B} \times \vec{C})\right] \hat{n}\, dS$$

$$\oint \vec{C} \cdot (d\vec{r} \times \vec{B}) = \int\int_S \left[(\vec{C} \cdot \vec{\nabla})\vec{B} - \vec{C}(\vec{\nabla} \cdot \vec{B})\right] \hat{n} dS$$

$$= \int\int_S \vec{C} \cdot \left[\vec{\nabla}(\vec{B} \cdot \hat{n})\right] ds - \int\int_S \vec{C} \cdot \left[\hat{n}(\vec{\nabla} \cdot \vec{B})\right] dS$$

or, $$\vec{C} \cdot \oint d\vec{r} \times \vec{B} = \vec{C} \cdot \int\int_S \left[\vec{\nabla}(\vec{B} \cdot \hat{n}) - \hat{n}(\vec{\nabla} \cdot \vec{B})\right] dS$$

$$= \vec{C} \cdot \int\int_S (\hat{n} \times \vec{\nabla}) \times \vec{B}\, dS$$

Since $\vec{C}$ is an arbitrary constant vector, we have

$$\oint d\vec{r} \times \vec{B} = \int\int_S (\hat{n} \times \vec{\nabla}) \times \vec{B}\, dS$$

DEDUCTION 7.3.1

Stokes' theorem states that

$$\oint_C \vec{F} \cdot d\vec{r} = \int\int_S (\vec{\nabla} \times \vec{F}) \cdot \hat{n}\, dS = \int\int_{\triangle S} (\vec{\nabla} \times \vec{F}) \cdot \hat{n}\ dS$$

where $\triangle \mathcal{S}$ is the surface bounded by the simple closed curve $\mathcal{C}$.

Let P be any point of $\triangle \mathcal{S}$, not on $\mathcal{C}$ and $\hat{n}$ be a unit normal to $\triangle \mathcal{S}$ and P. So, by the mean value theorem of integral calculus, we have

$$\int\int_{\triangle S} (\vec{\nabla} \times \vec{F}) \cdot \hat{n}\, dS = \left[(\nabla \times \vec{F})_P + \varepsilon\right] \triangle S$$

where $(\nabla \times \vec{F})_P$ is the value of $(\nabla \times \vec{F})$ (independent of $\triangle \mathcal{S}$) at the point P and ε is function of $\triangle \mathcal{S}$ such that $\varepsilon \longrightarrow 0$ as $\triangle \mathcal{S} \longrightarrow 0$. Then

$$(\nabla \times \vec{F})_P + \varepsilon = \frac{1}{\triangle S} \oint_C \vec{F} \cdot d\vec{r}$$

$$\text{or,} \qquad (\nabla \times \vec{F})_P = \lim_{\triangle S \longrightarrow 0} \oint_C \vec{F} \cdot d\vec{r}$$

Example 7.3.9

Evaluate the line integral $\oint_C \frac{x\,dy - y\,dx}{x^2 + y^2}$, when the origin is included in the region bounded by the curve C.

Solution

The required region is shown in Fig. 7.32. The given line integral takes the form $\oint_C \vec{F} \cdot d\vec{r}$, where $\vec{F} = \frac{1}{x^2+y^2}(-y\hat{\imath} + x\hat{\jmath}) = -\frac{y}{r}\,\hat{\imath} + \frac{x}{r}\,\hat{\jmath}$. Using the definition of curl, we get

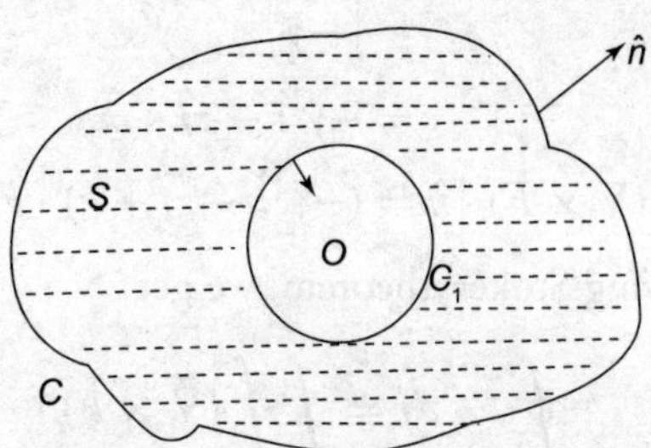

Figure 7.32 Region of integration

$$\vec{\nabla} \times \vec{F} = \left[\frac{\partial}{\partial x}\hat{\imath} + \frac{\partial}{\partial y}\hat{\jmath} + \frac{\partial}{\partial z}\hat{k}\right] \times \frac{1}{x^2+y^2}(-y\hat{\imath} + x\hat{\jmath})$$

$$= \frac{y^2 - x^2 + x^2 - y^2}{(x^2+y^2)^2}\,\hat{k} = 0, \quad \text{provided } x^2 + y^2 \neq 0$$

Using Stokes' theorem (Eq. 7.35), we get

$$\oint_C \vec{F} \cdot d\vec{r} = \int\int_S (\vec{\nabla} \times \vec{F}) \cdot \hat{n}\, dS = 0$$

when the origin is not included within S. If S includes the origin, then we enclose it in a positively oriented circle of radius a. Thus, the origin is not included in the region bounded by C and C_1. So, using the above result, we get

$$\oint_{C+C_1} \vec{F} \cdot d\vec{r} = \oint_C \vec{F} \cdot d\vec{r} + \oint_{C_1} \vec{F} \cdot d\vec{r} = 0$$

$$\Rightarrow \oint_C \vec{F} \cdot d\vec{r} = -\oint_{C_1} \vec{F} \cdot d\vec{r} = -\int \frac{1}{x^2+y^2}(-y\hat{\imath} + x\hat{\jmath}) \cdot (dx\,\hat{\imath} + dy\,\hat{\jmath})$$

$$= -\frac{1}{a^2}\int (x\,dy - y\,dx) = \frac{1}{a^2}\,2\pi\, a^2 = 2\pi$$

as the orientation of C and C_1 are opposite.

Example 7.3.10

Let C be the boundary of a triangle with vertices $(0, 1, 0)$, $(1, 0, 0)$ and $(2, 1, 0)$. If $\vec{F} = -y\hat{\imath} + y^2 z\hat{\jmath} + zx\hat{k}$, then use Stokes' theorem to evaluate the line integral $\oint_C \vec{F} \cdot d\vec{r}$, when C is traversed counter-clockwise when viewed from above.

Solution

In Fig. 7.33, C is the boundary of the triangle with vertices $(0,1,0)$, $(1,0,0)$ and $(2,1,0)$. For the surface bounded by the triangle, $\hat{n} = \hat{k}$. Since $\vec{F} = -y\hat{i} + y^2 z\hat{j} + zx\hat{k}$, using the definition of curl, we get

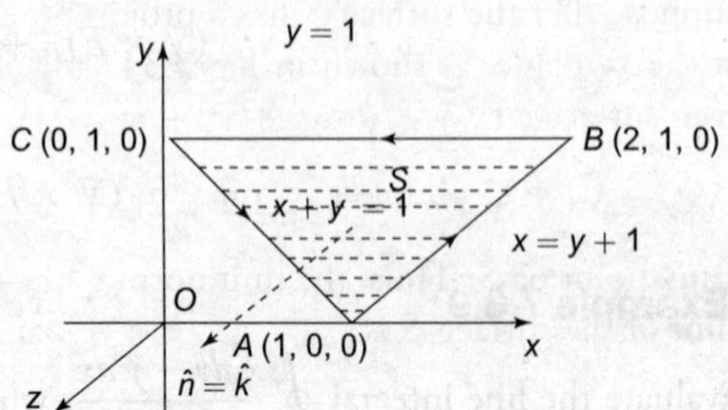

Figure 7.33 Region of integration

$$\vec{\nabla} \times \vec{F} = \begin{vmatrix} \hat{i} & \hat{j} & \hat{k} \\ \dfrac{\partial}{\partial x} & \dfrac{\partial}{\partial y} & \dfrac{\partial}{\partial z} \\ -y & y^2 z & zx \end{vmatrix}$$

$$= -y^2\hat{i} - z\hat{j} + \hat{k}$$

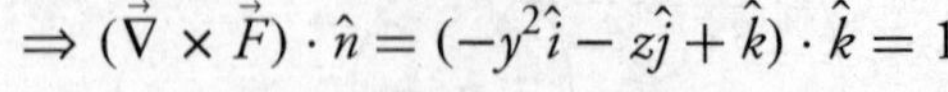

$$\Rightarrow (\vec{\nabla} \times \vec{F}) \cdot \hat{n} = (-y^2\hat{i} - z\hat{j} + \hat{k}) \cdot \hat{k} = 1$$

Applying Stokes' theorem, we get

$$\oint_C \vec{F} \cdot d\vec{r} = \iint_S (\vec{\nabla} \times \vec{F}) \cdot \hat{n}\, dS = \int_0^1 \left[\int_{1-y}^{y+1} 1\, dx\right] dy = 2\int_0^1 y\, dy = 1$$

Example 7.3.11

(i) Consider a constant vector field $\vec{F} = F_0\,\hat{k}$. Find any one of the many possible vectors $\vec{G}$, for which $\vec{\nabla} \times \vec{G} = \vec{F}$.

(ii) Using Stokes' theorem, evaluate the flux associated with the field $\vec{F}$ through the curved hemispherical surface defined by $x^2 + y^2 + z^2 = a^2, z > 0$.

Solution

Given a constant vector field $\vec{F} = F_0\,\hat{k}$. Let $\vec{G} = G_1\hat{i} + G_2\hat{j} + G_3\hat{k}$ be the required differentiable vector field. Using Eq. (5.55) as the definition of curl, we get

$$\vec{\nabla} \times \vec{G} = \begin{vmatrix} \hat{i} & \hat{j} & \hat{k} \\ \dfrac{\partial}{\partial x} & \dfrac{\partial}{\partial y} & \dfrac{\partial}{\partial z} \\ G_1 & G_2 & G_3 \end{vmatrix}$$

$$= \left(\frac{\partial G_3}{\partial y} - \frac{\partial G_2}{\partial z}\right)\hat{i} + \left(\frac{\partial G_1}{\partial z} - \frac{\partial G_3}{\partial x}\right)\hat{j} + \left(\frac{\partial G_2}{\partial x} - \frac{\partial G_1}{\partial y}\right)\hat{k}$$

$$= F_0\,\hat{k}, \text{ given}$$

Comparing the coefficients of $\hat{i}, \hat{j}$ and $\hat{k}$, we get

$$\frac{\partial G_3}{\partial y} - \frac{\partial G_2}{\partial z} = 0, \quad \frac{\partial G_1}{\partial z} - \frac{\partial G_3}{\partial x} = 0, \quad \frac{\partial G_2}{\partial x} - \frac{\partial G_1}{\partial y} = F_0$$

From the above equations, we come to the conclusion that one of the possible solutions for $\vec{G}$ is $F_0\, x\,\hat{j}$.

(ii) Here, the surface is above the xy plane (Fig. 7.34), so we have to calculate the given area in the xy plane. If A is the projected area on that xy plane, then $A = \{(x,y) : x^2 + y^2 = a^2, z = 0\}$.

Suppose that the surface S has a projection R_1 on the xy plane, as shown in Fig 7.34. For the given surface $\phi(x, y, z) = x^2 + y^2 + z^2 = a^2$

$$\vec{\nabla}\phi = \vec{\nabla}(x^2 + y^2 + z^2) = 2x\hat{i} + 2y\hat{j} + 2z\hat{k}$$

at any point of S. Thus, the unit normal to any point of the surface S is

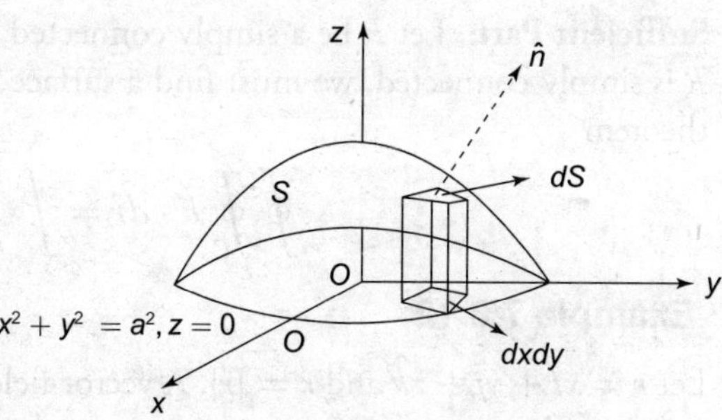

Figure 7.34 Region of integration

$$\hat{n} = \frac{1}{\sqrt{(2x)^2 + (2y)^2 + (2z)^2}}(2x\hat{i} + 2y\hat{j} + 2z\hat{k})$$

$$= \frac{x}{a}\hat{i} + \frac{y}{a}\hat{j} + \frac{z}{a}\hat{k}$$

$$\Rightarrow \quad \hat{n} \cdot \hat{k} = \left(\frac{x}{a}\hat{i} + \frac{y}{a}\hat{j} + \frac{z}{a}\hat{k}\right) \cdot \hat{k} = \frac{z}{a}$$

Using Stokes' theorem and Eq. (6.12), we obtain the flux associated with the vector field $\vec{G}$ through the curved hemispherical surface $x^2 + y^2 + z^2 = a^2, z > 0$ as

$$\oint_C \vec{G} \cdot d\vec{r} = \int\int_S (\vec{\nabla} \times \vec{G}) \cdot \hat{n}\, dS = \int\int_S \vec{F} \cdot \hat{n}\, dS = \int\int_S F_0\, \hat{k} \cdot \hat{n}\, dS$$

$$= F_0 \int_{-a}^{a} \left[\int_{-\sqrt{a^2-x^2}}^{\sqrt{a^2-x^2}} dy\right] dx$$

$$= 2F_0 \int_{-a}^{a} \sqrt{a^2 - x^2}\, dx = 4F_0 \left[\frac{x\sqrt{a^2 - x^2}}{2} + \frac{a^2}{2}\sin^{-1}\frac{x}{a}\right]_0^a$$

$$= 4\, F_0 \cdot \frac{a^2}{2} \cdot \frac{\pi}{2} = \pi\, F_0\, a^2$$

Theorem 7.6

A necessary and sufficient condition that $\oint \vec{F} \cdot d\vec{r} = 0$ for every closed curve C lying in a simply connected region R is that curl $\vec{F} = \vec{\nabla} \times \vec{F}$ is identically $\vec{0}$ (using Stokes' theorem).

Proof

Necessary Part: Let $\oint_C \vec{F} \cdot d\vec{r} = 0$ for every closed curve C lying in the simply connected region R and let, if possible, $\vec{\nabla} \times \vec{F} \neq \vec{0}$ at some point A in it. Taking $\vec{\nabla} \times \vec{F}$ as continuous, there must exist a region with A as an interior point where $\vec{\nabla} \times \vec{F} \neq 0$.

Let S be a surface contained in this region, whose outward drawn unit normal $\hat{n}$ at each point is in the same direction as $\vec{\nabla} \times \vec{F}$, i.e; $\vec{\nabla} \times \vec{F} = \lambda\hat{n}$: λ is a positive constant. Let C be a boundary of S. Then, by Stokes' theorem

$$\oint\oint_C \vec{F} \cdot d\vec{r} = \int\int_S (\vec{\nabla} \times \vec{F}) \cdot \hat{n}\, dS = \int\int_S \lambda\hat{n} \cdot \hat{n}\, dS = \lambda\, S > 0$$

This contradicts the hypothesis that $\oint\oint_C \vec{F} \cdot d\vec{r} = 0$ for every closed curve C. So we must have $\vec{\nabla} \times \vec{F} = \vec{0}$ everywhere in R.

Sufficient Part: Let R be a simply connected region and let $\vec{\nabla} \times \vec{F} = \vec{0}$ everywhere in R. Since R is simply connected, we must find a surface $\mathcal{S}$ in R having $\mathcal{C}$ as its boundary. Then, by Stokes' theorem

$$\oint \oint_{\mathcal{C}} \vec{F} \cdot d\vec{r} = \int \int_{\mathcal{S}} (\vec{\nabla} \times \vec{F}) \cdot \hat{n}\, dS = 0$$

Example 7.3.12

Let $\vec{r} = x\hat{i} + y\hat{j} + z\hat{k}$ and $r = |\vec{r}|$. A vector field is given by

$$\vec{F}\,(\vec{r}) = \begin{cases} \alpha(x\hat{j} - y\hat{i}); & I : x^2 + y^2 \le r_0^2 \\ \dfrac{r_0^2 \alpha}{x^2 + y^2}(x\hat{j} - y\hat{i}); & II : x^2 + y^2 > r_0^2 \end{cases}$$

where α and r_0 are constants. Find the

(i) curl of this field in both the regions.

(ii) line integral $\oint \vec{F} \cdot d\vec{l}$ along the closed semicircular path of radius $2r_0$, as shown in Fig. 7.35.

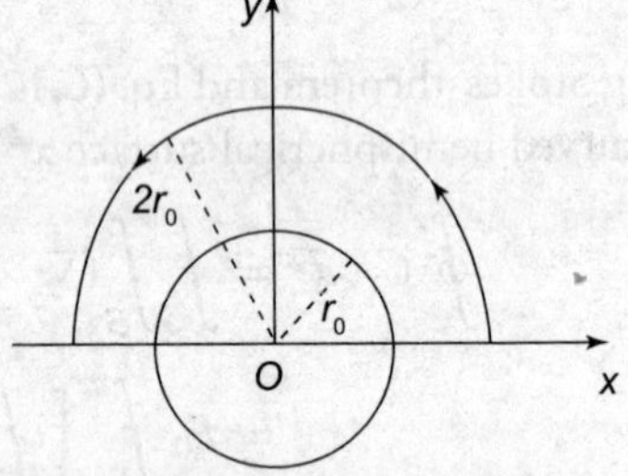

Figure 7.35 The region

Solution

(i) For the given vector field $\vec{F}$, using the definition of curl, we have (in region I: $x^2 + y^2 \le r_0^2$)

$$\vec{\nabla} \times \vec{F} = \begin{vmatrix} \hat{i} & \hat{j} & \hat{k} \\ \dfrac{\partial}{\partial x} & \dfrac{\partial}{\partial y} & \dfrac{\partial}{\partial z} \\ -\alpha\, y & \alpha\, x & 0 \end{vmatrix} = (\alpha + \alpha)\hat{k} = 2\alpha\ \hat{k}$$

Using region II: $x^2 + y^2 > r_0^2$, we have

$$\vec{\nabla} \times \vec{F} = \begin{vmatrix} \hat{i} & \hat{j} & \hat{k} \\ \dfrac{\partial}{\partial x} & \dfrac{\partial}{\partial y} & \dfrac{\partial}{\partial z} \\ -\dfrac{r_0^2 y\alpha}{r^2} & \dfrac{r_0^2 x\alpha}{r^2} & 0 \end{vmatrix} = \vec{0}$$

(ii) Now, using Stokes' theorem, we get

$$\begin{aligned} \oint \vec{F} \cdot d\vec{l} &= \int \int_{S} (\vec{\nabla} \times \vec{F}) \cdot d\vec{S} \\ &= \int \int_{I} (\vec{\nabla} \times \vec{F}) \cdot d\vec{S} + \int \int_{r_0^2 \le x^2 + y^2 \le 4r_0^2} (\vec{\nabla} \times \vec{F}) \cdot d\vec{S} \\ &= 2\alpha \int \int_{x^2 + y^2 \le r_0^2} dS + 0 = 2\,\alpha\,\pi\, r_0^2 \end{aligned}$$

Example 7.3.13

How much work is done when an object moves from $O \to P \to Q \to R \to O$ in a force field given by $\vec{F} = (x^2 - y^2)\hat{i} + 2xy\,\hat{j}$ along the rectangular path shown in Fig. 7.36. Find the answer by evaluating the line integral and also using Stokes' theorem.

Solution

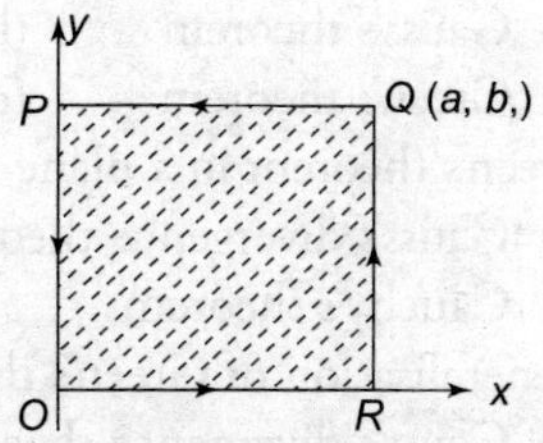

Figure 7.36 The region

Here, the given force field $\vec{F} = (x^2 - y^2)\hat{i} + 2xy\,\hat{j}$, so $\vec{F} \cdot d\vec{r} = (x^2 - y^2)dx + 2xy\,dy$. Now, along the line OP, $x = 0$ and $0 \le y \le b$, so

$$\int_{OP} \vec{F} \cdot d\vec{r} = \int_0^b 0\,dy = 0$$

Along the line PQ, $0 \le x \le a$ and $y = b$, so

$$\int_{PQ} \vec{F} \cdot d\vec{r} = \int_0^a (x^2 - b^2)\,dx = \frac{a^3}{3} - ab^2$$

Along the line QR, $x = a$ and y varies from b to 0, so

$$\int_{QR} \vec{F} \cdot d\vec{r} = \int_b^0 2ay\,dy = -ab^2$$

Along the line RO, x varies from a to 0 and $y = 0$, so

$$\int_{RO} \vec{F} \cdot d\vec{r} = \int_a^0 x^2\,dx = -a^3/3$$

Therefore, the work done is

$$W = \oint_C \vec{F} \cdot d\vec{r} = 0 + \frac{a^3}{3} - ab^2 - ab^2 - \frac{a^3}{3} = -2ab^2$$

Since $\vec{F} = (x^2 - y^2)\hat{i} + 2xy\,\hat{j}$, using the definition of curl, we get

$$\vec{\nabla} \times \vec{F} = \begin{vmatrix} \hat{i} & \hat{j} & \hat{k} \\ \dfrac{\partial}{\partial x} & \dfrac{\partial}{\partial y} & \dfrac{\partial}{\partial z} \\ x^2 - y^2 & 2xy & 0 \end{vmatrix} = (2y + 2y)\hat{k} = 4y\,\hat{k}$$

Here, $\hat{n} = -\hat{k}$, so using Stokes' theorem, we get

$$\oint_C \vec{F} \cdot d\vec{r} = \int_S (\vec{\nabla} \times \vec{F}) \cdot \hat{n} dS = -\int_0^a \left[\int_0^b y\,dy\right] dx$$

$$= -4 \int_0^a \left[\frac{y^2}{2}\right]_0^b dx = -2b^2 \int_0^a dx = -2ab^2$$

EXERCISES

Multiple Choice Questions

1. The relation between the line integral and the surface integral is explored by
   (a) Gauss's theorem (b) Stokes' theorem
   (c) Green's theorem (d) Bernoulli's theorem
2. Green's theorem in a plane is a special case of
   (a) Gauss's divergence theorem (b) Stokes' theorem
   (c) Cauchy's theorem (d) none of the above
3. Generalisation of Green's theorem in a plane leads to
   (a) Gauss's divergence theorem (b) Stokes' theorem
   (c) Cauchy's theorem (d) none of the above
4. R is a closed plane in a region, as shown by the shaded area in the figure below. Its boundary consists of the circles C_1 and C_2. If $P(x, y)$, $Q(x, y)$, $\frac{\partial P}{\partial y}$ and $\frac{\partial Q}{\partial x}$ are all continuous everywhere in R, then the statement of Green's theorem is (7.1). Which one of the following alternatives correctly describes the direction of integration along C?

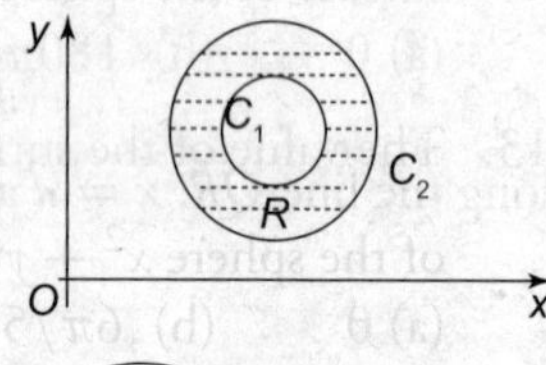

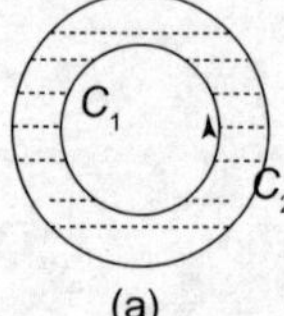

(a)

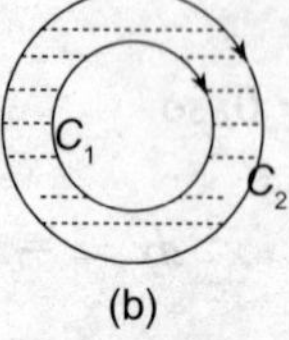

(b)

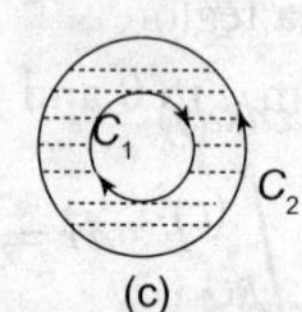

(c)

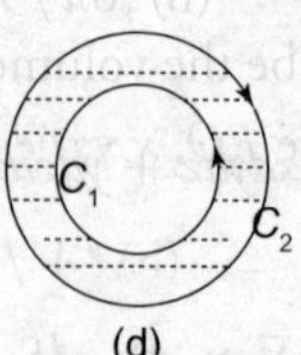

(d)

5. Let C be the circle $x^2 + y^2 = 1$ taken in the anti-clockwise sense. Then, the value of the integral $\oint_C (2xy^3 + y)dx + (3x^2y^2 + 2x)dy$ equals
   (a) 1 (b) $\pi/2$ (c) π (d) 0
6. The value of the surface integral $\int\int_S \vec{\nabla}(x^2 + y^2 + z^2)dS$, where S is any closed surface enclosing volume V is
   (a) V (b) $3V$ (c) $4V$ (d) $6V$
7. Gauss's divergence theorem is
   (a) $\int\int_S \vec{F} \cdot \hat{n}dS = \int\int\int_V \vec{\nabla} \cdot \vec{F}dV$ (b) $\int\int_S \vec{F} \cdot \hat{n}dS = \int\int\int_V \vec{\nabla} \times \vec{F}dV$
   (c) $\oint_C \vec{F} \cdot d\vec{r} = \int\int_S \vec{\nabla} \times \vec{F} \cdot \hat{n}dS$ (d) none of the above
8. Let C denote the boundary of the semi-circular disk $D = \{(x, y) \in \Re^2 : x^2 + y^2 \leq 1, y \geq 0\}$ and let $\phi(x, y) = x^2 + y$ for $(x, y) \in D$. If $\hat{n}$ is the outward unit normal to C, then the integral $\oint_C (\vec{\nabla}\phi) \cdot \hat{n}dS$, evaluated counter-clockwise over C is equal to
   (a) 0 (b) $\pi - 2$ (c) π (d) $\pi + 2$

9. Let $\vec{F} = 2xyze^{x^2}\hat{i} + ze^{x^2}\hat{j} + ye^{x^2}\hat{k}$ be the gradient of a scalar function. The value of $\oint_C \vec{F}\cdot d\vec{r}$ along the oriented path L from $(0,0,0)$ to $(1,0,2)$ and then to $(1,1,2)$ is
(a) 0 (b) $2e$ (c) e (d) e^2

10. The work done by the force $\vec{F} = 4y\hat{i} - 3xy\hat{j} + z^2\hat{k}$ in moving a particle over the circular path $x^2 + y^2 = 1, z = 0$ from $(1,0,0)$ to $(0,1,0)$ is
(a) $\pi + 1$ (b) $-\pi - 1$ (c) $-\pi + 1$ (d) $\pi - 1$

11. Let $\vec{F} = xy\hat{i} + y\hat{j} - yz\hat{k}$ denote the force field on a particle traversing the path L from $(0,0,0)$ to $(1,1,1)$ along the curve of intersection of the cylinder $y = x^2$ and the plane $z = x$. The work done by $\vec{F}$ is
(a) 0 (b) $1/2$ (c) $1/4$ (d) 1

12. The value of $\int\int_S \vec{F}\cdot\hat{n}dS$, when $\vec{F} = (2x+3z)\hat{i} - (xz+y)\hat{j} + (y^2+2z)\hat{k}$ and S is the surface of the sphere having its centre at $(3,-1,2)$ and radius 3 is
(a) 0 (b) 180π (c) 360π (d) 720π

13. The value of the surface integral $\int\int_S (x^3\,dydz + y^3\,dzdx + z^3\,dxdy)$, where S is the surface of the sphere $x^2 + y^2 + z^2 = 1$ is
(a) 0 (b) $6\pi/5$ (c) $12\pi/5$ (d) $18\pi/5$

14. Let V be the volume of a region bounded by a smooth closed surface S. Then
$\int\int (xdydz + ydzdx + zdxdy) =$
(a) V (b) $V/3$ (c) $3V$ (d) 0

15. $\int\int_S (\vec{\nabla}\times\vec{A})\cdot d\vec{S}$, where $\vec{A}$ is a vector, is equal to
(a) $\oint \vec{A}\cdot d\vec{l}$ (b) $\oint \vec{\nabla}\times\vec{\nabla}\times\vec{\nabla}A\cdot d\vec{l}$
(c) $\oint \vec{\nabla}\times\vec{A}\cdot d\vec{l}$ (d) $\int\int\int \vec{\nabla}\cdot\vec{A}dl$

16. Stokes' theorem connects
(a) a line integral and a surface integral (b) a surface integral and a volume integral
(c) a line integral and a volume integral
(d) the gradient of a function and its surface integral

17. Let $B = \{(x,y,z) : x,y,z \in \Re \text{ and } x^2 + y^2 + z^2 \le 4\}$. Let $v(x,y,z) = x\hat{i} + y\hat{j} + z\hat{k}$ be a vector-valued function defined in B. If $r^2 = x^2 + y^2 + z^2$, the value of the integral
$\int\int\int_B \vec{\nabla}\cdot(r^2 v(x,y,z))\,dV$ is
(a) 16π (b) 32π (c) 64π (d) 128π

18. Let $V = \left\{(x,y,z) \in \Re^3 : \frac{1}{4} \le x^2 + y^2 + z^2 \le 1\right\}$ and $\vec{F} = \frac{1}{(x^2+y^2+z^2)^2}(x\hat{i} + y\hat{j} + z\hat{k})$ for $(x,y,z) \in V$. Let $\hat{n}$ denote the outward unit normal vector to the boundary of V and S denote the part $\left\{(x,y,z) \in \Re^3 : x^2 + y^2 + z^2 = \frac{1}{4}\right\}$ of the boundary V. Then

$\int\int_S \vec{F} \cdot \hat{n}\, dS$ is equal to
(a) -8π (b) -4π (c) 4π (d) 8π

19. Stokes' theorem is
(a) $\oint_C \vec{F} \cdot \hat{n} dS = \int\int_S \hat{n} \cdot \text{ div } \vec{F} dS$ (b) $\oint_C \vec{F} \cdot \hat{n} dS = \int\int_S (\vec{\nabla} \times \vec{F}) \cdot \hat{n} dS$
(c) $\int\int_S \vec{F} \cdot \hat{n} dS = \int\int\int_V \vec{\nabla} \cdot \vec{F} dV$ (d) $\oint_C \vec{F} \cdot d\vec{r} = \int\int_S (\vec{\nabla} \times \vec{F}) \cdot \hat{n} dS$

20. The value of the integral $\int_{(0,0)}^{(\pi,2)} (6xy - y^2)\, dx + (3x^2 - 2xy)\, dy$, along the cycloid $x = t - \sin t$, $y = 1 - \cos t$ is
(a) $6\pi^2 - 4\pi$ (b) $6\pi^2 + 4\pi$ (c) $-6\pi^2 - 4\pi$ (d) $4\pi^2 - 6\pi$

21. Consider the hemisphere $x^2 + y^2 + (z-2)^2 = 9$, $2 \leq z < 5$ and the vector field $\vec{F}(x, y, z) = x\hat{i} + y\hat{j} + (z-2)\hat{k}$. The surface integral $\int\int_S \vec{F} \cdot \hat{n} dS$, evaluated over the hemisphere with $\hat{n}$ denoting the unit outward normal, is
(a) 9π (b) 27π (c) 54π (d) 162π

22. Suppose C_1 is the boundary of $\{(x, y) \in \mathcal{R}^2 : 0 \leq x \leq 1, 0 \leq y \leq 1\}$ and C_2 is the boundary of $\{(x, y) \in \mathcal{R}^2 : -1 \leq x \leq 0, -1 \leq y \leq 0\}$. Let $L_i = \oint_{C_i} xy^2\, dx + (x^2 y + 2x) dy$, $i = 1, 2$ be evaluated in the counter-clockwise direction. Then,
(a) $L_1 = 1, L_2 = 2$ (b) $L_1 = L_2 = 1$ (c) $L_1 = 2, L_2 = -2$ (d) $L_1 = L_2 = 2$

23. Let $B = \{(x, y, z) | x, y, z \in \Re \text{ and } x^2 + y^2 + z^2 \leq 4\}$. Let $V(x, y, z) = x\hat{i} + y\hat{j} + z\hat{k}$ be a vector-valued function defined on B. If $r^2 = x^2 + y^2 + z^2$, the value of the integral $\int\int\int_B \vec{\nabla} \cdot (r^2 V(x, y, z)) dV$ is
(a) 16π (b) 32π (c) 64π (d) 128π

24. Let $\vec{F}$ be a vector field defined on $\Re^2 - \{(0, 0)\}$ by $\vec{F}(x, y) = \frac{y}{x^2 + y^2}\hat{i} - \frac{x}{x^2 + y^2}\hat{j}$. Let $\gamma, \alpha : [0, 1] \rightarrow \Re^2$ be defined by $\gamma(t) = (8 \cos 2\pi t, 17 \sin 2\pi t)$ and $\alpha(t) = (26 \cos 2\pi t, -10 \sin 2\pi t)$. If $\int_\alpha \vec{F} \cdot d\vec{r} - \int_\gamma \vec{F} \cdot d\vec{r} = 2m\pi$, then m is
(a) 3 (b) 5 (c) 7 (d) 11

25. Let S be the surface bounded by the simple closed curve C and $\hat{n}$ be the unit outward drawn normal to it. Then $\oint_C \vec{r} \cdot d\vec{r}$ equals
(a) 0 (b) $2\pi S$ (c) πS^2 (d) $4\pi S^2$

26. Let S be the surface bounded by the simple closed curve C and $\hat{n}$ be the unit outward drawn normal to it. Then $\oint_C \vec{r} \times d\vec{r}$ equals
(a) $\int\int_S \hat{n}\, dS$ (b) $2\int\int_S \hat{n}\, dS$ (c) $3\int\int_S \hat{n}\, dS$ (d) $4\int\int_S \hat{n}\, dS$

27. The value of the line integral $\oint_C (2x - y^3)dx - xy\,dy$, where C is the boundary of the region enclosed by the circles $x^2 + y^2 = 1$ and $x^2 + y^2 = 9$ is
(a) 10π (b) 20π (c) 40π (d) 60π

28. The value of $\int\int_S x^2(x\,dy\,dz + y\,dx\,dz + z\,dx\,dy)$, where S is the surface of the cylinder $x^2 + y^2 = a^2$ bounded by $z = 0$ and $z = b$ is
(a) $a^4 b\pi/4$ (b) $3a^4 b\pi/4$ (c) $5a^4 b\pi/4$ (d) $7a^4 b\pi/4$

29. The value of $\int\int_S x^2\,dy\,dz + y^2\,dx\,dz + 2z(xy - x - y)\,dx\,dy)$, where S is the surface of the cube $0 \le x \le 1, 0 \le y \le 1, 0 \le z \le 1$ is
(a) 0 (b) 1/2 (c) 1/4 (d) 1/4

30. The value of $\int\int_S \frac{\vec{r} \cdot d\vec{S}}{r^3}$, where $\vec{r}$ is the position vector and S is a closed surface enclosing the origin is
(a) 0 (b) π (c) 4π (d) 8π

31. If $\hat{n}$ is the unit vector drawn outward normal to any closed surface of area S, then $\int\int\int_V \vec{\nabla} \cdot \hat{n}\,dV$ has the value
(a) S (b) $2S$ (c) $3S$ (d) 0

32. The volume integral $\int\int\int_V \frac{dV}{r^2}$ equals
(a) $\int\int_S \frac{\vec{r} \cdot \hat{n}}{r^2}\,dS$ (b) $\int\int_S \frac{\vec{r} \cdot \hat{n}}{r}\,dS$ (c) $\int\int_S \frac{\vec{r} \cdot \hat{n}}{r^3}\,dS$ (d) 0.

33. A vector $\vec{A}$ is always normal to a given closed surface S, then $\int\int\int_V \vec{\nabla} \times \vec{A}\,dV$ has the value
(a) $\int\int_S \vec{A} \cdot \hat{n}\,dS$ (b) $\int\int_S \vec{A}\,dS$ (c) $\oint_C \vec{A} \cdot d\vec{r}$ (d) 0

34. Consider a cylinder of height h and radius a, closed at both ends, centred at the origin. Let $x\hat{i} + y\hat{j} + z\hat{k}$ be the position vector and $\hat{n}$ a unit vector normal to the surface. The surface integral $\int\int_S \vec{r} \cdot \hat{n}\,dS$ over the closed surface of the cylinder is
(a) $2\pi a^2(a + h)$ (b) $3\pi a^2 h$ (c) $2\pi a^2 h$ (d) 0

35. The value of the surface integral $\int\int_S \vec{a} \cdot d\vec{S}$, where $\vec{a} = x\hat{i}$ and S is the surface of the hemisphere $x^2 + y^2 + z^2 = a^2$ with $z \ge 0$ is
(a) $\pi a^3/3$ (b) $2\pi a^3/3$ (c) πa^3 (d) $2\pi a^3$

36. Let $I = \oint_C (2x^2 + y^2)dx + e^y\,dy$, where C is the boundary (oriented anti-clockwise) of the region in the first quadrant bounded by $y = 0, x^2 + y^2 = 1$ and $x = 0$. The value of I is
(a) -1 (b) $-2/3$ (c) 2/3 (d) 1

37. Let $\mathcal{S}$ be the surface bounding the region $x^2 + y^2 \leq 1, x \geq 0, y > 0, |z| \leq 1$ and $\hat{n}$ be the unit outward normal to $\mathcal{S}$. Then $\int\int_S \left[\sin^2 x\hat{i} + 2y\hat{j} - z(1 + \sin 2x)\hat{k}\right] \cdot \hat{n}\, dS$ equals
(a) 1 (b) $\pi/2$ (c) π (d) 2π

Answers

1. b	2. b	3. a	4. d	5. c	6. d	7. a	8. b	9. b	10. b	11. b	12. b	13. c
14. c	15. a	16. a	17. d	18. a	19. d	20. a	21. c	22. d	23. d	24. c	25. a	
26. b	27. d	28. c	29. b	30. c	31. c	32. c	33. c	34. b	35. b	36. b	37. d	

Short Answer Questions

1. Express Green's theorem in the plane in vector form.
2. Show that the area bounded by a simple closed curve C is given by $\oint_C \dfrac{\lambda x\, dy + \mu\, y\, dx}{\lambda - \mu}$, where λ and μ are two unequal constants.
3. Using Green's theorem, interpret $\dfrac{1}{2}\oint_C x\, dy - y\, dx$.
4. If C is a simple closed curve in the xy plane, not enclosing the origin, show by using Green's theorem that $\vec{F} \cdot d\vec{r} = 0$, where $\vec{F} = \dfrac{1}{x^2 + y^2}(y\hat{i} - x\hat{j})$.
5. Using Green's theorem, find the area of the following regions enclosed by
(a) $y = x, y = -x, y = 4$ (b) $x^2 + y^2 = 1$ (c) $y^2 = 4x, y = x$
(d) the loop of the Folium of Descartes $x^3 + y^3 = 3axy, a > 0$
Ans: (a) 16 sq. units, (b) π sq. units, (c) $2\sqrt{2}$ sq. units, (d) $3a^2/2$ sq. units
6. Derive the divergence theorem for Green's theorem.
7. Find the area bounded by
(a) one arc of the cycloid $x = a(t - \sin t), y = a(1 - \cos t), a > 0$ *Ans*: $3\pi a^2$ sq. units
(b) the astroid $x^{2/3} + y^{2/3} = a^{2/3}, a > 0$ *Ans*: $3\pi a^2/8$ sq. units
(c) a loop of the four-leafed rose $r = 3 \sin 2\theta$ *Ans*: $9\pi/8$ sq. units
(d) both loops of the lemniscate $r^2 = a^2 \cos 2\theta$ *Ans*: a^2 sq. units
8. Let V be the region enclosed by the surface $\mathcal{S}$ and $\hat{n}$ be the unit outward drawn normal to it. Then prove the following:
(a) $\int\int\int_V \vec{\nabla}\phi\, dV = \int\int_S \phi\, \hat{n} dS$ (b) $\int\int_S (\vec{\nabla} \times \vec{A}) \cdot \hat{n} dS = 0$ *Hint*: div(curl $\vec{A}$) = 0
9. Use the divergence theorem to prove that
(a) $\int\int\int \vec{\nabla} \times \vec{b}\, dV = \int\int_S \hat{n} \times \vec{b}\, dS$, where V is the volume enclosed by the closed surface S
(b) $\int\int_S \vec{\nabla} r^2 \cdot d\vec{S} = 6V$
10. If the vector field $\vec{F}$ is the normal to the surface $\mathcal{S}$ bounding the region V, in which $\vec{F}$ has continuous first derivatives, then prove that $\int\int\int_V \vec{\nabla} \times \vec{F}\, dV = \vec{0}$.

11. For any closed surface $\mathcal{S}$, prove that $\iint_S d\vec{S} = \iint_S \hat{n} dS = \vec{0}$.
12. Let $\mathcal{S}$ be a closed surface enclosing a region V, having outward drawn unit normal $\hat{n}$. Prove that $\iiint_V \vec{\nabla} \cdot \hat{n}\, dV =$ area of $\mathcal{S}$.
13. If $\vec{\nabla}\phi = \vec{F}$ and $\nabla^2\phi = 0$, then show that for a closed surface $\mathcal{S}$ enclosing the volume V

$$\iiint_V \vec{F}^2\, dV = \iint_S \phi\, \vec{F} \cdot \hat{n}\, dS$$

14. If $\mathcal{S}$ be a closed surface enclosing a region V, prove that $\iint_S \vec{r} \cdot \hat{n} dS = 3V$, where $\vec{r}$ is the position vector of any point.
15. If $\hat{n}$ is the unit outward drawn normal to any closed surface of area S, show that

$$\iiint \text{div}\, \hat{n}\, dV = S$$

16. Prove that $\iiint_V \frac{dV}{r^2} = \iint_S \frac{\vec{r} \cdot \hat{n}}{r^2} dS$, where V is the volume of a region enclosed by the surface $\mathcal{S}$, $\hat{n}$ is the outward normal unit vector to $\mathcal{S}$.
17. If S be the surface bounded by the simple closed curve C and $\hat{n}$ be the unit outward drawn normal to it, then prove the following:

(a) $\oint_C \phi\, d\vec{r} = \iint_S \hat{n} \times \vec{\nabla}\phi\, dS$

(b) $\oint_C f(r)\, \vec{r} \cdot d\vec{r} = 0$, where $f(r)$ is a differentiable scalar point function

(c) $\oint_C \vec{r} \cdot d\vec{r} = 0$ (d) $\oint_C \phi\, \vec{\nabla}\psi \cdot d\vec{r} = -\oint_C \psi\, \vec{\nabla}\phi \cdot d\vec{r}$

(e) $\oint_C \phi\, \vec{\nabla}\phi \cdot d\vec{r} = 0$ (f) $\iint_S \vec{r} \times \hat{n}\, dS = \iint_S \vec{r} \times d\vec{S} = \vec{0}$

(g) $\iint_S (\vec{\nabla}\phi \times \vec{\nabla}\psi) \cdot \hat{n}\, dS = \oint_C \phi \vec{\nabla}\psi \cdot d\vec{r}$ (h) $\oint_C \vec{r} \times d\vec{r} = 2\iint_S \hat{n}\, dS$

(i) $\iint_S (\hat{n} \times \vec{\nabla}) f\, dS = \oint_C f\, d\vec{r}$ (j) $\iint_S \left\{(\hat{n} \times \vec{\nabla}) \times \vec{A}\right\} dS = -\oint_C \vec{A} \times d\vec{r}$

(k) $\iint_S \left[\phi \vec{\nabla} \times \vec{A} + \vec{\nabla}\phi \times \vec{A}\right] \cdot \hat{n}\, dS = \oint_C \phi \vec{A} \cdot d\vec{r}$

(*Hint*: With proper choice of the vector, apply Stokes' theorem)

18. If $\vec{b} = \vec{\nabla} \times \vec{a}$, prove that $\iint_S \vec{b} \cdot \hat{n}\, dS = 0$ for any closed surface $\mathcal{S}$.
19. Prove that $\iint_S r^5\, \hat{n}\, dS = \iiint_V 5r^3\, \vec{r}\, dV$.
20. If ϕ is harmonic in V, then prove that

(a) $\iint_S \frac{d\phi}{dn} dS = 0$ (b) $\iint_S \phi\, \frac{d\phi}{dn} dS = \iiint_V |\vec{\nabla}\phi|^2\, dV$

(c) ϕ is constant in V, if $\frac{d\phi}{dn} = 0$ on $\mathcal{S}$

(d) $\phi = \psi + C$ in V, where C is a scalar constant, if the normal derivatives of both the harmonic functions are equal on S

21. If $\vec{F}$ be a closed surface, prove that $\int\int_S (\vec{\nabla} \times \vec{F}) \cdot d\vec{S} = 0$.

Long Answer Questions

1. Express Green's theorem in the plane in vector notation. Make a generalization and write down Stokes' theorem in space.
2. Verify Green's theorem in the plane for the line integral:

   (a) $\oint_C (x^2 - xy^2)dx + (y^2 - 2xy)dy$, where C is the square with vertices $(0,0)$, $(2,0)$, $(2,2)$ and $(0,2)$. Common value: 0

   (b) $\oint_C (x^2 - \cosh y)dx + (y + \sin x)dy$, where C is the square with vertices $(0,0)$, $(\pi, 0)$, $(\pi, 1)$, and $(0,1)$. Common value: 0
3. Show that the integral $\int_{(1,2)}^{(3,4)} (xy^2 + y^2)dx + (x^2y + 3xy^2)dy$ is independent of the path joining the points $(1,2)$ and $(3,4)$. Hence, evaluate the integral. *Ans*: 254
4. Apply Green's theorem in the plane to evaluate

   (a) $\oint_C (y - \sin x)dx + \cos x\, dy$, where C is the triangle enclosed by the lines $y = 0, x = \pi$ and $y = 2x/\pi$. *Ans*: $-\frac{\pi}{4} - \frac{2}{\pi}$

   (b) $\oint_C (2x^2 - y^2)dx + (x^2 + y^2)\, dy$, where C is the boundary of the surface enclosed by the x axis and the semicircle given by $y = \sqrt{1 - x^2}$.
5. Let $\vec{F} = x^2\hat{i} + (2xy + x)\hat{j} + z\hat{k}$. Let C be the circle $x^2 + y^2 = 1$ and S be the disk $x^2 + y^2 \leq 1$ within the plane $z = 0$.

   (a) Determine the flux of $\vec{F}$ out of S. (b) Determine the circulation of $\vec{F}$ around C.

   (c) Find the flux of $\vec{\nabla} \times \vec{F}$. Verify Stokes' theorem directly in this case.
6. Prove the following by using Green's theorem:

   (a) $\oint_C (3x + 4y)\, dx + (2x - 3y)\, dy = -8\pi$, where C is the circle of radius 2 with centre at the origin of the xy plane, and is traversed in the positive sense.

   (b) $\oint_C (x^2 + y^2)\, dx + 3xy^2\, dy = 12\pi$, where C is the circle of radius 2 with centre at the origin of the xy plane, and is traversed in the positive sense.
7. Evaluate $\int_{(1,0)}^{(-1,0)} \frac{-ydx + xdy}{x^2 + y^2}$ along the following paths:

   (a) straight line segments from $(1,0)$ to $(1,1)$, then to $(-1,1)$, then to $(-1,0)$. *Ans*: π

   (b) straight line segments from $(1,0)$ to $(1,-1)$, then to $(-1,-1)$, then to $(-1,0)$. *Ans*: $-\pi$

   Show that although $\frac{\partial Q}{\partial x} = \frac{\partial P}{\partial y}$, the line integral is dependent on the path joining $(1,0)$ to $(-1,0)$ and explain.

8. Verify Gauss's divergence theorem for:
   (a) $\vec{F} = (x^2 - yz)\hat{i} + (y^2 - zx)\hat{j} + (z^2 - xy)\hat{k}$ over the rectangular parallelepiped $0 \le x \le a, 0 \le y \le b, 0 \le z \le c$. Common value: $abc(a+b+c)$
   (b) $\vec{F} = x\hat{i} + y\hat{j} + (z^2 - 1)\hat{k}$ over the cylinder $x^2 + y^2 = 4$ bounded by the planes $z = 0$ and $z = 1$. Common value: 4π
   (c) $\vec{F} = 2xz\hat{i} + y^2\hat{j} + yz\hat{k}$ over the cube bounded by the planes $x = y = z = 0$, $x = y = z = 1$. Common value: $5/2$
9. Verify Gauss's divergence theorem for the function $\vec{F} = x^2\hat{i} + z\hat{j} + yz\hat{k}$ over the unit cube. Common value: $3/2$
10. If $\vec{OA} = a\,\hat{i}$, $\vec{OB} = a\,\hat{j}$, and $\vec{OC} = a\,\hat{k}$, form three conterminous edges of a cube and $\mathcal{S}$ denotes the surface of the cube, evaluate $\int\int_S \left[(x^3 - yz)\hat{i} - 2x^2y\,\hat{j} + 2\,\hat{k}\right] \cdot \hat{n}\,dS$, by expressing it to a volume integral. Also verify the result by direct evaluation of the surface integral.

    Ans: $a^5/3$
11. Prove the following by using Gauss's divergence theorem:
    (a) $\int\int_S (x\,dy\,dz + y\,dz\,dx + z\,dx\,dy) = 4\pi a^3$, where $\mathcal{S}$ is the surface of the sphere $x^2 + y^2 + z^2 = a^2$
    (b) $\int\int_S (x^2 dydz + y^2 dzdx + z^2 dxdy) = 3$, where $\mathcal{S} : 0 \le x \le 1, 0 \le y \le 1, 0 \le z \le 1$
    (c) $\int\int_S \left[(x^2 - yz)\hat{i} - 2x^2y\hat{j} + 2\hat{k}\right] \cdot \hat{n}\,dS = \frac{1}{3}a^5$, where $\mathcal{S} : 0 \le x \le a, 0 \le y \le a, 0 \le z \le a$
    (d) $\int\int_S \left[y^2z^2\hat{i} + z^2x^2\hat{j} + z^2y^2\hat{k}\right] \cdot \hat{n}\,dS = \frac{\pi}{12}$, where $\mathcal{S}$ is the part of the sphere $x^2 + y^2 + z^2 = 1$ above the xy plane and bounded by the plane
12. Use the divergence theorem to evaluate $\int\int_S \vec{F} \cdot \hat{n}\,dS$, where
    (a) $\vec{F} = 4xz\hat{i} + xyz^2\hat{j} + 3z\hat{k}$ and $\mathcal{S}$ is the entire surface of the region above the xy plane bounded by the cone $x^2 + y^2 = z^2$ and the plane $z = 4$. *Ans*: 320π
    (b) $\vec{F} = x\hat{i} + y\hat{j} + z^2\hat{k}$ and $\mathcal{S}$ denotes the closed surface bounded by the cone $x^2 + y^2 = z^2$ and the plane $z = 4$. *Ans*: $128\pi/3$
    (c) $\vec{F} = 4xz\hat{i} - y^2\hat{j} + yz\hat{k}$, where $\mathcal{S}$ is the surface of the cube bounded by $x = 0, x = 1$; $y = 0, y = 1$ and $z = 0, z = 1$. *Ans*: $3/2$
    (d) $\vec{F} = e^x\hat{i} + e^y\hat{j} + e^z\hat{k}$ and S is the closed surface of the cube $|x| \le 1, |y| \le 1, |z| \le 1$. *Ans*: $12(1 - \frac{1}{e})$

    where $\hat{n}$ is the outward drawn unit normal to the surface $\mathcal{S}$.
13. Calculate the volume of the surface integral $\int\int_S \vec{F} \cdot \hat{n}\,dS$, where $\mathcal{S}$ is the surface of the sphere $x^2 + y^2 + z^2 = 4$, and $\hat{n}$ is the unit outward normal to the sphere and $\vec{F} = x\hat{i} + y\hat{j} + z\hat{k}$.

    Ans: 16π

14. Evaluate the surface integral $\int\int_S (\vec{\nabla} \times \vec{F}) \cdot \hat{n}\, dS$, where

(a) $\vec{F} = (x^2 + y - 4)\hat{\imath} + 3xy\hat{\jmath} + (2xz + z^2)\hat{k}$, where S is the surface of the hemisphere $x^2 + y^2 + z^2 = 16$ above the xy plane. *Ans*: -16π

(b) $\vec{F} = (x - z)\hat{\imath} + (x^3 + yz)\hat{\jmath} - 3xy^2\hat{k}$, where S is the surface of the cone $z = a - \sqrt{x^2 + y^2}$ above the xy plane. *Ans*: 12π

(c) $\vec{F} = (y - z + 2)\hat{\imath} + (yz + 4)\hat{\jmath} - xz\hat{k}$, where S is the surface of the cube $x = y = z = 0$, $x = y = z = 2$ above the xy plane. *Ans*: -4

(d) $\vec{F} = (y^2 + z^2 - x^2)\hat{\imath} + (z^2 + x^2 - y^2)\hat{\jmath} + (x^2 + y^2 - z^2)\hat{k}$, over the portion of the surface $x^2 + y^2 - 2ax + az = 0$ above the plane $z = 0$ and also verify Stokes's theorem.

15. Compute the surface integral $\int\int_S (ax^2 + by^2 + cz^2)dS$ over the sphere $x^2 + y^2 + z^2 = 1$. *Ans*: $4(a + b + c)\pi/3$

16. Verify Stokes' theorem for the vector functions:

(a) $\vec{F} = (y - x + 2)\,\hat{\imath} + (yz + 4)\,\hat{\jmath} - xz\hat{k}$, where S is the surface of the cube $x = 0, y = 0, z = 0, x = 2, y = 2, z = 2$ above the xy plane. Common value: -4

(b) $\vec{F} = x\,\hat{\imath} + z^2\,\hat{\jmath} + y^2\hat{k}$, over the plane surface $x + y + z = 1$ lying in the positive octant. Common value: 0

(c) $\vec{F} = 3y\,\hat{\imath} - xz\,\hat{\jmath} + y^2z\hat{k}$, where S is the surface of the paraboloid $x^2 + y^2 = 2z$ bounded by $z = 2$.

(d) $\vec{F} = (x^2 + y^2)\,\hat{\imath} - 2xy\,\hat{\jmath}$, taken rounded the rectangle bounded by $x = \pm a, y = 0, y = b$. Common value: $-4ab^2$

17. Consider a vector $\vec{F} = -4yx^2\hat{\imath} - 3y^2\hat{\jmath}$.

(a) Calculate the line integral $\int \vec{F} \cdot d\vec{r}$ from point $P \to O$ along the path $P \to Q \to R \to O$, as shown in the figure below: *Ans*: $-1/3$

(b) Using Stokes' theorem appropriately, calculate $\int \vec{F} \cdot d\vec{r}$ for the same path $P \to Q \to R \to O$. *Ans*: $-1/3$

18. Evaluate by using Stokes' theorem:

(a) $\oint_C \sin z\, dx - \cos x\, dy + \sin y\, dz$, where C is the boundary of the rectangle $0 \le x \le \pi, 0 \le y \le 1, z = 3$. *Ans*: 2

(b) $\oint_C y\, dx + (1 - 2z)x\, dy - xy\, dz$, where C is the circle $x^2 + y^2 = a^2, z = 0$. *Ans*: 0

(c) $\oint_C zy\, dx + zx\, dy + xy\, dz$, where C is the circle $x^2 + y^2 = 1, z = y^2$. *Ans*: 0

(d) $\oint_C e^x\, dx + 2y\, dy - dz$, where C is the circle $x^2 + y^2 = 1, z = 2$. *Ans*: 0

(e) $\oint_C (ydx + zdy + xdz)$ where C is the curve of intersection of $x^2 + y^2 + z^2 = a^2$ and $x + z = a$. *Ans*: $-\pi a^2/\sqrt{2}$

(f) $\oint_C y^2\,dx + z^2\,dy + x^2\,dz$, where C is the triangular closed path joining the points $(0,0,0)$, $(0,a,0)$ and $(0,0,a)$ by transfering the integral to surface integral. *Ans*: $a^3/3$

19. Evaluate the surface integral $\int\int_S (\vec{\nabla} \times \vec{F}) \cdot \hat{n}\, dS$ by using Stokes' theorem:

(a) $\vec{F} = (x^2 + y - 4)\hat{\imath} + 3xy\hat{\jmath} + (2xz + z^2)\hat{k}$, where $\mathcal{S}$ is the surface of the paraboloid $z = 4 - (x^2 + y^2)$ above the xy plane. *Ans*: -4π

(b) $\vec{F} = y\hat{\imath} + (x - 2xz)\hat{\jmath} - xy\hat{k}$, where $\mathcal{S}$ is the surface of the sphere $x^2 + y^2 + z^2 = a^2$ above the xy plane. *Ans*: 0

(c) $\vec{F} = (2x^2 + yz)\hat{\imath} + (y^2 + zx)\hat{\jmath} + (3z^2 + xy)\hat{k}$, where $\mathcal{S}$ is the surface of the region bounded by $x = 0, y = 0, z = 0$ and $x + y + z = 1$. *Ans*: 0

20. By using Stokes' theorem, evaluate the surface integral $\int\int_S \vec{F} \cdot \hat{n}\, dS$, where $\vec{F} = -2z\,\hat{\jmath} + 2y\,\hat{k}$ and $\mathcal{S}$ is the surface of the hemisphere $x^2 + y^2 + z^2 = 16$ above the xy plane. *Ans*: -16π

21. Evaluate the line integral $\oint_C \vec{F} \cdot d\vec{r}$ by Stokes' theorem, where $\vec{F} = y^2\hat{\imath} + x^2\hat{\jmath} - (x+z)\hat{k}$ and C is boundary of the triangle with vertices at $(0,0,0)$, $(1,0,0)$ and $(0,1,0)$. *Ans*: 0

Bibliography

[1] Chakravorty, JG and Ghosh, PR. 2013. *Vector Analysis: Vector Algebra & Vector Calculus*. UN Dhur & Sons Private Limited: Kolkata, India.

[2] Chatterjee, U and Chatterjee, N. 2010. *Vector and Tensor Analysis: Theory and Applications*. Academic Publishers: Kolkata, India.

[3] Davis, HF. 1963. *Introduction to Vector Analysis*. Allyn & Bacon: Boston, USA.

[4] Eliezer, CJ. 1961. *Concise Vector Analysis*. The Macmillan Company: New York, USA.

[5] Gupta, R. 2005. *Golden Vector Algebra*. Laxmi Publications: India.

[6] Hey, HG. 1953. *Vector and Tensor Analysis*. Dover Publications: New York, USA.

[7] Lindgren, BW. 1964. *Vector Calculus*. The Macmillan Company: New York, USA.

[8] Matthews, Paul C. 2000. *Vector Calculus*. Springer Undergraduate Mathematics Series. Springer Books: USA.

[9] Morrison, JB. 1965. *Principle of Vector Analysis*. Academic Press: New York, USA.

[10] Narayan, S and Mittal, PK. 2005. *A Textbook of Vector Calculus*. S Chand & Company Private Limited: India.

[11] Nayak, PK. 2010. *A Textbook of Tensor Calculus and Differential Geometry*. Prentice-Hall India Learning Private Limited: New Delhi, India.

[12] Nayak, PK. 2016. *A Textbook of Mechanics*. Narosa Publishing House Private Limited: India.

[13] Philips, HB. 1933. *Vector Analysis*. John Wiley & Sons: New York, USA.

[14] Spiegel, MR, Lipschutz, S and Spellman, D. 2009. *Vector Analysis, Second Edition* (Schaum's Outline Series). McGraw Hill: New York, USA.

[15] Spiegel, MR. 1959. *Vector Calculus* (Schaum's Outline Series), SI Edition. McGraw-Hill: New York, USA.

[16] Vasishtha, AR. 2014. *Vector Algebra*. Krishna Prakashan Media Private Limited: Meerut, India.

Index